Understanding the Fundamental Constituents of Matter

THE SUBNUCLEAR SERIES

Series Editor: **ANTONINO ZICHICHI**
European Physical Society
Geneva, Switzerland

1. 1963 STRONG, ELECTROMAGNETIC, AND WEAK INTERACTIONS
2. 1964 SYMMETRIES IN ELEMENTARY PARTICLE PHYSICS
3. 1965 RECENT DEVELOPMENTS IN PARTICLE SYMMETRIES
4. 1966 STRONG AND WEAK INTERACTIONS
5. 1967 HADRONS AND THEIR INTERACTIONS
6. 1968 THEORY AND PHENOMENOLOGY IN PARTICLE PHYSICS
7. 1969 SUBNUCLEAR PHENOMENA
8. 1970 ELEMENTARY PROCESSES AT HIGH ENERGY
9. 1971 PROPERTIES OF THE FUNDAMENTAL INTERACTIONS
10. 1972 HIGHLIGHTS IN PARTICLE PHYSICS
11. 1973 LAWS OF HADRONIC MATTER
12. 1974 LEPTON AND HADRON STRUCTURE
13. 1975 NEW PHENOMENA IN SUBNUCLEAR PHYSICS
14. 1976 UNDERSTANDING THE FUNDAMENTAL CONSTITUENTS OF MATTER

Volume 1 was published by W. A. Benjamin, Inc., New York; 2-8 and 11-12 by Academic Press, New York and London; 9-10 by Editrice Compositori, Bologna; 13-14 by Plenum Press, New York and London.

Understanding the Fundamental Constituents of Matter

Edited by

Antonino Zichichi
European Physical Society
Geneva, Switzerland

PLENUM PRESS · NEW YORK AND LONDON

Library of Congress Cataloging in Publication Data

International School of Subnuclear Physics, Erice, Italy, 1976.
Understanding the fundamental constituents of matter.

(The Subnuclear series; 14)
Includes index.
1. Particles (Nuclear physics)–Congresses. I. Zichichi, Antonino. II. Sicily (Region) III. Rehovot, Israel. Weizmann Institute of Science. IV. Title. V. Series.
QC793.I555 1976 539.7'21 78-2898

ISBN-13: 978-1-4684-0933-8 e-ISBN-13: 978-1-4684-0931-4
DOI: 10.1007/978-1-4684-0931-4

Proceedings of the 1976 International School of Subnuclear Physics (NATO-MPI-MRST Advanced Study Institute) held in Erice, Trapani, Sicily, July 23–August 8, 1976 and sponsored by the Sicilian Regional Government and the Weizmann Institute of Science

Softcover reprint of the hardcover 1st edition 1978
A Division of Plenum Publishing Corporation
227 West 17th Street, New York, N.Y. 10011

PREFACE

During July and August of 1976 a group of 90 physicists from 56 laboratories in 21 countries met in Erice for the 14th Course of the International School of Subnuclear Physics. The countries represented were Argentina, Australia, Austria, Belgium, Denmark, the Federal Republic of Germany, France, the German Democratic Republic, Greece, Israel, Italy, Japan, Mexico, Nigeria, Norway, Sweden, the United Kingdom, the United States of America, Vietnam, and Yugoslavia. The School was sponsored by the Italian Ministry of Public Education (MPI), the Italian Ministry of Scientific and Technological Research (MRST), the North Atlantic Treaty Organization (NATO), the Regional Sicilian Government (ERS), and the Weizmann Institute of Science.

The program of the School was mainly devoted to the elucidation and discussion of the progress achieved in the theoretical and experimental understanding of the fundamental constituents of matter. On the theoretical front we had a series of remarkable lecturers (C. N. Yang, S. Weinberg, G. C. Wick) attempting a description of finite size particles. Another group of lecturers covered such topics as the understanding of the new particles (H. J. Lipkin), whether or not jets really exist (E. Lillethun), and the unexpected A-dependence of massive dileptons produced in high-energy proton—nucleus collisions (J. W. Cronin). Two other outstanding questions were covered by E. Leader and G. Preparata respectively: whether strong interactions are still within the Regge framework, and if it is really possible to master strong interactions. A. J. S. Smith convinced everybody that a large fraction of single inclusive lepton production in hadronic interactions can be accounted for by pair production. The highlights of the School were the ($\mu^- e^+ K^\circ$) events presented by W. F. Fry. The program was completed by an excellent series of review lectures on the more classical field of Subnuclear Physics.

I hope the reader will enjoy this book as much as the students enjoyed attending the lectures and the discussion sessions, which are the most attractive features of the School. Thanks to the work of the Scientific Secretaries the discussions have been reproduced as faithfully as possible. At various stages of my work I have enjoyed the collaboration of many friends whose contributions have been extremely important for the School and are highly appreciated. I thank them most warmly. A final acknowledgement to all those who, in Erice, Bologna and Geneva helped me on so many occasions and to whom I feel very much indebted.

A. Zichichi
January, 1977
Geneva

CONTENTS

THEORETICAL LECTURES

REVIEW LECTURES

SEMINARS ON SPECIALIZED TOPICS

CLOSING LECTURE

CRITICAL PHENOMENA FOR FIELD THEORISTS

Steven Weinberg

Lyman Laboratory of Physics, Harvard University

Cambridge, Massachusetts 02138

1. INTRODUCTION

Many of us who are not habitually concerned with problems in statistical physics have gradually been becoming aware of dramatic progress in that field. The mystery surrounding the phenomenon of second-order phase transitions seems to have lifted, and theorists now seem to be able to explain all sorts of scaling laws associated with these transitions, and even (more or less) to calculate the "critical exponents" of the scaling laws.[1] Furthermore, the methods used to solve these problems appear to have a profound connection with the methods of field theory — one overhears talk of "renormalization group equations", "infrared divergences", "ultraviolet cut-offs", and so on. It is natural to conclude that field theorists have a lot to learn from their statistical brethren.

For this reason, I started a year or so ago to try to learn the modern theory of critical phenomena. It has not been easy. On one hand, there are a number of authors who use a language that is unfamiliar to field theorists, involving concepts (like block spins, lattice spacings, etc.) that refer specifically to crystal lattices. It is not so hard to rewrite the formulas in a continuum language, but the physical insight is harder to translate. On the other hand, there is a school of theorists who follow very closely the formalism of quantum field theory. This makes the theory even harder to understand, because many of the formal devices they use (like coupling-constant and field renormalization) were motivated in field theory by the need to deal with ultraviolet divergences, a problem that has little to do with critical phenomena. Above all, one wonders how renormalizable field theories, involving only a finite number of interaction terms, can have anything to do with the effective Hamiltonians of classical statistical mechanics, which must surely involve terms of unlimited complexity.

These lectures will present what I have been able to glean of the theory of critical phenomena. After a brief review of the field-theoretic formalism of statistical mechanics in Section 2, the qualitative theory will be described in Sections 3 - 7, and quantitative methods will be introduced in Sections 8 - 11. It hardly needs to be said that almost none of the theory I describe in these sections is originally due to me. (For detailed references, consult the reviews listed at the back of these notes.[1]) The only material which may possibly have originated with me is the proof of the invariance of the eigenvalues at a fixed point in Section 7; the "one-loop equations" presented in Section 8; and the use of renormalization-group methods to deal with the Bloch-Nordsieck problem in Section 3. Even here, I would not be surprised to be informed by a kind reader that some or all of this material already exists in the published literature.

In the last section I will try to draw some lessons for field theory from our study of critical phenomena. The formalism used in studying critical phenomena guarantees that physical quantities are cut-off independent for all theories, renormalizable or not. What then determines which of the infinite variety of possible Lagrangians in field theory is physically acceptable? Is renormalizability necessary? How does one handle phenomena like gravitation, where symmetries seem to rule out any renormalizable theory? Some tentative answers are offered, but the questions remain open.

2. STATISTICAL MECHANICS AND FIELD THEORY

This section will present a very condensed review of the field-theoretic formalism of statistical mechanics. I want especially to explain why it is that the statistical physicists who study critical phenomena can live in a three-dimensional world, unlike field theorists, who need to work in four space-time dimensions. Most of you probably know all about this, but at least this discussion will serve to fix our notation.

The aim of quantum field theory is to calculate S-matrix elements. However, at a finite temperature there is no such thing as an S-matrix; a finite temperature means that space is filled with debris like black-body radiation, so any particle that participates in a collision is scattered again and again before it gets out to infinity. Instead of S-matrix elements, one wants to calculate a partition function

$$Q = \mathrm{Tr}\exp(-H/\theta) \tag{2.1}$$

(H is the Hamiltonian and θ is the temperature in energy units, with Boltzmann's constant equal to unity.) Of course, H may depend on all sorts of external currents, in which case Q is a complicated functional of these currents, not a mere number. The "temperature Green's functions" are the derivatives of $\ell n\, Q$ with respect to these currents.

To calculate Q, one separates H into a free-particle term H_0 and an interaction term V, and uses the thermodynamic version of the Dyson formula:

$$e^{H_0/\theta}\, e^{-H/\theta} = \sum_{n=0}^{\infty} \frac{(-1)^n}{n!} \int_0^{1/\theta} d\tau_1 \ldots d\tau_n\, T_\tau\{V(\tau_1)\ldots V(\tau_n)\} \tag{2.2}$$

where $V(\tau)$ is the "interaction-representation" operator

$$V(\tau) \equiv e^{H_0\tau}\, V\, e^{-H_0\tau} \tag{2.3}$$

and T_τ is the operator which orders the operators in order of decreasing τ-arguments. (To prove this formula, simply differentiate with respect to $1/\theta$.) It follows immediately that the partition function is

$$Q = \sum_{n=0}^{\infty} \frac{(-1)^n}{n!} \int_0^{1/\theta} d\tau_1 \ldots d\tau_n\, \mathrm{Tr}\left[e^{-H_0/\theta}\, T_\tau\{V(\tau_1)\ldots V(\tau_n)\}\right] \tag{2.4}$$

Both H_0 and V can be expressed in terms of creation and annihilation operators as usual. (This is done even when there are conservation laws which prevent actual particle creation or annihilation, because it is by far the most convenient way of incorporating the correct cluster-decomposition properties in the theory.) Usually these operators appear in H_0 and V in the form of various fields, linear combinations $A_i(\vec{x})$ with simple translation properties. In the interaction representation, $V(\tau)$ is then a functional of the interaction representation "fields"

$$A_i(\vec{x},\tau) \equiv e^{H_0\tau}\, A_i(\vec{x})\, e^{-H_0\tau} \tag{2.5}$$

(The variable $\vec{x}$ may be continuous, as in field theory, or discrete, as on a lattice.) We can apply Wick's theorem to Eq. (2.4), and derive a set of Feynman rules which allow us to calculate Q order-by-order in perturbation theory, just as in quantum field theory. (Details can be found in the book of Fetter and Walecka.[2]) The contraction of two fields $A_1(\vec{x}_1\tau_1)$ and $A_2(\vec{x}_2\tau_2)$ yields a factor

$$\Delta_{12}(\vec{x}_1-\vec{x}_2,\tau_1-\tau_2) = \mathrm{Tr}\left[e^{-H_0/\theta}\, T_\tau\{A_1(\vec{x}_1\tau_1)A_2(\vec{x}_2\tau_2)\}\right] \tag{2.6}$$

with T_τ now defined with an extra minus sign for $\tau_2 > \tau_1$ when A_1 and A_2 are fermion field operators.

The only important difference between this formalism and that of quantum field theory is that we are now integrating the "time" $i\tau$ over imaginary values from 0 to i/θ, instead of over real values from $-\infty$ to $+\infty$. In consequence, we can express the propagators as Fourier integrals over momenta but Fourier sums over energies:

$$\Delta_{12}(\vec{x}_1-\vec{x}_2,\tau_1-\tau_2) = \int d^3p \sum_\omega \Delta_{12}(\vec{p},\omega)e^{i\vec{p}\cdot(\vec{x}_1-\vec{x}_2)}\ e^{-i\omega(\tau_1-\tau_2)} \tag{2.7}$$

with ω restricted to positive or negative integer multiples of $\pi\theta$. (On a lattice, the $\vec{p}$-integrals are cut off at momenta of the order of an inverse lattice spacing, while in a field theory they run over all $\vec{p}$.) Furthermore, the Green's functions have a remarkable periodicity property: for τ in the range $0 < \tau < 1/\theta$, we have

$$\begin{aligned}
\Delta_{12}(\vec{x},\tau) &\equiv \mathrm{Tr}\left[e^{-H_0/\theta}\ T_\tau\{A_1(\vec{x},\tfrac{1}{\theta})\ A_2(0,\tfrac{1}{\theta}-\tau)\}\right] \\
&= \mathrm{Tr}\left[e^{-H_0/\theta}\ e^{H_0/\theta}\ A_1(\vec{x})\ e^{-H_0/\theta}\ e^{H_0(\frac{1}{\theta}-\tau)}\ A_2(0)\ e^{-H_0(\frac{1}{\theta}-\tau)}\right] \\
&= \mathrm{Tr}\left[A_1(\vec{x})\ e^{-H_0\tau}\ A_2(0)\ e^{-H_0(\frac{1}{\theta}-\tau)}\right] \\
&= \mathrm{Tr}\left[A_2(0)\ e^{-H_0/\theta}\ e^{H_0\tau}\ A_1(\vec{x})\ e^{-H_0\tau}\right] \\
&= \mathrm{Tr}\left[e^{-H_0/\theta}\ A_2(0,\tfrac{1}{\theta})\ A_1(\vec{x},\tau)\right] \\
&\equiv \pm\Delta_{12}(\vec{x},\tau-\tfrac{1}{\theta}) \qquad \begin{cases}+\ \text{bosons}\\ -\ \text{fermions}\end{cases}
\end{aligned} \tag{2.8}$$

It follows that the sum in Eq. (2.7) runs only over ω-values with

$$\omega = \begin{cases}\pi\theta \times \text{even integer} & \text{(bosons)}\\ \pi\theta \times \text{odd integer} & \text{(fermions)}\end{cases} \tag{2.9}$$

To calculate Q, we add up all diagrams with no external lines. The momentum-space rules for these diagrams are the same as for the vacuum amplitude in quantum field theory, except that every internal energy is replaced with a quantity $i\omega$ satisfying the quantization conditions (2.9), and all energy integrals are replaced with ω sums:

$$p^0 \to i\omega \tag{2.10}$$

$$\int d^4p \rightarrow 2i\pi\theta\int d^3p\sum_{\omega} \qquad (2.11)$$

$$\delta^4(p-p') \rightarrow (2i\pi\theta)^{-1}\ \delta^3(\vec{p}-\vec{p}\,')\delta_{\omega\omega'} \qquad (2.12)$$

For instance, the Green's function for a scalar boson of mass m is now

$$\frac{-i}{(2\pi)^4}\ \Delta(\vec{p},\omega) = \frac{-i}{(2\pi)^4}\ \frac{1}{\vec{p}^{\,2}+\omega^2+m^2} \qquad (2.13)$$

(I should perhaps add here that Q is given by the sum of all diagrams, while $\ell n\ Q$ is given by the sum of all connected diagrams. The average value of any physical quantity A can be calculated by introducing a term JA in V and noting that

$$\langle A\rangle = \frac{\mathrm{Tr}\{A\,e^{-H/\theta}\}}{\mathrm{Tr}\{\ e^{-H/\theta}\ \}} = -\ \theta\ \frac{\partial}{\partial J}\ \ell n\ Q$$

The nth derivative of $\ell n\ Q$ with respect to external currents is given by the sum of all connected diagrams with n external lines of appropriate types. For systems with a large volume Ω, the quantity $\ell n\ Q$ will contain a factor

$$\delta^3(0) = (2\pi)^{-3}\,\Omega$$

and derivatives of $(\ell n\ Q)/\Omega$ with respect to various external currents will give the densities of the physical quantities coupled to these currents.)

A second-order phase transition occurs when one of the renormalized boson masses of the theory vanishes; the value to which the temperature must be lowered to make the boson mass vanish is known as the critical temperature. (The renormalized mass, or inverse correlation length, is defined in terms of the total inverse propagator at $\vec{p}=\omega=0$, including all "radiative" corrections. It is therefore a function of the temperature.) Our experience in field theory makes it familiar that a vanishing boson mass signals a smooth change in the properties of the physical states, including their symmetries. This is to be contrasted with a first-order phase transition, in which the properties of physical systems change abruptly. For instance, water at sea level pressure undergoes a first-order phase transition, with finite latent heat, at the boiling point, but water at the critical pressure and magnets in zero external field undergo second-order transitions at the critical temperatures.

It is in a second-order phase transition, when one of the boson masses of the theory vanishes, that the quantization of virtual energies has its most important effect.

A quantum field theory at zero temperature would at most have logarithmic infrared divergences, unless the Hamiltonian contained super-renormalizable ϕ^3 interactions. On the other hand, at finite temperature we have only a three-dimensional momentum-space volume element d^3p instead of a four-dimensional element $d^3p\,d\omega$ available to cut down the degree of infrared divergences, and the Feynman integrals will therefore have power instead of logarithmic divergences. (As far as the three-dimensional integrals with $\omega = 0$ are concerned, even ϕ^4 interactions are superrenormalizable.) We therefore expect that at finite temperature there will be dramatic long-range correlations when any of the boson masses approaches zero.

The infrared divergences arise solely from the blowing up of the $\omega = 0$ terms in the sums over boson energies as $\vec{p} \to 0$; in all other terms the finite value of ω acts as an infrared cut-off. [See Eq. (2.13).] Therefore, in order to study the long-range effects which occur when one of the boson masses approaches zero, it is very convenient to emphasize only the $\omega = 0$ terms, by burying all other terms in the sums over ω in "black boxes". That is, we construct an effective three-dimensional field theory, in which the propagators consist purely of the $\omega = 0$ terms in (2.7), with all terms having $\omega \neq 0$ absorbed into the coupling constants of the effective Hamiltonian. Of course, this effective Hamiltonian will consist not just of the simple interactions that may have been originally present in V, but of an infinite number of interactions, all with temperature-dependent coefficients. (For instance, in a ϕ^4 field theory, the effective Hamiltonian will contain ϕ^{2n} terms produced by loops with n corners, in which all the internal boson lines have $\omega \neq 0$.) As long as we keep track of all these interactions, the three-dimensional effective field theory is supposed to be fully equivalent to the original four-dimensional finite-temperature formalism.

In general, we would not expect to know very much about the temperature-dependence of the parameters in the effective Hamiltonian. The one thing we do know is that these parameters are given by sums over diagrams with $\omega \neq 0$. Therefore, the effective Hamiltonian itself does not feel the influence of the infrared divergences, and the parameters in this Hamiltonian are expected to be analytic functions of temperature even near the critical temperatures. Fortunately, as we shall see, this is all the information we need for most purposes.

In renormalizable theories with weak couplings and high temperature, it is possible to say a good deal more about the structure of the effective Hamiltonian.[3] (The material in the balance of this section is offered to readers who want to see some concrete formulas for coefficients in the effective Hamiltonian. It is not needed as a basis for the rest of the lectures.) If the couplings are weak, loop graphs are generally small, and most of the coefficients in the effective Hamiltonian simply have approximately the value they would have had in the original zero-temperature theory. However, some of the ω-sums in these loop graphs may contribute powers of the temperature to the coefficients in the effective Hamiltonian, and if the temperature is sufficiently high, these powers of temperature can compensate for powers of the coupling. For temperatures which are much larger than any masses or momenta, the temperature dependence of a one-loop diagram of dimensionality D will be simply θ^D; here D is just the degree of divergence of the theory at zero temperature. Each loop also introduces a coupling-constant factor, say f. Hence the leading contribution to the effective Hamiltonian when θ is large and f is small will come from diagrams in which each loop is as divergent as possible. The worst divergences in renormalizable theories are the quadratic divergences in scalar boson propagators, with $D=2$. Furthermore, the only graphs in which every loop is quadratically divergent consist of just a string of one-loop insertions in a scalar boson propagator. Hence, we can obtain the effective Hamiltonian to lowest order in f but to all orders in $f\theta^2$ by simply calculating the one-loop corrections to the scalar self-energies.

For instance, consider a theory involving a number of scalar fields ϕ_i, with an interaction

$$V = \frac{1}{24}\int d^3x\, f_{ijk\ell}\,\phi_i\,\phi_j\,\phi_k\,\phi_\ell$$

The one-loop correction to the boson mass matrix in the effective Hamiltonian is

$$\Delta M^2_{ij}(\theta) = \left(\frac{i}{2(2\pi)^4}\right)\left(2i\pi\theta\right)\int d^3p \sum_{\substack{n=-\infty \\ n\neq 0}}^{\infty} f_{ijk\ell}\left(\vec{p}^{\,2}+M^2+4n^2\theta^2\pi^2\right)^{-1}_{k\ell}$$

We are only interested here in the leading terms when θ is much greater than the zero-temperature mass M, so we can rewrite this as

$$\Delta M^2_{ij}(\theta) = \frac{-\pi\theta}{(2\pi)^4}\, f_{ijkk}\int d^3p \sum_{\substack{n=-\infty \\ n\neq 0}}^{\infty}\left(\vec{p}^{\,2}+4n^2\theta^2\pi^2\right)^{-1}$$

It is straightforward to calculate that for large θ

$$\Delta M^2_{ij}(\theta) \to \frac{-1}{24}\theta^2\, f_{ijkk}$$

There are also terms of first and zeroth order in θ which are respectively linearly and quadratically divergent, but we neglect these because we now only are keeping terms in ΔM^2 of order $f\theta^2$. Thus, to all orders in $f\theta^2$ but lowest order in f, the mass matrix in the effective Hamiltonian takes the form

$$M^2_{ij}(\theta) \simeq M^2_{ij} - \frac{1}{24}\theta^2 f_{ijkk}$$

We expect the actual phase transitions to occur somewhere near the temperatures where one of the eigenvalues of this matrix vanishes. As expected, the mass matrix $M^2_{ij}(\theta)$ is perfectly analytic near any critical temperature.

Similar calculations have been carried out for arbitrary renormalizable gauge theories.[3] The general result is that

$$M^2_{ij}(\theta) \simeq M^2_{ij} - \frac{1}{24}\theta^2\left\{f_{ijkk} + 6(\Theta_\alpha\Theta_\alpha)_{ij} + \mathrm{Tr}[\Gamma_i\gamma_4\Gamma_j\gamma_4]\right\}$$

where Θ_α are the representations of the gauge generators on the scalar fields (including gauge coupling constant factors) and the Γ_i are the matrices in the Yukawa coupling $\bar{\psi}\Gamma_i\psi\phi_i$ of the scalars and spinors.

It must be stressed that these formulas only provide an approximation to the "bare" mass matrix appearing in the effective Hamiltonian. The "renormalized mass" (or inverse correlation length) involves a sum over higher-order corrections, which become more and more important near the critical temperature. In fact, as we shall see, the renormalized mass has a dependence on temperature very different from that found here.

3. THE FLOATING CUT-OFF

The lesson of the last section is that critical phenomena are conveniently described in terms of an effective Euclidean three-dimensional field theory with temperature-dependent interactions of arbitrary complexity. We can now forget the four-dimensional theory from which we started, and deal with the three-dimensional effective theory on its own terms.

We are interested in what happens when one of the boson masses M in the effective Lagrangian becomes small. Because the theory is three-dimensional, the infrared divergences become so serious in this limit that perturbation theory necessarily breaks down, no matter how weak the coupling we started with may be. For instance, if the theory contains an interaction $u\phi^4$, then any graph with L loops constructed solely from this interaction will contain a factor $(\theta u)^L$. [See Eqs. (2.11) and (2.12).] But it follows then by simple dimensional reasoning that when all momenta are of the order of the effective boson mass M, the loop integrals must also contribute a factor M^{-L}. (In fact, this is true for sufficiently large L even when the external momenta are much larger than M.)

Hence the effective coupling constant contributed by each additional loop is $\theta u/M$. No matter how small u may be, this becomes large when M is sufficiently small, and perturbation theory becomes useless.

What can we do about this? The problem arises from very small virtual momenta of order M, but the effective Hamiltonian we are working with involves all momenta, or at least all momenta less than an inverse lattice spacing. Is it possible that we are working with an inappropriate effective Hamiltonian?

Suppose we choose the effective Hamiltonian to emphasize the degrees of freedom that are really important near a phase transition, by introducing a momentum-space cut-off Λ, and integrating only over momenta $\vec{p}$ with $|\vec{p}| < \Lambda$. (Eventually Λ will be allowed to go to zero.) We do not want the introduction of the cut-off to change the physics, so the effective couplings must all be chosen to depend on the cut-off, in such a way as to leave the partition function Λ-independent. This condition imposes a set of differential equations on the couplings in the effective Hamiltonian, known as the renormalization-group equations.

The idea of a floating cut-off, introduced in such a way as not to change the physics, is actually very old. For an early example, let's return briefly to four dimensions, and consider the classic problem of Bloch and Nordsieck,[4] the radiative corrections to scattering of an electron by an external Coulomb potential. In each order of perturbation theory, we encounter infrared divergences of the form $\ln^n m_\gamma$, where m_γ is a fictitious photon mass introduced as an infrared cut-off. We all know that these infrared divergences are cancelled by real photon emission, but let's forget that for a moment, and ask what is the behaviour of the purely elastic matrix element as $m_\gamma \to 0$? To answer this, we introduce an ultraviolet cut-off Λ on the photon momenta, and use an effective Hamiltonian, in which the interactions are Λ-dependent black boxes which include all effects of "hard" photons with momenta $|q| > \Lambda$. (To define the cut-off, we can either simply take $|q|$ as the non-invariant square $\vec{q}^2 + q^{0\,2}$, or use a regulator of mass Λ.) It is assumed that the photon mass m_γ is much smaller than the smallest energy E characteristic of the scattering process. (Here E is m_e, or the initial momentum $|\vec{p}_i|$, or the final momentum $|\vec{p}_f|$, whichever is smaller.) We take Λ in the range

$$m_\gamma << \Lambda << E \tag{3.1}$$

It is easy to see that the only graphs which can produce leading singularities in m_γ are those containing a single interaction of the form

$$\int \overline{\psi}(p')\Gamma_\Lambda(p',p)\psi(p)\,d^4 p\,d^4 p' \tag{3.2}$$

plus any number of emissions and absorptions of soft photons with $|q| < \Lambda$ from the incoming or outgoing electron line. (See Fig. 1.)

Furthermore, since $\Lambda << E$, the external electron momenta never get very far from the initial and final mass-shell values p_f and p_i (with $p_f^2 = p_i^2 = -m_e^2$). It follows that the matrix element takes the form

$$M(p_f,p_i,m_e,m_\gamma) = F(p_f,p_i,m_e,\frac{m_\gamma}{\Lambda})M_\Lambda(p_f,p_i,m_e) \tag{3.3}$$

where M_Λ is the mass-shell matrix element $\bar{u}_f \Gamma_\Lambda u_i$ of Γ_Λ. Note that

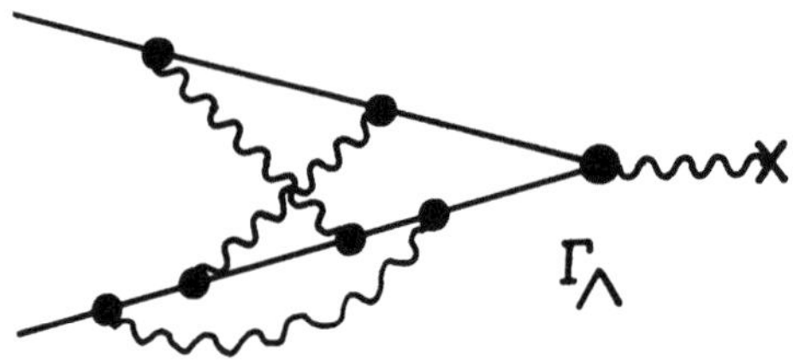

Figure 1. A typical graph which can produce leading singularities in m_γ. Here wavy lines are soft photons; straight lines are electrons near their mass shell; and darkened circles represent sums over graphs involving hard photons and electrons far from their mass shell.

M_Λ does not depend on m_γ and F depends on m_γ and Λ only in the ratio m_γ/Λ, because we took $\Lambda << E$. For instance, the lowest-order radiative correction function may be calculated directly as

$$F = 1 - A\,\ell n\frac{\Lambda}{m_\gamma} \tag{3.4}$$

where

$$A \equiv \frac{\alpha}{2\pi\beta}\,\ell n\left(\frac{1+\beta}{1-\beta}\right) \tag{3.5}$$

$$\beta \equiv \left|1 - m_e^4/(p_i \cdot p_f)^2\right|^{\frac{1}{2}} \tag{3.6}$$

Now, it is actually pretty easy to calculate F to all orders in α, and even to sum the series. Suppose however that we were all combinatoric cripples, and were unable to carry out the calculation directly. The "renormalization group" would do the job for us. We know that the matrix element M must be independent of the floating cut-off, so the derivative of Eq. (3.3) with respect to Λ gives

$$0 = -\frac{m_\gamma}{\Lambda^2}\frac{\partial F}{\partial(m_\gamma/\Lambda)}M_\Lambda + F\frac{\partial M_\Lambda}{\partial\Lambda}$$

or in other words

$$\Lambda\frac{\partial M_\Lambda}{\partial\Lambda} = \frac{\partial\,\ell n\,F}{\partial\,\ell n(m_\gamma/\Lambda)}M_\Lambda \tag{3.7}$$

But M_Λ does not depend on m_γ for $m_\gamma << E$, so the derivative on the right-hand side must be independent of m_γ/Λ.

$$\frac{\partial \ln F(p_f, p_i, m_e, m_\gamma/\Lambda)}{\partial \ln (m_\gamma/\Lambda)} = A(p_f, p_i, m_e) \tag{3.8}$$

and therefore F is just a power of m_γ/Λ

$$F \propto (m_\gamma/\Lambda)^A \tag{3.9}$$

in agreement with the lowest-order result (3.4).

It can be shown in precisely the same way that the infrared divergences in any QED matrix element always sum up to a power of m_γ. (However, the results are different if the electron mass is zero; for instance, the lowest-order correction involves $\ln^2 m_\gamma$ instead of $\ln m_\gamma$. The reason that our renormalization-group argument breaks down for massless QED is that the matrix element M_Λ in this case is not independent of m_γ, even though it only involves "hard" photons with $|q| \geq \Lambda >> m_\gamma$; this is because a massless photon can produce an infrared divergence for any value of $|q|$, when it is emitted parallel to the initial or final massless electron. Similar remarks apply to non-Abelian gauge theories.)

If m_γ were so small that $\ln(E/m_\gamma)$ was of order 137, the original perturbation theory in powers of $\alpha \ln(E/m_\gamma)$ would have been useless. However, by introducing a cut-off Λ with $\Lambda >> m_\gamma$ but $\alpha \ln \Lambda/m_\gamma << 1$, we can use perturbation theory to calculate the function F as accurately as we like. True, we cannot then also calculate the matrix element M_Λ, because $\alpha \ln E/\Lambda$ is of order unity, but all the m_γ dependence is in F, and this we can calculate. The moral is clear: when it is infrared effects that invalidate perturbation theory, the introduction of a floating cut-off may not restore perturbation theory, but it does allow us to say useful things about the infrared effects themselves.

4. RENORMALIZATION-GROUP EQUATIONS

The observations of the last two sections lead us to consider the theory of one or more boson fields in three dimensions, with a floating cut-off Λ, and with cut-off dependent and temperature-dependent interactions of arbitrary complexity, chosen subject to the condition that the physics be cut-off independent. We will take a single scalar field $\phi(\vec{p})$ for simplicity, but will work in a Euclidean space with arbitrary dimensionality d, for reasons which will become clear below. The effective Hamiltonian may be written

$$H_\Lambda/\theta = \sum_n \frac{(2\pi)^d}{n!}\int d^d p_1 \ldots d^d p_n \, u_n(\vec{p}_1 \ldots \vec{p}_n;\Lambda)\phi(\vec{p}_1)\ldots\phi(\vec{p}_n) \tag{4.1}$$

We do not explicitly show the temperature-dependence of the u_n; temperature is now just one of the many parameters on which the real symmetric coupling functions u_n may depend.

Each of the u_n contains a momentum-conservation delta function

$$u_n(\vec{p}_1 \ldots \vec{p}_n;\Lambda) = \delta^d(\vec{p}_1 + \ldots + \vec{p}_n)\bar{u}_n(\vec{p}_1 \ldots \vec{p}_n;\Lambda) \tag{4.2}$$

The usual clustering properties allow each of the $\bar{u}_n$ to be expanded as a power series in the momenta; the coefficients in these power series are our coupling parameters.

In particular, the function u_2 may be written

$$u_2(\vec{p}_1,\vec{p}_2;\Lambda) = G^{-1}(\vec{p}_1;\Lambda)\delta^3(\vec{p}_1+\vec{p}_2) \tag{4.3}$$

We can think of the partition function and Green's functions as being given by a sum of Feynman diagrams with propagators

$$(2\pi)^{-d} G(\vec{p};\Lambda)\ \theta(\Lambda - |\vec{p}|) \tag{4.4}$$

and with vertices $-(2\pi)^d u_3, -(2\pi)^d u_4, \ldots$. [According to Eqs. (2.11) and (2.12), the propagators and vertices in the finite-temperature graphs in d+1-dimensional space-time are associated with factors

$$2i\pi \times -i(2\pi)^{-d-1} = (2\pi)^{-d}$$

$$(2i\pi)^{-1} \times -i(2\pi)^{d+1} = -(2\pi)^d$$

respectively.] The function θ is taken here as the usual step-function, but our discussion could be easily adapted to deal with a smoother cut-off function.

We are interested in the behaviour of the Green's functions in the infrared limit, when all the momenta are scaled to zero together. If the coupling parameters were all dimensionless and Λ-independent this would be a trivial problem, because Λ would be the only dimensional quantity in the theory, and we demand that the Green's functions are Λ-independent. In this case, as all momenta are scaled together to zero, the Green's functions would simply scale with their naive dimensionality. Of course, life is not so simple, but we try to use dimensional analysis for all it's worth.

To this end, let us define a new dimensionless momentum

$$\vec{\ell} \equiv \vec{p}/\Lambda \tag{4.5}$$

for which the cut-off is

$$|\vec{\ell}| \leq 1 \tag{4.6}$$

Also, we define a new field

$$\chi(\vec{\ell}) \equiv a(\Lambda)\phi(\vec{p}) \tag{4.7}$$

with a(Λ) a constant to be chosen below. We can write the Hamiltonian in terms of new coupling functions

$$H/\theta = (2\pi)^d \sum_n \frac{1}{n!} \int d^d \ell_1 \ldots d^d \ell_n \, g_n(\vec{\ell}_1 \ldots \vec{\ell}_n;\Lambda)\chi(\vec{\ell}_1)\ldots\chi(\vec{\ell}_n) \tag{4.8}$$

with

$$g_n(\vec{\ell}_1 \ldots \vec{\ell}_n;\Lambda) \equiv a(\Lambda)^{-n} \Lambda^{nd} u_n(\vec{p}_1 \ldots \vec{p}_n;\Lambda) \tag{4.9}$$

Again, we factor out a delta function

$$g_n(\vec{\ell}_1 \ldots \vec{\ell}_n;\Lambda) = \delta^d(\vec{\ell}_1 + \ldots + \vec{\ell}_n)\bar{g}_n(\vec{\ell}_1 \ldots \vec{\ell}_n;\Lambda) \tag{4.10}$$

and find that

$$\bar{g}_n(\vec{\ell}_1 \ldots \vec{\ell}_n;\Lambda) = a(\Lambda)^{-n} \Lambda^{nd-d} \bar{u}_n(\vec{p}_1 \ldots \vec{p}_n;\Lambda) \tag{4.11}$$

In choosing the field scale factor a(Λ), we recall that our aim is to study the infrared behaviour of the theory. For this purpose, we must consider the low-momentum behaviour of the propagator. Before re-scaling, we would in general have

$$\bar{u}_2(\vec{p},-\vec{p};\Lambda) \to m^2(\Lambda) + \vec{p}^2 Z^{-1}(\Lambda) + \ldots \tag{4.12}$$

and the re-scaled inverse propagator therefore has the behaviour

$$\bar{g}_2(\vec{p};\Lambda) \to a(\Lambda)^{-2} \Lambda^d [m^2(\Lambda) + \vec{\ell}^2 \Lambda^2 Z^{-1}(\Lambda) + \ldots] \tag{4.13}$$

It is a very great simplification to choose a(Λ) so that the coefficient of $\vec{\ell}^2$ here is a Λ-independent constant. (We will see in the following sections the price that would have to be paid if we made any other choice.) While we are at it, we may as well pick a(Λ) so that this constant is unity, i.e.,

$$a(\Lambda) = \Lambda^{1+d/2} Z^{-1/2}(\Lambda) \tag{4.14}$$

With this definition, $\bar{g}_2$ is dimensionless, and since H_Λ/θ is dimensionless, the field χ and all the coupling functions $g_n(\vec{\ell}_1 \ldots \vec{\ell}_n;\Lambda)$ are also dimensionless.

Using (4.14) in (4.11), the coupling functions are now related by

$$\overline{g}_n(\vec{\ell}_1 \ldots \vec{\ell}_n;\Lambda) = \Lambda^{\frac{nd}{2}-n-d} Z^{n/2}(\Lambda)\overline{u}_n(\vec{p}_1 \ldots \vec{p}_n;\Lambda) \tag{4.15}$$

If u_i is the coefficient of a term of order N in the expansion of $\overline{u}_n$ in powers of $\vec{p}$, and g_i is the corresponding term in the expansion of $\overline{g}_n$, then

$$g_i = Z(\Lambda)^{n/2} \Lambda^{nd/2 -d-n+N} u_i \tag{4.16}$$

In particular, the mass has $n = 2$, $N = 0$, so the corresponding rescaled coupling is

$$\overline{g}_2(0) = Z(\Lambda)\Lambda^{-2} m^2 \tag{4.17}$$

The g_i [including $\overline{g}_2(0)$] comprise the dimensionless coupling parameters of the theory.

We demand that the Green's functions of the theory [for the <u>original</u> fields $\phi(\vec{p})$] should not depend on Λ. This imposes on the couplings u_i a set of differential equations, giving $du_i/d\Lambda$ in terms of u and Λ. These equations can then be rewritten in terms of the dimensionless couplings g_i, and must on dimensional grounds take the form

$$\Lambda \frac{d}{d\Lambda} g_i(\Lambda) = \beta_i[g(\Lambda)] \tag{4.18}$$

(The calculation of the β_i will be taken up in Section 8.) Note that β_i cannot depend directly on the temperature or similar parameters, because the Green's functions depend on temperature, etc. only through the $g(\Lambda)$; hence the condition that the Green's functions be Λ-independent must be expressed in terms of the g_i themselves. That is, the temperature and similar parameters enter the theory only as "initial conditions", determining the values of the $g(\Lambda)$ at some arbitrary point $\Lambda = \Lambda_0$. Since the β do not depend on any dimensional parameters except Λ, and the β's are dimensionless, they also cannot depend on Λ, except through the dimensionless couplings $g_i(\Lambda)$. The β_i are in general non-zero for all interactions, so even if we started with some simple (e.g. renormalizable) theory, the introduction of a cut-off would force us to include in the effective Hamiltonian all conceivable couplings consistent with the symmetries of the theory.

In addition, the renormalization constants $Z(\Lambda)$ or $a(\Lambda)$ satisfy renormalization group equations, which must be linear and homogeneous. [If $Z(\Lambda)$ is a solution, so must $\frac{1}{2} Z(\Lambda)$ be, because we could have started with a field $\phi/\sqrt{2}$ instead of ϕ.] On dimensional grounds, this equation must then take the form

$$\Lambda \frac{d}{d\Lambda} Z(\Lambda) = Z(\Lambda)\gamma[g(\Lambda)] \tag{4.19}$$

Now let's apply this formalism to the Green's functions of the theory. If $\overline{C}_E[\vec{p}_1 \ldots \vec{p}_E; u(\Lambda), \Lambda]$ is a connected (not amputated or IPI) Green's function with E external lines carrying momenta $\vec{p}_1 \ldots \vec{p}_E$, then the corresponding Green's function in the re-scaled theory is

$$\overline{\Gamma}_E[\vec{\ell}_1 \ldots \vec{\ell}_E; g(\Lambda)] = a(\Lambda)^E \Lambda^{-d} \; \overline{C}_E[\vec{p}_1 \ldots \vec{p}_E; u(\Lambda), \Lambda]$$

$$= \Lambda^D Z(\Lambda)^{-E/2} \overline{C}_E[\vec{p}_1 \ldots \vec{p}_E; u(\Lambda), \Lambda] \tag{4.20}$$

where

$$D = E + \frac{Ed}{2} - d \tag{4.21}$$

As always, a bar over Γ_E or C_E indicates that a delta function $\delta^d(\vec{\ell}_1 + \ldots)$ or $\delta^d(\vec{p}_1 + \ldots)$ has been factored out of these Green's functions; this is the origin of the factor Λ^{-d} in (4.20).

We do not include an explicit Λ-dependence in Γ_E, because this Green's function is dimensionless, and there are no dimensional parameters other than Λ on which it could depend. On the other hand, for any given theory, the Green's function C_E must (at fixed $\vec{p}$) have an explicit Λ-dependence which compensates for its dependence on Λ through the coupling parameter $u(\Lambda)$:

$$\overline{C}_E[\vec{p}_1 \ldots \vec{p}_E; u(\Lambda'), \Lambda'] = \overline{C}_E[\vec{p}_1 \ldots \vec{p}_E; u(\Lambda), \Lambda] \tag{4.22}$$

This is what we mean by the physics being Λ-independent.

The renormalization-group equation for the Green's functions can now be derived by using (4.20) to express $\overline{\Gamma}_E$ in terms of $\overline{C}_E$, then using (4.22) to change the cut-off, and then using (4.20) again to express $\overline{C}_E$ back in terms of $\overline{\Gamma}_E$. For an arbitrary momentum scale parameter κ, we have

$$\overline{\Gamma}_E[\kappa\vec{\ell}_1 \ldots \kappa\vec{\ell}_E; g(\Lambda)] = Z(\Lambda)^{-E/2} \Lambda^D \overline{C}_E[\Lambda\kappa\vec{\ell}_1 \ldots; u(\Lambda), \Lambda]$$

$$= Z(\Lambda)^{-E/2} \Lambda^D \overline{C}_E[\Lambda\kappa\vec{\ell}_1 \ldots; u(\kappa\Lambda), \kappa\Lambda]$$

$$= [Z(\kappa\Lambda)/Z(\Lambda)]^{E/2} \kappa^{-D} \; \overline{\Gamma}_E[\vec{\ell}_1 \ldots \vec{\ell}_E; g(\kappa\Lambda)] \tag{4.23}$$

It is convenient to suppress the Λ-dependence, writing

$$g_i(\kappa\Lambda) \equiv g_{i\kappa} \tag{4.24}$$

We can regard $g_{i\kappa}$ as the solution of the equation

$$\kappa \frac{d\, g_{i\kappa}}{d\kappa} = \beta_i(g_\kappa) \tag{4.25}$$

with initial condition

$$g_{i\kappa} = g_i(\Lambda) \qquad \text{for } \kappa = 1 \tag{4.26}$$

Equation (4.23) can be rewritten (using 4.19) as

$$\overline{\Gamma}_E\left(\kappa\, \vec{\ell}_1 \ldots ; g(\Lambda)\right) = \exp\left(\frac{E}{2}\int_1^\kappa \gamma(g_\kappa)\frac{d\kappa}{\kappa}\right)\kappa^{-D}\; \overline{\Gamma}_E\left(\vec{\ell}_1 \ldots ; g_\kappa\right) \tag{4.27}$$

We see in particular that the behaviour of the Green's function for low momenta is determined by the solution of Eq. (4.25) in the limit $\kappa \to 0$.

Each particular physical theory with a particular value of the temperature is represented by a <u>trajectory</u> in coupling-constant space, along which (4.25) is satisfied. Different points on a given trajectory do not represent different theories, but only different Hamiltonians, corresponding to different cut-offs.

5. FIXED POINTS AND SCALING LAWS

In general, we would not expect the solutions of the renormalization-group equations to have any particularly simple behaviour for $\Lambda \to 0$. For instance, if $m^2(\Lambda)$ does not vanish as $\Lambda \to 0$, then (4.17) suggests that $\overline{g}_2(0)$ would blow up like Λ^{-2}. In order to keep the physics Λ-independent, the coupling constants $g_i(\Lambda)$ would then also have to blow up for $\Lambda \to 0$. However, it might be that for some special trajectories, $m^2(\Lambda)$ vanishes for $\Lambda \to 0$, in which case the various dimensionless couplings $g_i(\Lambda)$ might all remain well-behaved for $\Lambda \to 0$. We are going to see that the trajectories for which this happens are just those corresponding to the critical temperatures of the theory.

The simplest kind of non-singular behaviour is for the $g_i(\Lambda)$ to approach fixed values g_i^* for $\Lambda \to 0$. According to Eq. (4.18), this would require that, for all i,

$$\beta_i(g^*) = 0 \tag{5.1}$$

If a trajectory leads to such a fixed point for $\kappa \to 0$, then in this limit Eq. (4.27) gives

$$\overline{\Gamma}_E[\kappa\, \vec{\ell}_1 \ldots \kappa\, \vec{\ell}_E ; g(\Lambda)] \propto \kappa^{-D_E^*} \tag{5.2}$$

where

$$D_E^* = D - \frac{E}{2}\gamma(g^*) = E[\frac{d}{2} + 1 - \frac{1}{2}\gamma(g^*)] - d \tag{5.3}$$

In particular, the two-point function behaves like

$$\overline{\Gamma}_2 \propto \kappa^{-2+\gamma(g^*)} \tag{5.4}$$

This result is conventionally written in the form

$$\overline{\Gamma}_2 \propto \kappa^{-2+\eta} \tag{5.5}$$

and we see that the "critical exponent" η is

$$\eta = \gamma(g^*) \tag{5.6}$$

The renormalized mass m_R (or inverse-correlation length $1/\xi$) is defined by

$$m_R^2 \equiv \lim_{\kappa\to 0}\left[\overline{\Gamma}_2 \Big/ \frac{d^2\overline{\Gamma}_2}{d\kappa^2}\right] \tag{5.7}$$

and (5.4) shows that this vanishes, just as we expect at a phase transition.

Even if there is a point g^* at which β vanishes, not all trajectories hit this point. In order to see what is involved, consider trajectories that pass close to g^*. The renormalization-group equation (4.25) can then be linearized

$$\kappa\frac{d}{d\kappa}(g_{i\kappa} - g_i^*) = \sum_j M_{ij}(g_{j\kappa} - g_j^*) \tag{5.8}$$

$$M_{ij} \equiv \left[\frac{\partial\beta_i(g)}{\partial g_j}\right]_{g=g^*} \tag{5.9}$$

The solution can be written (barring degeneracies) as

$$g_{i\kappa} - g_i^* = \sum_\ell c_\ell\, e_i^{(\ell)}\, \kappa^{\lambda_\ell} \tag{5.10}$$

where λ_ℓ is the ℓ-th eigenvalue of M, and $e^{(\ell)}$ is the corresponding eigenvector (with a fixed but arbitrary normalization)

$$\sum_j M_{ij}\, e_j^{(\ell)} = \lambda_\ell\, e_i^{(\ell)} \tag{5.11}$$

The eigenvectors are classified as <u>infrared-attractive</u> or <u>infrared-repulsive</u>, according as $\lambda_\ell > 0$ or $\lambda_\ell < 0$. (The case $\lambda_\ell = 0$ is a nuisance, and will not be considered here.) Clearly, the condition for a trajectory actually to hit the fixed point is that $c_\ell = 0$ for all infrared-repulsive eigenvectors $e^{(\ell)}$. The number of parameters which have to be adjusted to achieve this is just the number of infrared-repulsive eigenvectors. For a phase transition of the usual type, in which there is just one parameter (the temperature) that must be adjusted to achieve a vanishing renormalized mass, there must be just one repulsive eigenvector. We will see in Section 11 whether this is actually likely to be the case.

This approach yields useful information even when we are not precisely at the critical temperature. In general, we would know almost nothing about the temperature-dependence of the coupling parameters. However, a coupling function $u_n(\vec{p}_1 \ldots \vec{p}_n;\Lambda)$ may be thought of as a black box in which are buried all effects of virtual particles with momenta greater than Λ (or energies $\omega \neq 0$; see Sec. 2) so it is immume to infrared effects caused by virtual particles with very low momenta. That is, unlike the Green's functions, the coupling parameters ought to be smooth functions of the temperature, even near a critical temperature.

If there is just one repulsive eigenvector $e^{(R)}$, and if θ_c is a critical temperature at which the trajectory hits the fixed point, then the coefficient of this eigenvector in (5.10) must vanish at $\theta = \theta_c$:

$$c_R(\theta_c) = 0 \tag{5.12}$$

Therefore, by the above smoothness argument, we expect that for θ near θ_c,

$$c_R \propto (\theta - \theta_c) \tag{5.13}$$

As long as the g_κ are sufficiently close to g^*, the couplings are functions only of

$$g_{i\kappa} - g_i^* \propto (\theta - \theta_c)\kappa^{-1/\nu} \tag{5.14}$$

where ν is another critical exponent, defined in terms of the single repulsive eigenvalue λ_R by

$$\nu = -1/\lambda_R > 0 \tag{5.15}$$

From Eq. (4.27), we find that the Green's function with E external lines has infrared behaviour

$$\overline{\Gamma}_E = \kappa^{-D_E^*} f_E[(\theta - \theta_c)\kappa^{-1/\nu}] \tag{5.16}$$

or equivalently

$$\overline{\Gamma}_E = (\theta - \theta_c)^{-\nu D_E^*} F_E[\kappa(\theta - \theta_c)^{-\nu}] \tag{5.17}$$

with D_E^* given by (5.3) and (5.6) as

$$D_E^* = E\left(\frac{d}{2} + 1 - \frac{\eta}{2}\right) - d \tag{5.18}$$

and f_E and F_E unknown functions of a single variable. Equation

(5.2) can be regarded as a special case of the result (5.16), for $\theta = \theta_c$.

It must be stressed that whether or not we are able actually to calculate the critical exponents η and ν, Eq. (5.16) or (5.17) contains a remarkable quantity of information. Who would guess that the Green's functions would take such a simple form, with scaling parameters η and ν that do not depend on the number of external lines?

One case of special interest is provided by the limit $\kappa \to 0$, in which all momenta scale to zero together. For $\theta \neq \theta_c$ there are no infrared divergences, so we expect F_E in Eq. (5.17) to be analytic at $\kappa = 0$. We can write the two-point function in this limit as

$$\overline{\Gamma}_2(\vec{k},-\vec{k}) = \overline{\Gamma}_2(0,0)\,[1 + \xi^2\,\vec{k}^2 + \ldots]$$

with ξ a "correlation length". Comparison with Eq. (5.17) shows that for $\theta \to \theta_c$, the correlation length exhibits the scaling behaviour

$$\xi \propto (\theta - \theta_c)^{-\nu} \tag{5.19}$$

This is the way ν is usually defined. Note that the "renormalized mass" $1/\xi$ does vanish for $\theta \to \theta_c$, showing again that this really is a second-order phase transition.

As a special subcase, consider the partition function itself. As we have already remarked, $\ln Q$ is the connected Green's function with no external lines. Since we divided by a δ-function in defining $\overline{\Gamma}_E$, the function $\overline{\Gamma}_0$ is just the free energy density

$$\overline{\Gamma}_0 = \ln Q/\Omega \tag{5.20}$$

where Ω is the volume of the system. Because there are no external lines, $\overline{\Gamma}_0$ cannot depend on a momentum scale κ, so (5.17) gives

$$\frac{\ln Q}{\Omega} \propto (\theta - \theta_c)^{-\nu D_0^*} = (\theta - \theta_c)^{+\nu d} \tag{5.21}$$

The total energy density per unit volume is then

$$u = -\frac{\partial}{\partial(1/\theta)}\left(\frac{\ln Q}{\Omega}\right) \propto (\theta - \theta_c)^{\nu d - 1} \tag{5.22}$$

Finally, the specific heat per unit volume is

$$c = \frac{\partial u}{\partial \theta} \propto (\theta - \theta_c)^{-\alpha} \tag{5.23}$$

where α is yet another critical exponent

$$\alpha = 2 - \nu d \tag{5.24}$$

6. EXTERNAL-FIELD PROBLEMS

We can also apply the same formalism to study the effect of external fields. Suppose we add a "magnetic" perturbation to the Hamiltonian

$$H_\Lambda \to H_\Lambda + h\,\phi(\vec{k}=0) \tag{6.1}$$

The effect is to change the free energy density to

$$W(h) = \frac{\ell n\, Q}{\Omega} + \sum_{E=1}^{\infty} \overline{C}_E(0,0,\ldots 0)\,\frac{h^E}{E!} \tag{6.2}$$

But for zero external momentum, Eq. (5.17) gives

$$\overline{C}_E(0,0,\ldots 0) \propto \overline{\Gamma}_E(0,0,\ldots 0) \propto (\theta-\theta_c)^{-\nu D_E^*} = (\theta-\theta_c)^{\nu d}\,(\theta-\theta_c)^{-\nu(\frac{d}{2}+1-\frac{\eta}{2})E} \tag{6.3}$$

Hence the free energy density takes the form

$$W(h) = (\theta-\theta_c)^{\nu d}\, S\left[h(\theta-\theta_c)^{-\nu(\frac{d}{2}+1-\frac{\eta}{2})}\right] \tag{6.4}$$

with S some unknown function of a single variable. The "magnetization" $\langle\phi\rangle \equiv M$ is defined by the condition that

$$M = -\,\theta\,\frac{\partial}{\partial h}\,W(h) \propto (\theta-\theta_c)^{\nu(\frac{d}{2}-1+\frac{\eta}{2})}\, S'\left[h(\theta-\theta_c)^{-\nu(\frac{d}{2}+1-\frac{\eta}{2})}\right] \tag{6.5}$$

In particular, it may be possible to have a spontaneous magnetization: $M \neq 0$ for $h = 0$. In this case, Eq. (6.5) gives

$$M \propto (\theta-\theta_c)^{\beta} \tag{6.6}$$

where β is another critical exponent

$$\beta = \nu\left(\frac{d}{2} - 1 + \frac{\eta}{2}\right) \tag{6.7}$$

More generally, the value of the external field h required to produce any given magnetization is given by Eq. (6.5) as

$$h = (\theta-\theta_c)^{\nu(\frac{d}{2}+1-\frac{\eta}{2})} \times \text{function of } M(\theta-\theta_c)^{-\beta} \tag{6.8}$$

or equivalently

$$h = M^{\delta} \times \text{function of } M(\theta-\theta_c)^{-\beta} \tag{6.9}$$

where δ is one more (the last!) critical index

$$\delta = \nu(d/2 + 1 - \eta/2)/\beta = \frac{(d/2 + 1 - \eta/2)}{(d/2 - 1 + \eta/2)} \tag{6.10}$$

In field theory, it is very convenient to work with a potential, defined like the Gibbs free energy as a Legendre transform

$$V(M) \equiv W(h) - hM \tag{6.11}$$

From the above results, it is easy to see that

$$V(M) \propto M^{\delta+1} \times \text{function of } M(\theta-\theta_c)^{-\beta} \tag{6.12}$$

7. INVARIANCE OF THE EIGENVALUES: CUT-OFFS VS. RENORMALIZATION

The discussion in Sections 5 and 6 has shown that critical phenomena are to a great extent governed by the number and the values of the repulsive eigenvalues of the matrix (5.9). This naturally raises the question whether the eigenvalues are invariant to possible re-definitions of the coupling parameters or the cut-off procedure. In particular, would we obtain the same eigenvalues if we assumed the theory was renormalizable, and used a floating renormalization point (as in the original work of Gell-Mann and Low[5]) instead of a floating cut-off?

Suppose we introduce a new set of dimensionless coupling parameters $\tilde{g}_a(\tilde{\Lambda})$. They are defined in terms of a dimensional quantity $\tilde{\Lambda}$, which may be a new kind of cut-off (perhaps smoother than a θ-function) or a floating renormalization point. If the theory is non-renormalizable, or if $\tilde{\Lambda}$ is some sort of cut-off, there are as many $\tilde{g}_a$ as g_i. If the theory is renormalizable and $\tilde{\Lambda}$ is a floating renormalization point there are only a finite number of $\tilde{g}_a$ parameters, and we are restricted to a finite-dimensional surface of trajectories.

The old dimensionless coupling parameters g_i can only be functions of the new parameters $\tilde{g}_a$ and of the dimensionless ratio $\Lambda/\tilde{\Lambda}$:

$$g_i(\Lambda) = g_i[\tilde{g}(\tilde{\Lambda}), \tilde{\Lambda}/\Lambda] \tag{7.1}$$

Further, the value of the old coupling parameters cannot depend on how we choose the new quantity $\tilde{\Lambda}$, so

$$0 = \tilde{\Lambda}\frac{d}{d\tilde{\Lambda}}\, g(\Lambda) = \sum_a \frac{\partial g_i}{\partial \tilde{g}_a}\tilde{\beta}_a + \frac{\tilde{\Lambda}}{\Lambda}\frac{\partial g_i}{\partial(\tilde{\Lambda}/\Lambda)} \tag{7.2}$$

where $\tilde{\beta}$ is the new beta function

$$\tilde{\beta}_a(\tilde{g}) \equiv \tilde{\Lambda}\,\frac{d\tilde{g}_a(\tilde{\Lambda})}{d\tilde{\Lambda}} \tag{7.3}$$

The old beta function is then

$$\beta_i \equiv \Lambda\,\frac{dg_i}{d\Lambda} = -\,\frac{\partial g_i}{\partial(\tilde{\Lambda}/\Lambda)}\,\frac{\tilde{\Lambda}}{\Lambda}$$

or using (7.2)

$$\beta_i(g) = \sum_a \frac{\partial g_i}{\partial \tilde{g}_a}\,\tilde{\beta}_a(\tilde{g}) \tag{7.4}$$

That is, β transforms like a covariant vector in the space of coordinate parameters. One immediate consequence is the invariance of fixed points: if $\tilde{\beta}$ vanishes at $\tilde{g}_a = \tilde{g}_a^*$, then β vanishes at the corresponding point $g_i = g_i^*$. Now, how does the matrix (5.9) transform? From (7.4), we have immediately

$$\sum_j \frac{\partial \beta_i}{\partial g_j}\,\frac{\partial g_j}{\partial \tilde{g}_b} = \sum_a \left[\frac{\partial^2 g_i}{\partial \tilde{g}_a \partial \tilde{g}_b}\,\tilde{\beta}_a + \frac{\partial g_i}{\partial \tilde{g}_a}\,\frac{\partial \tilde{\beta}_a}{\partial \tilde{g}_b}\right] \tag{7.5}$$

This is moderately complicated (derivatives of vectors are not generally tensors) but it simplifies at a fixed point, where the first term on the right vanishes. At such a point, (7.5) reads

$$\sum_j M_{ij}\,S_{jb} = \sum_a S_{ia}\,\tilde{M}_{ab} \tag{7.6}$$

where

$$M_{ij} \equiv \Big[\partial\beta_i(g)/\partial g_j\Big]_{g=g^*} \tag{7.7}$$

$$\tilde{M}_{ab} \equiv \Big[\partial\tilde{\beta}_a(\tilde{g})/\partial\tilde{g}_b\Big]_{\tilde{g}=\tilde{g}^*} \tag{7.8}$$

$$S_{ia} = \Big[\partial g_i/\partial\tilde{g}_a\Big]_{\tilde{g}=\tilde{g}^*} \tag{7.9}$$

It follows then that if $\tilde{e}$ is an eigenvector of $\tilde{M}$ with eigenvalue λ,

$$\sum_b \tilde{M}_{ab}\,\tilde{e}_b = \lambda\,\tilde{e}_a \tag{7.10}$$

then there is a corresponding eigenvector of M

$$e_i = \sum_a S_{ia} \tilde{e}_a \tag{7.11}$$

with the same eigenvalue

$$\sum_j M_{ij} e_j = \lambda e_i \tag{7.12}$$

The eigenvalues are therefore independent of the cut-off procedure. Also, if we constrained the theory to be renormalizable, and calculated $\tilde{\beta}$ using a floating renormalization point instead of a floating cut-off, the eigenvalues we would obtain would be some subset of the eigenvalues obtained with a cut-off. The use of renormalizable field theories in statistical mechanics does not give all the eigenvalues of M, but those it gives, it gives correctly. As it happens, one of the eigenvalues of $\tilde{M}$ in a renormalizable ϕ^4 theory is the repulsive eigenvalue λ_R, so renormalizable theories can be used to calculate the critical exponent ν, as done by Brézin *et al.*[1]

8. THE ONE-LOOP EQUATIONS

It is remarkable that the functions $\beta_i(g)$ can be calculated exactly in terms of one-loop diagrams. The derivation is lengthy, and will be outlined in an Appendix. Here, we just give the results.

The one-loop equations are written in terms of a modified rescaled coupling function $\bar{g}_n(\vec{\ell}_1 \ldots \vec{\ell}_n;\Lambda)$. [These are not precisely the same as the coupling functions discussed up to now - the relation is explained in the Appendix. This modification will not be important in using the one-loop equations to study critical phenomena.] Let $-(2\pi)^d \bar{L}_n(\vec{\ell}_1 \ldots \vec{\ell}_n;\mu,\Lambda)$ be the sum of all one-loop, one-particle irreducible graphs, with n external lines carrying outgoing rescaled momenta $\vec{\ell}_1 \ldots \vec{\ell}_n$. These graphs are constructed with vertices

$$-(2\pi)^d \, \delta^d(\vec{\ell}_1+\ldots+\vec{\ell}_n)\bar{g}_n(\vec{\ell}_1\ldots\vec{\ell}_n;\Lambda) \tag{8.1}$$

and propagators with an <u>infrared</u> cut-off

$$(2\pi)^{-d} \, \theta(|\vec{\ell}|-\mu)\Delta(\vec{\ell};\Lambda) \tag{8.2}$$

where

$$\Delta(\vec{\ell};\Lambda) \equiv \bar{g}_2^{-1}(\vec{\ell},-\vec{\ell};\Lambda) \tag{8.3}$$

Propagators are not included on the external lines of $\bar{L}_n$, and a delta-function $\delta(\vec{\ell}_1+\ldots+\vec{\ell}_n)$ is factored out, as indicated by the bar over L. The one-loop equations take the form

$$\Lambda\frac{d}{d\Lambda}\,\bar{g}_n(\vec{\ell}_1\ldots\vec{\ell}_{n-1},-\vec{\ell}_1-\ldots-\vec{\ell}_{n-1};\Lambda)$$

$$= \left[\frac{nd}{2}-d-n+\frac{n\gamma}{2}+\sum_{r=1}^{n-1}\vec{\ell}_r\cdot\frac{\partial}{\partial\vec{\ell}_r}\right]\bar{g}_n(\vec{\ell}_1\ldots;\Lambda)$$

$$+\left[\frac{\partial}{\partial\mu}\,\bar{L}_n(\vec{\ell}_1\ldots\vec{\ell}_{n-1},-\vec{\ell}_1-\ldots-\vec{\ell}_{n-1};\mu,\Lambda)\right]_{\mu=1} \qquad (8.4)$$

For instance, let us suppose that our theory is invariant under a transformation $\phi \to -\phi$, so that all odd couplings are eliminated. For $n = 2$ and $n = 4$, the one-loop integrals are

$$-(2\pi)^d\;\bar{L}_2(\vec{\ell}_1,-\vec{\ell}_1;\mu,\Lambda) \equiv -\tfrac{1}{2}\int d^d\ell\;\theta(|\vec{\ell}|-\mu)\Delta(\vec{\ell};\Lambda)\bar{g}_4(\vec{\ell},-\vec{\ell},\vec{\ell}_1,-\vec{\ell}_1;\Lambda) \qquad (8.5)$$

and

$$-(2\pi)^d\;\bar{L}_4(\vec{\ell}_1\ldots\vec{\ell}_4;\mu,\Lambda) \equiv -\tfrac{1}{2}\int d^d\ell\;\theta(|\vec{\ell}|-\mu)\Delta(\vec{\ell};\Lambda)\bar{g}_6(\vec{\ell},-\vec{\ell},\vec{\ell}_1,\ldots\vec{\ell}_4;\Lambda)$$

$$+\tfrac{1}{2}\int d^d\ell\,d^d\ell'\;\theta(|\vec{\ell}|-\mu)\theta(|\vec{\ell}|-\mu')\Delta(\vec{\ell};\Lambda)\Delta(\vec{\ell}';\Lambda)$$

$$\times[\bar{g}_4(\vec{\ell},-\vec{\ell}',\vec{\ell}_1,\vec{\ell}_2;\Lambda)\bar{g}_4(\vec{\ell}',-\vec{\ell},\vec{\ell}_3,\vec{\ell}_4;\Lambda)\,\delta(\vec{\ell}_1+\vec{\ell}_2+\vec{\ell}-\vec{\ell}')$$

$$+\;2\text{ permutations}] \qquad (8.6)$$

where $\vec{\ell}_4 \equiv -\vec{\ell}_1-\vec{\ell}_2-\vec{\ell}_3$. (See Figure 2.) In consequence, the one-loop equations for $\bar{g}_2 \equiv \Delta^{-1}$ and $\bar{g}_4$ take the form

$$\Lambda\frac{d}{d\Lambda}\Delta^{-1}(\vec{\ell}_1;\Lambda) = \left[\gamma-2+\vec{\ell}_1\cdot\frac{\partial}{\partial\vec{\ell}_1}\right]\Delta^{-1}(\vec{\ell}_1;\Lambda)$$

$$-\tfrac{1}{2}(2\pi)^{-d}\int d^d\ell\;\delta(|\vec{\ell}|-1)\Delta(\vec{\ell};\Lambda)\bar{g}_4(\vec{\ell},-\vec{\ell},\vec{\ell}_1,-\vec{\ell}_1;\Lambda) \qquad (8.7)$$

and

$$\Lambda\frac{d}{d\Lambda}\bar{g}_4(\vec{\ell}_1\ldots\vec{\ell}_4;\Lambda) = \left[d-4+2\gamma+\sum_{r=1}^{3}\ell_r\cdot\frac{\partial}{\partial\vec{\ell}_r}\right]\bar{g}_4(\vec{\ell}_1,\ldots,\vec{\ell}_4;\Lambda)$$

$$-\tfrac{1}{2}(2\pi)^{-d}\int d^d\ell\,\delta(|\vec{\ell}|-1)\Delta(\vec{\ell};\Lambda)\bar{g}_6(\vec{\ell},-\vec{\ell},\vec{\ell}_1\ldots\vec{\ell}_4;\Lambda)$$

$$+\tfrac{1}{2}(2\pi)^{-d}\int d^d\ell\,d^d\ell'\{\delta(|\vec{\ell}|-1)\theta(|\vec{\ell}'|-1)+\delta(|\vec{\ell}'|-1)\theta(|\vec{\ell}|-1)\}$$

$$\times\Delta(\vec{\ell};\Lambda)\Delta(\vec{\ell}';\Lambda)\Big[\bar{g}_4(\vec{\ell},-\vec{\ell}',\vec{\ell}_1,\vec{\ell}_2;\Lambda)\bar{g}_4(\vec{\ell}',-\vec{\ell},\vec{\ell}_3,\vec{\ell}_4)\,\delta^d(\vec{\ell}-\vec{\ell}'+\vec{\ell}_1+\vec{\ell}_2)$$

$$+\bar{g}_4(\vec{\ell},-\vec{\ell}',\vec{\ell}_1,\vec{\ell}_3;\Lambda)\bar{g}_4(\vec{\ell}',-\vec{\ell},\vec{\ell}_2,\vec{\ell}_4)\,\delta^d(\vec{\ell}-\vec{\ell}'+\vec{\ell}_1+\vec{\ell}_3)$$

$$+\bar{g}_4(\vec{\ell},-\vec{\ell}',\vec{\ell}_1,\vec{\ell}_4;\Lambda)\bar{g}\;(\vec{\ell}',-\vec{\ell},\vec{\ell}_2,\vec{\ell}_3)\,\delta^d(\vec{\ell}-\vec{\ell}'+\vec{\ell}_1+\vec{\ell}_4)\Big] \qquad (8.8)$$

Since the fields are normalized so that the coefficient of $\vec{\ell}_1^2$ in the

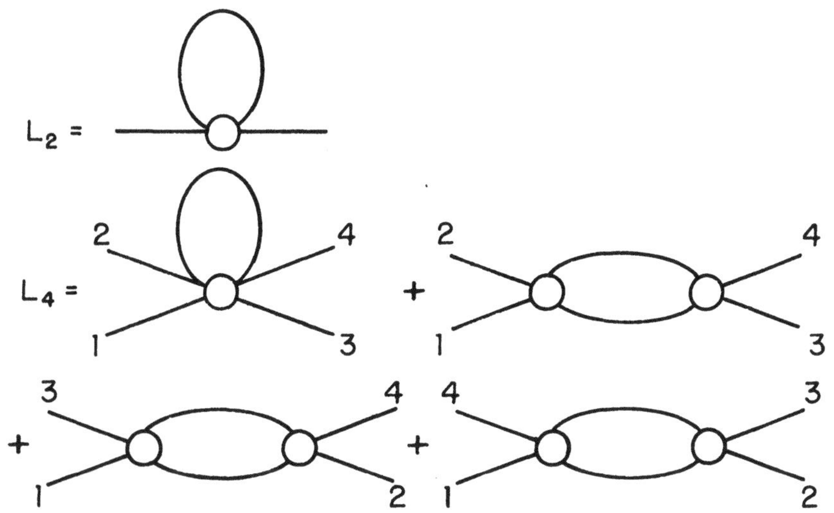

Figure 2. Graphs for the one-loop functions L_2 and L_4. Here internal lines represent the cut-off propagators (8.2), while the small circles represent the interactions g_n.

power series expansion of $\Delta^{-1}(\vec{\ell}_1;\Lambda)$ is unity, Eq. (8.7) yields a formula for γ:

$$\gamma = + \tfrac{1}{2}(2\pi)^{-d}\left[\frac{d^2}{d\vec{\ell}_1^{\,2}}\right]_{\vec{\ell}_1=0} \int d^d \ell \, \delta(|\vec{\ell}| - 1)\Delta(\vec{\ell};\Lambda)\overline{g}_4(\vec{\ell},-\vec{\ell},\vec{\ell}_1,-\vec{\ell}_1;\Lambda) \tag{8.9}$$

Note that the integrals in all these equations are taken over a closed d-1-dimensional surface, and are therefore automatically finite.

These equations are all exact. The "mistake" we make in dropping graphs with more than one loop is cancelled by the "mistake" we make in differentiating only the θ's, not the $\overline{g}$'s. However, the fact that we have exact equations in closed form does not mean that we can derive an exact solution. The equation for $\overline{g}_2$ involves $\overline{g}_4$; the equation for $\overline{g}_4$ involves $\overline{g}_6$; and so on. Only by using some sort of perturbation theory can we get useful results.

9. THE GAUSSIAN FIXED POINT

A fixed point is characterized by a set of coupling functions Δ^*; $\bar{g}_4^*$, $\bar{g}_6^*$,... for which the right-hand side of Eq. (8.4) vanishes. There is one obvious such fixed point, with

$$\Delta^{*-1}(\vec{\ell}) = \vec{\ell}^2$$

$$\bar{g}_n^*(\vec{\ell}_1 \ldots \vec{\ell}_n) = 0 \qquad \text{(for } n > 2) \qquad (9.1)$$

[Equation (8.9) gives $\gamma = 0$.] This is known as the Gaussian fixed point, because it corresponds to a free massless field theory, for which field distributions are Gaussians. The Gaussian fixed point turns out to be relatively unimportant for statistical mechanics, but the reasons why this is so are enlightening. Also, the Gaussian fixed point will play an important part for us in our discussion of renormalizability in Section 12.

To find the eigenvalues of the M-matrix at the Gaussian fixed point, we note that there is a simple basis in which the matrix is triangular. (That is, the matrix has only zero elements above the main diagonal.) Observe that there is no term in $\Lambda d\bar{g}_n/d\Lambda$ linear in $\bar{g}_m$ for $m < n$, while the term linear in $\bar{g}_n$ is simply

$$\left[\frac{nd}{2} - d - n + \sum_{r=1}^{n-1} \vec{\ell}_r \cdot \frac{\partial}{\partial \vec{\ell}_r}\right] \bar{g}_n \qquad (9.2)$$

Hence M is triangular in a basis in which the vectors correspond to interactions i with definite numbers n_i of fields and N_i of momentum-factors. The eigenvalues of a triangular matrix are just its diagonal elements, so (9.2) shows that the eigenvalues at the Gaussian fixed point are the numbers

$$D_i = \frac{n_i d}{2} - d - n_i + N_i \qquad (9.3)$$

We recognize that D_i is just the "dimensionality" of renormalization theory: an interaction is just super-renormalizable if $D < 0$, just renormalizable if $D = 0$, and not renormalizable if $D > 0$.

For $d > 2$, the value of D_i increases as either n_i or N_i is increased. Here are the eigenvalues for the interactions with the lowest values of n_i and N_i:

Interaction	n	N	D
ϕ^2	2	0	-2
$\phi\Box^2\phi$	2	2	0
$\phi\Box^4\phi, (\Box^2\phi)^2$	2	4	2
- - -			
ϕ^4	4	0	d-4
$\phi^3\Box^2\phi$	4	2	d-2
- - -			
ϕ^6	6	0	2d-6
- - -			

We see that for d = 3, there are two repulsive eigenvectors with $D < 0$, corresponding to the super-renormalizable interactions ϕ^2 and ϕ^4, and one marginal eigenvector with $D = 0$, corresponding to the renormalizable interaction ϕ^6. (We do not include $\phi\Box^2\phi$ as an "interaction", because its coefficient is fixed by our field renormalization convention.) Even for $3 < d < 4$, there are two repulsive eigenvectors of M. Hence in order for the Hamiltonian to approach the Gaussian fixed point as $\Lambda \to 0$ it would be necessary to adjust at least two free parameters in the Hamiltonian to eliminate the components along both repulsive directions, not just the temperature. This is what is called a tricritical fixed point.

Tricritical fixed points do occur in nature, as for instance in antiferromagnets and He^3-He^4 mixtures. However, the second-order phase transitions considered here (as in ferromagnets at zero field, or water at the critical pressure) occur at a critical value of just one parameter, the temperature, so they have nothing to do with a tricritical fixed point, like the Gaussian fixed point. We must look for some other kind of fixed point.

10. THE WILSON-FISHER FIXED POINT

We have already noted that the renormalization-group equations cannot generally be solved without some sort of perturbative expansion. However, perturbation theory would not generally be expected to work at all near any fixed point except the Gaussian fixed point. The β-function for an interaction i involving n_i

factors of ϕ and N_i factors of momentum is given by Eq. (8.4) as

$$\beta_i = D_i g_i + \text{loop terms} \tag{10.1}$$

where

$$D_i = \frac{n_i d}{2} - d - n_i + N_i \tag{10.2}$$

and the "loop terms" arise from the γ and L terms in (8.4). At a fixed point g^*, all β_i must vanish, so the first-order terms $D_i g_i^*$ must cancel the higher-order loop terms. In general, this would not be possible for small non-zero values of the g_i^*.

However, suppose that one of the g_i, say g_I, has a very small dimensionality D_I, while $D_{II}, D_{III}, \ldots$ are all of order unity or greater. If we tentatively suppose that g_I is small, then we can find functions $g_{II}^*(g_I)$, $g_{III}^*(g_I)$, ... as power series solutions of the equations

$$0 = \beta_{II}(g_I, g_{II}^*, g_{III}^*, \ldots) = \beta_{III}(g_I, g_{II}^*, g_{III}^*, \ldots) = \ldots \tag{10.3}$$

These power series can then be inserted in the "loop terms" for β_I, generating a series in powers of g_I:

$$\beta_I(g_I, g_{II}^*(g_I), g_{III}^*(g_I), \ldots) = D_I g_I + a g_I^2 + b g_I^3 + \ldots \tag{10.4}$$

For small D_I, the equation $\beta_I = 0$ has a perturbative solution

$$g_I^* = -\frac{D_I}{a} + \frac{b D_I^3}{a^3} - \ldots \tag{10.5}$$

Note incidentally that this solution for g^* may be useful even if D_I is not so small, provided that the coefficient a is sufficiently large; this is what actually seems to make the method work in practice.

Of course, for any integral number d of spatial dimensions, each dimensionality (10.2) is an integer or half-integer. Therefore, in order to use the above expansions, we have work in a mythical world, in which d is nearly but not quite equal to a whole number.

To see how this works in practice, we return again to our standard example, of a single scalar field ϕ, with a symmetry under the transformation $\phi \to -\phi$ to eliminate the interactions odd in ϕ. Inspection of the table in Section 9 shows that when d is near 4, the dimensionality of the ϕ^4 interaction is small, while all other interactions have dimensionalities of order unity or greater. (We exclude $\phi \Box^2 \phi$ from the list of interactions, because its coupling is fixed by our field-renormalization convention to have the value unity.) Since we are really interested here in $d = 3$, we will therefore set

$$d = 4 - \varepsilon, \qquad 0 < \varepsilon \ll 1 \tag{10.6}$$

and hope that the expansion in powers of ε gives good results even at $\varepsilon = 1$.

As we have seen, the first step in finding the fixed point would be to express all other couplings in terms of the coupling g_I of the ϕ^4 interaction. It is easy to see that, apart from the ϕ^4 (and $\phi\Box^2\phi$) interaction, the coupling parameter $g_i^*(g_I)$ of any interaction with n_i factors of the field ϕ will have a power series expansion in g_I which begins with a term of order

$$g_i^*(g_I) \sim g_I^{n_i/2} \tag{10.7}$$

To check this, we suppose that it is true, and see if it gives consistent results at a fixed point. Equation (10.7) shows that $g_2(\vec{p},\vec{p}\,')$ is of order g_I, except for the zeroth order term $\vec{p}^{\,2}\delta(\vec{p}+\vec{p}\,')$ which is fixed by our field-normalization conventions to have coefficient unity. Hence the propagator takes the form

$$\Delta(\vec{\ell}) \simeq \frac{1}{\vec{\ell}^{\,2}} + O(g_I) \tag{10.8}$$

Also, Eq. (10.7) shows that $\overline{g}_4(\vec{\ell}_1\vec{\ell}_2\vec{\ell}_3\vec{\ell}_4)$ is of order g_I^2, except for the $\vec{\ell}$-independent term, which is of course equal to g_I by definition. A one-loop graph in β_i that is built entirely out of ϕ^4 couplings will have $n_i/2$ vertices, and hence be of order $g_I^{n_i/2}$, in agreement with the order of the term $D_i g_i$ in β_i as given by Eq. (10.7). These terms therefore can cancel at a fixed point satisfying (10.7). (About the γ term, see below.) On the other hand, if we construct the loop from r ϕ^4 vertices and s other vertices with $m_1, m_2, \ldots, m_s$ external lines apiece, then we must have

$$n_i = m_1 + m_2 + \ldots + m_s - 2s + 2r$$

so the number of powers of g_I in the graph will be

$$\tfrac{1}{2}(m_1 + m_2 + \ldots + m_s) + r = \tfrac{1}{2}n_i + s$$

which is greater than $\frac{1}{2}n_i$. This incidentally shows that in lowest order, the one-loop graphs for β_i can be calculated using ϕ^4 couplings only, a fact that will prove useful later.

Finally, we must give special attention to the $\phi\Box^2\phi$ "interaction". This has $D = 0$ (for all d) and accordingly Eq. (8.9) gives the fixed-point condition for this coupling as

$$\gamma^*(g_I) = O(g_I^2)$$

(The loop terms here are of order g_I^2 rather than g_I, because the one-loop graph constructed from a ϕ^4 interaction is just a constant; it therefore contributes to β for the ϕ^2 interaction, but not to the $\phi\Box^2\phi$ or higher-derivative interactions.) The fact that γ^* is of order g_I^2 justifies the neglect of γ in the estimates made above. Note that if we had not renormalized our fields, we would not have the term γ available to cancel the one-loop terms in the β-function for the $\phi\Box^2\phi$ interaction, and in consequence there could not be a fixed point of the type considered here. This is the whole motivation for field-renormalization in statistical mechanics - to make a fixed point possible.

Now we are in a position to do some actual calculations. From Eq. (8.8), we see that the β-function for the ϕ^4 interaction is exactly given by

$$\begin{aligned}\beta_I &= [2\gamma-\varepsilon]g_I \\ &-\tfrac{1}{2}(2\pi)^{-d}\int d^d\ell\,\delta(|\vec{\ell}|-1)\Delta(\vec{\ell};\Lambda)\overline{g}_6(\vec{\ell},-\vec{\ell},0,0,0,0;\Lambda) \\ &+\tfrac{3}{2}(2\pi)^{-d}\int d^d\ell\,\delta(|\vec{\ell}|-1)\Delta^2(\vec{\ell};\Lambda)\overline{g}_4^2(\vec{\ell},-\vec{\ell},0,0;\Lambda)\end{aligned} \tag{10.9}$$

[The two terms in the curly brackets in (8.8) add up to just $\delta(|\vec{\ell}|-1)$. The factor 3 arises because there are three equal terms in the last integral in (8.8).] We have seen that when the fixed-point equations are used to express all the other couplings in terms of g_I, it turns out that

$$\gamma = O(g_I^2)$$

$$\overline{g}_6 = O(g_I^3)$$

$$\overline{g}_4 = g_I + O(g_I^2)$$

$$\Delta^{-1} = \vec{\ell}^2 + O(g_I)$$

Therefore in lowest order, we can simply drop γ and $\overline{g}_6$, set $\Delta = 1/\vec{\ell}^2$, and replace $\overline{g}_4$ with the constant g_I, so that (10.9) gives

$$\beta_I(g_I) = -\varepsilon g_I + \tfrac{3}{2}(2\pi)^{-d}S_d\, g_I^2 + O(g_I^3) \tag{10.10}$$

where S_d is the area of a unit sphere in d dimensions. Solving the equation $\beta_I = 0$ then gives the Wilson-Fisher fixed point

$$g_I^* = + \frac{2\varepsilon}{3} S_4^{-1} (2\pi)^{+4} + O(\varepsilon^2) \tag{10.11}$$

11. CALCULATION OF THE EIGENVALUES

We have found a fixed point in 4-ε dimensions with couplings g^* of order ε or less. Now we shall use perturbation theory to calculate the eigenvalues of the matrix $\partial \beta_i / \partial g_j$ at this fixed point.

From Eq. (10.1), we see that

$$M_{ij} \equiv \left(\partial \beta_i / \partial g_j \right)_{g=g^*} = D_i \, \delta_{ij} + \Delta M_{ij} \tag{11.1}$$

with ΔM_{ij} of order ε or less. In zeroth order, the jth component of the ith eigenvector of M is just δ_{ij}, so first-order perturbation theory gives the eigenvectors of M as the "expectation value"

$$\lambda_i \simeq D_i + \Delta M_{ii} = \left(\frac{\partial \beta_i}{\partial g_i} \right)_{g=g^*} \tag{11.2}$$

to order ε.

It is convenient to consider three special cases:

(a) Any interaction with $D_i \ll -\varepsilon < 0$ corresponds to a negative eigenvalue of M, and hence to an infrared-repulsive eigenvector. In our standard example of a single real scalar field with a $\phi \to -\phi$ symmetry, there is just one of these eigenvectors, corresponding to the ϕ^2 interaction, with D = -2.

(b) Any interaction with $D_i \gg \varepsilon > 0$ corresponds to a positive eigenvalue of M, and hence to an infrared-attractive eigenvector. There are an infinite number of these.

(c) Any interaction with $|D_i|$ of order ε corresponds to a borderline eigenvalue of M, which might be positive or negative, depending on the value of the interaction term ΔM. In our standard example, there is just one of these, the interaction ϕ^4, with $D_I = -\varepsilon$. (Recall that $\phi \Box^2 \phi$ is not counted as an interaction.) This is repulsive at the Gaussian fixed point, but we cannot tell whether it is repulsive or attractive at the Wilson-Fisher fixed point without taking interactions into account. From Eq. (10.10), we have

$$\lambda_I = \left(\frac{\partial \beta_I}{\partial g_I} \right)_{g^*} = -\varepsilon + 3(2\pi)^{-d} S_d \, g_I^* + O(g_I^{*2})$$

or, using (10.11),

$$\lambda_I = +\varepsilon + O(\varepsilon^2) \tag{11.3}$$

We see that this eigenvector is infrared-attractive, not repulsive, so there is just one repulsive eigenvector, as required for an ordinary second-order phase transition. Now let us calculate the critical index ν to first order in ε. As shown in Section 5,

$$\nu = -1/\lambda_{II} \tag{11.4}$$

where λ_{II} is taken as the single repulsive eigenvalue of M. This eigenvalue is just the one mentioned in case (a) above, corresponding to the interaction ϕ^2. The β-function for this interaction is given exactly by the one-loop equation (8.7):

$$\beta_{II} = (-2+\gamma)g_{II} - \tfrac{1}{2}(2\pi)^{-d}\int d^d \ell\, \delta(|\vec{\ell}|-1)\Delta(\vec{\ell};\Lambda)\bar{g}_4(\vec{\ell},-\vec{\ell},0,0;\Lambda) \tag{11.5}$$

The coupling g_{II} is defined as the constant term in the inverse propagator Δ^{-1}, so that

$$\frac{\partial\beta_{II}}{\partial g_{II}} = -2 + \gamma + \tfrac{1}{2}(2\pi)^{-d}\int d^d \ell\, \delta(|\vec{\ell}|-1)\Delta^2(\vec{\ell};\Lambda)\bar{g}_4(\vec{\ell},-\vec{\ell},0,0;\Lambda)$$

So far, this is exact.

Now let us use the ε-expansion. At the fixed point, γ is of order ε^2; $\bar{g}_4$ equals g_I^* plus terms of order ε^2; Δ equals $1/\ell^2$ plus terms of order ε; so that to order ε,

$$\lambda_{II} = \left(\frac{\partial\beta_{II}}{\partial g_{II}}\right)_{g=g^*} = -2 + \tfrac{1}{2}(2\pi)^{-d} S_d\, g_I^*$$

or, using (10.11)

$$\lambda_{II} = -2 + \frac{\varepsilon}{3} + O(\varepsilon^2) \tag{11.6}$$

The critical index (11.4) is then

$$\nu = \frac{1}{2} + \frac{\varepsilon}{12} + O(\varepsilon^2) \tag{11.7}$$

Note the factor 1/12; this makes the correction to the lowest-order value of ν rather small even for $\varepsilon = 1$. If we neglect higher-order terms, then (10.7) gives $\nu = 0.5875$ for $\varepsilon = 1$; the experimental value is in the neighborhood of 0.6 to 0.7.

12. RENORMALIZABILITY AND ALTERNATIVES

As I indicated at the beginning, the purpose of these lectures is to present the theory of critical phenomena to field theorists who want to apply this machinery in quantum field theory. We now turn to one of these applications. From now on, our efforts are directed at phenomena in $d = 4$ dimensions, though we will again have to consider formal variations in dimensionality. However, we will continue to work in a Euclidean rather than a Minkowskian space, leaving the analytic continuation to the physical region to be dealt with separately.

A good deal of modern elementary particle theory is based on the assumption that nature is described by a renormalizable quantum field theory. However, the "floating cut-off" formalism described in these lectures raises serious questions about the physical significance of the renormalizability requirement. In the formalism described here, the effective Hamiltonian depends on the cut-off in such a way that the physics (i.e., the set of Euclidean Green's functions) is cut-off-independent for all theories, not only renormalizable theories. In this formalism, a renormalizable theory merely corresponds to a subset of trajectories (characterized by a few renormalized coupling constants) for which all but a few of the couplings vanish (at least in perturbation theory) as $\Lambda \to \infty$. Why should the real world correspond to such trajectories?

It might be argued that renormalizable field theories are distinguished because they have only a finite number of free parameters. This seems like a rather unphysical requirement. We can eliminate all free parameters by demanding that at $\Lambda = 1$ MeV all dimensionless coupling constants g_i have the value 37. What is needed is not uniqueness itself, but a rationale for uniqueness.

Nor is experiment much help here. Non-renormalizable quantum field theories always involve a mass scale, such as $F_\pi \simeq 190$ MeV for chiral dynamics; $G_F^{-1/2} \simeq 300$ GeV for the Fermi theory of weak interaction; and $G_{NEWTON}^{-1/2} \simeq 2 \cdot 10^{19}$ GeV for general relativity. At energies which are much smaller than the characteristic mass scale, a non-renormalizable theory will look as if it were renormalizable. (If symmetries do not allow any renormalizable interactions, it will look like a free field theory.) The experimental success of quantum electrodynamics only shows that any non-renormalizable interactions have a characteristic scale larger than a few GeV. The success (so far) of renormalizable gauge theories of weak and electromagnetic interactions only indicates that any non-renormalizable interactions have a characteristic scale greater than 300 GeV. We need theoretical guidance to tell us whether physics will continue to look renormalizable at really high energies, like 10^{19} GeV. And if not, then what does pick out the trajectory corresponding to the real world from the infinite number of possible theories?

I know of only one promising approach to this problem. Randomly chosen quantum field theories tend to develop unphysical singularities if extended to sufficiently high energies. (In the Euclidean region, any singularity is unphysical.) For instance, suppose some coupling constant obeys the very simple renormalization-group equation

$$\Lambda \frac{dg(\Lambda)}{d\Lambda} = a\, g(\Lambda)^2 \qquad (a > 0)$$

The solution is, for arbitrary κ,

$$g(\kappa\Lambda) = \frac{g(\Lambda)}{1 - a\, g(\Lambda) \ln \kappa}$$

For $g(\Lambda) > 0$, this develops a singularity at $\ln \kappa = 1/a\, g(\Lambda)$, and Eq. (4.27) then suggests that some singularity occurs in Green's functions at finite Euclidean momenta. Nature must pick out trajectories which avoid singularities of this type.

One way to satisfy this requirement is to demand that the Hamiltonian lie on a trajectory which hits a fixed point for $\Lambda \to \infty$. Equation (4.27) shows that in this case, the Green's functions simply behave like powers of κ when the momentum scale κ goes to infinity, and do not develop unphysical singularities. [In particular, this is the case for our example above if we start with a negative coupling, $g(\Lambda) < 0$.] Theories with this property, that the trajectory hit a fixed point for $\Lambda \to \infty$, will be called asymptotically safe.

Of course, in order to be asymptotically safe, the trajectory near the fixed point must have no components along eigenvectors of the M-matrix, that are ultraviolet-repulsive. (Since Λ now goes to infinity instead of zero, ultraviolet-repulsive eigenvectors are those with positive eigenvalues.) But in all cases that have been studied, it turns out that there are only a finite number of ultraviolet-attractive eigenvectors (i.e., negative eigenvalues), and all the rest are ultraviolet-repulsive. Therefore, the demand that a theory be asymptotically safe imposes an infinite number of constraints on the coupling parameters, leaving only a finite number of free parameters, i.e., the components of the trajectory at the fixed point along the finite number of ultraviolet-attractive eigenvectors. Thus, asymptotic safety can provide a rationale for picking physically acceptable quantum field theories, which may either explain renormalizability, or else replace it.

To see how this works in practice, let us consider the theory of a single real scalar field in five dimensions. Let us ask whether or not it is possible for such a theory to achieve asymptotic safety by hitting the Gaussian fixed point. As we saw in Section 9, the eigenvalues at the Gaussian fixed point are simply equal to the dimensionalities of the various interactions: for each interaction with n ϕ-factors and N momentum factors, there is an eigenvalue given by (9.3), which for d = 5 reads

$$D = -5 + \frac{3}{2} n + N$$

If we do not impose the symmetry under $\phi \to -\phi$, then there are 2 ultraviolet-attractive eigenvectors, corresponding to the super-renormalizable interactions ϕ^2, with $D = -2$, and ϕ^3, with $D = -\frac{1}{2}$. (As usual, we do not include $\phi\Box^2\phi$ as an interaction, and we do not include a ϕ interaction because such an interaction can always be eliminated by shifting ϕ by a constant.) Thus there is a two-parameter set of trajectories which hit the Gaussian fixed point, and are therefore asymptotically safe.

It is easy to see that this two-dimensional surface of trajectories simply corresponds to the super-renormalizable theories of a scalar field in five dimensions. (Working in these super-renormalizable theories, we can calculate the matrix of derivatives of the Gell-Mann - Low beta function[5] at the Gaussian fixed point, and check that the eigenvalues of this matrix are -2 and $-\frac{1}{2}$.) The two parameters needed to describe the trajectories which hit the Gaussian fixed point are just the renormalized mass and ϕ^3 coupling. Thus in this case, asymptotic safety is achieved by requiring renormalizability. More generally, the effective Hamiltonian will approach the Gaussian fixed point for $\Lambda \to \infty$ if it corresponds to a super-renormalizable theory or an asymptotically free renormalizable theory.

However, it is not clear that nature really does choose trajectories which hit the Gaussian fixed point for $\Lambda \to \infty$. The notorious problem is gravitation: no one has been able to think of a satisfactory theory of gravitation which is renormalizable. Is it possible that nature achieves asymptotic safety by aiming the trajectories of the effective Hamiltonian at some fixed point other than the Gaussian fixed point?

As an example of what is possible, let's return to our example of a scalar field in five dimensions, but let's now impose the symmetry under the transformation $\phi \to -\phi$. This plays a role here similar to that of general covariance in the theory of gravitation - it eliminates the only interaction ϕ^3 that is renormalizable. There still is one ultraviolet-attractive eigenvector, corresponding to the "interaction" ϕ^2, but a trajectory that reaches the Gaussian fixed point along this direction simply corresponds to a free field theory. An interacting asymptotically safe theory with this symmetry must hit some other fixed point.

What about the Wilson-Fisher fixed point? We are really interested (in this example) in five-dimensional space, but in order to do calculations, let's work in $4+\epsilon$ dimensions. The eigenvalues can then be calculated from the results of Sec. 11, by simply changing ϵ to $-\epsilon$. We see that there are two ultraviolet-attractive eigenvectors at the Wilson-Fisher point, with eigenvalues

$$\lambda_I = -\varepsilon + O(\varepsilon^2)$$

$$\lambda_{II} = -2 - \frac{\varepsilon}{3} + O(\varepsilon^2)$$

and an infinite number of ultraviolet-repulsive eigenvectors, with positive eigenvalues. If we assume that the eigenvalues do not change sign for $0 < \varepsilon \leq 1$, we can conclude that there are just two ultraviolet-attractive eigenvectors in five dimensions. With this assumption, there is a two-parameter set of asymptotically safe field theories in five dimensions associated with the Wilson-Fisher fixed point. These theories are <u>not</u> renormalizable in the usual sense - the symmetry under $\phi \to -\phi$ rules out the possibility of <u>any</u> renormalizable theory in five dimensions. However, they are interacting theories with no unphysical singularities at high energy.

(In fact, this theory does have unphysical features, but of a different kind. Changing ε to $-\varepsilon$ in Eq. (10.11) shows that the ϕ^4 coupling constant at the Wilson-Fisher fixed point is negative in $4+\varepsilon$ dimensions. This means that the potential (6.11) goes to $-\infty$ for $|\langle\phi\rangle| \to \infty$, so the energy is not bounded below. However, this is an accident of this particular model. Any theory will have non-trivial ultraviolet-attractive fixed points in $d+\varepsilon$ dimensions if it is renormalizable and asymptotically free in d dimensions. It is well known that asymptotic freedom can only be achieved for a ϕ^4 theory in four dimensions by giving the coupling constant an unphysical negative value. However, there are plenty of other asymptotically free renormalizable theories, such as chiral dynamics in two dimensions[6] and non-Abelian gauge theories in four dimensions, which are not plagued by negative-energy problems.)

The asymptotically safe theories of a scalar field in five dimensions are characterized by two free parameters: one dimensionless parameter picks out a particular trajectory in the two-dimensional surface of trajectories which hit the Wilson-Fisher fixed point, and one parameter with the dimensions of mass gives the value of the cut-off Λ at which any given point along this trajectory is reached. In order to perfect the analogy with general relativity, we can eliminate the first parameter by demanding that the theory have zero boson mass. This is done simply by assuming that the theory lies on a trajectory which not only hits the Wilson-Fisher fixed point for $\Lambda \to \infty$, but also hits the Gaussian fixed point for $\Lambda \to 0$. (To see that there is likely to be such a trajectory, note that in $d > 4$ dimensions the symmetry $\phi \to -\phi$ eliminates all infrared-repulsive eigenvectors at the Gaussian fixed point except for the one corresponding to the "interaction" ϕ^2. Hence it is only necessary to adjust one dimensionless parameter to hit the Gaussian fixed point for $\Lambda \to 0$, and we do have one free parameter at our disposal.) This theory is then described by a single free parameter, with the dimensions of a mass, just like general relativity.

Perhaps gravitation works this way. It may be that general covariance rules out any renormalizable theory of gravitation in four dimensions, just as the symmetry under $\phi \to -\phi$ rules out any renormalizable theory of a scalar field in 5 dimensions. Nevertheless, nature may achieve asymptotic safety anyway, by picking out trajectories which hit some fixed point other than the Gaussian fixed point for $\Lambda \to \infty$. For this to be possible, there must be some non-trivial fixed point with at least one ultraviolet-attractive eigenvector.

Unfortunately, it is not so easy to check that this is so. General relativity becomes formally renormalizable in $d = 2$ dimensions, so one might try to study the fixed points in four dimensions by working in $d = 2 + \varepsilon$ dimensions and expanding in ε. However, general relativity actually becomes a trivial theory in two dimensions (because $R_{\mu\nu} - \frac{1}{2} g_{\mu\nu} R$ vanishes identically) and it is not clear how to expand around $d = 2$. Work on this is in progress.

In any case, I think it is wrong to hope that we will learn how to make sense out of arbitrary non-renormalizable field theories. It would be a disaster for theoretical physics if it were found that the infinite variety of physical theories with all possible couplings were all equally acceptable. We would then have no guide in understanding how nature picks out the particular theory that describes our world. What we need instead is some principle, like asymptotic safety, which picks out a very limited class of physically acceptable theories - perhaps renormalizable, perhaps not.

Acknowledgments

I am very grateful to Paul Martin and David Nelson for frequent enlightening discussions on the theory of critical phenomena throughout the last year. I would not have been able to prepare these lectures without their help in learning this subject. I also wish to thank Edouard Brezin, Sidney Coleman, Leo Kadanoff, Philippe Nozieres, Kenneth Wilson, and Edward Witten for useful conversations, and Antonio Zichichi for his kind hospitality at Erice.

Appendix

DERIVATION OF THE ONE-LOOP EQUATIONS

We will now derive the renormalization-group equations satisfied by the coupling functions. For the present, it will be convenient to work with the original functions $u_n(\vec{p}_1 \ldots \vec{p}_n;\Lambda)$ rather than the re-scaled functions $g_n(\vec{\ell}_1 \ldots \vec{\ell}_n;\Lambda)$. It will also be convenient to treat the quadratic coupling u_2 as much as possible like any other kind of interaction. In order to accomplish this, let us write the original function $u_{2,OLD}$ appearing in the Hamiltonian as

$$u_{2,OLD}(\vec{p}_1,\vec{p}_2;\Lambda) \equiv u_{2,NEW}(\vec{p}_1,\vec{p}_2;\Lambda) + G^{-1}_{NEW}(\vec{p})\delta^d(\vec{p}_1+\vec{p}_2) \quad (A.1)$$

where G_{NEW} is arbitrary but <u>Λ-independent</u>. (We can take G_{NEW} of the form $(a\vec{p}^2+b)^{-1}$, but this is not necessary.) The propagator can then be taken as

$$(2\pi)^{-d}G_{NEW}(\vec{p})\theta(\Lambda-|\vec{p}|) \quad (A.2)$$

and $u_{2,NEW}$ is now regarded as just another interaction. From now on, we drop the label "NEW": G and u_2 are to be understood as G_{NEW} and $u_{2,NEW}$ until further notice.

We consider the set of all connected Green's functions $C(\vec{p}_1 \ldots \vec{p}_n)$, with outgoing momenta $\vec{p}_1 \ldots \vec{p}_n$. These functions are calculated with vertices $-(2\pi)^d u_2$, $-(2\pi)^d u_4$, etc., and with propagators $(2\pi)^{-d} G(\vec{p})\theta(\Lambda-|\vec{p}|)$, except that the θ-function is omitted on external lines. (We do <u>not</u> assume that the external momenta are below the cut-off.) Our demand on the couplings is that these Green's functions be independent of the cut-off:

$$\frac{d}{d\Lambda} C(\vec{p}_1 \ldots \vec{p}_n) = 0 \quad (A.3)$$

for all n and all momenta. When we differentiate C with respect to Λ, we encounter terms of three different kinds, shown in Figure 3:

(a) The derivative $d/d\Lambda$ may act on one of the Λ-dependent coupling functions in the graphs for C. Such terms may be written

$$-(2\pi)^d \sum_m \int d^dk_1 \ldots d^dk_m \left[\frac{d}{d\Lambda} u_m(\vec{k}_1 \ldots \vec{k}_m;\Lambda)\right] \Gamma_\Lambda^{(mn)}(\vec{k}_1 \ldots \vec{k}_m|\vec{p}_1 \ldots \vec{p}_n)$$

Here $\Gamma_\Lambda^{(mn)}(\vec{k}_1 \ldots \vec{k}_m|\vec{p}_1 \ldots \vec{p}_n)$ is a sum of graphs having m external lines carrying incoming momenta $\vec{k}_1 \ldots \vec{k}_m$ and n external lines carrying outgoing momenta $\vec{p}_1 \ldots \vec{p}_n$. It is defined with cut-off propagators for the external k-lines but propagators without cut-offs on the p-lines, except that if a k-line turns into a p-line with no interaction, there is just one propagator without cut-off. Since we

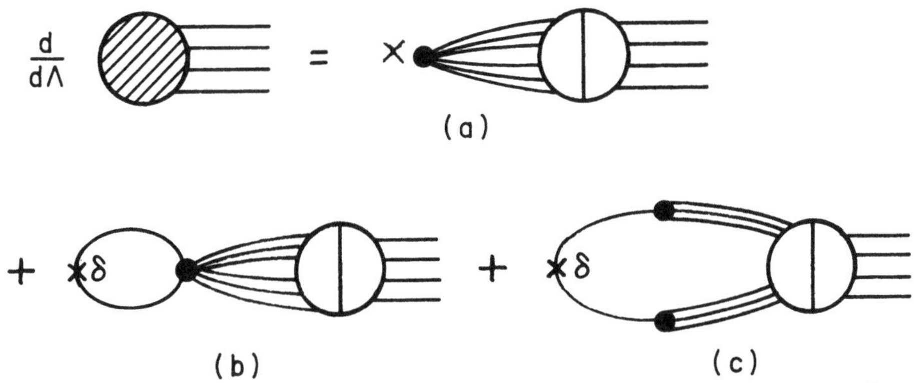

Figure 3. An example of the variation of a connected Green's function with cut-off. Here the shaded circle represents the total connected Green's function with four external lines; the circle with a vertical bar represents the sum of all graphs such that every line leaving on the right is connected to at least one line entering on the left; the darkened circles represent various interactions u_n; the x in the first term on the right represents the derivative of u_n with respect to Λ; and the x in the following two terms represents the replacement of a cut-off θ-function by a δ-function $\delta(|\vec{q}|-\Lambda)$. For the sake of clarity, the only graphs shown are those with six lines entering the barred circle from the left.

want to include all connected diagrams in C, Γ_Λ consists only of the graphs in which each p-line is connected by some path to at least one k-line, so that the interaction u_m can connect the whole graph together. Apart from this proviso, $\Gamma_\Lambda^{(mn)}$ may contain disconnected parts.

(b) The derivative $d/d\Lambda$ may act directly on one of the cut-off functions $\theta(\Lambda-|\vec{q}|)$ associated with any one of the internal lines of C. The two ends of this internal line may terminate either in the same vertex, or in two different vertices. If they terminate in the same vertex, there must be some other set of $m \geq 2$ lines with outgoing momenta $\vec{k}_1\ldots\vec{k}_m$ attached to this vertex. Also, since C is connected, each p-line will be connected by some path to at least one of the k-lines. The contribution of such graphs to $dC/d\Lambda$ is then

$$-\tfrac{1}{2}\sum_{m=2}^{\infty}\int d^d q\, d^d k_1 \ldots d^d k_m\, \delta(\Lambda-|\vec{q}|)\Delta(\vec{q})\, u_{m+2}(\vec{q},-\vec{q},\vec{k}_1\ldots\vec{k}_m)\Gamma_\Lambda^{(mn)}(\vec{k}_1\ldots\vec{k}_m|\vec{p}_1\ldots\vec{p}_n)$$

(c) Finally, when $d/d\Lambda$ acts on the cut-off function associated with an internal line which ends in two different vertices, there will be two sets of lines with outgoing momenta $\vec{k}'_1 \ldots \vec{k}'_r$ and $\vec{k}''_1 \ldots \vec{k}''_s$ respectively, attached to these two vertices. Again, since C is connected, each $\vec{p}$-line is connected by some path to at least one $\vec{k}'$- or $\vec{k}''$-line. The contribution of such graphs to $dC/d\Lambda$ is then

$$\tfrac{1}{2}(2\pi)^d \sum_{r,s} \frac{(r+s)!}{r!s!} \int d^d q\, d^d k'_1 \ldots d^d k'_r\, d^d k''_1 \ldots d^d k''_s\, \delta(\Lambda - |\vec{q}|)$$

$$\times\, G(\vec{q})\, u_{r+1}(\vec{q}, \vec{k}'_1 \ldots \vec{k}'_r;\Lambda)\, u_{s+1}(-\vec{q}, \vec{k}''_1 \ldots \vec{k}''_s;\Lambda)$$

$$\times\, \Gamma_\Lambda^{(r+s,n)}(\vec{k}'_1 \ldots \vec{k}'_r\, \vec{k}''_1 \ldots \vec{k}''_s | \vec{p}_1 \ldots \vec{p}_n)$$

We can now put this all together, and write the result as

$$\frac{d}{d\Lambda} C(\vec{p} \ldots \vec{p}_n) = -(2\pi)^d \sum_{m=2}^{\infty} \int d^d \vec{k}_1 \ldots d^d \vec{k}_m B_\Lambda(\vec{k}_1 \ldots \vec{k}_m) \Gamma_\Lambda^{(mn)}(\vec{k}_1 \ldots \vec{k}_m | \vec{p}_1 \ldots \vec{p}_n) \tag{A.4}$$

where

$$B_\Lambda(\vec{k}_1 \ldots \vec{k}_m) \equiv \frac{d}{d\Lambda} u_m(\vec{k}_1 \ldots \vec{k}_m;\Lambda)$$

$$+\tfrac{1}{2}(2\pi)^{-d} \int d^d q\, \delta(\Lambda - |\vec{q}|)\, G(\vec{q})\, u_{m+2}(\vec{q}, -\vec{q}, \vec{k}_1, \ldots \vec{k}_m;\Lambda)$$

$$-\tfrac{1}{2} \sum_{k \to k' \& k''} \int d^4 q\, \delta(\Lambda - |\vec{q}|)\, G(\vec{q})\, u_{r+1}(\vec{q}, \vec{k}'_1 \ldots \vec{k}'_r;\Lambda)\, u_{s+1}(-\vec{q}, \vec{k}''_1 \ldots \vec{k}''_s;\Lambda) \tag{A.5}$$

The sum in the last term runs over all values of r and s with $r+s = m$, and over all $m!/r!s!$ ways of partitioning the momenta $\vec{k}_1 \ldots \vec{k}_m$ into subsets $\vec{k}'_1 \ldots \vec{k}'_r$ and $\vec{k}''_1 \ldots \vec{k}''_s$. We see that for C to be independent of Λ, it is sufficient that

$$B_\Lambda(\vec{k}_1 \ldots \vec{k}_m) = 0 \tag{A.6}$$

for all m and all $\vec{k}_1 \ldots \vec{k}_m$. This is one form of renormalization-group equation. (See Figure 4.) We note further that the kernel $\Gamma^{(mn)}(\vec{k}_1 \ldots \vec{k}_m | \vec{p}_1 \ldots \vec{p}_n)$ generally will have an inverse, because in the absence of interactions it is simply proportional to

$$\delta_{mn}\, \delta(\vec{k}_1 - \vec{p}_1) \ldots \delta(\vec{k}_m - \vec{p}_m)\, G(\vec{k}_1) \ldots G(\vec{k}_m)$$

Therefore we expect (A.6) to be necessary as well as sufficient.

[Equation (A.6) is similar to a set of equations derived in quite a different way by Wegner and Houghton.[7] They differ in that G appears instead of u_2^{-1}, and that the sum in the last term includes

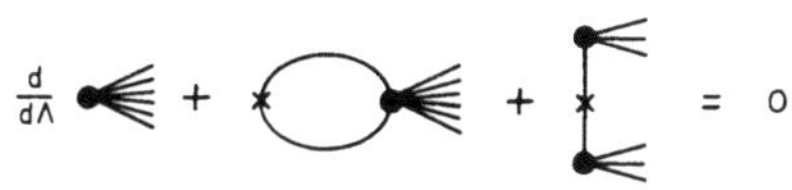

Figure 4. The Wegner-Houghton equations for u_6. These are derived here by demanding that the Λ-derivative in Figure 3 vanish.

terms with r or s equal to one. The difference arises because our cut-off procedure is different for external lines.]

These equations have a well-known drawback: the last term in Eq. (A.5) is highly discontinuous in the external momenta, contributing only when the total momentum of some subset with an odd number of lines has an absolute value equal to Λ. Also, these equations mix terms with different numbers of loops, so they are not very convenient for use in iteration schemes. To avoid these problems, we introduce a new set of coupling functions.

The reason for the discontinuities in the last term in (A.5) is obvious: in calculating Green's functions we must include tree graphs, whose internal lines contain cut-off functions $\theta(\Lambda-|\vec{q}|)$. Even though all external lines of such graphs lie below the cut-off Λ, the total momentum of subsets of the external lines may be above this cut-off, and the contribution of such graphs will contain discontinuities in the external momenta. The total Green's functions certainly do not contain such discontinuities (they are Λ-independent!) so in order to cancel them out the u_n must contain Λ-dependent discontinuities of some sort. These are automatically generated by the last term in Eq. (A.5).

The diagnosis immediately suggests a cure. Let us introduce a new set of coupling functions $v_n(\vec{k}_1 \ldots \vec{k}_n;\Lambda)$, by the prescription that $-(2\pi)^d u_n$ is to be constructed as a sum of all possible tree graphs, with vertex functions $-(2\pi)^d v_n$, and with internal-line factors

$$(2\pi)^{-d} G(\vec{q})\theta(|\vec{q}|-\Lambda) \tag{A.7}$$

When we take the derivative of u_n with respect to Λ, we then encounter terms of two different types (see Figure 5):

(a) The derivative $d/d\Lambda$ may act on one of the cut-off functions $\theta(|\vec{q}|-\Lambda)$ associated with an internal line of the tree, changing it to $-\delta(|\vec{q}|-\Lambda)$. The two ends of this internal line must be connected to separate trees built out of v-vertices.

(b) The derivative $d/d\Lambda$ may act on one of the vertex factors, say a v_m. Since the whole graph is a tree, each of the m lines attached to this vertex must itself be connected to a separate tree built out of v-vertices and propagators (A.7).

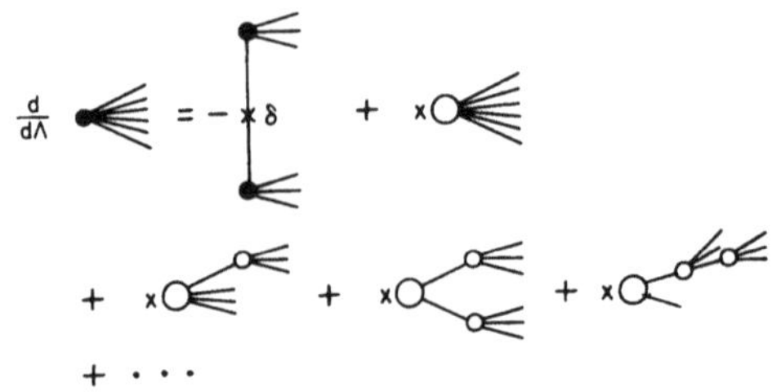

Figure 5. Λ-derivative of a u_n-function expressed in terms of Λ-derivatives of v_n-functions. Here dark circles represent u_n-functions; light circles represent v_n-functions; the x in the first term on the right represents the replacement of a cut-off θ-function with a δ-function; and the x's in the other terms represent Λ-derivatives acting on u_n-functions. When inserted into the equations of Figure 4, this yields the one-loop equations.

When we insert these results for $du_m/d\Lambda$ in Eq. (A.5), we find that term (a) cancels the last term in (A.5), so the sum of (b) and the second term of (A.5) must vanish.

Now, u_{m+2} in Eq. (A.5) is itself a sum of tree graphs built out of v-vertices. There must be a unique path from the line with momentum $\vec{q}$ to the line with momentum $-\vec{q}$, and each v-vertex along this path is attached to a tree built out of v-vertices. The v-vertices in term (b) are also connected to such trees, so the trees may be divided out. leaving us with

$$\begin{aligned}
0 = {} & \frac{d}{d\Lambda}\, v_m(\vec{k}_1 \ldots \vec{k}_m;\Lambda) \\
& + \tfrac{1}{2}(2\pi)^{-d}\int d^d q\, \delta(\Lambda-|\vec{q}|)G(\vec{q}) \sum_{\ell=1}^{\infty}\int d^d q_1 \ldots d^d q_{\ell-1}(-)^{\ell-1} \\
& \times \sum_{k\to k^{(1)}\ldots k^{(\ell)}} v_{m_1+2}(\vec{q},-\vec{q}_1,\vec{k}_1^{(1)}\ldots\vec{k}_{m_1}^{(1)};\Lambda)\,\theta(|\vec{q}_1|-\Lambda)G(\vec{q}_1) \\
& \times v_{m_2+2}(\vec{q}_1,-\vec{q}_2,\vec{k}_1^{(2)}\ldots\vec{k}_{m_2}^{(2)};\Lambda)\,\theta(|\vec{q}_2|-\Lambda)G(\vec{q}_2) \\
& \times \ldots\ v_{m_\ell+2}(\vec{q}_{\ell-1},-\vec{q},\vec{k}_1^{(\ell)}\ldots\vec{k}_{m_\ell}^{(\ell)};\Lambda)
\end{aligned} \tag{A.8}$$

The last sum is over all values of m_1, m_2,...,m_ℓ, with

$$m_1 + m_2 + \ldots + m_\ell = m$$

and over all $m!/m_1!m_2!\ldots m_\ell!$ ways of partitioning the m momenta into ℓ sets, with the j-th set consisting of momenta $k_1^{(j)}\ldots k_m^{(j)}$.

It should be noted that the last sum in (A.8) includes terms in which any of the m_j are zero. These can be trivially summed: their effect is simply to change $G(\vec{q})$ into

$$G'(\vec{q};\Lambda) = G(\vec{q}) - G(\vec{q})\overline{v}_2(\vec{q},-\vec{q};\Lambda)G(\vec{q}) + \ldots$$

$$= G(\vec{q})\,[1 + \overline{v}_2(\vec{q},-\vec{q};\Lambda)G(\vec{q})]^{-1}$$

where $\overline{v}_2$ is v_2 with the δ-function factored out:

$$v_2(\vec{q}_1,\vec{q}_2;\Lambda) = \overline{v}_2(\vec{q}_1,\vec{q}_2;\Lambda)\delta^d(\vec{q}_1+\vec{q}_2)$$

This leads us to introduce a new v_2 function

$$v_2'(\vec{q}_1,\vec{q}_2;\Lambda) \equiv G^{-1}(\vec{q}_1)\delta^d(\vec{q}_1+\vec{q}_2) + v_2(\vec{q}_1,\vec{q}_2;\Lambda)$$

which has the same Λ-derivative as v_2

$$\frac{dv_2'}{d\Lambda} = \frac{dv_2}{d\Lambda}$$

but which yields the corrected propagator

$$v_2'(\vec{q}_1,\vec{q}_2;\Lambda) = G'^{-1}(\vec{q}_1;\Lambda)\delta^d(\vec{q}_1+\vec{q}_2)$$

That is, Eq. (A.8) still holds if we replace G with G', v_2 with v_2', and restrict the sums so that $m_j > 0$ for all j. From now on we will drop the primes on v_2 and G, and interpret (A.8) to include only terms with $m_j > 0$.

We now make the transition to the re-scaled variables. Define new coupling functions:

$$g_n\left(\frac{\vec{p}_1}{\Lambda}\cdots\frac{\vec{p}_n}{\Lambda};\Lambda\right) \equiv v_n(\vec{p}_1\ldots\vec{p}_n;\Lambda)\Lambda^{nd/2-n}\,Z_\Lambda^{n/2} \qquad (A.9)$$

and a new propagator

$$\Delta\left(\frac{\vec{p}}{\Lambda};\Lambda\right) \equiv G(\vec{p};\Lambda)\Lambda^2 Z_\Lambda^{-1} \qquad (A.10)$$

so that

$$g_2(\vec{\ell}_1,\vec{\ell}_2;\Lambda) = \delta^d(\vec{\ell}_1+\vec{\ell}_2)\Delta^{-1}(\vec{\ell}_1;\Lambda) \qquad (A.11)$$

Inserting (A.9) and (A.10) in (A.8) gives

$$\left[\Lambda\frac{\partial}{\partial\Lambda} - \sum_{r=1}^{n}\vec{\ell}_r\cdot\frac{\partial}{\partial\vec{\ell}_r}\right]g_n(\vec{\ell}\ \ldots\vec{\ell}_n;\Lambda)$$

$$= \left(\frac{nd}{2} - n + \frac{n\gamma}{2}\right)g_n(\vec{\ell}_1\ldots\vec{\ell}_n;\Lambda)$$

$$- \tfrac{1}{2}(2\pi)^{-d}\int d^d\ell\,\delta(1-|\vec{\ell}|)\Delta(\vec{\ell};\Lambda)\sum_{k=1}^{\infty}\int d^d\ell_1\ldots d^d\ell_k(-)^{k-1}$$

$$\times \sum_{\ell\to\ell^{(1)}\ldots\ell^{(k)}} g_{m_1+2}(\vec{\ell},-\vec{\ell}_1,\vec{\ell}_1^{(1)}\ldots\vec{\ell}_{m_1}^{(1)};\Lambda)\theta(|\vec{\ell}_1|-1)\Delta(\vec{\ell}_1;\Lambda)$$
$$\times g_{m_2+2}(\vec{\ell}_1,-\vec{\ell}_2,\vec{\ell}_1^{(2)}\ldots\vec{\ell}_{m_2}^{(2)};\Lambda)\theta(|\vec{\ell}_2|-1)\Delta(\vec{\ell}_2;\Lambda)$$
$$\times \ldots g_{m_\ell+2}(\vec{\ell}_{k-1},-\vec{\ell},\vec{\ell}_1^{(k)}\ldots\vec{\ell}_{m_k}^{(k)};\Lambda) \tag{A.12}$$

After factoring out a momentum-conservation delta function, this becomes just the general one-loop equation (8.4).

REFERENCES

1. For surveys of the modern theory of critical phenomena, including references to the original literature, see the following reviews: K. G. Wilson and J. Kogut, Physics Reports 12C, No. 2 (1974); M. E. Fisher, Rev. Mod. Phys. 46, 597 (1974); E. Brézin, J. C. Le Guillou, and J. Zinn-Justin, in *Phase Transitions and Critical Phenomena*, ed. by C. Domb and M. S. Green (Academic Press, New York, 1975), Vol. VI; F. J. Wegner, in *Trends in Elementary Particle Theory* (Springer-Verlag, Berlin, 1975), p.171; K. Wilson, Rev. Mod. Phys. 47, 773 (1975); Shang-Keng Ma, *Modern Theory of Critical Phenomena* (W. A. Benjamin, Inc., Reading, Mass., 1976).

2. A. L. Fetter and J. D. Walecka, *Quantum Theory of Many-Particle Systems* (McGraw-Hill, Inc., New York, 1971), Chapter 7.

3. S. Weinberg, Phys. Rev. D9, 3357 (1974); L. Dolan and R. Jackiw, Phys. Rev. D9, 3320 (1974); D. A. Kirzhnits and A. D. Linde, Zh. Eksp. Teor. Fiz. 67, 1263 (1974); C. W. Bernard, Phys. Rev. D9, 3312 (1974).

4. F. Bloch and A. Nordsieck, Phys. Rev. 52, 54 (1937). For a version in modern notation, see S. Weinberg, Phys. Rev. 140, B516 (1965).

5. M. Gell-Mann and F. E. Low, Phys. Rev. 95, 1300 (1954).

6. The fixed point in the nonlinear σ-model in 2 + ε dimensions has been under intensive study lately; see W. A. Bardeen, B. W. Lee, and R. E. Shrock, Fermilab-Pub-76/33-THY, March 1976; E. Brézin, J. Zinn-Justin, and J. C. Le Guillou, Saclay preprints, May 1976. The motivation of these studies appears to be quite different from that described here.

7. F. J. Wegner and A. Houghton, Phys. Rev. A 8, 401 (1973).

DISCUSSIONS

CHAIRMAN: Prof. S. Weinberg

Scientific Secretary: F. Posner

DISCUSSION 1

- *FERBEL:*

Could you elaborate on the question of Higg's boson production relative to W production in hadronic collisions?

- *WEINBERG:*

Higg's production has been discussed in detail in a recent paper by Ellis, Gaillard, and Nanopoulos. I believe that they do the calculations you ask about, but I do not know the answer. I will make one point though: Higg's bosons' couplings are proportional to mass, so Higg's bosons tend to be emitted from internal lines of heavy particles. A good place to look for Higg's bosons therefore is in neutrino reactions, where they are emitted from the exchanged W line. Ellis, Gaillard, and Nanopoulos, and LoSecco have calculated the probability for Higg's production near threshold in neutrino-nucleon reactions to be about 10^{-5}.

- *FERBEL:*

If charm violating processes occur, might one expect final state correlations such as K^+K^+ or K^+e^+?

- *WEINBERG:*

I have been assuming that the neutral currents do conserve charm. If there is a milliweak $\Delta C = 2$ neutral Higg's exchange, then D^0-$\bar{D}^0$ oscillations will be much faster than D^0 decay. The result would be that an incoherent mixture of D_1^0 and/or D_2^0 would be produced, each decaying equally into Ks or leptons of either charge.

- *PARSONS:*

How easy would it be to introduce more flavours into your model?

- *WEINBERG:*

This model needs four quarks because if there were more, this would be likely to introduce CP violation into the W-exchange process. To include a bottom quark b, one requires

$$\begin{pmatrix} u \\ d\alpha_1 + s\alpha_2 + b\alpha_3 \end{pmatrix}$$

as the appropriate multiplet for weak interactions where

$$\alpha_1^2 + \alpha_2^2 + \alpha_3^2 = 1 \ .$$

The experimental data on β decay provides information on u → d. The experimental data on Λ decay provides information on u → s. An over-all normalization can be fixed by μ decay. This sets tight limits on α_1 and α_2, so that we must have $|\alpha_3| \lesssim 10^{-3}$. Such small numbers in the theory are rather unsatisfactory.

It is, however, possible to introduce extra doublets into the theory which do not mix with u, d, s, and c. These can be used to cancel anomalies due to heavy leptons. The real problem would be if neutrinos were found to cause transitions between u and b quarks, for instance, thus indicating a non-zero mixing angle α_3.

- *CRONIN:*

In order to be sure of your ideas about the nature of CP violation, it is really insufficient to have the electric dipole moment of the neutron come out $\sim 2 \times 10^{-24}$ e-cm. Thus one really needs to see the CP-violating effects of the Higg's scalar. What is the nature of the CP-violating effects of the Higg's scalar, if it could be isolated?

- *WEINBERG:*

Let me first answer the question I thought you were going to ask: What experimental evidence could be used to check that Higg's exchange is responsible for CP non-conservation? Since Higg's bosons couple essentially to mass, we would expect CP-violating amplitudes of order 10^{-3} in $K_{\mu 3}$ decay, but no corresponding effect in β decay.

As to the way that CP violations would show up in Higg's boson decay, I have not worked it out. It would be a good exercise for a student. The Higg's decays should violate CP strongly.

- *LEADER:*

You have offered us a beautiful and natural mechanism for producing a small number, namely 10^{-3}, in amplitude. However, you rely for this on a knowledge of the quark and the Higg's masses. Could you explain what it is that gives you such confidence in your knowledge of these masses?

- *WEINBERG:*

As regards the quark masses, I am just going along with all the standard ideas on quarks and constituent models. As for the Higg's mass, it is expected to be of order $\sqrt{f} \times 300$ GeV, where f is the ϕ^4 coupling constant. If f is of order α, as generally supposed, then the Higg's mass is of the order of the intermediate vector boson mass. Linde and I have recently shown that there is an effective lower bound of order α^2. In the simplest $SU(2) \times U(1)$ model, the Higg's mass is greater than 3.72 GeV.

- *POSNER:*

A rather elementary question: Why does a CP and P non-conserving milliweak interaction imply a detectable electric dipole moment for the neutron?

- *WEINBERG:*

If CP is broken, but not P, then there is no moment. If the neutron is really two d and one u quarks, then the moment equals $^4/_3$ (the moment of the d) - $^1/_3$ (the moment of the u). The moments arise from the virtual processes

$$u \to H^+ + s \text{ (or d)} \to u$$

$$d \to H^- + c \text{ (or u)} \to d \ .$$

- *MARCIANO:*

Is the statement that the CP violation is due to the H^+ propagator and not the W propagator, a gauge-dependent statement? Could you clarify your statement that you feel that there will be CP violation even for the case $\phi_1 = \phi_2$?

- *WEINBERG:*

The sum of the W-propagator and the part of the H propagator corresponding to a Goldstone boson is gauge invariant and conserves CP. The remaining part of the H propagator is then also gauge invariant, and violates CP. Thus the amount of CP violation is gauge invariant.

Even if $\phi_1 = \phi_2$, there can be a CP violation in the interaction of these Higg's bosons with other Higg's bosons that cannot couple to quarks. I have not tried to calculate such effects.

- *FREEDMAN:*

What classes of theories permit Higg's bosons which do not violate CP? How do these couplings differ from those discussed in this morning's talk? What, if any, characteristics are different between the two, mass, etc., and can one incorporate both in a single model?

- *WEINBERG:*

Such theories are certainly possible. One can always impose CP on L. Then Crs would have to be real, or have phases that could be made real. However, for a finite range of parameters there will still be a spontaneous breaking of CP.

DISCUSSION 2

- *YOON:*

Does the fact that critical phenomena in boson systems has its origin in the infrared behaviour of the system reflect the physical picture of phase transitions arising from long-range correlations? How does one understand critical phenomena in fermion systems where there are no infrared divergences?

- *WEINBERG:*

Second-order phase transitions can occur in systems composed purely of fermions, because the fermions can have bosonic collective excitations. That is a large subject, which I will not go into in these lectures.

- *ALVAREZ:*

In ϕ^4 quantum field theory, the two-point function is ultraviolet quadratically divergent. In the finite temperature theory, the three momentum integral is logarithmically divergent. When the energy sum is performed, do the logarithms sum up to power behaviour?

- *WEINBERG:*

Of course, they had better, because the divergences must be the same at finite temperature as at zero temperature. The Lagrangian does not know what the temperature is, so the counter-terms available to absorb infinities are necessarily temperature-independent. However, it is difficult to do the energy sums as you suggest after doing the momentum integrals. I have found it much easier to do the energy sum first by converting it into a contour integral, and *then* do the momentum integral.

- *PAULI:*

In the statistical mechanics formalism, if you take the fourth derivative of the partition function, you get a Feynman-like graph with four external legs. We know what this corresponds to in quantum field theory *vis-a-vis* the S matrix and differential cross-sections. We also know is field theory how to include symmetries. What physical and formal mathematical changes occur when we do this in a statistical mechanical form at finite temperatures?

- *WEINBERG:*

Symmetries like rotation or isospin invariance govern the temperature Green's functions just as they govern the S matrix or the Green's functions in quantum field theory. The only symmetry that is really fouled up by a finite temperature is Lorentz, or Galilean, invariance.

- *KLEINERT:*

If you pass the critical point of a ferromagnet, however, do the 4-point Green's functions not change symmetry?

- *WEINBERG:*

No, the Green's functions as I have defined them have the full symmetry of the underlying theory, i.e. of the Hamiltonian. The symmetry-breaking affects the various expectation values calculated using these Green's functions.

- *POSNER:*

A useful and powerful formulation of quantum field theory is by means of functional integrals. Are there any particular pit-falls in dealing with statistical mechanics by functional techniques? In field theory functional integrals, one rotates from Minkowskian to Euclidean space to avoid ambiguities. In statistical mechanics one has iw instead of p^0. How does this affect the functional integrals of statistical mechanics and their possible ambiguities?

- *WEINBERG:*

Your classmate at Harvard, Claude Bernard, has worked out a very nice formulation of statistical mechanics by using functional integrals. He has derived Feynman rules at a finite temperature in a gauge theory.

Because we are now in Euclidean space, the troubles with cut-offs due to the metric are gone. Things work very well.

- *MARCIANO:*

What plays the role of an infrared regulator in this formalism?

- *WEINBERG:*

The partition function is not well-defined at the critical temperature. The temperature θ regulates these divergences since for $\theta \neq \theta_c$, the partition function Q is finite and well-defined.

DISCUSSION 3 *(Scientific Secretary: O. Alvarez)*

- *MARCIANO:*

What does the renormalization group have to do with infrared divergences?

- *WEINBERG:*

It is unfortunate that Wilson, probably out of modesty, called these equations renormalization group equations, because of the formal appearance to the equation of Gell-Mann and Low. The Wilson work concerns infrared behaviour while Gell-Mann and Low considered ultraviolet behaviour. The "floating renormalization point" introduced by Gell-Mann and Low has a formal similarity to the floating cut-off used by Wilson, but the purpose is entirely different.

- *GARCIA:*

Following your discussion of the Bloch-Nordsieck problem, can you find a similar approach for coupled massless fields?

- *WEINBERG:*

I believe there is a renormalization group argument, but I have not been able to complete it.

- *GARCIA:*

I do not clearly see the connection between your use of the renormalization group and the idea of "thinning" out of the degrees of freedom as used by Wilson and Kadanoff.

- *WEINBERG:*

This is an example of the trouble with translating solid state language into the language of field theorists. Kadanoff's idea of block spins, or changing the lattice spacing by integer multiples would be equivalent to changing the cut-off by integer multiples. In quantum field theory we usually change the cut-off continuously, but we could change it by discrete steps.

- *PAULI:*

When you compared the work of Brezin et al. with the work of the Wilson school, you stated that every eigenvalue of $\tilde{M}$ is an eigenvalue of M, but not all eigenvalues of M are eigenvalues of $\tilde{M}$. How do you know that Brezin's method will give the repulsive and the "important" eigenvalues of M?

- *WEINBERG:*

The only way to check this is to calculate the eigenvalues, using an approximation scheme such as the ε expansion. Brezin et al. work in $4 - \varepsilon$ dimensions, and find one attractive and one repulsive eigenvalue. They cannot show, within a strictly renormalizable theory, that all ignored eigenvalues are attractive, but this seems reasonable, and can be shown by introducing non-renormalizable perturbations.

- *PAULI:*

You stated that if you have more than two repulsive eigenvalues in three dimensions, the ε expansion is necessary. Why can you not vary temperature and say magnetic field? Are there alternatives to the ε expansion?

- *WEINBERG:*

I must have been unclear. We believe that ordinary second-order phase transitions are associated with fixed points that have only one repulsive eigenvector, because these transitions can be brought about by adjusting only one free parameter, the temperature. Even with a magnetic field, the Gaussian fixed point cannot describe an ordinary second-order phase transition, because introduction of the field destroys the $\phi \to -\phi$ symmetry, and there are therefore *three* repulsive eigenvectors at the Gaussian fixed point.

One alternative to the ε expansion is an expansion in 1/N, where N is the number of fields. I will not discuss this in these lectures.

- *FREEDMAN:*

We learned that in three dimensions it was not possible to do perturbation theory about the Gaussian fixed point because of two repulsive eigenvectors. In order to circumvent this we go to $4 - \varepsilon$ dimensions where only one repulsive eigenvector exists. Although we can perform our calculations now, what can we learn about the physics in our original three dimensional problem?

- *WEINBERG:*

The hope is that the eigenvalues do not change sign between $\varepsilon \ll 1$ and $\varepsilon = 1$. The qualitative features of the physics will then remain.

- *PHAM QUANG HUNG:*

To which physical situations corresponds the limit $\lambda = 0$?

- *WEINBERG:*

The Gaussian fixed point has a zero eigenvalue corresponding to the interaction ϕ^6. However, it is only possible to reach this fixed point if two parameters are adjusted to eliminate components of the trajectory along the two repulsive eigenvectors which correspond to the ϕ^2 and ϕ^4 interactions.

MONOPOLES AND FIBER BUNDLES

Chen Ning Yang

State University of New York

Physics Department, Stony Brook, N.Y. 11794

Magnetic Monopole and Need to Introduce Sections

The magnetic monopole is the magnetic charge. While the idea of magnetic monopoles must have been discussed in classical physics early in the history of electricity and magnetism, modern discussions date back to 1931 in the important paper of Dirac[1] in which he pointed out that magnetic monopoles in quantum mechanics exhibit some extra and subtle features. In particular, with the existence of a magnetic monopole of strength g, electric charges and magnetic charges must necessarily be quantized, in quantum mechanics. We shall give a new derivation of this result in a few minutes.

If one wants to describe the wave function of an electron in the field of a magnetic monopole, it is necessary to find the vector potential $\vec{A}$ around the monopole. Dirac chose a vector potential which has a string of singularities. The necessity of such a string of singularities is obvious if we prove the following theorem[2].

Theorem. Consider a magnetic monopole of strength $g \neq 0$ at the origin and consider a sphere of radius R around the origin. There does not exist a vector potential $\vec{A}$ for the monopole magnetic field which is singularity free on the sphere. This theorem can be easily proved in the following way. If there were a singularity free $\vec{A}$ we consider the loop integral

$$\oint A_\mu dx^\mu$$

around a parallel on the sphere as indicated in Figure 1. By

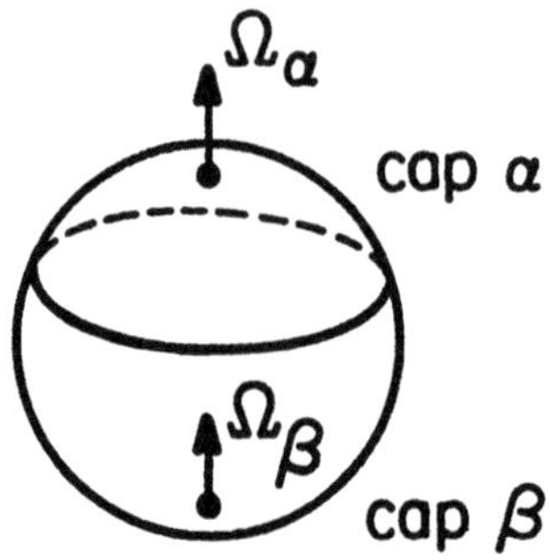

FIGURE 1. A sphere of radius R with a magnetic monopole at its center. The parallel divides the sphere into two caps α and β.

Stoke's theorem this loop integral is equal to the total magnetic flux through the cap alpha:

$$\oint A_\mu dx^\mu = \Omega_\alpha . \tag{1}$$

Similarily we can apply Stoke's theorem to cap β obtaining

$$\oint A_\mu dx^\mu = \Omega_\beta .. \tag{2}$$

Here Ω_α and Ω_β are the total upward magnetic flux through the caps α and β, both of which are bordered by the parallel. Subtracting these two equations we obtain

$$0 = \Omega_A - \Omega_B, \tag{3}$$

which is equal to the total flux <u>out</u> of the sphere, which in turn is equal to $4\pi g \neq 0$. We have thus reached a contradiction.

Having proved this theorem, we observe that R is arbitrary. Thus one concludes that there must be a string of singularities or strings of singularities in the vector potential to describe the monopole field. Yet we know that the magnetic field around the monopole is singularity free. This suggests that the string of singularities is not a real physical difficulty. Indeed the situation is reminiscent of the problem that one faces when one wants to find a parametrization of the surface of the globe. The coordinate system that we usually use, the latitude and the longitude, is not singularity free. It has singularities at the north pole and at the south pole. Yet the surface of the globe is evidently without singularities. We deal with this situation usually in something like the way illustrated in figure 2. We consider a rubber sheet with nicely defined coordinates and stretch and wrap it down onto the globe so that it covers more than the northern hemisphere. Similarly,

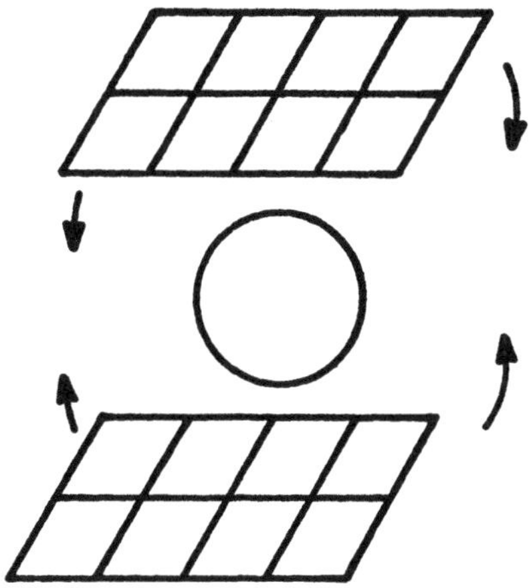

FIGURE 2. Method of parametrizing the globe.

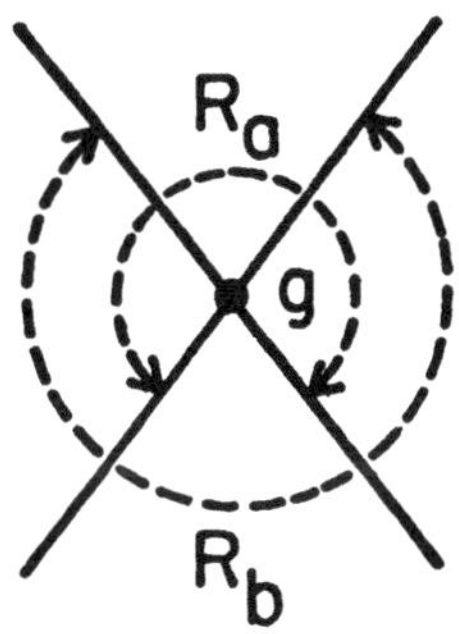

FIGURE 3. Division of space outside of monopole g into overlapping regions R_a and R_b.

we consider another rubber sheet with nicely defined coordinates and stretch and wrap it upwards so that it covers more than the southern hemisphere. We now have a double system of coordinates to describe the points on the globe. The description is analytic in the domain covered by each sheet, if we had done no violence in the stretching and wrapping. In the overlapping region covered by both sheets, one has two coordinate systems which are transformable into each other by an analytic non-vanishing Jacobian. This double coordinate system is an entirely satisfactory way to parametrize the globe.

Following this idea we shall now try to exorcise the string of singularities in the monopole problem by dividing space into two regions. We shall call the points outside of the origin, above the lower cone in figure 3, region R_a. Similarly, we shall call the points outside of the origin, under the upper cone, R_b.

The union of these two regions gives all points outside of the origin. In R_a we shall choose a vector potential for which there is only one non-vanishing component of A, the azimuthal component:

$$(A_r)_a = (A_\theta)_a = 0, \quad (A_\phi)_a = \frac{g}{r \sin\theta}(1 - \cos\theta), \tag{4}$$

It is important to notice that this vector potential has no singularities anywhere in R_a. Similarly in R_b we choose the vector potential

$$(A_r)_b = (A_\theta)_a = 0, \quad (A_\phi)_b = \frac{-g}{r \sin\theta}(1 - \cos\theta), \tag{5}$$

which has no singularities in R_b. It is simple to prove that the curl of either of these two potentials give correctly the magnetic field of the monopole.

In the region of overlap, since both of the two sets of vector potentials share the same curl, the difference between them must be curlless and therefore must be a gradient. Indeed a simple calculation shows

$$(A_\mu)_a - (A_\mu)_b = \partial_\mu \alpha, \text{ where } \alpha = 2g\phi \tag{6}$$

where ϕ is the azimuthal angle. The Schrodinger equation for an electron in the monopole field is thus

$$\frac{1}{2m}(p-eA_a)^2\psi_a + V\psi_a = E\psi_a, \qquad \text{in } R_a,$$

$$\frac{1}{2m}(p-eA_b)^2\psi_b + V\psi_b = E\psi_b, \qquad \text{in } R_b,$$

where ψ_a and ψ_b are respectively the wave functions in the two regions. The fact that the two vector potentials in these two equations are different by a gradient tells us, by the well known gauge principle, that ψ_a and ψ_b are related by a phase factor transformation

$$\psi_a = S\psi_b, \quad S = \exp(ie\alpha), \tag{7}$$

or

$$\psi_a = [\exp(2iq\phi)]\psi_b, \quad q = eg. \tag{8}$$

Around the equator which is entirely in R_a, ψ_a is single valued. Similarly, since the equator is also entirely in R_b, ψ_b is single valued around the equator. Therefore, S must return to its original value when one goes around the equator. That implies Dirac's quantization condition:

$$2q = \text{integer}. \tag{9}$$

Hilbert Space of Sections

Two ψ's, ψ_a and ψ_b, in R_a, and R_b respectively, that satisfy the condition of transition (8) in the overlap region, is called a section by the mathematicians. We see that around a monopole the electron wave function is a section and not an ordinary function. We shall call these wave sections.

Different wave sections (belonging to different energies, for example), clearly satisfy the same condition of transition (8) with the same q. Thus we need to develop[3] the concept of a Hilbert space of sections. To do this we define the scalar product of two sections ξ,η (for the same q) by

$$(\eta,\xi) \equiv \int \eta^* \xi d^3r. \tag{10}$$

(The question of convergence at $r = 0$ and $r = \infty$ is ignored here). Notice that in the overlap

$$(\eta_a)^* \xi_a = (\eta_b)^* \xi_b \tag{11}$$

so that (10) is well defined.

It is clear that if ξ is a section, then $x\xi$ is also a section, since

$$x\xi_a = S(x\xi_b).$$

Thus x is an operator in the Hilbert space of sections. Similarly we prove that the components of $(\vec{p}-e\vec{A})$ are operators, but those of $\vec{p}$ are not. Furthermore $\vec{x}$ and $\vec{p}-e\vec{A}$ are both Hermitian.

Following Fierz[4] we shall now try to construct angular momentum operators. Define

$$\vec{L} = \vec{r} \times (\vec{p} - e\vec{A}) - \frac{q\vec{r}}{r}. \tag{12}$$

It is clear that L_x, L_y, L_z are Hermitian operators on the Hilbert space of sections. The following commutation rules can be easily verified:

$$\begin{aligned} &[L_x,x] = 0, \quad [L_x,y] = iz, \quad [L_x,z] = -iy, \\ &[L_x,p_x - eA_x] = 0, \quad [L_x,p_y - eA_y] = i(p_z - eA_z), \\ &[L_x,p_z - eA_z] = - i(p_y - eA_y). \end{aligned} \tag{13}$$

It follows from these that

$$[L_x,L_y] = iL_z, \text{ etc.} \tag{14}$$

Eq. (13), together with its consequence (14), show that L_x, L_y, L_z are the angular momentum operators. We emphasize that neither the Hilbert space, nor these operators, possess any "singularities". (The singularities of A_a and A_b are not real singularities because they occur outside of R_a and R_b, respectively.)

Monopole harmonics $Y_{q,\ell,m}$

Since $[r^2, \vec{L}] = 0$, we can diagonalize r^2 and study operator $\vec{L}$ for fixed r^2. I.e. we shall study sections of the form

$$\delta(r^2 - r_0^2)\xi,$$

where ξ is a section dependent only on angular coordinates θ and ϕ. $\vec{L}$ operates then on "angular sections".

Eq. (14) shows that $[L^2, L_z] = 0$. Simultaneous diagonalization produces the familiar multiplets with eigenvalues $\ell(\ell + 1)$ and m

$$L^2 Y_{q,\ell,m} = \ell(\ell+1) Y_{q,\ell,m}; \quad L_z Y_{q,\ell,m} = m Y_{q,\ell,m}, \tag{15}$$

where $\ell = 0, \frac{1}{2}, 1, \ldots$ and, for each value of ℓ, m ranges from $-\ell$ to $+\ell$ in integral steps of increment. The $Y_{q,\ell,m}$ are the eigensections 3 which we shall call monopole harmonics. We shall show later that the allowed values of ℓ and m are

$$\begin{aligned} \ell &= |q|, \ |q| + 1, \ |q| + 2, \ldots, \\ m &= -\ell, -\ell + 1, \ldots, \ell, \end{aligned} \tag{16}$$

and that each of these ℓ, m combinations occur exactly once. We shall choose each Y normalized so that

$$\int_0^{\pi} \sin\theta \, d\theta \int_0^{2\pi} |Y_{q,\ell,m}|^2 d\phi = 1. \tag{17}$$

(Notice that in R_{ab}, $|(Y_{q,\ell,m})_a|^2 = |(Y_{q,\ell,m})_b|^2$.) Different $Y_{q,\ell,m}$ for a fixed q) are orthogonal, a fact one easily proves in the usual way from (15). We shall choose the phases of $Y_{q,\ell,m}$ such that the matrix elements of L_z, L_y, L_z between the Y's conform to the convention adopted in ch. 2 of Edmonds' book[5]. In particular

$$(L_x + iL_y) Y_{q,\ell,m} = (\ell-m)^{\frac{1}{2}} (\ell+m+1)^{\frac{1}{2}} Y_{q,\ell,m+1}. \tag{18}$$

These monopole harmonics will be explicitly exhibited. Each is analytic. That is, $(Y_{q,\ell,m})_a$ is analytic in R_a and $(Y_{q,\ell,m})_b$ is analytic in R_b. The set of all monopole harmonics for a fixed q forms a complete set of sections, as we shall see.

Explicit expressions for $Y_{q,\ell,m}$

Stating from (12) one easily verifies

$$L^2 = [r \times (p - eA)]^2 + q^2, \tag{19}$$

$$mY_{q,\ell,m} = L_z Y_{q,\ell,m} = (-i\partial_\phi - q)Y_{q,\ell,m}, \text{ in } R_a,$$

$$mY_{q,\ell,m} = L_z Y_{q,\ell,m} = (-i\partial_\phi + q)Y_{q,\ell,m}, \text{ in } R_b. \tag{20}$$

Eq. (20) shows that

$$Y_{q,\ell,m} = \Theta_{q,\ell,m}(\theta)e^{i(m+q)\phi} \quad \text{in } R_a,$$

$$Y_{q,\ell,m} = \Theta_{q,\ell,m}(\theta)e^{i(m-q)\phi} \quad \text{in } R_b. \tag{21}$$

The condition for a section shows that $[\Theta_{q,\ell,m}(\theta)]_a = [\Theta_{q,\ell,m}(\theta)]_b$ in the overlap. They are, in fact, the same function. Apply (19) to $Y_{q,\ell,m}$. An explicit evaluation of the operator $[rx(p- eA)]^2$ acting on $Y_{q,\ell,m}$ gives

$$[\ell(\ell + 1) - q^2]\Theta_{q,\ell,m} = \left[-\frac{1}{\sin\theta}\frac{\partial}{\partial\theta}\sin\theta\frac{\partial}{\partial\theta} + \frac{1}{\sin^2\theta}(m + q\cos\theta)^2\right]\Theta_{q,\ell,m}. \tag{22}$$

Writing $\cos\theta = x$, this gives

$$[\ell(\ell + 1) - q^2]\Theta = -(1-x^2)\Theta'' + 2x\Theta' + \frac{1}{1-x^2}(m + qx)^2\Theta,$$

$$-1 \leq x \leq 1, \tag{23}$$

where prime means differentiation with respect to x. This equation can be treated in the usual way, through analyzing the indical equations at $x = \pm 1$. We shall, however, pursue a different method which yields the normalization constant and phase factor automatically.

Before proceeding we note that since Y is single valued in each region, (21) shows that

$$m - q = \text{integer}.$$

Thus

$$\ell - q = \text{integer}. \tag{24}$$

Now (19) shows that

$$\ell(\ell + 1) \geqq q^2. \tag{25}$$

Eqs. (24) and (25) show that the allowed values of ℓ are among those given in (16).

We shall now show that each value of ℓ in (16) is allowed, by constructing, for each of them, the explicit function $\Theta_{q,\ell,m}$:

$$\Theta_{q,\ell,-\ell} = N_{q,\ell} \sqrt{1-x}^{\,\ell-q} \sqrt{1+x}^{\,\ell+q}, \quad \ell - |q| = \text{integer} \geqq 0, \tag{26}$$

where

$$N_{q,\ell} = \left| \frac{(2\ell+1)!}{4\pi(2^{2\ell})(\ell-q)!(\ell+q)!} \right|^{\frac{1}{2}} > 0. \tag{27}$$

To show this one substitutes (26) into (23) and verifies that the latter is satisfied. The factor $N_{q,\ell}$ is inserted so that $Y_{q,\ell,-\ell}$ is normalized in the sense of (17).

Repeated application of (18) onto the monopole harmonics $Y_{q,\ell,-\ell}$ (given by (21) and (26)) leads to, (for ℓ,m satisfying (16)) the explicit expression for $Y_{q,\ell,m}$ given[3] below. (As stated above, this method leads to automatically normalized $Y_{q,\ell,m}$ starting from normalized $Y_{q,\ell,-\ell}$).

$$(Y_{q,\ell,m})_a = M_{q,\ell,m}(1-x)^{\alpha/2}(1+x)^{\beta/2} P_n^{\alpha,\beta}(x) e^{i(m+q)\phi},$$

$$(Y_{q,\ell,m})_b = (Y_{q,\ell,m})_a e^{-2iq\phi}, \tag{28}$$

where

$$\alpha = -q - m, \qquad \beta = q - m, \qquad n = \ell + m, \qquad x = \cos\theta, \tag{29}$$

$$M_{q,\ell,m} = 2^m \left| \frac{2\ell+1}{4\pi} \frac{(\ell-m)!(\ell+m)!}{(\ell-q)!(\ell+q)!} \right|^{\frac{1}{2}}, \tag{30}$$

and $P_n^{\alpha,\beta}(x)$ are the Jacobi polynomials,

$$P_n^{\alpha,\beta}(x) = \frac{(-1)^n}{2^n n!}(1-x)^{-\alpha}(1+x)^{-\beta} \frac{d^n}{dx^n}[(1-x)^{\alpha+n}(1+x)^{\beta+n}], \tag{31}$$

which are defined if

$$n, n + \alpha, n + \beta \text{ and } n + \alpha + \beta \text{ are all integers} \geqq 0. \tag{32}$$

Completeness of Monopole Harmonics

For a given q(q may be negative) the set of $Y_{q,\ell,m}$ with ℓ,m satisfying (16) form a complete set of orthonormal sections. I.e. every continuous section (i.e. a section satisfying (9), with ξ_a and ξ_b being continuous in R_a and R_b) can be expanded as a series

$$\sum_{\ell,m} a_{\ell,m} Y_{q,\ell}$$

Proof: $Y_{q,\ell,m}$ can be expressed[3] in terms of $P_\nu^{|\alpha|,|\beta|}(x)$. Now for fixed q = integer or half-integer, and q + m = integer, there are four possible cases:

$$\alpha \geq 0,\ \beta \geq 0,\ \text{so that } -m \geq |q| \quad \text{and} \quad \nu = \ell + m, \tag{33}$$

$$\alpha \geq 0,\ \beta \leq 0,\ \text{so that } |m| \leq -q,\ q \leq 0 \quad \text{and } \nu = \ell + q, \tag{34}$$

$$\alpha \leq 0,\ \beta \geq 0,\ \text{so that } |m| \leq q,\ q > 0 \text{ and } \nu = \ell - q, \tag{35}$$

$$\alpha \leq 0,\ \beta \leq 0,\ \text{so that } m \geq |q| \quad \text{and } \nu = \ell - m. \tag{36}$$

In case (33), the allowed values of ℓ, according to (16), are $\ell = |m|, |m| + 1, \ldots$ which are precisely

$$\nu = 0, 1, 2, \ldots . \tag{37}$$

In case (34), the allowed values of ℓ according to (16) are $\ell = q, -q + 1, \ldots$ which are also precisely (37). Continuing this way we conclude that given q = integer or half-integer, q + m = integer, the allowed values of ℓ according to (16) are always precisely those given by (37).

Now for fixed $|\alpha|$, $|\beta|$, the Jacobi polynomials $P_\nu^{|\alpha||\beta|}$,(ν=0,1, 2,...) form a complete set. The exponential functions $e^{i\phi(m+q)^\nu}$, m+q = all integers) also form a complete set. It can be proved from these results that $Y_{q,\ell,m}$ forms a complete set of section for fixed q.

Examples and Analyticity of $Y_{q,\ell,m}$

For the case q = 0, $\alpha = \beta$, and (31) shows that

$$P_{\ell+m}^{-m,-m} = \frac{(-1)^m}{2m} \frac{\ell!}{(\ell+m)!} (1-x^2)^{m/2} P_\ell^m . \tag{38}$$

where P^{ℓ}_{m} is the associated Legendre function. Substitution of (38) into (28) shows that

$Y_{0,\ell,m}$ = usual spherical harmonics $Y_{\ell,m}$.

We tabulate in table 1 a few of the monopole harmonics for $q = \frac{1}{2}, 1$, these examples illustrate the fact that $Y_{q,\ell,m}$ is analytic everywhere. I.e., $(Y_{q,\ell,m})_a$ is analytic in R_a and $(Y_{q,\ell,m})_b$ is analytic in R_b. For example, $(Y_{\frac{1}{2}\frac{1}{2}\frac{1}{2}})_a$ is clearly analytic in R_a, which includes the point $\theta=0$, and

$$(Y_{\frac{1}{2}\frac{1}{2}\frac{1}{2}})_b = \sqrt{1 - \cos\theta} / \sqrt{4\pi} \tag{39}$$

is clearly analytic in R_b which includes the point $\theta=\pi$.

Schrodinger Equation

It is simple to show by explicit evaluation, and with the aid of (19) that

$$(p - eA)^2 = -\frac{1}{r^2}\frac{\partial}{\partial r}\left(r^2 \frac{\partial}{\partial r}\right) + \frac{1}{r^2}[r \times (p - eA)]^2$$

$$= -\frac{1}{r^2}\frac{\partial}{\partial r}\left(r^2 \frac{\partial}{\partial r}\right) + \frac{1}{r^2}[L^2 - q^2]. \tag{40}$$

The Hamiltonian thus commutes with L^2 and L_z. Hence in solving for eigenfunctions of H we can choose specific eigenvalues for L^2 and L_z. I.e. we take

$$\psi = R(r)Y_{q,\ell,m}, \tag{41}$$

obtaining

$$\left[-\frac{1}{2mr^2}\frac{\partial}{\partial r}\left(r^2 \frac{\partial}{\partial r}\right) + \frac{\ell(\ell+1)-q^2}{mr^2} + V - E\right]R = 0. \tag{42}$$

For the case that V = 0 this equation was solved by Tamm who found that R is a Bessel function, if $E>0$,

$$R = \frac{1}{\sqrt{kr}} J_\mu(kr), \tag{43}$$

where

$$\mu = \sqrt{\ell(\ell+1) - q^2 + \tfrac{1}{4}} = \sqrt{(\ell+\tfrac{1}{2})^2 - q^2} > 0,$$

$$k = \sqrt{2mE} \ . \tag{44}$$

If $E \leq 0$, (42) has no meaningful solution.

Table 1

Examples of $\sqrt{4\pi}Y$ in a region

q	ℓ	m	$(\sqrt{4\pi}Y_{q,\ell,m})_a$
$\frac{1}{2}$	$\frac{1}{2}$	$\frac{1}{2}$	$-e^{i\phi}\sqrt{1-x}$
	$\frac{1}{2}$	$-\frac{1}{2}$	$e^{0}\sqrt{1+x}$
	$\frac{3}{2}$	$\frac{3}{2}$	$\sqrt{3/2}\,e^{2i\phi}\sqrt{1+x}(1-x)$
	$\frac{3}{2}$	$\frac{1}{2}$	$-\sqrt{\tfrac{1}{2}}\,e^{i\phi}\sqrt{1-x}(1+3x)$
	$\frac{3}{2}$	$-\frac{1}{2}$	$-\sqrt{\tfrac{1}{2}}\,e^{0}\sqrt{1+x}(1-3x)$
	$\frac{3}{2}$	$-\frac{3}{2}$	$\sqrt{3/2}\,e^{-i\phi}\sqrt{1-x}(1+x)$
1	1	1	$\sqrt{3/4}\,e^{2i\phi}(1-x)$
	1	0	$-\sqrt{3/2}\,e^{i\phi}\sqrt{1-x^2}$
	1	-1	$\sqrt{3/4}\,e^{0}(1+x)$

$x = \cos\theta$. To obtain $Y_{q,\ell,m}$ in R_b apply (8).

Dirac Equation

Using the monopole harmonics discussed above, one can also discuss the motion of a Dirac electron in the field of a magnetic monopole. This was done in references 6 and 7 where bound states were found.

Remarks

(A) It is important to realize that the above-described way of using $(A)_a$ and $(A)_b$ together to describe the magnetic field of a monopole has an additional advantage: It gives the magnetic field H correctly everywhere. In older papers one oftentimes took a single A with a string of singularities. Since by definition

$$\nabla \cdot (\nabla \times A) = 0,$$

the magnetic field described by $\nabla \times A$ must have continuous flux lines. Thus its flux lines consist of the dotted lines of Figure 4, plus the bundle of lines described by the solid line, so as to make the net flux at the origin zero. Thus, $\nabla \times A$ does not correctly describe the magnetic field of the monopole, a point already emphasized by Wentzel.[8]

(B) For ordinary spherical harmonics there are a number of important theorems such as the spherical harmonics addition theorem, the decomposition of products of spherical harmonics using Clebsch-Gordon coefficients, etc. These theorems can be[9] generalized to monopole harmonics.

(C) It is instructive to go back to the reasoning concerning Figure 1 and try to repeat the steps for the combined A_a, A_b description of the magnetic field. Choose the parallel to be the equator. Then

$$\oint (A_\mu)_a dx^\mu = \Omega_\alpha$$

$$\oint (A_\mu)_b dx^b = \Omega_\beta .$$

Thus $$4\pi g = \Omega_a - \Omega_\beta = \oint [(A_\mu)_\alpha - (A_\mu)_\beta] dx^\mu$$

which is, by (6), equal to the increment of α around the equator i.e. $2g(2\pi) = 4\pi g$.

We have arrived at an identity. My reason for going through this simple argument is that it embodies exactly the gist of the proof of the famous Gauss-Bonnet-Allendoerfer-Weil-Chern theorem and the later Chern-Weil theorem which play seminal roles in contemporary mathematics. As a matter of fact, gauge field, of which electromagnetism is the simplest example, is conceptually identical to some mathematical concepts of fiber bundle theory. Table 2 gives[2] a translation table for the terminologies used by physicists on the one hand and mathematicians on the other. We notice that in particular Dirac's monopole quantization (9) is identical to the mathematical concept of classification of U(1) bundles according to the first Chern class.

Table 2. Translation of Terminology

Gauge Field terminology	Bundle terminology
gauge (or global gauge)	principal coordinate bundle
gauge type	principal fiber bundle
gauge potential b_μ^k	connection on a principal fiber bundle
S Eq.(8)	transition function
phase factor ϕ_{QP}	parallel displacement
field strength $f_{\mu\nu}^k$	curvature
source[a] J_μ^k	?
electromagnetism	connection on a U_1 bundle
isotopic spin gauge field	connection on a SU_2 bundle
Dirac's monopole quantization	classification of U_1 bundle according to first Chern class
electromagnetism without monopole	connection on a trivial U_1 bundle
electromagnetism with monopole	connection on a nontrivial U_1 bundle

[a] I.e., electric source

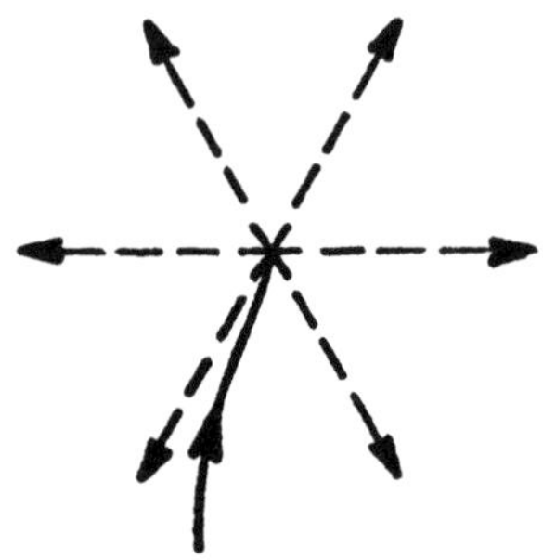

FIGURE 4. Magnetic Flux lines due to A. Since $\nabla\cdot(\nabla \times A) = 0$, flux lines are everywhere continuous. Hence there is "return flux" along solid line.

FIGURE 5. Examples of a trivial and a nontrivial fiber bundle.

The last two entries of the table identifies electromagnetism with and without magnetic monopoles with connections on trivial and nontrivial U(1) bundles. We can gain some understanding of these facts by looking at (i) a paper loop and (ii) a Moebius strip (Figure 5). If they are cut along the dotted lines, each would break into two pieces. Looking at the resultant pieces we cannot differentiate between cases (i) and (ii). The two cases are different only in the way the "resultant pieces" are put together. In case (ii), a twist of one of the "resultant pieces" is necessary. Thus case (i) corresponds to

$$\psi_a = S\psi_b, \text{ where } S = 1, \text{ (no twist);}$$

and case (ii) corresponds to ($q \neq 0$),

$$\psi_a = S\psi_b, \text{ where } S = \exp(2iq\phi), \text{ (twist).}$$

A bundle where the transition function S is necessarily different from 1 is called nontrivial, because a twist is needed. Hence electromagnetism with a megnetic monopole is nontrivial [cf.(7) and (8)].

FOOTNOTES

[1] P.A.M. Dirac, Proc. Roy. Soc. A133, 60 (1931).

[2] Tai Tsun Wu and Chen Ning Yang, Phys. Rev. D12, 3845 (1975).

[3] Tai Tsun Wu and Chen Ning Yang, Nuclear Phys. B107, 365 (1976).

[4] M. Fierz, Helv. Phys. Acta 17, 27 (1944).

[5] A.R. Edmonds, Angular Momentum in Quantum Mechanics (Princeton,1960).

[6] Yoichi Kazama, Chen Ning Yang and Alfred S. Goldhaber, to appear in Phys. Rev. D.

[7] Yoichi Kazama and Chen Ning Yang, to appear in Phys. Rev. D.

[8] G. Wentzel, Prog. Theor. Phys. Suppl. 37-38, 163 (1966).

[9] Tai Tsun Wu and Chen Ning Yang - to be published.

DISCUSSIONS

CHAIRMAN: Prof. C.N. Yang

Scientific Secretaries: N. Parsons and B. Jancewicz

DISCUSSION 1

- *PHAM QUANG:*

Could you comment on the differences between Dirac's quantization relation and Schwinger's quantization relation which has twice the value of Dirac?

- *YANG:*

We agree with Dirac's quantization relation for g and found no reason for Schwinger's. Schwinger required two strings for reasons of additional symmetry. In our approach, *no* strings are required at all, provided we keep the vector potentials A_a^μ and A_b^μ in the allowed regions R_a and R_b, respectively. One can deal with the singularities in the forbidden regions arbitrarily. In our opinion, the quantization rule of Schwinger is groundless.

- *BERLAD:*

Can one construct operators which will cause transitions between states of different q?

- *YANG:*

Maybe, but one does not know how to do it and be physically meaningful.

- *KLEINERT:*

From a purely group theoretical approach, such operators are easily constructed. Your Y_{qlm} for fixed q form a representation of

O(3.1). The quantum number q seems to be analogous to the helicity of a rotating top so that the angular momentum, of course, has to be greater than the intrinsic helicity of such a top. The quantum number q seems to fill precisely the same role here. For different values of q, the Y_{qlm} seem to form representations of a larger group, for example O(4.1), in which you can construct these raising and lowering operators.

- *YANG:*

You may be right. The electron-monopole problem is certainly more complex than that described by two coordinates θ,ϕ for fixed r, since the electromagnetic field has a momentum density distribution and therefore has inertia. A complete clarification of this type of problem is yet to be made.

- *GARCIA:*

Could you clarify the motivation for introducing the extra term in the expression for the angular momentum commutation relations in the presence of a singular potential?

- *YANG:*

It was first shown by Saha that if you have a monopole of strength g and an electron charge e, then everywhere in space you have crossed $\bar{E}$ and $\bar{H}$ fields. You will have a non-zero Poynting vector

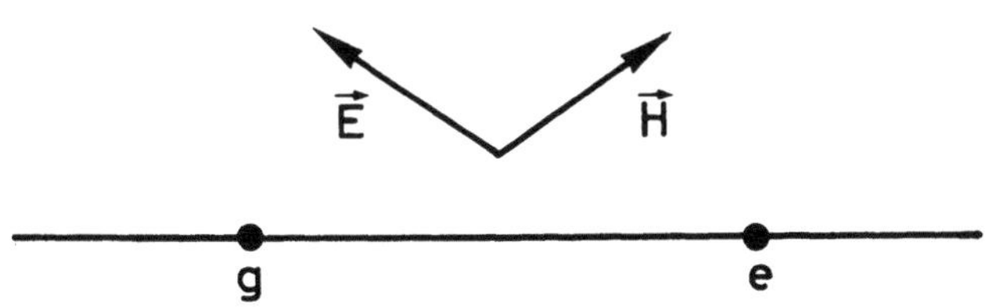

The system has cylindrical symmetry about the line joining the electron and the monopole. Hence the Poynting vector, and the field momentum, points in an azimuthal direction with respect to the axis of symmetry, giving rise to an angular momentum $-q\vec{r}/r$ from the electron to the monopole.

On the other hand, Fierz, in 1944, observed that the term $-q\vec{r}/r$ is needed to obtain the correct commutation relations for the total angular momentum. Both points of view should be equivalent in a field theory of electrons and monopoles in interaction with the electromagnetic field. However, such a field theory is still to be worked out.

There is, however, a problem in the paper of Fierz owing to boundary conditions which were not definable due to string singularities. In our section approach, we have "opened up" the singularities instead of bundling them up into a string. In this way, we are able to handle them rigorously.

- *JONES:*

In the Dirac approach, the total flux leaving a sphere surrounding a monopole is zero because all the flux comes back in along the string, even though you cannot see where the string is. In your approach, is the total flux leaving the charge equal to $4\pi g$?

- *YANG:*

Yes. There is, in our scheme, no "return flux" at all. In Dirac's scheme, the vector potential has zero curl. This inevitably leads to a divergenceless magnetic field. In other words, there must be a return flux, as you said. This return flux in the past was supposed to be cancelled by the "Dirac veto", and led to much confusion. In our approach, this confusion is removed from the beginning by not considering a string of "return" flux at all.

- *LIPKIN:*

To describe one monopole, you have divided space into two regions. Would you need more regions to describe systems with several monopoles?

- *YANG:*

One needs more regions when one has more monopoles of whatever sign.

- *LIPKIN:*

Is there a simple relation between the number of regions and the wrapping number discussed in Wick's lectures?

- *WICK:*

You need only two regions for one monopole, no matter what the value of $q = ge$ is. Therefore, I do not think there is a connection with my wrapping number. However, there may be a somewhat different way of formulating the question, to that it makes sense.

- *WIGNER:*

If one writes down the ordinary Dirac equation for two oppositely charged monopoles, the Hamiltonian is not self-adjoint -- it shows a mathematical pathology as discussed by von Neumann. Does this difficulty manifest itself also in quantum field theory, and if so, how?

- *YANG:*

This same difficulty exists. However, the difficulty does not imply that quantum field theory equations for magnetic monopoles are necessarily meaningless. It merely means that one cannot interpret the "static approximation". I believe the situation is similar to the case for ordinary QED in the thirties, or even now: the equations for QED are probably correct, but have to be understood correctly. With monopoles included, the renormalization theory, particularly in view of the large value of g, bec mes a completely open question.

- *WIGNER:*

One thing that I am not clear about is whether a theory with a coupling constant of the order of $g \sim 137$ can be useful because of the difficulties of convergence.

- *YANG:*

I agree, but there is one thing I would like to add. It is also not known whether QED forms a convergent series in α; in fact, there are many guesses that it does not converge in α.

- *WIGNER:*

That is true, but QED is still useful; whether it would still be useful with α equal to 137 rather than 1/137 is less clear.

- *YANG:*

Yes, I agree completely; it is 137^2 less clear!

- *WIGNER:*

What are the experimental indications for the existence of the monopole?

- *YANG:*

Last summer, 1975, Price and collaborators published a paper reporting on experimental evidence for a magnetic monopole. The report generated many discussions. It is generally regarded now as not conclusive.

- *SOHNIUS:*

Is the particular form of the Fierz term gauge invariant?

- *YANG:*

Yes, any $\underline{A}$ describing a magnetic monopole would lead to the same $-q\bar{r}/r$ in $\bar{L}$.

Let me raise the following question: Can Dirac's monopole idea be generalized to the case of SU(2) gauge fields?

The answer to this question depends on its precise interpretation. I believe the most important characteristics of Dirac's monopole is that it is the *source* of magnetic flux. Thus one should look for a generalization of the following type. What SU(2) gauge field has a point source of some "generalized flux"? It turns out that in five-dimensional Euclidean flat space or on a four-dimensional Euclidean sphere, there are such generalizations, but not in lower space dimensions. Furthermore, in these two cases, you get only two monopoles nnd not an infinite number as Dirac's quantization rule allows.

- *MARCIANO:*

Is your Yang-Mills SU(2) magnetic monopole a point source magnetic monopole with infinite energy? If so, then why can this not exist in 3 + 1 dimensions?

- *YANG:*

It is a point magnetic monopole with infinite energy. For lower dimensions, such as in four-dimensional flat space, generalization do not exist which have a point source of a "generalized flux".

DISCUSSION 2

- *PARSONS:*

You defined a quantity α^{μ} by $A^{\mu}_{a} - A^{\mu}_{b}$. Presumably, this quantity is only defined in the intersection of R_a and R_b. Is that correct?

- *YANG:*

Yes. You can try to continue α_{μ} into R_a or R_b, but it will have singularities in these regions.

- *PARSONS:*

Presumably, then, that is why you can allow the two conditions to apply of having a zero four-dimensional curl of α and a non-zero loop integral without inconsistency?

- *YANG:*

Yes, that is correct. The fact that we have a zero curl means that one can deform the four-dimensional loop in any way and not alter the value of the integral provided it always remains in R_{ab}. Because of this, the loop is not contractable to zero and the integral is equal to a space-time independent numerical constant. This means that we have a non-trivial fibre bundle associated with the monopole.

- *JANCEWICZ:*

I did not understand completely your picture:

How many dimensions does the cone have? How many dimensions has the loop surrounding it?

- *YANG:*

The surface of the cone is a three-dimensional manifold and the loop is a one-dimensional loop. If we consider the cone at *fixed time*, the surface is two dimensional. As the time is varied, the apex of the cone will move in a time-like direction so the surface of the cone will trace out a three-dimensional surface in four-dimensional space-time. The loop is just a one-dimensional loop. One can draw it at one time and then one is able to distort it in four-dimensional space-time so that different parts of the loop are at different times. Nevertheless, one cannot shrink it to a point.

- *MARCIANO:*

Can your formalism be trivially extended to include dyons?

- *YANG:*

Probably, but we have not done this. I would not say it is trivial. Nothing is quite trivial since the joint idea is new to us.

- *MARCIANO:*

Can you comment on the soliton type of monopole, such as the t'Hooft and Polyakov type of soliton?

- *YANG:*

If you have a pure non-Abelian gauge field, not interacting with anything else; I would think that if you had a solution in this, it would be most interesting, but such a solution has not been found without singularities. Wu and I found three solutions about nine years ago, but all of them are singular at the origin. The t'Hooft and Polyakov solution consisted in coupling the gauge field with an isovector scalar field. In that case, they found that you can have an analytic solution.

Let me make a further comment about solutions with pure gauge fields not interacting with anything else. I have said that you cannot have an analytic solution without singularities. That is not exactly correct. No-one has found analytic solutions if one confines oneself to compact gauge groups. But, if one allows oneself the luxury of non-compact gauge groups, one can write down analytic solutions with no singularities. Wu and myself have found such a static solution for a gauge field of the non-compact Lorentz group which has just been published in *Physical Review*. For such a non-compact group, one can find a simple solution with the following properties: i) it is static; ii) it is everywhere analytic; and iii) the field goes to zero at infinity. We believe that there is a theorem, which we have not been able to prove, which says that unless the group is non-compact, one cannot have such a solution. This theorem was supposed to have been proved in last year's lecture notes by Coleman. But the proof breaks down, because it is based on a fallacious scaling argument.

- *VINCIARELLI:*

There exist examples of *bona fide* monopole-solitons within a "pure" gauge theory based on compact gauge groups, such as SU(2). In this case, the SU(2) gauge invariance must be broken down to U(1) by the interaction of explicit mass terms for two of the three gauge fields. The U(1) monopole which one obtains has a finite energy in spite of the presence of a physical pole at the location of the point-like magnetic charge.

- *YANG:*

This solution seems to be in a different ball park, as it possesses a singularity at the origin. I feel that a solution without such a singularity would be more interesting, because of the question of convergence. A convergent solution may indeed point to the direction in which physics must move, but, unfortunately, this seems to be impossible at present.

- *BUCCELLA:*

Up to now, there is no convincing evidence for magnetic charges. I wonder whether one can associate the magnetic charge with the unobserved colour degrees of freedom in the same way as electric charge is associated with the observed flavour degrees of freedom?

- *YANG:*

This is an interesting question. I have not thought about it.

- *PAULI:*

What are SU(2), or SU(3) or SU(2) × U(1), etc., fibre bundles?

- *YANG:*

SU(2) bundles can be considered as generalizations of U(1) bundles. It is a new topic for physicists and has not yet been very much studied.

You divide space-time into regions and try to match the SU(2) gauge fields in one region with those in another region in the overlap of the two regions by a non-Abelian gauge transformation. If the gauge transformation cannot be deformed to the identity transformation then you have a non-trivial fibre bundle. The mathematical generalization from U(1) has been greatly studied by mathematicians. The only known non-trivial generalizations in the physics literature are due to Polyakov, reported in *Physics Letters*, October 1975, and the cases discussed by me earlier. In all these cases, you need a more complicated space than four-dimensional space-time in order to accommodate the complicated matching between the regions.

- *GARCIA:*

You made the remark that the condition $A_{\mu\mu} = 0$ is unnecessary in the theory of monopoles and electrons. In the quantization of ordinary electrodynamics, the subsidiary condition has an important role in restricting the gauge freedom. What is the situation in the quantized monopole theory?

- *YANG:*

The situation is quite similar to that in the quantization of electrodynamics. We think that one can quantize the theory if one wishes to make a specific choice of gauge.

- *PHAM QUANG:*

You talk about a theory of electrons, positrons, and monopoles. What should be the spin of the monopole? Is it a fermion or boson?

- *YANG:*

That theory is not yet complete. Tu, Wu, and I are working on it. We believe the spin could be 0 or ½ or higher. For integral spin, it would be a boson; for spin-½, a fermion, as usual.

- *ROSSI:*

You spoke of the monopole-electron system as having some properties analogous to those of a spinning top. Besides that, the equations you have written to assure the integrability of the action in the overlap region recall some general relativistic expressions which appear in the study of non-irrotational manifolds, and, in that context, are referred to the gravitational potentials.

Are there any connections between these properties of the gauge fields and the properties of the gravitational potential? In other words, can we look at the gravitational field as a gauge field, and vice-versa? Can we look at gauge properties as linked to geometrical properties of space?

- *YANG:*

It seems very likely that gravitation is, in some sense, a gauge field, but how it is a gauge field is not yet clear. I discussed this topic in *Physical Review Letters* 33, *445, 1974*. The discussion is, however, still incomplete.

DISCUSSION 3 *(Scientific Secretaries: A. Garcia and B. Yoon)*

- *TELEGDI:*

I think that it is not generally known that the work of Gibbs was essentially paralleled in the first two or three papers of Einstein's, where he reinvented the canonical and grand canonical ensembles.

A second remark: if I remember correctly, Einstein also knew about De Broglie's paper at the time that he received Bose's paper. Once you had the light waves and you could do particle counting with the photons, the jump to the real gas with the matter waves was maybe made a bit easier by this idea of De Broglie's. I think he should get some credit.

- *YANG:*

About the first remark, I have only recently learned that Einstein had duplicated a lot of what Gibbs wrote in 1901. As to the second point, certainly De Broglie's contribution was extremely important, but everything must have been in a fog at the time. It required great insight and boldness, so characteristic of Einstein's thinking, to marry two things together and reach the right conclusion.

- *WIGNER:*

You mentioned the great accomplishment of Fermi's to have provided Fermi statistics. I was in Berlin at that time and I know that Bose's article was an enormous surprise and Einstein was delighted with it. The idea that particles behaved in the opposite was was, to Einstein, quite obvious. It was interesting to see that this explained what is called Duder's paradox. But, that there is an opposite behaviour of particles due to Pauli's exclusion principle was, in Berlin, taken for granted. This shows that different groups of physicists take different things for granted and are astonished by different things.

- *YANG:*

I think this is a very interesting point and I would like to pursue it a bit further. A year and a half ago, when I was writing this article and was particularly thinking about Fermi-Dirac statistics, I had a conversation with Professor Wigner at Rockefeller University. Then I remarked to Professor Wigner that, perhaps, Fermi-Dirac statistics is the greatest contribution of Fermi's in physics proper, with which he immediately disagreed. I was completely taken by surprise, specially when he asserted that the greatest contribution of Fermi's was the theory of β-decay. I was surprised because Fermi-Dirac statistics is, in many senses, essentially a final story; β-decay theory, although it is a great piece of work which produced a profound impact, is not a final theory. In retrospect, as a final judgement of its position in the history of physics, I think we must put Fermi-Dirac statistics first. I would like now to ask Professor Wigner whether he agrees with this assessment.

- *WIGNER:*

You will forgive me, but I am afraid I do not. At that time we all believed, perhaps foolishly, that there were electrons in the nucleus. I calculated myself the kinetic energy of the electrons and it did turn out to be large but not terribly so. Then came the discovery of nitrogen. The nitrogen nuclei obeyed Bose statistics, and this was terrible; the nitrogen nucleus clearly consisted of 14 protons and seven electrons and we did not know what to do. Fermi's paper had the idea of the creation of particles, which was at that time a great surprise, even if it is now quite natural because we all create particles with a^+ and similar operators.

- *YANG:*

What you say illuminates the situation to me but there are other aspects of it that are opaque to us at this time because we have seen too much of what happened later. The creation and a-nihilation operators were familiar to you in the late twenties,and, furthermore, Dirac derived the correspondance principle of atomic transitions using quantum field theory. So creation and annihilation operators were already used.

- *WIGNER:*

We had operators for the creation of particles but the fact that particles were really created and that neutrinos were created at the same time -- well, we just did not think of this.

- *YANG:*

Is it the case, therefore, that by the late twenties, it was recognized that photons could be created and annihilated, but a material particle could not be created and annihilated, despite your paper with Jordan?

- *WIGNER:*

Yes, that is the way our thoughts went.

- *TELEGDI:*

If you look in Pauli's Handbuch article, it seems that the a, a^+ formalism was considered a luxury, a trick, because then one could, by cheating, obtain most of the results on photon emission, it was very close to classical physics. So, it appeared that these operators were weird mathematical techniques one did not obviously need. When he heard of Pauli's idea of the neutrino, Fermi recognized that there was a wonderful application to try the usefulness of the formalism.

DISCUSSION 4

- *WILKIE:*

I was always told that there is an ambiguity in the definition of form factors. As you use one specific choice of form factors, Sachs, as a measure of the matter distribution, can you give the physical reasons for making this choice?

- *YANG:*

In the limit that the mass $\to \infty$, but the magnetic moment remains finite, e.g. for a hydrogen atom with proton mass = ∞ and no proton spins, the Sachs form factors, G_E and G_M, have definite geometrical meanings. If the mass is $< \infty$, these geometrical meanings become fuzzy because of the recoil velocity, not recoil momentum, which is not bothersome.

- *LEADER:*

Although I do not believe that anyone really predicted the growth of σ_{tot}, it should be noted that on the basis of cosmic ray multiplicities, Heisenberg suggested in a published paper that σ_{tot} would grow like $(\log s)^2$ long before the work of Cheng and Wu.

- *YANG:*

I did not mention Heisenber's work because, in my opinion, his guess was based on the physics so far back in time that it really has no relationship with our present picture of high-energy collisions.

- *LEADER:*

You have given a nice heuristic derivation of the possibility of having both the density ρ, and the current j, play a role in hadron-hadron scattering. However, as early as 1969 Martinis et al., suggested that the natural generalization of $\rho \otimes \rho$ should be $j_\mu j^\mu$, and they examined the consequences in detail. Also, in the past few years, the Marseilles group of Bourrely, Sofer and Wray have numerically investigated models based on an eikonal proportional to $j_\mu j^\mu$ and have studied the resulting spin effects.

- *YANG:*

I am unfamilair with Martinis' work, I recently talked with Bourrely and Sofer. Their spin effects are quite different from ours.

- *LEADER:*

I think that the absence of a second dip in the recent ISR data on elastic pp scattering at large $|t|$ will effectively eliminate models based on taking $\rho_{hadronic}$ proportional to ρ_{em}. All such models predict a sharp second dip inside the region covered by this experiment.

- *YANG:*

I want to make two remarks. First, the ISR result in the $t \sim 7\ (GeV/c)^2$ region is statistically not weighty. It certainly has much fewer counts than the preliminary FNAL data.

Secondly, I would strongly disagree with your statement that if the second dip is not found it would "eliminate" models based on $\rho_{hadron} \propto \rho_{charge}$. "Eliminate" is a strong word. If a parameterless fit works for a region of 10^{10} and runs into trouble at smaller values, I would hesitate to immediately eliminate it. That is based on simple physical sense.

- *LEADER:*

You rely on the growth of σ_{tot} to produce spin effects. This would imply that in the large region, 30 GeV/c $\leq p_T \leq$ 100 GeV/c, where σ_{tot} is flat, that no spin effects should be seen. Is this a prediction of your picture? Moreover, if you insist on p^{α} growth, then you certainly cannot neglect the real part of the amplitude.

- *YANG:*

I will comment on the second question first. Increase of σ_T carries implications about the real part of the amplitude. When one aims for an improvement of the theory, one would put it in.

About the first question, it is clear that at low energies there *are* other effects that have spin dependence. Wu and I are working on the problem of how to make a "two-component" approach to the spin effects in the intermediate region. But, we do not know of a unique approach. Without the "low-energy" effects, flat σ_T would, in our picture, lead to an R parameter = 0, because there would then be no right-left difference of blackness.

- *LEADER:*

You justify your approach by considering scattering from a nucleus. However, your method breaks down for the scattering of two nuclei with high spin states. In your picture, you rely on the fact that sy, the spin component perpendicular to the scattering plane, cannot change. This will not be true for higher spins. Worse yet, for each value of sy there will be many different spin states possible for the nuclei. Thus your eikonalization procedure will fail. It therefore disturbs me that you appeal to the nuclear picture to support your theory.

- *YANG:*

It does not disturb me at all. For higher spin targets and beam, the analysis is necessarily complicated. Brown and his collaborators are working on this problem.

Let me emphasize thet we do not "justify" anything. The concept of hadronic matter current is clear to us, although not precisely defined. The nuclear discussion is (a) for its own interest and (b) to convince people of the usefulness of the hadronic matter current idea.

- *GOURDIN:*

I suppose that in your approach, the matter density ρ, and the current density $\vec{j}$, are members of a Lorentz four-vector. However, you identify these quantities which are frame dependent, with the form factors, which are, by definition, Lorentz invariant scalars. Therefore, such an identification has to be done in a particular frame of reference. What happens in other frames? For instance, what are the corrections in going from a proton at rest to a fast-moving proton?

- *YANG:*

In my answer to Doctor Wilkie's question, I already touched upon this point. I now repeat: if the recoil velocity is $\neq 0$, the physical picture becomes fuzzy. Thus, for $t/M_p^2 > 1$, there are troubles with recoil velocity which we do not now know how to deal with.

- *BERLAD:*

In 1968, working on my M.Sc., I tried to follow up on Professor Yang's idea of the "coherent droplet model" by giving an invariant formulation of the absorption phase shift. One starts from two time-independent matter densities, no spin included, in the two respective rest frames of the colliding particles. These densities are then boosted to a common frame, and, hence, two currents are obtained. The absorptive phase shift is then constructed by contracting the two vectors $j_\mu^{(1)} j_\mu^{(2)}$. As a result, one gets back Professor Yang's results, multiplied by a simple kinematical factor, i.e.

$$\rho_1 \otimes \rho_2 \to \frac{1}{V_{\text{relative}}} \rho_1 \otimes \rho_2$$

where

$$V_{\text{relative}} = \frac{\beta_1 + \beta_2}{L + \beta_1 \beta_2} , \qquad \beta = \frac{v}{c} .$$

This factor, of course, goes to 1 as the energy grows; and Professor Yang's formula is recovered. This paper was rejected by *Il Nuovo Cimento* because it was not interesting enough. *(This last remark has been added by special request of Professor Zichichi.)*

- *WEILL:*

The "geometrical" picture that you gave for hadron-hadron scattering is a very accurate description of proton-proton scattering. How does it depend on the quark content of the projectile? For example, if one scatters an Σ or Ξ off a proton, does your model predict a new value for the slope of the differential cross-section?

- *YANG:*

We cannot make predictions of that type until we are given information about hadronic matter form factors of Σ or Ξ.

- *ZICHICHI:*

The pion and kaon form factors have been measured in the time-like region. How does your calculation fit the time-like data?

- *YANG:*

Chen, in a paper in *The Physical Review, 1975,* computed with the geometrical picture from $\pi^{\pm}p$ and $K^{\pm}p$ scattering, the hadronic form factors of $\pi^{\pm}$ and $K^{\pm}$. No extrapolation to the time-like region has been attempted.

- *TELEGDI:*

In your talk, you have shown a graph for R as a function of t for different scattering processes; why was $K^{-}p$ very different from $K^{+}p$?

- *YANG:*

Phenomenologically, $K^{+}p$ total cross-section increases faster than $K^{-}p$ at FNAL energies.

- *TING:*

Can you comment on what happens if you use a photon beam?

- *YANG:*

We have not considered it.

- *LIPKIN:*

The parameter α used is really d/dp (log σ_{tot}). Is this what you actually use, rather than a power fit to the data over an extended region? If you are using both $\sigma(K^{-}p)$ and $\sigma(K^{+}p)$, you should use $\sigma(\bar{p}p)$ as well as $\sigma(pp)$. If the difference between particle and antiparticle scattering is a low-energy phenomenon outside the scope of your analysis, you should disregard both $\sigma(K^{-}p)$ and $\sigma(\bar{p}p)$. If you think that your model works also for the mechanism responsible for the particle-particle and particle-antiparticle difference, you should consider all cross-sections. There is no qualitative difference between $\sigma(\bar{p}p)-\sigma(pp)$ and $\sigma(K^{-}p)-\sigma(K^{+}p)$. The reason why $\sigma(\bar{p}p)$ is still decreasing at 200 GeV while $\sigma(K^{-}p)$ is increasing is irrelevant; it is because $\sigma(K\ p)$ increases faster than $\sigma(pp)$.

- *YANG:*

Wu and I are working on some two-component picture.

- *BUCCELLA:*

The hypothesis that the electric charge density is proportional to the matter density seems to be disproved by the behaviour of $F_2^n(x)/F_2^p(x)$ near x = 1. The behaviour implies, in the framework of the quark-parton model, that the p_0 and the n_0 momentum distribution in the proton are different.

- *YANG:*

We stated in our paper that for the proportionality question we are confused about the neutron form factors.

- *SAKURAI:*

I understand that your hadronic density distributions are isoscalar, i.e., the same for the proton and the neutron; yet you identify your hadronic density with the proton charge density, which is the sum of the isovector density and the isoscalar density. Perhaps you would say that we should not use the neutron density because there is a delicate cancellation in the neutron case. But, in the neutron magnetic density, there is no delicate cancellation! $G_\mu^{(n)}$ is large and its sign is opposite to that of $G_\mu^{(p)}$. It appears totally arbitrary to me that you identify the hadronic magnetic density with the proton magnetic density. If you identify it with the neutron density, your prediction for R changes sign. If you identify it with the isoscalar magnetic density, you will presumably predict a very small value for R.

- *YANG:*

Let me first emphasize that we do not have the same confidence in the proportionality hypothesis as we have in the main points of our paper. We therefore do not predict any R value. We try to estimate it with the proportionality hypothesis. Your discussion is logically correct; however, we venture to fix our attention on the protons which, we believe, are more homogenized.

- *GOURDIN:*

Do you use proton form factors which are a superposition of I = 0 and I = 1 quantities for the deuteron target which has I = 0? In other words, have you solved the neutron problem?

- *YANG:*

We are confused by the neutron problem.

- *KLEINERT:*

Does anything go wrong if you prescribe using isoscalar form factors?

- *YANG:*

We have more information with the proton and good agreement with the V-independent part of the theory with the proton data. Thus, we stay with the proton. We are confused about the neutron.

THREE LECTURES ON SOLITONS* †

G. C. Wick

Columbia University

Lecture I

It has been known for a long time, that certain non-linear two-dimensional** wave equations possess "soliton" solutions, describing solitary waves, that travel without changing shape or size. The speed of the waves, however, may depend on their size. For illustration purposes, I shall use the equation:

$$\partial^2\phi/\partial t^2 - \partial^2\phi/\partial x^2 + \operatorname{Sin}\phi = 0 \qquad \text{(S.G.)}$$

which has a long history[1]. Clearly it has stable static solutions ϕ = const (where const $= 0, \pm 2\pi, \pm 4\pi, \ldots$) which will be regarded as fully equivalent states of rest of the system. This is legitimate if we view ϕ as an angle, or as the phase of a complex $z = e^{i\phi}$. Then the position $\phi = 0$ is identical with $\phi = 2\pi n$. It is now very easy to find solutions of the S.G. equation of the form

$$\phi(x, t) = f(x - vt) \qquad (1.1)$$

* I apologize if, for the sake of clarity, one finds in these lectures some repetition of matters already covered in S. Coleman's "classical lumps and their quantum descendants," lectures held at this school in 1975. I have kept overlap with these lectures (quoted as Coleman in the following) to a minimum, I hope, by: a) a different emphasis on subjects b) using recent material and c) referring to Coleman for more detail (especially in my second lecture) whenever this was possible without interrupting the flow of the main argument.

** i.e., in one space- and one time-dimension.

† Research supported in part by U.S. Energy Research & Development Adm.

In fact, assume $|v| < 1$, let $\gamma = (1-v^2)^{-\frac{1}{2}}$ and

$$f(\xi) = 4 \text{ Arc tan } [e^{\pm\gamma\xi + \text{const}}] \tag{1.2}$$

Then (1.1) is a solution. For a given time t, let us plot $\partial\phi/\partial x$ vs. x, see fig. 1.

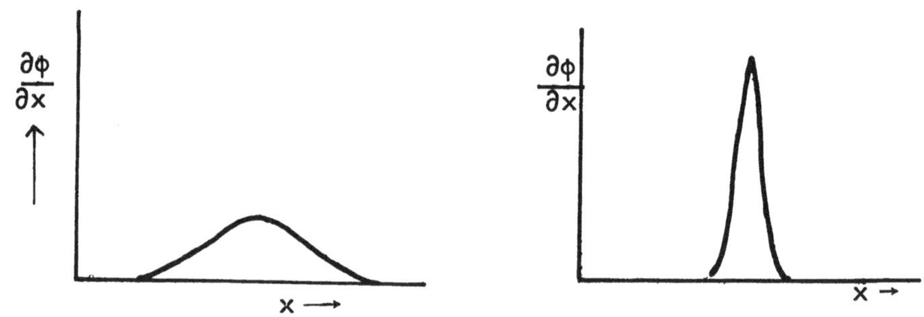

A slow soliton A fast soliton

Fig. 1

We obtain a peak, that will move with constant velocity v without changing shape, as t varies. The positive peak corresponds to the + sign in Eq. (1.2); the - sign gives a negative peak ("antisoliton") . In either case the peak indicates where the energy of the disturbance is located. The width of the peak tends to a finite limit as $v \to 0$ (slow solitons); the peak is narrower and higher for faster moving solitons[2].

Another venerable example is the Korteweg de Vries equation[3] of classical hydrodynamics. It describes waves in a canal; its solitons move all in the same direction, with speeds depending on the size of the soliton: bigger solitons move faster.

Since these wave equations are non-linear the superposition principle does not apply in general. We can, of course, obtain by superposition the asymptotic form (as $t \to \infty$) of the solution corresponding to the collision between two solitons. As soon as the two soliton peaks overlap noticeably, however, the solution will differ from the sum of the two one-soliton solutions. The surprising fact is that, nevertheless, a closed analytic expression for the solution at all times can be given. (This result extends further to n-soliton collisions!) We demonstrate this in the case of the (S. G.) equation.

The calculation will also disclose a rather deep reason for the result. It shows that, given a solution $\phi(x, t)$ of the second order equation (S. G.), we can get another solution $\psi(x, t)$ by solving a system of two first order equations:

$$\frac{\partial\psi}{\partial x} = -\frac{\partial\phi}{\partial t} + \lambda \sin\frac{\psi+\phi}{2} + \frac{1}{\lambda}\sin\frac{\psi-\phi}{2}$$

$$\frac{\partial\psi}{\partial t} = -\frac{\partial\phi}{\partial x} - \lambda \sin\frac{\psi+\phi}{2} + \frac{1}{\lambda}\sin\frac{\psi-\phi}{2} \qquad (1.3)$$

where λ is a constant. We can calculate $\frac{\partial}{\partial t}\frac{\partial\psi}{\partial x}$ from the first equation, and $\frac{\partial}{\partial x}\frac{\partial\psi}{\partial t}$ from the second equation and the result must, of course, be equal. This is in fact the integrability condition of the system (1.3), when ψ is the unknown. It turns out (as the reader can easily verify) that the condition is satisfied precisely if ϕ is a solution of (S.G.)!

In this case, ψ exists. But then notice that Eq.s (3) remain unchanged if we interchange ψ and ϕ and change λ to $-\lambda$[4]; then the calculation we have just sketched also shows that ψ is a solution of (S.G.).

The gist of the argument is then the following: Eq.s (1.3) "define" a transformation[5]

$$\psi = B_\lambda \phi \qquad (1.3')$$

of a solution ϕ of (S.G.) into a new solution ψ. Incidentally B_λ belongs to a general type known in mathematics as "Bäcklund transformations."

Integrating (1.3), when ϕ is given, is much easier than hunting for solutions of the original 2nd order equation. The process can be repeated generating always new solutions. Setting $\phi = 0$ in (1.3), to begin with, the ψ one gets is the one-soliton solution, Eq.s (1.1) (1.2). In the next step, using this ψ as the next ϕ, one generates a solution which describes a collision between two solitons, ...and so on.

There is a remarkable "commutability theorem" of Bianchi[6] which states that the result of two successive Bäcklund transformations is independent of the order $B_{\lambda_2} B_{\lambda_1} \phi = B_{\lambda_1} B_{\lambda_2} \phi$. In proving this theorem, Bianchi obtains as a by-product a remarkable result. Let $\psi_1 = B_{\lambda_1}\phi$, respectively $\psi_2 = B_{\lambda_2}\phi$, be the result of the first

Bäcklund transformation (for $\lambda = \lambda_1$ or λ_2). Bianchi proves by direct verification (it is a little tricky) that the function ψ deffned implicitly by the equation

$$\mathrm{Tan}\frac{1}{4}\ (\psi - \phi) = k\ \mathrm{Tan}\ \frac{1}{4}\ (\psi_1 - \psi_2) \tag{1.4}$$

where k is a suitable antisymmetric function of λ_1 and λ_2, satisfies the Bäcklund conditions (1.3) with $\phi = \phi_1$, $\lambda = \lambda_2$ (or alternatively $\phi = \phi_2$, and $\lambda = \lambda_1$) on the right-hand side. This, of course, proves the commutability property, but we are more interested in another consequence[7] of Eq. (1.4): it gives ψ in finite form; in other words after the first step, no further integrations of the system(1.3) are required ! This method is to be recommended, if one wishes to gain some first hand knowledge of solitons, without delving too deeply into the subject. There are more powerful methods, but their study requires much greater effort.

The explicit form of the two-soliton-(or of the soliton-antisoliton-) solution is given in many papers[8] and it is easy to verify the following remarkable property: after coalescing for a while, the two solitons reemerge and separate with their shapes and sizes (and speeds) completely unchanged, almost as if they had gone through one-another without interacting. This is very remarkable, since it is not required by any of the obvious conservation theorems[9]. Energy, for instance, could be transferred to other degrees of freedom (inelastic collision) but this does not happen. It is this result, that has attracked the most attention, and since several non-linear equations exhibit similar phenomena, it has become customary in the Applied Mathematics literature to include these peculiar properties in the definition of solitons.

The particle-physicist may be interested in solitons, insofar as they constitute stable extended objects, that could conceivably serve as interesting models for certain particles. In the traditional scheme of "second quantization" a wave theory generates particle-like states only after quantization. The "elementary" particles, that is, appear as "quanta" of the linearized form of the wave equation *; in subsequent approximations, other particle-like states may appear as bound states containing two or more elementary particles. Solitons, however, already exist at the classical level, a circumstance that opens up new possibilities for the description of particles in a non-linear field theory[10].

* In particular their mass (if it is not zero) is related to the parameters of the wave equation via Planck's constant in the familiar way.

It is amusing, that in a certain sense, after more than 60 years speculation about the nature of elementary particles has come back full circle to ideas that were first explored by Gustav Mie[11]. Needless to say, the setting is quite different, and the classical extended objects are viewed only as a possible first approximation to states that, we hope, will also exist in the quantized form of the theory.

I must now point out, however, that it is extremely unlikely that any non-linear wave equation in ordinary 3-dimensional space will possess solitons with the very special properties we have mentioned above. We must therefore either enlarge the definition, or eventually drop the name soliton altogether, as some authors propose. I shall go on using soliton in these lectures.

I wish to describe now a particularly simple example[12] of a wave equation in ordinary 3-dimensional space with "soliton" solutions in the general sense mentioned above. In setting up such an equation, one has to by-pass a well-known scaling argument[13] which shows that a three dimensional soliton can only exist in a theory endowed with some additional features, such as a conservation law. The reader may wish to compare the present example to earlier models of "extended objects", which contain fermions. In those models it is the conservation law of the "number of fermions" that helps to stabilize the system. Here, for simplicity, we stick to bosons. The conservation law is obtained by assuming that the Lagrangean contains, besides a real scalar field A, also a complex field ψ. The terms containing ψ are bilinear in ψ and its complex conjugate ψ^*. Specifically the Lagrangian density is:

$$L = -\frac{1}{2}\partial_u A\,\partial^\mu A - \partial_\mu\psi^*\,\partial^\mu\psi - V(A,\psi,\psi^*) \qquad (1.5)$$

The first two terms have the familiar Klein-Gordon form. The potential term:

$$V(A,\psi,\psi^*) = \frac{1}{8}(A^2-1)^2 + k^2A^2\psi^*\psi \qquad (1.6)$$

is chosen so as to have an absolute minimum at $\psi = 0$, $A^2 = 1$. This we shall call the vacuum configuration[14]. Any solution (of the field equations) with a definite energy must have fields tending to their vacuum values as $|\vec{x}| \to \infty$ (i.e. $\psi \to 0$ and $A \to +1$, say).

The conservation law arises, of course, from invariance with respect to the (global) gauge transformation $\psi \to e^{i\alpha}\psi$ where α is an arbitrary real constant. The conserved "charge" is

$$Q = i\int\left\{\psi^*\frac{\partial\psi}{\partial t} - \frac{\partial\psi^*}{\partial t}\psi\right\} d^3x \qquad (1.7)$$

A static Solution (A and ψ time-independent) will have $Q = 0$. Our solitons, however, will be solutions of minimum energy under the constraint, that Q has a given non zero value. One can show, by familiar arguments (we shall not do it) that the usual "separation" of the time dependence can be assumed for these solutions, and that in fact A may be assumed time independent, while $\psi(\vec{x},t) = \frac{1}{\sqrt{2}} e^{-i\omega t} B(\vec{x})$. Furthermore we may assume B real. Now $A(\vec{x})$ and $B(\vec{x})$ obey a set of two coupled partial differential equations, which are easily obtained from the Lagrangian (1.5). For the sake of simplicity we shall consider only spherically symmetric solutions, i.e. we shall write $A = A(r)$, $B = B(r)$, $r = |\vec{x}|$. The equations are then:

$$\frac{d^2A}{dr^2} + \frac{2}{r}\frac{dA}{dr} - k^2B^2A - \frac{1}{2}(A^2-1)A = 0 \tag{1.8a}$$

$$\frac{d^2B}{dr^2} + \frac{2}{r}\frac{dB}{dr} + \left[\omega^2 - k^2A^2\right]B = 0 \tag{1.8b}$$

For such solutions, furthermore, we have:

$$Q = 4\pi\omega \int_0^\infty B^2 r^2 dr \tag{1.9}$$

We notice that these equations allow the transformation $\omega \to -\omega$, $Q \to -Q$. From now on we shall assume that ω, and hence Q, is positive.

It is easy to see, that in a finite energy solution $A^2 \to 1$ and $B \to 0$ at infinity, so that also Q is finite. Now a qualitative insight into the nature of the solution is obtained if we assume that in some way the function $A(r)$ is known, and look upon (1.8b) as a radial Schrödinger equation in which B is the wave function, ω^2 is the energy parameter, and $k^2A^2(r)$ the potential. Since the potential $\to k^2$ at ∞, the energy ω^2 must be $< k^2$. Only in this case can $B(r)$ tend to zero (exponentially!) as $r \to \infty$. Clearly the condition: "Q is finite" is the usual square-integrability condition for a bound state wave function. No bound state will exist, however, unless there is an inner region where the value of the potential is lower than at ∞. Thus we may imagine $k^2A^2(r)$ to be, roughly, as sketched in Fig. 2.

Fig. 2

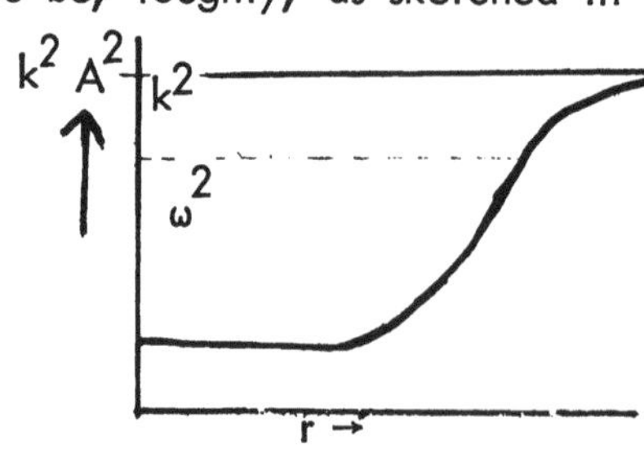

Then, of course, ω^2 must in addition, be one of the discrete eigenvalues corresponding to this potential. The wave function B(r) is also sketched; it is large inside the potential and has a small tail outside. All this is quite convincing, if we remember that the solution must have minimum energy, when we look at the potential term in the Lagrangian, Eq. (1.6). Outside the potential well, $A^2 \approx 1$ and the first term is negligible; but the second term is also small, because ψ, i.e. B is small. On the other hand, inside, where A^2 is smaller, ψ is allowed to take larger values. Of course, once we have determined B(r) from (1.8) for a given A(r) we have hardly solved the real problem. Let us consider, however, the limiting case when Q, Eq. (1.9) is very large. This implies that B(r) inside the well is very large. In this case the second term in the potential (1.6) will be very large inside the well unless A is very close to zero. Thus to keep the energy down as much as possible it seems desirable to have a rather abrupt transition from an inner region (the "potential well") where $A \approx 0$ and B is larger, to an outer region when $A \approx 1$ and B is negligible One can show that a variational method based on trial functions A(r) and B(r) consistent with these expectations is a good approximation when Q is large. The radius R of the well, which is adjusted to yield minimum energy for a given Q, has an optimal value which varies like $Q^{\frac{1}{4}}$. This is then the "size" of the soliton; its energy is proportional to $Q^{\frac{3}{4}}$.

In order to decide whether these solitons are really stable, we must examine whether their energy cannot be transferred to other degrees of freedom. The space surrounding a soliton is, as we know, in the normal vacuum configuration ($\psi = 0$ and $A = +1$, say). Could some of the energy and charge of the soliton be transferred to this region in the form of weak oscillations of the fields A and ψ about the vacuum values? These oscillations are easy to study; they are of two types. There are "pure A - waves" (with $\psi = 0$; they carry no charge and are of no interest in this connection) and there are ψ waves which the authors call "free meson" waves. These waves carry charge and the relation between E and Q is $E = kQ$ for long wavelength[12]. Clearly, if Q is suficiently large, the energy of a soliton ($E \simeq Q^{3/4}$) is less than that of free meson waves carrying the same charge. The emission of free mesons is energetically impossible. We have considered long wavelengths only, but for short wavelengths the situation is even more unfavorable. Please notice that the conservation of the charge Q plays an essential role in the stability consideration !

There are many other interesting points about this example of three-dimensional solitons, which are studied carefully in the original paper. But time is pressing on, and we must turn to a different subject.

Lecture II

We shall now devote some attention to the so-called "topological" solitons. Despite the esoteric name, the stability of these objects is due to quite ordinary energetic factors, like potential barriers and energy differences. The name refers only to the possibility of employing a certain technique, to infer the existence of infinite barriers from qualitative considerations about the continuity of functions or "maps."

All the theories envisaged here contain a gauge-field and possess "degeneracy of the vacuum." Assuming some familiarity with these notions as background, I hope to present a reasonably self-contained discussion of the general ideas involved and of their relation to unfamiliar (to physicists) but classical mathematical theories. In order to get a good grasp of the subject, however, a physicist should look at many different examples. I won't have time for this; even less shall I attempt an exhaustive discussion and classification of all possibilities[15].

The general procedure will be to show, that the set of accessible, finite energy, field configurations of the system decomposes into disconnected components or sectors: there is no continuous path from one sector to another, without going through configurations of infinite potential energy. The following treatment is entirely at the classical level; nevertheless, it may be pertinent to a quantised version of the theory for the following reason. By field-configuration we mean a set of values of field components as functions in space at a given time.

Now assume that these components are commuting variables in the quantised theory[16] (the analogue of Lagrange's generalized coordinates $q_1, q_2, \ldots, q_n$ for an ordinary dynamical system); in this case the study of classical configurations is also a study of the functional space, in which the Schrödinger functional is defined[17].

The following structure is general enough for our purposes. The theory has a local Lagrangian containing a gauge-field $A_\mu(x)$ and a field $\phi(x)$ with values in some abstract n-dimensional* real vector space Φ. With regard to proper Lorentz transformations the components of ϕ are scalars. The Lagrangian is invariant with respect to a continuous group G, which acts as an orthogonal group on the vector ϕ. Leaving out more general possibilities[18] we identify G with this group; simplifying further, we assume G is the group SO(n). (It could be a subgroup).

The action of an element g of G is denoted :

* i.e. ϕ (x) has n real components $\phi_1(x), \ldots ; \phi_n(x)$.

$\phi \mapsto g\phi$. We assume not only invariance with respect to G as a global symmetry, but also with respect to the (much larger) group of gauge transformations, of the form: $\phi(x) \mapsto g(x)\,\phi(x)$. As is well-known, this requires an accompanying transformation of the gauge field and a particular form of the terms in the Lagrangian containing derivatives*. We shall not discuss all this here. We must concern ourselves, however, with the remaining term in the Lagrangian, the "potential" $V(\phi)$, which because of our assumption ($G = SO(n)$) must be a function only of the square of the vector ϕ . For consistency, it must be bounded from below; we may assume its minimum value to be zero (subtract a constant if necessary).

Vacuum Values: In an ordinary theory, $V(\phi)$ has an absolute minimum at $\phi = 0$ and $A_\mu(x) \equiv 0$, $\phi(x) \equiv 0$ is a solution of the field equations, corresponding to an absolute minimum of the total energy. We may call this solution the (classical) vacuum state. In the theories we are studying, $V(\phi)$ attains its minimum value on an infinite set M of values of the field ϕ [19]. A closer look at the field equations shows that, in this case also, $A_\mu(x) = 0$ yields a solution[20] . $\phi(x)$, however, is not zero but may be set equal to any constant value ϕ belonging to the set M . Owing to the assumption: $G = SO(n)$ this set is an S^{n-1} (an (n-1)-dimensional sphere) of equation:

$$\phi_1^2 + \phi_2^2 + \ldots\ldots + \phi_n^2 = \text{const} \;(\neq 0) \qquad (2.1)$$

Notice that, if G were only a subgroup of SO(n), the set M could be a submanifold of the sphere (2.1). Barring "accidental" degeneracy, the set would consist of all vectors of the form $\phi = g\,\phi^0$ (g any $\in$ G) where ϕ^0 is some fixed vector. Just to remember this more general possibility, we shall keep referring to the set of values of ϕ, where $V(\phi)$ attains its absolute minimum, as the manifold M (this also avoids confusion with other spheres we have to mention).

Boundary Values: We now turn our attention to excited states, with finite excitation energy. Since the kinetic part of the energy density is positive definite, and the "gradient terms" also[21], the term $\int V(\phi)\, d^3x$ in the energy must be finite by its own virtue. Hence the condition : "$V(\phi)$ must tend to zero at infinity" or:

"In a finite energy configuration $\phi(\vec{x})$ must tend to a vacuum value (a value on M when $|\vec{x}| \to \infty$." At first sight one would assume that this limiting vacuum value is a constant, independent of direction[22]. The configuration would then differ from one of the zero-

*i.e. derivatives must appear as "covariant" derivatives.

energy vacuum configuarations only in a finite region. $V(\phi)$, however, will also vanish at infinity if we consider (using polar coordinates r, θ, ϕ) the more general assumption:

$$\lim_{r\to\infty} \phi(r,\theta,\phi) = f(\theta,\phi) \qquad (2.2)$$

as long as $f(\theta,\phi)$ lies on M. We can describe f as a "mapping of the boundary at ∞ into M."

A finite energy configuration must also satisfy other conditions, for instance the space-integral of the "gradient terms" must also be finite. Here the gauge field plays a crucial role, but we shall simply refer to Coleman. The upshot of the discussion is, that everything is all right as long as f is smooth, or at any rate continuous. This will do for the following discussion.

Sectors

We must now examine the question, whether any two field-configurations, $\phi = u(\vec{x})$ and $\phi = v(x)$, say, can be connected by a continuous path (or deformation or homotopy). I shall use homotopy since it is the technical term now in common use. Two configurations u and v are said to belong to the same homotopy class (we shall write $u \sim v$) if they can be embedded into a common one-parameter continuous family of configurations. At first sight, there is only one homotopy class, since the linear interpolation $\phi_\lambda(x) = (1-\lambda)\, u(x) + \lambda v(x)$ apparently supplies the desired family. The reason this does not work is the boundary condition, Eq. (2.2). The boundary value of ϕ_λ is $(1-\lambda) f + \lambda g$, where f and g are the boundary values of u and v respectively; this is not a value on M if $\lambda \neq 0$ or 1, see Fig. 3.

Fig. 3

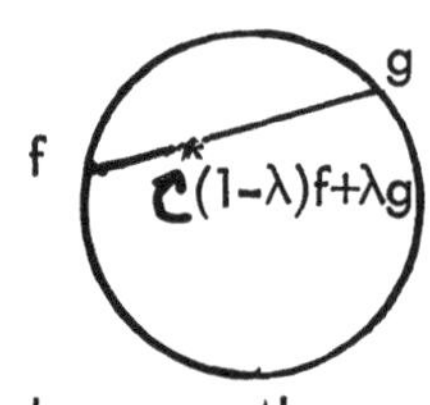

Owing to this, there can be more than one homotopy class.

To proceed further, it may be convenient to introduce the following (now standard) notation. An arrow connecting two spaces: $X \xrightarrow{f} Y$ denotes a continuous function $f(x)$ defined for all $x \in X$, with values in the space Y. The formulation of a homotopy problem always specifies that all functions concerned are from some fixed "domain" X to a fixed "range" Y. For instance our configurations $\phi(\vec{x})$ are arrows like $E \xrightarrow{\phi} \Phi$, where E is ordinary space and Φ is the vector space of values of the field ϕ, whereas the boundary values $f(\theta,\phi)$ of Eq. (2.2) are properly described as functions: $B \xrightarrow{f} m$, where B is the boundary at infinity. M, the manifold of vacuum values, is of course, a subspace of Φ.*

We can easily surmise (and will verify later) that the set of homotopy classes will depend on the domain and range of the functions concerned. It also helps to remember that our homotopy problem is of a special type: it concerns not just functions $E \longrightarrow \Phi$, but functions subject to a restriction on their boundary values, Eq. (2.2). Obviously any homotopy between configurations $(u \sim v)$ implies a homotopy $f \sim g$ between their boundary values. If we can classify maps $B \longrightarrow M$ into homotopy classes (which is, of course, in principle a different problem[23]) we will have learned something about our configurations. If there is more than one class of maps $B \longrightarrow M$, then there must be also several classes (or sectors) in the space of finite energy configurations. Consider then the sector containing the configurations with constant boundary values, hence in particular those with ϕ = const. everywhere. That is to say: the (degenerate) vacuum states of the theory are in this sector. Now consider any other sector; a minimum energy configuration in it must be stable, since it cannot make a transition to the sector containing the vacuum[24]. It is here that we expect to find (and will find in concrete examples, see also Coleman) solitons (or maybe in higher sectors configurations containing several solitons). This is the qualitative conclusion we were aiming at.

Field Theories in 2-, 3- (or 4-) Dimensional Space. Homotopy Groups

Let us have a brief look at the kind of problem we have been led to. First notice that although $B = S^2$ is physically the most interesting

* For the reader who is not repelled by "commutative diagrams": one can summarize all these relations by the diagram in Fig.4.

Fig.4

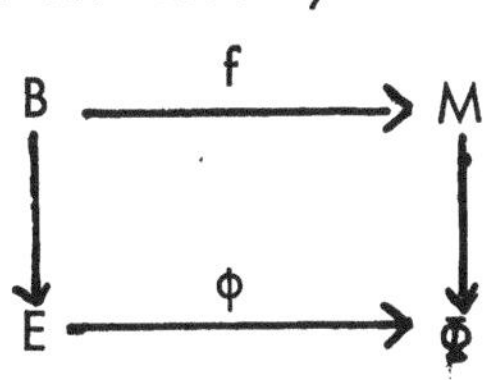

Here ordinary space E is turned into a compact space by including "points at infinity" amongst its elements. (This is natural, since the existence of the limit in Eq. (2.2) is then just a continuity requirement on ϕ). The vertical arrows are just "inclusion maps." The diagram then summarizes a) Eq. (2.2) b) $f(\theta,\phi) \in M$ c) all the continuity requirements we have stated.

case, there is also some interest in two-dimensional configurations, as in the examples studied by Nielsen and Olesen (see Coleman), with $B = S'$ (a circle) or in configurations for a field theory in four-dimensional Euclidian space, with $B = S^3$ (see the instanton in our third lecture). Thus, generalizing a bit, our problem is to study and classify maps $S^m \longrightarrow M$, where $m = 1, 2, 3, \ldots$. This is a very famous problem or rather family of problems, as a glance at the history of mathematics in our century will show.

In 1895 Poincaré, in order to classify closed paths on a connected manifold M (in other words, maps $S^1 \longrightarrow M$!) chose to consider only paths starting from, and ending at, some fixed point on the manifold. The set of homotopy classes of such paths is denoted $\pi_1(M)$ and is Poincaré's famous fundamental group. A group structure is assigned to the set by a) defining the inverse of a path f as the path obtained by running along f in the opposite direction, b) defining the product fg of two paths f and g as the closed path obtained by running along f and g in succession*, and c) showing that those concepts can be transferred unambiguously to the homotopy classes of paths. Finally one shows that d) the class of the "constant path", i.e. the class of all paths that can be shrunk to a point, plays the rule of unit element. A simply connected manifold has, of course, a trivial fundamental group (it consists of just one element).

Now the question for us is: suppose M is the vacuum manifold of a theory, and we study two-dimensional configurations $(B = S')$. Can we disregard Poincaré's fixed point constraint, and assert that $\pi_1(M)$ is identical with the set of classes we are seeking? It turns out that the answer is yes, if the fundamental group is abelian[25] (as is always the case when M is a group-manifold). In this case, therefore $\pi_1(M)$ is in one-to-one corrpondence with the "sectors" when $B = S'$ (two-dimensional space). The simplest example of non-trivial group π_1 is given by the case: M itself is a circle. The group $\pi_1(S')$, as one shows quite easily, is the infinite cyclic group; equivalently, the homotopy classes of maps $S' \to S'$ are characterized by an integer $p = 0, \pm 1, \pm 2, \ldots$.

Moving on to higher dimensions, a famous topological theorem is Brower's result, in 1912, that homotopy classes of maps $S^2 \longrightarrow S^2$ are characterized likewise by an integer (sometimes called the "wrapping number"). This means for instance that in three-dimensional space a

* It is, of course, important here that the endpoint of f concides with the starting point of g.

gauge theory with $G = SO(3)$ and $M = S^2$ has an infinity of sectors. An example of soliton occurring in such a theory is 't Hooft's monopole solution, also described independently by Polyakov. Maps $S^2 \to S^2$ can be most simply thought of as maps of a sphere into itself. Trivial maps (with "wrapping number" equal to zero) are homotopic to constant maps (mapping the whole sphere into a single point). If the image of a map leaves some part of S^2 uncovered, it can be shrunk continuously to a point; the map is then trivial. The identity map, which maps every point into itself, has wrapping number = + 1 and is not trivial. The reader will easily check that the boundary value of the Higgs field in the monopole solution is such a map, and this is why the monopole is a "topological soliton."

Around 1930 H. Hopf obtained some important results on spheres of higher dimensionality. He showed that the maps $S^m \to S^n$ $(m < n)$ are always homotopically trivial; and generalized Brower's result by showing that for $m = n$ the set of homotopy classes is always the same (it is characterized by a "wrapping number"). Finally he obtained some surprising results for values $m > n$, for instance $S^3 \to S^2$ (see also 3rd Lecture). In order to summarize some results, we adopt the language of higher homotopy groups, an invention of Witold Hurewicz.

In analogy with Poincaré's definition, we consider maps $S^m \to M$ $(m = 2, 3, \ldots)$ and again we single out a special point on M, and we only consider maps such that the image of the North Pole (say) of S^m is the special point. "Trivial" maps are then homotopic to the "constant" map, which maps the whole S^m into the special point. $\pi_m(M)$ is then defined as the set of homotopy classes of the maps just defined; a group structure can be assigned to π_m (see Discussion session) and in this case $(m \geq 2)$ it turns out that the group is always abelian. Just as in Poincaré's case, the introduction of a group structure is useful: it puts a powerful tool in our hands. What we are really interested in is the set of homotopy classes without the fixed point constraint. In many cases the fixed point makes no difference[26)] and we shall proceed on that assumption.

The determination of the groups $\pi_m(S^n)$ $(m > n)$ has been tackled by many authors, with surprising results, some of which are summarized in Table ... . It is often a very tough problem, and the general case is still unsolved.

TABLE
Some Examples of Manifolds
With Given Homotopy Groups $\pi_p(M)$

π_p	p	SPHERES	UNITARY[1] GROUPS	ORTHOGONAL[1] GROUPS
0	1	SU(n) (spanning all three columns)		
	2	ALL COMPACT LIE GROUPS[2] (spanning all three columns)		
	> 1	S'		
	all[3]	$S^n (n > p)$		
Z	1		U(n)	SO(2)
	all[3]	S^p		
	3	S^2	U(n) and SU(n) $n \geq 2$	SO(n) $n = 3$ or (4) $n \geq 5$
Z_2[5]	1			O(n) and SO(n) $(n \geq 3)$
	≥ 4	S^{p-1}		

Remarks
(1) The S in SU(n) or SO(n) means unimodular (i.e. with determinant = 1) .
(2) This includes S' which is identical with U(1) or SO(2) (as group manifolds), and similarly S^3, identical with SU(2) .
(3) All means $p = 1, 2, \ldots$.
(4) The exception, SO(4), has $\pi_3 = Z \otimes Z$.
(5) Z_2 is the (one and only) group of order 2.

Z, of course, is the infinite cyclic group, or the additive group of integers, and 0 is the trivial group.

Lecture III

We shall now describe an example[27] of soliton in a pure Yang-Mills theory, that is a theory in which the only field is a (non-abelian) gauge-field B_μ. The example is known as the "Instanton" and is of interest in connection with more elaborate models, in which other fields are coupled to the gauge-field[28]. This part of the question, however, is much more complicated and we shall not be concerned with it. I hope, nevertheless, that your curiosity will carry you further than I can go today.

By pretending that the usual $x_4 = ict$ of relativity theory is a <u>real</u> variable we turn Minkowski space into a four-dimensional Euclidean space with a positive-definite metric $x^2 = \sum_{i=1}^{4} x_i^2$. This is where the Instanton lives. Aside from this, the field equations have the standard Yang-Mills form, with gauge-group $G = SU(2)$. Denoting by σ_a $(a = 1,2,3)$, the customary Pauli matrices, the generators $T_a = \frac{1}{2}\sigma_a$ of the group satisfy commutation relations

$$[T_a, T_b] = i\epsilon_{abc} T_c \tag{3.1}$$

The components B^a_μ of the gauge-field four-vector carry an index a (= 1,2,3), which disappears when we turn them into matrices: $B_\mu = -i B^a_\mu T_a$ (summation over a is, of course, understood). Likewise, the gauge-field components $G_{\mu\nu}$ are matrices: $G_{\mu\nu} = -i G^a_{\mu\nu} T_a$ defined by:

$$G_{\mu\nu} = \partial_\mu B_\nu - \partial_\nu B_\mu + \gamma[B_\mu, B_\nu] \tag{3.2}$$

where γ is a coupling constant. The Lagrangian matrix density[29]: $-\frac{1}{4}\sum_{\mu\nu}(G_{\mu\nu})^2$ yields the field equations:

$$\partial_\mu G_{\mu\nu} + \gamma[B_\mu, G_{\mu\nu}] = 0 \tag{3.3}$$

These equations are satisfied trivially not only by $B_\mu = 0$, but also by any B_μ such that $G_{\mu\nu} = 0$. Let us recall that an element of the group of gauge transformations is a function $g(x)$ with values in the group G. The corresponding transformation of the B - field (in matrix form) is:

$$B_\mu(x) \to g^{-1}(x)\, B_\mu(x) g(x) + \frac{1}{\gamma} g^{-1}(x)\, \partial_\mu g(x) \tag{3.4}$$

so that, if the original B-field is zero, the gauge transformation <u>generates</u> a field:

$$B_\mu(x) = \frac{1}{\gamma} g^{-1}(x)\, \partial_\mu g(x) \tag{3.5}$$

The G-field, however, remains zero[30] as one can also verify directly from (3.2)(3.5). The energy density of such a solution is, of course, also zero. One assumes that such a state is physically indistinguishable from $B_\mu \equiv 0$, and we shall call it a vacuum state.

The Instanton is a non trivial solution of the field equations, which tends asymptotically to a vacuum state as the point x tends to infinity. An especially simple form of this solution is mentioned by the authors[27] towards the end of their paper. In order to describe it, let us first define a particular gauge-function $g(x)$. Recall that an element u of SU(2) is of the form

$$u = u_4\, 1 + i\, \vec{u}\cdot\vec{\sigma} \tag{3.6}$$

where $(u_4, \vec{u})$ is a unit four vector:

$$u_1^2 + u_2^2 + u_3^2 + u_4^2 = 1 \tag{3.7}$$

To define $g(x)$ we must define (3.6) as a function of x, i.e. define four functions $u_\mu(x)$ such that (3.7) is identically satisfied. We chose the simplest Ansatz:

$$u_\mu = x_\mu R^{-1} \; ; \; R = \left(\sum_{\mu=1}^{4} x_\mu^2\right)^{\frac{1}{2}} \tag{3.8}$$

Having defined $g(x)$, we now set:

$$B_\mu(x) = \frac{1}{\gamma} \frac{R^2}{R^2 + \lambda^2} g^{-1}(x)\, \partial_\mu g(x) \tag{3.9}$$

Before we examine the interesting topological properties of this solution[31], let us verify that it is indeed a solution. We first evaluate (3.9), with $g(x) = u$, and u defined by (3.6) (3.9). An elementary calculation gives:

$$\gamma B_\mu(x) = i(R^2 + \lambda^2)^{-1}\, x_\nu\, \sigma_{\nu\mu} \tag{3.10}$$

where the symbols $\sigma_{\nu\mu}$ are defined by the equations

$$\sigma_{\nu\mu} = -\sigma_{\mu\nu} \; ; \; \sigma_{ab} = \epsilon_{abc}\, \sigma_c \; ; \; \sigma_{4a} = \sigma_a \tag{3.11}$$

where a, b, c are indices running from 1 to 3[32]. One easily derives commutation laws for the $\sigma_{\mu\nu}$ symbols, and it is then a simple matter to calculate:

$$\gamma G_{\mu\nu}(x) = 2i\lambda^2 (R^2 + \lambda^2)^{-2} \sigma_{\mu\nu} \tag{3.12}$$

and to verify then the field equations (3.3). Let us notice in passing the duality relations $\sigma_{12} = -\sigma_{34}$, ... etc., so that the dual tensor $\tilde{G}_{\mu\nu} = \frac{1}{2}\epsilon_{\mu\nu\alpha\beta} G_{\alpha\beta}$ obeys the simple relation $\tilde{G}_{\mu\nu} = -G_{\mu\nu}$. The whole structure of the solution (3.10) (3.12) is in fact remarkably simple.

Let us now understand why this solution is a "topological" soliton. Let us first notice that in the limit $R \to \infty$ the B-field of Eq.(3.9) is indeed of the form (3.5); that is to say, it satisfies the general condition that the physical state of the field must be indistinguishable from the vacuum on any finite portion of the boundary B at infinity. (Notice incidentally how rapidly the components of the G-field go to zero: $G_{\mu\nu} \sim R^{-4}$). Notice also that here we have no scalar fields, there is no degeneracy of the vacuum in the special sense of our earlier discussion; nevertheless an analogous degeneracy arises from the possibility of gauge transformations. In particular we can ask whether the Instanton solution (3.9) can be gauge transformed to another form in which the vacuum state at infinity is a normal vacuum, with B-field (and not just the G-field) equal to zero. By this we mean more precisely a B-field that tends to zero at infinity faster than R^{-1}. We can state this requirement in another way, which makes it clearer why R^{-1} is not "fast enough." At large distances from the origin, neglecting terms of the first (or higher) order in λ^2/R^2, the field defined by (3.9) (3.10) is indeed (as we noticed already) of the form (3.5); that is to say the difference between this field and the normal vacuum ($B_\mu = 0$) is entirely attributable to the gauge transformation $g(x)$. With $g(x)$ defined by Eq.s (3.6) (3.8) there is, however, something quite peculiar about this "vacuum state at infinity" in the sense that $g(x)$ tends to a limit as $R \to \infty$, but this limit depends on the direction ! The slow (1/R) decrease of the B-field at infinity is indeed a direct manifestation of that dependence. We now repeat our question: can the Instanton solution be gauge transformed[33] to another form, such that for large R the B-field corresponds, in the Eq.(3.5) to a gauge function which is constant at infinity? On any restricted portion of the boundary B this is obviously possible, but globally, that is simultaneously over the whole boundary, the answer is No.

The reason (you guessed it) is topological. Notice firstly that the boundary value of the gauge function $g(x)$ defined by (3.6) (3.8) is a map $B \to G$ (= SU(2)). Since both manifolds are three-dimensional spheres, we are reminded of the Hopf classification of maps $S^3 \to S^3$. Treating the two manifolds as if they were the same sphere, we will describe (3.8) as the identity map of an S^3 onto itself, which as we know is non-trivial, that is non-homotopic to a constant map. (It has "wrapping number" equal to one, not zero).

Why is this relevant? Consider any map $B \xrightarrow{f} G$ and suppose f can be extended to a continuous function f(x) defined over the whole space inside B. It is easy to see that f(x) supplies a homotopy of the map $B \xrightarrow{f} G$ to a constant map (just consider the values of f(x) on concentric spheres of smaller and smaller radius; as $R \to 0$, the limit is the constant map with value f(0)). It is further easy to see that if B_μ is gauge transformed, as indicated in footnote[33], the gauge function that generates the transformed field B'_μ is f(x) g(x) and if the answer to the stated question were Yes, it would mean that f(x)g(x) = const on B. This is impossible since f(x) is "trivial" and g(x) is not.

Let me conclude this very brief description of the Instanton with a couple of remarks, that may be of help to you, if you wish to do some further reading on the subject.

The first remark concerns the expression of indices or "wrapping numbers" in terms of integrals. The first famous example of such an integral is due to Kronecker[34]. A four-dimensional integral expression for the index of the Instanton is given in the quoted paper by Belavin and o.; the integrand is then recognized as the divergence of a current[35] and transformed into an integral on the boundary at infinity. If you engage in calculations of this kind, you probably ought to know that quite considerable simplifications (and also a greater insight into what you are doing) can be achieved if you adopt the notations of the calculus of exterior forms. One way or another, the end product of the calculation is that the three-dimensional integral reduces to the invariant volume of the group space (see also the discussion session).

The second and final remark is that the boundary values of the gauge function g(x), that we have studied, can be easily transformed into a gauge function in ordinary, three dimensional space E_3 by a stereographic projection of $B(= S^3)$ into ordinary Euclidean space. If the North-Pole (i.e. $u_4 = 1$, $\vec{u} = 0$) is used as the projection center, that is if it goes to infinity in the projection, we then obtain a map $E_3 \longrightarrow SU(2)$ (a gauge function in ordinary space) such that the group element tends to 1 at infinity. The static B-field generated by this gauge function is then a vacuum-state, but a vacuum of a peculiar kind. If we try to gauge transform this state to an ordinary vacuum ($B_\mu = 0$) by means of gauge functions that have a constant limit at infinity, we discover that we cannot do that. The reason is the same as before.

Whether this has a deeper significance, in connection with the "structure of the vacuum" as some known AA. maintain is perhaps still a subject for discussion. I can only refer you to the papers cited in footnote[28) for further reading.

References

(1) See e.g. Footnote 3 in Coleman. See also W. Blaschke "Differentialgeometrie" Vol. I. p. 207 (Chelsea Publ. Co. N. York, 1967)

(2) This is a consequence of the Lorentz-contraction. Eq. (S. G.) is, of course, relativistic.

(3) See for example, G. B. Whitham, "Linear and Nonlinear Waves" J. Wiley and Sons 1974, Chapters 13 and 17.

(4) Another transformation is $(\phi, \psi, \lambda, x) \rightarrow (-\psi, \phi, \lambda^{-1}, -x)$.

(5) Strictly speaking ψ is defined only up to an integration constant. The notation (1.3') must be used with some caution.

(6) See Bianchi's textbook cited by Coleman. The theorem assumes, of course, a suitable choice of integration constants. (See earlier remarks !)

(7) G. L. Lamb, Physics Letters 25A, 181.

(8) See e.g. N. Christ, and T. D. Lee, Phys. Rev. D11, 1606 (1975); R. F. Dashen, B. Hasslacher and A. Neveu Phys. Rev. D11, 3424 (1975); J. L. Gervais and A. Jevicki, (1976) CCNY-HEP-76/2 preprint. L. D. Faddev, L. A. Takhtajan, T. M. P. 21, 1046 (1975)

(9) See e.g. B. Yoon, Phys. Rev. D13, 3440 (1976).

(10) T. Skyrme, Proc. Roy. Soc. A247, 260 (1958); A262, 237 (1961)

(11) G. Mie Ann. Phys. 37, 511 (1912) 39, 1 (1912) 40, 1 (1913); See also W. Pauli "Theory of Relativity" (Pergamon Press 1958) Part V S64.

(12) R. Friedberg, T. D. Lee and A. Sirlin, Phys. Rev. D13, 2739 (1976). The fields A and ψ in the original paper differ from those used here by proportionality constants. We omit an overall factor g^{-2} in the Lagrangian. This factor is irrelevant at the classical level.

(13) Coleman (loc. cit.) refers to this argument as "Derrick's theorem" comp. G.H. Derrick Journ. Math. Phys. 5, 1252 (1964). See also J. Goldstone and R. Jackiw, Phys. Rev. D11, 1486 (1975)

(14) There are, of course, two possible vacuum values for the field A (= ± 1). This "degeneracy," however, will not play any significant role in the following.

(15) As indicated earlier, the omissions are deliberate. The reader will find much of the needed complementary information in Coleman, Section 3. It should be possible to read the following, however, independently of this reference.

(16) A discussion of solitons involving Fermions (anticommuting fields) is beyond the scope of these lectures

(17) Our viewpoint is somewhat different from Coleman's. He claims to be discussing a set of initial conditions (this includes velocities).

(18) In general G can act in the space of ϕ through a representation D(G). The representation need not be faithful. We shall also occasionally assume that ϕ is a complex vector, and G is SU(n).

(19) In the S.G. theory we have encountered an example of infinite but denumerable set of minima of $V(\phi)$, i.e. $\phi = 0, \pm 2\pi, \pm 4\pi, \ldots$ etc.

(20) Other, physically equivalent, solutions with $A_\mu \neq 0$ are obtained by gauge-transformations. See Third Lecture.

(21) We gloss over questions connected with A_0 components of the gauge field, see Coleman.

(22) In ordinary field theories we have, of course, no other choice. That interesting solutions arise if we allow a more general behavior, Eq. (2.2) was pointed out by H. Nielsen and P. Olesen Nucl. Phys. B61, 45 (1973).

(23) That the two problems are nevertheless equivalent, when E is ordinary space and B its boundary, is intuitively obvious, but we shall not prove it.

(24) This conclusion presumably extends to the quantised theory, since there is an infinite potential barrier to be crossed.

(25) There are quite simple cases of manifolds with non-abelian fundamental group, e.g. the surface of a doughnut with two holes.

(26) A sufficient condition for this has been stated by S. Eilenberg, Fund. Math. 32, 167 (1939). It is satisfied if M is a sphere. We cannot go into this, however. See also: N. Steenrod "The Topology of Fibre Bundles" S 16.5 .

(27) A. A. Belavin, A.M. Polyakov, A.S. Schwartz and Yu. S. Tyupkin, Phys. L. 59B, 85 (1975).

(28) G.'t Hooft Phys. Rev. L. 37, 8(1976) and Nucl. Phys. to be published. R. Jackiw and C. Rebbi, Phys. Rev. D14,517(1976), and to be published. C.G. Callan, R.F. Dashen, D.J. Gross, to be published.

(29) To obtain the customary expression for the Lagrangian take twice the trace. Remember $\mathrm{Tr}\,(T_a T_b) = \frac{1}{2}\,\delta_{ab}$.

(30) Unlike (3.4) the transformation law for the G-field:

$$G_{\mu\nu}(x) \rightarrow g^{-1}(x)\, G_{\mu\nu}(x)\, g(x)$$

does not contain an inhomogeneous terms

(31) Please notice, although g(x) is ambiguously defined at the origin and its derivatives are singular, the solution (3.9) (3.10) is regular everywhere.

(32) See also 't Hooft's $\bar{\eta}$ symbol. One can write $\sigma_{\mu\nu} = \bar{\eta}_{a\mu\nu}\sigma_a$. See especially Eq. (A.4) and (A.13) in 't Hooft's paper.

(33) In pedantic detail: is there a gauge function f(x), well defined and non singular over the whole Euclidean four-space such that $B'_\mu = f^{-1} Bf + \frac{1}{\gamma} f^{-1}\partial_\mu f$ has the required properties?

(34) Interesting references may be found in J. Arafune, P.G.O. Freund and C.J. Goebel Journ. Math Phys. 16, 433 (1975)

(35) See also W. Bardeen Nucl. Phys. B75, 246 (1974)

DISCUSSIONS

CHAIRMAN: Prof. G.C. Wick

Scientific Secretary: P. Rossi

DISCUSSION 1

- *ROSSI:*

Can we study models containing fermions, such as the SLAC bag, with the same techniques as in the charged-boson case, in the case of large charge quantum number or in the case of higher states?

- *WICK:*

The anticommutativity of fermion fields, of course, requires, in principle, different techniques. There is, however, some ingenious work by Neveu and others in which, by eliminating the fermion variables, one is able to handle the problem by the usual technique.

- *GARCIA:*

Does the Bäcklund transformation have a one-parameter group structure? If this is so, could one learn anything about the solutions of the Sine-Gordon equation by studying the representations of such a group?

- *WICK:*

By combining an infinitesimal Bäcklund transformation with an infinitesimal translation, one can construct the generator of a one-parameter group which, however, does not coincide with the set of "finite" Bäcklund transformations. One does not get, unfortunately, anything very exciting, since it is contained as a special case in the general results of Faddeev.

- *MARCIANO:*

Are there uniqueness proofs which allow one to determine when all solutions to coupled non-linear equations have been found? For example, has it been done for a simple case like the Sine-Gordon equation?

- *WICK:*

I do not know of general proofs. One would have to examine one model at a time.

In the case of the Sine-Gordon equation, of course, the problem has been thoroughly explored by Faddeev and others, and I think one knows the properties of the system very well.

I assume you are asking about the classical, that is non-quantized aspect of the problem.

- *PAULI:*

Could you illustrate roughly how soliton solutions could give an oscillating bag and how possibly bag fissioning could be described through non-linear wave equations?

- *WICK:*

Let me try to answer especially the second part of your question. There are well-known methods to study small oscillations around an equilibrium configuration. On the other hand, I do not know of any serious attempt to calculate bag fissioning. This would probably require an enormous amount of numerical work, especially if you consider that the interesting problems would be in three-dimensional space.

- *PARSONS:*

You mentioned two categories of solitons: the mathematical physicists' "collisionless" solitons and the topological type. Does the Sine-Gordon "kink" type of soliton belong to both categories?

- *WICK:*

To be more precise, I distinguished three different types. The Sine-Gordon kink belongs simultaneously to two of these, that is it is topological and it is collisionless.

A slight modification of the potential will give rise to a very similar "kink", which is no longer collisionless, however. In other words, it no longer satisfies the mathematicians' definition of soliton, as the numerical investigations of Ablowitz have shown.

DISCUSSION 2

- *POSNER:*

I would like to know more about Hopf's work on the mappings of S^m into S^n. Last year at Erice, Coleman used a theorem of Cartan's to show that certain models could not possess monopole solutions. What is the connection between Hopf's work and Cartan's theorem? Are there any other physical consequences of Hopf's work?

- *WICK:*

Hopf showed that all maps $S^m \to S^n$, where $m < n$, are trivial, that is homotopic to a constant map. When $n = m$, the equivalence classes are in one-to-one correspondence with the integers. The case $m > n$ is much more difficult and is not completely solved even now. Hopf gave some famous examples, such as a map $S^3 \to S^2$, which is, in essence, what physicists do when they associate a direction in space to a normalized spinor or to an element of SU(2); as is well known, the group space of SU(2) is homeomorphic to S^3. The theorem of Cartan that Coleman used is the statement that the second homotopy group of a Lie group is always trivial. There is no simple connection between this and the results of Hopf -- of course, they belong to the same chapter of mathematics.

The work on magnetic monopoles and on instantons are examples of physical problems in which topology may be relevant.

- *LEADER:*

Restricting ourselves just to one-dimensional equations, say, of the type of the wave equation with some function $f(\phi)$ replacing $\sin \phi$, do the mathematicians know the conditions which $f(\phi)$ must satisfy in order that there exist "collisionless" soliton solutions?

- *WICK:*

I am not aware of a general theorem that will tell you *a priori* whether a wave equation possesses "collisionless" solitons.

- *ROSSI:*

What are the relations between the homotopy groups of a space and its connection properties?

- *WICK:*

There are, of course, very deep connections. The simplest one is the obvious statement of the equivalence between saying that a space is simply connected and saying that the group π_1 is trivial.

DISCUSSION 3 *(Scientific Secretary: P. Sodano)*

- *PARSONS:*

In the paper by Polyakov et al. there was mention of the solitons providing long-range effects and having a relevance to the infrared problem. Do you have any comments on this? Could you generalize the Polyakov solitons to SU(3)?

- *WICK:*

To your first question, my answer is no. To your second question, I choose the case of SU(2) as an example of the topological part I wanted to illustrate. Undoubtedly, there are other examples. In particular, the stability of the instnaton depends on the existence of non-trivial elements of the third homotopy group π_3, and this is true for Lie groups other than SU(2). However, I have not looked at these cases in detail.

- *ROSSI:*

What about quantum solitons?

- *WICK:*

The quantization of the classical solitons is, in a certain sense, analogous to the well-known W.K.B. method. It was discussed last year by Coleman. For a different method, I must refer you to the paper by Christ and Lee.

- *ROSSI:*

Topological solitons are often associated to spontaneous symmetry breaking. Is it possible to study quantum solitons in Lagrangians which do not possess classical ones, but show the phenomenon of spontaneous symmetry breaking by defining an effective Lagrangian in the one loop approximation and obtaining Lagrangian equations of motion?

In other terms, can the variational equation for effective action be used, not only to define the broken vacuum, but also for obtaining non-constant solutions which can be dynamic topological soliton solutions?

- *WICK:*

In principle, I see no objection to trying to do what you say. As far as I know, it has not been done.

- *POSNER:*

Last year at Erice, Coleman, in his lumps lectures, discussed the patching together of two widely separated lumps. The process involved the gauging of the invididual solutions at large distances to a constant almost everywhere. The possibility of ambiguity increased greatly as the number of space dimensions went from three to two. Is there also possible ambiguity in the procedure you discussed of generating two soliton solutions in one time and one space dimension? You have discussed magnetic monopole solutions. Would you elaborate on an antimonopole solution? Is there any possible ambiguity in considering monopole-antimonopole solutions?

- *WICK:*

In the procedure which I discussed of generating two soliton solutions, there is no ambiguity. The possibility of ambiguity arises perhaps in the case of topological solitons.

A magnetic antimonopole solution can be obtained from a monopole by changing the sign of the Higg's field. A monopole-antimonopole pair has a Higg's field at infinity which is homotopically equivalent to the normal vacuum.

- *WILKIE:*

What is the equivalent for extended objects in relativistic quantum field theory? In the non-extended case, it is impossible to construct a local number operator; that is,that if one looks closely at an electron, then one sees it, by the very act of looking closely, as surrounded by virtual electron-positron pairs.

- *WICK:*

It is possible, if one wishes, to describe a soliton. For example, the kink solution of the Sine-Gordon equation, as a coherent state, involving the excitation of one of the normal vacuum states of the theory, such as the state with $\langle\phi\rangle_{vac} = 0$ everywhere. The number of elementary excitations, mesons, in the coherent state is totally undefined.

- *JARVIS:*

You gave the construction of the first homotopy group π_1, the Poincaré group, and the group operation. Could you also give the definition of the group operation for π_2, π_3, and so on?

- *WICK:*

There are various ways of defining the higher homotopy groups π_2, π_3, etc. of a space X. First one chooses in the space a distinguished point X_0. One then considers maps of spheres S^2, S^3, etc. into the space, subject to the restriction that a certain point on the sphere, say the north pole, is always mapped into X_0. This restriction is maintained throughout, for instance, in defining homotopy between maps. Two maps are equivalent if they are homotopic; that is to say the class of all maps $S^2 \to X$ is subdivided into equivalence classes. These classes are the elements of the group π_2. Analogously, maps $S^3 \to X$ yield the group π_3, etc. For a connected space, it does not matter much which point X_0 is chosen as distinguished point.

In order to define the group operation, one may proceed as follows: consider, for instance, π_2. Instead of a sphere S^2, consider a square I^2, e.g. the set of all points (t_1,t_2) with $0 \le t_1 \le 1$; $0 \le t_2 \le 1$. One can obviously map I^2 into S^2 in such a way that the boundary of I^2 is mapped into the north pole and the interior of I^2 is mapped homeomorphically, that is one-to-one and continuously both ways, onto the sphere minus the north pole. We now replace the maps $S^2 \to X$ of the earlier definitions by maps $I^2 \to X$, such that the boundary of I^2 is mapped onto X_0. The composition is now defined as follows: suppose f and g are two such maps, i.e. $f(t_1,t_2)$ and $g(t_1,t_2)$ are functions of two variables t_1, t_2 satisfying conditions you can easily state. We squeeze the map f into the lower half of I^2 by defining

$$F(t_1,t_2) = f(2t_1,t_2) \qquad 0 \le t_1 \le \tfrac{1}{2}, \quad 0 \le t_2 \le 1 .$$

Likewise, we squeeze g into the upper half of I^2 by

$$F(t_1,t_2) = g(2t_1-1,t_2) \qquad \tfrac{1}{2} \le t_1 \le 1, \quad 0 \le t_2 \le 1 .$$

The two equations now define F over the whole square, and you can easily see that the two definitions match on the dividing line $t_1 = \frac{1}{2}$. The map $F : I^2 \to X$ satisfied all the conditions stated earlier. It is not hard to show that the homotopy equivalence class to which F belongs only depends on the equivalence classes to which f and g belong. This defined the group operation. It is also not hard to see that the operation is commutative, i.e. π_2 is always Abelian -- π_1 is not always Abelian! π_3 is defined analogously by considering a cube I^3, and so on.

- *GARCIA:*

If I remember correctly, in 't Hooft's paper it is stated that any Yang-Mills theory with a gauge group having a compact covering group has a monopole solution. I am not sure if you have already explained this in a different language. Could you clarify the statement?

- *KLEINERT:*

This depends on the choice of the Higg's field. If you take an isospin ½ field, all vector mesons become massive and no monopole can exist.

- *JANCEWICZ:*

You said during the lecture that g(X) is not homotopic to a constant. One should rather show that $B_\mu(X) = g^{-1}\partial_\mu g$ is not homotopic to a constant.

- *WICK:*

There is no interesting homotopy question concerning $B_\mu(X)$. The possible vacuum values of B_μ are arbitrary. As a result, $B_\mu(X)$ is always homotopic to a constant.

- *McPHERSON:*

Expand upon the use of exterior forms.

- *WICK:*

The general rules of the calculus of exterior forms are really quite simple and more useful than most physicists realize. Allow me to demonstrate this on an example.

In the discussion of the instanton by Belavin and others, as mentioned in my lecture, there occurs the statement that a certain integral, extended to the boundary of the instanton, must be an integer. It is further stated that this is so because the expression under the integral sign

$$\Omega = \frac{1}{24\pi^2} \mathrm{Tr}(g^{-1}\, dg \wedge g^{-1}\, dg \wedge g^{-1}\, dg) \tag{1}$$

is the invariant volume element in group space. Let us examine the second statement first.

I recall the notation: g is a matrix belonging to a three parameter Lie group G, specifically, SU(2); and g + dg is another element of G differing from g by an infinitesimal displacement. The produce under the trace sign in (1) is therefore a matrix product.

$$\sum_{k,e} (g^{-1}\, dg)_{ik}\, (g^{-1}\, dg)_{ke}\, (g^{-1}\, dg)_{eJ} \tag{1'}$$

but also, as I have indicated by the wedge symbols $..\Lambda..\Lambda$, an *exterior product* of linear forms. Since G = SU(2) is a three parameter group, I can write

$$g^{-1}\,dg = \sum_{r=1} a_r(u)\,du_r \tag{2}$$

I shall use the familiar parametrization

$$g = u_0 + i(u_1\sigma_x + u_2\sigma_y + u_3\sigma_z) \tag{3}$$

where

$$u_0^2 + u_1^2 + u_2^2 + u_3^2 = 1 \tag{4}$$

so that du_0 is linearly dependent on du_1, du_2, and du_3, and can be eliminated as I have done in Eq. (2). Each $a_r(u)$ is, of course, a 2 × 2 matrix with elements $(a_r)_{ik}$, that is $g^{-1}\,dg$ has four matrix elements $(\omega)_{ik}$ (i,k = 1,2) which are linear forms in du_1, du_2, and du_3. The basic rule is that these differentials must be treated as the generator units of a Grassman algebra, i.e. $du_1 \Lambda du_2 = -du_2 \Lambda du_1$, etc. There is only one unit of the third order, namely du du du and in Eq. (1) must be a multiple of this. After some calculations, starting from (3) one gets:

$$\Omega = \frac{1}{2\pi^2}(u_0\,du_1 \Lambda du_2 \Lambda du_3 - u_1\,du_0 \Lambda du_2 \Lambda du_3 + u_2\,du_0 \Lambda du_1 \Lambda du_3 - u_3\,du_0 \Lambda du_1 \Lambda du_2)\ . \tag{5}$$

This expression is nice and symmetric, but in order to demonstrate what I first said, I eliminate du_0 and obtain

$$\Omega = \frac{1}{2\pi^2}\,u_0^{-1}\,du_1 \Lambda du_2 \Lambda du_3\ . \tag{5'}$$

The essential remark is now that this form is left- and right-invariant. I recall that a left-translation is a mapping defined by $g \to hg$, where h is a fixed element of the group $G \to G$. This entails $dg \to hdg$, and $g^{-1}\,dg \to (g^{-1}h^{-1})h\,dg = g^{-1}\,dg$. Hence, each of the linear forms is left-invariant, and this applies consequently also to Ω.

Under a *right-translation* $g^{-1}\,dg$ is *not* invariant. From $g \to gh$ we get: $g^{-1}\,dg \to (h^{-1}g^{-1})dg\,h = h^{-1}(g^{-1}\,dg)h$. When we form the triple product of Eq.'s (1) and (1′) and *take the trace*, the h drops out again and Ω is found to be left-invariant. As is well known, these properties define the invariant volume element up to a proportionality constant. The constant in (1) is fixed by the requirement that the total volume be equal to one. This completes my comments on the second statement of Belavin and others.

Turning to the first statement, let us now assume that the unitary matrix g, see Eq. (3), is a function g(X) of a point $X = (X_0, X_1, X_2, X_3)$ in a Euclidean four-space. Specifically, assume X lies on a large sphere S^3:

$$X_0^2 + X_1^2 + X_2^2 + X_3^2 = R^2 \tag{6}$$

which is the boundary of the region containing the instanton. The assumption is, then, that the parameters u_r in (3) are given as functions $u_r = f_r(X)$ $(r = 0, \ldots, 3)$ satisfying (4) identically, when X lies on S^3.

This functional dependence then transforms the form Ω into a three-form in the X variables:

$$\Omega \to \Omega' = F(x)dx_1 \wedge dx_2 \wedge dx_3 \tag{7}$$

The statement is then, that $\int \Omega'$ *extended to the sphere* S^3 of Eq..(6) is equal to a positive or negative integer. Without going into much detail, I will sketch a possible proof. It is easy to see that the form Ω of Eq. (5) or (5′), can be expressed, in many ways, as the *exterior* differential of a two-form: $\Omega = d\omega$. For example, we can choose:

$$\omega = \frac{1}{2\pi^2} f(u_0)\ (u_1\ du_2 \wedge du_3 + u_2\ du_3 \wedge du_1 + u_3\ du_1 \wedge du_2) \tag{8}$$

provided f satisfies the differential equation:

$$(1 - u_0^2)f' - 3u_0\ f + 1 = 0\ .$$

A solution is

$$f(u_0) = \tfrac{1}{2}\ (1 - u_0)^{-3/2} \arccos u_0 - u_0(1 - u_0)^{-1}\ . \tag{9}$$

This is, of course singular at one point in group space, namely $u = (-1,0,0,0)$. If this value does not occur amongst values $u_i = f_i(x) \times \theta\delta^3$, then the two-form ω' obtained from ω,ω' can be approximated by transforming to x-variables the expression

$$\omega = \frac{1}{4\pi}\ (\vec{u}^2)^{-3/2}\ (u_1\ du_2 \wedge du_3 + u_2\ du_3 \wedge du_1 + u_3\ du_1 \wedge du_2)\ .$$

It is then easy to see that the contribution of each singularity to $\int_{\partial\varepsilon}\omega'$ is a positive or negative integer, which proves the assertion.

One final remark: the equations $u_i = f_i(x)$ may be said to define a mapping $S^3 \to G$, and since $G = SU(2)$ is again an S^3, as one sees from Eq. (3) and (4), what we have studied is one famous Hopf mapping $S^n \to S^m$, more precisely, the case $n = m = 3$. We have, of course, made assumptions about the map f which are quite unnecessary; mere continuity is sufficient, but the proof then requires quite different tools.

CAN WE EVER UNDERSTAND HADRONIC MATTER ? A PROPOSAL

Giuliano PREPARATA
CERN - Geneva, Switzerland

The aim of these lectures is to discuss and clarify some important questions concerning a recently proposed quark theory of hadronic matter [1)-3)].

While the introduction and the discussion of quark spin effects, internal quantum numbers and three-quark systems is necessary to turn very theoretical proposal into a realistic theory of hadrons, it is nevertheless crucial at this stage to ask ourselves whether it is possible to make this general framework agree with some necessary general requirements. And to accomplish this we do not need to specify the otherwise important properties like spin, and internal quantum numbers.

The two main questions which I shall try to answer in these lectures are the proper description of off-shell effects and an adequate definition of current operators on the space of physical states' wave functions.

In previous work [1)-3)], the importance of these problems was adequately stressed, but the answer to them was left somewhat open, needing a detailed analysis which had not yet been carried out.

I am quite aware that some of the conclusions I will reach may have to be reassessed in the future, in the face of a deeper and more accurate analysis. However, I have not refrained from discussing these points now, just to stimulate discussion on very important issues which are not peculiar to this approach, but rather generally invest theories

Quark confinement, i.e., the impossibility of observing quarks as real physical states, is introduced through the requirement :

$$\psi(p;x) = 0 \qquad \text{for} \quad x \notin R^4(p) , \tag{1.2}$$

where $R^4(p)$ is a certain bounded space time domain (bag) *), depending on the momentum p of our state, whose structure will be specified shortly.

To determine the quark "motion" inside the bag $R^4(p)$ we need some differential equations for the wf (1.1). The simplest choice appears to be

$$\begin{aligned} -\Box_1 \psi(p;x_1,x_2) &= m^2 \psi(p;x_1,x_2) \\ -\Box_2 \psi(p;x_1,x_2) &= m^2 \psi(p;x_1,x_2) \end{aligned} \tag{1.3}$$

for $x \in R^4(p)$. However, as discussed at length elsewhere [1),3)], when one converts (1.3) into equations for $\psi(p,x)$; i.e.,

$$\begin{aligned} \left(\frac{p}{2} - i\frac{\partial}{\partial x}\right)^2 \psi(p;x) &= m^2 \psi(p;x) \\ \left(\frac{p}{2} + i\frac{\partial}{\partial x}\right)^2 \psi(p;x) &= m^2 \psi(p;x) \end{aligned} \tag{1.4}$$

or, equivalently

$$\left(\frac{p^2}{4} - \Box\right) \psi(p;x) = m^2 \psi(p;x) \tag{1.5a}$$

$$p \cdot \frac{\partial}{\partial x} \psi(p;x) = 0 \tag{1.5b}$$

one encounters the difficulty that (1.5b) cannot be matched with the solution (1.2) outside the bag, due to its first order character. The simplest way out of this difficulty is to modify (1.5b) to

$$\left(p \cdot \frac{\partial}{\partial x}\right)^2 \psi(p;x) = -p^2 \omega_0^2 \psi(p;x) , \tag{1.5c}$$

with the further constraint that only that solution of (1.5c) should be kept which corresponds to the minimum eigenvalue ω_0. Our equations (1.5a) and (1.5c) will only be compatible with those bags $R^4(p)$ which in the rest frame, $p \equiv (M,\vec{0})$, factorize into a space bag (R_s) and a time bag (R_t), i.e.,

$$R^4(M,\vec{0}) \equiv \begin{cases} |\vec{x}| \leq R_s(M) \\ |t| \leq R_t(M) \end{cases} \quad . \tag{1.6}$$

where the quark degree of freedom is unobservable (i.e., "confined"). We can convince ourselves that the analyticity, the off-shell behaviour and the currents' structure of a theory where the fundamental fields are confined, must drastically differ from conventional quantum field theory, upon which much of our intuition is based. It is precisely the unavailability of well-established reference points which makes an investigation of such questions so difficult. In order to voyage as safely as possible in such quicksands, it is necessary that we check our answers by means of a number of calculations of processes of fundamental importance.

For this reason extensive calculations will be outlined on such processes as high energy scattering, deep inelastic phenomena and production of hadrons at large p_T. The very satisfactory picture one thus obtains can be taken as an indication that we may be moving along a sensible track.

The plan of these lectures is as follows. In Section 1 the basics of our approach shall be briefly reviewed ; Section 2 contains a more detailed discussion of the calculational rules, especially of the extension of our rules in the case of complex loop momenta. Section 3 takes up the question of how to normalize the physical states, and therefore of how to introduce currents in a physically meaningful way. High energy hadron scattering and the problem of Reggeization will be discussed in Sections 4 and 5. Deep inelastic phenomena are treated in Section 6, while hadron production at high p_T is the object of Section 7. The final part will be devoted to the conclusions.

1. - THE PHYSICAL HADRON STATES

A. - The meson $(\bar{q}q)$ wave function and its equations of motion

The simplest description of a meson state is given by the

---

*) The reader may notice some superficial similarities between this approach and the so-called MIT bag [4].

wave function (wf)

$$\psi(p;x_1,x_2) = e^{ip\frac{x_1+x_2}{2}} \psi(p;x) \tag{1.1}$$

where p is the total four-momentum of the $q\bar{q}$ system and $x_{1,2}$ are the four-co-ordinates of the $q\bar{q}$ pair. Note that (1.1) is a trivial consequence of translational invariance.

In the rest frame (1.5a) and (1.5c) become respectively

$$\left[\frac{M^2}{4} - \frac{\partial^2}{\partial t^2} + \vec{\nabla}^2\right] \psi_M(\vec{x},t) = m^2 \psi_M(\vec{x},t) \tag{1.7}$$

and

$$\frac{\partial^2}{\partial t^2} \psi_M(\vec{x},t) = -\omega_0^2(M)\, \psi_M(\vec{x},t) . \tag{1.8}$$

The solutions of these equations which vanish at the boundary of the region $R^4(M,\vec{0})$ given in (2.6) are easily found to be

$$\psi_{n\ell m}(M;\vec{x},t) = N_{n\ell}(M) \cos\omega_0 t \; j_\ell\left[\sqrt{\frac{M^2}{4}+\omega_0^2-m^2}\,|\vec{x}|\right] Y_\ell^m(\theta,\varphi) \tag{1.9}$$

with the eigenvalue conditions :

$$\omega_0(M) = \frac{\pi}{2R_t(M)} \tag{1.10}$$

and

$$\left[M^2 + 4\omega_0^2(M) - 4m^2\right]^{1/2} = \frac{2\beta_{n\ell}}{R_s(M)} . \tag{1.11}$$

$N_{n\ell}(M)$ is a normalization factor to be determined later, $j_\ell(z)$ is the spherical Bessel function of order ℓ, $\beta_{n\ell}$ its nth positive zero, and $Y_\ell^m(\theta,\varphi)$ is the normalized spherical harmonic. Note that (1.10) and (1.11) will give us a spectrum labelled by ℓ, the angular momentum, and n, the radial quantum number, only after we have specified the dependence of R_s and R_t on the mass M of the meson state. The important massive quark model (MQM) criterion of universal quark mass behaviour of quark Green's functions [5] yields, as discussed elsewhere [1)-3)],

$$R_s , R_t \underset{M^2 \text{ large}}{\longrightarrow} R^2 M \qquad (1.12)$$

which when inserted into (1.10) and (1.11) provides a set of exchange degenerate Regge trajectories which are almost linear and parallel. The magnitude of R^2 in (1.12) is fixed by the slope of our Regge trajectories : for $\alpha' \cong 1\ \text{GeV}^{-2}$ one obtains $R^2 \simeq \pi\ \text{GeV}^{-2}$. We end this subsection by recording the general structure of the high mass mesons wf in momentum space (see Fig. 1)

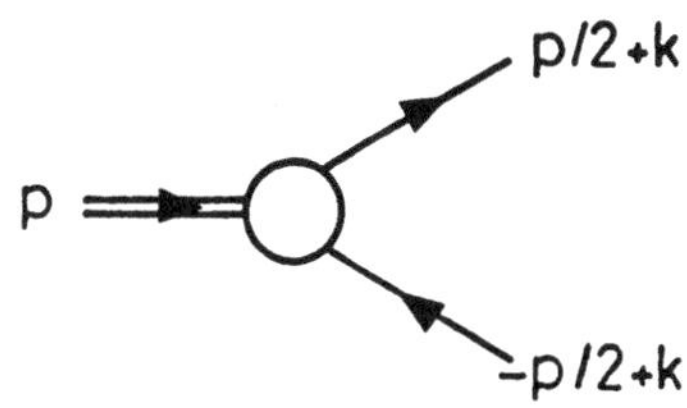

Figure 1 The momentum space wave function

$$\psi_{n\ell m}(p,k) \simeq (2\pi)^2 \frac{2\pi}{R^2} \delta_{R^2}\left(\frac{p^2}{4}+k^2-m^2\right) \delta_{R^2}(pk)\, Y_\ell^m(\Omega_k) \qquad (1.13)$$

where Ω_k is the direction of the three vector

$$\hat{k}_i = \left[\Lambda^{-1}(\vec{p})\, k\right]_i \qquad (1.14)$$

$\Lambda^{-1}(\vec{p})$ being the Lorentz transformation bringing our state to rest. The function $\delta_{R^2}(z)$ is a "fat δ function", characterized by the following two properties :

$$\int_{-\infty}^{+\infty} dz\, \delta_{R^2}(z) = 1 \qquad (1.15)$$

$$\delta_{R^2}(0) = \frac{R^2}{\pi} , \qquad (1.16)$$

in the limit $R^2 \to \infty$ $\delta_{R^2}(z)$ tends to a Dirac δ function. A very simple $\delta_{R^2}(z)$, particularly useful in the calculations, is given by

$$\delta_{R^2}(z) = \frac{1}{\pi} \frac{\sin R^2 z}{z} . \qquad (1.17)$$

Finally, in (1.13) the factor $N_{n\ell}$ has been chosen in such a way that

$$\int \frac{d^4k}{(2\pi)^4} |\psi_{n\ell m}(p,k)|^2 = 1 \quad ; \qquad (1.18)$$

one should keep in mind that this particular choice is only suggested by convenience and has no physical motivation. A physically meaningful normalization can only be given by introducing the electric current ; this shall be discussed in Section 3.

B. - Decoupling the high ℓ states. The fire sausage

We have just derived a meson spectrum consisting of approximately linear and parallel Regge trajectories. On one hand this looks very promising, but on the other it may lead us into serious trouble if quarks do not decouple the meson states whose angular momentum ℓ is higher than a value ℓ_0 given by

$$\ell_0 = \frac{M}{2} R_\perp(M) \qquad (1.19)$$

where $R_\perp(M)$ can increase at most logarithmically with the mass of the state.

The reason for this is quite simple ; it stems from the fact that if the strong interactions are to have a range which increases at most logarithmically with energy, the maximum impact parameter of a $q\bar{q}$ collisiom must also be so limited. Calling $R_\perp(M)$ such maximum value, (1.19) follows immediately.

Now, the quark coupling to the meson states is uniquely determined by the wf, and we can easily check that if the normalization is given by (1.13) at a given mass M all angular momentum states (up to $\ell_{max} = (R^2M^2)/\pi$) are coupled with comparable strength. In our previous work [1),3)] an inadequate definition of the electromagnetic current lead us to a normalized wf simply proportional to (1.13), and to the consequent necessity to decree away the troublesome states. We shall see later that from a physically reasonable electromagnetic current we can derive this decoupling and we shall also be able to give a very physical characterization of $R_\perp(M)$ appearing in (1.19).

Given that at a large mass M only those states are coupled for which $\ell \leq \ell_0$, we can get a nice intuitive picture of the space time structure of the typical meson state of mass M. We can in fact construct a set of coherent states characterized by the direction $\Omega_0 \equiv (\theta_0, \varphi_0)$, by taking the following linear combination of the wave functions (2.13)

$$|\Omega_0, M\rangle = \frac{1}{\ell_0^2} \sum_{\ell=0}^{\ell_0} \sum_{m=-\ell}^{+\ell} Y_\ell^{-m}(\Omega_0)\, \psi_{n\ell m}(p,k) \,. \tag{1.20}$$

Notice that from the asymptotic relation

$$M^2 \simeq \frac{\pi}{R^2}(2n+\ell) \tag{1.21}$$

in (1.20) n is not an independent index. Going back to configuration space these states correspond to cylindrical space domain (see Fig. 2), which we shall call fire sausage (FS), of height $R_\ell \simeq R^2 M$ and width $R_\perp(M)$, containing stationary quark waves moving with very small divergences $[\delta \simeq (R_\perp(M))/ /(R^2 M)]$ along the direction Ω_0.

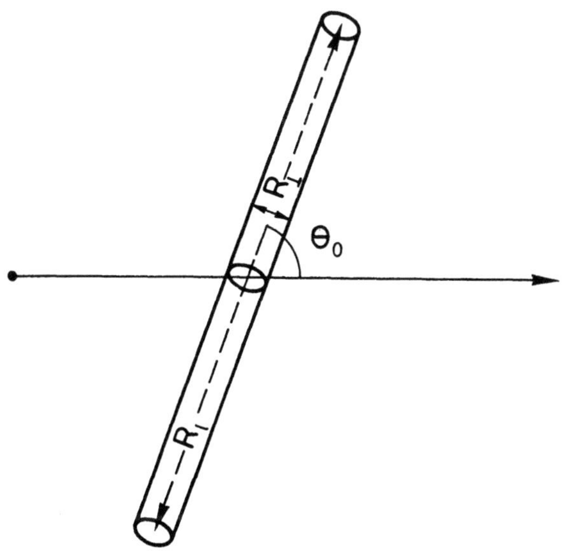

Figure 2 The space structure of a fire sausage

On introducing an interaction among three bags along the lines we shall discuss in the next section, we can study the decay properties of such a peculiar hadronic structure. This is a short summary of the main results of such a study [2] :

i) the strongly preferred decay configuration of the FS of mass M is a linear decay chain (Fig. 3) in which the initial FS cascades down emitting at each step a low mass state (a) (π) and a lower mass FS oriented around the initial direction ;

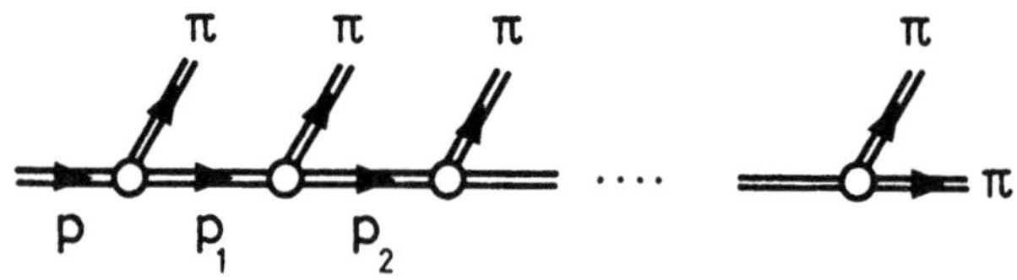

Figure 3 The linear decay chain of an excited $q\bar{q}$ system

ii) calling $\Gamma(M)$ the width of FS of mass M, we get asymptotically

$$M\Gamma(M) \longrightarrow \lambda \frac{2\pi}{R^2} \tag{1.22}$$

where λ depends on the coupling among bags ;

iii) the one-particle yield exhibits strong interaction scaling, i.e.,

$$2E \frac{d^3N}{dp^3} \simeq f(x;p_T) \tag{1.23}$$

where $x = 2p_\ell/M$ and p_ℓ and $p_\perp$ are the particle momenta longitudinal and transverse with respect to the FS direction Ω_0 ;

iv) the distribution in transverse momentum in (1.23) is strongly cut-off with an average value given roughly by

$$\langle p_\perp^2 \rangle \simeq \frac{\pi}{R_\perp^2(M)} \quad ; \tag{1.24}$$

v) when $x \to 0$ $f(x,p_\perp)$ in (1.23) tends to a non-zero limit, as a consequence the decay multiplicity of the FS increases logarithmically with its mass M :

$$\langle n \rangle = c \log M^2 + \text{const} \tag{1.25}$$

where c is given by

$$c = \int d^2 p_\perp \, f(0, p_\perp) \; .$$

These results suggest that what is experimentally observed, from hadronic jets in e^+e^- annihilation to usual peripheral high energy hadron interactions, originate from the decay of a small number of FS's which get produced in a variety of ways, depending, as we shall see later on, on the particular process one considers.

C. - The $q\bar{q}$ Green's function. The Regge states

From the bag states considered so far we can construct the $q\bar{q}$ Green's function following the usual procedure (see Fig. 4)

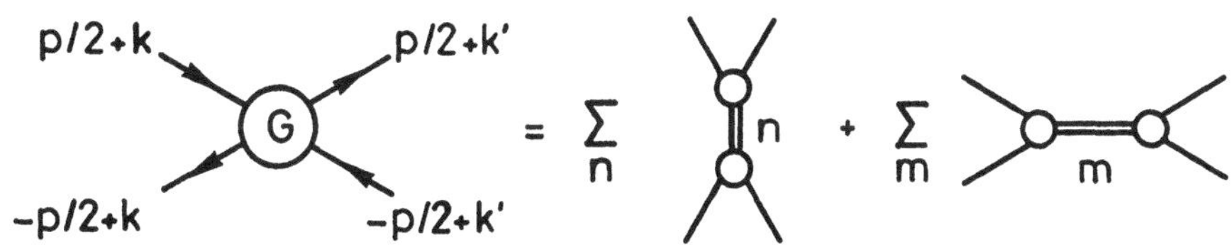

Figure 4 The diagrams making up the $q\bar{q}$ Green's function

In order to have crossing invariance G must be the sum of two pieces,

$$G = G_s + G_t \tag{1.26}$$

representing the contributions of bag states in the s and in the t channel. Note that the u channel, carrying non-zero triality, must be free from singularities. Let us study the G_s contribution first ; we write

$$G_s = \sum_{n\ell m} \frac{\psi_{n\ell m}(p,k)\,\psi^*_{n\ell m}(p,k')}{s-m^2_{n\ell}} \tag{1.27}$$

where ψ is a physically normalized wf (see Section 3), and $s = p^2$ is the square of the centre-of-mass energy. We are interested in the high energy behaviour of G_s. Making use of (1.13), we can write

$$G_s = (2\pi)^4 \left(\frac{2\pi}{R^2}\right)^2 \delta_{R^2}\left(\frac{m_1^2+m_2^2}{2}-m^2\right) \delta_{R^2}\left(\frac{m_1^2-m_2^2}{2}\right)$$
$$\delta_{R^2}\left(\frac{m_1'^2+m_2'^2}{2}-m^2\right) \delta_{R^2}\left(\frac{m_1'^2-m_2'^2}{2}\right) G_s(s,t) \tag{1.28}$$

where

$$G_s(s,t) = \sum_{n\ell} \frac{1}{Z_{n\ell}} \frac{(2\ell+1)\,P_\ell(\cos\theta_s)}{s-m^2_{n\ell}+i\,(m\Gamma)_{n\ell}}\;, \tag{1.29}$$

$m^2_{1,2} = ((p/2)\pm k)^2$, $m'^2_{1,2} = ((p/2)\pm k')^2$, $Z_{n\ell}$ is the factor needed to correctly normalize the wf and $\cos\theta_s \cong (1+(2t/s))$. Note that in (1.29) we have added to the dispersive denominator an imaginary part which, according to the discussion of the previous section [see (1.22)] is asymptotically constant. We shall now calculate the imaginary part $\mathrm{Im}\,G_s(s,t)$ with the approximation $(1/Z_{n\ell}) = (2/Z_0)\Theta(\ell_0-\ell)$ *) where ℓ_0 is given by (1.19). From (1.29) we easily get [1)-3)]

$$\mathrm{Im}\,G_s(s,t) \to \frac{R^2}{2Z_0} \sum_{\ell=0}^{\frac{\sqrt{s}}{2}R_\perp(s)} (2\ell+1)P_\ell(\cos\theta_s) \simeq \frac{sR^2R_\perp^2}{2Z_0} \frac{J_1(R_\perp\sqrt{-t})}{R_\perp\sqrt{-t}} . \tag{1.30}$$

exhibiting the "primeval pomeron" behaviour of the massive quark model [5)]. As for the G_t contribution, we have for large s :

$$G_t \simeq (2\pi)^4 \left(\frac{2\pi}{R^2}\right)^2 \delta_{R^2}\left(\frac{m_1^2+m_1'^2}{2}-m^2\right) \delta_{R^2}\left(\frac{m_1^2-m_1'^2}{2}\right)$$
$$\delta_{R^2}\left(\frac{m_2^2+m_2'^2}{2}-m^2\right) \delta_{R^2}\left(\frac{m_2^2-m_2'^2}{2}\right) G_t(s,t) \tag{1.31}$$

---

*) This approximation is adequate only for small t. At large t one must be more careful as will be shown in Section 7.

where

$$G_t(s,t) = \sum_{n\ell} \frac{(2\ell+1)\, P_\ell(-\cos\theta_t)}{Z_{n\ell}\,(t-m^2_{n\ell})} , \qquad (1.32)$$

and $\cos\theta_t = (1 + (2s/t-4m^2))$. By a Sommerfeld-Watson transform we convert (1.32) into

$$G_t(s,t) = \sum_{n=1}^{\infty} \frac{\pi\alpha_n'(t)\,[2\alpha_n(t)+1]}{\sin\pi\alpha_n(t)} \frac{1}{Z_n(\alpha_n)} P_{\alpha_n(t)}\left(1+\frac{2s}{t-4m^2}\right) \qquad (1.33)$$

which is asymptotically Regge behaved :

$$G_t(s,t) \underset{s\to\infty}{\longrightarrow} \frac{\pi\alpha_1'(t)\,(2\alpha_1(t)+1)}{\sin\pi\alpha_1(t)} \frac{1}{Z_1(\alpha_1)} \frac{\Gamma(1+2\alpha_1)}{\Gamma(1+\alpha_1)^2} \left(\frac{-s}{4m^2}\right)^{\alpha_1(t)} \qquad (1.34)$$

Thus the Regge behaviour of $G_t(s,t)$ is controlled by a set of trajectories $\alpha_0(t)$ corresponding precisely to the bag states. These trajectories are exchange degenerate, a crucial property in order to ensure the absence of physical states (i.e., a discontinuity for $s < 0$) corresponding to unobserved non-zero triality states. This means that at sufficiently high $s > 0$, when we look at the discontinuities in the s channel we do not find only bag states but also a continuum associated with the Regge pole. These states must be looked upon as genuine physical states of a character different from simple bag states ; we shall call them "Regge states". They will fill up the final states with π's in a way similar to the bag states. The only important difference between bag and Regge states at high energy is that the former builds up diffractive scattering while the latter generates normal Regge behaviour.

2. - HADRON SCATTERING AMPLITUDES

A. - Implementing unitarity in a perturbative way

In the preceding section we have explored the consequences of some simple general ideas on the structure and the spectrum of meson systems. We shall now take up the problem of how can a precise framework for hadronic interactions be introduced. As discussed elsewhere [1)-3)] this can be done along the lines of the massive quark model [5)], and it can be shown that it amounts to a diagrammatical expansion of the S matrix which parallels that of a $\lambda\varphi^3$ perturbation theory. Actually further development of these ideas has shown the incompleteness of such a scheme and the necessity for a piece of the type $\mu\varphi^4$ (see Section 4). Enlarging accordingly our scheme, the parallelism between the S matrix expansion and the fictitious $\lambda\varphi^3 + \mu\varphi^4$ perturbation theory is spelled out by the following transcription (see Fig. 5)

i) φ propagator $\rightleftarrows$ $q\bar{q}$ Green's function
ii) φ^3 vertex $\rightleftarrows$ irreducible six-point kernel V_6
iii) φ^4 vertex $\rightleftarrows$ irreducible eight-point kernel V_8

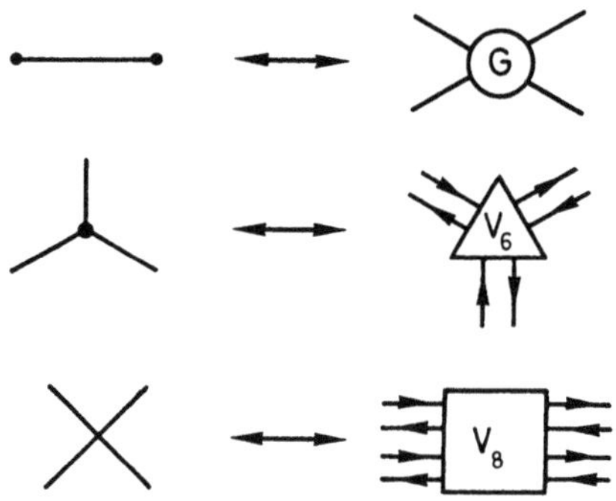

Figure 5 Dictionary leading from $\lambda\varphi^3 + \mu\varphi^4$ to our perturbative expansion

To any given order in λ and μ, which should then have the rôle of relevant expansion parameters, we can draw the diagrams of $(\lambda\varphi^3 + \mu\varphi^4)$ and then proceed to substitute the φ propagators with $q\bar{q}$ Green's functions, the φ^3 with the irreducible six-point kernel V_6, and the φ^4 with the kernel V_8.

Thus the whole construction of unitary hadron scattering amplitudes boils down to determining the structure of G, V_6, V_8 and the rules governing their juxtaposition. The structure of G has been discussed in the preceding section, the forms of V_6 and V_8 will be discussed in the next paragraphs.

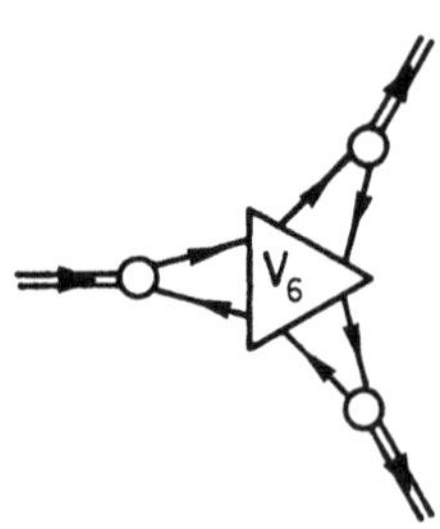

Figure 6 The coupling of three-bag states

B. - The irreducible kernel V_6, the coupling of three bags and its extension to complex momenta

Let us look at the coupling of three-bag states (Fig. 6)

V_6 determines the probability amplitude for the quarks of one bag, say p_1, to disappear from it and reappear as the constituents of a two-bag system (p_2,p_3). According to a geometrical picture of the interaction this amplitude should be zero whenever the space-time regions spanned by the three bags do not overlap. A way of achieving this is to write

$$V_6(x_1,\dots,y_3) = \mu^6\ \delta^4(x_1-x_2)\,\delta^4(y_2-y_3)\,\delta^4(x_3-y_1) + \text{perm.} \tag{2.1}$$

where μ^2 is a parameter, with dimension of a $(\text{mass})^2$, playing the rôle of a coupling constant. It can be interpreted as the amplitude for a quark to tunnel from one bag to another. According to (2.1) the three-bag coupling is given in momentum space by the diagrams in Fig. 7, i.e., by the expression $[p_i \equiv (p_i, n_i \ell_i m_i)]$:

$$a(p_1\to p_2+p_3) = \frac{\mu^6}{Z_1^{1/2} Z_2^{1/2} Z_3^{1/2}} \int \frac{d^4k}{(2\pi)^4}\, \psi_{n_1\ell_1 m_1}(p_1,k)\, \psi^*_{n_2\ell_2 m_2}\!\left(p_2; k+\frac{p_3}{2}\right) \psi^*_{n_3\ell_3 m_3}\!\left(p_3; k-\frac{p_2}{2}\right)$$
$$+ (p_2 \leftrightarrow p_3) \tag{2.2}$$

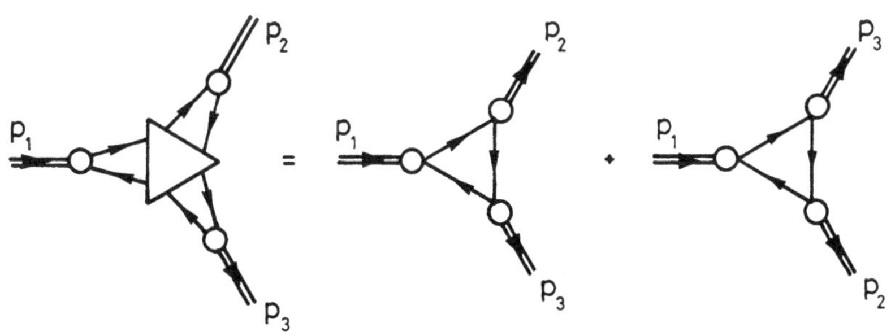

Figure 7 The diagrams describing the three-bag coupling

Substituting the approximate expression (1.13) correctly normalized one gets (we set for convenience $m^2 = 0$)

$$a(p_1\to p_2+p_3) \simeq \mu^6 \frac{(2\pi)^2}{(Z_1 Z_2 Z_3)^{1/2}} \left(\frac{2\pi}{R^2}\right)^3 \int d^4k\ \ \delta_{R^2}\!\left(\frac{p_1^2}{4}+k^2\right)$$
$$\times\ \delta_{R^2}(p_1 k)\ \delta_{R^2}\!\left[\frac{p_2^2}{4}+\left(k+\frac{p_3}{2}\right)^2\right] \delta_{R^2}\!\left[p_2\cdot\left(k+\frac{p_3}{2}\right)\right] \delta_{R^2}\!\left[\frac{p_3^2}{4}+\left(k-\frac{p_2}{2}\right)^2\right] \tag{2.3}$$
$$\times\ \delta_{R^2}\!\left[p_3\left(k-\frac{p_2}{2}\right)\right] Y_{\ell_1}^{m_1}(\Omega_k)\, \bar{Y}_{\ell_2}^{m_2}(\Omega_{k+p_3/2})\, \bar{Y}_{\ell_3}^{m_3}(\Omega_{k-p_2/2}) + (p_2 \leftrightarrow p_3).$$

This quite formidable expression can be evaluated, to lowest order in $1/R^2$, by making use of the δ like properties of the "fat δ function" δ_{R^2}.

We get quite easily :

$$a(p_1 \to p_2+p_3) \simeq \frac{\mu^6 (2\pi)^2}{4(Z_1Z_2Z_3)^{1/2}} \left(\frac{2\pi}{R^2}\right)^3 \int_{-1}^{+1} dz \int_0^{2\pi} d\varphi \; Y_{\ell_1}^{m_1}(\theta,\varphi)$$
$$\times \; Y_{\ell_2}^{-m_2}(\Omega_{k+\frac{p}{2}}) Y_{\ell_3}^{-m_3}(\Omega_{k-\frac{p}{2}}) \; \delta^4_{R^2}\left[\frac{1}{4}(m_1^2-m_2^2-m_3^2) + \frac{1}{4}\lambda^{1/2}(m_1^2,m_2^2,m_3^2) Z_{\vec{k}\vec{p}}\right] \quad (2.4)$$

$z_{\vec{k}\vec{p}}$ is the cosine of the angle between the vector $\vec{k}$ and the decay momentum $\vec{p} = 1/2m\,\lambda(m_1^2,m_2^2,m_3^2)$, where $\lambda(a,b,c) = a^2+b^2+c^2-2ab-2ac-2bc$ is the well-known triangular function, which is positive provided we are in the physical decay region $[m_1 \geq (m_2+m_3)]$. Notice that (2.4) is well behaved for $\lambda \geq 0$, but for $\lambda < 0$ $\delta^4_{R^2}$ develops exponentially exploding tails and we should try and define such a coupling in a way which does not show this disease. This will be done shortly, but before doing it we would like to analyze the structure of (2.4).

The presence of the term $\delta^4_{R^2}$ makes the integrand strongly peaked for $z_{\vec{k}\vec{p}} = 1$, and we get immediately

$$a(p_1 \to p_2+p_3) \simeq \frac{\mu^6}{(Z_1Z_2Z_3)^{1/2}} (2\pi)^2 \left(\frac{\pi}{R^2}\right)^3 \left[(2\ell_2+1)(2\ell_3+1)\right]^{1/2} \delta_{h_2 0}\,\delta_{h_3 0}\, Y_{\ell_1}^{m_1}(\theta,\varphi)$$
$$\times \int_{-1}^{+1} dz \; \delta^4_{R^2}(a-bz) \quad (2.5)$$

where θ,φ are the polar co-ordinates of the decay momentum $\vec{p}$, and

$$a = \frac{1}{4}(m_1^2-m_2^2-m_3^2)$$
$$b = \frac{1}{4}\left[(m_1^2-m_2^2-m_3^2)^2 - 4m_2^2m_3^2\right]^{1/2}. \quad (2.6)$$

Thus (2.5) and (2.6) tell us that the matrix element strongly favours the decay of the highly excited state p_1 into a light meson (say a π) and another high mass state. This is the reason why the decay of a FS proceeds through a linear chain, a fact which has been stressed in the previous section.

Let us now turn to the case when $\lambda < 0$. Its interest lies in the fact that, in order to compute virtual effects coming from bag exchanges we shall need to know how the three-bag vertex behaves in the unphysical region. To see what happens in Appendix 1 we report a calculation of a simplified form of the vertex by means of Gaussian integration over the loop momentum. The disease for $\lambda < 0$ manifests itself by the presence in the final expression of a logarithmically divergent piece. We cure this by a subtraction procedure, and then determine the subtraction by the requirement of continuity at $\lambda = 0$, the boundary of the physical region. As a result [see (A1.8) and (A1.9)], we get the following prescription : whenever a complex momentum flows through a quark line the related "fat δ function" is to be defined as (a,b real)

$$\delta_{R^2}(a+ib) = \frac{1}{2\pi}\int_0^{R^2} d\alpha \cos(\alpha a)\, e^{-|b|\alpha} . \tag{2.7}$$

Thus for $\lambda < 0$, the vertex is given by (2.4), where the expression

$$\delta^4_{R^2}\left[\frac{m_1^2-m_2^2-m_3^2}{4} + i\,\frac{|\lambda|^{1/2}}{4} z_{\vec{k}\vec{p}}\right]$$

is to be interpreted according to (2.7).

C. - The kernel V_8 and the four-bag coupling

The ideas of the preceding paragraph can be immediately carried over to the description of the kernel V_8 and the four-bag coupling (Fig. 8). Thus we write

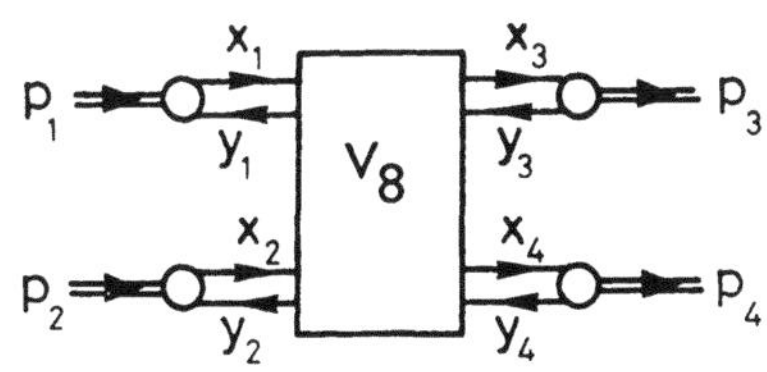

Figure 8 The four-bag coupling

$$V_8(x_1,\dots y_4) = \rho^8\, \delta^4(x_1-x_3)\,\delta^4(y_3-x_4)\,\delta^4(y_4-y_2)\,\delta^4(x_2-y_1) \tag{2.8}$$
$$+ \text{ permutations}$$

ρ^2 is a mass parameter which should be equal to μ^2, if our interpretation of μ^2 as a quark tunnelling amplitude is correct. From (2.8) the four-bag coupling in momentum space is given by the diagrams in Fig. 9.

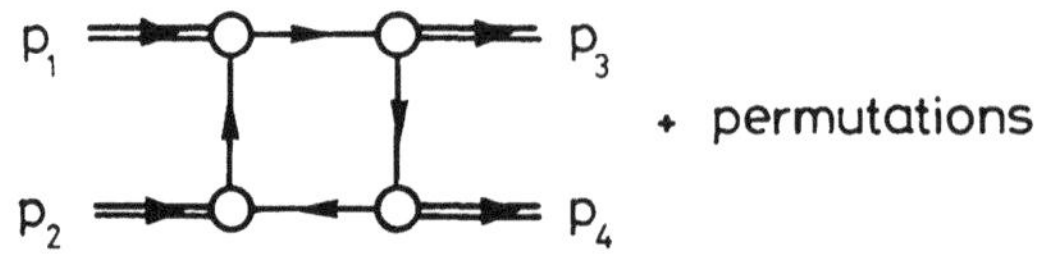

Figure 9 The four-bag coupling diagrams

D. - Graphical rules. Off-shell effects

The hadrons' S matrix is therefore given by a collection of diagrams which can be reduced to a network of quark lines joining quark Green's functions and hadron wave functions. The rules to evaluate any such diagram are the following *) :

i) each incoming bag of momentum p_i and relative quark momentum k_i is introduced through the properly normalized wf $\psi(p_i,k_i)$;

ii) each outgoing bag state of momentum p_f and relative quark momentum k_f is given by $\psi^*(p_f,k_f)$;

iii) each "exchanged" bag state (see Fig. 10) of momentum p_ℓ and relative momentum k_ℓ is given by $(i^\ell)\psi(p_\ell,k_\ell)$, where ℓ is the angular momentum of the state ;

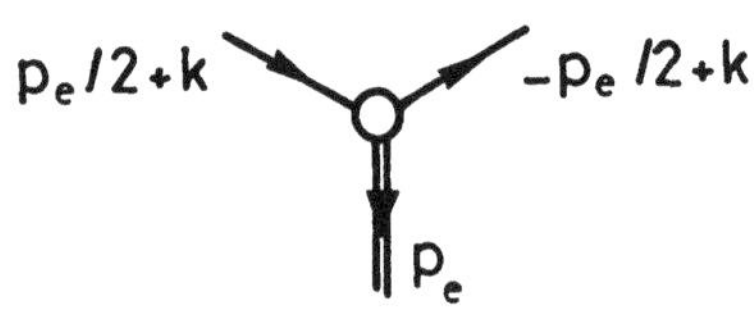

Figure 10 An exchanged bag state

iv) for each quark line we must multiply by μ^2 ;

v) four-momentum at each two-quark hadron vertex must be conserved ;

vi) for each loop we have an integration $d^4\ell/(2\pi)^4$;

vii) $\bar{q}q$ Green's functions are totally connected and their discontinuities correspond to physical hadron states.

These rules, however, are not unambiguous. In order to lift the ambiguity we must specify the procedure by which off-shell effects are to be calculated. To be concrete, let us take the simple Born diagram in "π-π scattering" (Fig. 11). According to our rules we are led to a

---

*) For the reasons previously mentioned these rules turn out to be slightly different from the ones given in Refs. 1)-3).

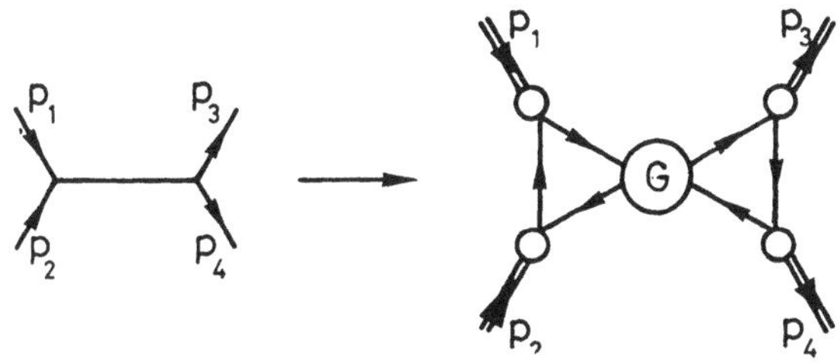

Figure 11 The lowest order diagram in $\pi\pi$ scattering and its quark counterpart.

two-loop integral of four wf's and one $\bar{q}q$ Green's function. The question now is how to calculate the Green's function to be inserted in the two-loop calculation. There are two procedures which give in principle different results *).

1) Calculate first the Green's function as a function of s and t at fixed quark masses, i.e., put the imaginary part (1.30) into an s dispersion relation at fixed t. Insert then such a Green's function into the two-loop integral.

2) Calculate the s discontinuity of the whole diagram first and then insert it in a fixed t dispersion relation.

While, obviously, the two procedures give the same imaginary part (on-shell), the real parts in principle differ drastically. It is clear that in general only procedure 2 is in agreement with usual dispersion relations, and therefore should be followed throughout. In more complicated situations we shall have to perform a thorough analysis of the correct analyticity properties of the amplitudes under consideration before we can decide about the correct calculational procedure.

3. - CURRENTS AND NORMALIZATIONS

A. - A simple way to introduce currents

Let us now suppose that our quarks, still Lorentz scalars, form an isospin doublet. The states one can build out of a $q\bar{q}$ pair will then carry isospin 1 and 0, and their wave functions are given by ($\alpha = 0,\dots,3$; $i,j = 1,2$)

---

*) In the MQM framework 5) procedure 2 was implicitly assumed.

---

*) This was done in Refs. 1)-3).

$$(\psi^{\alpha}_{n\ell m})^{j}_{i} = (\tau^{\alpha})^{j}_{i} \; \psi_{n\ell m}(p,k) \tag{3.1}$$

where $\psi_{n\ell m}(p,k)$ is the wave function (1.9), and for simplicity the isotriplet and the isosinglet have been assumed degenerate. The simplest and most natural way *) to introduce an isospin current is to write for the current quark vertex operator (see Fig. 12) :

$$\mathrm{disc}_{q^2}[\Gamma^{\beta}_{\mu}(q,k)]^{j}_{i} = \left(\frac{\tau^{\beta}}{2}\right)^{j}_{i} \sum_{n} (2\pi)\delta(q^2-m^2_{n1})\,\mu^6 \int \frac{dk'}{(2\pi)^4}\, 2k'_{\mu}\,\psi_{n1}(q,k')\,\psi^{*}_{n1}(q,k) \tag{3.2}$$

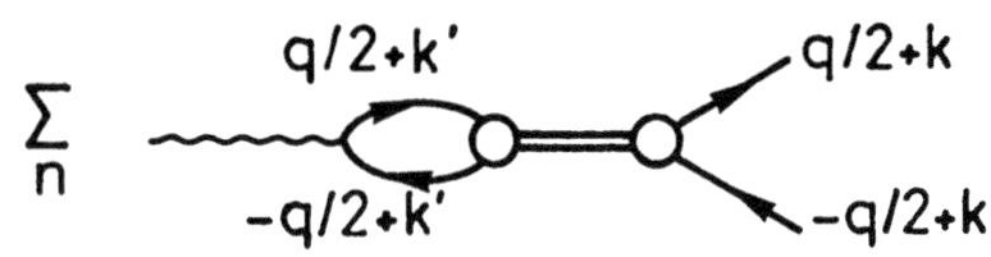

Figure 12 The vector meson dominated current-quark vertex operator

Any current matrix element will be calculated by first folding (3.2) with the relevant wave functions and then getting the real part by dispersing in q^2. This definition coincides in an obvious manner with generalized vector meson dominance [6], whose phenomenological success is well known. Using (1.5c) we can check that the current is conserved, i.e.,

$$q^{\mu}\,\mathrm{disc}_{q^2}[\Gamma^{\beta}_{\mu}(q,k)]^{j}_{i} = 0 \tag{3.3}$$

In Appendix 2 a calculation of the form factor discontinuity is reported for the generic state of angular momentum ℓ and mass M. Upon inserting the discontinuity in the q^2 dispersion relation we obtain for the charge form factor :

$$\langle p\ell\beta | J^{\alpha}_{\mu}(0) | p'\ell\gamma\rangle = (p+p')_{\mu}\, \mathrm{Tr}\left(\frac{\tau^{\alpha}}{2}[\tau^{\beta},\tau^{\gamma}]\right) F^{ch}_{M\ell}(q^2) \tag{3.4}$$

where for $M \gg m_1 = m_\rho$

$$F^{ch}_{M\ell}(q^2) \simeq \frac{2\mu^6}{R^2 Z_\rho Z_\ell(M)} \frac{1}{m_\rho M} \frac{1}{m^2_\rho - q^2}\, P_\ell\left[\sqrt{1+\frac{m^2_\rho}{M^2}}\right] \tag{3.5}$$

The normalization condition $F^{ch}_{M\ell}(0) = 1$ leads to :

$$Z_\ell(M) \simeq \frac{2\mu^6}{R^2 Z_\rho} \frac{1}{m_\rho^3 M} P_\ell\left[\sqrt{1+\frac{m_\rho^2}{M^2}}\right] . \tag{3.6}$$

For low partial waves $[\ell \lesssim (m_\rho/M)]$ asymptotically one has :

$$Z_\ell(M) \underset{M\to\infty}{\longrightarrow} \frac{c}{M} . \tag{3.7}$$

If we go back to Section 1, we see that (3.7) wrecks the nice concept of the "primeval Pomeron" [see Eq. (1.3)], and with it the successful phenomenology of the MQM. This result is therefore unacceptable, and the only way for us to salvage the situation is to admit that (3.2) does not represent the whole isospin current.

The vanishing of $Z_\ell(M)$ when M becomes very large can be understood if we think that the currents can measure the isospin content of our hadrons only through their hadronic fluctuations, the vector mesons. Any massive vector meson can sample the isospin charge only in space region of the order of its Compton wavelength $1/m_V$.

On the other hand the space volumes of our states increase linearly with their mass, and massive vector mesons can only "see" a fraction of the charge of high mass states which decrease with the mass M in agreement in (3.7).

B. - Adding a long range piece to the currents

How can we cure the pathological result (3.7) ? The only way out of this difficulty seems to admit that the vector dominated piece is only one contribution to the isospin currents, and that the full current contains a direct coupling to the quarks comprising a hadronic state.

The idea that currents can couple directly to quarks is certainly not new and arises quite naturally *). However, it is not widely appreciated that it is far from an innocent and harmless hypothesis. In fact if we try to implement it, we can see right away that all our (unproven) ideas about form factors and their analyticity properties are incorrect.

---

*) For instance in the parton approach of Feynman and Bjorken [7].

**) See for example the discussion in Ref. 8).

Suppose, in fact, we know how to introduce this additional direct coupling of the current to the quarks, we will end up with calculating diagrams of the type in Fig. 13. Such diagrams are quite familiar to nuclear physicists, it suffices to recall the calculation of the deuteron form factor. There the q^2 dependence arises from the well-known anomalous thresholds which are sensitive to the structure of the wave function. Also there is no trouble with dispersion relations due to the fact that anomalous singularities do arise from the configuration where all lines are on the "mass shell" **). When we carry this picture

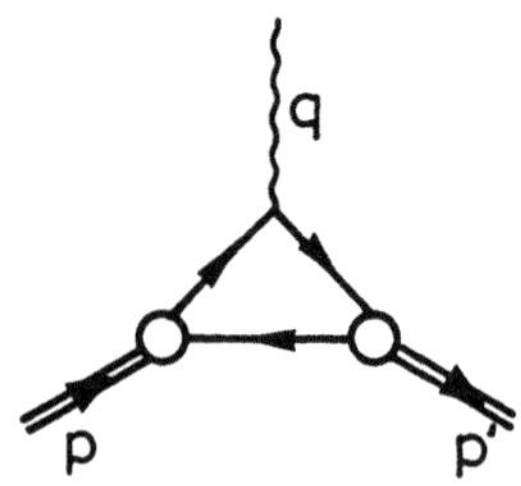

Figure 13 The direct current quark coupling contribution to the hadron form factor

over to quarks, this does not make sense any more ; and we must face the serious possibility that our ideas about form factors and their analyticity have to be revised.

I am aware that this step is a momentous one and may lead the whole theory into disaster, but I have found no other way out. On the other hand this new piece of the currents might be just what we need to account for the seemingly abundant production of photons in high energy collision [9].

I propose that in addition to (3.2) the isospin current has a "direct" quark contribution, $\hat{J}^a_\mu$, and the matrix element between any two bag states is given by

$$\langle p_1,\alpha | \hat{J}^\beta_\mu(x) | p_2,\gamma\rangle = Z_0 \, \mathrm{Tr}\left(\frac{\Gamma^\beta}{2}[\tau^\alpha,\tau^\gamma]\right) \int dy \, \psi^*(p_2;x,y) \overleftrightarrow{\frac{\partial}{\partial x_\mu}} \psi(p_1;x,y) \qquad (3.8)$$

where Z_0 is a constant carrying the dimension of a mass squared. In order to ascertain that the new current piece is indeed conserved we must utilize (1.5a) which is equivalent to

$$-\frac{1}{2}(\Box_x+\Box_y)\,\psi(p;x,y) = m^2\psi(p;xy) \quad ; \tag{3.9}$$

and the vanishing of ψ at the boundary.

From (3.8) we can readily evaluate the charge form factor for any state of mass M and angular momentum ℓ, in particular we get

$$\hat{F}^{ch}_{M\ell}(0) = \frac{Z_0}{2Z_\ell(M)} \,. \tag{3.10}$$

Putting (3.6) and (3.10) together one gets the normalization condition :

$$Z_\ell(M) \simeq \frac{Z_0}{2} + \frac{2\mu^6}{Z_\rho M m_\rho^3}\, P_\ell\left[\sqrt{1+\frac{m_\rho^2}{M^2}}\right] . \tag{3.11}$$

This result not only reinstates the "primeval Pomeron" behaviour, in fact for $\ell \lesssim (M/m_\rho)$ at large M

$$Z_\ell(M) \simeq \frac{Z_0}{2} \quad ,$$

but for $\ell \gg M/m_\rho$ $Z_\ell(M)$ increases like $(1/M)e^{+\ell(m_\rho/M)}$, and this implie that the high angular momenta decouple from the two-quark Green's function exactly in the way which was discussed in Section 1. One can estimate that (3.11) corresponds to the approximation made in Section 1 :

$$\frac{1}{Z_\ell(M)} = \frac{2}{Z_0}\,\theta(\ell_0-\ell) \tag{3.12}$$

with

$$\ell_0 = \frac{M}{2}R_\perp(M)$$
$$R_\perp = \frac{1}{m_\rho}\log\left(\frac{M^2}{M_0^2}\right) \quad , \tag{3.13}$$

where M_0 is a mass whose explicit expression depends on the constants appearing in (3.11).

It is worth noticing that the "primeval Pomeron" acquires [see (3.13) and (1.30)] a logarithmic behaviour as well as a shrinkage. A discussion of how this behaviour affects scaling and its possible breaking will be presented in Section 6.

Thus we see that while, on one hand, the original MQM picture [5)] is basically confirmed, on the other, one sees emerging new interesting structures which get the whole approach closer to Nature. We end this section by adding an item to our list of calculational rules in Section 2 :

viii) currents are given by the sum of two contributions which in the fictitious φ theory are given by the diagram in Fig. 14. The "vector dominated" piece is given by the quark diagram in Fig. 15, while the "direct" piece is described by the diagram in Fig. 13, multiplied by the normalization constant Z_0.

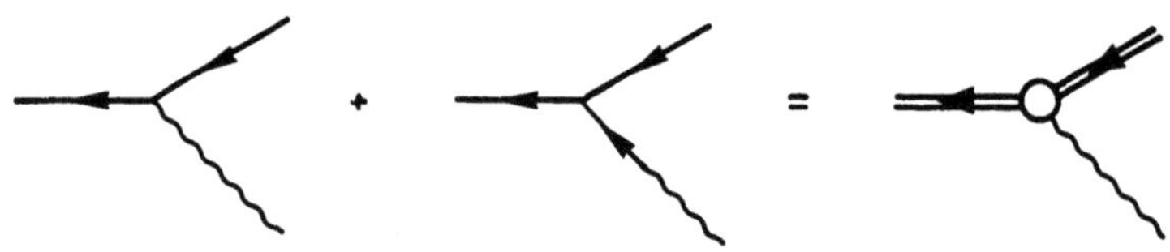

Figure 14 The current coupling in the fictitious φ theory

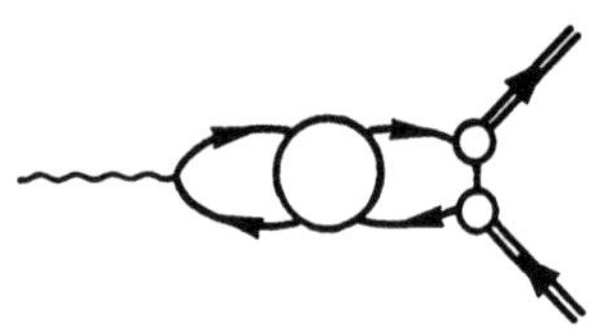

Figure 15 The "vector dominated" piece of the current

4. - HADRON-HADRON SCATTERING TO LOWEST ORDER

We begin in this section a series of calculations of several interesting hadronic processes. As emphasized in the Introduction the importance of these calculations lies in the fact that from them we should get a clear idea about the real meaning of the rules of Section 2. Even though the phenomenologically crucial quark spin has no place in all this, the reader should appreciate that many of these results, when properly transposed, are indeed relevant for the actual hadronic world.

A. - The production of resonances and of a single FS

The φ diagram and its quark correspondent is reported in

Fig. 16, where the external hadron shall be always taken as the ground $q\bar{q}$ state.

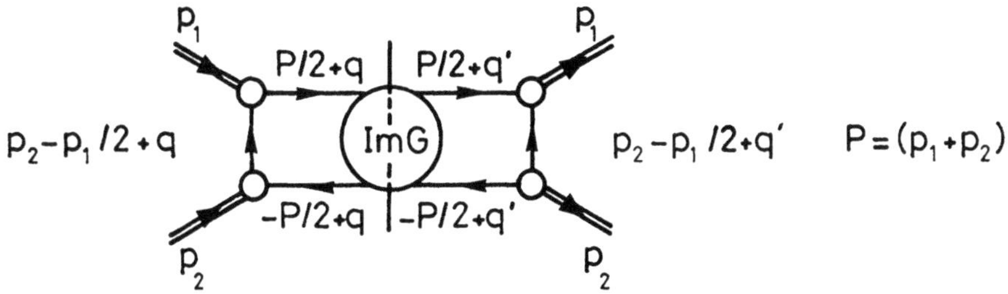

Figure 16 The hadron-hadron scattering amplitude to lowest order

The scattering amplitude at low energy will have poles in s corresponding to the positions of low mass resonances. At high s we calculate (see Appendix 3)

$$\mathrm{Im}\, a_s(s,t) \longrightarrow \frac{c}{s} , \qquad (4.1)$$

which leads to a resonance cross-section decreasing like $1/s^2$. Thus single FS production in high energy hadron-hadron collisions is strongly suppressed. If, instead of FS production, we are interested in the production of Regge states, a calculation similar to the previous one shows that such production cross-section goes as $s^{-5/2}$.

B. - Bag exchange. The question of Reggeization

The amplitude we want to compute now is the one shown in Fig. 17.

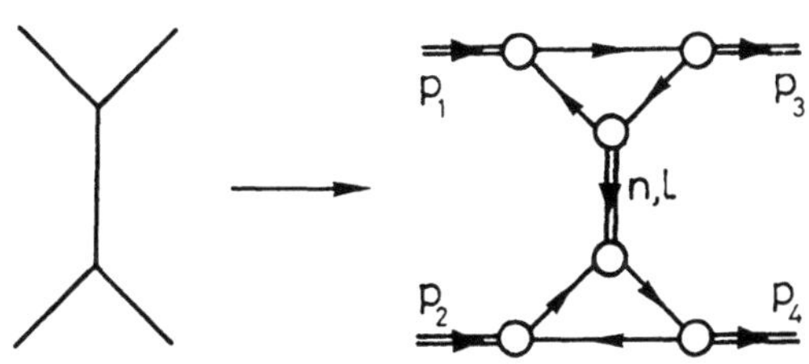

Figure 17 The bag exchange contribution to lowest order

Proceeding in a standard way we write

$$a_t = \sum_{n\ell} g_{n\ell}^2 \frac{(2\ell+1)(-1)^\ell P_\ell(\cos\theta_t)}{t - m_{n\ell}^2} \tag{4.2}$$

where $g_{n\ell}$ is the coupling constant of the excited state $n\ell$ to the ground states, and

$$\cos\theta_t = 1 + \frac{2s}{t} . \tag{4.3}$$

Equation (4.2) is nothing but the model of Van Hove [10], and it is well known to lead to Regge behaviour. Furthermore the amplitude a_t exhibits for s positive and bigger than a certain s_0 a cut coming from the divergence of the series defined by (4.2). But what is the meaning of this cut ? From Fig. 17, it is immediately clear that it cannot correspond to a physical state, but rather to a state consisting of a Regge state and a pair of non-interacting (spectators) quarks. Thus (4.2) and (4.3) lead us into the fatal difficulty of yielding a discontinuity which cannot correspond to an observable physical state. Is there any way out ? Well, let us go back to (4.3), $\cos\theta_t$ is unambiguously defined only for $t = m_{n\ell}^2$ (i.e., on-shell) ; when we go off-shell $(t \neq m_{n\ell}^2)$ (4.3) is a possible extrapolation, but far from unique. We shall show that if rather than via (4.3) we extrapolate according to

$$\cos\theta_t = 1 + \frac{2s}{m_{n\ell}^2} \tag{4.4}$$

our trouble will disappear and the amplitude defined by (4.2) and (4.4) is actually analytic in s. To see this we need only recall that our particles lie on Regge trajectories $m_{n\ell}^2 \simeq a_n + \alpha'\ell$, and that when ℓ increases so does $m_{n\ell}^2$, which makes the convergence of (4.2) for s fixed a trivial matter to prove.

In Appendix 4 the behaviour of a_t for $s \to \pm\infty$ and t fixed and negative is estimated to be :

$$a_t \xrightarrow[s\to+\infty]{} \frac{a(t)}{s^{3/2}} \tag{4.5}$$

$$\xrightarrow[s\to-\infty]{} \frac{b(t)}{s^2} , \tag{4.6}$$

which makes it legitimate to neglect this contribution at high energy, except in some special cases for $t \simeq 0$ where non-Reggeized π exchange may play an important rôle.

To conclude, the main point of this section is that, in defining the rules to compute "off-shell" behaviours, we should be guided only by the criterion of physical interpretability of the amplitude discontinuities arising from our rules. This criterion excludes the interpretation of the diagram in Fig. 17 as a Van Hove model, and forces us to find an alternative solution. It is utterly non-trivial that such solution exists in the definition (4.4) of the meaning of $\cos\theta_t$, and that the amplitude so defined does not appreciably contribute to the high energy behaviour of hadron-hadron scattering [see (4.5) and (4.6)]. In the language of dispersion relations (4.4) corresponds to writing a t dispersion relation keeping s and not $\cos\theta_t$ fixed ; this strongly suggests that in calculating "off-shell" behaviours the correct procedure to follow will always be to disperse in the variable we want to continue, while keeping <u>all other independent Lorentz scalar fixed</u>.

5. - HIGH ENERGY HADRON-HADRON SCATTERING : THE POMERON AND REGGE BEHAVIOUR

We have seen in the preceding section that the lowest order diagram contribution to high energy hadron-hadron scattering is negligible. Here we want to discuss how Pomeranchuk and Regge behaviour may emerge from our approach. Once we have disposed of single FS production, as an unlikely high energy process [see Eq. (4.1)], the perturbative approach described in Section 2 leads us to consider the production of two FS's as the dominant high energy process. This is actually what will be shown in this section ; but before entering in the midst of calculations I would like to recall that in previous work [3] the belief was held that a diagram like the one reported in Fig. 18 would lead to the experimentally observed Pomeron and Regge behaviours.

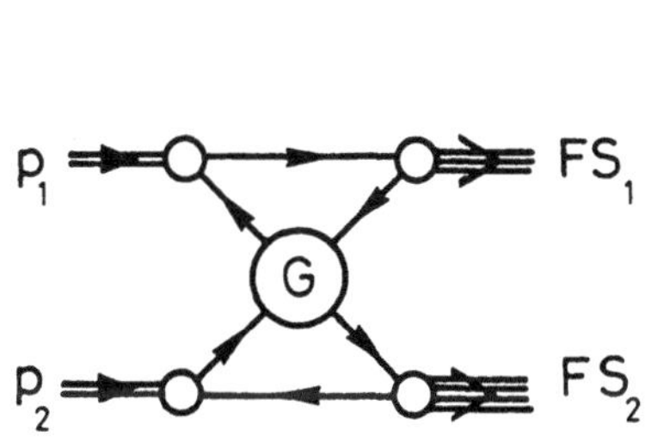

Figure 18 The two FS production diagram generated by the iteration of the φ^3 coupling

However, carrying out a calculation similar to the one reported in Appendix 4, this belief turns out to be unfounded, and one must conclude that one cannot get an acceptable high energy behaviour for the two FS's production processes through the "bag exchange" of Fig. 18.

As anticipated in Section 2, we must have a bag coupling scheme which in addition to the three-bag vertex allows also a <u>direct</u> four-bag coupling (see Fig. 19).

A. - The four-bag coupling and the Pomeron

The s discontinuity of the two-body scattering amplitude is given by the diagram in Fig. 19.

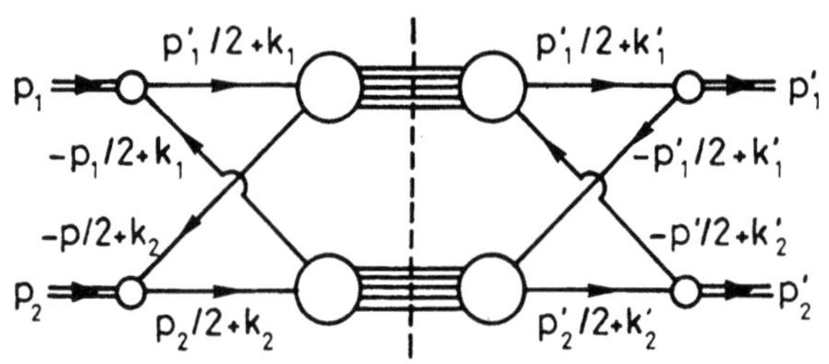

$$k_1-k_2=k_1'-k_2'$$
$$s=(p_1+p_2)^2=(p_1'+p_2')^2$$
$$k=(p_1-p_2)^2=(p_2'-p_1')^2$$

Figure 19 The Pomeron diagram

The calculation is completely standard : we write down the wf's and the Green's functions in the approximate forms (1.13) and (1.30), respectively. We parametrize all momenta à la Sudakov :

$$q=\alpha p_1+\beta p_2+q_\perp$$

and carry out all the integrals in the longitudinal variables $(\alpha,\beta,\ldots)$ by means of the "fat" δ functions and obtain :

$$\mathrm{disc}_s\, \mathcal{A} \simeq 2^5 \mu^{16} \left(\frac{2}{Z_0}\right)^4 \left(\frac{2\pi}{R^2}\right)^2 R_\perp^2 s \int_{-1/2}^{+1/2} d\alpha \int_{-1/2}^{+1/2} d\beta \, \frac{(\frac{1}{2}-\alpha)(\frac{1}{2}-\beta)}{(\frac{1}{2}+\alpha)(\frac{1}{2}+\beta)}$$

$$\times \int d^2\vec{k}_1 \int d^2\vec{k}_2 \int d^2\vec{k}_1' \; \delta^2_{R^2}\left(\frac{\vec{k}_1^{\,2}}{\frac{1}{2}+\alpha}\right) \delta^2_{R^2}\left(\frac{\vec{k}_2^{\,2}}{\frac{1}{2}+\beta}\right)$$

$$\times \; \delta^2_{R^2}\left[\frac{\vec{p}^{\,2}/4+\vec{k}_1'^{\,2}}{\frac{1}{2}+\alpha} + \frac{\vec{p}\cdot\vec{k}_1'}{\frac{1}{2}+\alpha}(2\alpha)\right] \delta^2_{R^2}\left[\frac{\frac{\vec{p}^{\,2}}{4}+(\vec{k}_1'-\vec{k}_1+\vec{k}_2)^2}{\frac{1}{2}+\beta} + \frac{\vec{p}\cdot(\vec{k}_1-\vec{k}_2-\vec{k}_1')}{\frac{1}{2}+\beta}\,2\beta\right]$$

$$\times \; \frac{J_1(R_\perp|\vec{k}_1-\vec{k}_1'-\vec{p}/2|)}{R_\perp|\vec{k}_1-\vec{k}_1'-\vec{p}/2|} \; \frac{J_1(R_\perp|\vec{k}_1-\vec{k}_1'+\vec{p}/2|)}{R_\perp|\vec{k}_1-\vec{k}_1'+\vec{p}/2|} \qquad (5.1)$$

where $\vec{k}_{1,\ldots}$ are two-dimensional vectors and $t = -\vec{p}^{\,2}$. By using the approximations (A3.5) and (A3.6) we can perform the three two-dimensional ingegrals, and get :

$$\mathrm{disc}_s\, \mathcal{A} \simeq \frac{9}{2} \left(\frac{R_\perp^2\mu^2}{\pi}\right)^4 \left(\frac{2\mu^2}{Z_0}\right)^4 \pi R_\perp^2 s \, \exp\left(\frac{R_\perp^2}{16} t\right) . \qquad (5.2)$$

We have indeed obtained a diffractive amplitude. From (3.13) we see that the total hadron-hadron cross-section goes asymptotically like $(\log s)^2$ and that the diffraction peak shrinks with energy. Our calculation, therefore, confirms that the origin of diffraction in hadron collisions is the production of two fire sausages through the four-bag coupling mechanism. Such a diffractive behaviour lends itself to a geometrical interpretation in terms of a grey disc with a radius expanding logarithmically with energy.

To get the full amplitude we follow our general strategy and write a dispersion relation in s with t fixed (5.2) being twice the imaginary part. The fact that the amplitude thus obtained is already a quite good approximation to the real high energy amplitude, does nothing but confirm that the expansion of the S matrix envisaged in Section 2 is a meaningful one.

B. - The Regge behaviour

To obtain the observed non-leading Regge behaviour, we need simply identify one of the blobs in Fig. 19 with the discontinuity of the part of the Green's function G_t, which according to (1.34) gives rise asymptotically to Regge behaviour. Setting

$$\mathrm{disc}_s\, G_t(s,t) \rightarrow c \left(\frac{s}{s_0}\right)^{\alpha(t)}$$

and performing a calculation analogous to the one reported above, we easily obtain the non-leading piece of $\mathrm{disc}_s a$ as :

$$\mathrm{disc}_s\, a \simeq K \left|\frac{s}{s_0}\right|^{\alpha(t)} . \tag{5.3}$$

This result may appear surprising ; we have in fact obtained unadulterated Regge behaviour in spite of the complicated structure of the diagram leading to (5.3). We would have rather expected a Regge cut type behaviour. The reason for this is that at very high energy the t distribution of $\mathrm{disc}G_s$ shrinks faster ($\sim\log^2 s$) than that of $\mathrm{disc}G_t$, thus dominating the covolution between the two Green's functions. Putting (5.3) into a fixed t s dispersion relation reproduces the correct Regge phase in the full amplitude. We can therefore conclude that in our approach Regge behaviour does not arise from the diagram in Fig. 17 but from the one in Fig. 19. However, amusingly enough, due to the peculiar structure of the Green's function G_s, everything goes as if we were calculating precisely the incriminated diagram of Fig. 17.

6. - THE SCALING PHENOMENON

After having successfully passed the (qualitative) test of hadronic scattering at high energy and recovered all essential experimental features, we must ask ourselves whether such a nice picture survives another crucial test : the deep inelastic phenomena. In this section our theory will demonstrate its healthy status in this respect also. In order not to encumber the notation the case of scalar currents and scalar targets shall be considered throughout. A full account of the spin structure will be possible only after we have introduced the quark spin degree of freedom.

A. - "e^+e^- annihilation" into hadrons

From the discussion of the currents' structure carried out in Section 3 the coupling of the current to a generic hadronic system is made up of two distinct contributions : the "vector dominated" one and the "direct" contribution. The scalar current, whose effects will be calculated here, will accordingly be given by :

$$J = J^{VD} + J^{Direct} \; . \tag{6.1}$$

In calculating the e^+e^- annihilation process into hadrons we must therefore consider the two diagrams in Fig. 20 ; notice that we are neglecting interference terms between the two current pieces. That this is legitimate shall become apparent in the following

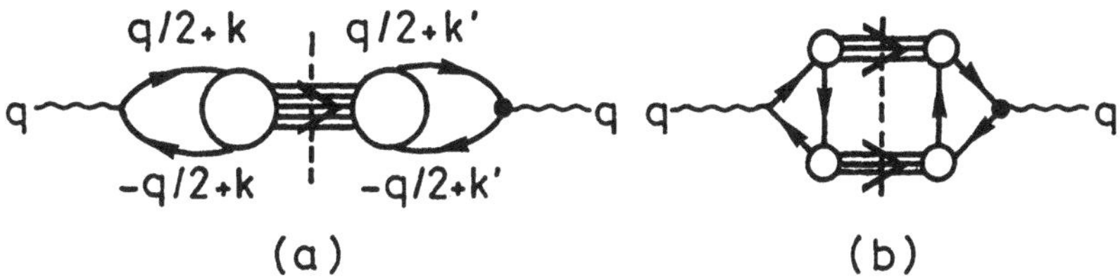

Figure 20 The diagrams contributing to "e^+e^- annihilation"

(a) - The vector dominated contribution

The high energy behaviour of the imaginary part ρ_a of the vacuum polarization function is given by

$$\rho_a(q^2) \to \frac{1}{2} \int \frac{d^4k}{(2\pi)^4} \int \frac{d^4k'}{(2\pi)^4} \, \mathrm{disc}_s \, G_s \; . \tag{6.2}$$

According to (1.28) and (1.30), (6.2) becomes :

$$\rho_a(q^2) \to \frac{1}{2} \frac{1}{(2\pi)^4} \int d^4k \int d^4k' \; \frac{R^2 R_\perp^2}{8\pi Z_0} \cdot q^2 \; \frac{J_1\left(R_\perp \sqrt{(k-k')^2}\right)}{R_\perp \sqrt{(k-k')^2}}$$

$$\times \left(\frac{2\pi}{R^2}\right)^2 \delta_{R^2}(qk) \, \delta_{R^2}(qk') \, \delta_{R^2}\left(\frac{q^2}{4}+k^2\right) \delta_{R^2}\left(\frac{q^2}{4}+k'^2\right) . \tag{6.3}$$

The k' integration is easily made by use of the identity (in the sense of distribution theory) [5] :

$$\delta_{R^2}(a^2)\,\delta_{R^2}(b^2)\,\frac{J_1(R_\perp\sqrt{(a-b)^2})}{R_\perp\sqrt{(a-b)^2}} \to \frac{\pi}{R_\perp^2}\,\delta_{R^2}(a)\int d\alpha\,\delta^4(b-\alpha a)\,. \tag{6.4}$$

Thus we get

$$\begin{aligned}\rho_a(q^2) &\to \frac{1}{16\pi^2}\,\frac{1}{R^2 Z_0}\int d^4k\;\delta_{R^2}(qk)\,\delta_{R^2}\left(\frac{q^2}{4}+k^2\right)\\ &= \frac{1}{16\pi}\,\frac{1}{R^2 Z_0} + O\left(\frac{1}{\sqrt{q^2}}\right),\end{aligned} \tag{6.5}$$

where the non-leading term originates from the Regge behaviour of the G_t piece of the Green's function. In order to appreciate the result (6.5) it is necessary to calculate the "canonical contribution" reported in Fig. 21 ; a trivial calculation yields immediately :

$$\rho^{can}(q^2) \longrightarrow \frac{1}{16\pi}\,, \tag{6.6}$$

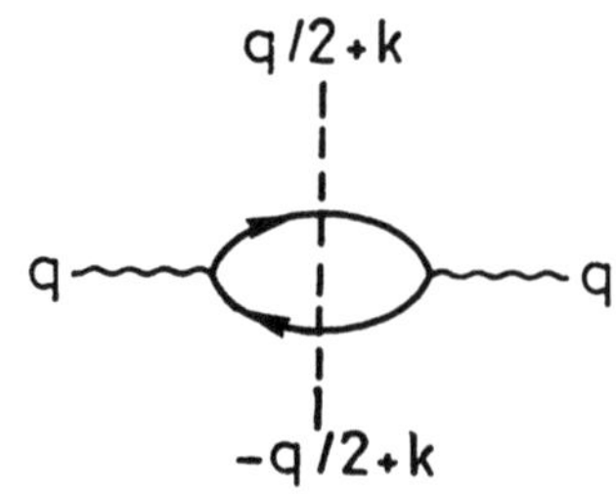

Figure 21 The canonical contribution to e+e- annihilation

i.e., the imaginary part of the vacuum polarization function for large q^2 goes to a constant limit given numerically by (6.6). Introducing a familiar notation we can define R_a, the ratio between the hadronic and the leptonic annihilation, and obtain

$$R_a = \frac{\rho_a(q^2)}{\rho^{can}(q^2)} \rightarrow \frac{1}{R^2 Z_o} + O\left(\frac{1}{\sqrt{q^2}}\right). \qquad (6.7)$$

Thus we see that the hadronic production cross-section has a pointlike, scaling behaviour and the ratio R_a tends to a constant which is in principle different from one, being determined by two crucial hadronic parameters like R^2 and Z_o. This is at variance with the general belief that this ratio should be one.

(b) - The direct contribution

We now must turn to the contribution of the direct current coupling. Recalling the discussion of Section 2B, we need consider only the configuration where a low mass hadron recoils against a high mass FS.

Proceeding in a, by now, standard fashion we find that also this term scales and for R_b we obtain the unexpectedly large numerical result :

$$R_b \simeq 2. \qquad (6.8)$$

The fact that R_b does not depend on the hadronic parameters can be understood by the cancellation between the Z_o^2 terms arising from the current coupling and the Z_o^{-2} term appearing in the denominator coming from the normalization. No particular weight, however, should be given to the particular numerical results (6.7) and (6.8), what is quite important is that both contributions (a) and (b) give rise to a scaling behaviour as observed at high energy in the SPEAR experiments.

What about the one-particle distributions ? If you recall the discussion in Section 1B, you can convince yourself that Bjorken scaling holds for the "vector dominated" contribution due to the scaling properties of the decay mechanism of the FS. As for the "direct" contribution we also get a Bjorken scaling as can be checked without any particular trouble.

B. - "Deep inelastic scattering"

To lowest order we have the two contributions reported in Fig. 22.

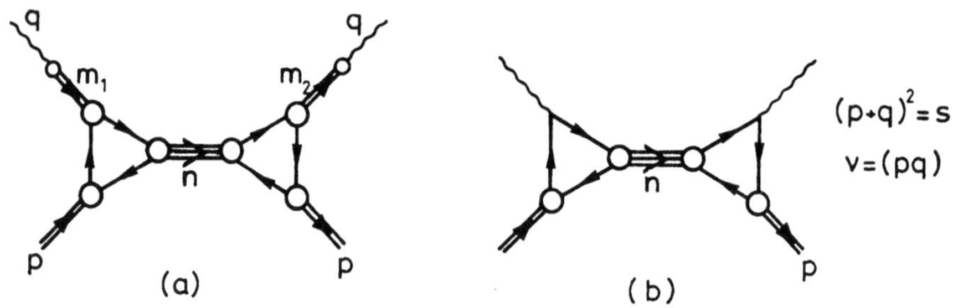

Figure 22 The two lowest order contributions to "deep inelastic scattering"

The evaluation of (a) presents the problem of defining the off-shell extrapolation of the vector mesons m_1 and m_2. The procedure we shall follow, as previously discussed, is to evaluate first the discontinuity $W_a(s;m_1^2,m_2^2)$ of our diagram by keeping m_1 and m_2 fixed. To get then the physical discontinuity $W_a(s;q^2)$ we shall disperse in m_1^2 and m_2^2 by keeping s for fixed, i.e.,

$$W_a(s;q^2)=\sum_{m_1,m_2} g^{Jm_1} g^{Jm_2} \frac{W(s;m_1^2,m_2^2)}{(m_1^2-q^2)(m_2^2-q^2)} \tag{6.9}$$

where

$$\begin{aligned} g^{Jm} &= \left(\frac{2}{Z_0}\right)^{1/2} \int \frac{d^4k}{(2\pi)^2}\left(\frac{2\pi}{R^2}\right)\frac{1}{\sqrt{4\pi}} \quad \delta_{R^2}(qk)\,\delta_{R^2}\left(\frac{m^2}{4}+k^2\right) \\ &\simeq \left(\frac{2}{\pi Z_0}\right)^{1/2} \frac{1}{4R^2} \end{aligned} \tag{6.10}$$

is the current-hadron coupling constant. From the calculation reported in Appendix 5, we get [see (A5.6)]

$$W_a(s;m_1^2 m_2^2) \simeq \left(\frac{2\mu^4}{Z_0}\right)^3 (2\pi)^4 \left(\frac{4\pi}{R^2}\right)^3 \frac{3\pi R^2}{|s-m_1^2|}$$
$$\times \; \delta^4_{R^2}\left[\frac{\mu_0^2 s}{4}\frac{(m_1^2-m_2^2)^2}{(s-m_1^2)(s-m_2^2)}\right] \frac{R_\perp^2}{R_\perp^2+\frac{8R^2}{\pi}\left(1-\frac{m_1^2+m_2^2}{2s}\right)} . \qquad (6.11)$$

The very interesting feature of this result consists in the peaking of $W_a(s;m_1^2,m_2^2)$ for $m_1^2=m_2^2$, exhibited by the "fat δ". Thus in this approach the so-called "diagonal" approximation of generalized vector dominance [11) arises in a very natural way. Moreover, this same mechanism will be always operative in suppressing the interferences between the "direct" and the "hadron dominated" pieces of the currents ; a fact already used in the preceding paragraph.

Inserting (6.10) and (6.11) into (6.9) we have $(x = Q^2/2\nu)$

$$W_a(\nu,Q^2) \simeq \frac{1}{2\nu} F_a(x) , \qquad (6.12)$$

where

$$F_a(x) = \frac{\mu'^2}{Z_0^4}(2\pi)^2\, 3R^2\,(1-x)\int_0^\infty dx_1 \int_0^\infty dx_2 \frac{1}{|1-x_1|}\, e^{-\frac{2R^2}{\pi}\mu_0^2\frac{(x_1-x_2)^2}{|x_1-1|^2|x_2-1|}}$$
$$\times \frac{1}{(x+x_1(1-x))(x+x_2(1-x))}\;\frac{R_\perp^2}{R_\perp^2+\frac{8}{\pi}R^2\left(1-\frac{x_1+x_2}{2}\right)} . \qquad (6.13)$$

The features of $F_a(x)$ worth noticing are :

$$F_a(x) \xrightarrow[x\to 0]{} \text{const} \qquad (6.14)$$

and

$$F_a(x) \xrightarrow[x\to 1]{} (1-x) . \qquad (6.15)$$

What does all this mean ? Let us look at the free field diagram (Fig. 23). We immediately calculate

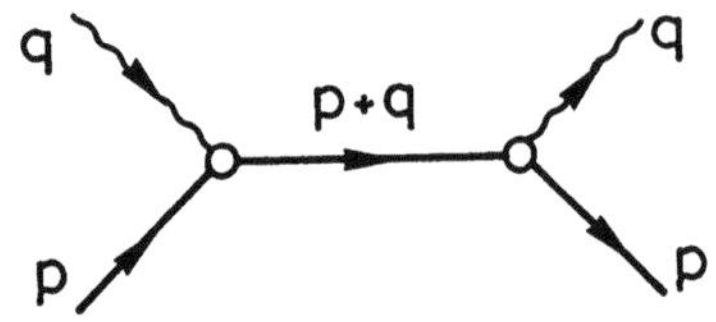

Figure 23 The free field contribution to "deep inelastic scattering"

$$W^{can}(\nu,Q^2) = \frac{1}{2\nu} F^{can}(x) \tag{6.16}$$

where

$$F^{can}(x) = (2\pi)\,\delta(1-x)\,. \tag{6.17}$$

Thus our results (6.12) and (6.13) show scaling with a "structure function" $F_a(x)$ which is highly non-trivial. Equation (6.12) is the leading behaviour ; if we add the Regge behaved piece $\text{disc}_s G_t$ we obtain a non-scaling contribution of the form

$$W_a'(\nu,Q^2) = \frac{1}{(2\nu)^{3/2}} F_a'(x) \tag{6.18}$$

with

$$F_a'(x) \underset{x\to 1}{\longrightarrow} (1-x)^{1/2}\,. \tag{6.19}$$

This piece could play an important rôle [5] in the observed pattern of breaking of scaling behaviour near $x=1$.

We now turn to the contribution (b) ; the relevant calculation is reported in the Appendix 6, which shows that we have a behaviour which simulates scaling

$$W_b(\nu,Q^2) = \frac{1}{2\nu} F_b(x,\nu) \tag{6.20}$$

with

$$F_b(x,\nu) = \frac{12 R_\perp^2 (1-x)}{R_\perp^2(1-x) + \frac{s}{\pi} R^2} . \tag{6.21}$$

In fact, due to the logarithmic dependence of $R_\perp^2$ on s, (6.21) exhibits a peculiar scaling breaking behaviour near $x = 1$. For x small $F_b(x,\nu)$ tends like $F_a(x)$ to a constant.

We want to ask the question whether W_a and W_b are enough to take care of the scaling phenomenon in deep inelastic scattering. Phenomenologically we know that the data are compatible with a "Regge behaviour" of scaling functions near $x = 0$. In our scalar case such a behaviour would mean for the scaling function $F(x)$:

$$F(x) \underset{x \to 0}{\longrightarrow} x^{-\alpha(0)-1} , \tag{6.22}$$

and setting $\alpha(0) = 1$ for the leading Pomeron behaviour we get $F(x) \sim x^{-2}$. The constant limit reached by both $F_a(x)$ and $F_b(x)$ obviously shows that such terms must be unimportant for $x \to 0$, and it behoves upon us to show that we do have a contribution which exhibits a behaviour such as (6.22). The lowest order candidate for this is the Pomeron diagram in Fig. 24, which we shall compute by following a procedure analogous to the one employed for $W_a(s,Q^2)$. The calculation of $W_P(\nu,Q^2)$ is reported in Appendix 7. The result exhibits again scaling behaviour but this time

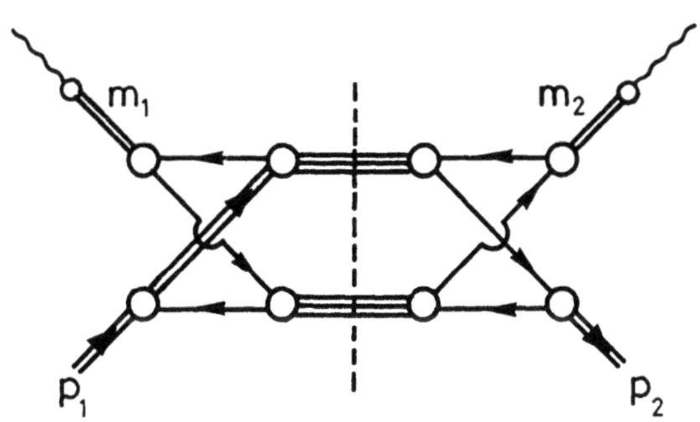

Figure 24 The Pomeron contribution to "deep inelastic scattering"

$$F_P(x) \underset{x\to 0}{\longrightarrow} \frac{1}{x^2}(\log x)^2 , \qquad (6.23)$$

in complete agreement with our expectations and (5.2). On the other hand, when $x \to 1$ $F_P(x)$ clearly vanishes much faster than either F_a and F_b.

This concludes our analysis of deep inelastic phenomena. It is almost superfluous to stress that in this approach we could in principle discuss several other topics like form factors, structure of final states and massive μ pairs production. However, carrying out such an extensive programme will be really meaningful only after we have introduced in a realistic way the quark internal degrees of freedom : SU(3), SU(4)?,..., and spin.

7. - LARGE p_T

It is by turning our attention to this very interesting piece of physics, which has been unravelled for the first time at the CERN ISR that we conclude our analysis. Again no extensive discussion will be carried out of these processes, and we shall simply concentrate on the leading contribution to large p_T particle production in the kinematical region for which $x_T = (2P_T/\sqrt{s}) \ll 1$.

As already discussed in the framework of the MQM we can isolate two distinct contributions : A) the totally inclusive FS decay, and B) the semi-inclusive decay at large angles. We shall see that the results previously obtained in the MQM framework will be naturally reproduced.

A. - The totally inclusive FS decay at large angle

At very high energy production of large p_T hadrons can occur according to the following mechanism : a $q\bar{q}$ pair produced along the collision direction according to the usual mechanism (see Fig. 19) can propagate and emerge at large angle and yield along such direction a corresponding FS which by its canonical decay (see Section 1B) produces hadrons at large angle and consequently large p_T.

To evaluate the amplitude for the pair to emerge at large angle we must sort out the large t behaviour of the Green's function G_s given in (1.29). Going to the impact parameter representation and making use of the results of Section 3 [see (3.11)] we write :

$$\mathrm{Im}\, G_s(s,t) = \frac{1}{2} R^2 s \int_0^\infty b\, db \frac{J_0(b\sqrt{-t})}{\left(1 + \frac{a}{\sqrt{s}} e^{b\frac{m_\rho}{2}}\right)} \tag{7.1}$$

and defining the new variable $x = b\sqrt{-t}$ we have

$$\mathrm{Im}\, G_s(s,t) = \frac{R^2 s}{2(-t)} \int_0^\infty x\, dx \frac{J_0(x)}{1 + \frac{a}{\sqrt{s}} e^{\frac{m_\rho}{2} \frac{x}{(-t)^{1/2}}}} . \tag{7.2}$$

Expanding the exponential for large t we obtain :

$$\mathrm{Im}\, G_s(s,t) \longrightarrow \frac{R^2 s}{2(-t)} \left[\frac{-a m_\rho}{\sqrt{-st}} + O\left(\frac{1}{\sqrt{t}}\right)\right] , \tag{7.3}$$

and feeding $\mathrm{Im}\, G_s(s,t)$ in a dispersion relation we get asymptotically a purely imaginary amplitude *) :

$$G_s(s,t) \to i \frac{R^2 s^{1/2}}{(-t)^{3/2}} \left(\frac{-a m_\rho}{2}\right) + O\left(\frac{1}{\sqrt{s}}, \frac{1}{\sqrt{t}}\right) . \tag{7.4}$$

To compute the production of large p_T hadrons through this mechanism, use will be made of the following formula :

$$2E \frac{d^3\sigma}{dp^3} \simeq \int d^4k \frac{d^4\sigma}{dk^4} \delta^4\left(k - \frac{p}{x}\right) f(x) , \tag{7.5}$$

where $d^4\sigma/dk^4$ is the production cross-section of the FS as a function of the vector k which can be taken as the four-momentum of one of its quarks, $x = 2|p|/M$ is the usual Feynman variable, and M is the FS mass.

From the results of Section 5 and (7.4) we can write the following expression for the FS production cross-section :

$$\frac{d^4\sigma}{dk^4} \sim 18\, \sigma_{TOT} \int_{\frac{4p_\perp^2}{s}}^{1} \frac{dM^2}{s} \int_0^{1-\frac{M^2}{s}} \frac{d\xi}{\sqrt{\frac{M^2}{s} + \frac{\xi^2}{4}}} \frac{M^2}{s}\left(1 - |\xi| - \frac{M^2}{s}\right)$$

$$\times \left(\frac{a m_\rho R^2}{2}\right)^2 \frac{1}{M(k_0 - k_3)^2} \frac{1}{\left(\frac{\xi}{2} - \sqrt{\frac{M^2}{s} + \frac{\xi^2}{4}}\right)^3} , \tag{7.6}$$

---

*) Actually a more careful evaluation yields in addition a real term going like $1/s$; but this will not change any of the following conclusions.

and inserting it into (7.5), we get :

$$2E\frac{d^3\sigma}{dp^3} \simeq \frac{18}{s^2}\sigma_{TOT}\int_{\frac{4p_\perp^2}{s}}^{1}\frac{dM^2}{s}\int_0^{1-M^2/s}\frac{d\xi}{\sqrt{\frac{M^2}{s}+\frac{\xi^2}{4}}}$$

$$\frac{M^2}{s}\left(1-|\xi|-\frac{M^2}{s}\right)\left(\frac{R^2 a m_\rho}{2}\right)^2\frac{8s}{M^2(1-\cos\theta)^3}f\left(\frac{2P}{\sqrt{s}}\left(\frac{s}{M^2}\right)^{1/2}\right) \quad . \qquad (7.7)$$

Thus asymptotically we get the "scaling" $1/p_T^4$ behaviour. The meaning of this result and its experimental evidence are fully discussed elsewhere [5)-13)]. However here we must recall that such a behaviour should be valid only at very high energies and very small x_T ; in an intermediate situation like the one at ISR the leading rôle should be played by the semi-inclusive decay of the FS, which will be now calculated.

B. - The semi-inclusive FS decay at large angle

We must now compute the diagram in Fig. 25, which describes a mechanism for producing large p_T hadrons which at not too high energies is much more efficient than the canonical fire sausage decay.

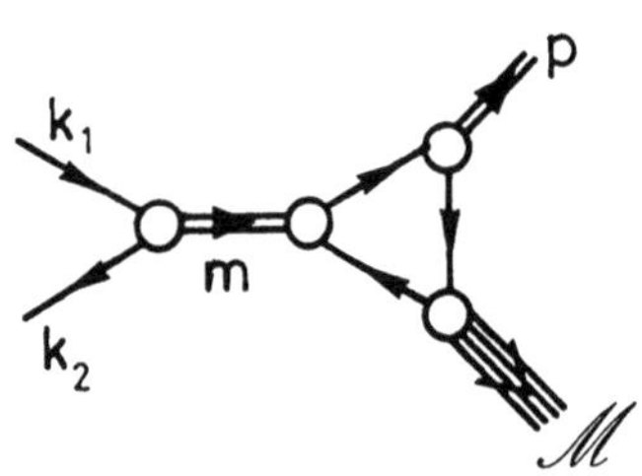

Figure 25 The semi-inclusive FS decay

Fixing $t=(k_1-p)^2$ and $\mathcal{M}^2$ we calculate our diagram by dispersing in $(k_1+k_2)^2=M^2$. Using Im G_S as given by (7.3), we have

$$A(\mathcal{M}^2, M^2, t) \simeq \text{const}\int_0^\infty\frac{dm^2}{m^2-M^2+i\epsilon}\,\frac{m}{(-t)^{3/2}}\,\frac{1}{|m^2-\mathcal{M}^2|} \quad . \qquad (7.8)$$

where the last factor comes from the loop integration [see, for instance, (A5.6)]. Scaling the variables by M^2, we have :

$$\mathcal{A}(\mathcal{M}^2, M^2, t) \rightarrow \text{const} \frac{1}{(-t)^{3/2} s^{1/2}} \frac{1}{1 - \frac{\mathcal{M}^2}{M^2}} . \tag{7.9}$$

To get the final answer we must take (7.9), square it, sum over all states $\mathcal{M}$ [which gives a factor $R_1^2(\mathcal{M}^2/4)$] and finally multiply everything by the q pair distribution in a way analogous to the one employed in the preceding paragraph. Thus :

$$2E \frac{d^3\sigma}{dp^3} \sim \frac{1}{s^4} \frac{1}{(1-\cos\theta)^3} \left(\frac{\sqrt{s}}{2p}\right)^5 \int_{\frac{4p_\perp^2}{s}}^{1} \frac{dM^2}{s} \int_0^{1-\frac{M^2}{s}} \frac{d\xi}{\sqrt{\frac{M^2}{s} + \frac{\xi^2}{4}}}$$

$$\frac{M^2}{s}\left(1 - |\xi| - \frac{M^2}{s}\right) \left(\frac{M^2}{s}\right)^{-3/2} \left(1 - \frac{2p}{\sqrt{s}} \left(\frac{s}{M^2}\right)^{1/2}\right) . \tag{7.10}$$

which reproduces the result of the MQM, and the experimentally observed p_T^{-8} behaviour. We do not expect that the introduction of spin will change these results.

CONCLUSIONS

It seems appropriate to conclude by trying to assess where does the proposed approach stand. Let me emphasize once again that any conclusion one can extract from this work can only be qualitative due to the neglect of a realistic set of internal quantum numbers. What has then been achieved ?

Starting from a minimally simple description of meson states in terms of the quark co-ordinates we have obtained an infinite number of meson states having well-defined space-time properties. The spectrum is quite closely described by a set of parallel linear and exchange degenerate Regge trajectories.

Having constructed our meson states we have attacked the problem of introducing an interaction among them following the ideas of the massive quark model. The picture emerging is very simple and quite appealing, however, it does not give us yet the possibility to calculate physical processes. In order to do this we must introduce current operators which will give a physical normalization to our states, i.e., through the correct specification of the electric charge. At this point, something very distressing happens ; if we want to define the current operators through vector dominance the normalization, which requires an extension of the computational rules to deal with complex momenta (see Appendix 1), is found to be totally unacceptable. This prompts us to reconsider the definition of current operators, and to propose a new definition which takes also into account the possibility for current operators to couple directly to quark lines. This proposal is a very momentuous one, it entails a drastic change of our views about the analytic structure of form factors, which may help us in explaining some puzzling photon behaviours in high energy collisions.

The previous distress now turns itself into gratification, because not only do we recover the nice properties of the massive quark model, but also we obtain gratis the damping of the states with angular momentum exceeding $\ell_o = (M/2)R_\perp$, where $R_\perp$ increases logarithmically with M.

At this stage the framework is precise enough so that we can carry out calculations of several high energy processes. The question of Reggeization is taken up in Section 4 and it is shown that the diagram usually associated with Regge behaviour (Fig. 17) does not Reggeize but rather falls quite fast at high energy, thus presenting no problem with unitarity. The high energy behaviour of two-body scattering, discussed in Section 5, shows the relation of the Pomeron with the production of two fire sausages, as well as its peculiar geometrical nature. A similar behaviour is also shown by the normal Regge term.

Deep inelastic physics is amenable to a similar kind of analysis, which yields Bjorken scaling behaviour, and a physically appealing distinction between "valence" quarks and "sea" quarks.

Finally, large p_T physics is analyzed. The massive quark model picture led to the expectation that the "scaling" p_T^{-4} behaviour is delayed by kinematical effects, so that even at ISR energies the dominant behaviour is still p_T^{-8}. This is completely confirmed by the results of Section 7.

I believe that the quark theory of hadronic phenomena recently proposed is showing a remarkable fitness for attacking the basic problems of hadrodynamics. It will not take long before we shall be able to answer the question whether these ideas can lead us to the long sought mastering of strong interactions.

APPENDIX 1

In this Appendix we shall calculate the simplified overlap integral $(p_1 = p_2 + p_3 \;;\; m_i^2 = p_i^2)$

$$I(m_1^2, m_2^2, m_3^2) = \int d^4k \; \delta_{R^2}(k^2)\, \delta_{R^2}((k-p_2)^2)\, \delta_{R^2}((k-p_1)^2) \tag{A1.1}$$

where

$$\delta_{R^2}(k^2) = \frac{1}{\pi}\, \frac{\sin R^2 k^2}{k^2} = \frac{1}{\pi} \int_0^{R^2} d\alpha \cos \alpha k^2$$

In the physical region $(\lambda = \lambda(m_1^2; m_2^2, m_3^2) \geq 0)$ we can immediately calculate to lowest order in R^2 :

$$I \simeq \pi \int_{-1}^{+1} dz \; \delta_{R^2}\left[\frac{m_1^2 - m_2^2 - m_3^2}{2} - \frac{1}{2}\lambda^{1/2} z\right] . \tag{A1.2}$$

We are now going to calculate (A1.1) in a different way which will give us the possibility to extend our result to the complex momentum case $\lambda < 0$. We write

$$I = \frac{1}{\pi^3} \int_{-R^2}^{+R^2} d\alpha_1 \int_{-R^2}^{+R^2} d\alpha_2 \int_{-R^2}^{+R^2} d\alpha_3 \int d^4k \; e^{i[\alpha_1 k^2 + \alpha_2 (k-p_2)^2 + \alpha_3 (k-p_1)^2]}$$

$$= \frac{1}{i\pi} \int_{-R^2}^{+R^2} d\alpha_1 \int_{-R^2}^{+R^2} d\alpha_2 \int_{-R^2}^{+R^2} d\alpha_3 \; \frac{\varepsilon(\alpha_1+\alpha_2+\alpha_3)}{(\alpha_1+\alpha_2+\alpha_3)^2} \exp\left\{ i\, \frac{\alpha_2\alpha_3 m_3^2 + \alpha_1\alpha_3 m_2^2 + \alpha_1\alpha_2 m_1^2}{\alpha_1+\alpha_2+\alpha_3} \right\} , \tag{A1.3}$$

where the second line has been obtained through Gaussian integration. On changing variables $\alpha_1 \to \alpha = \alpha_1 + \alpha_2 + \alpha_3$, and neglecting higher order terms in R^2, we can perform a Gaussian integration over α_3 to get

$$I \simeq \frac{1}{2i\,\pi^{1/2}} \int_{-R^2}^{+R^2} d\alpha_2\, e^{-i\alpha_2 \frac{m_1^2-m_2^2-m_3^2}{2}} \int_{-R^2}^{+R^2} \frac{d\alpha\, \varepsilon(\alpha)}{|\alpha|^{3/2}}\, e^{-i\frac{\pi}{4}\varepsilon(\alpha)} \exp\left[\frac{i}{4}\left(\alpha m_1^2 + \frac{\alpha_3^2\lambda}{\alpha}\right)\right]. \qquad (A1.4)$$

To leading order in R^2, by integrating over α we obtain :

$$I \simeq \frac{1}{2i} \int_{-R^2}^{R^2} d\alpha_2\, e^{-i\alpha_2 \frac{m_1^2-m_2^2-m_3^2}{2}} \left\{ \frac{e^{i(\alpha_2^2\lambda)^{1/2}}}{(\alpha_2^2\lambda)^{1/2}} - c.c. \right\} \qquad (A1.5)$$

For $\lambda \geq 0$,

$$I(\lambda \geq 0) \simeq \pi \int_{-1}^{+1} dz\, \delta_{R^2}\left[\frac{m_1^2-m_2^2-m_3^2}{2} - \frac{\lambda^{1/2}}{2} z\right] \qquad (A1.6)$$

in agreement with (A1.2) ; while for $\lambda < 0$ we have :

$$I(\lambda<0) \simeq -\int_0^{R^2} d\alpha_2\; 2\cos\left(\alpha_2 \frac{m_1^2-m_2^2-m_3^2}{2}\right) \frac{e^{-\alpha_2|\lambda|^{1/2}}}{\alpha_2|\lambda|^{1/2}}, \qquad (A1.7)$$

which diverges logarithmically. We cure this disease by subtracting the singular part of the integrand at $\alpha_2 = 0$. The subtraction term can then be determined by imposing the natural requirement that at the boundary of the physical region $(\lambda = 0)$ the two functions coincide. Thus, instead of (A1.2) we get <u>uniquely</u> :

$$I(\lambda<0) \simeq + \int_0^{R^2} d\alpha\; 2\cos\left(\alpha \frac{m_1^2-m_2^2-m_3^2}{2}\right)\left[\frac{1-e^{-\alpha|\lambda|^{1/2}}}{\alpha|\lambda|^{1/2}}\right] \qquad (A1.8)$$

One can interpret this result by saying that whenever a complex momentum flows through a quark line the correct extension of the fat δ function in this situation is given by (a,b real) :

$$\delta_{R^2}(a+ib) = \frac{1}{2\pi} \int_0^{R^2} d\alpha\, e^{-|b|\alpha} \cos(\alpha a). \qquad (A1.9)$$

APPENDIX 2

In this Appendix we shall carry out an approximate calculation of the charge form factor of an arbitrary bag of mass M and angular momentum ℓ.

We begin with the contribution to the discontinuity $\mathrm{disc}_{q^2} < M\ell|J_\mu(0)|M\ell >|_i$ of the ith vector meson. Using the wave function given by (1.13) and the rules of Section 3

$$\mathrm{disc}_{q^2} \frac{1}{2\ell+1} \sum_m \langle p,\ell m\beta | J_\mu^\alpha(0) | p',\ell m\gamma\rangle\Big|_i = \mathrm{Tr}\left(\frac{\tau^\alpha}{2}[\tau^\beta,\tau^\gamma]\right) (2\pi)\,\delta(q^2-m_i^2)$$

$$\mu^6 \left(\frac{2\pi}{R^2}\right)^4 \frac{1}{(4\pi)^2} \frac{1}{Z_i Z_\ell(M)} \left(g_{\mu\nu} - \frac{q_\mu q_\nu}{m_i^2}\right) \int dk\, (p'+2k)_\mu$$

$$\delta_{R^2}(pk)\, \delta_{R^2}\left(\frac{p^2}{4}+k^2\right) \delta_{R^2}\left(p'\left(k-\frac{q}{2}\right)\right) \delta_{R^2}\left(\frac{p'^2}{4}+\left(k-\frac{q}{2}\right)^2\right)$$

$$\delta_{R^2}\left(q\left(k+\frac{p'}{2}\right)\right) \delta_{R^2}\left(\frac{q^2}{4}+\left(k+\frac{p'}{2}\right)^2\right) P_\ell(\cos\hat{\theta}) \,, \tag{A2.1}$$

where $\cos\hat{\theta}$ is the angle between the vector $\vec{k}$ and the vector defined by (1.14) relative to the wf $\psi_\ell(p',k-(q/2))$. Notice that we have chosen to evaluate (A2.1) in the rest frame of the state $|p,\ell m>$. Proceeding in the usual way we obtain :

$$\mathrm{disc}_{q^2} \frac{1}{2\ell+1} \sum_m \langle p,\ell m| J_\mu(0) | p',\ell m\rangle\Big|_i \simeq (2\pi)\,\delta(q^2-m_i^2) \frac{\mu^6}{(4\pi)^2}\left(\frac{2\pi}{R^2}\right)^4 \frac{1}{Z_i Z_\ell(M)}$$

$$(p+p')_\mu \int \frac{d\Omega_k}{4}\, \delta_{R^2}^4\left(\frac{m_i^2}{4}+\frac{\lambda^{1/2}}{4}\cos\theta_{pk}\right) P_\ell(\cos(\theta-\theta')) \tag{A2.2}$$

where $\lambda(M^2,M^2,m_i^2) = m_i^2(m_i^2 - 4M^2)$ is the triangular function determining the magnitude of the centre-of-mass momentum $\vec{p}$ through the relation

$$|\vec{p}| = \frac{1}{2M}\lambda^{1/2}(M^2,M^2,m_i^2) \,, \tag{A2.3}$$

and $\cos\theta_{pk}$ is the angle between $\vec{p}$ and the vector $\vec{k}$. By using (1.14) we calculate

$$\cos\theta' = \frac{2M^2-m_i^2}{2M^2} \frac{\frac{|\vec{p}|}{2}+\frac{M}{2}\cos\theta_{pk}-\frac{|\vec{p}|m_i^2}{4M^2}}{\left[\frac{M^2-m_i^2}{4}+\frac{|\vec{p}|M}{2}\cos\theta_{pk}+\frac{1}{M^2}\left(\frac{m_i^2}{4}-\frac{|\vec{p}|M}{2}\cos\theta_{pk}\right)^2\right]^{1/2}} . \quad \text{(A2.4)}$$

In view of the fast decrease of the integrand of (A2.2) with the mass of the vector meson m_i, we need only consider the contribution of the first vector meson (i.e., $i=1$), which we call ρ with mass m_ρ. For $M^2 > > m_\rho^2/4$, λ is negative and according to (A2.3), $|\vec{p}|$ becomes purely imaginary. Upon substituting (2.7) for the δ_{R^2} function appearing in (A2.2), we notice that the integrand is strongly peaked for $\cos\theta_{pk}=0$, and inserting this value in (A2.4) we obtain approximately :

$$\frac{1}{2\ell+1}\sum_m \mathrm{disc}_q = \langle p\ell m\beta|\, J_\mu^\alpha(0)\, |p'\ell m\gamma\rangle\Big|_1 \simeq (2\pi)\,\delta(q^2-m_\rho^2)\;(p+p')_\mu\; \mathrm{Tr}\left(\frac{\tau^\alpha}{2}[\tau^\beta,\tau^\gamma]\right)\; C_\ell^{(\rho)}(M) \quad \text{(A2.5)}$$

where

$$C_\ell^{(\rho)}(M) \simeq \frac{\mu^6}{Z_\rho Z_\ell(M)}\left(\frac{2\pi}{R^2}\right)^4 \frac{1}{8\pi}\int_{-1}^{+1} dz\; \delta^4_{R^2}\left(-\frac{m_\rho^2}{4}+i\frac{|\lambda|^{1/2}}{4}z\right) P_\ell\left(\sin\theta'(z=0)\right) \quad \text{(A2.6)}$$

which for M large becomes

$$C_\ell^{(\rho)}(M) \xrightarrow[M\ \text{large}]{} \frac{4\mu^6}{R^2 Z_\rho Z_\ell(M)}\,\frac{1}{m_\rho M}\, P_\ell\left(\sqrt{1+\frac{m_\rho^2}{M^2}}\right) . \quad \text{(A2.7)}$$

APPENDIX 3

Here we shall carry out the calculation of the diagram appearing in Fig. 16. Following the steps which lead to (2.5), the imaginary part of the scattering amplitude is given by :

$$\mathrm{Im}\, a \simeq \pi^4 \mu^{12} \left(\frac{Z}{Z_0}\right)^3 \left(\frac{2\pi}{R^2}\right)^6 \int_{-1}^{+1} dz \int_0^{2\pi} d\varphi \int_{-1}^{+1} dz' \int_0^{2\pi} d\varphi' \times$$

$$\times \sum_{\ell m n}^{\ell_0} \delta^4_{R^2}\left(\frac{m^2_{n\ell}}{4}(1-z)\right) \delta^4_{R^2}\left(\frac{m^2_{n\ell}}{4}(1-z')\right) \frac{Y^m_\ell(\theta,\varphi) Y^{-m}_\ell(\theta',\varphi')\,(m\Gamma)_{n\ell}}{(s-m^2_{n\ell})^2 + (m\Gamma)^2_{n\ell}} \tag{A3.1}$$

We use the well-known addition theorem to perform the sum over m :

$$\sum_{m=-\ell}^{\ell} Y^m_\ell(\theta,\varphi) Y^{-m}_\ell(\theta',\varphi') = \frac{2\ell+1}{4\pi} P_\ell(\cos\hat{\theta}) \tag{A3.2}$$

where

$$\cos\hat{\theta} = \cos\theta\cos\theta' + \sin\theta\sin\theta'\cos(\varphi-\varphi') .$$

For large s (A3.1) becomes

$$\mathrm{Im}\, a \simeq \pi^4 \mu^{12} \left(\frac{Z}{Z_0}\right)^3 \left(\frac{2\pi}{R^2}\right)^6 \frac{R^2}{2} \int_{-1}^{+1} dz \int_0^{2\pi} d\varphi \int_{-1}^{+1} dz' \int_0^{2\pi} d\varphi'$$

$$\times \frac{sR_\perp^2}{2} \frac{J_1(R_\perp\sqrt{-t})}{R_\perp\sqrt{-t}} \delta^4_{R^2}\left(\frac{s}{4}(1-z)\right) \delta^4_{R^2}\left(\frac{s}{4}(1-z')\right) , \tag{A3.3}$$

where we have used, for small t, the asymptotic formula :

$$\sum_{\ell=0}^{\ell_0} (2\ell+1) P_\ell(\cos\theta) = 2\ell_0^2 \frac{J_1(R_\perp\sqrt{-t})}{R_\perp\sqrt{-t}} . \tag{A3.4}$$

The steep peaking of the $\delta^4_{R^2}$ function allows us to integrate quite easily over z', φ' and φ, and obtain :

$$\mathrm{Im}\,a \simeq \pi^4 \left(\frac{4\mu^4}{Z_0}\right)^3 R^2 R_\perp^2 \left(\frac{2\pi}{R^2}\right)^3 (2\pi)^2 \int_{-1}^{+1} dz \, \frac{J_1\left(R_\perp\sqrt{\frac{s}{2}(1-z)}\right)}{R_\perp\sqrt{\frac{s}{2}(1-z)}} \, \delta^4_{R^2}\left(\frac{s}{4}(1-z)\right) . \tag{A3.5}$$

The integral in (3.5) can be easily carried out by use of the following approximations (valid for small z)

$$\frac{J_1(z)}{z} \simeq \frac{1}{2} \, e^{-\frac{z^2}{8}} \tag{A3.6}$$

and

$$\delta^4_{R^2}(z) \simeq \left(\frac{R^2}{\pi}\right)^4 e^{-\frac{2R^2}{\pi}|z|} . \tag{A3.7}$$

We obtain finally :

$$\mathrm{Im}\,a \simeq (2\pi)^5 \left(\frac{4\mu^2}{Z_0}\right)^3 (4\mu^2 R^2)^2 \, \frac{4\mu^2}{s} \, \frac{8\pi^2 R_\perp^2}{\pi R_\perp^2 + 8R^2} . \tag{A3.8}$$

APPENDIX 4

We want to calculate the high energy behaviour of a_t given by (4.2) according to the prescription (4.4) :

$$a_t = \sum_{n\ell} g^2_{n\ell} \frac{(2\ell+1)\ (-1)^\ell}{t-m^2_{n\ell}} P_\ell \left(1+\frac{2s}{m^2_{n\ell}}\right), \tag{A4.1}$$

where the coupling $g_{n\ell}$ is easily calculated from (2.5) ; one gets :

$$\begin{aligned} g^2_{n\ell} &\simeq \frac{c}{m^2_{n\ell}} \qquad \text{for } \ell \leq \ell_0 = \frac{m_{n\ell}}{2} R_\perp \\ g^2_{n\ell} &\simeq 0 \qquad \text{for } \ell > \ell_0 . \end{aligned} \tag{A4.2}$$

Keeping ℓ fixed, the sum over n can be converted approximately into an integral over m^2, according to

$$\sum_n \rightarrow \frac{R^2}{2\pi} \int_{4\ell^2/R_\perp^2}^{\infty} dm^2 \tag{A4.3}$$

where the approximate spectrum equation

$$m^2_{n\ell} \simeq \frac{\pi}{R^2}(2n+\ell)$$

has been used.

Thus we write

$$a_t \simeq c^2 \sum_{\ell=0}^{\infty} (2\ell+1) \frac{R^2}{2\pi} (-1)^\ell \int_{\frac{4\ell^2}{R_\perp^2}}^{\infty} \frac{dm^2}{m^4} \frac{1}{m^2-t} P_\ell \left(1+\frac{2s}{m^2}\right) \tag{A4.4}$$

which, defining $x=\sqrt{|s|}/m$, becomes

$$a_t \sim \frac{R^2c^2}{2\pi} \frac{1}{|s|^2} \sum_{\ell=0}^{\infty} (2\ell+1)(-1)^\ell \int_0^{\frac{\sqrt{|s|}R_\perp}{2\ell}} \frac{dx\ x^3}{1-\frac{t}{|s|}x^2} P_\ell (1 \pm 2x^2) \tag{A4.5}$$

where in the P_ℓ argument the upper sign stands for positive s, and the lower for negative s.

To perform the ℓ sum more easily we introduce in the m^2 integration the smooth cut-off $e^{-2\ell x(sR_\perp^2)^{-\frac{1}{2}}}$ rather than the sharp $\Theta((\sqrt{s}R_\perp/2\ell)-x)$; and we write

$$a_t \simeq \frac{R^2 c}{(2\pi)|s|^2} \sum_{\ell=0}^{\infty} (2\ell+1)(-1)^\ell \int_0^{\sqrt{|s|}/\mu} \frac{dx \; x^3}{1-\frac{t}{|s|}x^2} \, e^{-x\frac{2\ell}{\sqrt{|s|}R_\perp}} \, P_\ell(1 \pm 2x^2) , \qquad \text{(A4.6)}$$

where μ is the lowest mass in the hadron spectrum. To perform the ℓ sum we would like to use the identity

$$\Sigma(\zeta,\xi) = \sum_{\ell=0}^{\infty} (2\ell+1)(-1)^\ell \zeta^\ell P_\ell(\xi) = \frac{1-\zeta^2}{[1+\zeta^2+2\xi\zeta]^{3/2}} \qquad \text{(A4.7)}$$

which is valid, however, only for $|\xi| \leq 1$. In view of the fact that a_t is <u>real</u> and <u>analytic</u> in s due to the convergence of the series, for $|\xi| > 1$ we extrapolate (A4.9) by taking the average

$$\bar{\Sigma}(\zeta,\xi) = \frac{1}{2}\left(\Sigma(\zeta,\xi+i\epsilon) + \Sigma(\zeta,\xi-i\epsilon)\right) . \qquad \text{(A4.8)}$$

On setting

$$\zeta = e^{-\frac{2x}{R_\perp\sqrt{|s|}}} , \quad \xi = (1 \pm 2x^2) ,$$

and inserting (A4.8) into (A4.6), we get

$$a_t \xrightarrow[s\to+\infty]{} \frac{1}{|s|^{3/2}} \qquad \text{(A4.9)}$$

$$a_t \xrightarrow[s\to-\infty]{} \frac{1}{|s|^2} . \qquad \text{(A4.10)}$$

APPENDIX 5

By means of the identity (6.4) we can dispose of one of the two loops and reduce our diagram to a simple four-bag overlap :

$$W_a(s;m_1^2,m_2^2) \simeq \mu^{12}\left(\frac{2}{Z_0}\right)^3\left(\frac{2\pi}{R^2}\right)^6 \pi R^2 (2\pi)^4 \frac{R_\perp^2}{R_\perp^2+\frac{8}{\pi}R^2\left(1-\frac{m_1^2+m_2^2}{s}\right)}$$

$$\int d^4k\, \delta^4_{R^2}\left(q_1k+\frac{s-m_1^2-\mu_0^2}{4}\right)\delta^4_{R^2}\left(q_2k+\frac{s-m_2^2-\mu_0^2}{4}\right) \qquad \text{(A5.1)}$$

where the factor $R_\perp^2/\ldots$ has the same origin as in (A3.7), and $\mu_0^2 = p_1^2 = p_2^2$ is the mass of the target. In the centre-of-mass frame the momenta of the targets p_1 and p_2 such that $(p_1-p_2)^2 = t = 0$ are given by :

$$p_1 = \left(\frac{s+\mu_0^2-m_1^2}{2\sqrt{s}}, 0, 0, |\vec{p}_1|\right)$$

$$|\vec{p}_1| \simeq \frac{1}{2\sqrt{s}}\left[(s-m_1^2-\mu_0^2)^2-4m_1^2\mu_0^2\right]^{1/2} \qquad \text{(A5.2)}$$

$$p_2 = \left(\frac{s+\mu_0^2-m_2^2}{2\sqrt{s}}, |\vec{p}_2|\sin\theta, 0, |\vec{p}_2|\cos\theta\right)$$

$$|\vec{p}_2| = \frac{1}{2\sqrt{s}}\left[(s-m_2^2-\mu_0^2)^2-4\mu_0^2m_2^2\right]^{1/2} \qquad \text{(A5.3)}$$

where

$$\cos\theta \simeq 1+\mu_0^2 s\frac{(m_1^2-m_2^2)^2}{(s-m_1^2)^2(s-m_2^2)^2} \quad ; \qquad \text{(A5.4)}$$

notice that for $m_1 \neq m_2$ $\cos\theta > 1$; i.e., $t = 0$ is in the unphysical region.

Integrating over k_0 and $|\vec{k}|$ by means of the δ_{R^2} functions in (A5.1), we obtain

$$W_a(s;m_1^2,m_2^2) \simeq \mu^{12}\left(\frac{2}{Z_0}\right)^3\left(\frac{2\pi}{R^2}\right)^6(2\pi)^4\;\frac{\pi R^2}{4}\;\frac{R_\perp^2}{R_\perp^2+\frac{8}{\pi}R^2\left(1-\frac{m_1^2+m_2^2}{2s}\right)}$$

$$\int_{-1}^{+1}dz\int_0^{2\pi}d\varphi\;\delta^4_{R^2}\left[\frac{s-m_1^2-\mu_0^2}{4}+\frac{\sqrt{s}}{2}|\vec{p}_1|\,z\right]\delta^4_{R^2}\left[\frac{s-m_2^2-\mu_0^2}{4}+\frac{\sqrt{s}}{2}|\vec{p}_2|\left(\cos\theta+\sin\theta\sqrt{1-z^2}\times\cos\varphi\right)\right] \quad (A5.5)$$

For $|s-m_1^2| \gtrsim 1/R^2$ the integral can be approximately evaluated as :

$$W_a(s;m_1^2,m_2^2) \simeq \mu^{12}\left(\frac{2}{Z_0}\right)^3\left(\frac{2\pi}{R^2}\right)^3(2\pi)^4\;\frac{\pi^2R^2}{|s-m_1^2|}$$

$$\frac{R_\perp^2}{R_\perp^2+\frac{8R^2}{\pi}\left(1-\frac{m_1^2+m_2^2}{s}\right)}\;\delta^4_{R^2}\left(\frac{\mu_0^2 s}{4}\;\frac{(m_1^2-m_2^2)}{(s-m_1^2)(s-m_2^2)}\right). \quad (A5.6)$$

APPENDIX 6

The diagram in Fig. 22b is given by the expression :

$$W_b(s,Q^2) = Z_0^2 \left(\frac{2}{Z_0}\right)^2 \left(\frac{\pi}{R^2}\right)^4 \frac{R^2}{2} \frac{1}{(4\pi)^2} \sum_{\ell m}^{\frac{R_\perp\sqrt{s}}{2}} \int_{-1}^{+1} dz \int_0^{2\pi} d\varphi \; Y_\ell^m(\theta,\varphi)$$

$$\times \; \delta^4_{R^2}\left(\frac{s+Q^2}{4}(1-z)\right) \int_{-1}^{+1} dz' \int_0^{2\pi} d\varphi' \; Y_\ell^{-m}(\theta',\varphi') \; \delta^4_{R^2}\left(\frac{s+Q^2}{4}(1-z)\right) . \qquad \text{(A6.1)}$$

Using the addition theorem for spherical harmonics we have :

$$W_b(s,Q^2) = \frac{1}{(4\pi)^2} \frac{R^2}{8} \left(\frac{2\pi}{R^2}\right)^4 \sum_{\ell=0}^{\frac{R_\perp\sqrt{s}}{2}} \int_0^{2\pi} d\varphi \int_0^{2\pi} d\varphi' \int_{-1}^{+1} dz \int_{-1}^{+1} dz' \; (2\ell+1)$$

$$P_\ell(\cos\hat{\theta}) \; \delta^4_{R^2}\left(\frac{s+Q^2}{4}(1-z)\right) \; \delta^4_{R^2}\left(\frac{s+Q^2}{4}(1-z')\right) , \qquad \text{(A6.2)}$$

where

$$\cos\hat{\theta} = zz' + (1-z^2)^{1/2}(1-z'^2)^{1/2} \cos(\varphi-\varphi') .$$

We now sum over all partial waves up to $R_\perp(\sqrt{s}/2)$ and using (1.30), we get :

$$W_b(s,Q^2) = \frac{R^2}{8} \frac{1}{(4\pi)^2} \left(\frac{2\pi}{R^2}\right)^4 \frac{R_\perp^2 s}{4} \int_0^{2\pi} d\varphi \int_0^{2\pi} d\varphi' \int_{-1}^{+1} dz \int_{-1}^{+1} dz' .$$

$$\times \; \frac{2 J_1\left(R_\perp\sqrt{\frac{s}{2}(1-\cos\hat{\theta})}\right)}{R_\perp\sqrt{\frac{s}{2}(1-\cos\hat{\theta})}} \; \delta^4_{R^2}\left(\frac{s+Q^2}{4}(1-z)\right) \delta^4_{R^2}\left(\frac{s+Q^2}{4}(1-z')\right) . \qquad \text{(A6.3)}$$

By using the approximations (A3.5) and (A3.6), and after integrating over z', we obtain

$$W_b(s,Q^2) = \frac{\pi}{4} \frac{R_\perp^2 s}{(s+Q^2)} \int_{-1}^{+1} dz\, e^{-\frac{R_\perp^2}{16} s(1-z)}\, e^{-\frac{R^2}{2\pi}(s+Q^2)(1-z)} \tag{A6.4}$$

and making the last trivial integration we get :

$$W_b(s,Q^2) \simeq \frac{12}{s+Q^2} \frac{R_\perp^2 s}{R_\perp^2 s + \frac{8}{\pi} R^2 (s+Q^2)} . \tag{A6.5}$$

This fills the gap left in (A5.1), by showing the necessity of introducing there the factor

$$\frac{R_\perp^2 s}{R_\perp^2 s + \frac{8}{\pi} R^2 (s+Q^2)} .$$

APPENDIX 7

Using the identity (6.4) we can reduce the calculation of $W_p(s,Q^2)$ to the evaluation of the diagram in Fig. 26.

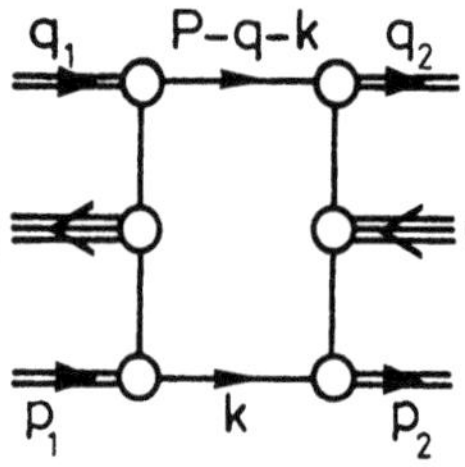

$$\times \frac{\pi^2}{R^2 Z_0} \delta_{R^2}(k^2)\, \delta_{R^2}[(P-q-k)^2] \qquad \begin{aligned} &P=(q_1+p_1)=(q_2+p_2) \\ &P^2=s \\ &\nu_1=(q_1 q) \\ &\nu_2=(p_1 q) \end{aligned}$$

Figure 26 The diagram equivalent to Fig. 24

We shall proceed as follows : first compute $(d^4W_p/dq^4)(s;m_1^2,m_2^2,q)$, then disperse in m_1^2, m_2^2, by keeping s, $p_1q=\nu_2$, $q_1q=\nu_1$ and q^2 fixed and finally integrate over d^4q to get $W_p(s,Q^2)$.

i) we write :

$$\frac{dW_p}{dq}(s;m_1^2,m_2^2,q) \simeq \frac{\pi^3 \mu^{16}}{4R^2 Z_0} \left(\frac{2}{Z_0}\right)^3 \left(\frac{2\pi}{R^2}\right)^6 \frac{R^2 R_1^2}{8}\, q^2$$
$$\times \int d^4k\, \delta_{R^2}(k^2)\, \delta_{R^2}(s+q^2-2\nu_1-2\nu_2+2qk)\, \delta^2_{R^2}(-2p_1k) \qquad \text{(A7.1)}$$
$$\times\ \delta^2_{R^2}(-2p_2k)\, \delta^2_{R^2}(q^2-2\nu_2+2qk)\, \delta^2_{R^2}(q^2-2q(p_2-k)) .$$

Defining p_1 and p_2 as in (A5.2) and (A5.3), we notice the kinematical relations :

$$q^2+q_\perp^2 = \frac{1}{s}\left(2\nu_1\nu_2\frac{\sqrt{s}}{|\vec{p}_1|} - \nu_2^2\frac{m_1^2}{|\vec{p}_1|^2}\right) \tag{A7.2}$$

$$q_0 = \frac{1}{\sqrt{s}}(\nu_1+\nu_2) \tag{A7.3}$$

$$q_3 = \frac{1}{\sqrt{s}}\left(\frac{s+m_1^2}{2p_1\sqrt{s}}\,\nu_2 - \nu_1\right). \tag{A7.4}$$

We integrate over $k_\perp$, and k_3 by using $\delta_{R^2}(k^2)$ and $\delta^2_{R^2}(-2p_1k)$ and obtain

$$\begin{aligned}\frac{dW_P}{dq} \simeq {} & \frac{\mu^{16}\pi^3}{4R^2Z_0}\left(\frac{2}{Z_0}\right)^3\left(\frac{2\pi}{R^2}\right)^6\frac{R_\perp^2R^2}{8}\,q^2\,\frac{R^2}{\pi}\,\frac{1}{2|\vec{p}_1|}\,\times\\ & \times\int dk_0\;\delta^2_{R^2}\left(\frac{k_0}{\sqrt{s}}\,\mu_0^2\,\frac{s(m_1^2-m_2^2)^2}{(s-m_1^2)^2(s-m_2^2)}\right)\;\delta^2_{R^2}\left(\frac{(m_1^2-m_2^2)\,2\nu_2}{(s-m_1^2)}\right)\\ & \times\,\delta_{R^2}\left(s+q^2-2\nu_1-2\nu_2-2k_0(\sqrt{s}-q_0-q_3)\right)\;\delta^2_{R^2}\left(2\nu_1-s+2k_0\sqrt{s}\right),\end{aligned} \tag{A7.5}$$

and integrating over k_0 by means of $\delta_{R^2}[s+q^2-2\nu_1-2\nu_2-2k_0(\sqrt{s}-q_0-q_3)]$ we obtain :

$$\begin{aligned}\frac{dW_P}{dq} \simeq {} & \frac{\pi^3\mu^{16}}{4R^2Z_0}\left(\frac{2}{Z_0}\right)^3\left(\frac{2\pi}{R^2}\right)^6\frac{R^2R_\perp^2}{16}\,q^2\,\frac{R^2}{s-m_1^2-2\nu_2}\\ & \times\,\delta^2_{R^2}\left(\frac{s+q^2-2\nu_1-2\nu_2}{s-m_1^2-2\nu_2}\,\frac{\mu_0^2\,(m_1^2-m_2^2)^2}{2(s-m_1^2)(s-m_2^2)}\right)\;\delta^2_{R^2}\left(\frac{2\nu_2}{s-m_1^2}(m_1^2-m_2^2)\right)\\ & \times\,\delta^2_{R^2}\left(\frac{(s-m_1^2)q^2+2\nu_2m_1^2-4\nu_1\nu_2}{s-m_1^2-2\nu_2}\right).\end{aligned} \tag{A7.6}$$

ii) We now disperse over m_1^2 and m_2^2 and multiply by the current-hadron couplings (6.10), and obtain :

$$\frac{dW_P}{dq}(q;s,Q^2) = \frac{\mu^{16}}{32 Z_0} \left(\frac{2}{Z_0}\right)^3 \left(\frac{2\pi}{R^2}\right)^6 \frac{R^4 R_\perp^2}{16} q^2$$
$$\times \frac{1}{s} \int_0^\infty dx_1 \int_0^\infty dx_2 \frac{1}{\left(1-x_1-\frac{2q_3}{s}\right)} \frac{1}{\left(Q^2/s+x_1\right)\left(Q^2/s+x_2\right)} \delta^2_{R^2}\left(2\nu_2 \frac{x_1-x_2}{1-x_1}\right)$$
$$\times \delta^2_{R^2}\left[\frac{1+q^2/s-\frac{2\nu_1}{s}-\frac{2\nu_2}{s}}{1-x_1-\frac{2\nu_2}{s}} \frac{\mu_0^2 (x_1-x_2)^2}{2(1-x_1)(1-x_2)}\right] \delta^2_{R^2}\left(\frac{q^2\left(Q^2/s+x_1\right)+q_\perp^2\left(1+Q^2/s\right)+2\nu_2 \frac{2\nu_2/s}{s+Q^2}}{1-x_1-\frac{2\nu_2}{s}}\right).$$

(A7.7)

iii) We finally integrate over d^4q to obtain $W_P(s,Q^2)$. We have, neglecting Q^2/s

$$W_P(s,Q^2) \to \frac{\mu^{16}}{Z_0^4} \frac{\pi}{2} \frac{R^6 R_\perp^2}{s} \int \frac{dq_3}{2q_0} \int dq^2 \frac{q^2}{2\nu_2} \left(\frac{s}{Q^2}\right) e^{-\frac{R^2}{\pi}\left(q^2 \frac{Q^2}{s}\right)}$$
$$= \frac{\mu^{16}}{4Z_0^4} \frac{\pi^2}{s} \left(\frac{s}{Q^2}\right)^2 R_\perp^2 R^6$$

(A7.8)

Thus we have obtained for $W_P(s,Q^2)$ a scaling behaviour

$$W_P(s,Q^2) = \frac{1}{2\nu} F_P(x) \tag{A7.9}$$

with

$$F_P(x) \underset{x\to 0}{\longrightarrow} \frac{1}{x^2} \log^2 x \ . \tag{A7.10}$$

REFERENCES

1) G. Preparata and N. Craigie - Nuclear Phys. B102, 478 (1976).

2) N. Craigie and G. Preparata - Nuclear Phys. B102, 497 (1976).

3) G. Preparata - in Proceedings 1975 Erice Summer School (A. Zichichi, ed.) to be published.

4) A. Chodos, R.L. Jaffe, K. Johnson, C.B. Thorn and V.F. Weisskopf - Phys.Rev. D9, 3471 (1974).

5) G. Preparata - in Proceedings 1974 Erice Summer School (A. Zichichi, ed.) Academic Press, New York (1975), p. 54.

6) A. Bramon, E. Etim and M. Greco - Phys.Letters 41B, 507 (1972).

7) R.P. Feynman - "The Photon Hadron Interaction", W.A. Benjamin, New York (1972) ;

J. Bjorken - Proceedings Summer Institute on Particle Physics, SLAC Report 167, Stanford (1967), Vol. 1, p. 1.

8) R.J. Eden, P.V. Landshoff, D.I. Olive and J.C. Polkinghorne - "The Analytic S Matrix", Cambridge University Press (1966).

9) P. Darriulat et al. - "Large Transverse Momentum Photons from High Energy Proton-Proton Collisions", submitted to Nuclear Phys.

10) L. Van Hove - Phys.Letters 24B, 183 (1967).

11) M. Greco - Nuclear Phys. B63, 398 (1973) ;

H. Fraas, B. Read and D. Schildknecht - Nuclear Phys. B86, 346 (1975).

12) R.E. Taylor - Proceedings of the 1975 Lepton-Photon Symposium, Stanford University, SLAC (1976).

13) G. Preparata and G. Rossi - "Looking at Final States : I. The Large p_T Phenomena" CERN Preprint TH. 2163 (1976).

D I S C U S S I O N

CHAIRMAN: Prof. G. Preparata

Scientific Secretary: M. Falcioni

DISCUSSION

- *BERLAD:*

How is the baryon constructed?

- *PREPARATA:*

The baryon has not yet been constructed. In this case, one has two relative internal coordinates and if one imposes appropriate boundary conditions and differential equations on each of the three coordinates -- quarks -- that compose it, one might hope to get a unique description of the baryon.

- *BERLAD:*

How could one get away with SU(3) triality states without introducing colour and glue?

- *PREPARATA:*

It is of equivalent theoretical reliability to introduce triality zero *ad hoc* and assume unseen colour and glue to produce it. In the present approach, I try to avoid as much of unobserved quantities as possible; thus colour and glue will not be used.

- *DE LA TORRE:*

You told us that quarks are not fields or elementary constitutents of hadrons; however, in your diagrams the lines look very much like propagators of quark fields. I would like to ask you what is the exact meaning of your diagrams?

- *PREPARATA:*

The diagrams drawn represent the Fourier transform of the overlap of the wave functions which, in turn, correspond to the transition probability as indicated in the diagram.

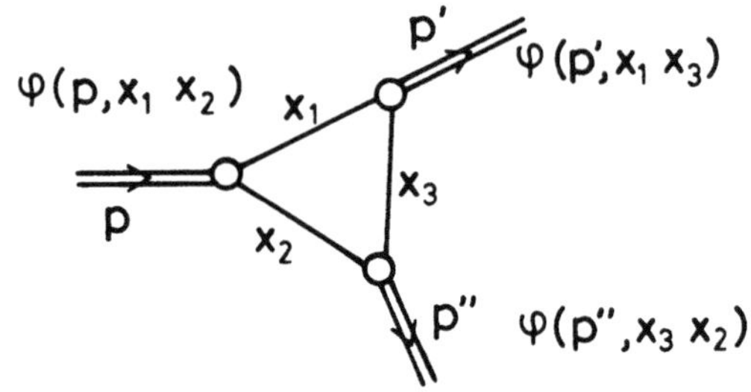

- *KLEINERT:*

Why do you only include three- and four-bag couplings and not higher ones?

- *PREPARATA:*

It is a question of power counting in this perturbation theory: If I do not go to five- six- etc.- bag couplings, I get a renormalizable theory, and I do not get bad behaviour at very high frequencies -- just for the same reason you take ϕ^3, ϕ^4 and not ϕ^5, ϕ^6, and so on.

- *PAULI:*

Could you please compare and contrast your work on fire sausages and strong interaction dynamics with that of the CIM (Constituent Interchange Model) dimensional counting rules of Brodsky, Blankenbecler, Farrar and Gunion.

- *PREPARATA:*

We do not need to appeal immediately to fire sausages and can go back just to the massive quark model. CIM is good in fitting data but lacks physical basis. CIM has some diseases as other models with the production of objects with zero triality. With bags, we avoid their intepretational troubles. Similar behaviour comes again from free quarks, both are related to the primeval Pomeron which gives power counting rules and one obtains similar but slightly different results to CIM. This is more physical and ties in a nice way to the bare Pomeron.

- *PAULI:*

Can your theory be applied to explain the prompt production of leptons in hadron-hadron collisions, such as $pp \to \mu^{\pm}X$, $pN \to e^{\pm}X$? If so, what sort of results do you obtain and how do they agree with data?

- *PREPARATA:*

Prompt production of leptons is still mysterious. The combination of many effects can fit data as suggested by Lederman and others:

i) hyperon + strange particle leptonic decay
ii) vector meson-photon couplings
iii) new particle (J)
iv) electromagnetic lepton production:

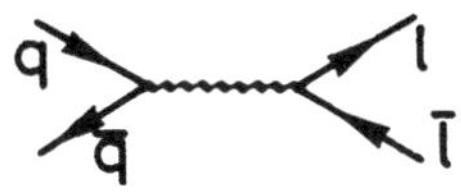

v) new lepton or other new particles .

Looking simply at the lepton-pion ratio may not be the best way of understanding the subject. Small p_T leptons may come from direct production by photons coupling to quark lines in the fire sausage model -- fire sausages sometimes look like Hertz dipoles -- which can radiate high-energy photons which, in turn, can give rise to prompt leptons.

- *ETIM:*

Would you elaborate on the compelling reasons for introducing two kinds of couplings of hadrons to the photon?

- *PREPARATA:*

First, the sampling of the charge content of the hadron by the photon through its coupling via vector mesons involves space time volumes of order $(1/m_\rho)$ and that is much less than the volume of a fire sausage. The direct photon coupling to a quark pair can probe unlimited volumes. The two couplings together therefore probe the whole fire sausage and can be normalized to give the correct charge of the firesausage at zero momentum transfer.

- *FREEDMAN:*

Should not any extended model of hadrons enjoy the same success in explaining Zweig's rule, as long as over-lap between bags is necessary for hadron-hadron interaction?

- *PREPARATA:*

I do not know of any model formulated in this way which can "explain" the Zweig rule. Let me explain how my approach can do this. First, for the case of the ϕ, the corrections to the Zweig rule come from unitarity through the self energy diagram

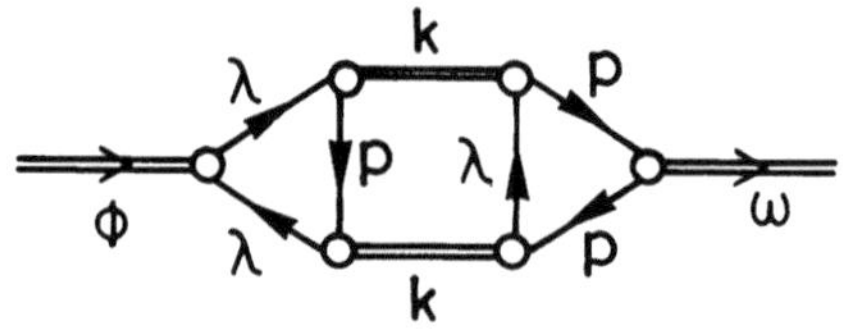

and the fact that this is small -- $\sim 10^{-1}$ the direct amplitude -- is supporting evidence for the "weakness" of strong interactions. In the case of the ψ, supposing, as many do, that it is a $c\bar{c}$ system, we have the analogous diagram

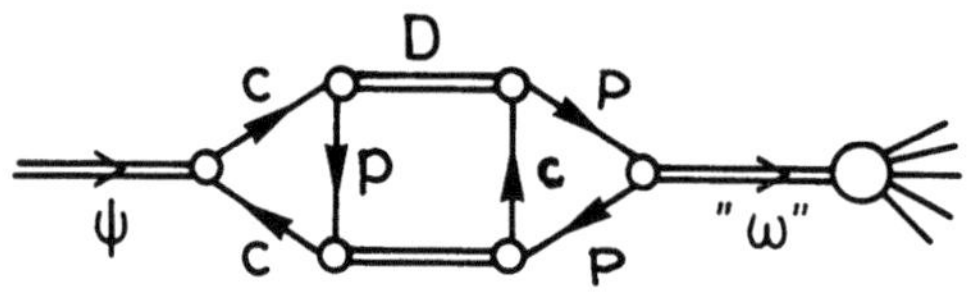

and we must see whether we get an extra factor 10^{-1} needed to explain the extraordinarily small $\Gamma_{tot} \simeq 50$ eV. It is quite straightforward to show that this factor comes out quite naturally and is approximately given by

$$\sim \exp - \left(\frac{m_c^2 - m_\lambda^2}{(\text{Gev})^2} \right) \simeq \frac{1}{10}$$

where $m_c \simeq 1.5$ GeV and $m_x \simeq 0.3$ GeV. The reason for this factor stems from the fact that the wave functions are quite strongly peaked around the "mass" of the quarks. In the last loop in the right, the integration runs, for kinematical reasons, over a negative value of the "mass" of the c quark; and one is only drawing on the small tail.

- *ETIM:*

Is the perturbation coupling constant in your model small because effectively, a large part of the interaction has been used in manufacturing your bags?

- *PREPARATA:*

Yes. You can look at the situation that way.

- *PARSONS:*

If strong interactions are weak, would you not expect the indirect contribution via vector mesons to be small compared with the direct one in the deep inelastic processes?

- *PREPARATA:*

In the indirect type of scattering, one is scattering off an extended fire sausage type of structure, which for simple geometrical reasons has a large overlap with the incoming photon or neutrino.

CAN PEDESTRIANS UNDERSTAND THE NEW PARTICLES?*

Harry J. Lipkin

Weizmann Institute of Science, Rehovot, Israel
Argonne National Laboratory, Argonne, Illinois 60439
Fermi National Accelerator Laboratory, Batavia, Illinois 60510

I. INTRODUCTION

What is so special about the new particles and why is there such excitement? So many old particles have been known for a long time that the discovery of an additional particle has created very little excitement. An unbiased observer from another field would conclude "seen one particle, seen them all." However, the new particles were immediately seen to be peculiar and interesting because they are very narrow states at high excitation and nobody still understands why these states are so narrow. The only argument given supporting the narrowness is based on the Okubo-Zweig-Iizuka rule, but nobody understands the OZI rule even for the old particles, where many interesting open questions still remain. There must be interesting physics in this rule worthy of further theoretical and experimental investigation. The major part of these talks is devoted to interesting questions regarding the theoretical validity and possible experimental tasks of the OZI rule.

In trying to explain to some of my nuclear colleagues why these new particles are so exciting and interesting, I first tell them they are narrow resonances that appear at much too high excitation for their narrowness. The nuclear physicist says, "We know about those things. We have them too; isobaric analog[1] states". Then I explain that they were completely unexpected. No one thought that they were going to be there. He says, "Of course. Theorists didn't expect isobaric analog resonances either. But as soon as they were found, they had the explanation".

*Supported in part by the Israel Commission for Basic Research and the U.S.E.R.D.A., Division of Physical Research.

Then I say, "That's the difference. As soon as the new particles were found, all the theorists came out with explanations, but all of them are wrong. And they still don't understand why the particles are so narrow." As soon as the new particles were found theorists dug into their old files and tried to show that their old theories really predicted these particles. One theorist actually quoted the reference to an old paper where he claimed he had predicted these particles. One of our nasty graduate students actually looked up the reference and gleefully circulated the abstract around the department. The abstract said that this was the only paper explaining the new weak neutral currents without requiring the existence of new particles for which there was no experimental evidence. The status of the new particles is well described by the following quotation:

> "I have no data as yet. It is a capital mistake to theorize before one has data. Insensibly one begins to twist facts to suit theories, instead of theories to suit facts."
>
> ....A. Conan Doyle
> 'The Adventures of Sherlock Holmes'

Niels Bohr developed his model of the atom on the basis of the experimental data of the Balmer series. Much hard work by many people was then needed to get modern quantum mechanics started. But today's theorists are trying to develop a theory for the new particles as beautiful as modern quantum mechanics when they don't even have the Balmer series.

During the time that this talk was being prepared and even during the Erice school itself, new data were being accumulated indicating that the new particles were those theoretically predicted by the charm scheme in which an additional fourth "charmed" quark[2] is added to the three conventional members of the quark triplet. All this charm spectroscopy was known long before the discovery of the new particles. Searches for charm have been suggested for a long time. Yet nobody suggested that SLAC search for very narrow resonances in electron-positron annihilation in the 3-GeV range. It is instructive to examine why nobody suggested such a search.

Vector meson states constructed from a charmed quark-antiquark pair were predicted long before the discovery of the new particles but they were not expected to have narrow widths. Since their decays into ordinary uncharmed states was known to violate the Okubo[3]-Zweig[4]-Iizuka[5] rule, these states were expected to be narrower than conventional uncharmed states at this mass. But there was no reliable theory underlying the OZI rule and its breaking and no theoretical calculation predicting the strength of OZI violation. The only clue was experimental OZI violating decays into nonstrange hadrons of mesons consisting of a strange quark-antiquark pair e.g., the $\phi \rightarrow \rho\pi$ decay. These indicated that OZI suppression factors were one order of magnitude, possibly two but certainly not more. This would still leave a large width for a state at 3 GeV with many open channels. Such a state would not easily be seen as a resonance in electron-positron annihilation.

Thus even if the charm model is correct, one crucial step is missing in the description of the new particles and responsible for the failure to predict their discovery. This missing link is understanding the OZI rule and why the suppression factor is very much larger for the new particles than for forbidden old-particle transitions. This question is still open and considered in detail in these talks. Some indications, but no conclusive answers, are given, but answers should not be expected from this talk. They say that when one asks a Jew a question he answers by answering with another question and when he is asked why he always answers a question by asking another question, he answers "why not?". I shall raise many questions in this talk but I shall not answer them. Instead I will raise more questions. I hope that pursuing the answers to these questions will lead to even more interesting questions and to a better understanding of hadron physics.

II. THE SU(6) BANDWAGON

How can pedestrians understand the new particles when we still have so much trouble understanding the old particles? We still don't understand why the old hadron spectrum has been fit very successfully by an SU(6) symmetry scheme[6] which suggests that hadrons are built from elementary objects called quarks with spin 1/2 and three flavors. If hadrons are made of quarks, and the forces are independent of charge, strangeness and spin, all of these six states are equivalent and transformations among them generate an SU(6) symmetry. Particles can then be classified into SU(6) multiplets. The lowest-lying mesons and baryons fit very beautifully into two SU(6) supermultiplets, the baryons in a 56-plet, the mesons in a 35-plet and a singlet. Since the SU(6) scheme was proposed more new evidence has been found for additional 56-plets, 35-plets and 70-plets. Rosner's review[7] at the 1974 London conference listed the known SU(6) multiplets as a "Michelin Guide" in which he gave four stars, three stars, and two stars to the multiplets, depending on how well they were established experimentally. But where are the quarks?

Now we have a new SU(6), which I call the Sicilian SU(6) because it was invented by a Sicilian (with some help from another island). Arima and Iachello[8] suggest that we should not stop with the 56-plet obtained by putting three basic building blocks having six states in a totally symmetric configuration. Why not try the

252-plet or 1287-plet, obtained by using five or eight building blocks? Arima and Iachello have played the standard game of building an SU(6) supermultiplet degenerate in the symmetry limit, removing the degeneracy by using a simple ansatz for symmetry breaking and obtaining a mass formula which they compare with experiment. Figure 2.1 shows a typical hadron spectrum obtained in this way from the 1287-plet. Figure 2.2 shows a comparison with experiment of the lowest states in the 252-plet and the spectrum of hadrons with baryon number 170 and electric charge 68. Further spectra and comparisons with experiment are shown in Figs. 2.3 and 2.4.

This SU(6) bandwagon is very amusing: Particle physicists build particles from a fundamental building block with six possible states and introduce an SU(6) symmetry. But nobody has found any quarks, and more data are needed to see whether this symmetry is really there and particles are really made out of quarks.

Now the nuclear physicists have jumped on the SU(6) bandwagon. If particles can be made out of a sextet of objects that are not really there, maybe nuclei are too. But the Arima-Iachello nuclear SU(6) model is not based on elementary fermion quarks. Their building blocks have six states, but they are bosons, one with spin zero and one with spin two. Everyone knows that there are no elementary bosons in the nucleus, but the agreement with experiment shown in Figs. 2.2, 2.3 and 2.4 is just as impressive as the quark model fits to particle data. Perhaps the bosons in the nucleus are just as real or unreal as the quarks in the particles.

Why do nuclei look like composite systems of $S = 0$ and $S = 2$ bosons which nobody has seen? Why do hadrons look like composite systems of spin-1/2-three-flavored quarks which nobody has seen? Are these bosons or quarks confined? Are they in a bag? Or are they simply not there? Perhaps there is an underlying substructure which makes nuclei behave as if they were made of bosons in certain experiments and makes hadrons behave as if they were made out of quarks.

But bosons in the nucleus is not really as crazy as it sounds. We know that nuclei are made of neutrons and protons. If they pair in some fashion they may behave somehow like bosons. Before the BCS theory there were suggestions that Bose condensation of electron pairs was responsible for superconductivity. Then BCS showed how a proper treatment gave not only the properties that look like bosons but also the important differences between fermion-pairs and real bosons. So maybe there is something in this boson model of the nucleus and we are waiting for the right theoretical description.

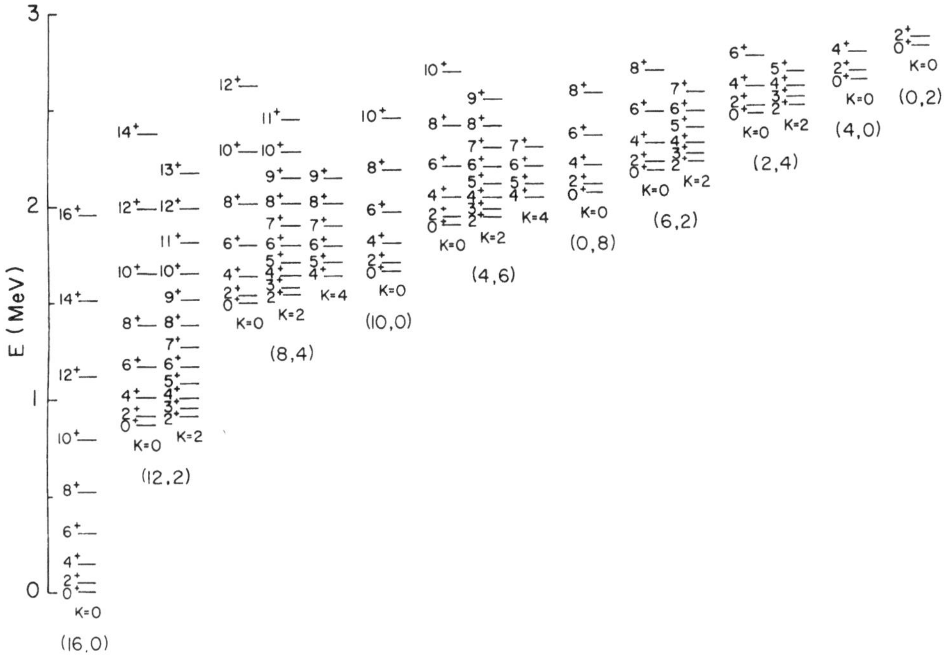

Fig. 2.1. Typical hadron spectrum from 1287-plet.

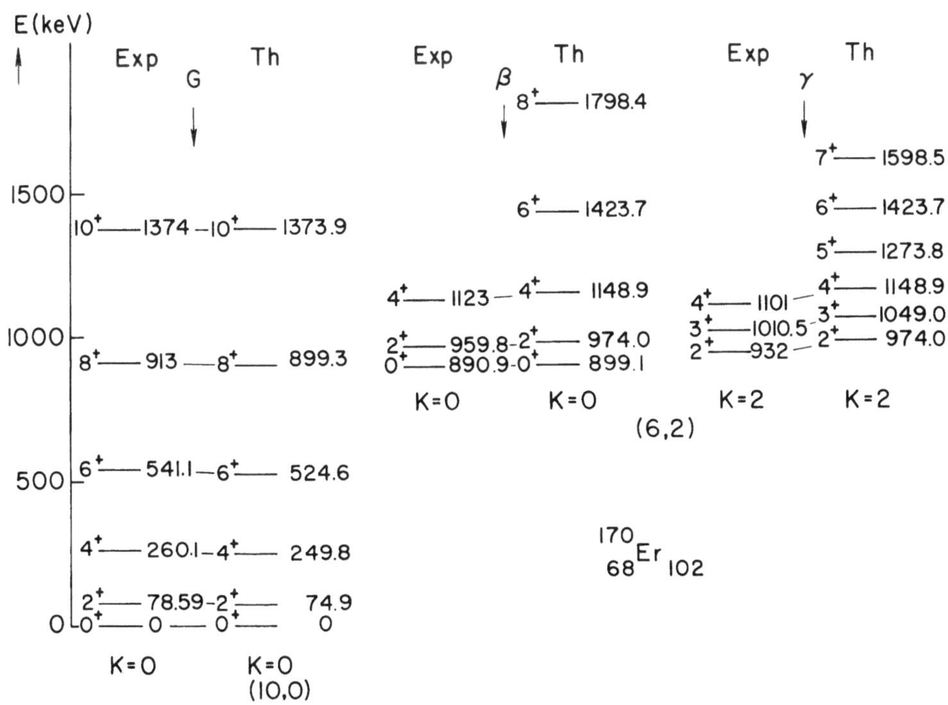

Fig. 2.2. Comparison of 252-plet spectrum with experiment.

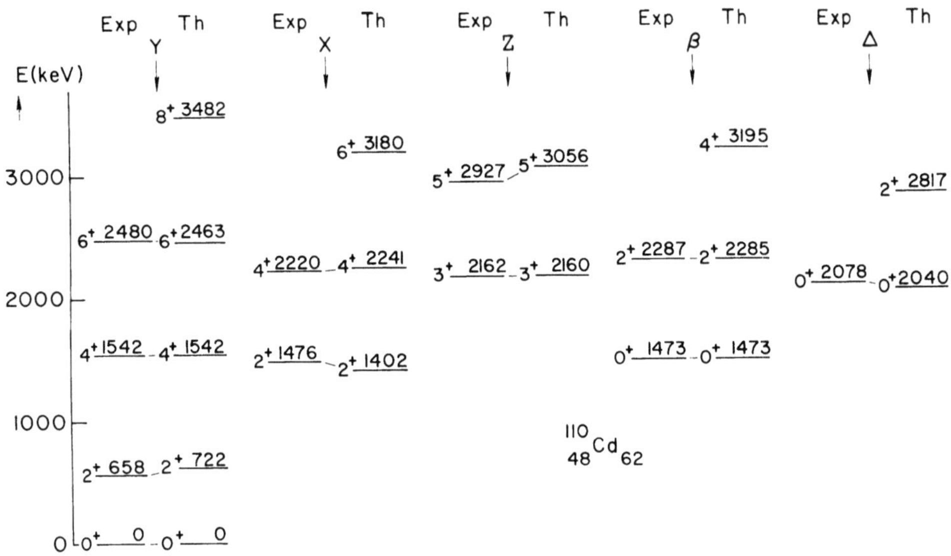

Fig. 2.3. Boson model fit to the ^{110}Cd vibrational spectrum.

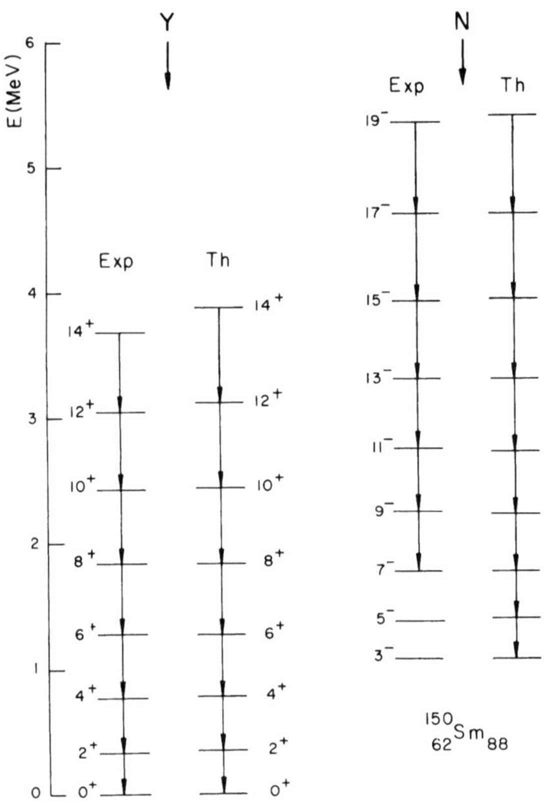

Fig. 2.4. Boson model fit to the ^{150}Sm vibrational spectrum.

Let us now discuss the nuclear SU(6) and Figs. 2.1–2.4 in slightly more detail. Figure 2.1 shows a typical SU(6) supermultiplet with the energy spectrum obtained when SU(6) is broken by a particular type of boson-boson interaction that preserves the subgroup SU(3). The spectrum looks very much like that found in the nuclear collective model with a ground-state rotational band and beta and gamma vibrations. The states are labelled by the quantum numbers of the SU(3) classification. Data are fitted in a large group of nuclei in the rare earth and transuranium regions where these rotational and vibrational spectra occur. The example of ^{170}Er shown in Fig. 2.2 has experimental energy levels which fit two SU(3) multiplets which are just the lowest two multiplets found in a single SU(6) supermultiplet.

In particle physics, the SU(3) subgroup of SU(6) gives a good classification for baryon states, but the SU(4) group works better for mesons. For nuclei the SU(3) subgroup of SU(6) gives a good description of rotational nuclei; while the SU(5) subgroup works better for vibrational nuclei. Figures 2.3 and 2.4 show typical vibrational spectra and some fits to vibrational nuclei.

III. THE SEARCH FOR NEW DEGREES OF FREEDOM

What are these new particles? They indicate some new degree of freedom, but what is it? At the Palermo conference, Cabbibo[9] presented the charm approach as analogous to the search for the planet Neptune where other data on irregularities in the orbit of Uranus had indicated something must be there. My guide to any search is two key questions[10]: 1) who needs it? and 2) who cares if it is not found? In the case of the planet Neptune anybody who believed Newton's descriptions of the motions of the planets knew that something had to be there. It would have been very serious if nothing was found to produce the observed irregularities in the orbit of Uranus. But if all the new additional particles that are suggested by new theories are not found the theorists will find new excuses for their absence and change the theory a bit to explain it. As a guide to the search for new degrees of freedom, it is instructive to recall the search for a higher symmetry[11,12] that eventually turned out to be SU(3). It begins with isospin, which is SU(2) and strangeness, which is U(1). The correct higher symmetry SU(3) which included SU(2) and U(1), was found by an eight-year journey, in which all possible wrong symmetries were tried first. When they finally found SU(3) they called it the eight-fold way because it took them eight years to find it.

Why did it take so long? Why did they try everything else because before they tried SU(3) instead of noting that $SU(2) + U(1) \subset SU(3)$ is as simple as 2+1 = 3. Physicists are not stupid; the reason that they could not see that 2 + 1 = 3 was because they did not know that they had two. In those days isospin was believed to be a rotation in a three-dimensional space like ordinary spin, described by isospin operators similar to angular momentum operators. The natural candidates for a higher symmetry to include isospin rotations were rotations in four, five, six, seven and eight dimensions. None of these worked because the algebra of the group of three dimensional rotations is accidentally isomorphic to the algebra of two-dimensional unitary transformations and isospin is really SU(2), not O(3). The two-dimensional Hilbert space of proton and neutron states and the transformations of protons and neutrons into one another have no relation to any physical three-dimensional space.

Beyond three dimensions there is no longer this isomorphism between rotations and unitary transformations. Thus theorists could not get anywhere by extrapolating what they already had. They had to learn something new, but they did not realize it. At the Princeton Institute for Advanced Study many now famous theoretical particle physicists did not bother going to Guilio Racah's famous lectures on Group Theory and Spectroscopy [13] because they did not think unitary groups had anything to do with particle physics.

Now the pendulum has swung to the other direction. We know all about SU(n). We begin with n basic building blocks and define unitary transformation among these objects to make an SU(n) symmetry. Now that something beyond SU(3) is needed, theorists play the same games with SU(n) instead of rotations: SU(4), SU(5), SU(6) and so on. If we keep it up, we will get to SU(ROSENFELD) where n is the number of entries in the Rosenfeld table of particles and all particles are classified in the fundamental representation.

This reminds me of an explanation I heard from a physicist who works in the field of controlled thermonuclear reactions about the difference between CTR and particle accelerators. The particle physicist builds an accelerator. It works and he is happy and does some physics with it. After a while he realizes that to progress further he needs a bigger accelerator. He gets money to build it, builds it, it works and he is happy and does some physics. After a while he realizes that he needs an even bigger accelerator, etc. etc. The CTR man builds a machine which does not work and he is unhappy. Then he decides that if he had more money he could build a bigger machine that might work. He gets the money, builds it, it does not work and he is unhappy. He then decides again that if he had more money he could build a bigger machine which might work etc. etc.

The quark model[14] started with the idea that everything is made from three fundamental building blocks. The experimentalists looked for the quarks and did not find them. So the theorists said maybe there are more quarks. It started from three, it has gone up to four, nine, twelve, etc. The current popular colored six-quark model has 18 quarks. But still nobody is finding them. So, perhaps what we need is not to keep adding more of the things that we know, maybe we need to learn something new. In other words, maybe the quark is something like three-dimensional isospace, a useful realization of the symmetry at a certain stage which enables us to do calculations very nicely. But it freezes our intuition in the wrong direction and thus hides the new things that we may have to learn to advance to the next stage.

IV. THE PRESENT STATUS OF THE NEW PARTICLES AND THE OZI RULE

Let me now briefly review the status of the new particles. They are narrow resonances discovered at SLAC as peaks in the cross section for $e^+ - e^-$ annihilation into hadrons and at Brookhaven as peaks in the mass spectrum of $e^+ - e^-$ pairs in nucleon-nucleon collisions produced with hadrons. In the SLAC experiment the mass of the ψ is about 3000 MeV, the instrumental resolution is about 1 MeV, and the width of the ψ is even smaller. The peak is already smeared by the instrumental resolution by an order of magnitude, and would be smeared further by magnet drift unless the magnetic field is kept stable to better than 1 part in 3000. Such a narrow resonance was not expected by theory and there is no point in getting that much stability in all the apparatus if it is not needed. So the particles were discovered by accident. The first J or ψ that was discovered had a mass of 3100, the same quantum numbers as the photon (spin 1, odd parity and odd charge conjugation), and decays into hadrons. Soon afterwards, a whole family of particles were found having the same quantum numbers as the photon. This does not mean that most of the new particles have the same quantum numbers as the photon; it is just that an $e^+ - e^-$ colliding beam experiment excites most strongly those states having the same quantum numbers as the photon.

Once these are found one asks what they might be; something completely new, a new kind of weak or semi-weak boson, a new kind of quark-antiquark pair, having a new quantum number like charm, or some hadron with a new internal degree of freedom. With too many possibilities and not enough experimental data, we can try to look for general properties and draw conclusions which are not too model dependent. It seems fairly clear that the production is electromagnetic, because it all fits together with what is known of electromagnetic production via a single virtual photon. The particles have the quantum numbers of the photon— if they are produced by some other mechanism there is no reason to pick out 1^- states in particular.

The most peculiar property of these particles is the inconsistency between their production and decay. They are produced electromagnetically by $e^+ - e^-$ collisions presumably through a photon, and with a production cross section comparable to that for the ordinary vector mesons ρ, ω and ϕ. But they decay into hadrons with a very narrow width. The characteristic widths for vector mesons are 100 MeV for strong decays like $\rho \to 2\pi$ and 1 MeV for electromagnetic decays like $\omega \to \pi\gamma$. This is consistent with the picture that 100 MeV is a strong width, and electromagnetic widths are first order in α and down by a factor of 100. The J/ψ decay into hadrons is of the order of 100 kilovolts while decays into hadrons $+\gamma$ are very small. Thus both hadronic and electromagnetic decays are down at least three orders of magnitude from the expected decays of an ordinary hadron.

Relatively large electromagnetic production and suppressed decays into hadrons suggest that the J/ψ might not be a hadron at all, but a composite state of some new kind of fermion-antifermion pair, coupled to the photon because it has electric charge, but not coupled to hadrons. That does not work either, because the coupling to hadrons is actually too strong. This coupling is conveniently measured by a quantity called R defined by the relation[15]

$$R = \frac{\sigma(e^+e^- \to \text{hadrons})}{\sigma(e^+e^- \to \mu^+\mu^-)} . \qquad (4.1)$$

Production of $\mu^+\mu^-$ pairs is known from quantum electrodynamics and experiments agree with QED predictions. Thus the $\mu^+\mu^-$ cross section provides the scale for measuring other processes. In parton models where hadron production occurs via the production of an intermediate state of a quark-antiquark pair by the virtual photon, the ratio R depends on the number of different kinds of quarks and their electric charges[16]

$$R = \sum_i Q_i^2 \qquad (4.2)$$

where Q_i is the charge of the quark of type i and the sum is over all types of quarks. Experiments show that R in the vicinity of the J/ψ but off resonance is of the order of two to five, but inside the resonance it rises to 25.

The J/ψ is produced electromagnetically with a strength comparable to the ρ, ω or ϕ. It is more strongly coupled than the photon to the hadrons but decays much more slowly than it should. So in some sense its coupling to hadrons is too small but it is also too big.

I conclude this review with a more detailed picture of why this narrowness problem is such a nuisance. There are two problems. The widths for both electromagnetic and strong decays are too narrow. We first consider the electromagnetic problem. The paradox is that no simple selection rule can forbid the decay of J/ψ into something else plus a photon. Any selection rule based on a conservation law that forbids electromagnetic decay also forbids electromagnetic production. We know that in $e^+ - e^-$ annilihation both the J/ψ and ω are produced with comparable cross sections. Thus the transition matrix elements for the two processes are comparable

$$\langle\psi|J_{em}|0> \sim <\omega|J_{em}|0\rangle \tag{4.3}$$

where J_{em} denotes the electromagnetic current. Thus the J/ψ cannot have a peculiar eignevalue of a new quantum number, which is conserved in electromagnetic interactions. Therefore, we cannot forbid by symmetry the decay of a J/ψ to a photon and some hadron states having the same quantum numbers as the vacuum for all conserved quantities. Since 3100 MeV can make many pions there are many such states. But experimentally we know that the sum of the squares of these matrix elements over all possible states is still very much smaller than the squared matrix element for the $\omega \to \pi\gamma$ decay. That is the electromagnetic trouble.

For the strong decays, a selection rule is possible. There are "generalized color models", with color symmetry in which all ordinary hadrons are color singlets.[14,17] If the ψ is not a color singlet, its decay into ordinary singlet hadrons is forbidden, but the photon can excite it because the photon need not be a color singlet. But in all such models the electromagnetic trouble is still there because the ψ is allowed to decay into color singlet hadrons plus a photon. Thus generalized color models solve the strong decay trouble but cannot explain the electromagnetic trouble.

There is a dynamical selection rule which may be relevant known as the Okubo-Zweig-Iizuka rule. We know from ordinary hadron physics that the $\phi \to \rho\pi$ and $f' \to 2\pi$ decays are suppressed. The initial states of both decays contain a strange quark-antiquark pair and the final states contain no strange quarks. Some dynamical principle suppresses the transition in which a strange quark-antiquark pair annihilates and only nonstrange quarks are produced. This suggests that if the ψ is a new kind of quark-antiquark pair like a charmed pair there might be a similar principle preventing a charmed pair from disappearing. But this selection rule can only hold in Born approximation. It cannot be rigorous in higher order because a succession of transitions which satisfy the selection rule can produce a forbidden transition. For example both the ϕ which consists only of strange quarks and the ω which consists only of nonstrange quarks are coupled to the $K\bar{K}$ state. So the ϕ can go to a nonstrange final state via the intermediate $K\bar{K}$ and ω states.

Thus the basic question of why the new particles are so narrow is still not understood. There are all kinds of hand waving explanations that may be right but are still unconvincing. It may very well be that the real structure in the new particles is completely different from anything that is being considered today.

V. THE OKUBO-ZWEIG-IIZUKA RULE

One of the principal open problems in trying to understand the old as well as the new particles is the Zweig rule, or the Okubo-Zweig-Iizuka rule, as it is now commonly called. The Okubo ansatz,[3] which antedated not only Zweig[4] and Iizuka[5] but even the quark model, applied to nonet couplings and gave all results for the three-meson vertex later obtained by Zweig and Iizuka from quark diagrams. The quark-line rules of Zweig and Iizuka define one possible generalization of the Okubo ansatz for four-point functions and more complicated vertices, but this generalization is not unique. Okubo has pointed out other possible generalizations[18] that may be relevant to experiment. Previous papers have separated the "cookbook rules" of Zweig and Iizuka and the Okubo ansatz, since the validity of the Okubo ansatz for three-meson couplings is experimentally well established, whereas the particular ZI generalization to more complicated vertices has not yet been conclusively tested. These notes follow the present common usage of giving Okubo proper credit for his pioneering work by using the name OZI rule. This leaves some ambiguity in its definition for four-point and higher couplings as discussed in detail below.

5.1. Pedagogical Examples and Some Basic Questions

The OZI rule has entered the folklore of particle physics without any clear theoretical understanding or justification. At the present time nobody really understands it, and anyone who claims to should not be believed. Investigating the OZI rule for the old particles raises many interesting questions[19,20] which may lead to a better understanding of strong interactions as well as giving additional insight into the experimentally observed suppression of new particle decays attributed to the OZI rule. We follow an iconoclastic approach emphasizing embarrassing questions with no simple answers which might lead to fruitful lines of investigation.

The most common applications of the OZI rule are selection rules forbidding the couplings of the ϕ and f' mesons to nonstrange mesons and nucleons

$$g_{\phi\rho\pi} \ll g_{\omega\rho\pi} \tag{5.1a}$$

$$g_{f'\pi\pi} \ll g_{f\pi\pi} \tag{5.1b}$$

$$g_{N\bar{N}\phi} \ll g_{N\bar{N}\omega} \tag{5.1c}$$

$$g_{N\bar{N}f'} \ll g_{N\bar{N}f} \tag{5.1d}$$

where the couplings on the left-hand sides of the inequalities are forbidden and those on the right-hand side are allowed. Since the selection rule is not exact, the degree of suppression is expressed quantitatively by comparing the corresponding forbidden and allowed couplings appearing in these inequalities.

The selection rules can also be formulated in terms of two-body reaction cross sections,

$$\sigma(\pi^- p \to \phi n) \ll \sigma(\pi^- p \to \omega n) \tag{5.1e}$$

$$\sigma(\pi^- p) \to f'n) \ll \sigma(\pi^- p \to fn) \ . \tag{5.1f}$$

Although these agree very well with experiment, no consistent theoretical or phenomenological model explains them without raising paradoxes and contradictions. There is also no theoretical indication of how good the selection rule should be in different processes; i.e. no description of the breaking mechanism.

A principal difficulty to be overcome in any theoretical formulation is that a succession of transitions all allowed by the OZI rule can lead to one which is forbidden. For example, all forbidden couplings (5.1) can proceed through an intermediate $K\bar{K}$ state via the following transition amplitudes observed experimentally and allowed by the OZI rule.

$$T(\phi \to K\bar{K}) \neq 0 \tag{5.2a}$$

$$T(f' \to K\bar{K}) \neq 0 \tag{5.2b}$$

$$T(K\bar{K} \to \rho\pi) \neq 0 \tag{5.2c}$$

$$T(K\bar{K} \to \pi\pi) \neq 0 \tag{5.2d}$$

$$T(K\bar{K} \to N\bar{N}) \neq 0 \ . \tag{5.2e}$$

All the selection rules can thus be broken by the following allowed higher-order transitions

$$\phi \to K\bar{K} \to \rho\pi \qquad (5.3a)$$

$$f' \to K\bar{K} \to \pi\pi \qquad (5.3b)$$

$$\phi \to K\bar{K} \to N\bar{N} \qquad (5.3c)$$

$$f' \to K\bar{K} \to N\bar{N} \ . \qquad (5.3d)$$

If the OZI rule holds only to first order in strong interactions, much greater violations are expected than those experimentally observed. Some mechanism for reducing these violations seems to be present. One possibility is a cancellation of the violating amplitudes (5.3) by other amplitudes, as occurs in the case of symmetry selection rules. This is described in detail below.

The essential features of many problems arising in applications of the OZI rule are illustrated in the following examples.

Consider the decays of vector mesons into two pseudoscalar mesons. The following decays are all allowed by the OZI rule and observed experimentally.

$$\Gamma(\rho \to 2\pi) \neq 0 \qquad (5.4a)$$

$$\Gamma(K^* \to K\pi) \neq 0 \qquad (5.4b)$$

$$\Gamma(\phi \to K\bar{K}) \neq 0 \ . \qquad (5.4c)$$

The decay

$$\Gamma(\phi \to 2\pi) \approx 0 \qquad (5.4d)$$

is forbidden by the OZI rule and experiments are consistent with zero decay rate. Thus, the VPP decays (5.4) all agree with the OZI rule.

But the decay

$$\Gamma(\omega \to 2\pi) \approx 0 \qquad (5.4e)$$

is allowed by the OZI rule and is also observed experimentally to be very weak. This apparent contradiction is resolved by noting that the decays (5.4d) and (5.4e) are forbidden by G parity. The OZI rule is thus completely irrelevant to VPP decays. A similar situation obtains for tensor-vector-pseudoscalar decays, where the

$f' \to \rho\pi$ decay forbidden by the OZI rule is also forbidden by G parity.

This example shows that experimentally-observed suppression of a transition forbidden by the OZI rule does not necessarily provide evidence for the validity of the rule. The transitions may be forbidden for other reasons.

The VPP example also raises two questions with interesting implications for the general case.

1. The Doubly Forbidden Question . Since both the $\omega \to 2\pi$ and $\phi \to 2\pi$ decays are forbidden by G parity but the $\phi \to 2\pi$ decay is also forbidden by OZI rule, is the doubly forbidden $\phi \to 2\pi$ decay weaker than the $\omega \to 2\pi$ decay?

2. The Higher Order Paradox. The scattering amplitude $T(K^+K^- \to \pi^+\pi^-)$ is allowed by both G parity and the OZI rule. Thus the $\phi \to 2\pi$ decay could take place as a two-step transition in which both steps are allowed

$$\phi \to K^+K^- \to \pi^+\pi^- \; . \qquad (5.5a)$$

How is this transition inhibited in a theory of strong interactions where there is no small parameter to make second-order transitions weaker than first order?

The answers to these questions are simple in this trivial case and very illuminating for more interesting non-trivial cases.

1. Double Forbiddeness. There is no simple answer to this question. A transition already otherwise forbidden can be additionally suppressed by the OZI rule only if the dynamical process which breaks the other selection rule also respects the OZI rule. In the case of the VPP decays the G-parity selection rule is broken by electromagnetic transitions which violate the OZI rule. The OZI violating transition

$$\phi \to \gamma \to 2\pi \qquad (5.4d)$$

and the OZI conserving transition

$$\omega \to \gamma \to 2\pi \qquad (5.4e)$$

have couplings of the same order of magnitude. The OZI conserving transition (5.4e) may be favored by kinematic factors if the $\gamma \to 2\pi$ transition is dominated by the ρ pole, but this has no simple connection to the OZI rule.

2. The Higher-Order Paradox. The answer is that the allowed transition (5.5a) is exactly cancelled by the transition

$$\phi \rightarrow K^0\overline{K^0} \rightarrow \pi^+\pi^- \, . \tag{5.5b}$$

This cancellation is characteristic of transitions rigorously forbidden by a conservation law. The conservation law, in this case G conservation, must hold to all orders. G-violating contributions to the transition amplitude can arise from particular intermediate states which are not eigenstates of G. These contributions must be cancelled by other contributions in the sum over all intermediate states related to these states by the G transformation.

Note that this cancellation (5.5) does not occur in the non-trivial selection rules (5.1) allowed by G parity. The contributions to the higher-order transitions (5.3) from the K^+K^- and $K^0\overline{K^0}$ intermediate states have the same phase and cannot cancel. Thus, any cancellation must come from some other state.

An instructive example of a symmetry selection rule broken in higher order is the SU(3) and charge conjugation selection rule forbidding the transition[21]

$$e^+ + e^- \rightarrow \gamma \rightarrow K^0 + \overline{K^0} \, . \tag{5.6}$$

However, this reaction is observed experimentally near the mass of the ϕ,

$$e^+ + e^- \rightarrow \gamma \rightarrow \phi \rightarrow K^0 + \overline{K^0} \, . \tag{5.7a}$$

All the individual transitions are allowed by charge conjugation and SU(3). If charge conjugation and SU(3) are exact symmetries this contribution (5.7a) to the amplitude (5.6) must be cancelled by other contributions. In the SU(3) limit the ρ and ω are degenerate with the ϕ and the amplitude for the reaction (5.7a) is exactly cancelled by the amplitudes for the reactions

$$e^+ + e^- \rightarrow \gamma \rightarrow \rho^0 \rightarrow K^0 + \overline{K^0} \tag{5.7b}$$

$$e^+ + e^- \rightarrow \gamma \rightarrow \omega \rightarrow K^0 + \overline{K^0}. \tag{5.7c}$$

In the real world the ρ, ω and ϕ are not degenerate and no such cancellation occurs at the ϕ peak. The SU(3)-violating reaction (5.6) thus occurs just as strongly as the corresponding SU(3) conserving transition to the K^+K^- final state.

The implications of these examples for the OZI rule are clear. Higher-order OZI violations can be suppressed by cancellations from different immediate states. But such cancellations require a degeneracy of the relevant intermediate states. Without exact degeneracy the OZI rule will be broken like the SU(3) selection rule forbidding the transition (5.6). But no physical state is degenerate with the $K\bar{K}$ intermediate state occurring in the transitions (5.3). Thus these cancellations can at best be only approximate.

We now pose a number of interesting questions which have no simple answers today.[20]

1. What is the theoretical basis of the OZI rule? Can it be formulated with predictive power to give strengths of forbidden transitions at least on the phenomenological level? Could there be a description analogous to the Cabibbo description of strangeness violation in weak interactions where one or more parameters analogous to Cabibbo angles describe the relative strengths of OZI-conserving and OZI-violating transitions?

2. What is the experimental evidence for the OZI rule? How many of the so-called OZI selection rules also follow from other considerations like G-parity conservation and therefore do not really test the OZI rule? How much unprocessed or easily-available data could be used to test the OZI rule?

3. Can the OZI rule be exact in some symmetry limit like SU(3)? Can it be formulated as a conservation law?

4. How is the rule broken and by how much? Can the rule be kept exact for vertices with all the breaking introduced in properties of external particles and propagators for virtual states, as in the conventional SU(3) phenomenology?

5. Where does the rule apply? To baryons as well as mesons? To the not ideally mixed pseudoscalar nonet as well as ideally mixed nonets? To the new as well as the old particles? Is it better in some cases and worse in others? How is it formulated for multiparticle vertices?

6. What is the relation of the OZI rule to ideally mixed nonets? Can the breaking be described as entirely due to deviations from ideal mixing? This would allow the deviation angle to play the role of a Cabibbo angle in a phenomenological description with symmetric vertices and breaking only in mixing angles for physical particles.

7. A cascade of OZI-allowed transitions can produce a forbidden transition. How are these higher-order transitions suppressed?

8. How are the large suppression factors observed in the new particles explained? If they are entirely due to the OZI rule then the breaking must be much smaller than SU(3) breaking.

9. Are thresholds important? Higher-order transitions which violate the OZI rule are possible for the ϕ and f' decays via the physical $K\bar{K}$ state with both kaons on shell. But no such physical states exist for the ψ and ψ' decays, which are below the threshold of the analogous $D\bar{D}$ state. Will the OZI rule be better for the ψ and ψ' which cannot proceed by a cascade of allowed on-mass-shell transitions than it is for the old particles or the higher ψ's above the $D\bar{D}$ threshold?

5.2. Symmetry, Dynamics, Mixing and the K_1-K_2 Analogy

There are two possible approaches to explaining the OZI selection rules (5.1), symmetry and dynamics. All these rules apply to processes involving mixed meson nonets, where the SU(3) breaking and mixing of singlet and octet states plays a crucial role. When the couplings and amplitudes for the processes (5.1) are expressed in terms of the unbroken SU(3) singlet and octet states, the two corresponding quantities are of the same order of magnitude and nothing vanishes.

$$g_{\omega_1\rho\pi} \neq 0 \neq g_{\omega_8\rho\pi} \quad (5.8a)$$

$$g_{f_1\pi\pi} \neq 0 \neq g_{f_8\pi\pi} \quad (5.8b)$$

$$g_{N\bar{N}\omega_1} \neq 0 \neq g_{N\bar{N}\omega_8} \quad (5.8c)$$

$$g_{N\bar{N}f_1} \neq 0 \neq g_{N\bar{N}f_8} \quad (5.8d)$$

$$\sigma(\pi^- p \to \omega_1 n) \neq 0 \neq \sigma(\pi^- p \to \omega_8 n) \quad (5.8e)$$

$$\sigma(\pi^- p \to f_1 n) \neq 0 \neq \sigma(\pi^- p \to f_8 n) \; . \quad (5.8f)$$

Thus the selection rules (5.1) apparently arise from a mysterious mixing mechanism which chooses the physical states to be just the right linear combinations of the SU(3) singlet and octet eigenstates so that the singlet and octet contributions to the processes (5.1) exactly cancel for one of the physical states. There must be a better way to understand this.

Each cancellation of singlet-octet contributions required for the selection rule depends upon two parameters not constrained by SU(3): (1) the ratio of the singlet and octet couplings, (2) the mixing angle of the meson nonet which determines the ratio of singlet to octet in the physical meson. These two parameters appear very different in character. A specific ratio of singlet-to-octet couplings suggests a higher symmetry beyond SU(3) which relates amplitudes unrelated by SU(3). A mixing of singlet and octet states to give physical states having different masses requires SU(3) breaking.

In the symmetry approach one looks for a higher symmetry which classifies the full nonet of mesons in a single supermultiplet and describes singlet and octet couplings by a single coupling. The mixing angles of the physical states could have a simple description if the symmetry-breaking mechanism breaks SU(3) but conserves another subgroup of the higher symmetry which does not commute with SU(3) and mixes SU(3) eigenstates. This other subgroup would then predict new conservation laws not found in SU(3) and could give the selection rules (5.1). One example is an SU(6) description[6] which breaks both SU(6) and SU(3) while conserving the subgroup SU(4) × SU(2). But such symmetry approaches have all had troubles.[14,22] The basic difficulty is that a conservation law rigorously forbidding the transitions (5.1) must either forbid the higher order transitions (5.3) or cancel them exactly with other transitions. They cannot be forbidden without also forbidding some of the experimentally observed amplitudes (5.2). They cannot be cancelled exactly without introducing additional channels whose intermediate states are exactly degenerate with the kaon pair states, and such degeneracies do not exist.

The dynamical approach looks for dynamical models which naturally give the mixing that decouples one eigenstate.

The neutral kaon system provides an instructive analogy for describing the mixing and selection rule problem and illustrates both the symmetry and dynamical approaches. We begin with the eigenstates K^o and $\overline{K^o}$ of a symmetry, strangeness, that is broken in the decay process. Symmetry breaking determines new eigenstates which are linear combinations of the strangeness eigenstates and could be described by a mixing angle

$$|K_A\rangle = |K^o\rangle \cos\theta + \overline{|K^o\rangle} \sin\theta \qquad (5.10a)$$

$$|K_B\rangle = -|K^o\rangle \sin\theta + |\overline{K^o}\rangle \cos\theta \,. \tag{5.10b}$$

Both the K^o and $\overline{K}^o$ states are coupled to the 2π decay channel. Consider the particular value of the mixing angle defined by the relation

$$\tan\theta = \frac{\langle 2\pi|T|\overline{K^o}\rangle}{\langle 2\pi|T|K^o\rangle} \,. \tag{5.11}$$

Then we can write

$$\langle 2\pi|T|K^o\rangle = T\cos\theta \tag{5.12a}$$

$$\langle 2\pi|T|\overline{K^o}\rangle = T\sin\theta \tag{5.12b}$$

where T is defined by the relation

$$T = \{ |\langle 2\pi|T|K^o\rangle|^2 + |\langle 2\pi|T|\overline{K^o}\rangle|^2 \}^{1/2} . \tag{5.12c}$$

Substituting Eqs. (5.10) into Eqs. (5.12) we obtain the transition matrix elements for the 2π decay in the K_A, K_B basis

$$\langle 2\pi|T|K_A\rangle = T \tag{5.13a}$$

$$\langle 2\pi|T|K_B\rangle = 0 \,. \tag{5.13b}$$

Thus if both the K^o and $\overline{K^o}$ are coupled to the 2π decay mode, a mixing angle can always be found which decouples one state from the 2π system. The physical mixing angle is chosen by diagonalization of the mass matrix. If experimentally one of the two kaon eigenstates is decoupled from the 2π system, one can turn the question around. Instead of asking why one of the eigenstates is decoupled from the 2π system one can ask why diagonalizing the mass matrix chooses the particular state decoupled from the 2π system to be an eigenstate, or why nature chooses the mixing angle given by Eq. (5.11). We examine two possible answers to this question, one based on symmetry and one based on dynamics.

1. Symmetry. If an additional symmetry remains unbroken it can give rise to the selection rule. For example, if CP is conserved in the kaon decays, then the eigenstates of the mass matrix must be eigenstates of CP and the mixing angle θ must

be equal to $\pm 45^{\circ}$. Since the 2π state is an eigenstate of CP the 2π decay is forbidden by CP conservation for the 1 kaon eigenstate of CP with the wrong eigenvalue.

2. Dynamics. If the symmetry is broken by the decay itself and there is no other symmetry breaking, the decoupling is automatic. Assume, for example, that only the $K \to 2\pi$ coupling breaks strangeness conservation. Then all breaking and mixing come from the loop diagram shown in Fig. 5.1a. The mass matrix is a 2×2 matrix whose elements are given by computing this loop diagram. But Eqs. (5.13) show that the loop diagram is diagonal in the K_A, K_B basis, where all matrix elements vanish except the diagonal matrix element for the state K_A.

Thus we see that two completely independent mechanisms naturally decouple one of the eigenstates of the neutral kaon system from the 2π decay mode. If the conditions required for either mechanism hold exactly the decay of one neutral kaon state into two pions is forbidden. In the real world, both conditions are very good approximations but not exact. CP is conserved to a good approximation, but is still violated. The 2π decay mode is the dominant strangness-violating decay mode of the neutral kaon system but it is not the only decay mode. Additional loop diagrams involving other states contribute to the mass matrix. Thus, both conditions are slightly violated and the neutral kaon system contains a long-lived state which decays weakly into two pions.

5.3 A Dynamical Model for Selection Rules at the SU(3) Level

The first attempt to explain the selection rule (5.1a) took the point of view[23] of the K_1-K_2 analogy and searched for a dynamical symmetry breaking mechanism which naturally chose the decoupled state as an eigenstate of the mass matrix. The approach begins in the nonet symmetry limit where all nine states of the vector or tensor nonet are degenerate and breaks the symmetry by coupling to decay channels. This removes the degeneracy and mixes the two states via transitions through decay channels as shown by the loop diagrams of Fig. 5.1. If a single loop diagram gives the

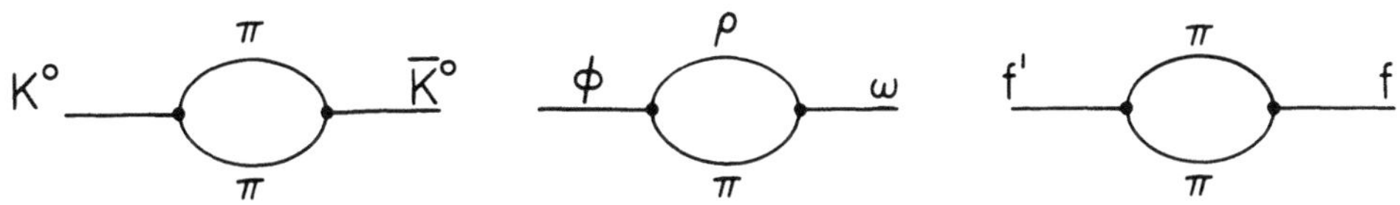

Fig. 5.1. Loop diagrams.

only contribution to the symmetry breaking the diagonalization of the mass matrix is trivial and leaves one of the two eigenstates decoupled from that decay channel. Thus, just as the K_2 is decoupled from the 2π channel the ϕ and f' are decoupled from the $\rho\pi$ and $\pi\pi$ channels if these are the only decay modes. The presence of other decay modes complicates the diagonalization but their effect is small if a single mode is dominant as in the case of the 2π mode in K^o decay. The dominant decay mode then remains decoupled from one of the two eigenstates to a very good approximation.

For the ϕ and f' decays the final states involving the other members of the pseudoscalar and vector octets must be considered together with the $\rho\pi$ and $\pi\pi$ modes. However, in the SU(3) symmetry limit these loop diagrams conserve SU(3) and cannot mix singlet and octet states. Mixing can arise only from symmetry breaking. The conventional description of symmetry breaking assumes that the vertices in loop diagrams satisfy SU(3) symmetry and breaks SU(3) by using physical non-degenerate masses for particles in the propagators. Because the dominant breaking effect in the masses is the low mass of the pion relative to the K and η, the dominant symmetry breaking effects in the loop diagrams come from the $\rho\pi$ and $\pi\pi$ channels. Thus, a dynamical model in which the symmetry is broken by loop diagrams gives the meson selection rules (5.1a) and (5.1b). The particular model considered for the vector mesons gave a natural suppression of the $\phi \to \rho\pi$ decay and a mass formula which fit the vector meson masses.

The loop diagram model does not give the baryon selection rules (5.1c) and (5.1d) in any simple way. It also gives no indication why the particular linear combinations of singlet and octet states which satisfy the selection rules (5.1a) and (5.1b) should also happen to satisfy the baryon coupling selection rules (5.1c) and (5.1d). Furthermore although the mass formula obtained gave a non-trivial fit to the data, the very simple experimental mass spectrum with ρ-ω and f-A2 degeneracy and equal ρ - K^* - Φ and A2 - K^* - f' spacings is not obtained naturally in this model and is fit by adjusting a free parameter. Thus, despite its initial promise the loop diagram model does not provide a satisfactory description of the selection rules.

5.4 The Quark Line Selection Rules

A crucial mystery at the SU(3) level is the difference between the meson couplings (5.1a) and (5.1b) and the baryon couplings (5.1c) and (5.1d). Baryon selection rules having the same SU(3) couplings as the meson selection rules (5.1a) and (5.1b) would decouple the Σ rather than the nucleon from the ϕ and f', since the Σ and π both have zero hypercharge and occupy corresponding positions in the baryon and meson octets. The simple unified statement of the selection rules (5.1) is in terms of strangeness, rather than hypercharge. But hypercharge is simple at the SU(3) level, while strangeness is not, since it depends on baryon number which is outside SU(3).

In the quark model the meson and baryon octets are very different because mesons are quark-antiquark pairs while baryons are three-quark states, and the very different SU(3) couplings arise naturally. There does not seem to be any simple description of the baryon selection rules (5.1c) and (5.1d) without invoking a quark-like structure for the baryons in which they are composed of three fundamental SU(3) triplets. The Zweig-Iizuka formulation with quark diagrams provides a simple unified description of the meson vertex selection rules (5.1a) and (5.1b), the baryon vertex selection rules (5.1c) and (5.1d), the choice as eigenstates of the mass matrix of just those particular linear combinations of singlet and octet which satisfy the selection rules, and the simple nonet mass spectrum.

The quark picture begins with a degenerate meson nonet and breaks the nonet degeneracy by a mass difference between strange and nonstrange quarks. This gives the so-called "ideal mixing" which chooses as eigenstates those mixtures of singlet and octet states corresponding to a pure strange quark-antiquark pair and a pure nonstrange quark-antiquark pair, and gives a simple mass formula with a mass splitting proportional to the number of strange quarks. The selection rule is simply stated by drawing quark line diagrams for the three-point vertex functions as in Figs. 5.2 and postulating that only the connected diagrams Fig. 5.2a and Fig. 5.2c are allowed while the disconnected diagrams Figs. 5.2b and Fig. 5.2d are forbidden. All the couplings (5.1) are forbidden since the ϕ and f' both consist only of strange quarks while the remaining particles consist only of nonstrange quarks. Since the quantum numbers of the quark remain the same on a given line, the strange quark lines begin and end on the ϕ or f' and are completely disconnected from the nonstrange quark lines. Thus, the couplings (5.1) are described by forbidden diagrams, Figs. 5.2b and 5.2d.

However, exactly the same meson selection rules (5.1a) and (5.1b) are obtainable from other approaches[14,19,20] without invoking connected and disconnected quark diagrams. A mathematically equivalent description for the three-meson coupling was first proposed by Okubo as a nonet coupling ansatz. The same selection rule was obtained by Alexander et al.[24] from the Levin-Frankfurt approach[25] in which any hadron transition involves a change in the state of only one active quark in the hadron while the remaining quarks are spectators. If pion emission is described as a single

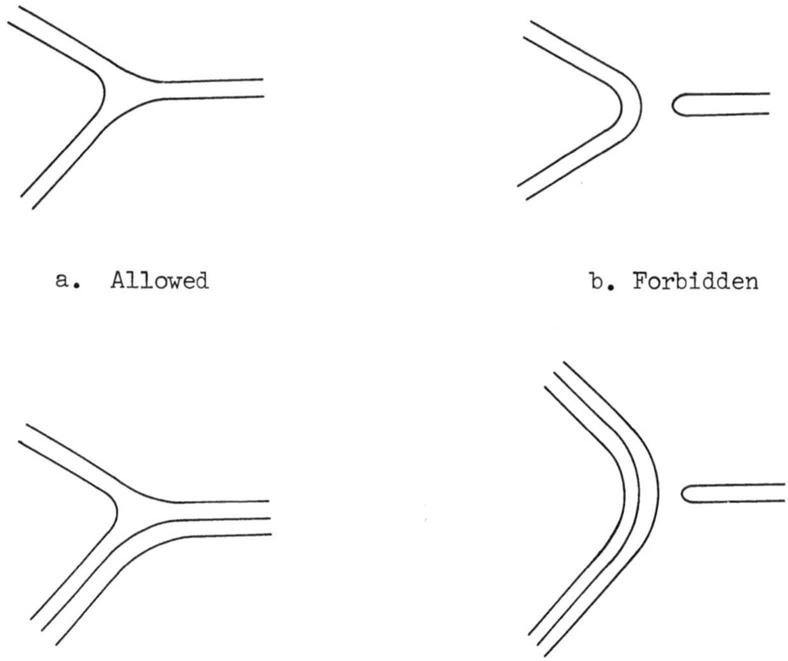

a. Allowed b. Forbidden

c. Allowed d. Forbidden

Fig. 5.2. Quark diagrams for three-point functions.

quark transition only nonstrange quarks can emit pions and conserve isospin. Thus, a state like the ϕ or f' which contains only strange quarks cannot decay by pion emission. This argument can be stated more precisely in the language of PCAC and the Melosh transformation.[26] The PCAC prescription relates the amplitude for a pionic decay to the matrix elements of the axial charge operator Q_5 between the initial state and the final state remaining after pion emission. The Melosh prescription postulates that Q_5 is a single quark operator. Since a single quark operator cannot change a strange quark-antiquark pair into a nonstrange pair,

$$\langle \phi | Q_5 | \rho \rangle = 0 \tag{5.14a}$$

$$\langle f' | Q_5 | \pi \rangle = 0 \, . \tag{5.14b}$$

The PCAC prescription then gives the selection rules (5.1a) and (5.1b).

None of these alternative approaches for the meson selection rules (5.1a) and (5.1b) are applicable to the baryon selection rules (5.1c) and (5.1d). The baryon couplings are much more complicated than meson couplings at the phenomenological SU(3) level. The octet three-meson coupling is constrained by charge conjugation to be pure F and pure D depending on the charge conjugation properties of the mesons. The singlet meson coupling is forbidden when the octet coupling is pure F. For baryons no such restrictions exist. The D/F ratio is a free parameter and the singlet coupling is always allowed even if the octet is pure F. Furthermore, an additional spin degree of freedom is present in the baryon coupling. There are two independent couplings corresponding to helicity flip and nonflip at the baryon vertex.

For the three meson couplings the OZI rule follows automatically from G conservation for those channels where charge conjugation invariance requires the antisymmetric F type SU(3) coupling. For example, the $\phi \to \pi\pi$ decay is forbidden while $\phi \to K^*\bar{K}$ is allowed. For cases like (5.1a) and (5.1b) where the symmetric D type octet coupling is required and the singlet is allowed one additional constraint on the couplings is needed to obtain the selection rules; namely a particular value for the ratio of the singlet octet couplings. The correct value of this singlet to octet ratio is naturally obtained from the Okubo-Levin-Frankfurt and Melosh approaches.

For the baryon case, the singlet to octet coupling ratio needed to obtain the selection rules (5.1c) and (5.1d) depend upon the D/F ratio. These ratios are different for the spin flip and nonflip transitions. Thus very complicated constraints on the couplings are needed in order to obtain the baryon selection rules. The only simple description so far has been the Zweig-Iizuka formulation with connected and disconnected quark diagrams.

The baryon selection rules (5.1c) and (5.1d) thus provide the best unambiguous tests of the ZI rule and its breaking. One can envision a hierarchy in which the rule holds best when several mechanisms reinforce one another. The smallest breaking effects would occur in pionic meson transitions where the PCAC derivation with eqs. (5.14) still holds while the other mechanisms which give rise to the ZI rule in the other cases are broken. The strongest breaking effects would occur in the baryon vertex and meson vertices not involving pions would lie somewhere between these two extremes.

The extension to more complicated vertices of the quark-line selection rules for three-point functions is not unique, as has been pointed out by Okubo.[18] The forbidden diagrams of Figs. 5.2b and 5.2d can be characterized either as "disconnected diagrams," which allow external particles to be separated without breaking quark lines, or as "hairpin diagrams" in which one external meson has its two lines connected together rather than joining lines to other particles. Forbidding all disconnected diagrams and forbidding all hairpin diagrams are equivalent for three-point functions. However, for four-point and higher functions there are disconnected diagrams which are not hairpin diagrams, multiply disconnected diagrams and diagrams with more than one hairpin, as shown in Fig. 5.3. Without some fundamental theory for the OZI rule it is not clear whether disconnected diagrams which are not hairpin diagrams are forbidden just as much as hairpin diagrams, and whether multiply disconnected diagrams or multiple hairpin diagrams are forbidden more than corresponding single diagrams.

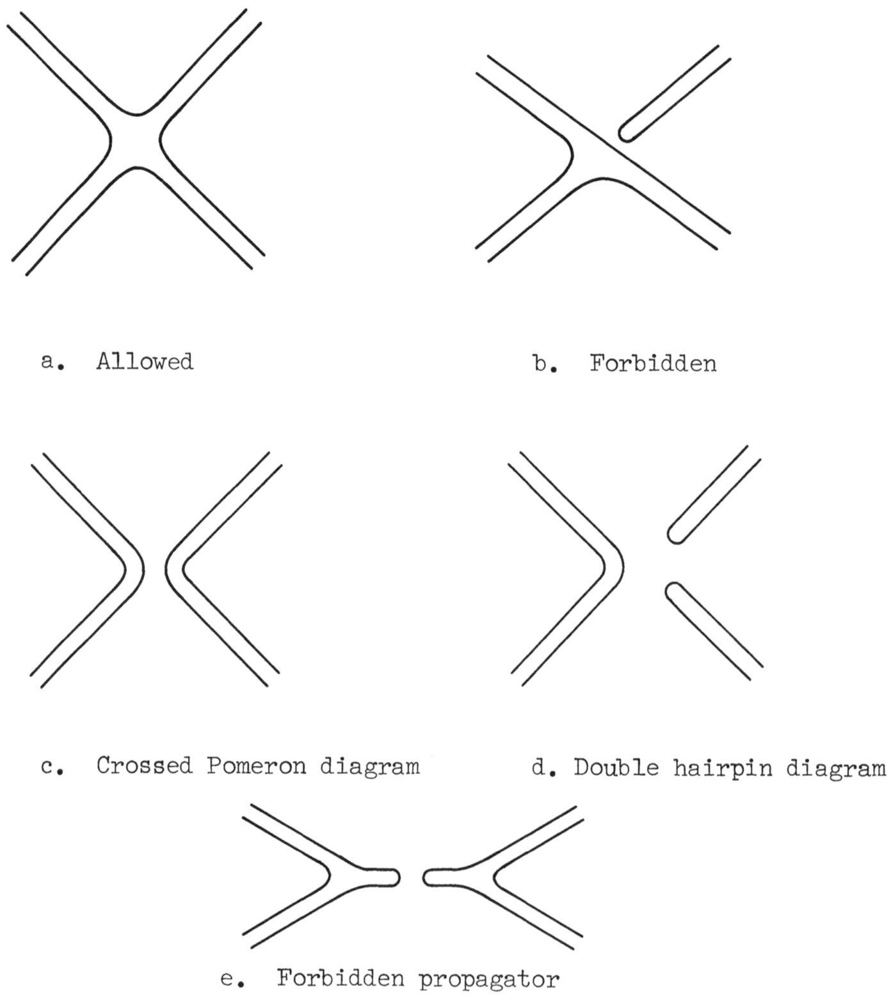

Fig. 5.3. Quark diagrams for three-point functions.

The four-point function of Fig. 5.3a is clearly allowed and that of Fig. 5.3b is clearly forbidden by any version of the OZI rule. But Figs. 5.3c and 5.3d are ambiguous. The simplest example of a disconnected diagram which is not a hairpin diagram is shown in Fig. 5.3c and could describe the process $\pi\pi \to \phi\phi$, or elastic $\phi\pi$ scattering in the cross channel. The elastic amplitude must have a Pomeron contribution, as in ϕN scattering. This "crossed Pomeron diagram" is forbidden in the Zweig-Iizuka formalism but is not forbidden if only hairpin diagrams are forbidden.

Double hairpin diagrams like that of Fig. 5.3d can occur when charm and strangeness are both present, because the two hairpins could describe a strange and a charmed quark-antiquark pair. One example of a process described by such a diagram is the decay $J/\psi \to \phi\pi\pi$ in the charm model for the J/ψ. In this model all decays of the J/ψ into normal hadrons are forbidden by the OZI rule in any formulation because one hairpin is required for the J/ψ. Whether the two-hairpin decays are more forbidden or not has been questioned.[18] The most recent experimental results on the $\phi\pi\pi$ decay mode suggest that such decays are indeed more forbidden,[27] but further data are necessary before any gendral conclusions can be drawn.

One possible approach to the generalization of the OZI rule to more complicated vertices is to build everything from three-point functions which satisfy the OZI rule. Inconsistencies arise in this approach because of the higher order paradox in which combinations of OZI-allowed transitions can produce an OZI-forbidden transition, as in Eqs. (5.3). There is also the question of possible OZI violations in the propagators of particles appearing as internal lines in the diagrams, as indicated in Fig. 5.3e. These points are discussed in detail below.

5.5 The Higher Order and Unitarity Paradoxes

We now examine in more detail the violation of the OZI rule by the transitions (5.3) in which two OZI-conserving amplitudes (5.2) combine to produce an OZI-violating transition. All these transitions are from an initial state containing only strange quarks to a final state containing only nonstrange quarks via the intermediate state of a kaon pair. The kaon plays a crucial ambivalent role. Since it contains one strange and one nonstrange quark, it couples equally to strange and nonstrange systems and can go either way. A kaon pair state contains one strange and one nonstrange quark-antiquark pair. It can therefore be created from a strange pair by the creation of a non-strange pair or vice versa. The kaon pair state thus links two kinds of states between which transitions are forbidden by the OZI rule.

The quark diagram Fig. 5.4 for the forbidden transition (5.3b) illustrates the essential features of the paradox. Viewed as a single topological diagram it is indeed disconnected and can be separated into two disconnected hairpin diagrams. But when it is separated into two individual transitions, each half is connected. Connecting the two diagrams together results in a topological disconnected diagram because of the twist in the quark and antiquark lines in the kaon intermediate state.

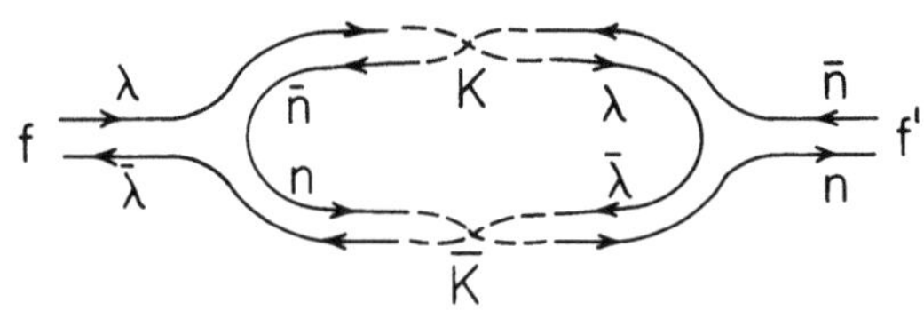

Fig. 5.4. OZI Violation via twisted diagram

Thus, to save the OZI rule the connection of allowed diagrams by a "twisted propagator" must somehow be forbidden. But a twisted propagator has physical meaning only if there is additional information in a kaon pair state to specify "which way it is twisted"; i.e., whether it originally came from a strange or a nonstrange system. Some memory of the origin of the pair is necessary to prevent the nonstrange decay of a pair which originated in a strange system. But a physical kaon pair state has no such memory. A kaon pair produced from a nonstrange system is indistinguishable from a pair produced from a strange system.

The transition (5.3) thus must exist in any consistent scheme which incorporates the OZI rule. Saving the rule requires additional transitions via other intermediate states which exactly cancel these amplitudes. Such cancellations do in fact occur in dual resonance models where twists in diagrams denote changes of the relative phase of the contributions of such intermediate states. But the degeneracy requirement discussed above poses difficulties. This is discussed in detail below.

The same higher order paradox appears in an S-matrix formulation as unitarity violation. Consider for example D wave $\pi\pi$ and $K\bar{K}$ scattering in the vicinity of the f' pole treated as a system with two coupled channels. The OZI selection rule (5.1b) requires the f' to appear as a pole only in the $K\bar{K}$ channel but not in the $\pi\pi$ channel. But the eigenstates of the 2 × 2 S matrix are not the $\pi\pi$ and $K\bar{K}$ states but mixtures of the two, because the $\pi\pi \to K\bar{K}$ transition amplitude (5.2d) does not vanish. Thus, both eigenstates of the S matrix contain a $\pi\pi$ component and any pole appearing in the S matrix must have a nonvanishing coupling to the $\pi\pi$ channel. Complete decoupling of a pole from the $\pi\pi$ channel is possible only when the $\pi\pi$ and $K\bar{K}$ channels are completely decoupled from one another.

This can be seen explicitly by writing the unitarity equation

$$\mathrm{Im}\langle K\bar{K}|T|\pi\pi\rangle = \langle K\bar{K}|T^{\dagger}|K\bar{K}\rangle\langle K\bar{K}|T|\pi\pi\rangle + \langle K\bar{K}|T^{\dagger}|\pi\pi\rangle\langle\pi\pi|T|\pi\pi\rangle \,. \tag{5.15}$$

Since there are only two channels the unitarity sum has only two terms. If the OZI rule holds and the f' pole is decoupled from the $\pi\pi$ channel, the f' pole appears only in a single term in Eq. (5.15), namely the first term in the right-hand side. Thus the OZI rule is inconsistent with unitarity in this two-channel model.

The way out of this paradox is to include more than the $\pi\pi$ and $K\bar{K}$ channels. Additional channels can introduce new terms in the unitarity sum of Eq. (5.15) which can cancel the term with the f' pole and enable the decoupling of the f' from the $\pi\pi$ channel without violating unitarity. Again the paradox is resolved by canceling the transitions via some other set of intermediate states.

Some symmetry scheme or dynamical model is needed to choose which additional intermediate states cancel the $K\bar{K}$ contribution. There are several possibilities. SU(3) symmetry suggests that the full pseudoscalar octet be included with the additional $\eta\eta$ intermediate state. Nonet symmetry requires the $\eta'\eta'$ and $\eta\eta'$ states as well. SU(6) symmetry suggests that vector and pseudoscalar mesons be treated together with the inclusion of intermediate states involving vector mesons. Duality and exchange degeneracy suggest that the vector and tensor mesons which lie on degenerate trajectories must be included together. Exactly how these conflicting suggestions can be resolved is an open question.

5.6 A Simple SU(3) Model

We now show how the higher order and unitarity paradoxes can be resolved in a simple way in the framework of SU(3) symmetry by including the $\eta\eta$ channel together with the $K\bar{K}$ channel. The model is not relevant to the physical particles, but the manner in which it avoids the difficulty of the higher order transition (5.3b) and the associated problems of the unitarity of the S-matrix is instructive. In the nonet symmetry limit there is always a particular linear combination of the two isoscalar tensor mesons for which the couplings of the two components to the 2π channel cancel one another exactly. We consider a model in which a small SU(3) symmetry breaking with appropriate properties splits the masses and leaves the decoupled states as an eigenstate.

We denote by f_1 and f_8 the two isoscalar tensor mesons classified in the SU(3) symmetry limit in the singlet and octet representations. We assume an unmixed octet of pseudoscalar mesons. There are three possible two-pseudoscalar decay modes for these tensor mesons namely, $\pi\pi$, $K\bar{K}$ and $\eta\eta$. The branching ratios for the f_1 and f_8 into these three decay modes are determined uniquely by SU(3), but the relative strengths of the f_1 and f_8 couplings are not determined

$$\sqrt{8/3}\,\langle\pi\pi|f_1\rangle = \sqrt{8}\,\langle\eta\eta|f_1\rangle = \sqrt{2}\,\langle K\bar{K}|f_1\rangle = \sqrt{\gamma_1} \tag{5.16a}$$

$$\sqrt{5/3}\,\langle\pi\pi|f_8\rangle = -\sqrt{5}\,\langle\eta\eta|f_8\rangle = -\sqrt{5}\,\langle K\bar{K}|f_8\rangle = \sqrt{\gamma_8} \tag{5.16b}$$

where γ_1 and γ_8 are reduced total widths for the f_1 and f_8 with phase space factors removed

$$\gamma_1 = |\langle\pi\pi|f_1\rangle|^2 + |\langle\eta\eta|f_1\rangle|^2 + |\langle K\bar{K}|f_1\rangle|^2 \tag{5.17a}$$

$$\gamma_8 = |\langle\pi\pi|f_8\rangle|^2 + |\langle\eta\eta|f_8\rangle|^2 + |\langle K\bar{K}\ f_8\rangle|^2. \tag{5.17b}$$

If $\gamma_1 \neq \gamma_8$, the states f_1 and f_8 have different lifetimes and cannot be mixed, as mixed states would not have simple exponential decays. To allow mixing of f_1 and f_8 in the symmetry limit we set their widths equal

$$\gamma_1 = \gamma_8. \tag{5.18}$$

The states f_1 and f_8 are now degenerate and any linear combinations can be chosen to give a basis of states with simple exponential decays. We choose a basis in which one state is completely decoupled from the 2π channel. From Eqs. (5.16) and (5.18) this basis is

$$|f\rangle = \sqrt{\frac{5}{13}}|f_1\rangle + \sqrt{\frac{8}{13}}|f_8\rangle \tag{5.19a}$$

$$|f'\rangle = \sqrt{\frac{8}{13}}|f_1\rangle - \sqrt{\frac{5}{13}}|f_8\rangle\ . \tag{5.19b}$$

We now assume that the symmetry breaking chooses these states as eigenstates of the mass matrix for some mysterious reason. The partial widths for the various decays of these states are given by

$$\langle \pi\pi | f \rangle = \sqrt{(39/40)\gamma} \tag{5.20a}$$

$$\langle \eta\eta | f \rangle = -\sqrt{(9/520)\gamma} \tag{5.20b}$$

$$\langle K\bar{K} | f \rangle = \sqrt{(1/130)\gamma} \tag{5.20c}$$

$$\langle \pi\pi | f' \rangle = 0 \tag{5.21a}$$

$$\langle \eta\eta | f' \rangle = \sqrt{(4/13)\gamma} \tag{5.21b}$$

$$\langle K\bar{K} | f' \rangle = \sqrt{(9/13)\gamma}. \tag{5.21c}$$

The state denoted by f' is decoupled from the 2π channel by construction. But the mixing angle described by Eqs. (5.19) is very different from the ideal mixing angle of the quark model. Note, however, that Eqs. (5.20) and (5.21) lead to the result

$$\langle (3K\bar{K} + 2\eta\eta) | f \rangle = 0 \tag{5.22a}$$

$$\langle (2K\bar{K} - 3\eta\eta) | f' \rangle = 0 \tag{5.22b}$$

$$\langle f' | K\bar{K} \rangle \langle K\bar{K} | f \rangle + \langle f' | \eta\eta \rangle \langle \eta\eta | f \rangle = 0. \tag{5.23}$$

Equation (5.23) shows that the higher order transition amplitude (5.3b) is canceled exactly by the analogous transition via the $\eta\eta$ intermediate state. Equations (5.22a) and (5.22b) show that the f and f' are coupled to two orthogonal linear combinations of the $K\bar{K}$ and $\eta\eta$ channels. The particular linear combination (5.20a) which is decoupled from the f is the eigenstate of the S-matrix whose amplitude has the f' pole. This pole does not appear in the other eigenstates of the S-matrix, which are decoupled from the f', namely, the 2π channel. Thus the selection rule (5.1a) can be rigorous without difficulties from the higher order transitions (5.3b) or the unitarity of the S matrix.

We now consider the conventional quark model formulation using ideal mixing and the OZI rule. This case is very different from the above SU(3) treatment where the cancellation (5.23) of the higher order transition amplitude (5.3b) depends crucially on the description of the η as a member of an unmixed octet. No such cancellation occurs in the case of an ideally mixed pseudoscalar nonet with couplings given by the OZI rule. We denote the pseudoscalar state consisting of a nonstrange quark-antiquark pair by η_n and the state of a strange quark-antiquark pair by η_s. The OZI rule then immediately gives

$$\langle\eta_n\eta_n|f'\rangle = \langle\eta_s\eta_s|f\rangle = \langle\eta_s\eta_n|f\rangle = \langle\eta_s\eta_n|f'\rangle = 0 \quad (5.24a)$$

$$\langle\eta_n\eta_n|f\rangle = \sqrt{2}\,\langle\eta_s\eta_s|f'\rangle = \sqrt{1/3}\,\langle\pi\pi|f\rangle$$

$$= \langle K\bar{K}|f\rangle = \sqrt{1/2}\,\langle K\bar{K}|f'\rangle \,. \quad (5.24b)$$

Thus if the coupling of the tensor mesons to the pseudoscalar nonet is described by the OZI rule with ideal mixing, there is only one pseudoscalar channel coupled to both f and f', the $K\bar{K}$ channel, and no additional channel is available to cancel the transition (5.3b).

For a more realistic treatment we consider pseudoscalars which are not ideally mixed,

$$\eta' = \eta_n \cos\theta_p + \eta_s \sin\theta_p \quad (5.25a)$$

$$\eta = -\eta_n \sin\theta_p + \eta_s \cos\theta_p \quad (5.25b)$$

where θ_p is the angle of deviation from ideal mixing. In this basis Eqs. (5.24a) and (5.24b) become

$$\frac{\langle\eta'\eta'|f\rangle}{\cos^2\theta_p} = \frac{\langle\eta\eta|f\rangle}{\sin^2\theta_p} = \frac{\sqrt{2}\langle\eta'\eta'|f'\rangle}{\sin^2\theta_p} = \frac{\sqrt{2}\langle\eta\eta|f'\rangle}{\cos^2\theta_p} =$$

$$= \frac{\langle\eta'\eta|f'\rangle}{\sin\theta\cos\theta} = -\frac{\langle\eta'\eta|f\rangle}{\sqrt{2}\sin\theta\cos\theta} = \frac{1}{\sqrt{3}}\langle\pi\pi|f\rangle$$

$$= \langle K\bar{K}|f\rangle = \frac{1}{\sqrt{2}}\langle K\bar{K}|f'\rangle. \quad (5.26)$$

Equations (5.26) give the transition amplitudes for the case where the couplings to ideally mixed states satisfy SU(3) symmetry and the OZI rule, but the physical pseudoscalar states are not ideally mixed.

From Eqs. (5.24b)

$$\langle f'|\eta\eta\rangle\langle\eta\eta|f\rangle = \frac{\sin^2(2\theta_p)}{8}\langle f'|K\bar{K}\rangle\langle K\bar{K}|f\rangle \tag{5.27a}$$

$$\langle f'|\eta'\eta'\rangle\langle\eta'\eta'|f\rangle = \frac{\sin^2(2\theta_p)}{8}\langle f'|K\bar{K}\rangle\langle K\bar{K}|f\rangle \tag{5.27b}$$

$$\langle f'|\eta\eta'\rangle\langle\eta\eta'|f\rangle = -\frac{\sin^2(2\theta_p)}{4}\langle f'|K\bar{K}\rangle\langle K\bar{K}|f\rangle. \tag{5.27c}$$

If the result (5.27) is summed over all channels with equal weighting the total contribution of η and η' decay modes vanishes, as expected from Eq. (5.24a). A finite contribution appears under the more realistic assumption that the $\eta\eta$ channel is dominant and others are neglected. However, this contribution has the same phase as the contribution from the $K\bar{K}$ intermediate state and cannot produce a cancellation. It is also much smaller than the $K\bar{K}$ contribution.

5.7 The Selection Rule in $SU(6)_W$ Symmetry

An early derivation of the selection rule forbidding the $\phi \to \rho\pi$ decay was based on $SU(6)_W$ symmetry.[6,28] This selection rule holds in any model which satisfies $SU(6)_W$. It is therefore of interest to examine the higher order transition (3a) and see how $SU(6)_W$ operates in this case. In $SU(6)_W$ the $K^*(890)$ is in the same super-multiplet with the kaon, and higher order transitions via all possible intermediate states involving one or two K^* mesons must also be considered. In the approximation where K and K^* are degenerate and all vertices are related by $SU(6)_W$ the contributions from the different K and K^* intermediate states all cancel and the decay $\phi \to \rho\pi$ is still forbidden. This cancellation is simply described by a conservation law. In the decay (5.1a) the outgoing ρ and the initial ϕ must all be in the same polarization state with $S_z = \pm 1$, since angular momentum and parity conservation forbids the transition for the states with $S_z = 0$. The transition from the initial state to the intermediate state of two strange mesons conserves separately the total W spins of strange and nonstrange quarks. The initial ϕ state has strange W-spin 1 and nonstrange W spin zero. Thus each strange meson in the intermediate state is a coherent linear superposition of a K and a K^* which is an eigenstate of the z component of the spin of the strange quark or antiquark, with the eigenvalue required to conserve the strange W spin. The spin directions of the two strange quarks are parallel. Thus, they cannot annihilate and conserve strange quark W spin.

The $SU(6)_W$ selection rule is thus follows simply from the conservation of the z component of the strange quark spin.

However, the transition (5.3a) is forbidden in a much simpler way by $SU(6)_W$, which also illustrates a basic weakness in $SU(6)_W$, which holds only for colinear processes. In $SU(6)_W$ all momenta are in the z direction and a $K\bar{K}$ state must have $S_z = J_z = 0$. Thus the $S_z = 0$ polarization state of the initial ϕ decays into the $K\bar{K}$ channel and the $S_z = \pm 1$ state decays into the $\rho\pi$ channel and angular momentum conservation forbids transitions between the two states. But this argument holds for the transition (5.3a) only if the momentum of the kaon pair in the intermediate state is in the same direction as the momentum of the final $\rho\pi$ state. Consider for example the decay of a ϕ in the state with $S_x = 0$ into a $\rho\pi$ final state with momentum in the z direction. The transition of a ϕ into an ω via an intermediate $K\bar{K}$ state with momenta in the x direction is allowed by $SU(6)_W$ because the ϕ and ω states with $S_x = 0$ are both states with $W = 0$ with respect to the x axis. However, the state of the ω with $S_x = 0$ is a linear combination of states with $S_z = \pm 1$ for which the $\rho\pi$ decay in the z direction is allowed.

The $SU(6)_W$ argument does not hold for the f' decay, because the outgoing pions are spinless and there can be no correlation between the quark spins of the final state and the quark spins in the initial state. In the decay (5.1b) a component in the initial f' wave function has the quark and antiquark spins antiparallel, and therefore has $W = 0$. They are allowed to annihilate without any angular momentum transfer, and the transition from the $W = 0$ component of the final two-pion state is allowed by $SU(6)_W$. Thus, $SU(6)_W$ cannot be used to give a general derivation of the OZI rule.

5.8 Ideal Mixing and Symmetry Cancellations

The peculiar role of ideal mixing and higher symmetries in the OZI rule is illustrated by the following example:

Consider the decay of a high K^{*-} resonance into three kaons via an intermediate nonstrange resonance M^o

$$K^{*-} \to K^- M^o \to K^- K\bar{K}, \tag{5.28}$$

where M^o is a resonance like the ρ^o, ω, f or A2 which consists only of nonstrange quarks.

There are two quark line diagrams for this decay shown in Fig. 5.5, using the $p\bar{p}$ and $n\bar{n}$ components of M^o. The $p\bar{p}$ diagram is connected, obeys the OZI rule and leads to the final state $K^-K^+K^-$. The $n\bar{n}$ diagram is disconnected, violates the OZI rule and leads to the final state $K^-K^0\bar{K}^0$. Thus if the OZI rule holds,

$$(K^* \to K^- M^0 \to K^- K^+ K^-) \text{ is allowed} \quad (5.29a)$$

$$(K^* \to K^- M^0 \to K^- K^0 \bar{K}^0) \text{ is forbidden .} \quad (5.29b)$$

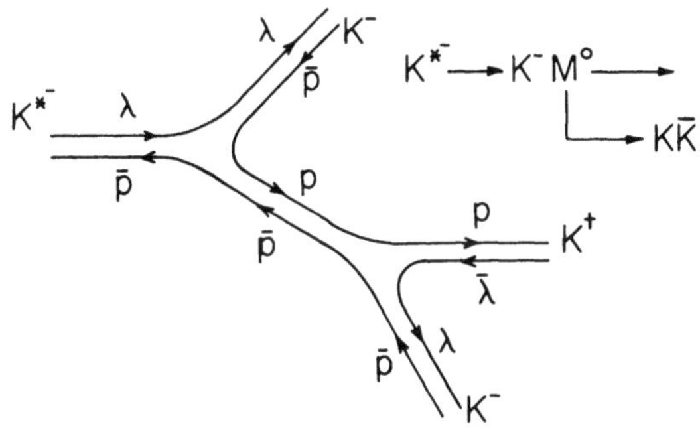

Fig. 5.5a. Allowed K^{*-} into charged mode.

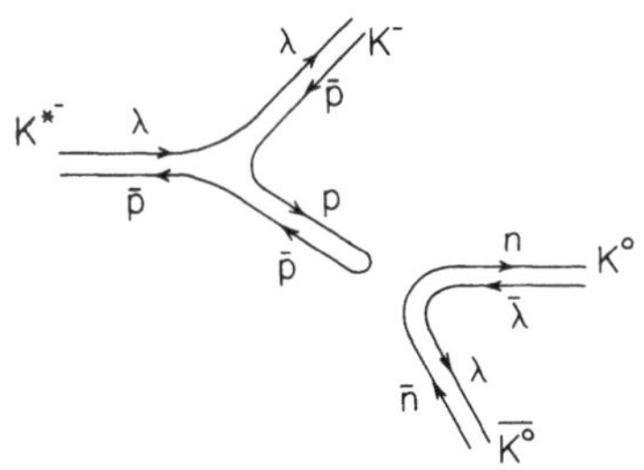

Fig. 5.5b. Forbidden K^{*-} decay into neutral mode.

But if the resonance M^0 has a definite isospin, the two transitions (5.29a) and (5.29b) must be equal by isospin invariance. Contradictions between the OZI rule and isospin invariance are avoided if the nonstrange meson spectrum consists of degenerate isospin doublets, like ρ and ω or f and A2. In that case the transition (5.29a) proceeds via the particular coherent linear combination of isovector and isoscalar particles which has the quark constitution $p\bar{p}$. The OZI rule is thus intimately related to the existence of the isospin doublets found in ideally mixed nonets.

If the M^o in the transitions (5.29) is not a member of an isospin doublet, the OZI rule is inconsistent with isospin invariance. This is the case if M^o is a π^o, which has no degenerate isoscalar partner. Although the π^o cannot appear as a physical resonance in the reactions (5.29) because of its low mass, it can appear as an exchanged particle in the analogous two-body scattering reactions

$$K^+ + K^- \rightarrow K^{*o} + \bar{K}^{*o} \tag{5.30a}$$

$$K^+ + K^- \rightarrow K^{*+} + K^{*-} \tag{5.30b}$$

$$K^o + K^- \rightarrow K^{*o} + K^{*-}. \tag{5.30c}$$

The quark diagrams for these reactions are shown in Fig. 5.6 .

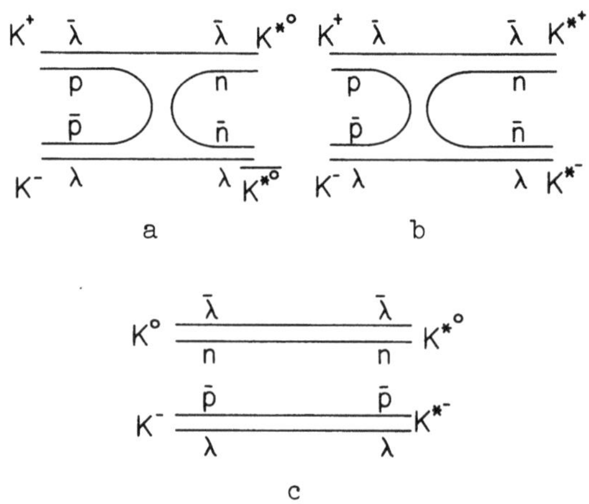

Fig. 5.6. Reactions allowed by pion exchange.

The charge exchange reaction (5.30a) is clearly allowed by the OZI rule and can go by pion exchange. The amplitudes for the pion exchange contribution to the reactions (5.30b) and (5.30c) are uniquely related to the charge exchange amplitude (5.30a) by isospin invariance. But the reaction (5.30b) is allowed by the OZI rule and the reaction (5.30c) is forbidden when only nonstrange quark exchange is considered. (The reaction (5.30c) is allowed by $\lambda\bar{\lambda}$ exchange but this is irrelevant to the present argument). The OZI rule could be saved from inconsistency with isospin invariance if a contribution from isoscalar exchange degenerate with pion exchange cancelled the pion exchange contribution to the reaction (5.30c). But no such isoscalar exists. Thus violations of the OZI rule might be expected in processes where pseudoscalar exchange plays a dominant role.

5.9 Cancellations and Degeneracies in Quark Line Models

Figure 5.7 shows the essential piece of the diagrams of Figures 5.5 and 5.6 which break the OZI rule, a transition between a $p\bar{p}$ and $n\bar{n}$ pair. This diagram exists in the propagator of any neutral isovector meson and will lead to OZI breaking unless it is cancelled by a contribution from a degenerate isoscalar partner. The diagrams of Fig. 5.4 describe the higher order transitions (5.3) which violate the OZI rule as the connection of two allowed diagrams by a twisted pair of lines. The essential piece of this diagram which breaks the OZI rule is shown in Fig. 5.8. This is a transition which interchanges the quark and antiquark lines.

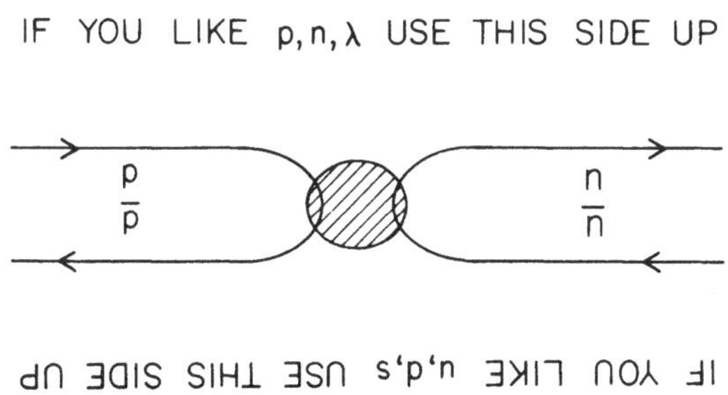

Fig. 5.7. OZI-Violating diagram required by isospin.

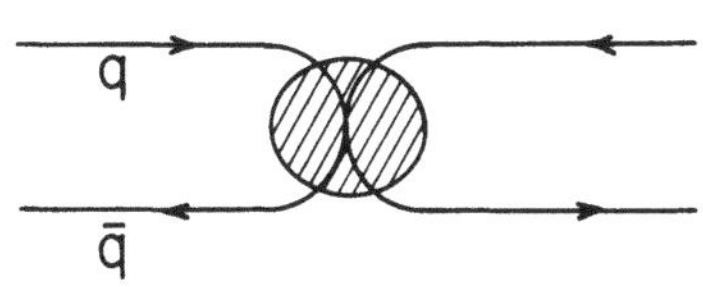

Fig. 5.8. OZI-Violating diagram required by charge conjugation.

The two basic OZI-violating diagrams, Fig. 5.7 and Fig. 5.8 have a very similar structure. Figure 5.7 describes a transition between two states related by isospin. It occurs naturally in the propagator of any particle which is an isospin eigenstate. Figure 5.8 describes a transition which interchanges quark and antiquark and occurs naturally in the propagator of any particle which is an eigenstate of charge conjugation. The OZI violation implied by the diagram of Fig. 5.7 can be avoided by an additional degeneracy of isoscalar and isovector particles. This allows the choice of a basis of states which are not isospin eigenstates and whose propagators do not include the diagram of

Fig. 5.7. Similarly, the OZI violation implied by Fig. 5.8 can be avoided by an additional degeneracy of particle states which are even and odd under charge conjugation. This allows the choice of a basis which are not eigenstates of charge conjugation and whose propagators do not include the diagram of Fig. 5.8. For the case where the quark and antiquark in Fig. 5.8 do not have the same internal symmetry quantum numbers, the relevant transformation is not charge conjugation but an appropriate combination like G parity of charge conjugation and an internal symmetry transformation. The conclusions are the same.

We can now specify the additional degeneracies essential for the validity of the OZI rule in higher order in models described by quark line diagrams. The states described by the left-hand and right-hand sides of the diagrams of Fig. 5.7 and Fig. 5.8 must be physical eigenstates which can propagate unchanged and in particular can avoid undergoing transitions indicated by Fig. 5.7 and Fig. 5.8. These states are eigenstates of quark number having a well-defined quark composition (either $p\bar{p}$ or $n\bar{n}$, but not a linear combination of them) and are linear combinations of states even and odd under charge conjugation (either $\bar{p}p$ or $p\bar{p}$ but not the linear combination of them which is a charge conjugation eigenstate). These states required by the OZI rule are not eigenstates of SU(3) and its isospin subgroup nor of charge conjugation. They can be physical eigenstates only if additional degeneracies are present beyond those imposed by these symmetries.

There must be ideal mixing of the SU(3) singlet and octet states so that the $\lambda\bar{\lambda}$ state remains an eigenstate and the $p\bar{p}$ and $n\bar{n}$ eigenstates which go into one another under isospin transformations are degenerate. When processes are described in terms of the isospin eigenstates as in Fig. 5.5 the amplitude for the forbidden diagram vanishes because of a cancellation between the contributions involving degenerate isoscalar and isovector states.

Charge conjugation degenerate doublets are required to eliminate the OZI violation due to twisted diagrams like Fig. 5.8. In duality and dual resonance models this degeneracy appears in Regge trajectories, rather than in individual particle states as exchange degeneracy of trajectories having opposite signature. In these formulations the two states on the left-hand and right-hand sides of Fig. 5.8 represent two different linear combinations of even signature and odd signature trajectories with equal magnitude and opposite phase. When the OZI-violating higher-order transitions (5.3) are described in the conventional basis using states which are eigenstates of charge conjugation the violating diagrams cancel one another in pairs involving states which behave oppositely under charge conjugation.

The peculiar relation between ideal mixing, exchange degeneracy and the OZI rule was noted in early treatments of duality using finite energy sum rules.[29] Two possible mechanisms for the breakdown of the necessary cancellations are immediately evident upon closer examination of the diagrams of Fig. 5.7 and Fig. 5.8.

The diagram of Fig. 5.7 will occur and break the OZI rule whenever a propagator appears for a state which is not ideally mixed, such as a pseudoscalar meson. One can expect the OZI rule to be violated in processes where there is a strong contribution from pseudoscalar exchange.

The cancellation of the diagram of Fig. 5.8 must break down because charge conjugation doubling exists only for Regge trajectories and not for individual physical states.[30] Thus, even if exchange degeneracy is exact the OZI-violating diagram of Fig. 5.8 is cancelled only in the kinematic region where Reggeization is a good approximation; i.e., where the scattering amplitude is well described by a Regge exchange rather than by one or two resonances.[31] This is clearly not the case for the higher-order transitions (5.3a) and (5.3b) where the mass is above the threshold for the intermediate $K\bar{K}$ state but below threshold for all other states on the kaon trajectory and its exchange degenerate partner. In general, one might say that contributions from high momentum intermediate states could be described in the Regge approximation and the desired cancellations from exchange degenerate pairs could occur. But at low momenta, where the large mass difference between individual resonances on exchange degenerate trajectories is significant, such cancellations should not be expected.

5.10 Quantitative Estimate of OZI Violation

Let us now attempt to estimate the violation of the OZI selection rule (5.1b) resulting from the higher order transition (5.3b). We consider the f-f' system analogous to the K_1-K_2 system and diagonalize a 2×2 mass matrix. We assume that the dominant portion of the mass splitting comes not from the loop diagram but from a quark mass term. We therefore use the ideally mixed basis in which the quark mass term is diagonal. The loop diagram contributes both a real part and an imaginary part to the mass matrix.

The real part is dominated by high momenta. There is no justification for considering only the contribution of the $K\bar{K}$ state rather than all states on the K and K^* Regge trajectories. Exchange degeneracy and Regge arguments suggest some kind of cancellation in these diagrams. But it is very difficult to estimate these cancellations quantitatively. One possibility is to consider the entire set of states on Regge trajectories and use duality and dual resonance models in order to calculate the contributions. Such calculations are beyond the scope of this paper.

The imaginary part, however, is dominated by the f and f' poles. It is therefore reasonable as a first approximation to consider only the $K\bar{K}$ intermediate state and neglect the contributions of higher strange meson resonances. Estimates of the imaginary part of the mass matrix are obtained by using the experimental partial widths for the decays. We therefore neglect the real part, which we cannot calculate anyway, and calculate the contribution of the imaginary part to the OZI violation. This gives a lower bound since the contribution of the real part cannot cancel that of the imaginary part.

In this formulation the violation of the OZI rule comes about as a result of a deviation from ideal mixing produced by the loop diagram. The relevant parameter which characterizes the mixing is the ratio of half the width of the f' to the f-f' mass splitting. For $\Gamma(f' \to K\bar{K}) = 40$ MeV and $m_{f'} - m_f = 240$ MeV this ratio is small and consistent with the small observed violation. This also implies that we can treat the deviation from ideal mixing as a first-order perturbation. Let us write

$$|f'\rangle = \cos\theta_{f'}|f_I'\rangle + \sin\theta_{f'}|f_I\rangle$$

where the subscript I denotes the ideally mixed states and $\theta_{f'}$ is the deviation of the f' mixing angle from ideal mixing. This is not necessarily equal to the f mixing angle because the eigenstates of a complex mass matrix are not necessarily orthogonal. However, only $\theta_{f'}$ is needed to calculate the OZI rule violation. In first-order perturbation theory $\sin\theta_{f'}$ is given by

$$\sin\theta_{f'} = \frac{\pi\langle f|T|K\bar{K}\rangle\langle K\bar{K}|T|f'\rangle\rho_F(f')}{\hbar(M_{f'} - M_f)} \tag{5.31a}$$

where $\rho_F(f')$ denotes the density of final $K\bar{K}$ states at the mass of the f'. This can be expressed in terms of the experimental width of the f' and the ratio of f and f' transition matrix elements

$$\sin\theta_{f'} = \frac{\Gamma(f' \to K\bar{K})}{2(M_{f'} - M_f)} \frac{\langle f|T|K\bar{K}\rangle}{\langle f'|T|K\bar{K}\rangle} \,. \tag{5.31b}$$

The amplitude of the OZI violating transition is then given by

$$\langle 2\pi|T|f'\rangle = \sin\theta_{f'} \langle 2\pi|T|f\rangle = \frac{\Gamma(f' \to K\bar{K})}{2(M_{f'} - M_f)} \frac{\langle f|T|K\bar{K}\rangle\langle 2\pi|T|f\rangle}{\langle f'|T|K\bar{K}\rangle} . \tag{5.31c}$$

To evaluate the expressions (5.31) we introduce the SU(3) relation between the transition amplitudes

$$\langle 2\pi|T|f\rangle = \sqrt{3}\, \langle K\bar{K}|T|f\rangle = \sqrt{3/2}\, \langle K\bar{K}|T|f'\rangle. \tag{5.32}$$

We now obtain

$$\sin\theta_{f'} = \frac{\Gamma(f' \to K\bar{K})}{2\sqrt{2}(M_{f'} - M_f)} \tag{5.33a}$$

$$|\langle 2\pi|T|f'\rangle|^2 = \frac{3\, \Gamma(f' \to K\bar{K})^2}{16(M_{f'} - M_f)^2} |\langle K\bar{K}|T|f'\rangle|^2 \, (p_\pi/p_k)^5 \tag{5.33b}$$

where the d-wave phase space factor $(p_\pi/p_k)^5$ is introduced to account for the difference between pion and kaon momenta in the decay of the f'. Substituting the experimental values $\Gamma(f' \to K\bar{K}) = 40$ MeV, $M_{f'} = 1514$ MeV, $M_f = 1270$ MeV we obtain $\sin\theta_{f'} = 0.06$ and

$$\left|\frac{\langle 2\pi|T|f'\rangle}{\langle K\bar{K}|T|f'\rangle}\right|^2 = 0.005\, (p_\pi/p_k)^5 = 0.02. \tag{5.34}$$

This result is in qualitative agreement with experiment.[32] Obtaining a better approximation is difficult because there are too many uncertainties in the additional effects which must be taken into account. In addition to the contribution from the real part there are additional contributions from channels like $\eta\eta$, $\eta'\eta'$ which have large uncertainties because of the deviation from ideal mixing. In the limit of ideal mixing these channels give no contribution as shown by Eqs. (5.24) since there is no state which is coupled both to the f and to the f'. However, the physical pseudoscalar mesons are approximately equal mixtures of η_n and η_s

and all three physical states contribute to the transition. The three contributions cancel exactly in the nonet symmetry limit where η and η' are degenerate and the ratio of the singlet and octet couplings is given by the OZI rule. For the physical nondegenerate and nonideally mixed states the cancellation no longer occurs and the magnitude of the contribution is very sensitive to the mixing angle and to possible changes in the singlet to octet ratio. Thus, there does not seem to be a serious possibility of obtaining a better approximation than Eq. (5.34).

An alternative approach to including the higher order transition (5.3b) is to use an S-matrix formalism and calculate unitarity corrections. In the most naive approximation results similar to (5.34) are obtained. However, attempts to include more channels run into the same difficulties discussed above for the η and η'.

Our analysis of the $f' \to \pi\pi$ decay suggests that the observed violation of the OZI rule is due to the higher order transition via the open $K\bar{K}$ channel, and that there is no effective mechanism available for cancelling the contribution of this channel. The effect is small because it is characterized by a small parameter, the ratio of half the width of the f' to the f-f' mass difference. Whether the smallness of this parameter has any deep theoretical significance is not clear at this point. Two fundamental quantities having the dimensions of mass appear in this ratio, the characteristic width of strong decays and the mass difference between strange and nonstrance particles or between the strange and nonstrange quark. Both these quantities are generally considered to be of the order of 100 MeV and the ratio of the two is then of order unity.

However, in the particular case of the f' decay there are several factors of 2 present which conspire to provide a factor which is an order of magnitude. Two factors of 2 arise because the relevant parameters are half of the width and twice the energy difference between strange and nonstrange quarks. The width of the f' is 40 MeV rather than 100 MeV. As long as there is no fundamental theory which predicts the ratio of the strangeness mass difference to strong decay widths, there can be no explanation for why the numerical value of the parameter which characterizes OZI violations is small in this particular case.

It is amusing that the selection rule holds both in the limit of very small and very large values for this ratio of the widths to the mass splitting. In the other extreme case of the large value the mixing of the two states is completely dominated by the decay process and leads to the decoupling of one state via the K_1-K_2 mechanism.

In intermediate cases, where the mixing is not dominated by the decay process the K_1-K_2 mechanism still has some effect in reducing OZI violation from higher-order transitions. In the above analysis of mixing produced via the intermediate $K\bar{K}$ state in f' decay contributions from the intermediate $\pi\pi$ state were neglected. In first order these contributions vanish because the unperturbed f' does not couple to the $\pi\pi$ channel. This approximation is justified in this case by the small magnitude of the results (3.31-3.34). But when the mixing is larger and higher order effects must be considered the contribution of the $\pi\pi$ intermediate state introduces an effective "restoring force" opposing the mixing by the $K\bar{K}$ intermediate state and opposing the increased violation of the OZI rule.

A qualitative estimate of this restoring force effect is obtainable by introducing all the two-pseudoscalar meson intermediate states into the calculation. We assume that the 2×2 mass matrix M has the form:

$$M = \Delta m + L_s + L_\pi \tag{5.35}$$

where Δm is a quark mass term diagonal in the ideal mixing basis as in the naive mixing model, L_s represents the contribution of the loop diagram of Fig. 5.1c but with all octet pseudoscalar mesons included in the intermediate state and full SU(3) symmetry, and L_π denotes an additional contribution from the $\pi\pi$ intermediate state resulting from SU(3) symmetry breaking. We do not have values for the strengths of these terms from first principles, but give each one a strength parameter and see how the OZI rule is affected by their variation. In particular, we shall see that the L_π term indeed has the effect of a restoring force reducing OZI violation.

If $L_s = L_\pi = 0$, the only contribution to the mass matrix comes from Δm and gives ideal mixing and the OZI rule. If $L_s \neq 0$ but L_π is kept zero; i.e. SU(3) symmetry in the loop diagram, there is no longer ideal mixing, and the OZI violating decay $f' \to \pi\pi$ occurs. However, if Δm and L_s are fixed and the symmetry breaking term L_π is turned on, the OZI violation is reduced by this "restoring force". Table 5.1 shows values of the OZI violation, expressed as the ratio of the forbidden and allowed decay rates, $\Gamma(f' \to \pi\pi)/(\Gamma(f \to \pi\pi)$ for different values of L_π and L_s. The value of L_s is given in arbitrary units, and L_π is given in units of the $\pi\pi$ contribution to L_s; i.e. $L_\pi = 1$ means that the contributions from the $\pi\pi$ intermediate state in L_π and L_s are equal. Table 5.1 shows that restoring force effects can be quite appreciable. The significant quantity is the change in $\Gamma(f' \to \pi\pi)/(\Gamma(f \to \pi\pi)$ with increasing L_π. For example, when the loop diagram contribution is sufficiently large to give a 31%

violation of the OZI rule the addition of a symmetry breaking $\pi\pi$ loop of equal magnitude to the SU(3) symmetric $\pi\pi$ contribution reduces the violation to 7%.

Table 5.1

Effect of $K_1 - K_2$ Restoring Force Mechanism

Values of $\Gamma(f' \to \pi\pi)/\Gamma(f \to \pi\pi)$ as Functions of L_s and L_π

L_s \ L_π	0	1/2	1	3/2	2
1	.04	.03	.02	.016	.012
2	.11	.06	.04	.03	.02
3	.18	.09	.05	.035	.02
7	.31	.14	.07	.04	.03

5.11 Experimental Tests of the OZI Rule

The OZI rule has been found to be in qualitative agreement with experiment for selection rules forbidding the production of the ϕ and f' mesons.

Two types of further experimental information are needed as a guide to theoretical understanding of the underlying dynamics: 1) Quantitative results on the magnitude of the OZI-violating transitions with systematic comparisons of different processes, 2) Tests of the OZI rule for nonets like the pseuco-scalar nonet which are not ideally mixed.

If production of the ϕ and f' from nonstrange systems arises primarily from a small admixture of a nonstrange quark-antiquark pair into the wave function, as in the example of section 5.10, then all OZI-violating transitions will be expressible[19] in terms of the angle $\theta_{f'}$ and the analogous angle θ_ϕ

$$\frac{g^2_{\phi\rho\pi}}{g^2_{\omega\rho\pi}} = \frac{g^2_{N\bar{N}\phi}}{g^2_{N\bar{N}\omega}} = \frac{\sigma(\pi N \to \phi X)}{\sigma(\pi N \to \omega X)} = \frac{\sigma(NN \to \phi X)}{\sigma(NN \to \omega X)} = \frac{\sigma(K^- p \to Y\phi)_{back}}{\sigma(K^- p \to Y\omega)_{back}} = \tan^2\theta_\phi$$

(5.36a)

$$\frac{g^2_{f'\pi\pi}}{g^2_{f\pi\pi}} = \frac{g^2_{N\bar{N}f'}}{g^2_{N\bar{N}f}} = \frac{\sigma(\pi N \to f'X)}{\sigma(\pi N \to fX)} = \frac{\sigma(NN\to f'X)}{\sigma(NN\to fX)} = \frac{\sigma(K^-p\to Yf')_{back}}{\sigma(K^-p\to Yf)_{back}} = \tan^2\theta_f \tag{5.36b}$$

where X denotes any single or multiparticle hadron state which contains no strange particles, Y denotes any neutral hyperon or hyperon resonance and the subscript back denotes backward neutral meson production by baryon exchange.

If on the other hand, the violation in a given process comes from a specific higher order transition appropriate for that particular process, the relations (5.36) will not hold. One example of such a higher order transition which has been considered in detail is[30,31]

$$\pi^- p \to K^{*o}\Lambda \to \phi n \tag{5.37a}$$

$$\pi^- p \to K^{*o}\Lambda \to \omega n \ . \tag{5.37b}$$

The unitarity paradox has been formulated for the box diagrams corresponding to these reactions and has been shown to be related by SU(3) symmetry to the reactions

$$K^- p \to \rho^o\Lambda \to \pi^+\Sigma^- \tag{5.38a}$$

$$K^- p \to \rho^o\Lambda \to \pi^-\Sigma^+. \tag{5.38b}$$

The transition (5.37a) is forbidden by the OZI rule; the transition (5.38a) is forbidden in standard models because it involves an exotic exchange. Yet both steps in these second-order processes are allowed and are simply related respectively to the corresponding allowed processes (5.37b) and (5.38b). The relations between corresponding allowed and forbidden processes have been written as a unitarity sum involving the intermediate states in the box diagram

$$\text{Im}\ T_{AB} = \sum_i T_{Ai}\ T_{iB} \tag{5.39}$$

where A, B and i denote initial, final and intermediate states for any of the reactions (5.37) and (5.38). Note that for any intermediate state i, the right-hand sides of Eq. (5.39) are either equal for corresponding allowed and forbidden processes or differ only by a symmetry coefficient. The paradox is resolved by requiring cancelations in the unitarity sum for the forbidden processes, which are seen in quark-line formations to be represented by twisted diagrams analogous to Fig. 5.4.

Without a detailed model for all possible intermediate states, it is impossible to estimate how good the cancellations are. However, SU(3) symmetry has been used to relate the two forbidden processes, so that OZI breaking can be predicted with experimental data on exotic exchange amplitudes used an input.[30] Estimates of OZI violation have also been made by this box-diagram mechanism for low energies where the lowest mass intermediate state can be assumed to be dominant without cancellations from other channels.

Experimental results indicate that in some cases the mixing mechanism giving rise to Eqs. (5.36) are valid, while in other cases the box diagram description (5.37) may hold.[32,33] Further experiments would be of great interest.

Interference experiments have been useful in detecting OZI-violating transitions, since these measure a small amplitude directly, rather than the square. The forbidden f' production amplitude has been observed as interference with the tail of the allowed f amplitude in the reaction[32]

$$\pi^- p \rightarrow M^o n \rightarrow K^+ K^- n \qquad (5.40a)$$

where M^o denotes either the f or f'. Similar f-f' interference effects between the tail of an allowed f peak and a forbidden f' amplitude could be seen in the reactions

$$K^- p \rightarrow M^o \Lambda \rightarrow \pi^+ \pi^- \Lambda \qquad (5.40b)$$

$$K^- p \rightarrow \Lambda M_o \rightarrow \Lambda K^+ K^- \quad (\text{backward}) . \qquad (5.40c)$$

Comparison of the reactions (5.40a) and (5.40b) would be interesting since they are very similar, with the roles of the production and decay of the f' interchanged. The production of the f' is forbidden by the OZI rule in reaction (5.40a) and the decay allowed, while the decay is forbidden and the production allowed in (5.40b). The reaction (5.40c) tests the forbidden baryon vertex, rather than the forbidden meson vertex and would give insight on the relation between OZI violations for the two cases.

Further tests of the OZI rule are now possible[11,18] in the decays of the new particles into final states containing the ϕ and f'. These also enable tests of the generalizations of the OZI rule to multiparticle vertices and can settle questions regarding the crossed Pomeron and multiply forbidden diagrams of Fig. 5.3.

Tests of the OZI rule are more complicated for production of mesons which are not ideally mixed like the pseudoscalars because there is no clear selection rule forbidding the production of any physical particle. The pseudoscalar state analogous to the ϕ or f' whose production is forbidden by the OZI rule in πp reactions is a linear combination of the η and η'. However, the OZI rule for neutral meson production can also be tested by other relations which are not selection rules. Such relations were first derived by Alexander et al.[24] together with the selection rule forbidding ϕ production in πp reactions. An SU(3) rotation of the selection rule forbidding ϕ production with incident pions leads to the observation that an incident K^- which contains no n-type quarks or antiquarks cannot produce the $(n\bar{n})$ vector meson state. Since this state is not a physical meson but a linear combination of the ρ^o and ω states, the selection rule is expressible as an equality between ρ^o and ω production amplitudes, and similarly for the tensor mesons

$$\sigma(K^-p \rightarrow \omega Y) = \sigma(K^-p \rightarrow \rho^o Y), \tag{5.41a}$$

$$\sigma(K^-p \rightarrow fY) = \sigma(K^-p \rightarrow A_2 Y). \tag{5.41b}$$

An additional relation is also obtained from the additive quark model for each case,

$$\sigma(K^-p \rightarrow \phi Y) = \sigma(\pi^-p \rightarrow K^{*o} Y), \tag{5.42a}$$

$$\sigma(K^-p \rightarrow f'Y) = \sigma(\pi^-p \rightarrow K^{*o} Y). \tag{5.42b}$$

In the quark model derivation, these relations follow from the requirement that the meson transition be described as a single quark transition with the other quark being a spectator. All meson transitions in Eqs. (5.42) have the same quark transition $\bar{p} \rightarrow \bar{\lambda}$ and differ only in the quantum numbers of the spectator quark, which is a λ on the left-hand side and an n on the right-hand side. The same relations (5.42) arise in Regge exchange models which assume SU(3) symmetry, no exotic exchanges and the OZI rule.

The relations (5.41) and (5.42) hold for any meson nonet but are not directly applicable to the pseudoscalar mesons because they assume ideal mixing. However, they can be combined to give a sum rule which holds independent of mixing angle and can be applied to the pseudoscalars as well

$$\sigma(K^-p \to \omega Y) + \sigma(K^-p \to \phi Y) = \sigma(K^-p \to \rho^o Y) + \sigma(\pi^-p \to K^{*o}Y), \tag{5.43a}$$

$$\sigma(K^-p \to fY) + \sigma(K^-p \to f'Y) = \sigma(K^-p \to A_2{}^oY) + \sigma(\pi^-p \to K^{*o}Y). \tag{5.43b}$$

$$\sigma(K^-p \to \eta Y) + \sigma(K^-p \to \eta'Y) = \sigma(K^-p \to \pi^oY) + \sigma(\pi^-p \to K^oY). \tag{5.43c}$$

Additional sum rules independent of mixing angle were also obtained by Alexander et al. for pseudoscalar meson production,[24]

$$\sigma(\pi^-p \to \pi^on) + \sigma(\pi^-p \to \eta n) + \sigma(\pi^-p \to \eta'n)$$
$$= \sigma(K^+n \to K^op) + \sigma(K^-p \to \overline{K}^on), \tag{5.44a}$$

$$\sigma(\pi^+p \to \pi^o\Delta^{++}) + \sigma(\pi^++p \to \eta\Delta^{++}) + \sigma(\pi^+p \to \eta'\Delta^{++})$$
$$= 3\sigma(K^-p \to \overline{K}^o\Delta^o) + \sigma(K^+p \to K^o+\Delta^{++}). \tag{5.44b}$$

Analysis of experimental data at 3.9 GeV/c shows striking agreement with experiment for the relations (5.41a) and (5.42a) for fector meson production but strong disagreement with experiment[34] for the sum rule (5.43c) for pseudoscalar meson production where the left-hand side is 705 ± 91 μb and the right-hand side is 1121 ± 59 μb. There are also troubles with the sum rule (5.44a). The validity of the OZI rule for the pseudoscalar mesons thus remains unsettled. There is also the possibility that the conventional nonet classification does not apply to pseudoscalars. One possible explanation for the disagreement of the sum rules with experiment is to assume that the OZI rule holds, that the η is well represented by a state which is mainly a member of the same octet that contains the pion and kaon, but that the η' is a more complicated mixture involving radially excited octet configurations.[35]

Since the selection rule forbidding the production of the $(n\bar{n})$ states leads to equality of amplitudes for ρ^0 and ω production and similarly for f and A2, the equalities (5.41) include the phase detectable in $\rho\omega$ and fA_2 interference experiments and found to agree with the predictions.[19] This interference provides another test of the OZI rule in reactions like (5.41) where the production of the $(n\bar{n})$ state is forbidden; e.g.

$$e^+e^- \to K^+K^-\pi^+\pi^-, \qquad \psi,\psi' \to K^+K^-\pi^+\pi^-, \qquad (5.45a,b)$$

$$(q\bar{q}) \to K^+K^-\pi^+\pi^-, \qquad e^+e^- \to K^+K^-(K\bar{K}), \qquad (5.45c,d)$$

$$\psi,\psi' \to K^+K^-(K\bar{K}), \qquad (q\bar{q}) \to K^+K^-(K\bar{K}), \qquad (5.45e,f)$$

where $(q\bar{q})$ denotes any quark-antiquark meson state, including new particles. In (5.45a-c) the $\pi^+\pi^-$ mass spectrum should show the characteristic "peak-dip" $\rho\omega$-interference pattern, constructive on the low-energy side of the ω peak and destructive on the high-energy side. In the reactions (5.45d-f) the $K\bar{K}$ spectrum in the fA_2 region should show interference constructive in the K^+K^- decay mode and destructive in the $K^0\bar{K}^0$ mode.[11] Since the relative magnitudes and phases of the amplitudes follow from the OZI rule independent of kinematics, data at different energies and from several reactions can be combined to improve statistics.

5.12 Applications of the OZI Rule to the New Particles

The narrow widths of the new particles are attributed to the OZI rule, but there has been no reliable quantitative estimate of these widths from any theoretical model. There have been attempts to estimate the OZI violation for the new particles by using experimental data from the old particles as input. However, the analysis of the $f' \to \pi\pi$ decay in section 5.10 shows that this is unjustified. There is no simple way to apply these results to new particles which are states of new heavy quark-antiquark pairs. For these states the channel analogous to the $K\bar{K}$ channel involves pairs of mesons each containing one new heavy and one ordinary light quark. These channels are closed for the decays of the lowest-lying new meson states made of heavy quark-antiquark pairs. Thus, there is no possible higher order decay to hadron states made of light quarks via an intermediate state on the mass shell. The higher-order paradox arises only in virtual transitions to states of pairs of mesons carrying charm or some new quantum number and which must be off their mass shell. The calculation of these effects requires some underlying field theory as well as a knowledge of the spectrum and of the couplings of all these new particles in order to determine the effectiveness of various cancellation mechanisms. There is no simple estimate analogous to the one above for the $f' \rightarrow \pi\pi$ decay which is dominated by the on shell transitions. There is no way to extrapolate the observed violations in the f' and ϕ decays to these other states by assuming a dependence on masses, since the effect observed in the ϕ and f' cases can be entirely attributed to open channels on shell which do not exist for the new particles. Such mass extrapolations could give upper limits for OZI violation under the reasonable assumption that the effects of the off shell transitions must be smaller than the observed violations in the f' and ϕ cases. However, such extrapolations are very risky since the contributions of these off shell transitions are probably very sensitive to subtle cancellation mechanisms which may be very different for the old and the new particles.

Thus the experimental observation that the OZI rule must be much better for the new particles than for the old particles leads to no contradictions. But there is also no simple way to estimate the widths of the new particles from theoretical models for OZI breaking.

Another application of the OZI rule to the new particles has been in estimates of charmed particle production in experiments searching for these particles. It has been suggested that charmed particles might be found in associated production with the J/ψ because the production of the J/ψ without charmed particles violates the OZI rule.[36] This argument is incorrect, as can be seen from the analogous case of the ϕ and strangeness. Kaons are not frequently found in associated production with the ϕ, even though the production of the ϕ without accompanying kaons in pp collisions is forbidden by the OZI rule. That ϕ production is dominated by OZI violating mechanisms is still consistent with the OZI rule.[19]

The OZI rule can be tested only by comparing corresponding pairs of processes as in Eq. (5.36). Other comparisons are misleading, such as comparing ϕ production with and without kaon pairs in NN reactions. A proper comparison of these processes involves other dynamical considerations. This can be seen by writing

$$\frac{\sigma(\mathrm{NN}) \rightarrow \phi + K + \bar{K} + X)}{\sigma(\mathrm{NN} \rightarrow \phi + X)} = \frac{K \cdot S}{Z}, \tag{5.46}$$

where X contains no strange particles and the parameters K, S and Z are defined as

$$Z \equiv \frac{\sigma(NN \to \phi + X)}{\sigma(NN \to \omega + X)} , \qquad (5.47a)$$

$$S \equiv \frac{\sigma(NN \to \omega + K + \bar{K} + X)}{\sigma(NN \to \omega + X)} , \qquad (5.47b)$$

$$K \equiv \frac{\sigma(NN \to \phi + K + \bar{K} + X)}{\sigma(NN \to \omega + K + \bar{K} + X)} . \qquad (5.47c)$$

The parameter Z is the ratio of a corresponding pair of OZI-violating and OZI-conserving processes and is small in any model which suppresses OZI-violating transitions. The parameter S is also small because experiment shows that it is hard to produce kaon pairs. The quantity K is of the order unity since it relates two processes allowed by the OZI rule and differing only by the interchange of two members of the same vector nonet. The value of the ratio (5.40) is thus not determined by the OZI rule and requires additional dynamical input. It depends upon which of the two small quantities S and Z, is smaller, i.e. whether it is harder to violate the OZI rule or to produce a pair of strange particles. The available data indicate that the strange-particle-production factor S overwhelms the OZI-violation factor Z and that ϕ production is dominated by the OZI-violating transition without kaons[37] up to energies of at least 24 GeV.

A similar argument holds for the production of J/ψ particles with and without pairs of charmed particles in a charmonium model. By analogy with Eqs. (5.46) and (5.47)

$$\frac{\sigma(NN \to J/\psi + D + \bar{D} + X)}{\sigma(NN \to J/\psi + X)} = \frac{K_c \cdot Z_c}{C} , \qquad (5.48)$$

where X can now contain strange particles but no charmed particles and the parameters $K_c Z_c$ and C are defined by

$$Z_c \equiv \frac{\sigma(NN \to J/\psi + X)}{\sigma(NN \to \omega + X)} , \qquad (5.49a)$$

$$C \equiv \frac{\sigma(NN \to \omega + D + \bar{D} + X)}{\sigma(NN \to \omega + X)} , \qquad (5.49b)$$

$$K_c \equiv \frac{\sigma(NN \to J/\psi + D + \bar{D} + X)}{\sigma(NN \to \omega + D + \bar{D} + X)} \; . \tag{5.49c}$$

Again the ratio (5.48) depends on which is the smaller of two small quantities Z_c and C; i.e. whether it is harder to violate the OZI rule or to produce a pair of charmed particles. In addition the factor K_c which is of order unity in the SU(4) symmetry is probably also small. Present experiments suggest here also that charmed particle production is suppressed more than OZI violating transitions and that the J/ψ is produced primarily without charmed particles. This can also be seen in a model like Eq. (5.36) with Z_c a universal factor describing the suppression of all OZI-violating transitions involving a charmed quark-antiquark pair. A value of about 10^{-3}—10^{-4} for Z_c is obtained by comparing the width of the OZI-violating decay of the J into normal hadrons with expected widths of about 100 MeV for OZI-conserving decays. The observed total cross sections for J/ψ and ω production are consistent with this value of Z_c and Eq. (5.49a).

If the OZI rule were exact and all OZI-violating processes had zero cross section, all production of particles like the ϕ and the J/ψ would be via OZI-conserving reactions. However, once the OZI rule is broken the OZI-violating transitions are proportional to small but finite suppression factors as in Eqs. (5.47a) and (5.49a). Whether a given production process is dominated by OZI-conserving or OZI-violating transitions depends upon whether the OZI-conserving transition is suppressed more strongly than the OZI-suppression factor by a different dynamical mechanism, as in the reactions (5.46) and (5.48).

VI. WHY ARE NARROW CONTINUUM STATES INTERESTING?

The new particles promise new exciting opportunities for interesting research. To understand why they are so interesting it is instructive to compare them with the isobaric analog states. There are many common features. The isobaric-analog states were understood very shortly after their discovery, but they are still very interesting and have opened a new field in nuclear physics. For the same reason the new particles will still be very interesting even after we understand their structure. There is much discussion about these new particles and whether they should be called ψ or J. I see that Italians know all about the Psi (Italian socialist party).

Narrow states in the continuum are interesting because they combine the best features of the ordinary low-lying states and the ordinary continuum states. In any non-trivial spectroscopy there are models which are not exact but useful because they are reasonably good for the low-lying states where configurations are simple. At higher excitation the models get worse; the configurations are much more complicated, the density of states is higher, and there is much more mixing. The interactions neglected in the models become more unimportant at higher excitation, where the level density becomes greater and there are many configurations close together and easily mixed. Thus there is no simple theoretical model for the wave functions high in the continuum. On the other hand, high continuum states are very convenient for experiment because they have many open channels, can be excited in many ways, and provide a very rich source of experimental data. The isobaric analog states give the best features of both worlds. They are up in the continuum and have many possible open channels, many ways of excitation, and many things to study. But they also have a very simple structure very much like the low-lying states and simple theoretical models can be used. Any model good for the low-lying states is also just about as good for the narrow states in the continuum. This then opens a wide field for experimental tests and investigations of the various models.

How do we know that narrow states in the continuum have a simple structure? We know the structure of the isobaric analog resonances and can answer this question in detail. But the answer is similar for the new particles even though we do not understand their structure. In both cases we know that the structure of the states must be simple because they are simply produced. The isobaric analog state is strongly produced by a nuclear charge-exchange reaction at zero momentum transfer on a nuclear ground state; i.e. by changing a neutron into a proton without changing its momentum. Any state strongly produced by this very simple operation must differ from the ground state only by a simple elementary excitation and cannot be a state with twelve particles excited.

The new particles are made very strongly by electron-positron annihilation through one virtual photon. The transition matrix element connects the particle state with a vacuum by the operator of the electromagnetic current and is appreciable only if the particle has a very simple structure. For example, in simple quark or parton models, the electromagnetic current operator creates only single quark-antiquark pair. This tells us that the single pair part of the wave function must be appreciable in any strongly produced state. If it were mainly a state of thirty quarks and thirty antiquarks it would not have a very large matrix element for a single photon transition from the vacuum.

VII. WHAT ARE STRANGENESS AND BARYON NUMBER?

The new charm degree of freedom[2] provides a new quantum number like electric charge, strangeness and baryon number. But understanding charm is difficult when we still do not understand the old internal degrees of freedom.[38] We have some understanding of the role of electric charge in particle interactions and dynamics even though we do not understand why electric charge is quantized and universal. But our understanding of baryon number and strangeness is much weaker. There is no theory like quantum electrodynamics in which baryon number or strangeness appear as coupling constants defining the strengths of interactions. There is no formula analogous to the Rutherford formula for Coulomb scattering describing the dependence of strong interaction scattering on baryon number and strangeness.

A few phenomenological models and symmetries like the quark model and SU(3) symmetry give rough descriptions of the dependence of total cross sections on baryon number and strangeness. But these descriptions are highly inadequate and the difference between mesons and baryons and between strange and non-strange hadrons are not really understood. Furthermore, many of the models developed work in only one area of hadron physics and are incompatible with models used in other areas. For example, the quark model used in describing hadrons strong interactions is not the same as the quark model used in weak interactions.

Consider, for example, the description by conventional models of the difference between pion and kaon wave functions. The quark model says that both are made from a quark-antiquark pair.[14] But weak interaction quarkists explain the ratio of the $\pi \to \mu + \nu$ and $K \to \mu + \nu$ decay requiring the wave functions at the origin to be very different as described by Weisskopf-Van-Royen[39] formula

$$\frac{|\psi_K(0)|^2}{|\psi_\pi(0)|^2} = \frac{M_K}{M_\pi} . \tag{7.1}$$

Strong interaction quarkists say that the difference between pion and kaon wave functions is measured by the difference between their scattering cross sections on nucleons. These differ by less than 20%. Recent data at high energies show that πp and Kp differential cross sections approach equality with increasing momentum transfer. This suggests equality within 20% of the mean square radii of pion and kaon wave functions and nearly identical short distance behavior, in sharp contrast with the weak quarkist Eq. (7.1) .

The very precise experimental data[40] now available on pion, kaon and nucleon total cross sections give us some information about the difference between the interactions of strange and nonstrange particles with matter. Careful examination of the data show that this difference is very interesting but also very puzzling and not really understood.[41] Instead of the conventional plot of total cross sections versus laboratory momentum on a logarithmic scale, we show the systematics in a more interesting plot (Fig. 7.1) with a square root scale rather than a logarithmic scale for P_{lab} and with the total cross section multiplied by $\sqrt{P_{lab}}$. This is equivalent at these high energies to a plot against center-of-mass momentum of the imaginary part of the forward amplitude obtained from the total cross section by the optical theorem. Theoretical reasons why the curve of 7.1 is so much simpler than the standard plot follow from a two-component description of the cross sections with a Regge component varying as $s^{-1/2}$ and a pomeron component varying slowly as a function of energy. A more detailed discussion

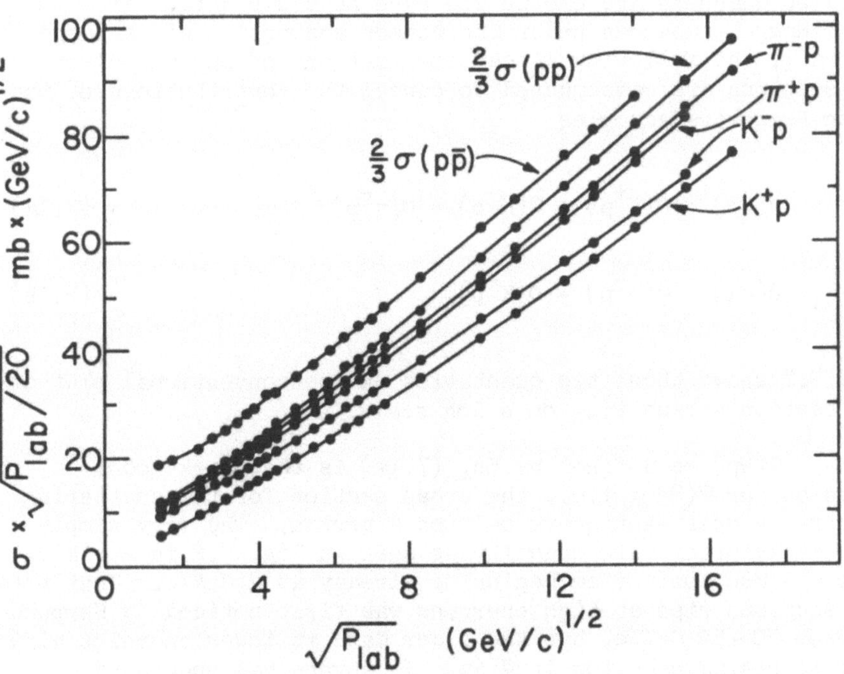

Fig. 7.1. $\sigma_{tot}\sqrt{P/20}$ vs. $\sqrt{P}$. Nucleon cross sections multiplied by 2/3.

is given elsewhere.[42] For our purposes this particular plot shows very clearly that there is a difference between strange and non-strange particles and that there are puzzles not explained by the quark model.

In Fig. 7.1 the nucleon-nucleon and nucleon-antinucleon cross sections are multiplied by a factor 2/3. The six quantities plotted are just those predicted to be equal asymptotically in the simple quark model with the pomeron component an SU(3) singlet coupled equally to pions and kaons and coupled to mesons and baryons by simple quark counting prescriptions. Figure 7.1 shows that these cross sections are indeed all equal at the 20% level. However, beyond this approximation of "seen one hadron, seen them all" the difference between the πp and the pp cross sections is seen to be strangely similar to the difference between the πp and Kp cross sections. The difference between mesons and baryons seems to be similar to the difference between nonstrange and strange mesons.

This regularity is shown more precisely by examining linear combinations of cross sections which have no Regge component and are therefore conventionally assumed to be pure pomeron. The K^+p and pp channels are exotic and have no contribution from the leading Regge exchanges under the common assumption of exchange degeneracy. The following linear combinations of meson-nucleon cross sections are constructed to cancel the contributions of the leading Regge trajectories

$$\sigma(\phi p) = \sigma(K^+p) + \sigma(K^-p) - \sigma(\pi^-p) \tag{7.1a}$$

$$\Delta(\pi K) = \sigma(\pi^-p) - \sigma(K^-p). \tag{7.1b}$$

Figure 7.2 shows these two quantities on the conventional plot of cross section versus P_{lab} on a log scale.

$\sigma(\phi p)$ as defined by Eq. (7.1a) is the quark model expression for $\sigma(\phi p)$; i.e., the cross section for the scattering of a strange quark-antiquark pair on a proton. The very simple energy behavior of this quantity as seen in Fig. 7.2 is striking. It shows a monotonic rise beginning already at 2 GeV/c. That total cross sections rise at high energies was first noticed by Serpukhov data from 20—50 GeV/c, but the older data at lower energies already show this rising behavior in $\sigma(\phi p)$. If anyone has suggested something particularly fundamental about this cross section for strange quarks on a nucleon before the Serpukhov data were available and concluded that its rising cross section indicated that all cross sections would eventually rise he would naturally

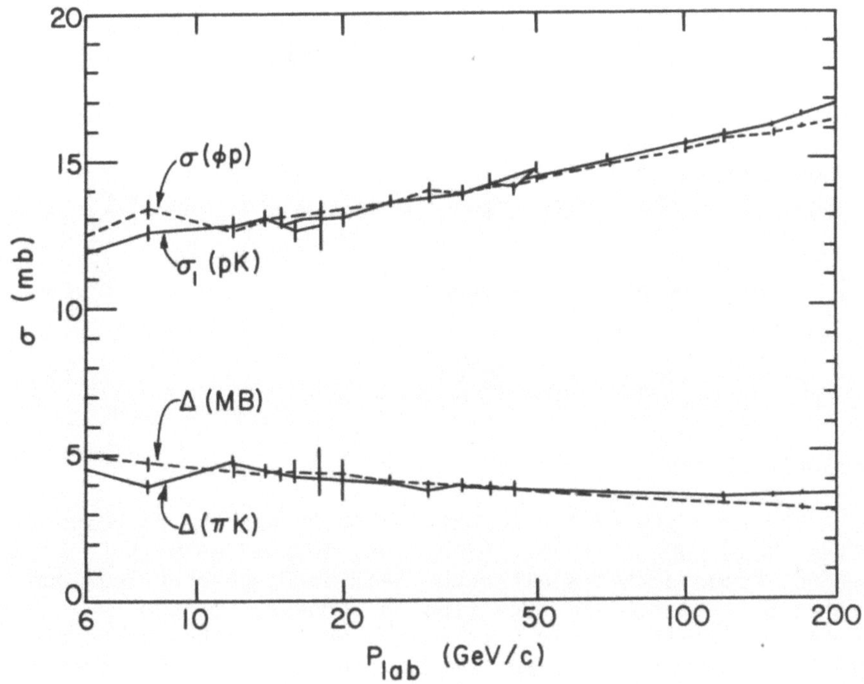

Fig. 7.2. Plots of Eqs. (7.1) and (7.2).

have been disregarded as crazy. But now that the whole picture up to 200 GeV/c is available we may conclude that there is indeed something simpler and more fundamental about the cross sections for strange quarks on a proton target. Understanding this simpler behavior may help us to understand the more complicated energy behavior of the other cross sections.

The quantity $\Delta(\pi K)$ defined by Eq. (7.1b) represents the difference in the scattering of a strange particle and a nonstrange particle on a proton target. In the quark model this is the difference between the scattering of a strange quark and a nonstrange quark on a proton target after the leading Regge contributions have been removed. This difference between strange and nonstrange also has a very simple energy behavior, decreasing constantly and very slowly (less than a factor of 2 over a range P_{lab} of two orders of magnitude). So far there is no good explanation for why strange and nonstrange mesons behave differently in just this way.

Since the two quantities (7.1) have no contribution from the leading Regge trajectories they represent something loosely called the pomeron. However, their energy behaviors are different from one another and also from that of the quantities $\sigma(K^+p)$ and $\sigma(pp)$ which should also be "pure pomeron." However the following linear combinations of $\sigma(K^+p)$ and $\sigma(pp)$ have exactly the same energy behavior as the meson-baryon linear combinations (7.1)

$$\sigma_1(pK) = \frac{3}{2}\,\sigma(K^+p) - \frac{1}{3}\,\sigma(pp) \tag{7.2a}$$

$$\Delta(MB) = \frac{1}{3}\,\sigma(pp) - \frac{1}{2}\,\sigma(K^+p). \tag{7.2b}$$

These quantities are also plotted in Fig. 7.2.

The equality of the quantities (7.2) and the corresponding quantities (7.1) suggest that the pomeron, defined as what is left in the total cross sections after the leading Regge contributions are removed by the standard prescription, consists of two components, one rising slowly with energy and the other decreasing slowly. The coefficients in Eq. (7.2) were not picked arbitrarily but were chosen by a particular model. In this model the rising component of the total cross section is assumed to satisfy the standard quark model recipe exactly.

$$\sigma_R(Kp) = \sigma_R(\pi p) = \frac{2}{3}\,\sigma_R(pp) = \frac{2}{3}\,\sigma_R(Yp) = \frac{2}{3}\,\sigma_R(\Xi p), \tag{7.3a}$$

where Y denotes a Λ or Σ hyperon. The falling component has been assumed to satisfy the following relation

$$\sigma_F(Kp) = \frac{1}{2}\,\sigma_F(\pi p) = \frac{2}{9}\,\sigma_F(pp) = \frac{1}{3}\,\sigma_F(Yp) = \frac{2}{3}\,\sigma_F(\Xi p). \tag{7.3b}$$

This particular behavior is suggested by a model in which the correction to a simple quark-counting recipe comes from a double exchange diagram involving a pomeron and an f coupled to the incident particle.[41]

We thus see unresolved problems in the total cross-section data associated with the questions of what is the difference between strange and nonstrange particles and what is the nature of the pomeron. Note that Eq. (3.1b) defines the difference between the scattering of a nonstrange quark and a strange quark while Eq. (7.2b) can be interpreted as the difference between the scattering of a quark in a baryon and a quark in a meson. The fact that the strange-nonstrange difference and the meson-baryon difference are equal and have the same energy behavior over such a wide range is a puzzle which may be explained by pomeron-f double exchange but may also indicate something deeper.

The cross section differences $\Delta(\pi K)$ and $\Delta(MB)$ are both predicted to vanish in the simple model where the pomeron is an SU(3) singlet which is coupled to the quark number and all the curves of Fig. 7.1 are equal. The deviation from the additive quark model ratio of 2/3 for meson to baryon scattering appears as a finite value of $\Delta(MB)$, the deviation of the pomeron coupling from an SU(3) singlet appears as a finite value of $\Delta(\pi K)$. One might ask whether both these decreasing quantities approach zero at high energies, so that the simple model would be valid in asymptopia.

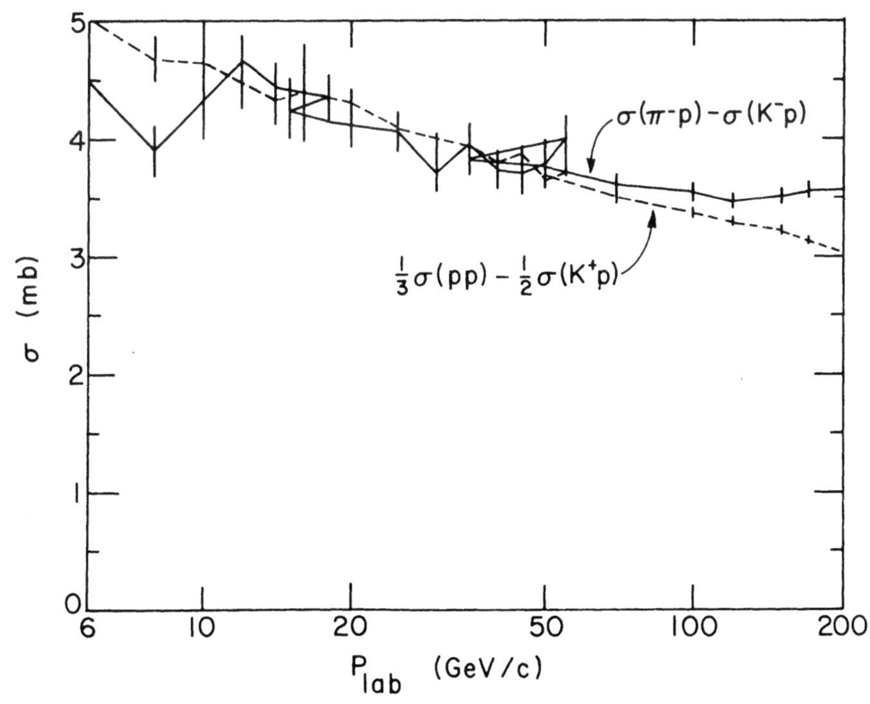

Fig. 7.3. Plots of $\Delta(\pi K)$ and $\Delta(MB)$ on an expanded scale.

A close look at the experimental plots on an expanded scale of $\Delta(\pi K)$ and $\Delta(MB)$ in Fig. 7.3 reveals a small difference in the behavior at the highest energies. The curve for $\Delta(\pi K)$ seems to be leveling off above 50 GeV/c, while that for $\Delta(MB)$ continues decreasing monotonically. This trend seems to continue in the one or two additional points available up to 280 GeV/c which are not plotted. Additional data up to 400 GeV/c should determine whether $\Delta(\pi K)$ has definitely stopped decreasing and is approaching a constant, while $\Delta(MB)$ is decreasing. If this is the case, then the additive quark model becomes good at high energies while pions and kaons continue to look different even at asymptopia and the SU(3) relation never becomes good. The equality observed between these two differences over the 6–200 GeV/c range and described by a two-component Pomeron ultimately breaks down at higher energies.

A search for similar systematics in elastic hadron scattering differential cross section data has led to new surprises and paradoxes.[38] With only differential cross section data available and no detailed amplitude analysis, it is convenient to define the quantity[41,43]

$$S(Hp) = [\frac{d\sigma}{dt}(\bar{H}p) + \frac{d\sigma}{dt}(Hp)]^{1/2} \qquad (7.4)$$

where H is any hadron. This quantity $S(Hp)$ is assumed to give a good approximation for the Pomeron contribution to the Hp scattering amplitude. With this assumption the simple additive quark model prediction that $\Delta(MB) = 0$ becomes

$$S(\pi p) = (2/3)S(pp) \qquad (7.5a)$$

when we use $S(\pi p)$ to represent a typical meson baryon cross section. The assumption that the Pomeron is a SU(3) singlet predicts $\Delta(\pi K) = 0$ and

$$S(\pi p) = S(Kp). \qquad (7.5b)$$

The two relations (7.5a) and (7.5b) describe the dependence of the scattering amplitude on baryon number and strangeness, respectively. The two component Pomeron model which relates the deviations from the two predictions (7.5a) and (7.5b) predicts the weaker sum rule

$$S(\pi p) = \frac{1}{2} S(Kp) + \frac{1}{3} S(pp). \qquad (7.6)$$

The experimental data[44] show that the weaker sum rule (7.6) is in much better agreement with experiment than the additive quark model prediction (7.5a). However, the SU(3) prediction which is not very good at t = 0 becomes better at larger values of t and becomes much better than the two component Pomeron prediction (7.6) or the additive quark model prediction (7.5a). Two examples of this comparison with experiment are given in Table 7.1. The same qualitative features are present in all the data.

Table 7.1

Tests of Additive Quark Model (AQM), Two-Component Pomeron (P2) and SU(3) Relations Between Differential Cross Sections. RHS/LHS of Eqs. (7.5a), (7.5b) and (7.6).

P = 100 GeV/c				P = 175 GeV/c			
t (GeV/c)	AQM (7.5a)	P2 (7.6)	SU(3) (7.5b)	t (GeV/c)	AQM (7.5a)	P2 (7.6)	SU(3) (7.5b)
0.0	1.2	1.0	0.84	0.0	1.1	0.97	0.84
-0.08	1.0	0.95	0.86	-0.08	0.98	0.92	0.85
-0.16	0.94	0.91	0.88	-0.16	0.89	0.88	0.86
-0.24	0.85	0.87	0.90	-0.24	0.81	0.84	0.88
-0.32	0.78	0.85	0.92	-0.32	0.74	0.81	0.89
-0.40	0.71	0.83	0.94	-0.40	C.68	0.79	0.90
-0.48	0.66	0.81	0.97	-0.48	0.63	0.77	0.92
-0.56	0.61	0.80	1.0	-0.56	0.58	0.76	0.93
-0.64	0.56	0.80	1.0	-0.64	0.54	0.74	0.95
-0.72	0.53	0.80	1.1	-0.72	0.50	0.73	0.96
-0.80	0.50	0.80	1.1	-0.80	0.47	0.72	0.98

The comparison with experiment of relations (7.5a) and (7.6) does not really add any new qualitative information. It is summed up by the observation that at the optical point the relation (7.5a) is not very good and the relation (7.6) is much better and that baryon-baryon cross sections decrease much more rapidly with t than meson-baryon cross sections. The behavior at the optical point is expected from the similar behavior of total cross sections. The high t behavior is expected since naive additive quark model predictions (7.5a) and (7.6) neglect differences between meson and baryon wave functions. These differences introduce additional form factors into the scattering amplitudes, which cause baryon amplitudes to decrease more rapidly with increasing t than meson amplitudes.

However, the improvement of the relation (7.5b) with increasing t comes as a complete surprise. One can ask why pions and kaons should look more alike[45] at high t than at low t. One might also ask whether the two are really approaching equality or whether there will be a cross over and that still at higher t the amplitude will differ in the opposite direction.

We thus seem to see a peculiar systematics in which the additive quark model becomes good at t = 0 and high s but not at high t, the SU(3)-symmetric pomeron becomes good at high t, but not at t = 0, even at high s, and the two-component pomeron description holds at t = 0 and s between 6 and 200 GeV/c, where there are discrepancies in both the additive quark model and the SU(3)-symmetric pomeron. Further data on total cross sections at higher energy, differential cross sections at higher momentum transfer, and hyperon total and differential cross sections everywhere will show whether these puzzling features are really in the data, and will provide clues for our understanding of the differences between strange and nonstrange particles and between mesons and baryons.

REFERENCES

[1] J. D. Anderson, C. Wong and J. W. McClure, Phys. Rev. 126 (1962) 2170; A. K Kerman, in "Nuclear Isospin" (J. D. Anderson, S. D. Bloom, J. Cerny and W. W. True, Eds.), p. 315, Academic Press, New York (1969).

[2] For a review of charm, see M. K. Gaillard, B. W. Lee and J. Rosner, Rev. Mod. Phys. 47 (1975) 277.

[3] S. Okubo, Phys. Letters 5 (1963) 165.

[4] G. Zweig, unpublished, 1964; and in Symmetries in Elementary Particle Physics (Academic Press, New York, 1965) p. 192.

[5] J. Iizuka, Supplement to Progress of Theoretical Phys. 37-38 (1966) 21.

[6] B. Sakita, Phys. Rev. 136 (1964) B1756; F. Gursey and L. A. Radicati, Phys. Rev. Letters 13 (1964) 173; A. Pais, Phys. Rev. Letters 13 (1964) 175.

[7] J. L. Rosner, in Proc. XVII Intern. Conf. on High Energy Physics, London (1974), ed. J. R. Smith, p. II-171, Physics Reports 11C (1974) 190.

[8]A. Arima and F. Iachello, Phys. Rev. Letters 35 (1975) 1069.

[9]N. Cabibbo, in Proc. of Intern. Conf. on High Energy Physics, Palermo, Italy, 1975.

[10]H. J. Lipkin, Symmetries and Resonances (Theoretical), in Particle Physics (Irvine Conference, 1971), edited by M. Bander, G.L. Shaw and D.Y. Wong (American Institute of Physics, New York, 1972), p. 30.

[11]H. J. Lipkin, Spectroscopy after the New Particles, in Proc. Intern. Conf. on High Energy Physics, Palermo, Italy, 1975.

[12]Y. Ne'eman, in Spectroscopic and Group Theoretical Methods in Physics, edited by F. Bloch et al., Wiley, New York (1968), p. 337.

[13]G. Racah, Group Theory and Spectroscopy, Ergebnisse der Naturwissenschaffen, 37 (1965) 28.

[14]H. J. Lipkin, Physics Reports 8C (1973) 173.

[15]S. J. Brodsky, Proc. of the 16th Int. Conf. on High Energy Physics, National Accelerator Laboratory, Batavia, Ill. (1972).

[16]J. Ellis, in Proc. XVII Intern. Conf. on High Energy Physics, London (1974) ed. J. R. Smith, p. II-171.

[17]O. W. Greenberg and C. A. Nelson, Color Models for Hadrons, to be published in Physics Reports.

[18]S. Okubo, Phys. Rev. D 13 (1976) 1994 and D 14 (1976) 108.

[19]H. J. Lipkin, Physics Letters 60B (1976) 371.

[20]H. J. Lipkin, Who Understands the Zweig-Iizuka Rule, to be published in the Proceedings of the 1975 Rencontre de Moriond.

[21]H. J. Lipkin, Phys. Rev. Letters 31 (1973) 656.

[22]H. J. Lipkin, Phys. Letters 9 (1964) 203; H. J. Lipkin and S. Meshkov, Phys. Rev. 143 (1966) 1269.

[23]A. Katz and H. J. Lipkin, Physics Letters 7 (1963) 44.

[24]G. Alexander, H. J. Lipkin and F. Scheck, Phys. Rev. Lett. 17 (1966) 412.

[25]E. M. Levin and L. L. Frankfurt, Zh. Eksp. i. Theor. Fiz-Pis'ma Redakt 2 (1965) 105 [JETP Lett. 2 (1965) 65].

[26]F. Gilman, M. Kugler and S. Meshkov, Phys. Rev. D9 (1974) 715; Phys. Letters 45B (1973) 481.

[27]B. Richter, private communication.

[28]H. J. Lipkin and S. Meshkov, Phys. Rev. Letters 14 (1965) 670.

[29]C. B. Chiu and J. Finkelstein, Phys. Letters 27B (1968) 576.

[30]C. Schmid, D. M. Webber and C. Sorensen, Nucl. Phys. B111 (1976) 317.

[31]E. L. Berger and C.Sorensen, Phys. Letters 62B (1976) 303.

[32]A. J. Pawlicki et al., Phys. Rev. Lett. 37 (1976) 971.

[33]D. Cohen et al., Argonne Preprint ANL-HEP-PR-76-64.

[34]J. Mott et al., Phys. Rev. Lett. 18 (1967) 355; M. Anguilar-Benitez et al., Phys. Rev. Letters 28 (1972) 574; Phys. Rev. D 6 (1972) 29.

[35]H. J. Lipkin, Fermilab Preprint-FERMILAB-Pub-76/94-THY.

[36]M. Binckley et al., Phys. Rev. Lett. 37 (1976) 58.

[37]V. Blobel et al., Phys. Letters 59B (1975) 88; H. Harari, Theoretical Implications of the New Particles, rapporteur talk at the Intern. Symp. on Lepton and Photon Interactions at High Energies, Stanford, California, August, 1975.

[38]H. J. Lipkin, "Why are Hyperons Interesting and Different from Nonstrange Baryons?", Particles and Fields 1975, Edited by H. J. Lubatti and P. M Mockett, University of Washington (Seattle) p. 352.

[39]R. Van Royen and V. F. Weisskopf, Nuovo Cimento 50A (1967) 617.

[40]E. Flaminio, J. D. Hansen, D. R. O. Morrison and N. Tovey, Compilation of Cross Sections. I. Proton Induced Reactions. II. Antiproton Induced Reactions, CERN/HERA 70-2 and 70-3; E. Bracci, J. P. Droulez, E. Flaminio, J. D. Hansen and D. R. O. Morrison, Compilation of Cross Sections. I. π^- and π^+ Induced Reactions. II. K^- and K^+ Induced Reactions, CERN/HERA 72-1 and 72-3; J B. Allaby, Yu. B. Bushnin, Yu. P. Gorin et al., Yad. Fiz. 12 (1970) 538; Phys. Lett. 30B (1969) 500; Yu. P. Gorin S. P. Denisov, S. V. Donskov et al., Yad. Fiz. 14 (1971); Phys. Lett. 36B (1971) 415; S. P. Denisov, S. V. Donskov, Yu. P. Gorin et al., Nucl. Phys. B65 (1973) 1; A. S. Carroll et al., Phys. Rev. Lett. 33 (1974) 927 and FERMILAB-Pub-75151-EXP.

[41]H. J. Lipkin, Nucl. Phys. B78 (1974) 381; Phys. Lett 56B (1975) 76; Phys. Rev. D 11 (1975) 1827.

[42]H. J. Lipkin, "What Are Total Cross Section Data Trying to Tell Us?," Proc. Int. Conf. on High Henergy Physics, Palermo, Italy, 23-28 June 1975.

[43]H. Harari, Ann. Phys. 63 (1971) 432; I. Ambats et al., Phys. Rev. D 9 (1974) 1179; V. Barger, K. Geer and F. Halzen, Nucl. Phys. B44 (1972) 475.

[44]C. W. Akerlof et al., Phys. Rev. Lett. 12 (1975) 631; Fermilab Single Arm Spectrometer Group, Phys. Rev. Lett. 35 (1975) 1195; G. Brandenburg et al., SLAC-PUB-1607 (1975). The data in Table I are taken from the Fermilab Single Arm Spectrometer group,

but data from all three experiments have been used to test relations (7.5) and (7.6) with similar results. The help of all three groups in making data available before publication and computing preliminary tests of relations (7.5) and (7.6) is gratefully acknowledged, and in particular correspondence and discussions with D.D. Yovanovitch, R. Diebold, D. Leith and J. Mikenberg.

[45] G. F. Chew and C. Rosenzweig, Nucl. Phys. B104 (1976) 290, propose a model in which SU(3) symmetry improves with increasing t.

DISCUSSIONS

CHAIRMAN: Prof. H. Lipkin

Scientific Secretary: M. Pauli

DISCUSSION 1

- *FALCIONI:*

Do you not think that deep inelastic scattering and neutrino experiments may support the idea of realistic spin-$\frac{1}{2}$ quarks?

- *LIPKIN:*

The behaviour to which you refer is limited to a definite low region of energy. We know nothing about the scaling at high energy. There is no indication that quarks are elementary and are not composite objects of still more fundamental particles.

- *FREEDMAN:*

When $\phi \to \rho\pi$ via $\phi \to K^+K^- \to \rho\pi$ and $\phi \to K^0\bar{K}^0 \to \rho\pi$, the strange quarks in the K's must annihilate to go into $\rho\pi$. Does this part of the decay not become suppressed by Zweig's rule?

- *LIPKIN:*

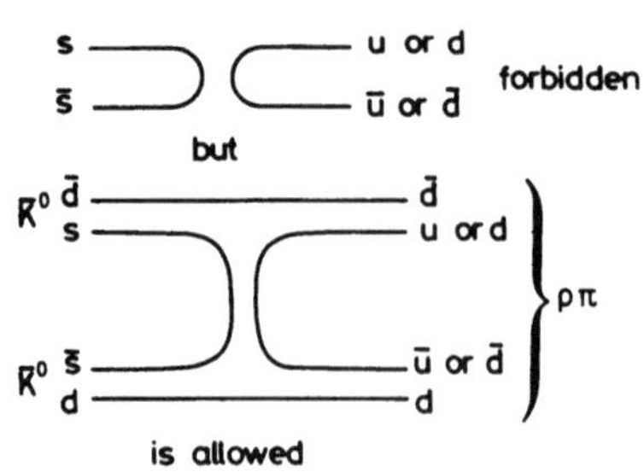

It is called a "duality diagram".
You can compose this into 3 point functions

No. The ZI rule does not suppress the annihilation of strange quarks from two different hadrons such as a kaon pair. Only the annihilation of a quark-antiquark pair from the same meson is suppressed.

- *MARCIANO:*

What does the SU(6) you discussed have to do with the SU(6) traditionally used in particle physics?

- *LIPKIN:*

Nothing. I thought the nuclear physicists can build beautiful models out of six building blocks -- yet no one takes these building blocks seriously. There may be a moral in that.

- *WICK:*

Is there a sharp formulation of the ZI rule that applies in general, rather than specific examples?

- *LIPKIN:*

The formulation of the ZI rule for three point functions is very sharp and clear. The dual resonance model provides a framework for extending the rule to n-point functions. One way to formulate the ZI rule for a general n-point function without drawing diagrams is the following: if the external particles can be divided into two groups such that there is no flavour of quark which appears in both groups, then the vertex is forbidden by the ZI rule. In the three-flavour model -- no charm -- this reduces to the following statement: if the external particles can be divided into two non-trivial groups such that one group contains only strange quarks and the other contains only non-strange quarks, the vertex is forbidden. It is easy to formulate this in the Born approximation -- but it is harder to formulate the ZI rule for higher order terms. Since these are strong interactions, we need to do this.

- *WICK:*

Could you give us an example?

- *LIPKIN:*

The simplest example is $\phi \to K\bar{K} \to \rho\pi$ because you know the $\phi, K, \bar{K}$ coupling. You could make this $K\bar{K} \to \omega$ and $\omega \to \rho\pi$. $\phi \to K\bar{K}$ and $\omega \to K\bar{K}$ are related by SU(3). If you know how to calculate this loop, then you can calculate the relationship between $\phi \to \rho\pi$ and $\omega \to \rho\pi$. There is no reason for this to be small unless there is another diagram

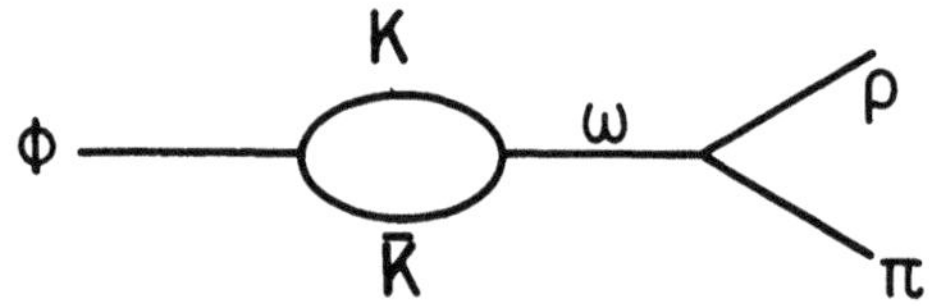

$\phi \to K\bar{K} \to \omega \to \rho\pi$
showing loop diagram

that cancels this one. There are a number of possibilities and this will be discussed in detail in my next talk.

- *FREEDMAN:*

Are there any other cases similar to that of $\phi \to K\bar{K} \to \rho\pi$ where the ZI rule is violated? Specifically, I mean where the two quarks annihilating from different hadrons in the intermediate state originated from the same initial hadron and are not a G-parity forbidden decay.

- *LIPKIN:*

There is the $f' \to K\bar{K} \to \pi\pi$ decay, and perhaps also $\psi \to D\bar{D} \to$ pions.

- *PAULI:*

How do you put G parity into an interacting theory of quarks and gluons so as to obtain the selection rules implied by G parity? Could the notions of G parity be generalized in such a way to give a dynamical explanation of Zweig's rule?

- *LIPKIN:*

G parity is a combination of isospin and charge conjugation. Any theory of quarks and gluons which is invariant under isospin and charge conjugation transformation automatically conserves G parity. There is no transformation or symmetry which can be put into a theory to give the ZI rule. The ZI rule is not formulated as a conservation law.

- *McPHERSON:*

Is there any reason why the non-planar diagrams must be cancelled in the same order? Could it not be cancelled by the same conspiracy of higher order diagrams?

- *LIPKIN:*

Not obviously. But all known formulations which provide such cancellations, in particular, dual resonance models with exchange degeneracy, have cancellations in the same order.

DISCUSSION 2

- *McPHERSON:*

In the discussion of the $KK^* \to KK^*$ Zweig rule situation as to the pseudoscalar intermediate exchanged state, how does the $\lambda\bar{\lambda}$ state behave?

- *LIPKIN:*

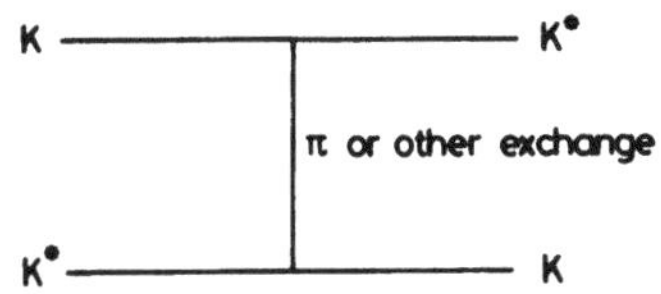

The $\lambda\bar{\lambda}$ state is always split from the $p\bar{p}$ and $n\bar{n}$ states by SU(3) symmetry breaking. Thus, $\lambda\bar{\lambda}$ exchange is always distinguishable from the other exchanges and does not affect the argument. There is no degenerate partner of the pion to cancel ZI rule violating transitions produced by pion exchange.

- *KLEINERT:*

Can the ZI rule explain the observation that the new narrow $p\bar{p} \to p\bar{p}$ resonance at 1930 MeV is not seen in $p\bar{p} \to n\bar{n}$?

- *LIPKIN:*

This has no direct connection with the ZI rule. Any resonance with definite isospin which appears in the $p\bar{p} \to p\bar{p}$ (elastic) amplitude must also appear in the $p\bar{p} \to n\bar{n}$ (charge exchange) amplitude. The experiment indicates that there must be degenerate isoscalar and isovector resonances like $\rho\omega$ or fA_2. The $p\bar{p}$ system is too complicated to provide a simple explanation of why these two resonances interfere in such a way as to exactly cancel the charge exchange reaction. There are just too many possibilities and too many free parameters to adjust. This makes experimental predictions meaningless.

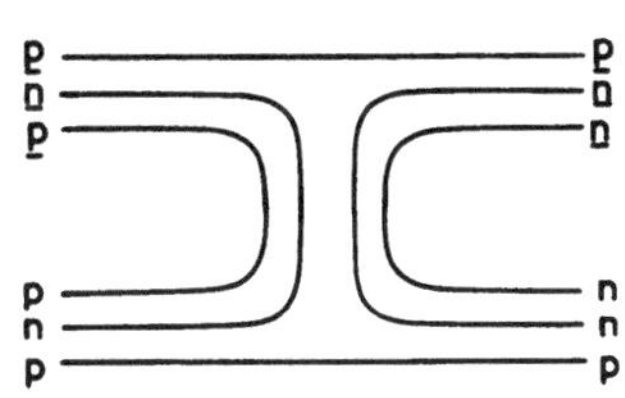

- *KLEINERT:*

Is not the re-arrangement collision the most important one where you do not have anything go across but just re-shuffle -- the six quarks come in, re-arrange, and go out again.

- *LIPKIN:*

Not necessarily.

- *FERBEL:*

You are so little above threshold that re-arrangement collisions are dominant.

- *LIPKIN:*

This would probably be true for elastic scattering which is commonly associated with a Pomeron exchange, but here you are dealing with an s-channel resonance. These are described in dual resonance models by quark-antiquark annihilations and not by a mere re-arrangement

- *WIGNER:*

Could you give the energy of the resonance?

- *KLEINERT:*

The energy above threshold is of order 100 MeV.

- *LITTENBERG:*

The mass is 1930 MeV, and the width, as determined by Carroll et al is about 9 ± 9 MeV. Montanet got a similar width with better errors which made it clear that it is not infinitely narrow. These two experiments had completely different systematic errors; and I think this is to be regarded as a well-founded object by now.

- *KLEINERT:*

It could be some exotic object which could have three quarks and three antiquarks in one bag.

- *ZICHICHI:*

I think we should not discuss this too much because the effect is three standard deviations. I would like to see some more convincing data.

- *PARSONS:*

Am I correct in assuming that the ZI rule combined with the assumption of exchange degeneracy produces all the inhibitions that the dual resonance model of Chan-Hong Mo et al. produces?

- *LIPKIN:*

One does not usually assume both the ZI rule and exchange degeneracy. One follows as a result of the other. Thus, for example, from the assumption of exchange degeneracy and the absence of exotics in all channels, Chiu and Finkelstein obtained the results of ideal mixing and the ZI rule.

- *FERBEL:*

At what level would the breaking of exchange degeneracy affect the ZI rule and the nice duality results? For example, 25% violation of, let us say $\alpha(0)$ with that observed at FNAL; would that hurt exchange degeneracy, exotics, and all those other nice ideas?

- *LIPKIN:*

This is a complicated and interesting problem. Different ways of breaking exchange degeneracy can have different effects on the ZI rule in different processes. The violations you mentioned involve non-strange trajectories. Their consequences would probably show up in other processes rather than in the violations of the ZI rule where the observed processes always involve strange quarks; e.g. the common selection rules for the ϕ and f' which contain only strange quarks. The relevant trajectories whose exchange degeneracy are necessary to prevent higher order violations of the ZI rule are the strange K, K^*, and Q trajectories. Experimental information is still weak on these.

- *KLEINERT:*

Is not the partner of the pion the B trajectory or the corresponding strange partner?

- *LIPKIN:*

The Q trajectories are the strange members of the SU(3) octets containing the B and the A_1 -- these are the trajectories required to be degenerate with the K trajectory.

DISCUSSION 3

- *WILKIE:*

Can you clarify for me the difference between the naive Zweig rule and the dual resonance model Zweig rule and what they have to say about $p\bar{p} \to \phi\phi$? Snnce $p\bar{p} \to \phi\phi$ is the t channel of the $\phi p \to \phi p$ scattering process which is dominated by Pomeron exchange experimentally, can one predict how it is going to behave?

- *LIPKIN:*

The naive Zweig rule says nothing about $p\bar{p} \to \phi\phi$ since it only concerns the annihilation of a quark with an antiquark from the same external particle. The generalized ZI rule for states with more than three particles forbids reactions like $p\bar{p} \to \phi\phi$ because the diagram describing it is disconnected. It is true that this is related by crossing to ϕp elastic scattering which is dominated by the Pomeron, but this seems to be irrelevant. The Pomeron gives a strong amplitude when it is exchanged in the t-channel, not the s-channel. Other examples of reactions, which are related to elastic scattering by crossing, but are found experimentally to be suppressed are: $p\bar{p} \to \Delta^-\bar{\Delta}^-$, $p\bar{p} \to \Sigma^-\bar{\Sigma}^-$, $p\bar{p} \to \Omega\bar{\Omega}$, and $p\bar{p} \to \Xi\bar{\Xi}$. These reactions are all suppressed because they cannot proceed via simple t-channel exchange. The fact that they have vacuum quantum numbers in the s-channel is irrelevant.

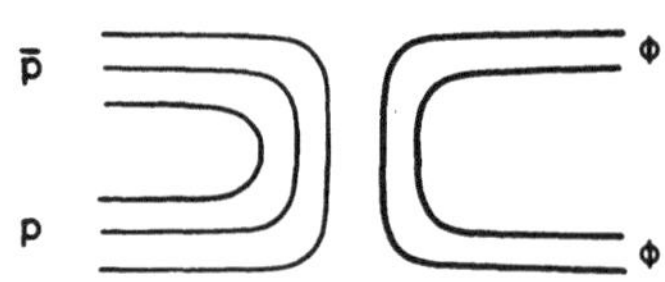

Diagram for $p\bar{p} \to \phi\phi$, forbidden by the generalized Zweig rule since it is disconnected

- *ETIM:*

You did not say much about radiative decays of mesons in connection with the Zweig rule. Would you briefly comment on this and on the discrepancies of these decays as predicted by the quark model with the results using the Primakoff effect, such as $\rho \to \pi\gamma$, $K^{0*} \to K^0\gamma$, and $\phi \to \eta\gamma$, and all these decays that show discrepancies with the quark model?

- *LIPKIN:*

All that the ZI rule says about radiative meson decays is to forbid $\phi \to \pi\gamma$, which seems to be all right. The discrepancies you mention involve difficulties of fitting the latest data with quark model or SU(3) relations which are not directly related to the ZI rule; in particular, the ratio of $\Gamma(\rho \to \pi\gamma)$ to $\Gamma(\omega \to \pi\gamma)$. This disagreement is very serious, since the relation between $\rho \to \pi\gamma$ and $\omega \to \pi\gamma$ is very much model independent. This can be seen by noting that the same vertex appears in both transitions in the vector dominance description, and that the mass difference between the ρ and the ω is negligible.

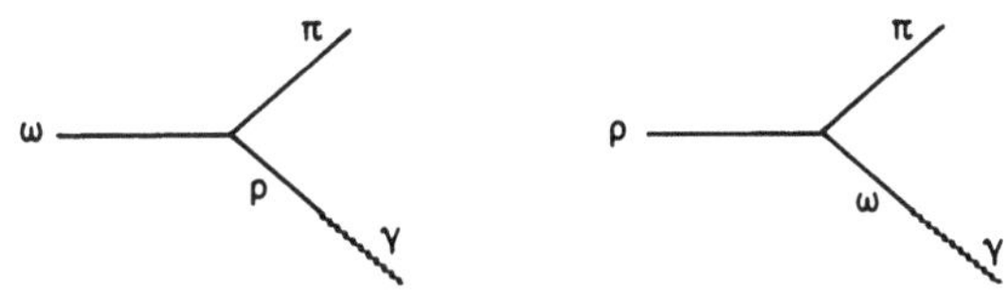

Vector meson dominance description of vector meson decay

- *WEILL:*

Is the Zweig rule applicable to multiparticle production? For an example, one could give $p\bar{p} \to$ many pions since when one draws the quark lines, there do not seem to be enough quark lines to produce many particles.

- *LIPKIN:*

Yes, it is very easy. You see that you cannot make a ϕ without having at least one additional kaon pair.

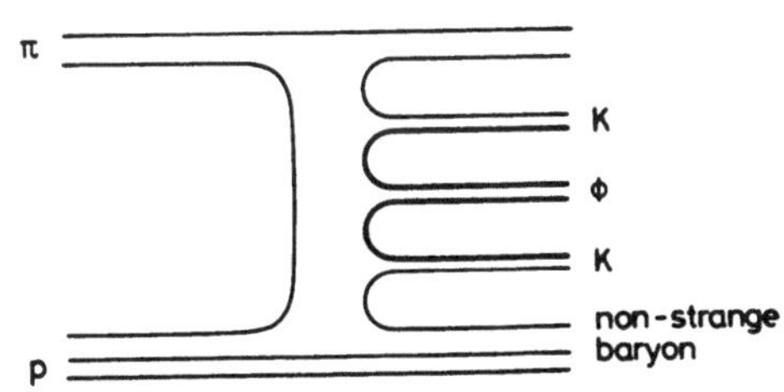

Possible graph in multiparticle production.
Note that a kaon pair must accompany a ϕ meson

- *KLEINERT:*

Consider $N\bar{N} \to \pi\pi$ or $N\bar{N} \to K\bar{K}$ at low energies. The production in the final state is suppressed more strongly than one would expect from SU(3) by a factor of ten.

- *LIPKIN:*

If you start with a non-strange initial state, it is always harder to make strange particles. $\pi p \to p\pi$ is greater than $\pi p \to K^*\Lambda$.

- *ZICHICHI:*

The $\pi\pi$ is 10^{-3} and the $K\bar{K}$ is how much?

- *LEADER:*

It is about a five- or ten-to-one ratio but it is also true for multiparticles.

- *KLEINERT:*

Yes, this is just an example.

- *LIPKIN:*

Quasi two-body annihilation seems to be limited by baryon exchange. The intercept difference between the Λ trajectory exchanged in $K\bar{K}$ production and the N trajectory exchanged in $\pi\pi$ production is sufficient to expalin this.

- *FERBEL:*

Is there any reason why the K production cross-section should be about a factor of ten smaller than that for pion production?

- *LIPKIN:*

In any Regge description where pions are produced by exchange of ρ or N trajectories while kaons are produced by exchange of strange trajectories, there is a natural suppression of kaon production because of the lower intercept of strange trajectories.

- *BERLAD:*

If one uses ideas of analyticity and absence of exotics in comparing $\pi p \rightarrow K\Sigma$ and πp charge exchange, one can explain quite well the strangeness exchange in $\pi p \rightarrow K\Sigma$ by SU(3) relations, relating it to πp charge exchange. There are no special numerical factors aside from SU(3) factors. I do not know about this for vector mesons, but I do know it for the pseudoscalars.

- *LIPKIN:*

If you look at this as a simple peripheral reaction, you always get a suppression simply by the intercept of the strange trajectory. Thus, if you look at high-energy inclusive things in any kinematic region, where you could use some kind of Regge formulation, you always find that producing a strange particle involves a strange trajectory somewhere; and the intercept will always knock down the amplitude compared to non-strange production.

- *ZICHICHI:*

This is one of the basic mysterious points which will be discussed next year at this school on the topics "The Whys of Subnuclear Physics" -- to say that strange quarks are coupled less than non--strange strange quarks, means to translate into the quark language the mystery of strangeness.

- *KLEINERT:*

The asymptotic freedom people have something to say about that.

- *ZICHICHI:*

This is very strange that they have anything to say. *(Laughter in audience.)* Asymptotic freedom exists independent of strangeness.

- *KLEINERT:*

Asymptotic freedom people always say that it is harder to make a pair with high masses such as a strange quark-antiquark pair with respect to a non-strange quark-antiquark pair because you need higher q^2 to make it, roughly speaking.

- *ZICHICHI:*

Do they know the strange quark mass?

- *LIPKIN:*

When I talk to the asymptotic freedom people, they say it is very nice for charmed things, but these light quarks, that includes strange and non-strange quarks, are all too light for them.

- *ZICHICHI:*

The mass difference between strange and non-strange quarks is very small if you believe in Kleinert -- who knows everything about quarks.

- *BUCCELLA:*

You can compute the imaginary part for $f' \to \pi\pi$ through unitarity, inserting the $K\bar{K}$ state. You also say that the off-shell contribution can give you the real part. Can you explain this?

- *LIPKIN:*

The real part has contributions from high momenta where there is no reason to include only the $K\bar{K}$ state. Once higher K^*'s are included, cancellations can occur which are model dependent. It is very difficult to make any definite statement. You could use some dual resonance model to sum over all trajectories. You need a definite model to tell you which things are going to contribute. Even the imaginary part is not so definite. There are transitions like $f' \to K\bar{K} \to f^{(R)} \to \pi\pi$, where $f^{(R)}$ is the Regge recurrence of the f. Veneziano has pointed out that the tail of the $f^{(R)}$ can give an appreciable contribution at the mass of the f'.

- *FREEDMAN:*

Recent spectroscopy calculations seem to give a strange quark mass of roughly 300 MeV and a non-strange quark mass of roughly 50 MeV. Threshold effects for model field theory calculations typically become important for q^2 greater than four times the mass squared. When strange quarks are absent in the initial state, would it not be more difficult to produce strange hadrons as compared to the production of non-strange hadrons?

- *LIPKIN:*

Experimentally, there is no evidence for such an effect. The simplest case is the $A_2 \to \rho\pi$, $K\bar{K}$, or $\eta\pi$ where you have SU(3) and SU(6) relations and you put in phase space and that works. There are no suppressions for making a kaon pair out of an initial non-strange pair other than phase space.

- *KLEINERT:*

What energies -- what values of q^2 -- are you referring to?

- *LIPKIN:*

Well this, of course, is at very low q^2. The production of ρ, ω, and ϕ by e^+e^- annihilation agrees with SU(3) predictions with no observed suppression of creation of strange quarks.

ARE STRONG INTERACTIONS STILL WITHIN THE REGGE FRAMEWORK?

ELLIOT LEADER

WESTFIELD COLLEGE, UNIVERSITY OF LONDON
LONDON NW3 7ST

0) PROLOGUE

This is the first time that I have had the pleasure of attending an Erice school. On many previous occasions, however, I have studied the circulars announcing the schools, when deciding whether or not to send a particular student, and I have to confess that there is one part of these posters that has always aroused my suspicions; and that is the little section headed "Poetic Touch", wherein the beauties and attractions of Erice are extolled, with references to Homer and other lustrous figures of Classical Antiquity.

You will appreciate that in this age of package holidays and advertising pressure, my reaction of suspicion is not unreasonable.

However, now that I am here I must freely admit that that paragraph is not only justified, it is, if anything, an understatement, and ever since setting foot in this enchanted village I have found myself overcome by a poetic urge, an irresistable desire to mimic Homer. You will forgive me, therefore, if I begin my lecture by showing you the fruits of this urge. I have to warn you, though, that there are two problems. Firstly my Classical Greek is simply dreadful and secondly my poetic ability is even worse. Nevertheless, here it is:-

POETIC TOUCH

HERE FOLLOWS A SAGA UN- GARNI
OF A STRANGE YOUNG LAD OF TRAPA(R)NI
WHOSE CHARM AND DETERMINATION
LED OFT' TO THE FUNIVA-STATION.

YET ONCE ON THE MOUNTAIN PICHCHI
OUR YOUTH, ANTONINO ZICHICHI,
TURNED HIS THOUGHTS FROM LOVE'S INTERACTIONS
TO BASIC HADRONIC REACTIONS.

IN THE HEART OF ERICE CELESTIAL
CAME AN IDEA SOME THOUGHT QUITE BESTIAL -
TURN THE CONVENT OF THE HOLIEST VISION
INTO A CENTRE FOR NUCLEAR FISSION.

AND LO! OUR PRESENCE HERE TODAY
GOES TO SHOW IN A DEFINITE WAY
WHAT RESULTS FROM A THOUGHT ICONOCLASTIC
WHEN CONCEIVED IN A MIND SO ELASTIC!

No sooner was my poetic urge satisfied than I found my thoughts turning to Philosophy. What would be needed to give a definitive answer to the question raised in the title of my lecture? How would one really decide whether "Regge Poles" are right or wrong? Historians and Philosophers of Science spend a great deal of time discussing how a theory is accepted or rejected or disproved. Their discussion is dominated by the rather naive concept of the "crucial experiment". Just how naive this is, is well illustrated by Regge Theory.

In truth:-

We do not wish to give up a theory.
We come to love it.
It becomes a part of our language
even if it is wrong.

Moreover:-

Theories are alive (and slippery).
They are Darwinian.
They adapt themselves to the experimental situation.

I shall deal today with a classic case of the "failed crucial experiment", the pion nucleon charge exchange reaction $\pi^- p \to \pi^0 n$. Not that it was a failed experiment. It was a superb experiment. But it failed to be crucial. However, since the concept of the crucial experiment is so unrealistic it is not surprising that the experiment has turned out to be exceedingly important and interesting.

Because of the great spectrum of interests and experience in the audience I should like to give a brief recapitulation of Regge Theory before getting down to the serious analysis.

I shall consider Regge Theory on two levels. About the first, the fundamental one, there will be no discussion. It contains ideas and concepts that have already appeared in Text Books on Quantum Mechanics.

The principal elements are:-

i) Continuation into the complex J-plane.
ii) Existence of poles at $J = \alpha(t)$.
iii) Relationship of these poles to the bound states and resonances of the system. A bound state or resonance of mass m and spin J occurs when $\alpha(m^2) = J$.
iv) Signature: Continuation from physical values of J must be done separately for even and odd values. Labelled by $\tau = \pm 1$.

These ideas are as solid as Quantum Mechanics itself. At an altogether different level are what I shall call the "derived" properties, which allow contact to be made between the above and the world of high energy scattering.

The ingredients are:

Analyticity
Crossing
Kinematic Accidents.

The result is that for a process

$$A + B \to C + D$$

with (C.M. energy)^2 = s and (four-momentum transfer)^2 = t the scattering amplitude has the asymptotic behaviour

$$f(s,t) \xrightarrow[t\ \text{fixed}]{s\to\infty} \xi(t)\ \beta(t)\ s^{\alpha(t)}$$

where

$\alpha(t)$ = Trajectory of Regge Pole in the crossed channel i.e. in the process $A\bar{C} \to \bar{B}D$

$\beta(t)$ = Residue function whose form is not specified by the theory.

$$\xi(t) = \text{Signature factor} = \frac{1+\tau e^{-i\pi\alpha}}{\sin(\pi\alpha)}$$

What are the $\alpha(t)$? They are labelled by internal Quantum Numbers corresponding to known particles or resonances e.g. ρ, ω, A_2, f etc. The most infamous of them is the Pomeron, with Quantum Numbers of the vacuum, invented specifically to "explain" elastic diffractive scattering. Historically, in the early 1960s the shrinkage of the pp differential cross section, $d\sigma/dt$, and its energy dependence led to the need for a trajectory

$$\alpha_{\mathbb{P}}(t) = 1 + t$$

where the value $\alpha_{\mathbb{P}}(0) = 1$ is of course correlated via the Optical Theorem with the then believed behaviour $\sigma_{TOT} \to$ constant. Looking at a present day picture of σ_{TOT} vs energy, one is tempted to ask what the $\mathbb{P}$ would be like if Regge Theory had been invented in 1975!

Historically the πN charge exchange reaction has been especially significant. Because the Quantum Numbers of the exchange are so restricted, namely to mesons with I=1 and G=+, and since the only particle with such Quantum Numbers was the ρ, it was expected that $\pi^- p \to \pi^0 n$ would display the classic features of Regge Pole exchange:

i) Shrinkage of $d\sigma/dt$: Since one has

$$\frac{d\sigma}{dt} \propto s^{2\alpha_\rho(t)-2}$$

and since for small t

$$\alpha(t) = \tfrac{1}{2} + \alpha'_\rho(t)$$

one expects

$$\frac{d\sigma}{dt} \propto e^{t\alpha'_\rho \log s}$$

which, bearing in mind that $\alpha'_\rho > 0$ and that $t < 0$, shrinks as s increases.

ii) Both the spin non-flip amplitude f_{++} and the spin-flip amplitude f_{+-} should have the same phase. One has

$$f_{++} \propto \xi_\rho(t)\ \beta_{++}(t)\ s^{\alpha_\rho(t)}$$

$$f_{+-} \propto \xi_\rho(t)\ \beta_{+-}(t)\ s^{\alpha_\rho(t)}$$

and the only non-real quantity, $\xi_\rho(t)$, is common to both. An immediate consequence is that the polarisation, which depends on interference between f_{++} and f_{+-}, will be zero. i.e.

$$P \propto \mathrm{Im}\ (f_{+-}{}^* f_{++}) = 0$$

As is well known (i) was beautifully confirmed by experiment, whereas for (ii) a significant non-zero polarisation was found!

How could one escape the dilemma?

A secondary trajectory called ρ' was postulated. It was to be a daughter of the ρ i.e. its trajectory would satisfy

$$\alpha_{\rho'}(t) = \alpha_\rho(t) - 1$$

Unfortunately this implies the existence of a vector meson at a mass of 1250 MeV, something which Nature has failed to provide us with.

A more attractive solution physically was provided by Nicolescu and myself. The polarisation has a zero at $t \simeq -1(\mathrm{GeV}/c)^2$. A natural explanation of this would be that at this point $\alpha_{\rho'} = \alpha_\rho$, thus making all phases equal at this point. Further, a study of the energy dependence of $\sigma_{\pi^- p} - \sigma_{\pi^+ p}$ indicated that at t=0 one should have $\alpha_\rho'(0) \simeq 0$. With two points on its trajectory fixed, assuming as usual a straight line, one found that one had a rather flat ρ' trajectory

$$\alpha_{\rho'}(t) \simeq 0 + \frac{1}{3} t$$

so that the implied vector meson would have quite a large mass, about 1700 MeV, and this is not ruled out experimentally. Thus points (i) and (ii) could be reconciled without grossly contradicting the known particle spectrum.

To summarize the situation on $\pi^- p \to \pi^0 n$ in the pre-FERMILAB days: A very good description of all data could be achieved using ρ and ρ' trajectories as specified above. We shall see, however, that the situation has changed as a result of the recent FERMILAB experiments. Indeed we shall suggest that once again this reaction may be playing a historical role and that it may be teaching us about entirely unforseen effects in exchange type reactions.

1) INTRODUCTION

The concept of Reggeisation of particle exchanges in dynamical calculations has been of enormous importance both from a phenomenological point of view and from the deeper point of view of the fundamental structure of strong interaction dynamics.

On the phenomenological side a major weakness of the theory has always been its virtually unconstrained flexibility, so that most attempts to pin it down or to subject it to critical tests have ended in abject failure, defeated by the sheer elusiveness of the theory. Despite this it cannot be denied that the Regge theory has provided a very convenient and useful framework in which to analyse reaction amplitudes both for elastic scattering and for inclusive multiparticle reactions. Moreover it has succeeded, on a qualitative level, in "explaining" and correlating a great many features of the data.

Regge theory, in its full generality, presumably requires the consideration of both poles and their accompanying cuts and the precise predictions emerging naturally reflect the level of sophistication achieved in handling these complicated effects. Thus many arguments have raged as to what precisely Regge theory does imply in a particular reaction, and failure to agree with experimental data is often only a reflection of the specific approach adopted in grappling with the cut complications.

In the present lectures we wish to suggest tentatively that there is growing evidence that in a rather absolute sense conventional Regge theory does not correctly describe the asymptotic behaviour of scattering amplitudes. By "absolute sense" we mean in a way transcending the details of a particular parametrisation of the poles or a specific approach to the handling of the associated cuts.

If this is correct then Regge theory is somehow or other an intermediate energy theory and not a truly asymptotic theory. It is easy enough to say this, but it is not at all clear how such a situation could arise at a fundamental dynamical level. Possibly it could reflect a new type of behaviour of matter at the very short distances involved in ultra-high energy experiments.

We shall argue that data emerging from Fermilab and from Serpukhov for exchange type reactions i.e. reactions in which there is an exchange of definite, non-vacuum quantum numbers, seem to show an energy dependence that is completely at variance with that expected from Regge theory with conventional poles and cuts. In particular there appear to exist exchange type amplitudes e.g. some with the quantum numbers of ρ A_2 that give rise to physical effects that do not vanish as $p_L \to \infty$. Although this is very startling and unexpected it should be remembered that recently we have learnt from the behaviour of σ_{pp} at the ISR that the symmetric, vacuum exchange, amplitude grows like slogs or $s(\log s)^2$ and is thus not Regge-behaved in the simple fashion expected of Pomeron exchange. Perhaps it is not too surprising then if we discover that anti-symmetric amplitudes are also non-Regge behaved at very high energies.

In the following we shall discuss in turn the reactions

$\pi^- p \to \pi^0 n$

$np \to pn$

and $\pi^\pm p \to p\pi^\pm$

We shall not go into the technical points but restrict ourselves to a qualitative description of the essential points. Full details and a complete set of references can be found refs. 1-3.

2) PION-NUCLEON CHARGE EXCHANGE SCATTERING

We consider the differential cross section $d\sigma/dt$ for small angle charge exchange scattering, $\pi^- p \to \pi^0 n$, and the difference of the total cross-sections for $\pi^+ p$ and $\pi^- p$. We define

$$\Delta\sigma \equiv \sigma_{\pi^- p} - \sigma_{\pi^+ p}$$

and use the standard notation A', B for the $\pi^- p \to \pi^0 n$ amplitudes. The optical theorem then reads

$$\mathrm{Im}\, A' \,(t=0) = \frac{p_L}{2} \Delta\sigma. \qquad (1)$$

To begin with, we shall study to what extent the data on $\Delta\sigma$ and on $d\sigma/dt|_{t=0}$ are compatible with various versions of standard Regge-pole theory.

i) Single, pure Regge pole.

a) If one parametrizes the imaginary part of A' at t=0 in the usual way

$$\text{Im } A'(t=0) = \beta s^{\alpha}, \tag{2}$$

and attempts to fit the data on $\Delta\sigma$ of all energies above p_L = 10 GeV/c, a very good fit can be achieved when the parameters take on the values

$$\beta = 106.6\ \mu b^{\frac{1}{2}},\ \alpha = 0.54.$$

Using the standard formula for the Regge signature factor, one takes the real part of A' at t=0 to be given by

$$\text{Re } A'\ (t=0) = \text{tg}\left(\frac{\pi\alpha}{2}\right)\ \text{Im } A'\ (t=0). \tag{3}$$

One then computes $d\sigma/dt|_{t=0}$ and compares it with the data. The fit is extremely poor. The χ^2 for the four Fermilab points alone is 122. The fit is shown by the solid curve in Fig. 1.

b) If one reverses the procedure, and first attempts to adjust the parameters so as to obtain a good fit to $d\sigma/dt|_{t=0}$ one now finds as best fit values

$$\beta = 184.5\ \mu b^{\frac{1}{2}},\quad \alpha = 0.42,$$

and not surprisingly, when one compares the expressions with the data on $\Delta\sigma$ one finds a poor fit both at low and high energies. The value of χ^2 is 76 for 16 points for the $\Delta\sigma$ fit. The fit is shown by the dashed curves in Fig. 1.

The results of the two methods of choosing best parameters indicate that, while either $\Delta\sigma$ or $d\sigma/dt|_{t=0}$ can be very well fitted by a single, pure Regge pole, the parameters required for the pole are significantly different in the two cases. Thus a single pole cannot provide a complete description of the physics at t=0.

ii) Pure ρ+ρ' Regge poles. Since the ρ alone is inadequate we attempt to fit to $\Delta\sigma$ and $d\sigma/dt|_{t=0}$ using a ρ and ρ' pole model. Old models based on ρ and ρ' do not fit the new data, so a reanalysis of the parameters is called for. We put

$$\text{Im } A'\ (t=0) = \beta_1\ s^{\alpha_1} + \beta_2 s^{\alpha_2}, \tag{4}$$

and try to fit all the data above 5 GeV/c, keeping α_ρ close to 0.5 as required for a linear trajectory to pass through the ρ and g mesons.

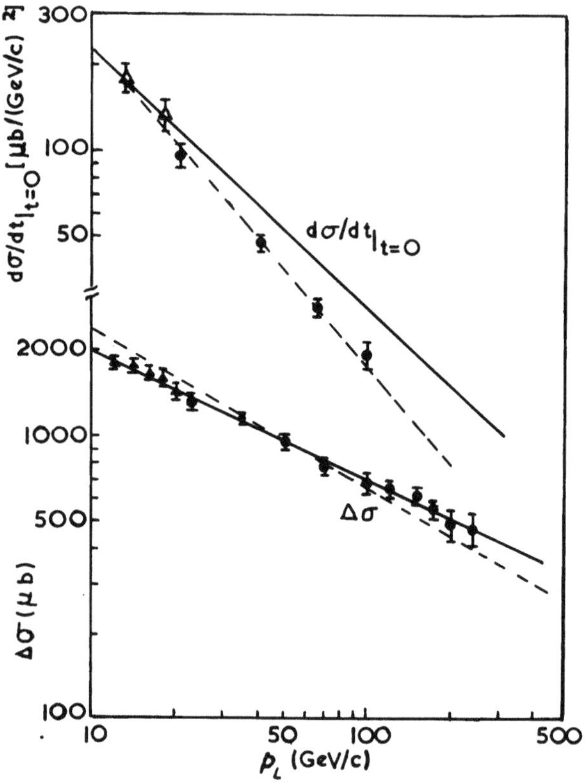

Fig. 1 - Indication of the mismatch between the ρ pole in the first case optimized to fit the energy variation of $\Delta\sigma$ (solid curve) and in the second case optimized to fit the energy variation of $d\sigma/dt|_{t=0}$ (dashed curve). (Data: Δ from ref. (4), ● from ref. (5), ▲ from ref. (6).

The χ^2 is quite good ($\simeq$ 1.3 per point) for the parameter values

$$\beta_1 = 124.5\ \mu b^{\frac{1}{2}}, \quad \alpha_1 = 0.49,$$

$$\beta_2 = -86.1\ \mu b^{\frac{1}{2}}, \quad \alpha_2 = -0.16,$$

but the shape of the curves is unsatisfactory at Fermilab energies. The bad shape at Fermilab energies is reflected in a bad χ^2 ($\simeq$ 2 per point) in the Fermilab region.

A good way to see the difficulty is to look at the behaviour of the experimental ratio

$$r = \frac{\text{Re } A'(t=0)}{\text{Im } A'(t=0)}$$

As seen in Fig. 2 neither ρ nor ρ + ρ' are able to follow the experimental trend.

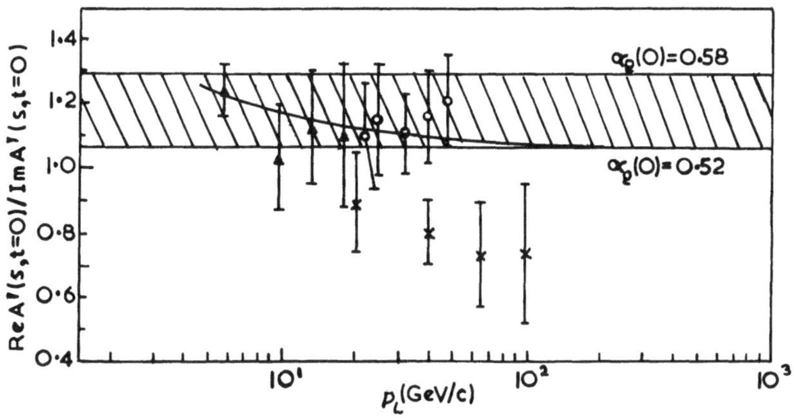

Fig. 2. - The experimental values of the ratio Re A'/Im A' at t=0 compared with the value of this ratio for i) a single ρ pole and ii) a ρ pole plus a ρ' pole (——), in both cases the pole parameters being constrained to fit the energy variation of $\Delta\sigma$.

We conclude that the ρ + ρ' model with a conventional ρ intercept cannot explain the data and that a new term whose real part is negative at Fermilab energies is required.

We have also tried fits in which α_ρ is left completely free. A good fit to the t=0 data can be achieved ($\alpha^2 \simeq 1.1$/point) with $\alpha_\rho = 0.44$. However,

a) the fit is systematically bad for $\Delta\sigma$ in the region $100 \leq p_L \leq 240$ GeV/c ($\chi^2 = 2$/point),

b) the ρ' in this description plays an exaggerated rôle. It is in no sense secondary to the ρ and one loses much of the physical interpretation based on the ρ as the dominant mechanism.

As a result, an attempt to use these t=0 parameter values in a study of the t-dependence leads to wildly unacceptable dσ/dt and polarizations.

We should also mention that other models such as ρ+ soft Regge cuts are very similar to the ρ + ρ' model and will suffer from the same deficiencies at Fermilab energies.

iii) Can dispersion relations influence the phase? Since all the above models fail as a result of a mismatch between the energy variation of their real and imaginary parts, or equivalently, as a result of their unacceptable phases, we wish to examine whether the whole problem has not arisen because of an injudicious use of asymptotic phases.

Remember that the usual "Regge phase" is in fact an asymptotic one, and that a Regge-behaved imaginary amplitude when fed into a dispersion relation will yield a real part differing from the "Regge real part" at low energies, but tending towards the Regge value as the energy increases.

For all the above models we assume a Regge-like behaviour in the whole complex plane, and then set up unsubtracted dispersion relations into which are fed the imaginary part of A' as given by each model. In no case does the calculated real part of A' differ significantly from the "Regge real part" beyond p_L = 10 GeV/c.

We conclude that the non-Regge behaviour observed is genuine and is not due to a misuse of the asymptotic phase formula.

iv) Qualitative properties of the new term at t=0. We have seen from the above analysis that, if we wish to retain the standard Regge terms, then we require in addition a new term with the properties

$$(5)\quad \begin{cases} \text{a)} & \text{Re (new term)} < 0 \text{ at Fermilab energies,} \\ \text{b)} & \text{Im (new term)} > 0 \text{ and increasing with energy at Fermilab.} \end{cases}$$

In fact it is also possible to have Im (new term) = 0 since the standard Regge-pole model can be optimized to provide an excellent fit to $\Delta\sigma$. It is important to note that analyticity does not allow us to use an analogous argument about fitting $d\sigma/dt|_{t=0}$ with a standard Regge form, and thereby to conclude that Re (new term) could be zero.

Thus, finally, our condition on the imaginary part is

$$\text{b')}\quad \text{Im (new term)} \geq 0. \tag{5'}$$

The new term cannot be of the form of a soft Regge cut since the relative signs of the real and imaginary parts would be wrong and also since the cut correction would decrease with energy, in contrast to the empirical requirements.

We could try an assumption that Im A'(t=0) has a term of the form $s^\alpha(\log s)^\beta$ with $\alpha > \frac{1}{2}$, $\beta > 0$ in it and with $|A'|$ bounded by $s^\alpha(\log s)^\beta$ for s complex and $|s| \to \infty$. In that case one would have asymptotically

$$A' \simeq \left(i + \mathrm{tg}\,\frac{\pi\alpha}{2}\right) s^\alpha(\log s)^\beta, \tag{6}$$

provided strictly $\alpha < 1$. We see that this is unacceptable since its real part has the same sign as its imaginary part. This conclusion is of course only exact asymptotically, but a detailed analysis using dispersion relations supports this conclusion at finite energies.

It is interesting to note that, if $\alpha = 1$, then with β at its maximum allowed value, i.e. $\beta = 1$, the asymptotic behaviour becomes

$$A'(t=0) \underset{\sim}{\sim} s[\log^2 s - i\pi \log s], \tag{7}$$

so that in this case Re A' and Im A' have the desired opposite sign as $s \to \infty$. It is noteworthy that this behaviour corresponds to a singularity at $J = 1$ in the complex-angular-momentum plane and has the maximum growth possible for an anti-symmetric amplitude.

It would also be possible to interpret the new term as arising from an odd signatured analogue of the pomeron, i.e. a Regge pole, which we shall refer to as the "odderon" and which has $\alpha(0) = 1$. This term would be purely real at $t = 0$ and would thus only play a rôle in $d\sigma/dt|_{t=0}$.

The above qualitative statements are supported by a careful and detailed study at t=0.

Thus we are led to contemplate the possibility of an exchange amplitude that behaves asymptotically something like s^{α} with $\alpha = 1$. Such a behaviour is quite outside the conventional picture of Regge exchanges. It seems to represent a totally new type of dynamical effect which only begins to show itself at ultra high energies.

If such an amplitude is present at t=0 one should ask what rôle it plays away from the forward direction. It turns out that there is not enough data to really pin down the properties for $t < 0$. However, an exceedingly good fit to the data can be obtained by making a very simple ansatz for the t-dependence of the new term. For example, taking a new term in A' of the form $\beta(t)\, s^{\alpha(t)}$ with $\alpha(0) = 1$ (the odderon picture) and an exponential t-dependence for $\beta(t)$ one finds the near perfect fits shown in Figs. 3, 4, 5.

In summary there is evidence in $\pi^- p \to \pi^0 n$ for something quite unexpected at high energies. However, the evidence is flimsy at this stage and more persuasive consequences must be sought before a truly convincing case can be established.

It is quite remarkable that despite the smallness of the new term a dramatic effect of its existence can be demonstrated. Although the magnitude of the new term is minute its phase is significantly different from the phases of other standard Regge contributions. Thus one might expect noticeable effects in an experimental quantity which is sensitive to interference. Indeed

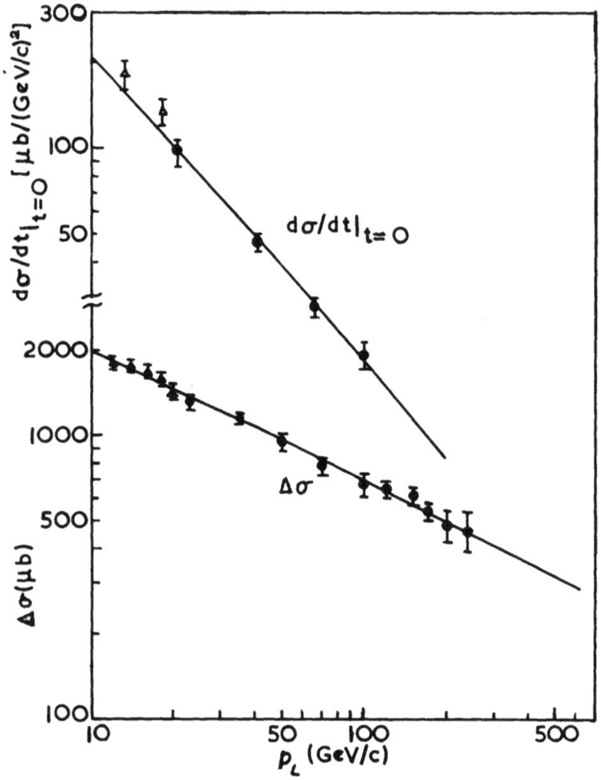

Fig. 3 - ρ+ρ'+odderon fit to Δσ and $d\sigma/dt|_{t=0}$

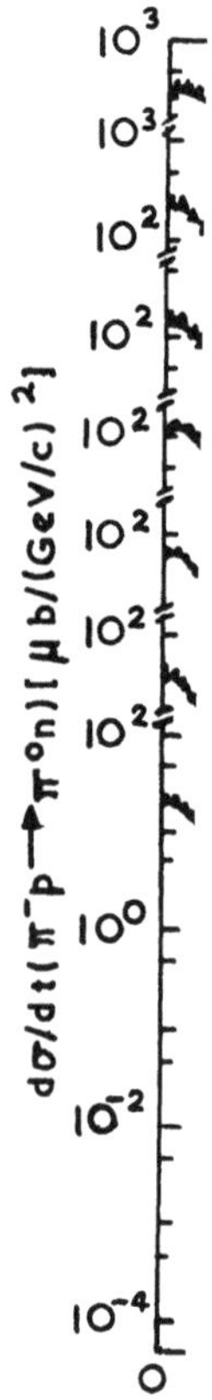

Fig. 4
ρ+ρ'+od

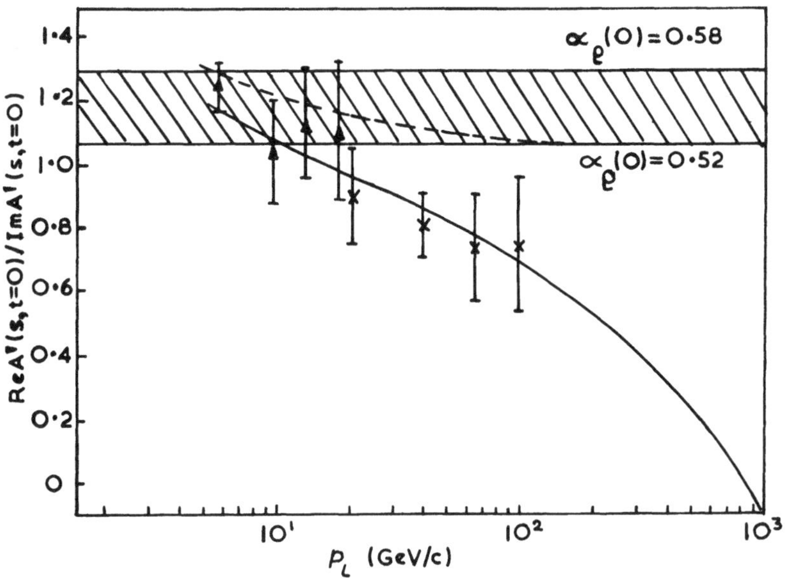

Fig. 5 - The ratio Re A'/Im A' at t=0 given by the $\rho+\rho'$ odderon model (——) compared with the data. The hatched band is the same as in Fig. 2.

one finds that the new term generates a huge and totally unexpected polarization at small t values in $\pi^-p \to \pi^0 n$ at Fermilab energies. The precise form of the polarization is not predictable far from t=0, but in the region of small t, say $-t \lesssim 0.5$, it is quite well determined. Just how surprising the predicted behaviour is can be seen in Fig. 6. The discovery of such an effect in the charge exchange polarization would constitute a most convincing confirmation of the unexpected asymptotic behaviour of the antisymmetric I = 1 exchange amplitude.

If the new asymptotic behaviour is a real effect one may or may not see a change in the energy dependence of $\Delta\sigma = \sigma_{\pi^-p} - \sigma_{\pi^+p}$. This is model dependent. However, one would certainly see a continuation of the strange behaviour of

$$r = \frac{\text{Re } A'(0)}{\text{Im } A'(0)}$$

shown in Fig. 2, with $r \to -\infty$ ultimately as $s \to \infty$, and as a consequence the value of $d\sigma/dt$ at t=0 would, for example, have dropped to about 2/3 of the value expected in conventional Regge theories at p_L = 300 GeV/c.

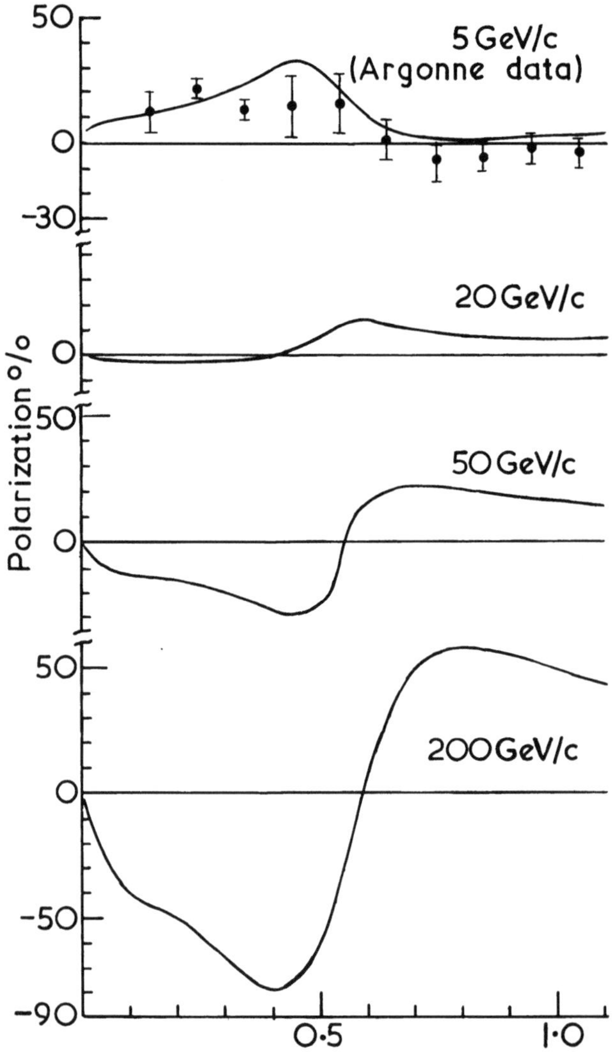

50
0
-30
5 GeV/c
(Argonne data)
20 GeV/c
0
Polarization %
50
50 GeV/c
0
50
200 GeV/c
0
-50
-90
0·5
1·0

Figure 6

3) NEUTRON-PROTON CHARGE EXCHANGE SCATTERING

In an analogous fashion we consider the variation with energy of the difference of total cross-sections $\sigma_{pp}-\sigma_{pn}$ and of the forward differential cross-section for $np \to pn$.

The nucleon-nucleon case differs from the meson nucleon case in two major aspects:-

i) It is extremely difficult to obtain a precise measurement of the cross-section difference. Experiments either use deuterium targets, in which case there is some uncertainty in the exact magnitude of the Glauber correction needed in order to extract σ_{np}, or they use neutron beams, in which case the statistical accuracy of the data has not been sufficiently accurate - at least not up to the present.

To make matters worse, different experimental groups have used different values of the "deuteron radius" when making their Glauber corrections, so that raw deuterium data on which there is perfect agreement between different groups ends up yielding significantly different values for σ_{np}. Clearly, a detailed study of the energy dependence of $\sigma_{pp} - \sigma_{np}$ will be very sensitive to such anomalies. An essential requirement therefore is a re-analysis of the raw deuterium data ensuring a compatible treatment of the Glauber correction in different experiments.

We have carried out such an analysis and all statements in the following will refer to the results of this analysis.

Also in the neutron beam case there is an unreasonable difficulty. The targets used for the np and pp measurements have been very different in length. In comparing σ_{pp} with σ_{pn} there is thus a danger of some uncontrollable systematic error.

Despite all these problems, some confidence in our conclusions is inspired by the fact that the results using neutron beam data are in fairly good agreement with those obtained from the re-analysed deuterium data.

ii) There is no unique relationship between $\sigma_{pp} - \sigma_{np}$ and $d\sigma/dt$ $(t=0)$ for $np \to pn$. This is because of the existence of three independent non-zero amplitudes at t=0 in nucleon-nucleon scattering. $d\sigma/dt$ $(t=0)$ is proportional to $|\phi_1|^2+|\phi_2|^2+|\phi_3|^2$. But $\sigma_{pp} - \sigma_{np}$ directly determines only Im $(\phi_1+\phi_3)$ whereas ϕ_2 which is dominated by π-exchange, is independent of the difference of total cross-sections. Thus a comparison of the energy variation of $\sigma_{pp} - \sigma_{np}$ and $d\sigma/dt$ $(t=0)$ for $np \to pn$ might not be so conclusive as for the analogous meson-nucleon case. We shall nevertheless see that fairly significant conclusions can be reached.

Let us begin by recalling the expected behaviour of $\sigma_{pp} - \sigma_{np}$ in a conventional model. Only exchanges with the quantum numbers of ρ and A_2 can contribute, so that one would have, at high energies,

$$\Delta_p \equiv \tfrac{1}{2}(\sigma_{pp} - \sigma_{np}) \sim \frac{1}{\sqrt{s}} \tag{8}$$

In fact Δ_p depends on $(\rho - A_2)$ so that in exchange degenerate models there is actually a cancellation between ρ and A_2 in (8) and one has

$$\Delta_p = 0 \tag{9}$$

at reasonably high energies.

Fig. 7 shows the published data on Δ_p using deuterium data and Fig. 8 shows the results of re-analysing the new data.

In Fig. 9 the neutron beam data can be seen to be very similar to the re-analysed deuterium data.

Looking at Figs. 8 and 9 we see that there appears to be a small non-vanishing component to Δ_p. If this effect is genuine then we are again facing the possibility of a most unexpected asymptotic behaviour for a quantum number exchange type amplitude.

Analysis of $\Delta_{\bar{p}} \equiv \frac{1}{2}(\sigma_{\bar{p}p} - \sigma_{\bar{p}n})$, which depends upon $(\rho + A_2)$, suggests that the new effect resides also in the A_2 type amplitude.

Taking these effects at face value and bearing in mind our study of meson-nucleon scattering, we are forced to conclude that both the symmetric (A_2) and anti-symmetric (ρ) $I=1$ exchange amplitudes are showing unexpected, non-Regge asymptotic behaviour.

However, even more so than in the meson-nucleon case, it is important to try to find some independent check on the existence of such behaviour which does not rely upon extremely precise measurements and the dangers inherent in the subtraction of similar large numbers. Again, as in the meson-nucleon case, despite the smallness of the new term one can find an experimental quantity wherein it would play a significant role and wherein it would thus produce quite unexpected and dramatic effects.

We have already mentioned that the forward differential cross-section for neutron proton charge exchange scattering depends upon three of the five nucleon-nucleon helicity amplitudes. One has

$$\frac{d\sigma}{dt}(t=0) \;\alpha\; \frac{1}{s^2}\{|\phi_1+\phi_3|^2 + |\phi_1-\phi_3|^2 + |\phi_5|^2\} \tag{10}$$

The imaginary part of $(\phi_1+\phi_3)$ is fed directly by $(\sigma_{pp} - \sigma_{np})$, while ϕ_2, (and only ϕ_2 at $t=0$) is fed by π exchange. It is a remarkable fact that up to the present, that is up to pre-Fermilab energies, the extraordinary spiked shape and the energy dependence of $d\sigma/dt$ for very small t have been in perfect agreement with π exchange coupled to a smooth coherent background having the same

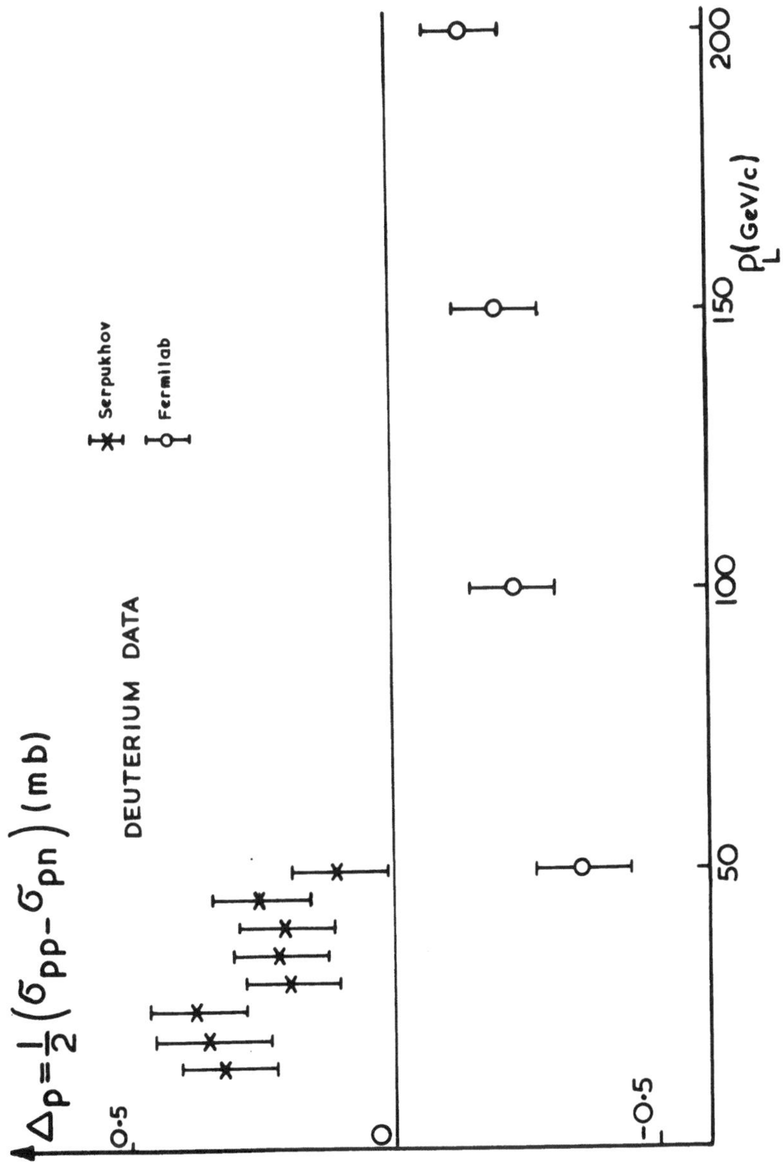
$\Delta p = \frac{1}{2}(\sigma_{pp} - \sigma_{pn})$ (mb)
DEUTERIUM DATA
Serpukhov
Fermilab
0·5
0
-0·5
50
100
150
200
p_L (GeV/c)

Figure 7

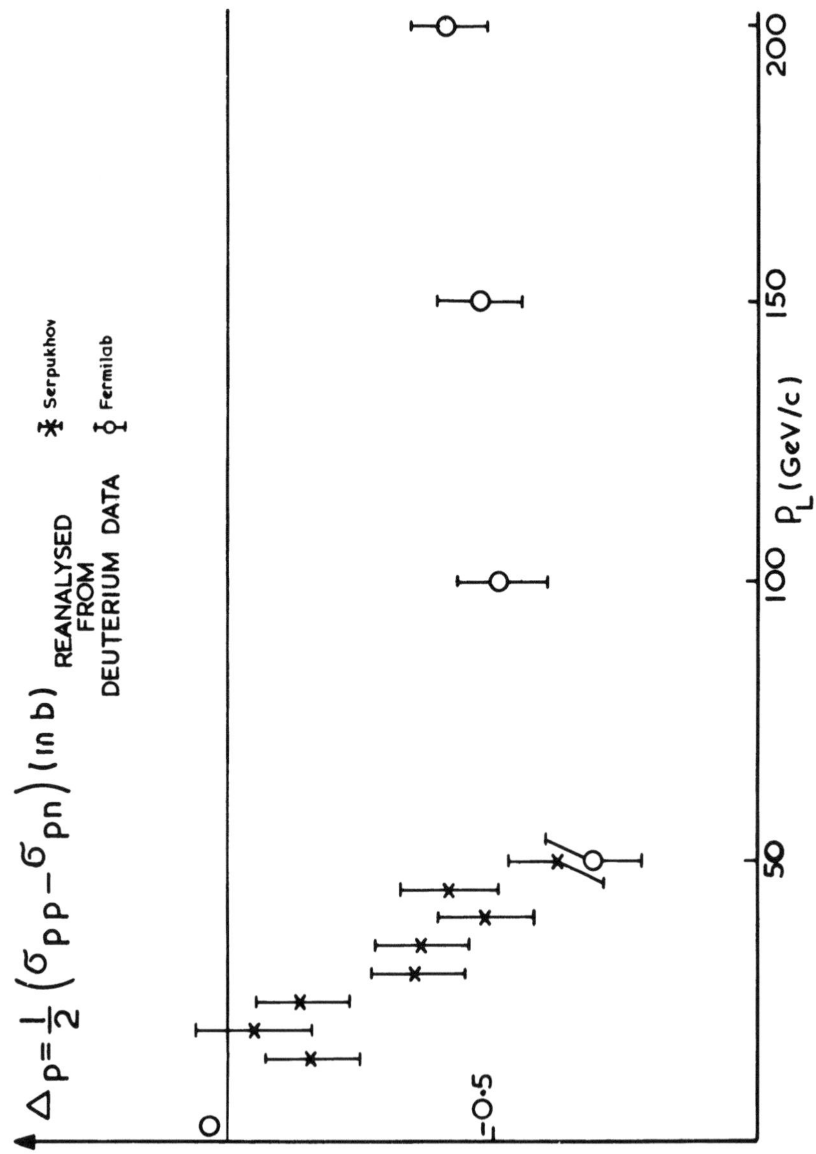
Δp = 1/2 (σpp − σpn) (in b)
REANALYSED FROM DEUTERIUM DATA
Serpukhov
Fermilab
0
-0·5
50
100
150
200
pL (GeV/c)

Figure 8

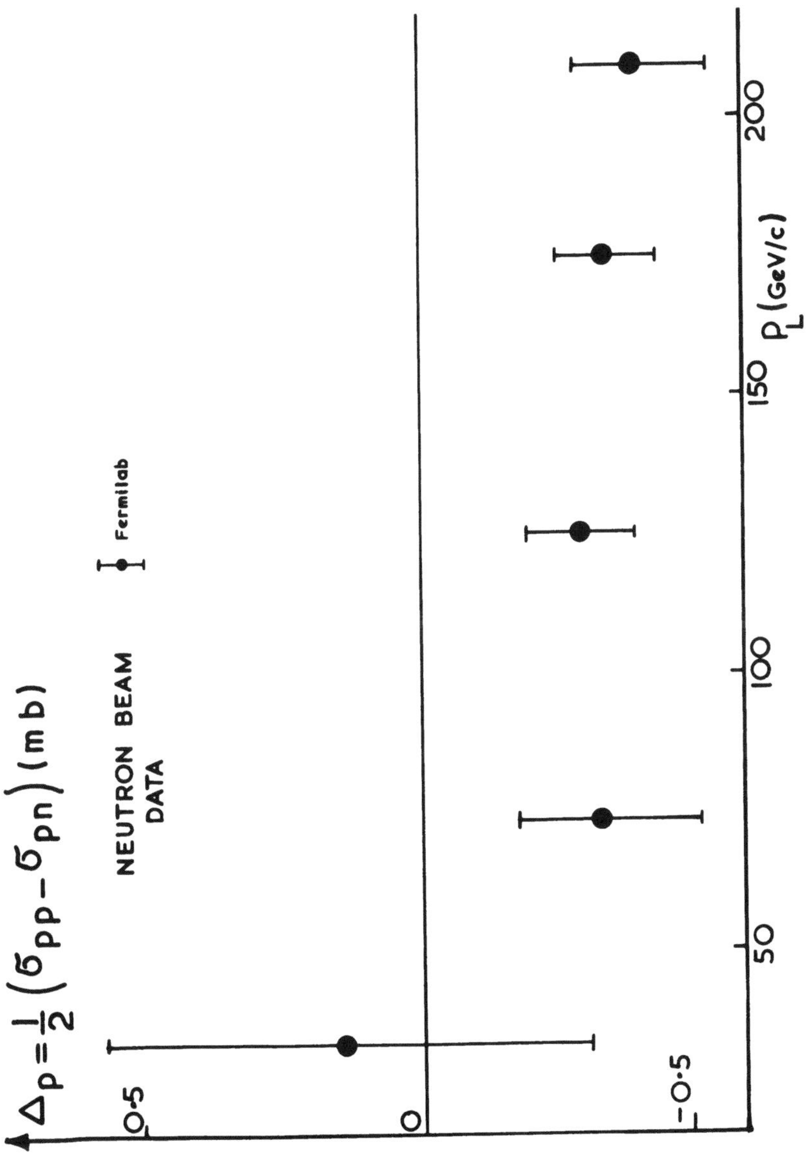

$\Delta_p = \frac{1}{2}(\sigma_{pp} - \sigma_{pn})$ (mb)
0·5
0
-0·5
NEUTRON BEAM DATA
Fermilab
50
100
150
200
P_L (GeV/c)

Figure 9

energy dependence as the π itself. No need for any contribution from $\phi_1+\phi_3$ has ever been found. At first sight this might suggest that $d\sigma/dt$ is not the place to look for consequences of the new asymptotic term. However, it must be remembered that ϕ_2 as given by π plus background is constant in energy so that $d\sigma/dt$ drops rapidly like $1/s^2$. At Fermilab energies this contribution has dropped to such a small value (less than 10 micro-barns per $(GeV/c)^{-2}$) that even a small magnitude of Δp can give a significant additional boost to $d\sigma/dt$. The exact numerical connection is as follows. The contribution to $d\sigma/dt$ (t=0) from a non-zero Δp is given by

$$\frac{d\sigma}{dt}\ (t=0)\ \text{in}\ \mu b.(GeV/c)^{-2} = 51.09\ [\sigma_{pp} - \sigma_{pn}\ \text{in mb}]^2 \tag{11}$$

Thus even $\frac{1}{2}$ mb produces about 12 $\mu b\ (GeV/c)^{-2}$! - a 100% effect!

Hence, if our interpretation is correct, if there really is a new asymptotic term present, then we must expect quite a dramatic change in the energy dependence of $d\sigma/dt$ (t=0) in $np \to pn$. There is no escape.

The situation with the data on neutron-proton charge exchange is tantalising. Fig. (10) shows an extrapolation of π + background contribution, which is in effect an extrapolation of the data below 50 GeV/c. A few experimental points below 50 GeV/c are shown as well as some preliminary data from Fermilab. It is remarkable that the Fermilab points fall significantly above the extrapolated curve.

However, the interpretation is complicated by the following. If $\sigma_{pp} - \sigma_{pn} \neq 0$ then no matter what happens to any other contributions one would have an exact mathematical lower bound,

$$\frac{d\sigma}{dt}\ (t=0) \geq 51.09\ [\sigma_{pp} - \sigma_{pn}]^2 \tag{12}$$

The value of the R.H.S. of eqn. (12) depends on whether we use the neutron beam data or the deuterium data. In Fig. 10 we see that the Fermilab data falls right between the two lower bounds shown. It violates the neutron beam lower bound and one is forced to conclude either that the neutron beam data is wrong or that the preliminary Fermilab charge exchange data is incorrect.

Thus despite the huge amplification which charge exchange produces in any new asymptotic term of Δp we cannot at present reach a clean conclusion. Suffice it to say that the data is pregnant with implication and that one awaits experimental confirmation with much exictement.

4) BACKWARD πp SCATTERING

We come now to what may well be the most startling example of unexpected asymptotic behaviour - the exchange amplitude with baryon number one that controls backward ($u \simeq 0$) scattering in $\pi^{\pm}p$ collisions.

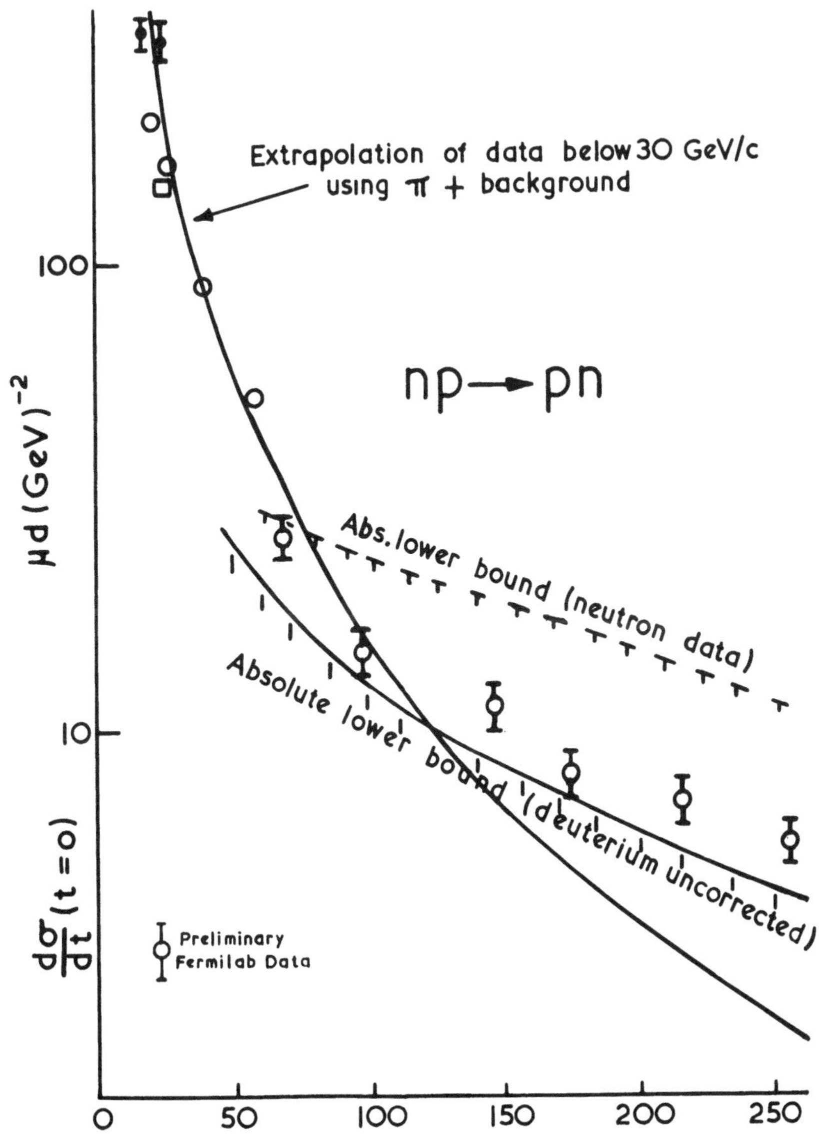
Extrapolation of data below 30 GeV/c
using π + background
np→pn
Abs. lower bound (neutron data)
Absolute lower bound (deuterium uncorrected)
Preliminary
Fermilab Data
$\frac{d\sigma}{dt}(t=0)$ μb(GeV)$^{-2}$
100
10
0
50
100
150
200
250

Figure 10

Before going on to considering the new results emerging from very high energy studies we must recall that baryon exchange reactions have generally been less well understood than meson exchange processes at typical medium energies (5-30 GeV/c). Thus in backward $\pi^{\pm}p$ scattering where the dominant exchanges are the N (nucleon) and Δ (N^*_{33}) trajectories there have always been certain difficulties of interpretation. The major features of the reactions, which are illustrated for p_L = 10 GeV/c in Fig. 11 are:

i) $\pi^+p \to p\pi^+$ has an extremely sharp backward peak and a deep minimum at $u \simeq -0.15$ (GeV/c)^2.

ii) $\pi^-p \to p\pi^-$ has neither a sharp peak nor a minimum.

iii) In the backward direction one has

$$\left.\frac{d\sigma}{du}\right|_{\pi^+p} \gtrsim 5 \times \left.\frac{d\sigma}{du}\right|_{\pi^-p} .$$

iv) The above behaviour remain qualitatively true up to 17 GeV/c.

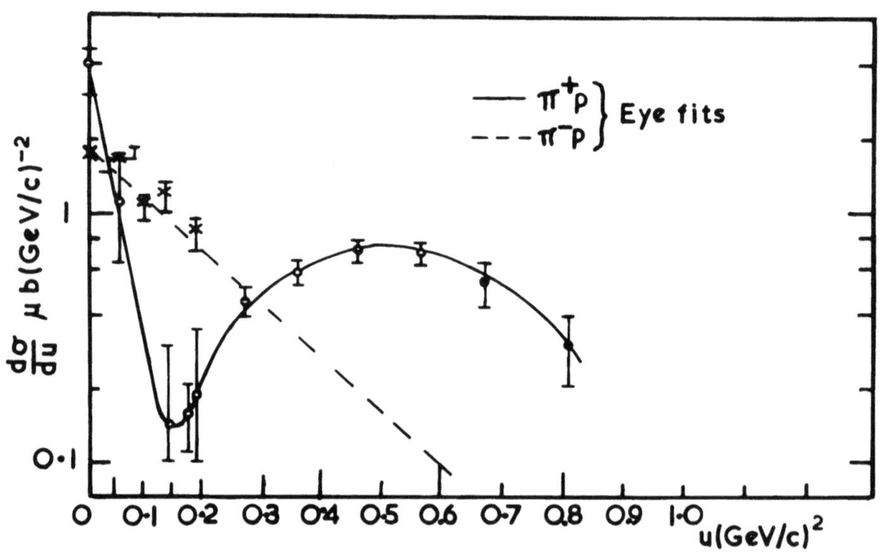

Fig. 11 - Data on backward $\pi^{\pm}p$ scattering at 10 GeV/c (Taken from ref. 7).

All these remarkable features appear to have a natural explanation in terms of the exchange of the N and Δ trajectories:-

i) Both N and Δ contribute to $\pi^+p \to p\pi^+$. N exchange dominantes, is responsible for the sharp peak, and the dip at $u \simeq -0.15$ is connected to the wrong-signature nonsense zero occurring at $\alpha_N(u) = -\frac{1}{2}$. This is in agreement with the linear trajectory

$$\alpha_N(u) = -0.36 + u$$

suggested by the Chew-Frautschi plot.

ii) Only Δ can contribute to $\pi^-p \to p\pi^-$; hence the absence of sharp peak or dip.

iii) The dominance of π^+p over π^-p reflects the dominance in the coupling strength of N over Δ.

iv) Although Δ exchange drops off more slowly with energy than N exchange does (one has $\alpha_\Delta(u) \simeq 0 + u$ from the Chew-Frautschi Plot) N continues to dominate up to 17 GeV/c and the main features are unaltered.

Unfortunately this very pretty picture is somewhat marred by the polarisation measurements which for $\pi^+p \to p\pi^+$ show a maximum at $u \simeq -0.15$ where the theory would predict a zero. There are also problems with the strength of the couplings needed to fit the cross-section data, but these may simply reflect inadequate methods of parametrisation.

However, the properties we wish to examine transcend these detailed questions. We shall argue that the very high energy data on $\pi^\pm p \to p\pi^\pm$ seem to be totally incompatible with this conventional picture.

In the first place as we go to higher energies the rôle of Δ exchange must become more important simply because $\alpha_\Delta(0) > \alpha_N(0)$. Ultimately it has to dominate completely, and at this point one must find several extraordinary results:-

i) $$\frac{d\sigma}{du}(\pi^-p \to p\pi^-) \simeq 9\,\frac{d\sigma}{du}(\pi^+p \to p\pi^+) \qquad (13)$$

because of the isospin properties.

ii) The peak in $\pi^+p \to p\pi^+$ becomes negligible.

iii) There is no longer a dip in $\pi^+p \to p\pi^+$ at $u = -0.15\ (\text{GeV/c})^2$.

All these results are absolutely contrary to the experimental data below 17 GeV/c and therefore stand as remarkable predictions of the conventional theory. In quantitative terms one expects the $\pi^-p \to p\pi^-$ cross-section to have caught up with $\pi^+p \to p\pi^+$ for u=0 at an energy of about 50 GeV.

Fermilab data is not yet available but data at 25 and 38 GeV/c from Serpukhov show no sign of the expected behaviour. The peak in

π^+p is just as sharp, and $d\sigma/du$ for π^+p is still about five times bigger than π^-p in the backward direction.

The overall decrease with energy also shows signs of slowing down, Fig. 12, but it is perhaps premature to try to estimate what the asymptotic trend will be.

However, a suggestion of a quite extraordinary behaviour has been inferred by Diu and Tchang using an indirect method of analysis. Basically they use partial wave dispersion relations to evaluate the πN $\ell=1$ partial waves in the region of s=0, where they are known from measurements of the πN phase-shifts and crossing. In particular the P_{11} phase shift varies very rapidly near s=0 and it is shown that in order to reproduce this behaviour it is necessary that the $I=\frac{1}{2}$ exchange amplitude increase strongly above p_L = 40 GeV/c. Interpreted in the language of

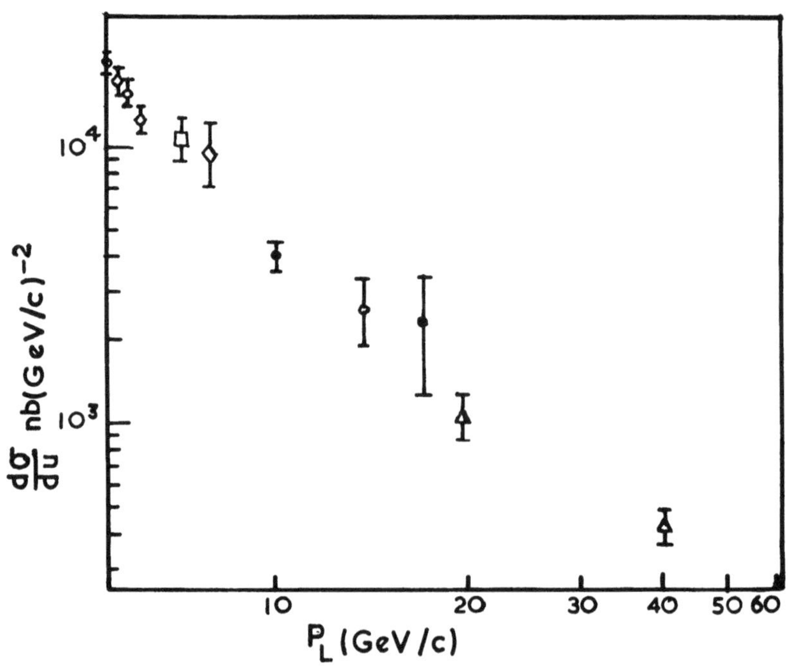

Fig. 12 - Energy variation of backward differential cross section for π^+p scattering. (Data from ref. 7)

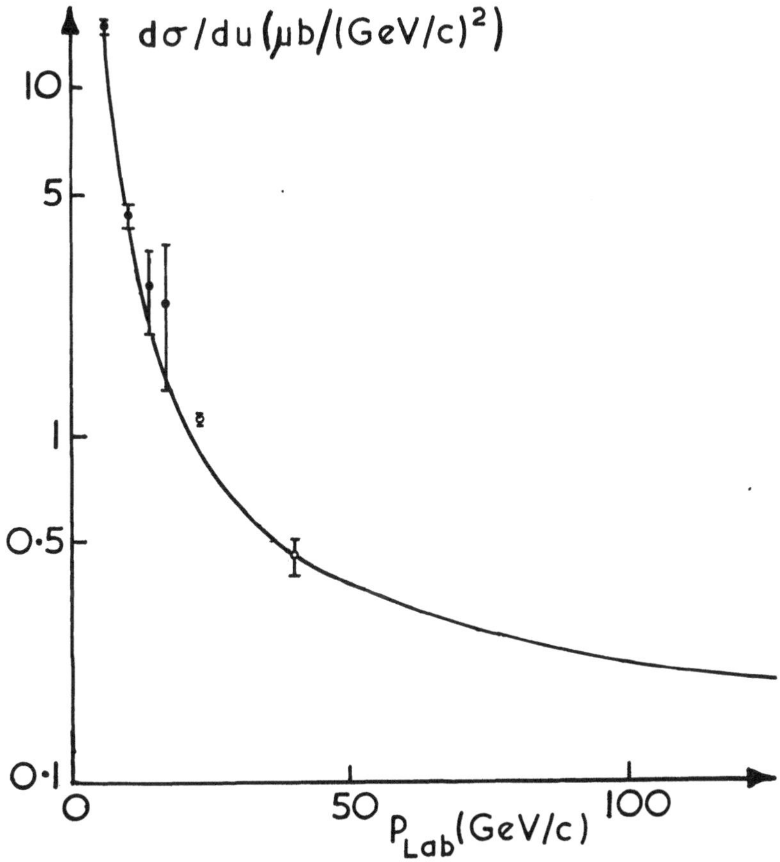

Fig. 13 - Possible high energy behaviour of backward π^+p differential cross section. (From ref. 3)

trajectories it requires

$$\alpha_{I=\frac{1}{2}}(0) \tilde{\sim} 1 \tag{14}$$

instead of $\alpha_{I=\frac{1}{2}} = \alpha_N = -0.36$.

If (14) is correct, and the isospin one half baryon exchange amplitude is dominated by an $\alpha \tilde{\sim} 1$, we shall have the unprecedented spectacle at Fermilab of constant backward differential cross sections for $\pi^{\pm}p \rightarrow p\pi^{\pm}$. Fig. 13 indicates how the slowing down of the energy variation shown in Fig. 12 might continue to take place at higher energies.

5) CONCLUSIONS

With the advent of ultra-high-energy machines we may be entering a realm in hadronic physics where the general ideas of Regge pole theory no longer hold. The ISR has already taught us that the pp total cross-section is growing at a rate which is difficult, and perhaps impossible, to accommodate within the Regge framework (though strong coupling Reggeon field theories may still hold the key). The early Fermilab results on $\pi^- p \rightarrow \pi^o n$ and on cross-section differences, $\sigma_{pp} - \sigma_{pn}$ in particular, suggest that also amplitudes with quantum number exchange, e.g. ρ and A_2 type, may be beginning to depart from the expectations based on the conventional Regge picture. Even baryon exchange, as shown by the energy trends in backward $\pi^{\pm}p$ scattering, may be behaving in a completely anomalous fashion as compared with the traditional Regge structure.

We thus face the disquieting, but exciting possibility that $\alpha_{effective}$ for ρ, A_2 and N exchange at ultra high energies is $\tilde{\sim} 1$; at least in two body reactions. The more conventional Regge description would then have to be seen somehow as appropriate only to an intermediate energy range and the "true" asymptotic behaviour awaiting our detailed investigation, at Fermilab and at SPS, may then be quite different.

All the experimental effects discussed above, and which are the basis for our suggestions, are small and difficult to measure. The consequences for our interpretation, on the other hand, are not. Herein, we have played the Devil's advocate but time will soon tell whether or not it is all just "Much Ado About Nothing".

REFERENCES

1) D. Joynson, E. Leader, B. Nicolescu and C. Lopez, Nuovo Cimento 30A, 345 (1975).

2) B. Diu and E. Leader, Nuovo Cimento 28A, 137 (1975); A. Bouquet, B. Diu, E. Leader and B. Nicolescu, Nuovo Cimento 31A, 411 (1976).

3) B. Diu and P. Tchang, Nuovo Cimento 32A, 185 (1976).

4) A. V. Stirling, P. Sonderegger, J. Kirz, P. Falk-Vairant, O. Guisan, C. Bruneton, P. Borgeaud, M. Yvert, J. P. Guillaud, C. Caverzasio and B. Amblard: Phys. Lett., 14, 763 (1965); P. Sonderegger, J. Kirz, O. Guisan, P. Falk-Vairant, C. Bruneton, P. Borgeaud, A. V. Stirling, C. Caverzasio, J. P. Guillaud, M. Yvert and B. Amblard: Phys. Lett., 20, 75 (1966).

5) A. V. Barnes, D. J. Mellema, A. V. Tollestrup, R. L. Walker, O. L. Dahl, R. A. Johnson, R. W. Keeney and M. Pripstein: California Institute of Technology preprint CALT-68-465 (1974); W. F. Baker, A. S. Carroll, I-H. Chiang, R. L. Cool, D. P. Eartly, O. Fackler, G. Giacomelli, T. F. Kycia, P. F. M. Koehler, K. K. Li, P. O. Mazur, P. Mockett, K. P. Pretzl, D. C. Rahm, R. Rubinstein and A. A. Wehmann: Phys. Rev. Lett., 33, 932 (1974).

6) K. J. Foley, R. S. Jones, S. J. Lindenbaum, W. A. Love, S. Ozaki, E. D. Platner, C. A. Quarles and E. H. Willen: Phys. Rev. Lett., 19, 330 (1967).

7) W. F. Baker, D. P. Eatly, P. Koehler, K. P. Pretzl, S. M. Pruss, R. Rubinstein, R. M. Kalbach and S. Mukhin, NAL Proposal No. 290, (1974).

DISCUSSION

CHAIRMAN: Prof. E. Leader

Scientific Secretary: T. Wilkie

DISCUSSION I

- *PARSONS:*

To remove the discrepancy between Regge theory and experiment, you have suggested a singularity at $J = 1$ for $t = 0$. Does this not imply the existence of a new zero-mass strongly interacting vector meson?

- *LEADER:*

It would be fun to have a new strongly-coupled photon but, unfortunately, experiment denies us this pleasure. We have to arrange that the contribution is non-singular at $t = 0$ by a decoupling mechanism analogous to a ghost-killing factor.

- *PARSONS:*

Is it obvious how to derive the energy dependence of these new terms? What happens to signature for instance?

- *LEADER:*

We show that the imaginary part of the new term is given by $s(\log s)^{\beta}$, with $0 \leq \beta \leq 1$. The two most interesting cases are the extremes $\beta = 0$ and $\beta = 1$. To show that $\beta = 1$ is the maximum possible value is tantamount to re-deriving the Froissart bound, but now for an antisymmetric amplitude. The phase then follows from the Phragmen-Lindelhof theorem. In the other extreme case $\beta = 0$, the singularity behaves like an odd signatured pomeron. We would have a signature factor $\xi = [1 - \exp(-i\pi\alpha)]/\sin \pi\alpha$, $(\tau = -1)$, and this would give the phase in the usual fashion. Thus the new term is real for $t = 0$, $A'_{NEW} = cs$, with c real.

- *BERLAD:*

Is there a non-relativistic potential scattering example of crossing trajectories?

- *LEADER:*

Yes, there are examples. There is a classic paper which discusses this problem. I really would not worry about colliding trajectories -- there is no theoretical difficulty.

- *BERLAD:*

In $\pi^{\pm}p \to \pi^{\pm}p$ there is a Coulomb contribution in the near forward direction. In the subtraction $\sigma_{\pi^+p} - \sigma_{\pi^-p}$ one *adds* these respective contributions. If they are not properly removed, one ends up with an energy independent term. Are you sure this possible contribution does not affect the high-energy tail of the reaction?

- *LEADER:*

The experimentalists always attempt to include the Coulomb corrections, but you are right to suggest that this could be a source of danger.

- *FERBEL:*

The correction can never be done too carefully.

- *LEADER:*

This problem was looked at by Carlitz at Chicago. It cannot explain the effects with which I am dealing.

- *TARNOPOLSKY:*

Some people think that the double Pomeron exchange process

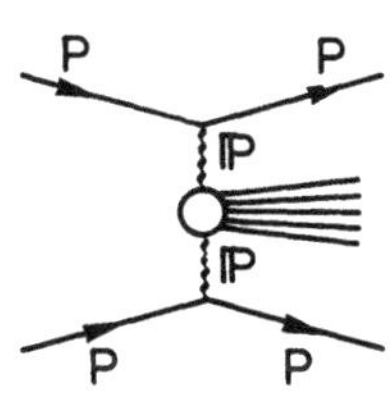

with small momentum transfers where a state of low rapidity is formed provides a stringent test of the Regge ideas? What is your opinion on this idea and the experimental situation?

- *LEADER:*

This is outside the topics which I covered in my talk. I am not an expert on Reggeon field theory, so I will say very little.

There were two approaches to this problem. In the weak field theory approach, one drew perturbative diagrams which just look like the eikonalization of a fundamental potential due to Pomeron exchange. This was never really successful. The alternative strong coupling approach started off with unphysical energy dependence $\sim s^{1+\varepsilon}$, $\varepsilon > 0$, which could then renormalize itself. One did get a rise in the total cross-section with energy, but at energies beyond the ISR.

The point is that Reggeon field theory is the study of the high energy behaviour of complicated Feynman graphs. At ISR energies, not all the sub-energies are large, and so one cannot just blindly Reggeize. Capella et al in Paris have adopted a more phenomenological approach and claim to have obtained a rise in σ_{tot} in accord with the data, but I would not like to comment, as I have not studied their work closely.

I have been so surprised at the diffraction data shown by Ferbel that I would not really like to comment on double Pomeron exchange processes. What I would like to know from Ferbel, however, is what the latest story is on double diffractive processes of the type

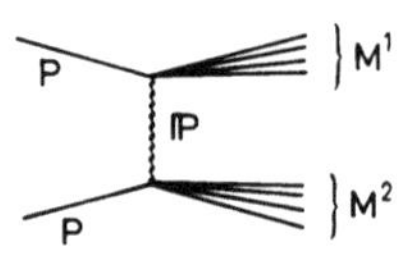

where M_1, M_2, and t are kept small

- *FERBEL:*

I cannot answer for Fermilab since there is no data for fixed M_1 and M_2. Because of poor statistics, what is done is to integrate over the masses. This is, I think, a dangerous procedure. There are some comments on Pomeron factorization from Schlein's group at IRS, again integrating over the masses. I do not know how to interpret this.

- *LEADER:*

What would be interesting would be to plot $(d^3\sigma)/(dM_1 dM_2 dt)$ at each energy, with M_1, M_2 in narrow bins, as a function $A \exp\left[b(M_1, M_2)t\right]$ and to check factorization by comparing with the elastic and with the single diffraction processes.

If the Pomeron factorizes, then for $M_1 \simeq M_2$, we should expect

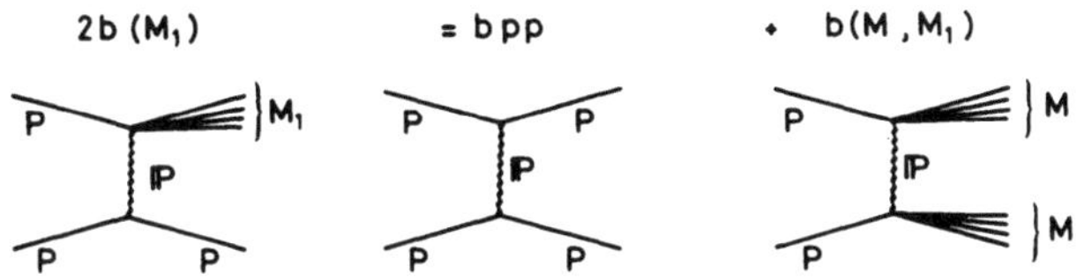

- *FERBEL:*

Using something similar to this, I get factorization to within the error bars, integrating over the masses -- but I think that due to this, the test would not be better than to 30% if the data were better.

- *LEADER:*

One should remember that all tests of Pomeron factorization work only to within 30%.

- *BUCCELLA:*

That you find this J = 1 singularity in both I = 1 meson exchange and in baryon exchange processes is perhaps a sign that the Regge language, which deals with the exchange of definite quantum numbers, is no longer appropriate. If you write the s-channel unitarity equation for elastic processes, then you have a coherent sum of contributions. For charge exchange, however, the contributions have different signs which tend to cancel. But there is no reason why the cancellations should switch off the J = 1 contribution.

- *LEADER:*

Yes, but the unitarity sum in the s-channel is very difficult at high energies. I stated, as part of my philosophical motivation at the beginning, that the coherent summation gives a larger answer $\sim s\,(\log s)^2$ than naive Regge theory. Perhaps, indeed, the cancellations in charge exchange only reduce this to O(s) behaviour.

- *LIPKIN:*

You have not included KN data in your analysis of isovector exchange. Nearly all models predict that the total cross-sections are related by

$$\sigma(pp) - \sigma(pn) = \sigma(K^+p) - \sigma(K^+n)$$

and

$$\sigma(\bar{p}p) - \sigma(\bar{p}n) = \sigma(K^-p) - \sigma(K^-n) \; .$$

Thus, any inconsistency in $\sigma(pp) - \sigma(pn)$ could be checked by looking at $\sigma(K^+p) - \sigma(K^+n)$ where the ambiguities in the Glauber corrections are much smaller. Are the KN data not consistent with

$$\sigma(K^+p) - \sigma(K^+n) = 0 \ ?$$

Is this consistent with the differences in $\sigma(pp) - \sigma(pn)$ which you find in the nucleon data?

- *LEADER:*

Our new result is not directly checked by looking at the KN system. The couplings could be different and we have no way to judge them except by appealing to SU(3). However, as long as the quark relations you mention continue to hold roughly, there cannot be any contradiction.

DISCUSSION 2

- *GOURDIN:*

Nucleon pole dominance in nucleon-deuterium inelastic scattering implies the well-known expression

$$\sigma(pd) = \sigma(pn) + \sigma(pp) - (\text{corrections}) \ ,$$

where the dominant contributions are shown by the diagrams

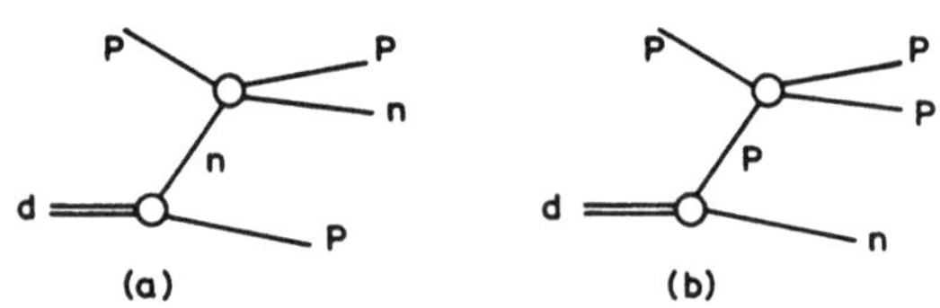

The difficulty is to extrapolate correctly at the nucleon pole, taking into account the deuteron structure. This procedure, which is suitable to isolate diagrams (a), can be checked by looking at diagram (b) which we can compare with the result given by a free proton target. The corresponding kinematical situation is a spectator neutron of very small kinetic energy; but if this were experimentally feasible, it would give more confidence on the way to compute the Glauber corrections.

- *LEADER:*

It is certainly a good idea to test one's ideas on Glauber corrections by attempting to derive the pp total cross-section from pd measurements. The diagrams, however, refer to specific channels and not to total cross-section measurements.

- *GOURDIN:*

This question concerns $(d\sigma/dt)_{t=0}$ in np → pn. At low energy, π + background seems to agree very will with experiment, but the contribution from $|\phi_1+\phi_3|^2$ due to $\Delta\sigma \neq 0$ is still present. Did you check that it is small compared to the previous contribution?

- *LEADER:*

Yes, the contribution to Im $(\phi_1+\phi_3)$ via the optical theorem is always present and cannot be cancelled. However, at lower energies, as at high energy, $\Delta\sigma$ is small. Compared with the contributions of π + background at these lower energies, its effect through Im $(\phi_1+\phi_3)$ is completely negligible. Do not forget that the contribution to $d\sigma/dt$ from π + background is decreasing as $1/s^2$. At p_L = 10 GeV/c, $d\sigma/dt \sim 700$ μb $(GeV/c)^{-2}$. At 200 GeV/c, π + background gives about 10 μb $(GeV/c)^{-2}$, and now the contribution from $\Delta\sigma$ is comparable and therefore very important.

- *BERLAD:*

What is the nature of the pion pole?

- *LEADER:*

Phenomenologically, since one is at such small t when talking about pion exchange, it does not matter whether one considers it as a fixed pole or a moving trajectory. On the other hand, your quastion at a fundamental level is an important one. One does not really know what kind of object the pion is. We once attempted to answer the question by looking at high order Feynman diagrams in photoproduction, but the results were so peculiar that we simply could not take them seriously.

- *PAULI:*

(A philosophical question from a non-Reggeist.)

Suppose the data show that Regge theory cannot work without a combination of baryonic Pomerons, and an obscene number of trajectories, and maybe some Ptolomeic epicycles, yet we cannot discard the fundamental idea of the complex J-plane. What should theoretical physicists do? Shoud we stick with S-matrix theory, wait for "the unified field theory", give up trying to systematize strong interactions?

- *LEADER:*

I have never been a fanatical Reggeist. In 1966, Slansky and I wrote a paper in Physics Review, "Critical tests for Regge pole theory", suggesting 13 ways of testing the theory critically in NN → NN. Unfortunately, the tests involved very difficult and esoteric spin measurements which were quite impossible at that time. What is very exciting is that the polarized beam and target at Argonne now make the tests experimentally feasible.

No-one has come up with any alternative to Regge theory in the last 15 years, nor has anyone produced a decent calculation of a residue function. People should try to get serious and reliable results in the theory -- then the tests would become really meaningful.

On the other hand, it may turn out that the situation in the complex J-plane really is exceedingly complicated. We cannot forget the J-plane. It is an established and unassailable concept in quantum mechanics, but nothing guarantees simplicity in that plane. That will be a pity, but we shall have to learn to live with it.

HADRONIZATION OF QUARK THEORIES

Hagen Kleinert

Institut für Theoretische Physik

Freie Universität Berlin, Berlin, Germany

ABSTRACT

Local quark gluon theories are converted into bilocal field theories via functional techniques. The new field quanta consist of all quark antiquark bound states in the ladder approximation. They are called "bare hadrons". Hadronic Feynman graphs are developed which strongly resemble dual diagrams. QED is a special case with the "bare hadrons" being positronium atoms. Photons couple to hadrons via intermediate vector mesons in a current-field identity . The new theory accommodates naturally bilocal currents measured in deep-inelastic e p scattering Also these couple via intermediate mesons.

In the limit of heavy gluon masses, the hadron fields become local and describe π, ρ, A_1, σ mesons in a chirally invariant Lagrangian (the "σ model"). Many interesting new relations are found between meson and quark properties such as $m_\rho^2 \approx 6M^2$ where M is the "true" non-strange quark mass after spontaneous breakdown of chiral symmetry. There is a simple formula linking these quark masses with the small "bare masses" of the Lagrangian. The quark masses also determine the vacuum expectations of scalar densities. These show an SU(3) breaking in the vacuum of $\approx$ - 16 %.

I. INTRODUCTION

In the attempt to understand strong interaction, two basically different theoretical approaches have been developed in the past years. One of them, the dual approach, is based on complete democracy among all strongly interacting particles. Within this approach, an elaborate set of rules assures the construction of certain lowest order vertex functions for any number of hadrons[1]. The other approach assumes the existence of a local field equation involving fundamental quarks bound together by vector gluons[2] Here strong interaction effects on electomagnetic and weak currents of hadrons can be analyzed in a straight-forward fashion without detailed dynamical computations[3]. Either approach has its weakness where the other is powerful. Dual models have, until now, given no access to currents while quark theories have left the problem of hadronic vertex function intractable. Not even an approximate bound state calculation is avaiable (except in 1 + 1 dimensions[4] or by substituting the field couplings by simple ad-hoc forces[5]).

At present there is hope that the problems connected with quark models are of a purely technical nature. A Lagrangian field theory of Yang-Mills type seems to have a good chance of defining a true fundamental theory of elementary particles. Dual models, on the other hand, seem to be of a more phenomenological character. Once the fundamental vertices are determined, it is difficult to find next corrections and to extend the prescriptions to what might be called a complete theory. If this could be done it would certainly have to be phrased in terms of local infinite-component or multi-local fields[6)].

It would be very pleasing if both models were, in fact, essentially equivalent both being different languages for one and the same underlying dynamics. In this case one could use one or the other depending on whether one wants to answer short- or long-distance questions concerning quarks.

In order to learn how a translation between the different languages might operate we shall consider, in these lectures, the simplified field theory in which quarks are colorless, have N flavours, and are held together by vector gluons of arbitrary mass μ. This theory incorporates several realistic features of strong interactions, for example current algebra and PCAC. Moreover, the case N=1 and $\mu=0$ includes ordinary quantum electrodynamics (Q.E.D.). This will provide a good deal of intuition as well as the possibility of a detailed test of our results.

We shall demonstrate how functional methods can be employed to transform the local quark gluon theory into a new completely equivalent field theory involving only bilocal fields. The new free field quanta coincide with quark-antiquark bound states when calculated by ladder exchanges only. They may be considered as "bare hadrons". Accordingly, the transition from the local quark- to the bilocal hadron-theory will be named "hadronization". In the special case of Q.E.D., "bare hadrons" are positronium atoms in ladder approximation.

The functional technique will ensure that bare hadrons have exactly the correct interactions among each other in order that hadronization preserves the equivalence to the original quark gluon theory. It is simple to establish the connection between classes of Feynman graphs involving quarks and gluons with single graphs involving hadrons. The topology of hadron graphs is the same as that of dual diagrams. It is interesting to observe the appearence of a current-hadron field identity for photons just as employed in phenomenological discussions of vector meson dominance. Moreover, since the theory is bilocal, this identity can be extended to bilocal currents which are measured in deeply inelastic electromagnetic and weak interactions.

The limit of a very heavy gluon mass can be hadronized most simply. Here the bilocal fields become local and describe only a few hadrons with the quantum numbers of σ, π, ρ, A_1 mesons. The Lagrangian coincides with the standard chirally invariant σ model which is known to account quite well for the low-energy aspects of meson physics. Here hadronization renders additional connection between quark and meson properties. It also makes transparent the connection between the very small bare quark masses (which describe the explicit breakdown of chiral symmetry) and the mechanical quark masses (which include the dynamic effects due to spontaneous symmetry violations).

II TECHNICAL PRELIMINARIES

Before we embark in our program we have to recall certain functional techniques[7]. They are generalizations of the basic Fresnel integrals, valid for all real A≠0

$$\int_{-\infty}^{\infty} \frac{d\xi}{\sqrt{2\pi i}}\, e^{\frac{i}{2}\xi A \xi} = A^{-\frac{1}{2}} \tag{2.1}$$

$$\int \frac{d\xi}{\sqrt{2\pi i}} \frac{d\xi^*}{\sqrt{2\pi i}}\, e^{i\xi^* A \xi} = A^{-1} \tag{2.2}$$

where the complex integral $\int d\xi\, d\xi^*$ stands symbolic for $\int_{-\infty}^{\infty} d\,\frac{\xi+\xi^*}{\sqrt{2}} \int_{-\infty}^{\infty} d\,\frac{\xi-\xi^*}{\sqrt{2}\,i}$. Quadratic completion in the exponent yields:

$$\int_{-\infty}^{\infty} \frac{d\xi}{\sqrt{2\pi i}}\, e^{i\left(\frac{1}{2}\xi A \xi + j\xi\right)} = A^{-\frac{1}{2}}\, e^{-\frac{i}{2} j A^{-1} j} \tag{2.3}$$

$$\int \frac{d\xi}{\sqrt{2\pi i}} \frac{d\xi^*}{\sqrt{2\pi i}}\, e^{i\left(\xi^* A \xi + j^*\xi + \xi^* j\right)} = A^{-1}\, e^{-i j^* A^{-1} j} \tag{2.4}$$

If ξ is a vector $\xi = (\xi_1, \ldots, \xi_n)$ and $\int \mathcal{D}\xi$ stands short for $\prod_k \int \frac{d\xi_k}{\sqrt{2\pi i}}$, then for non-singular symmetric or hermitian matrices A these formulas become

$$\int \mathcal{D}\xi\, e^{i\left(\frac{1}{2}\xi A \xi + j\xi\right)} = (\det A)^{-\frac{1}{2}}\, e^{-\frac{i}{2} j A^{-1} j} \tag{2.5}$$

$$\int \mathcal{D}\xi \mathcal{D}\xi \; e^{i(\xi^\dagger A \xi + j^\dagger \xi + \xi^\dagger j)} = (\det A)^{-1} e^{-i j^\dagger A^{-1} j} \qquad (2.6)$$

respectively, as can be seen immediately by diagonalizing A via an orthogonal or unitary transformation (which leaves the measures $\mathcal{D}\xi$ or $\mathcal{D}\xi\,\mathcal{D}\xi^\dagger$ invariant). If finally $\xi(x)$ is a function $\xi(x)$ and $\xi^\dagger A \xi$ is understood in the functional sense as

$$\xi^\dagger A \xi \equiv \int dx\,dy\; \xi^*(x)\, A(x,y)\, \xi(y) \qquad (2.7)$$

these integrals may still be defined by grating the x axis into finer and finer lattices of points $x_k = k\varepsilon$ with $k=0,\pm1,\pm2,\ldots$ and reducing the problem to the previous case via the vector components $\xi_k = \sqrt{\varepsilon}\,\xi(x_k)$.
For large matrices A, the calculation of the determinant is performed most simply by expanding

$$\det A = \exp \ln \det A = \exp \operatorname{tr} \ln A$$

$$(2.8)$$

$$= \exp \operatorname{tr} \ln[1+(A-1)] = \exp \operatorname{tr}\left\{A-1-\tfrac{1}{2}(A-1)^2+\tfrac{1}{3}(A-1)^3 \pm \ldots\right\}$$

This formula is directly applicable in the functional case if all sums over intermediate indices are replaced by the corresponding integrals, for example:

$$\mathrm{tr}\, A = \sum_k A_{kk} \rightarrow \int dx\, A(x,x)$$
$$\mathrm{tr}\, A^2 = \sum_{k,l} A_{kl} A_{lk} \rightarrow \int dx dy\, A(x,y) A(y,x) \tag{2.9}$$

The integrals (2.5) and (2.6) can also be extended to functions $\xi(x)$ with values in an anticommuting algebra (i.e. $\{\xi(x), \xi(y)\} = 0$). In this case the (det A) in equs. (2.5),(2.6) appears in the inverse forms $(\det A)^{1/2}$ and $(\det A)^{1}$, respectively.

With these preparations consider now a Lagrangian composed of fermion and boson fields and split into free and interacting part $\mathcal{L}(\psi, \bar\psi, \varphi) = \mathcal{L}_0 + \mathcal{L}_{int}$. All time ordered Green's functions can be obtained from the derivatives with respect to the external sources of the generating functional

$$Z[\eta, \bar\eta, j] = \mathrm{const} \langle 0 | T e^{i\int dx\, (\mathcal{L}_{int} + \bar\eta\psi + \bar\psi\eta + j\varphi)} | 0 \rangle \tag{2.10}$$

The fields in the exponent follow free equations of motion and $|0\rangle$ is the free-field vacuum. The constant is conventionally chosen to make $Z[0,0,0] = 1$, i.e.

$$\mathrm{const} = \left[\langle 0 | T e^{i\int dx\, \mathcal{L}_{int}(\psi, \bar\psi, \varphi)} | 0 \rangle \right]^{-1} \tag{2.11}$$

This normalization may always be enforced at the very end of any calculation such that $Z[\eta, \bar\eta, j]$ is only interesting as far as its functional dependence is concerned, modulo the irrelevant constant in front.

It is then straight-forward to show that $Z[\eta,\bar{\eta},j]$ can alternatively be computed via the Feynman path integral formula

$$Z[\eta,\bar{\eta},j] \propto \int \mathcal{D}\psi\mathcal{D}\bar{\psi}\mathcal{D}\varphi\, e^{i\int dx\,(\mathcal{L}_0(\psi,\bar{\psi},\varphi)+\mathcal{L}_{int}+\bar{\eta}\psi+\bar{\psi}\eta+j\varphi)} \tag{2.12}$$

Here the fields are no more operators but classical functions (with the mental reservation that classical Fermi fields are anticommuting objects). Notice that contrary to the operator formula (2.10) the <u>full</u> action appears in the exponent.

For simplicity, we demonstrate the equivalence only for one real scalar field $\varphi(x)$. The extension to other fields is immediate [7,8]. First note that it is sufficient to givethe proof for the free field case, i.e.

$$\begin{aligned} Z_0[j] &= \langle 0|T e^{i\int dx\, j(x)\varphi(x)}|0\rangle \\ &\propto \int \mathcal{D}\varphi\, e^{i\int dx\left(\frac{1}{2}\varphi(x)(-\Box_x-\mu^2)\varphi(x)+j(x)\varphi(x)\right)} \end{aligned} \tag{2.13}$$

For if it holds there a simple multiplication on both sides of (2.13) by the differential operator

$$e^{i\int dx\,\mathcal{L}_{int}\left(\frac{1}{i}\frac{\delta}{\delta j(x)}\right)} \tag{2.14}$$

would extend it to the interacting functionals (2.10) or (2.12). But (2.13) follows directly from Wick's theorem according to which any time ordered product of a free field can be expanded into a sum of normal products with all possible time ordered contractions. This statemen can be summarized in operator form valid for any functional $F[\varphi]$ of a free field $\varphi(x)$:

$$T\,F[\varphi] = e^{\frac{1}{2}\int dx\,dy\,\frac{\delta}{\delta\varphi(x)} D(x-y) \frac{\delta}{\delta\varphi(y)}} : F[\varphi]: \tag{2.15}$$

where $D(x-y)$ is the free-field propagator

$$D(x-y) = \int \frac{d^4q}{(2\pi)^4}\, e^{-iq(x-y)} \frac{i}{q^2-\mu^2+i\varepsilon} = \frac{i}{-\Box_x-\mu^2+i\varepsilon}\,\delta(x-y) \tag{2.16}$$

Applying this to (2.13) gives

$$\begin{aligned} Z_0[j] &= e^{\frac{1}{2}\int dx\,dy\,\frac{\delta}{\delta\varphi(x)} D(x-y)\frac{\delta}{\delta\varphi(y)}} \langle 0|:e^{i\int dx\, j(x)\varphi(x)}:|0\rangle \\ &= e^{-\frac{1}{2}\int dx\,dy\, j(x) D(x-y) j(y)} \langle 0|:e^{i\int dx\, j(x)\varphi(x)}:|0\rangle \\ &= e^{-\frac{1}{2}\int dx\,dy\, j(x) D(x-y) j(y)} \end{aligned} \tag{2.17}$$

The last part of the equation follows from the vanishing of all normal products of $\varphi(x)$ between vacuum states.

Now exactly the same result is obtained by performing the functional integral in (2.13) and using (2.5). The matrix A is in this case $A(x,y) = (-\Box_x - \mu^2)\delta(x-y)$, such that its inverse becomes the propagator $D(x-y)$:

$$A^{-1}(x,y) = \frac{1}{-\Box_x - \mu^2 + i\varepsilon}\,\delta(x-y) = -i\,D(x-y) \tag{2.18}$$

yielding again (2.17).

Notice that it is Wick's expansion which supplies the free part of the Lagrangian when going from the operator form (2.15) to the functional version (2.12).

III QUARK GLUON THEORY

Consider now a system of N quarks $\psi(x)$ held together by one gluon field $G^\nu(x)$ of mass μ via a Lagrangian

$$\mathcal{L}(x) = \bar\psi(x)(i\not\partial - m)\psi(x) + g\,\bar\psi(x)\gamma_\nu\psi(x)G^\nu(x) - \frac{1}{4}F_{\mu\nu}^2(x) + \frac{\mu^2}{2}G_\nu^2 \tag{3.1}$$

Here $F_{\mu\nu}$ is the usual curl $\partial_\mu G_\nu - \partial_\nu G_\mu$. In the special case in which N = 1, $\mu = 0$, and $g^2 = 4\pi\alpha$, this Lagrangian describes quantum electrodynamics. In other cases it may be considered as a model field theory which carries many interesting properties of strong interactions, for example approximate SU(3) symmetry, chiral SU(3) xSU(3) current algebra, PCAC, and scaling up to small corrections. Certainly this model will never be able to confine quarks, give symmetric baryon wave functions, and explain infinitely rising hadron trajectories. For this it would have to contain an additional, exactly conserved, color symmetry with $G^{\nu}(x)$ being its non-abelian gauge mesons. Before attempting to deal with this far more complicated situation we shall develop[10] our tools for the less realistic but much simpler model (3.1) without color.

The generating functional of all time ordered Green's function is

$$Z[\eta,\bar{\eta},j^{\nu}] \propto \int \mathcal{D}\psi\mathcal{D}\bar{\psi}\mathcal{D}G\, e^{i\int dx\,(\mathcal{L}+\bar{\psi}\eta+\bar{\eta}\psi+g\,j^{\nu}G_{\nu})} \tag{3.2}$$

The exponent is quadratic in $G^{\nu}(x)$, such that the functional integration over the gluon field can be performed[7,8] (using (2,5)) leaving:

$$Z[\eta,\bar{\eta},j^{\nu}] \propto \int \mathcal{D}\psi\mathcal{D}\bar{\psi}\, e^{i\mathcal{A}[\bar{\psi},\psi,\bar{\eta},\eta,j^{\nu}]} \tag{3.3}$$

where the action is

$$\mathcal{A}[\psi,\bar{\psi},\eta,\bar{\eta},j^{\nu}] = \int dx\,dy \Big\{ \big(\mathcal{L}(x)+\bar{\psi}(x)\eta(x)+\bar{\eta}(x)\psi(x)\big) \times\delta(x-y) - \frac{i}{2}g^{2}D(x-y)\big(\bar{\psi}(x)\gamma^{\nu}\psi(x)+j^{\nu}(x)\big)\big(\bar{\psi}(y)\gamma_{\nu}\psi(y)+j_{\nu}(y)\big)\Big\} \tag{3.4}$$

By employing the Fierz identity:

$$\gamma^{\nu}_{\alpha\beta} \otimes \gamma_{\nu\,\gamma\delta} = 1_{\alpha\delta} \otimes 1_{\gamma\beta} + (i\gamma_5)_{\alpha\delta} \otimes (i\gamma_5)_{\gamma\beta}$$

$$- \frac{1}{2} \gamma^{\nu}_{\alpha\delta} \otimes \gamma_{\nu\,\gamma\beta} - \frac{1}{2} (\gamma^{\nu}\gamma_5)_{\alpha\delta} \otimes (\gamma_{\nu}\gamma_5)_{\gamma\beta} \tag{3.5}$$

the quartic quark interaction term can be written in a different fashion

$$\frac{i}{2} g^2 D(x-y) \{ \bar{\psi}(x)\psi(y)\bar{\psi}(y)\psi(x) + \bar{\psi}(x) i\gamma_5 \psi(y) \bar{\psi}(y) i\gamma_5 \psi(x)$$

$$- \frac{1}{2} \bar{\psi}(x)\gamma^{\nu}\psi(y)\bar{\psi}(y)\gamma_{\nu}\psi(x) - \frac{1}{2}\bar{\psi}(x)\gamma^{\nu}\gamma_5\psi(y)\bar{\psi}(y)\gamma_{\nu}\gamma_5\psi(x)\}$$

$$\equiv \frac{i}{2} g^2 D(x-y) \bar{\psi}_{\alpha}(x) \psi_{\delta}(y) \xi_{\alpha\delta,\gamma\beta} \bar{\psi}_{\gamma}(y) \psi_{\beta}(x) \tag{3.6}$$

This is the point where our elimination of quark fields in favor of new bilocal fields can set in.

Let $S(x,y)$, $P(x,y)$, $V^{\nu}(x,y)$, $A^{\nu}(x,y)$ be a set of hermitian auxiliary fields, i.e.

$$S(x,y) = S(y,x) \;,\; P(x,y) = P^{*}(y,x) \;,\; etc \tag{3.7}$$

With these fields one can certainly construct the following functional identities[9)]

$$\int \mathcal{D}S(x,y)\, e^{-\frac{i}{2}\left|S(x,y)+ig^2 D(x-y)\,\bar{\psi}(y)\psi(x)\right|^2/ig^2 D(x-y)} = \text{const}$$

$$\int \mathcal{D}P(x,y)\, e^{-\frac{i}{2}\left|P(x,y)+ig^2 D(x-y)\,\bar{\psi}(y)i\gamma_5\psi(x)\right|^2/ig^2 D(x-y)} = \text{const}$$

$$\int \mathcal{D}V(x,y)\, e^{i\left|V^{\nu}(x,y)-\frac{i}{2}g^2 D(x-y)\,\bar{\psi}(y)\gamma^{\nu}\psi(x)\right|^2/ig^2 D(x-y)} = \text{const}$$

$$\int \mathcal{D}A(x,y)\, e^{i\left|A^{\nu}(x,y)-\frac{i}{2}g^2 D(x-y)\,\bar{\psi}(y)\gamma^{\nu}\gamma_5\psi(x)\right|^2/ig^2 D(x-y)} = \text{const} \qquad (3.8)$$

which are independent of the fields $\psi(x)$. If we now multiply $Z[\eta,\bar{\eta},j^{\nu}]$ in (3.3) by these constants and make use of (3.6), all quartic quark terms are seen to cancel. The generating functional becomes

$$Z[\eta,\bar{\eta},j^{\nu}] \propto \int \mathcal{D}\psi\,\mathcal{D}\bar{\psi}\,\mathcal{D}S\,\mathcal{D}P\,\mathcal{D}V\,\mathcal{D}A\, e^{i\tilde{\mathcal{A}}} \qquad (3.9)$$

where the new action $\tilde{\mathcal{A}}$ denotes the integral

$$\tilde{\mathcal{A}}[\psi,\bar{\psi},S,P,V,A,\eta,\bar{\eta},j^{\nu}] = \int dx\,dy\,\mathcal{L}(x,y) \qquad (3.10)$$

with the bilocal Lagrangian

$$\begin{aligned}\mathcal{L}(x,y) \equiv & \left\{\bar{\psi}(x)(i\not{\partial}-m)\psi(x)+\bar{\psi}(x)\eta(x)+\bar{\eta}(x)\psi(x)\right\}\delta^{(4)}(x-y) \\ & -\bar{\psi}(x)m(x,y)\psi(y) - \frac{i}{2}g^2 D(x-y)\,j^{\nu}(x)\,j_{\nu}(y) \qquad (3.11) \\ & -\left\{\tfrac{1}{2}|S|^2+\tfrac{1}{2}|P|^2-|V|^2-|A|^2\right\}/ig^2 D(x-y)\end{aligned}$$

Here m(x,y) has been introduced as an abbreviation for the combined field

$$m(x,y) \equiv S(x,y) + P(x,y)i\gamma_5 + \left(V^\nu(x,y) + \delta^{(4)}(x-y)\int dz\, ig^2 D(x-z) j^\nu(z)\right)\gamma_\nu + A^\nu(x,y)\gamma_\nu\gamma_5 \quad (3.12)$$

Due to (3.7), the matrix m(x,y) is self-adjoint in the sense

$$\left(\overline{m(x,y)}\right)_{\alpha\beta} \equiv \gamma_{0\,\alpha\alpha'}\left(m^*(x,y)^T\right)_{\alpha'\beta'}\gamma_{0\,\beta'\beta} = m_{\alpha\beta}(\,,x) \quad (3.13)$$

At this place it is worth remarking that the Lagrangian (3.11) shows its equivalence to the previous form (3.4) also quite directly. By virtue of the Euler Lagrange equations the fields S, P, V, A are seen to be dependent fields coinciding with the corresponding bilocal quark expressions

$$S(x,y) = -ig^2 D(x-y)\bar\psi(y)\psi(x)$$

$$P(x,y) = -ig^2 D(x-y)\bar\psi(y)i\gamma_5\psi(x)$$

$$V^\nu(x,y) = \frac{i}{2}g^2 D(x-y)\bar\psi(y)\gamma^\nu\psi(x) \quad (3.14)$$

$$A^\nu(x,y) = \frac{i}{2}g^2 D(x-y)\bar\psi(y)\gamma^\nu\gamma_5\psi(x)$$

Inserting these relations back into (3.11) reproduces (3.4). In the action (3.10), quark fields enter only in quadratic form such that they can be integrated according to formula (2.6) (in its fermionic version). The matrix A is in this case

$$A(x,y) = (i\not\partial - m)\,\delta^{(4)}(x-y) - m(x,y)$$

Hence $A^{-1}(x,y) \equiv -iG(x,y)$ becomes simply the Green's function of the equation

$$\int dy\, \left[(i\not\partial - m)\,\delta^{(4)}(x-y) - m(x,y)\right] G(y,z) = i\delta^{(4)}(x-z) \qquad (3.15)$$

With this notation, the quark integration brings the functional (3.9) to the form

$$Z[\eta,\bar\eta,j^\nu] \propto \int \mathcal{D}m(x,y)\, e^{i\mathcal{A}[m,\eta,\bar\eta,j^\nu]} \qquad (3.16)$$

with

$$\mathcal{A}[m,\eta,\bar\eta,j^\nu] = \int dx\,dy \Big\{ -i\,\mathrm{tr}\left(\ln iG^{-1}\right)(x,y)\,\delta(x-y) \qquad (3.17)$$

$$-\frac{1}{2}\,\mathrm{tr}\left(m(x,y)\,\xi^{-1} m(y,x)\right)/ig^2 D(x-y) + i\bar\eta(x)\,G(x,y)\,\eta(y)$$

$$-\frac{2}{D(0)} V^\nu(x,x)\,D(x-y)\,j_\nu(y) - ig^2 \frac{\delta^{(4)2}(x-y)}{D(0)} \int dz\,dz'\, D(z-x)\,D(y-z')\,j^\nu(z)\,j_\nu(z') \Big\}$$

Here we have introduced, for brevity, the notation $\mathcal{D}m(x,y) \equiv \mathcal{D}S\,\mathcal{D}P\,\mathcal{D}V\,\mathcal{D}A$. Notice that the effect of the matrix ξ^{-1} defined in equ. (3.6) is simply to devide the projections into S, P, V, A by 4, -4, -2, 2, respectively+). The trace refers only to Dirac indices.

+) Since $\frac{1}{4}\,1\otimes 1$, $-\frac{1}{4}\,(\gamma_5)\otimes(\gamma_5)$, $\frac{1}{4}\,\gamma^\nu\otimes\gamma_\nu$, $-\frac{1}{4}\,\gamma^\nu\gamma_5\otimes\gamma_\nu\gamma_5$ are the correcponding projection operators.

The new functional (3.16) is identical to the original one in equ. (3.2). As a consequence, a quantum theory based on the action (3.17) must be completely equivalent to the original quantized quark gluon theory.

A word is in order concerning the internal symmetry SU(N) among the N quarks (i=1,...N) under consideration. Since the gluon is an SU(N) singlet, the interaction in equ. (3.1) is $g\sum_{i=1}^{N}\bar{\psi}_i\gamma^\nu\psi_i G_\nu(x)$. In the Fierz transformed version (3.6) the indices i and j appear separated

$$\frac{i}{2}g^2 D(x-y)\bar{\psi}^j(x)\psi_i(y)\,\xi\,\bar{\psi}^i(y)\psi_j(x)$$

Hence in the presence of N quarks, the fields m(x,y) have to be thought of a matrices in SU(N) space $m(x,y)_i{}^j$. This carries over to the action with the traces including Dirac as well as SU(N) indices.

Let us now develop a quantum theory for the new action. In general, the field m(x,y) may oscillate around some constant non-zero vacuum expectation value $m_0\delta^{(4)}(x-y)$. It is convenient to subtract such a value from m(x,y) and introduce the field

$$m'(x,y) \equiv m(x,y) - m_0\delta^{(4)}(x,y) \tag{3.18}$$

With this and the definition

$$M \equiv \mathfrak{M} + m_0 \,, \tag{3.19}$$

equ. (3.15) can be rewritten as

$$\int dy \left[(i\not{\partial} - M)\delta^{(4)}(x-y) - m'(x,y) \right] G(y,z) = i\delta^{(4)}(x-z) \tag{3.20}$$

Now let us assume that the oscillations m'(x,y) are sufficiently small as to permit a perturbation expansion for $G(x,y)$:

$$G(x,y) = G_M(x,y) - i(G_M m' G_M)(x,y) - (G_M m' G_M m' G_M)(x,y) + \ldots \tag{3.21}$$

where $G_M(x,y)$ are the usual propagators of a free fermion of mass M.

$$G_M(x,y) \equiv G_M(x-y) \equiv \int \frac{d^4p}{(2\pi)^4} e^{-ipx} \frac{i}{\not{p} - M}$$

Using this expansion, the action (3.17) takes the form+)

$$\mathcal{A}[m', \eta, \bar{\eta}, j^\nu] = \mathcal{A}_1[m'] + \mathcal{A}_2[m'] + \mathcal{A}_{int}[m'] + \mathcal{A}_{ext}[m', \eta, \bar{\eta}, j^\nu] \tag{3.22}$$

with $\mathcal{A}_1$ denoting the term linear in the field m'(x,y)

$$\mathcal{A}_1[m'] \equiv \int dx dy \, tr \left[G_M(x-y) m'(x,y) - \xi^{-1} m'(x,y) m_0 \delta(x-y) / i g^2 D(x-y) \right] \tag{3.23}$$

+) A trivial additive constant has been dropped.

and $\mathcal{A}_2$ being quadratic in m'

$$\mathcal{A}_2[m'] \equiv \int dx\,dy\, \mathrm{tr}\left[\frac{i}{2} G_M m\, G_M m')(x,y) - \frac{1}{2}\xi^{-1} m'(x,y)\, m'(y,x)/ig^2 D(x-y)\right] \tag{3.24}$$

The term $\mathcal{A}_{int}[m']$ collects all remaining powers in m'

$$\mathcal{A}_{int}[m'] \equiv \int dx\, \mathrm{tr}\left[-\sum_{n=3}^{\infty} \frac{(-i)^{n+1}}{n} (G_M m')^n (x,x)\right] \tag{3.25}$$

The last piece $\mathcal{A}_{ext}$, finally, contains all interactions with the external sources

$$\mathcal{A}_{ext}[m', \eta, \eta, j^\nu] \equiv \int dx\,dy \Big\{ i\bar{\eta}(x) G(x,y) \eta(y)$$

$$- \frac{i}{2} g^2 D(x-y)\, j^\nu(x)\, j_\nu(y) \tag{3.26}$$

$$- \frac{2}{D(0)} V^\nu(x,x) D(x-y)\, j_\nu(y) - ig^2 \frac{\delta^{(4)}(0)\,\delta^{(4)}(x-y)}{D(0)} \int dz\,dz'\, D(z-x) D(y-z')\, j^\nu(z)\, j_\nu(z') \Big\}$$

For the quantization we shall adopt an interaction picture. As usual, the quadratic part of the action, $\mathcal{A}_2[m']$, serves for the construction of free-particle Hilbert space. According to the least action principle, the free equation of motion are obtained from $\delta \mathcal{A}_2[m']/\delta m'(x,y) = 0$ rendering

$$m'(x,y) = g^2 \xi\, D(x-y) \left(G_M\, m' G_M\right)(x,y) \tag{3.27}$$

Going to momentum space

$$m'(p_2,p_1) \equiv \int dx_2 dx_1 \, e^{i(x_2p_2 - x_1p_1)} m'(x_2,x_1)$$

and introducing relative and total momenta

$$P \equiv (p_2+p_1)/2 \quad , \quad q \equiv (p_2-p_1)/2$$

together with the notation

$$m'(P|q) \equiv m'(p_2,p_1)$$

the field equation becomes

$$m'(P|q) = \xi g^2 \int \frac{d^4P'}{(2\pi)^4} D(P'-P) G_M(P'+\tfrac{q}{2}) m'(P'|q) G_M(P'-\tfrac{q}{2})$$

(3.28)

In this form we easily recognize the Bethe Salpeter equation[11] in ladder approximation for the vertex functions of quark-antiquark bound states

$$\Gamma^H(P|q) \equiv N_H G_M(P+\tfrac{q}{2}) \int dz \, e^{iPz} \langle 0|T\,\psi(\tfrac{z}{2})\bar{\psi}(-\tfrac{z}{2})|0\rangle G_M(P-\tfrac{q}{2})$$

(3.29)

where N_H is some normalization factor.As a consequance our free field m'(x,y) can be expanded in a complete set of ladder bound state solutions. These are the bare quanta spanning the Hilbert space of the interaction picture. Because of their bound quark-antiquark nature, they will be called "bare hadrons". In the special case of QED, "quarks" are electrons and the bare hadrons" are positronium atoms.

For mathematical reasons it is convenient to solve (3.28) for fixed $q^2 \in (0, 4M^2)$ and all possible coupling constants g^2, to be called $g_H^2(q^2)$, i.e.

$$\Gamma^H(P|q) = \xi\, g_H^2(q^2) \int \frac{d^4P'}{(2\pi)^4} D(P-P') G_M(P'+\tfrac{q}{2}) \Gamma^H(P'|q) G_M(P'-\tfrac{q}{2}) \tag{3.30}$$

A useful normalization condition is

$$-i \int \frac{d^4P}{(2\pi)^4} \mathrm{tr}\left[G_M(P+\tfrac{q}{2}) \Gamma^{H'}(P|q) G_M(P-\tfrac{q}{2}) \bar{\Gamma}^{H}(P|q) \right] = \varepsilon^H \delta^{HH'} \tag{3.31}$$

Here we have allowed for a sign factor $\varepsilon^H(q)$ which cannot be absorbed in the normalization N_H of (3.29). It may take the values +1, -1 or zero. Then the expansion of the free field m'(x,y) in terms of hadron creation and annihilation operators $a_H^\dagger(q), a_H(q)$ can be written as

$$m'_{\alpha\beta}(x,y) = \int \frac{d^4q}{(2\pi)^4} \sum_H \delta^{(+)}(g_H^2(q^2) - g^2) \int \frac{d^4P}{(2\pi)^4} \left\{ e^{-i(q(x+y)/2 + P(x-y))} \Gamma^H(P|q) n_H a_H(q) \atop e^{i(q(x+y)/2 - P(x-y))} \bar{\Gamma}^H(P|-q) n_H^* a_H^\dagger(q) \right\} \tag{3.32}$$

where n_H are appropriate factors giving $a_H(q)$ the standard normalization

$$\left[a_H(q), a_{H'}^\dagger(q') \right] = (2\pi)^3 \delta^{(3)}(q-q') \, 2\omega^H(q) \varepsilon^H(q) \tag{3.33}$$

Now the sign factor $\varepsilon^{H}(q)$ appears at the norm of the hadronic state $a_H^+|0\rangle \equiv |H\rangle$. In general there will be many states with unphysical norms since the "bare hadrons" are produced by ladder diagrams only and may not be directly related to physical particles. This situation presents no fundamental difficulty. There are many interactions among bare hadrons which are capable of excluding unphysical states from the S-matrix. In fact, the equivalence of the hadronized theory to the healthy original quark gluon version is a guarantee for physical results (on shell).

The propagator of the free field m'(x,y) can be found most directly by adding an external disturbance to the free action

$$\mathcal{A}_2[m'] \to \mathcal{A}_2[m'] - \int dx\,dy\, \mathrm{tr}\,[m'(x,y)J(y,x)] \tag{3.34}$$

This current enters the equation of motion as

$$\begin{aligned} m'(x,y) &= \xi g^2 D(x-y)\left(G_M\, m' G_M\right)(x,y) \\ &\quad - \xi i g^2 D(x-y)\, J(x,y) \end{aligned} \tag{3.35}$$

The propagator $\mathcal{G}_{\alpha\beta,\alpha'\beta'}(xy,x'y') \equiv \overbrace{m'_{\alpha\beta}(x,y)\, m'_{\alpha'\beta'}(x',y')}$ is then defined as the solution of (3.35) for the δ-function disturbance

$$J_{\alpha\beta}(x,y) = i\,\delta(x-x')\,\delta(y-y')\,\delta_{\alpha\alpha'}\delta_{\beta\beta'} \tag{3.36}$$

It satisfies the inhomogeneous Bethe-Salpeter equation

$$\mathcal{G}_{\alpha,\beta,\alpha'\beta'}(x,y;x',y') = \xi_{\alpha\beta,\alpha'\beta'} D(x-y) \int d\bar{x}\, d\bar{y}\, G_M(x-\bar{x})_{\alpha\bar{\alpha}}$$

$$\mathcal{G}_{\bar{\alpha}\bar{\beta},\alpha'\beta'}(\bar{x},\bar{y};x',y') G_M(\bar{y}-y)_{\bar{\beta}\beta} + \xi_{\alpha\beta,\beta'\alpha'} g^2 D(x-y)\delta(x-x')\delta(y-y') \tag{3.37}$$

This is immediately recognized as the equation for the two-quark trasition matrix in ladder approximation (see equ. (A.19) in App. A .

An explicite representation of the Green's function in terms of the solutions $\Gamma^H(P|q)$ of the homogeneous equation (3.30) can now be given.

If $\mathcal{G}_{\alpha\beta,\alpha'\beta'}(P,P'|q)$ denotes the Fourier transform

$$(2\pi)^4 \delta^{(4)}(q-q') \mathcal{G}_{\alpha\beta,\alpha'\beta'}(P,P'|q) \tag{3.38}$$

$$\equiv \int dx\,dy\,dx'\,dy'\, e^{i[P(x-y)+q(x+y)/2-P'(x'-y')-q'(x'+y')/2]} \mathcal{G}_{\alpha\beta,\alpha'\beta'}(x,y;x',y')$$

it can be written as the sum over all hadron solutions:

$$\mathcal{G}_{\alpha\beta,\alpha'\beta'}(P,P'|q) = -ig^2 \sum_H \varepsilon_H(q) \frac{\Gamma^H_{\alpha\beta}(P|q)\,\bar{\Gamma}^H_{\beta'\alpha'}(P'|-q)}{q_H^2(q^2)-q^2} \tag{3.39}$$

where the sum comprises possible integrals over a continuous set of solutions. If quarks and gluons were scalars, the sum would be discrete for $q^2 \in (0, 4M^2)$ since the kernel of the integral equation (3.30) would be of the Fredholm type. A more detailed discussion is given in Appendix A. Here we only note that a power series expansion of the denominator

$$\mathcal{G}_{\alpha\beta,\alpha'\beta'}(P,P'|q) = -i \sum_{n=1}^{\infty} \sum_{H} \left(\frac{q^2}{g_H^2(q^2)} \right)^n \varepsilon_H(q) \Gamma^H_{\alpha\beta}(P|q) \bar{\Gamma}^H_{\beta'\alpha'}(P'|-q) \qquad (3.40)$$

renders explicit the exchange of one, two, three etc. gluons. Hence one additional gluon can be inserted (or removed) by multiplying (or dividing) (3.39) by a factor $q^2/g_H^2(q^2)$. This fact will be of use later on.

Seen microscopically in terms of quarks and gluons, the free hadron propagator (3.39) is given by the sum of ladders (see Fig. 1)

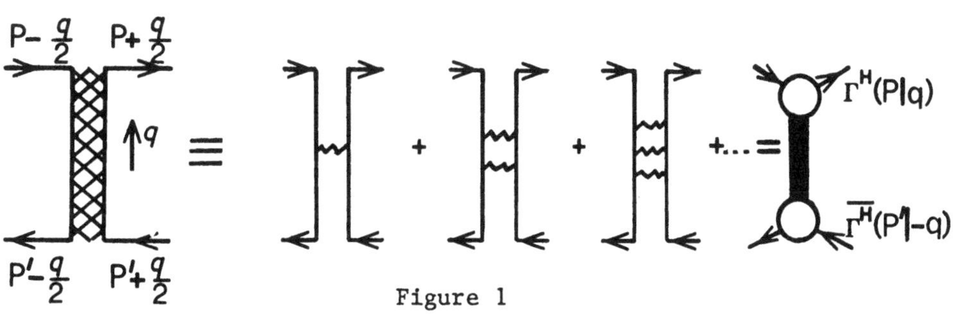

Figure 1

Graphically it will be represented by a wide band. In the last term of Fig 1 we have also given a visualisation of the expansion (3.39). Here the fat line denotes the propagator

$$\Delta_H(q) = -i\,\varepsilon_H(q)\,\frac{g^2}{g_H^2(q^2)-g^2} \tag{3.41}$$

while upper and lower bubbles stand for the Bethe Salpeter vertices $\Gamma^H(P|q)$ and $\bar{\Gamma}^H(P'|-q)$, respectively. This picture suggests another way of representing the new bilocal theory in terms of an infinite component hadron field depending only on the average position $X=(x+y)/2$ For this we simply expand the interacting field $m'(P|q)$ in terms of the complete set of free vertex function

$$m'(P|q) = \sum_H \Gamma^H(P|q)\,m_H(q) \tag{3.42}$$

Inserting this expansion into (3.22), the free action becomes directly

$$\mathcal{A}_2[m'] = \frac{1}{2}\int dX\, m_H(X)\left(1-g_H^2(q^2)/g^2\right)m_H(X) \tag{3.43}$$

implying the free propagator (3.41) for the field $m_H(X)$ With this understanding of the free part of the action we are now prepared to interprete the remaining pieces.

Consider first the linear part $\mathcal{A}_1[m']$. The first term in it can graphically be represented as whown in Fig. 2 . When attached to other hadrons it produces a tadpole correction.

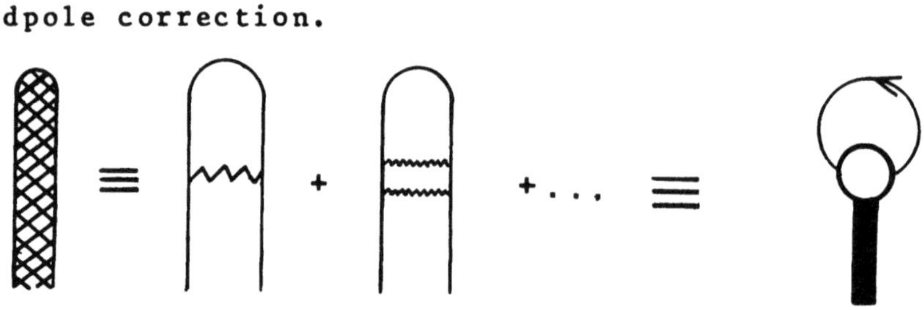

Figure 2

When interpreted within the underlying quark gluon picture, such a correction sums up all rainbow contributions to the quark propagator. Also the second term in $\mathcal{A}_1[m']$ has a straight-forward interpretation. First of all, the division by $\xi i g^2 D(x-y)$ has the effect of removing one rung from the ladder sum (such that the ladder starts with no rung, one rung, etc.) and creating two open quark legs. This can be seen directly from (3.30) and (3.39): Suppose a hadron line ends at the interaction $-\int dx dy \, \mathrm{tr}\,[m'(x,y)\,\xi^{-1} m_0]/ig^2 D(x-y)\,\delta(x-y)$

Then the factor $[\xi i g^2 D(x-y)]^{-1}$ applied to $\mathcal{G}(P,P'|q)$

gives (leaving out irrelevant indices)

$$[\xi g^2 D]^{-1}\mathcal{G} = -ig^2 \sum_H \varepsilon_H \frac{[\xi g^2 D]^{-1} \Gamma^H \bar{\Gamma}^H}{g_H{}^2(q^2) - g^2} \tag{3.43}$$

Using (3.30) this yields

$$= -ig^2 \sum_H \varepsilon_H \frac{g_H^2(q^2)}{g^2} \frac{(G_M \Gamma^H G_M)\bar{\Gamma}^H}{g_H{}^2(q^2) - g^2} \tag{3.44}$$

As discussed before, the factor $g_H^2(q^2)/g^2$ amounts to the removal of one rung. Multiplication by $-m_0$ and integration over $\int dP/(2\pi)^4$ yields the total contribution of this hadron graph

$$i m_0 \sum_H \frac{g_H^2(q^2)}{g_H^2(q^2)-g^2} \int \frac{d^4P}{(2\pi)^4} \, \mathrm{tr}\left[G_M(P+\tfrac{q}{2})\Gamma^H(P|q)G_M(P-\tfrac{q}{2})\right] \times \varepsilon_H(q)\bar{\Gamma}^H(P'|-q) \qquad (3.45)$$

As far as quarks and gluons are concerned, this amounts to the insertion of a mass term m_0 on top of a ladder graph with one rung removed (this being indicated by a slash in Fig. 3 .

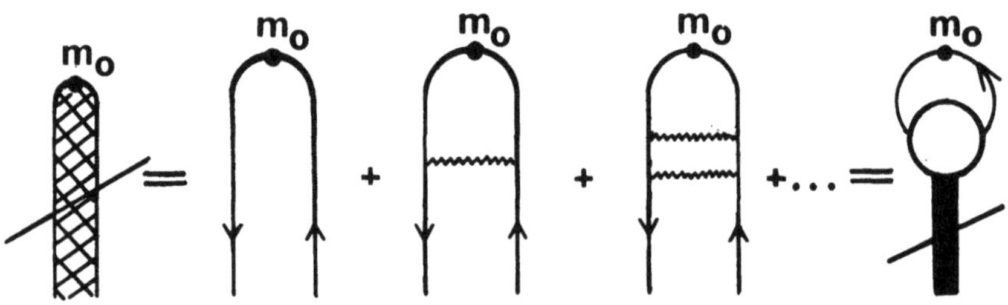

Figure 3

The quark gluon picture leads us to expect that m_0 must be a cutoff dependent quantity cancelling the logarithmic divergence in every upper loop of the ladder sum of Fig. 2 . Numerically, m_0 is most easily calculated by cancelling the infinite contributed by $\mathcal{A}_1[m]$ to the equation of motion (3.28). If we include $\mathcal{A}_1[m']$, this equation reads

$$m_0(2\pi)^4\delta^{(4)}(q) + m'(P|q) = \left[\xi i g^2 \int \frac{d^4P'}{(2\pi)^4} D(P-P')G(P')\right](2\pi)^4\delta^{(4)}(q) \qquad (3.46)$$

$$+ \xi g^2 \int \frac{d^4P'}{(2\pi)^4} D(P-P')G_M(P'+\tfrac{q}{2})m'(P'|q)G_M(P'-\tfrac{q}{2})$$

The first term on the right-hand side is exactly the usual self energy $\Sigma(P)(2\pi)^4\delta^{(4)}(q)$ in second order

$$\Sigma_{\alpha\beta}(P) \equiv -\xi_{\alpha\beta,\gamma\delta}\, i\int \frac{d^4P'}{(2\pi)^4}\frac{1}{(P-P')^2-\mu^2}\frac{1}{\not{P}'-M} \quad (3.47)$$

Normalizing $\Sigma(P)$ on mass shell one finds the usual expression

$$\Sigma(P) = \Sigma_0 + \Sigma_1(\not{P}-M) + \Sigma_R(\not{P}) \quad (3.48)$$

where Σ_R is the regularized self-energy.
The cutoff dependent term

$$\Sigma_0 = \frac{3}{4\pi}\frac{g^2}{4\pi} M\left(\log \Lambda^2/M^2 + \tfrac{1}{2}\right) \quad (3.49)$$

must be balanced by choosing $m_0 = -\Sigma_0$ on the left had side of (3.46). Also the second term Σ_1 is cutoff dependent:

$$\Sigma_1 = \frac{1}{4\pi}\frac{g^2}{4\pi}\left(\log \Lambda^2/M^2 + \frac{9}{2} + 2\log \mu^2/M^2\right) \quad (3.50)$$

and a renormalization is necessary to cancel this infinity. Most economic is the introduction of an appropriate wave function counter term $(Z_2-1)\bar{\Psi}(i\not{\partial}-M)$ in the original Lagrangian (3.1). Such a term would enter equ. (3.15) as

$$\int dy \left\{(i\not{\partial}-m)\delta^{(4)}(x-y) + (Z_2-1)(i\not{\partial}-M)\delta^{(4)}(x-y) - m(x,y)\right\} G(y,z)$$
$$= i\delta^{(4)}(x-z) \quad (3.51)$$

Instead of (3.18), m(x,y) should now be assumed to oscillate around $\left[m_0 + (Z_2^{-1}-1)(i\partial\!\!\!/-M)\right]\delta(x-y)$.
By defining a new $m'(x,y)$ via

$$m'(x,y) \equiv m(x,y) - \left[m_0 + (Z_2^{-1}-1)(i\partial\!\!\!/-M)\right]\delta^{(4)}(x-y), \tag{3.52}$$

the full action (3.22) is obtained exactly as before except for the linear part $\mathcal{A}$ in which the new wave function renormalization term enters together with m_0:

$$\mathcal{A}_1[m'] = \int dx\,dy\ \mathrm{tr}\Big\{ G_M(x-y)\,m'(x,y) \\ - \xi^{-1}\, m'(x,y)\left[m_0 + (Z_2^{-1}-1)(i\partial\!\!\!/-M)\right]\delta(x-y)/ig^2 D(x-y)\Big\} \tag{3.53}$$

By choosing

$$Z_2^{-1} - 1 = -\Sigma_1 \tag{3.54}$$

the cutoff dependent term Σ_1 is exactly compensated in the equation of motion (3.46). After this renormalization procedure, only the finite term $\Sigma_R(P)$ is left. The regularized action is

$$\mathcal{A}_1[m']_R = -\int dx\,dy\ \Sigma_R(x-y)\, m'(x,y)/ig^2 D(x-y) \tag{3.55}$$

Using the expansion (3.42), this can be rewritten as

$$\mathcal{A}_1[m']_R = \sum_H \int dX\, f_H(-\Box)\, m_H(X) \tag{3.56}$$

with

$$f_H(q^2) = i\int \frac{d^4p}{(2\pi)^4}\,\mathrm{tr}\left[\Sigma_R(p)G_M(p+\tfrac{q}{2})\Gamma^H(p|q)G_M(p-\tfrac{q}{2})\right]\frac{g_H^2(q^2)}{g^2} \tag{3.57}$$

By momentum conservation, the tadpole momentum always vanishes such that only $f_H(0)$ is needed eventually.

Let us now proceed to the discussion of the interaction part $\mathcal{A}_{int}[m']$ of equ. (3.25). Take as an example the term of the third order in m'. If a hadron line ends at every m', it can be represented graphically as shown in Fig. 4

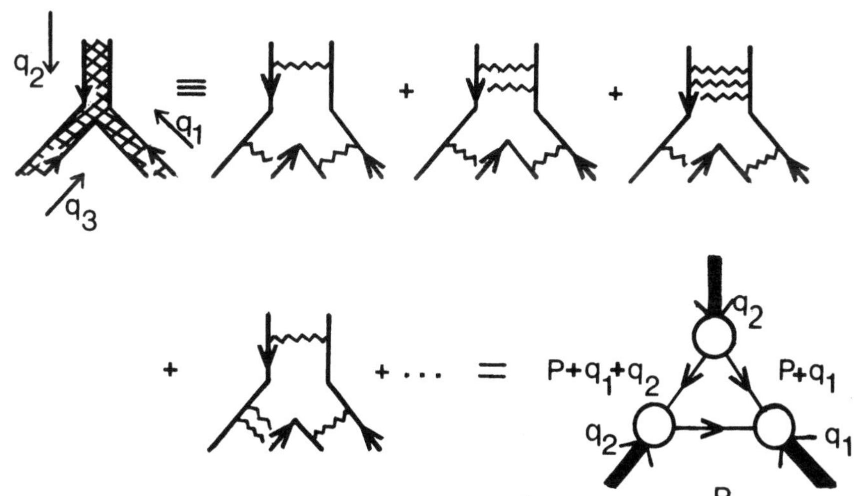

Figure 4

Employing the expansion (3.42), this interaction term can be rewritten as

$$A_{int}^{3\,hadr}[m'] = -\frac{1}{3}\sum_{H_1H_2H_3}\int \frac{d^4q_3}{(2\pi)^4}\frac{d^4q_2}{(2\pi)^4}\frac{d^4q_1}{(2\pi)^4}(2\pi)^4\delta(q_1+q_2+q_3)$$

$$\times\int\frac{d^4P}{(2\pi)^4}\,\mathrm{tr}\Big[\Gamma^{H_3}(P-\tfrac{q_3}{2}|q_3)G_M(P+q_1+q_2)\Gamma^{H_2}(P+q_1+\tfrac{q_2}{2}|q_2) \quad (3.58)$$

$$G_M(P+q_1)\Gamma^{H_1}(P+\tfrac{q_1}{2}|q_1)G_M(P)\Big]m_{H_3}(q_3)m_{H_2}(q_2)m_{H_1}(q_1)$$

$$=\frac{1}{3}\sum_{H_1H_2H_3}\int dX\, v_{H_3H_2H_1}(i\partial_X^{H_3}, i\partial_X^{H_2}, i\partial_X^{H_1})\, m_{H_3}(X)m_{H_2}(X)m_{H_1}(X)$$

with a vertex function $v_{H_3H_2H_1}(i\partial_X^{H_3}, i\partial_X^{H_2}, i\partial_X^{H_1})$ whose derivatives $\partial_X^{H_i}$ are to be applied only to the argument of the corresponding field $m_{H_i}(X)$. A corresponding formula holds for every power of m'.

Notice that the flow of the quark lines in every interaction is anticlockwise. When drawing up hadronic Feynman graphs it may sometimes be more convenient to draw a clockwise flow. A simple identity helps to write down directly the corresponding Feynman rules. Consider a graph for a three hadron interaction and cross the upper band downwards (see Fig. 5). The interaction appears now with the hadronic bands in anticyclic order, and the fermion lines in the hadron vertex flowing clockwise. This is topologically compensated by twisting every band once. Mathematically, this deformation displays the

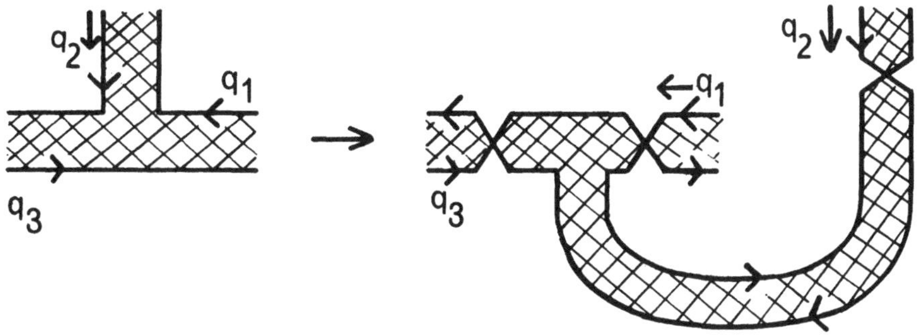

Figure 5

following identity of the vertex functions

$$v_{H_3 H_2 H_1}(q_3, q_2, q_1) = \eta_{H_3} \eta_{H_2} \eta_{H_1} v_{H_1 H_2 H_3}(q_1 q_2 q_3) \quad (3.59)$$

where the phase η_H denotes the charge parity of the hadron H. This phase may be absorbed in the propagator characterizing the twisted band.

The proof of this identity (3.59) is quite simple. Let C be the charge conjugation matrix. Then the vertices satisfy:

$$C \Gamma^H(P|q) C^{-1} = \eta_H \Gamma^H(-P|q)^T \quad (3.60)$$

Inserting now CC^{-1} between all factors in (3.58) and observing $C \gamma^\mu C^{-1} = -\gamma^{\mu T}$ one has

$$v_{H_3 H_2 H_1}(q_3, q_2, q_1) = -\eta_{H_3}\eta_{H_2}\eta_{H_1} \int \frac{dP}{(2\pi)^4} \operatorname{tr}\left\{ \Gamma^{H_3}\left(-P+\frac{q_3}{2}\Big|q_3\right)^T \right.$$

$$\left. \left(\frac{i}{-\not P-\not q_1-\not q_2-M}\right)^T \Gamma^{H_2}\left(-P-q_1-\frac{q_2}{2}\Big|q_2\right)^T \left(\frac{i}{-\not P-\not q_1-M}\right)^T \Gamma^{H_1}\left(-P-\frac{q_1}{2}\Big|q_1\right)^T \left(\frac{i}{-\not P-M}\right)^T \right\}$$

Taking the transpose inside the trace and changing the dummy variable P to $-P$, the vertices appear in anti-cyclic order and the right hand side coincides indeed with $\eta_{H_3}\eta_{H_2}\eta_{H_1} v_{H_1 H_2 H_3}(q_1, q_2, q_3)$. Twisted propagators are physically very important. They describe the strong rearrangement collisions of quarks and certain classes of cross-over gluon lines. Fig. 6 shows some twisted graphs together with thei quark gluon contents. In meson scattering rearrangement collisions (Fig 6a) have roughly the same coupling strength as direct (untwisted) exchanges. In QED, on the other hand, they provide for the main molecular binding forces.

The exchange of two twisted hadron lines (Fig 6b) seems to be an important part of diffraction scattering (Pomeron).

Two more examples are shown in Fig. 7. Notice that in the pseudoscalar channel these graphs incorporate the effect of the Adler triangle anomaly.

In this connection it is worth pointing out that all fundamental hadron vertices are planar graphs as far as the quark lines are concerned. Non-planar graphs are generated by building up loops involving twisted

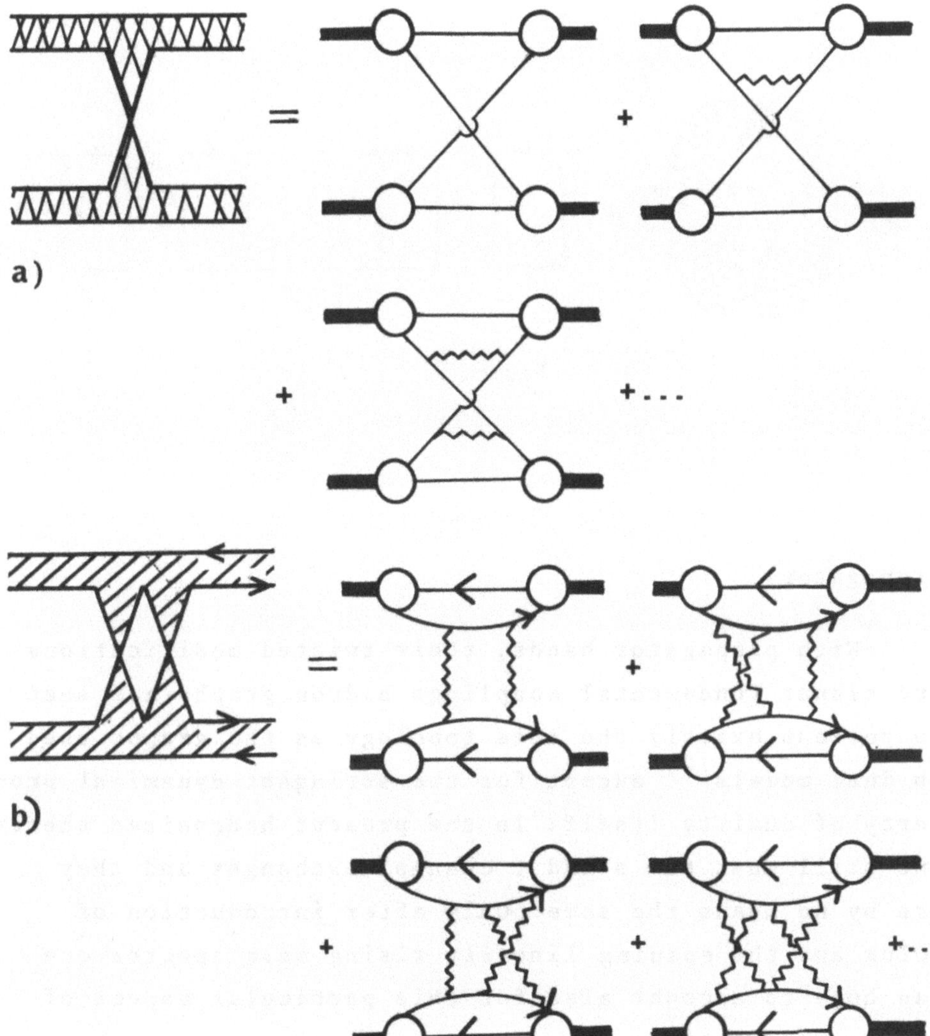

Figure 6

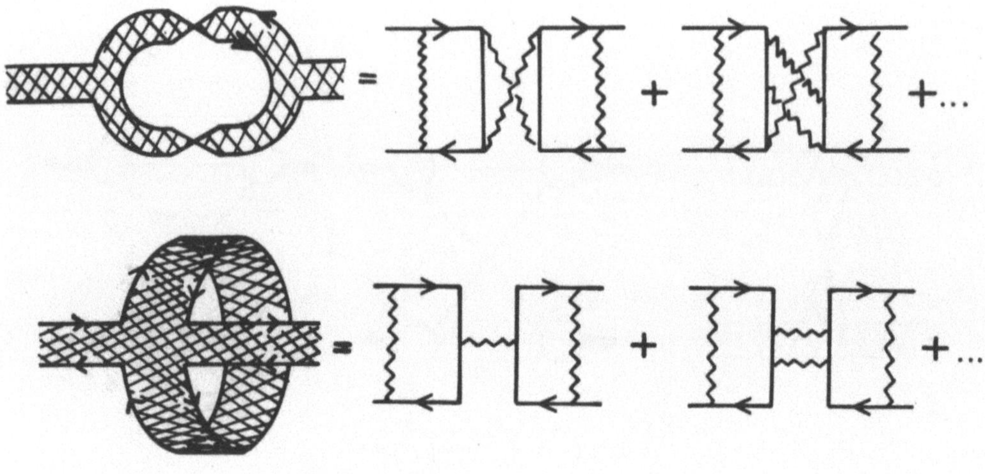

Figure 7

propagators.

With propagator bands, their twisted modifications and planar fundamental couplings hadron graphs are seen to possess exactly the same topology as the graphs used in dual models[12)] except for the stringent dynamical property of duality itself: In the present hadronized theory one still must sum s and t channel exchanges and they are by no means the same. Only after introduction of color and the ensuing linearly rising mass spectra one can hope to account also for this particular aspect of strong interactions.

The similarity in topology should be exploited for a model study of an important phenomenon of strong interactions: the Okubo, Zweig, and Iizuka rule. Obviously all hadron couplings derived by hadronization exactly respect this rule. All violations have to come from graphs of the so called cylinder type[13] (for example Fig 6b). If it is true that the topological expansion[12] is the correct basis for explaining this rule+),it may also provide the appropriate systematics for organizing the hadronized perturbation expansion.

Let us finally discuss the external souces. From $\mathcal{A}_{ext}$ in (3.36) we see that external fermion lines are connected via the full propagator G wich after expansion in powers of m' amounts to radiation of any number of hadrons (see Fig. 8)

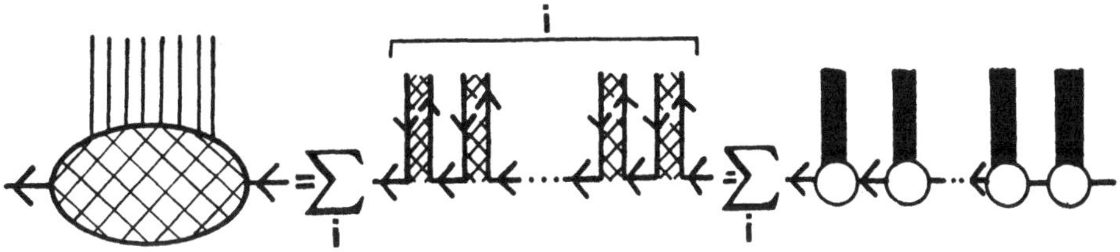

Figure 8

These hadrons then interact among each other as quantum fields. Diagrammatically, every bubble carries again a factor $\Gamma^{H}(P|q)$.

It has to be watched out that hadrons are always emitted to the right of each line. For example, the lowest order quark-quark scattering amplitude should initially be drawn as shown in Fig 9 in order to avoid phase errors due to twisted bands. Then the graphical rules yield directly the expression (3.39) as they should. Afterwards, arbitrary deformations can be performed if all twisted factors η_H are respected.

+) See the forth of Ref. 14).

External gluons interact with hadrons according to the third term in equ. (3.26)

$$-\frac{2}{g^2 D(0)} \int dx\,dy \; V^{\nu}(x,x)\, g^2 D(x-y)\, j_\nu(y) \tag{3.63}$$

Hence every external gluon enters the hadronic world only via an intermediate vector particle and there is a current field identity as has been postulated in phenomenological treatments of vector mesons (VMD). Here one finds a non-trivial coupling between the gluon and the vector mesons: As discussed before, the division by $g^2 D$ amounts to a removal of one rung from the ladder of the incoming hadron propagator and takes care of the direct coupling of the gluon to the quarks without the ladder corrections. This effect was shown to be accounted for a factor $g_H^2(q^2)/g^2$ in the propagator sum (3.39). Thus the direct coupling of the vector meson field $m_H(x)$ to an external gluon field $G^{ext}_\nu(x)$ can be written as:

$$g \sum_H \int \frac{d^4q}{(2\pi)^4} \int \frac{d^4P}{(2\pi)^4} \operatorname{tr}\left\{ \gamma^\nu G_M\!\left(P+\tfrac{q}{2}\right) \Gamma^H(P|q)\, G_M\!\left(P-\tfrac{q}{2}\right) \right\} \times \frac{g_H^2(q^2)}{g^2}\, m_H(q)\, G^{ext}_\nu(-q) \tag{3.64}$$

In a hadronic graph, the removal of one rung will be indicated by a slash. As an example, the lowest order contribution to the quark gluon form factor is illustrated in Fig 10. The slash guarantees the presence of the direct coupling. The free propagator of external gluon is given by the second term of equ. (3.26). The lowest radiative corrections consist in an intermediate slashed vector mesons (see Fig 11).

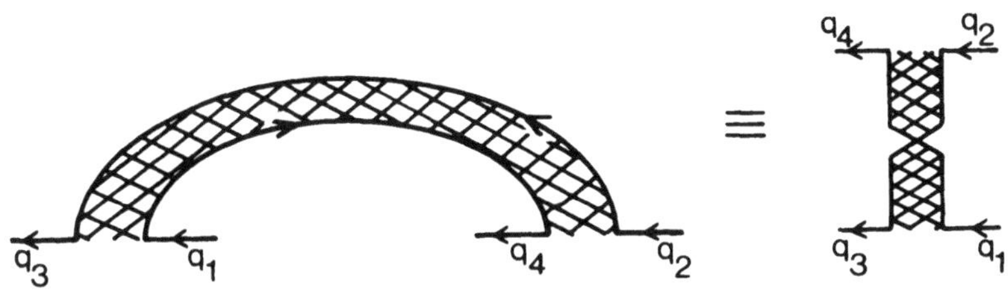

Figure 9

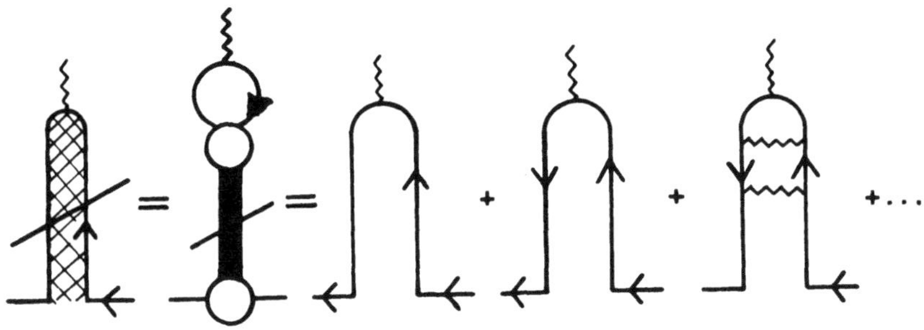

Figure 10

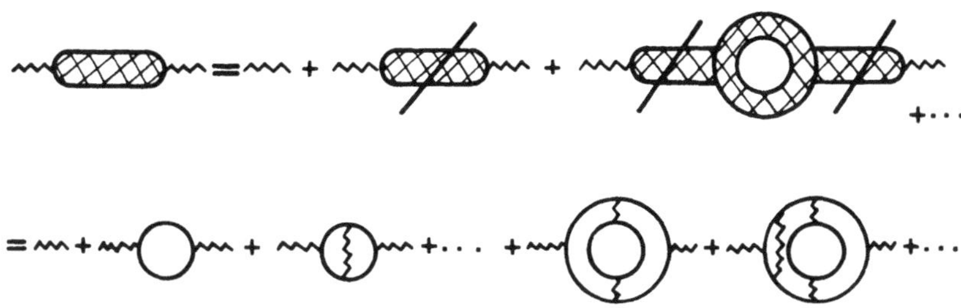

Figure 11

Here the slash is important to ensure the presence of one single quark loop.

The divergent last term in the external action (3.26) has no physical significance since it contributes only to the external gluon mass and can be cancelled by an appropriate counter term.

A final remark concerns the bilocal currents as measured in deep inelastic electron and neutrino scattering. These are vector currents of the type

$$j^{\nu}(x,y) = \bar{\psi}(x)\gamma^{\nu}\psi(y) \tag{3.65}$$

It is obvious, that also for bilocal currents there is a current-field identity with the bilocal field $V^{\mu}(x,y)$. In fact, if one would have added an external source term $C_{\nu}(x,y)$ in the quark action:

$$\Delta \mathcal{A}_{ext} \equiv \int dx dy\, \bar{\psi}(x)\gamma^{\nu}\psi(y)\, C_{\nu}(x,y) \tag{3.66}$$

this would appear in the hadronized version in the form

$$\Delta \mathcal{A}_{ext} = \int dx dy \frac{1}{ig^2 D(x-y)} V^{\nu}(x,y)\, C_{\nu}(x,y) \tag{3.67}$$

which proves our statement. Again, a rung has to be removed in order to allow for the pure quark contribution (see Fig 12)

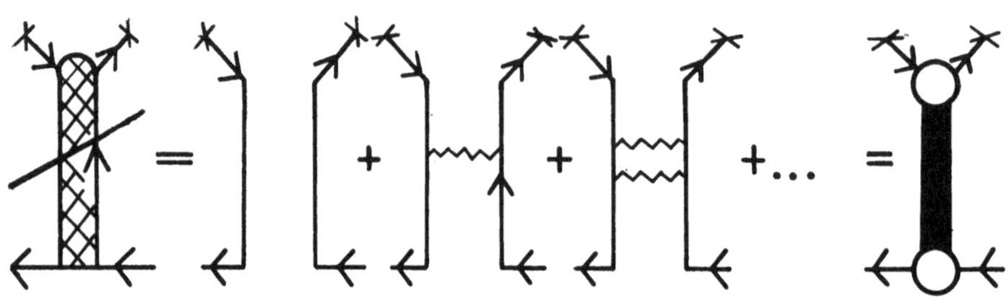

Figure 12

Bilocal currents carry direct information on the properties of Regge trajectories[15]. Therefore the present bilocal field theory seems to be the appropriate tool for the construction of a complete field theory of Reggeons[16], which is again equivalent to the original quark gluon theory. Technically, such a construction would proceed via analytic continuation of the propagators (3.39) in the angular momentum (and the principal quantum number) of the hadrons H. The result would be a "reggeonized" quark gluon theory. The corresponding Feynman graphs would guarantee unitary in all channels. Present attempts at such a theory enforces t channel unitarity only[17]. Also, they are asymptotically valid by construction and apparently have a chance of approximating nature only at energies unaccessible in the near future[+)].

IV. THE LIMIT OF HEAVY GLUONS

As an illustration of the hadronization procedure we now discuss in detail the limit of very heavy gluons[18,34]. Apart from its simplicity, this limit is quite attractive on physical grounds since it may yield a reasonable approximation to low energy meson interactions. This is suggested by the following arguments:

Suppose hadrodynamics follows a colored quark gluon theory. In this theory the color degree of freedom is very important for generating a potential between quarks rising at long distances which can explain the observed great number of high mass resonances. However, as far as low-energy interactions among the lowest lying mesons

---

+) See, for example, D. Amati and R. Jengo, Physics Letters B 54 (1974).

are concerned, color seems to be a rather superfluous luxury:

First, many fundamental aspects of strong interaction dynamics such as chiral SU(3) x SU(3) current algebra, PCAC (together with the low-energy theorems derived from both) and the approximate light cone algebra are independent of color.

Second, there is no statistics argument concerning the symmetry of the meson wave functions as there is for baryons[19].

Third, high-lying resonances are known to contribute very little in most dispersion relations of low-energy amplitudes. For example, the low-energy value of the isospin odd $\pi\pi$ scattering amplitude is given by a dispersion integral over the mesons ρ and σ with $\approx$90% accuracy[20]. Similarly, $\pi\rho$ scattering is saturated by the intermediate mesons π and A_1. By looking at all scattering combinations one can easily convince oneself that the resonances π, ρ, σ, A_1, form an approximately closed "subworld" of hadrons as far as dispersion relations are concerned. As a consequence, it would not at all be astonishing if the neglect of color in a quark gluon theory would not change the dynamics when restricting the attention to this hadronic "subworld"[+)]

+) There is one estimate concerning the electromagnetic decay of $\pi^0 \to \gamma\gamma$ which is based on short distance arguments and therefore depends on color[21]. However, the same decay can be estimated also via intermediate distance argmments, namely by using the coupling $\rho\omega\pi$ and vector meson dominance such that color does not come in.

The point is now that in the limit of a large gluon mass $\mu \to \infty$, exactly this restricted set of mesons appears as particles in the hadronized quark gluon theory (3.1) without color. Thus it might be considered as some approximation to the low-energy aspects of the colored version. Indeed, we shall see that the hadronized theory coincides exactly with the well known chirally invariant σ model. This model has proven in the past to be an appropriate tool for the rough description of low-energy meson physics[23]. Our derivation of the σ model via hadronization will render several new relations between meson and quark properties[18]. We shall at first confine ourselves to SU(2) quarks only, such that symmetry breaking may be neglected. The extension to broken SU(3) will be performed afterwards.

In order to start with the derivation observe that in the limit $\mu \to \infty$, the gluon propagator approaches a δ-function:

$$iD(x-y) \to \frac{1}{\mu^2}\delta(x-y) \tag{4.1}$$

The equation of motion (3.27) forces m'(x,y) to become a local field m'(x):

$$m'(x,y) \to m'(x)\delta(x-y) \tag{4.2}$$

which satisfies the free field equation

$$m'(x) = -i\frac{g^2}{\mu^2}\,\xi \int dy\, G_M(x-y)\, m'(y)\, G_M(y-x) \tag{4.3}$$

In the local limit, the action without external sources takes the form

$$\mathcal{A}\,[m'] = \int dx\, \mathrm{tr}\Big\{ G_M(x,x)\, m'(x) - \tfrac{1}{2}\big(G_M m' G_M m'\big)(x,x)$$

$$+ \sum_n \frac{(-i)^{n-1}}{n}\big(G_M m'\big)^n(x,x) - \frac{\mu^2}{g^2}\frac{1}{2\xi} m'(x)^2 - \frac{1}{\xi} m'(x)\, m_0 \Big\} \tag{4.4}$$

where $(G_M m' G_M m')(x,x)$ stands short for $\int dy\, G_M(x-y)\, m'(y)\, G_M(y-x)$ etc. As before in the general discussion, the constant m_0 is determined by the vanishing of the tadpole parts in (4.4) which amounts to balancing the constant contributions in the wave equation. Due to the singularity of $G_M(x-y)$ for $x \to y$ this condition has a meaning only if a cutoff is introduced such that $G_M(0)$ is finite:

$$[G_M(0)]_{\alpha\beta} = \int \frac{d^4p}{(2\pi)^4}\left[\frac{i}{\not p - M}\right]_{\alpha\beta} = M \int_0^{\Lambda} \frac{d^4p_E}{(2\pi)^4}\,(p_E^2 + M^2)^{-1}\,\delta_{\alpha\beta}$$

$$= M\,\frac{\pi^2}{(2\pi)^4}\left(\Lambda^2 - M^2 \log \Lambda^2/M^2\right)\delta_{\alpha\beta} \equiv M\,Q\,\delta_{\alpha\beta} \tag{4.5}$$

Here the dp^0 integration has been Wick-rotated by 90° such that the momentum $p^\mu = (p^0, \underset{\sim}{p})$ becomes $(ip^4, \underset{\sim}{p})$ with $p^4 \in (-\infty, \infty)$ along the integration path. The new real momentum $(p^4, \underset{\sim}{p})$ has been denoted by p_E^μ and its euclidean scalar product by $p_E^2 = p^{4\,2} + \underset{\sim}{p}^2 = -p^2$ The tadpoles can now be cancelled by setting m_0 equal

$$m_0 = 4\,\frac{g^2}{\mu^2}\,Q\,M \tag{4.6}$$

Remembering the relation to the bare quark mass $m_0 = M - \mathfrak{m}$, this determines the connection between the "true" quark mass M and the bare mass $\mathfrak{M}$ contained in the Lagrangian:

$$M = \mathfrak{M} + 4\frac{g^2}{\mu^2} Q M \tag{4.7}$$

Equation (4.8) is often called "gap equation" because of its analoguous appearance in the theory of superconductivity[24].

Consider now the free part $\mathcal{A}_2[m']$ of the action. Performing again a decomposition of type (3.12) but with the local field m'(x), it can be written in the form

$$\mathcal{A}_2[m'] = \int dx \, \mathrm{tr}_{SU(2)} \Big\{ \frac{1}{2} m_i'(x) \, \mathcal{J}_{ij}(i\partial) \, m_j(x) - \frac{\mu^2}{2g^2}\left(S^2_{(x)} + P^2_{(x)} - 2V^2(x) - 2A^2(x)\right)\Big\} \tag{4.8}$$

where $m_i'(x)$ (i=1,2,3,4) stands short for the fields+) $S(x)$, $P(x)$, $V(x)$, $A(x)$ and the trace runs only over internal SU(2) indices. The coefficients $\mathcal{J}_{ij}(q)$ are given by the integrals

$$\mathcal{J}_{ij}(q) \equiv -4 \int \frac{d^4P_E}{(2\pi)^4} \frac{1}{\left(P+\frac{q}{2}\right)_E^2 + M^2} \frac{1}{\left(P-\frac{q}{2}\right)_E^2 + M^2} t_{ij}(P|q) \tag{4.9}$$

where $t_{ij}(P|q)$ denotes the Dirac traces

$$t_{ij}(P|q) \equiv \frac{1}{4} \mathrm{tr}_{Dirac} \Big\{ \Gamma_i \left(\not P + \frac{\not q}{2} + M\right) \Gamma_j \left(\not P - \frac{\not q}{2} + M\right)\Big\} \tag{4.10}$$

+) The Lorentz indices of V^ν and A^ν fields are suppressed.

with Γ_i (i=1,2,3,4) abbreviating the standard Dirac covariants 1 , $i\gamma_5$, γ^ν , $\gamma^\nu\gamma_5$. The traces are displayed in Appendix A equ. (A.33). Some of them grow quadratically in P . The corresponding integrals $J_{ij}(q)$ are quadratically divergent for large cutoffs. The others diverge logarithmically. If one introduces the basic integral

$$L = \int \frac{d^4P_E}{(2\pi)^4} \frac{1}{(P_E^2+M^2)^2} = \frac{\pi^2}{(2\pi)^4}\left(\log \frac{\Lambda^2}{M^2} - 1\right) \qquad (4.11)$$

the divergent parts of $J_{ij}(q)$ are (see App. A)

$$J_{SS}(q) = Q + L\left(\frac{q^2}{2} - 2M^2\right)$$

$$J_{PP}(q) = Q + L\,\frac{q^2}{2} \quad ; \quad J_{PA^\nu} = -iLMq^\nu$$

$$J_{V^\mu V^\nu}(q) = -\frac{1}{3}(q^2 g^{\mu\nu} - q^\mu q^\nu)\,L \qquad (4.12)$$

$$J_{A^\mu A^\nu}(q) = -\frac{1}{3}\left[(q^2 g^{\mu\nu} - q^\mu q^\nu) - 6M^2 g^{\mu\nu}\right] L \;;\; J_{A^\mu P} = iLMq^\mu$$

with all other integrals vanishing.

If we neglect the finite contributions as compared with these divergent ones, the action $\mathcal{A}_2[m']$ is seen to corresponds to the local Lagrangian+)

+) Since $m(x)$ and $m'(x)$ differ only by a Dirac scalar constant $m_0 1_{\text{(4)}}$ there is no difference between primed and unprimed fields except for $S'(x) = S(x) - m_0$

$$\mathcal{L}(x) = \mathrm{tr}_{SU(2)} \Big\{ \tfrac{1}{2} S'(x)\Big[4Q - 2(\Box + 4M^2)L - \frac{\mu^2}{g^2}\Big] S'(x)$$
$$+ \tfrac{1}{2} P(x)\Big[4Q - 2\Box L - \frac{\mu^2}{g^2}\Big] P(x)$$
$$+ \tfrac{1}{2} V_\mu(x)\Big[\tfrac{4}{3}(\Box g^{\mu\nu} - \partial^\mu\partial^\nu)L + \frac{2\mu^2}{g^2}\Big] V_\nu(x)$$
$$+ \tfrac{1}{2} A_\mu(x)\Big[\tfrac{4}{3}(\Box g^{\mu\nu} - \partial^\mu\partial^\nu)L + 8M^2 g^{\mu\nu} L + 2\frac{\mu^2}{g^2}\Big] A_\nu(x)$$
$$+ 2ML\,\big(\partial_\mu P(x) A^\mu(x) + A^\mu(x)\partial_\mu P(x)\big) \Big\} \qquad (4.13)$$

If we respect the gap equation (4.7) in this Lagrangian, the quadratically divergent terms Q can be eliminated. The mixed terms can be removed by introducing a new field $\hat{A}^\mu(x)$ via

$$A^\mu(x) = \hat{A}^\mu(x) + \lambda\,\partial^\mu P \qquad (4.14)$$

and fixing λ as

$$\lambda = -3M/m_A^2 \qquad (4.15)$$

where m_A^2 stands short for

$$m_A^2 = m_V^2 + 6M^2 \qquad (4.16)$$

with

$$m_V^2 = 3\mu^2/(2g^2L) \qquad (4.17)$$

This substitution produces additional kinetic terms for the pseudoscalar fields which now appears with a factor

$$-\mathrm{tr}_{SU(2)}\big(P(x)\Box P(x)\big)\big(1 + \tfrac{2}{3}m_A^2\lambda^2 + 4M\lambda\big)L \qquad (4.18)$$
$$\triangleq \mathrm{tr}_{SU(2)}\big(\partial_\mu P(x)\,\partial^\mu P(x)\big)\, Z_P^{-1} L$$

Using (4.15), this renormalization factor becomes

$$Z_p^{-1} = 1 - 6M^2/m_A^2 \tag{4.19}$$

After this diagonalization, the Lagrangian reads

$$\mathcal{L}(x) = \mathrm{tr}_{SU(2)} \{ \partial_\mu S' \partial^\mu S' - (4M^2 + \tfrac{1}{3} m_V^2 \mathfrak{M}/M) S'^2$$
$$+ \partial_\mu P \partial^\mu P \, Z_p^{-1} - \tfrac{1}{3} m_V^2 \mathfrak{M}/M) P^2$$
$$- \tfrac{1}{3} F_V^{\mu\nu} F^V_{\mu\nu} + \tfrac{2}{3} m_V^2 V_\mu^2 \tag{4.20}$$
$$- \tfrac{1}{3} F_{\tilde{A}}^{\mu\nu} F^{\hat{A}}_{\mu\nu} + \tfrac{2}{3} m_A^2 \tilde{A}_\mu^2 \} \times L$$

where $F^{\mu\nu}_{V,\hat{A}}$ are the usual field tensors of vector and axial vector fields. The particle content of this free Lagrangian is now obvious. There are vector mesons of mass m_V^2, axial-vector mesons of mass m_A^2 and scalar and pseudoscalar mesons of mass

$$m_S^2 = 4M^2 + \tfrac{1}{3} m_V^2 \mathfrak{M}/M \tag{4.21}$$

$$m_P^2 = \qquad \tfrac{1}{3} m_V^2 \mathfrak{M}/M \; Z_P \tag{4.22}$$

With (4.17), the constant (4.19) can also be written as

$$Z_P^{-1} = m_V^2/m_A^2 \tag{4.23}$$

As we have argued before, there is a good chance that the fields P, V, S, A describe approximately the lowest lying mesons π, ρ, σ, A_1. Let us test this hypothesis as far as the masses are concerned. Since experimentally $m_{A_1}^2 \approx 2m_\rho^2$ the factor Z_π becomes ≈ 2.

Forthermore, equ. (4.16) determines the quark mass as:

$$6M^2 = m_{A_1}^2 - m_\rho^2 \quad ; \quad M \approx 310 \text{ MeV} \tag{4.24}$$

in good agreement with other estimates.[25] The small pion mass yields via (4.22)

$$m \approx 15 \text{ MeV} \tag{4.25}$$

Thus the bare quark mass has to be extremely small. Aslo this result has been obtained by many authors[26]. It is common to all models in which the smallness of the pion mass is related to the approximate conservation of the axial current (PCAC).

The scalar meson finally is predicted from (4.21) to have a mass

$$m_\sigma \approx 2M \sim 620 \text{ MeV} \tag{4.26}$$

This agrees well with the observed broad resonance in $\pi\pi$ scattering.[27)20)]

One disagreement with experiment appears in connection with the SU(2) singlet pseudoscalar mass (the η meson). According to (4.22) it should be degenerate with the pion. The resolution of this problem will be discussed later when the theory has been extended to SU(3).

After these first encouraging results we shall rename the fields P , V , S , A by the corresponding particle symbols

$$\sqrt{L}\, P \equiv \pi, \quad \sqrt{L}\, S' \equiv \sigma', \quad \sqrt{\tfrac{2}{3}L}\, V^\mu \equiv \rho^\mu, \quad \sqrt{\tfrac{2}{3}L}\, A^\mu \equiv A_1^\mu \tag{4.27}$$

where a normalization factor has been introduced in order to bring the kinetic terms in the Lagrangian to a conventional form.

A comment is in order concerning the appearance of a quadratic divergence in equations (4.7), (4.12). Such a strong divergence indicates, that the limiting procedure $\mu \to \infty$ of equ. (4.1) has been performed too carelessly. In fact, if one inserts (4.1) into the action (3.4), the theory becomes of the $(\bar{\psi}\psi)^2$ type and thus non-renormalizable. In order to keep the renormalizability while dealing with a large gluon mass $\mu^2 \gg M^2$ we actually have to watch out that the gluon mass stays always far below the cutoff: $\mu^2 \ll \Lambda^2$. Then the quadratic divergence becomes actually of the logarithmic type (compare (3.47)):

$$Q = \int \frac{d^4 p_E}{(2\pi)^4} \frac{1}{p_E^2+M^2} \frac{1}{p_E^2+\mu^2} = \frac{\pi^2}{(2\pi)^4} \frac{\mu^2}{\mu^2-M^2} \left[\mu^2 \log\left(\frac{\Lambda^2}{\mu^2}+1\right) - M^2 \log\left(\frac{\Lambda^2}{M^2}+1\right)\right] \tag{4.28}$$

(which in the careless limit $\mu^2 \to \infty$ reduces again to (4.5)). The logarithmic divergence (4.11) on the other hand becomes in this more careful treatment independent of the cutoff which is replaced by the large gluon mass

$$L = \int \frac{d^4 p_E}{(2\pi)^4} \frac{1}{(p_E^2+M^2)^2} \frac{1}{(p_E^2+\mu^2)} = \frac{\pi^2}{(2\pi)^4} \frac{\mu^2}{(\mu^2-M^2)^2} \tag{4.29}$$

$$\times \left[\mu^2 \log \frac{\mu^2}{M} + M^2 - \mu^2\right] \approx \frac{\pi^2}{(2\pi)^4} \log \frac{\mu^2}{M^2}$$

Hence all our results refer to a renormalizable theory if one reads both Q and L as logarithmic expression once in the cutoff and once in the gluon mass, respectively.

Let us now proceed to study the interaction terms. The n'th order contribution to the action is given by

$$\mathcal{A}_n[m'] = \frac{(-i)^{n-1}}{n} \int dx \; \mathrm{tr}\,(G_M\, m')^n \tag{4.30}$$

In momentum space this can be written as the one loop integral

$$\mathcal{A}_n[m'] = 4\,\frac{(-1)^{n-1}}{n} \int \frac{dq_n}{(2\pi)^4} \cdots \frac{dq_1}{(2\pi)^4} (2\pi)^4 \delta(q_n + \dots + q_1)$$

$$\int \frac{d^4P_E}{(2\pi)^4} \frac{1}{(P+q_n+\dots+q_1)_E^2+M^2} \cdots \frac{1}{(P+q_1)_E^2+M^2}\, t_{i_n\dots i_1}(P|q_{n-1},\dots,q_1) \tag{4.31}$$

$$\mathrm{tr}_{SU(2)}\left[m'_{i_n}(q_n)\cdot\dots\cdot m'_{i_1}(q_1)\right]$$

where $t_{i_n\dots i_1}(P|q_{n-1},\dots,q_1)$ is the generalization of the tensor (4.10)

$$t_{i_n\dots i_1}(P|q_{n-1},\dots,q_1) = \frac{1}{4}\,\mathrm{tr}\left[\Gamma_{i_n}(P+q_{n-1}+\dots+q_1+M)\Gamma_{i_{n-1}}\cdots \right. \tag{4.32}$$

$$\left. \Gamma_{i_2}(P+q_1+M)\Gamma_{i_1}(P+M)\right]$$

The result is hard to evaluate in general (except in a 1 + 1 dimensional space). With the approximation of a large cutoff one may however, neglect again all contributions which do not diverge. This considerably simplifies the results. Since $t_{i_n \dots i_1}(P|q_{n-1},\dots,q_1)$ are polynomials in P of order n, the integral is seen to converge for $n > 4$. For $n = 4$ there is a logarithmic divergence with only the leading momentum behaviour of $t_{i_n \dots i_1}$ contributing. For $n \leq 3$ also lower powers in momentum P of $t_{i_n \dots i_1}(P|q_{n-1},\dots,q_1)$ diverge logarithmically. A simple but somewhat tedious calculation of all the integrals (see App. B) yields the remaining terms in the Lagrangian. They can be written down in a most symmetric fashion by employing the unshifted fields+) $S(x) \equiv M + S'(x)$ rather than S', or in renormalized form

$$\sigma(x) = -\sqrt{L}\, M + \sigma'(x) \tag{4.33}$$

Then the Lagrangian reads

$$\begin{aligned}\mathcal{L}(x) = \mathrm{Tr}_{SU(2)} \Big\{ & [(D_\mu\sigma)^2 + (D_\mu\pi)^2] + 2M_0^2(\sigma^2+\pi^2) \\ & - \tfrac{2}{3}\gamma^2 [\sigma^4 + \pi^4 + 4\sigma^2\pi^2 - 2\sigma\pi\sigma\pi] - \tfrac{1}{2} F_{\mu\nu}^{V\,2} - \tfrac{1}{2} F_{\mu\nu}^{A\,2} \\ & + m_V^2 (V_\mu^2 + A_\mu^2) - \tfrac{2}{3} m_V^2 \sqrt{L}\, \mathfrak{m}\sigma \Big\}\end{aligned} \tag{4.34}$$

Here $D_\mu\sigma$ and $D_\mu\pi$ are the usual covariant derivatives:

$$\begin{aligned} D_\mu\sigma &= \partial_\mu\sigma - i\gamma[V_\mu\sigma] - \gamma\{A_\mu\pi\} \\ D_\mu\pi &= \partial_\mu\pi - i\gamma[V_\mu\pi] + \gamma\{A_\mu\sigma\} \end{aligned} \tag{4.35}$$

and $F_{\mu\nu}^V$, $F_{\mu\nu}^A$ are the covariant curls

+) Notice that with this notation $m(x) = m_0 + m'(x)$
$= (M - \mathfrak{m}) + m'(x) = -\mathfrak{m} + S + P i\gamma_5 + V^\mu\gamma_\mu + A^\mu\gamma_\mu\gamma_5$

$$F_{\mu\nu}^{V} = \partial_\mu V_\nu - \partial_\nu V_\mu - i\gamma\,[V_\mu, V_\nu] - i\gamma\,[A_\mu, A_\nu]$$

$$F_{\mu\nu}^{A} = \partial_\mu A_\nu - \partial_\nu A_\mu - i\gamma\,[V_\mu, A_\nu] - i\gamma\,[A_\mu, V_\nu] \tag{4.36}$$

The constant γ denotes

$$\gamma = \sqrt{\frac{3}{2L}} \tag{4.37}$$

It describes the direct coupling of the vector mesons to the currents, i.e. it coincides with the coupling conventionally denoted by γ_g . Here γ has its origin in the renormalization of the fields. The mass term stands short for

$$2M_0^2 \equiv 2M^2 - \frac{1}{3} m_V^2\, \mathfrak{m}/M \tag{4.38}$$

Actually, the so defined mass quantity has an intrinsic significance. This can be seen by deriving the Lagrangian in a different fashion from the beginning. consider the tadpole terms of the action

$$\mathcal{A}_1[m'] = \int dx\, \mathrm{tr} \left\{ G_M(x,x) m'(x) - \frac{1}{g} m'(x) m_0 \right\} \tag{4.39}$$

In the former treatment we have eliminated m_0 completely by giving the quarks a mass M satisfying the gap equation

$$M - \mathfrak{m} \equiv m_0 = 4 \frac{g^2}{\mu^2} Q M \tag{4.40}$$

Instead, we could have introduced an auxiliary mass M_0 satisfying the equation

$$M_0 = 4\frac{g^2}{\mu^2} Q_0 M_0 \tag{4.41}$$

where Q_0 is the same function of M_0 as Q is of M. The connection between this M_0 and the other masses is obtained by inserting $M = M_0 + \delta M$ into Q:

$$Q = Q_0 - 2M_0 \delta M (1 + \delta M / 2M_0) L \tag{4.42}$$

which holds exactly in δM with only small corrections for large cutoffs (notice that at this accuracy $L_0 = L$). Inserting this into (4.40) we find

$$\mathfrak{m} = 4\frac{g^2}{\mu^2} L\, 2M_0 M (1 + \delta M / 2M_0) \delta M \tag{4.43}$$

and using m_V^2 from

$$\mathfrak{m} = \frac{12}{m_V^2} M_0 M (1 + \delta M / 2M_0) \delta M \tag{4.44}$$

If now $m(x)$ is split in a different fashion

$$m(x) = \tilde{m}_0 + m''(x) \tag{4.45}$$

with a new $\tilde{m}_0 = M_0 - \mathfrak{m}$ then the propagator $G(x,y)$ would have an expansion

$$G(x,y) = G_{M_0}(x-y) - i(G_{M_0} m'' G_{M_0})(x,y) + \tag{4.46}$$

For this reason, the derivation of all Lagrangian terms yields exactly the same results as before only with m'', M_0, L_0, and Q_0 occurring rather than m', M, L and Q, respectively. There are only two differences: First, due to the gap equation (4.41), the scalar and pseudoscalar mass terms become $4M_0^2$ and 0 rather than (4.21),(4.22) second, the tadpole terms in this derivation do not cancel completely. Instead one finds from (4.39)

$$\mathcal{A}_1[m'] = \int dx \, \mathrm{tr}\{(4Q_0 M_0 - \frac{M_0 - \mathfrak{M}}{g^2/\mu^2}) m''(x)\}$$
$$= \int dx \, \frac{\mu^2}{g^2} \mathrm{tr}\{\mathfrak{M} \, m''(x)\} = \frac{2}{3} m_V^2 \int dx \, \mathrm{tr}\{\mathfrak{M} m''(x)\} \qquad (4.47)$$

These tad pole terms provide exactly the necessary additional shifts in the fields which are needed in order to bring the scalar and pseudoscalar masses from $4M_0^2$ and 0 to their correct values m_σ^2 and m_π^2 . The symmetric form (4.34) of the Lagrangian is again reached by introducing the original unprimed fields

$$S(x) \equiv M_0 + S''(x) \quad , \sigma = -\sqrt{L}\, M_0 + \sigma''$$

Then the mass term appears as an SU(3) x SU(3) invariant

$$2M_0^2(\sigma^2 + \pi^2).$$

+) With this substitution, the unprimed field S really coincides with the formerly introduced field S since now

$$m(x) = (M_0 - \mathfrak{M}) + S''(x) + P(x) i\gamma_5 + \dots$$
$$= -\mathfrak{M} + S(x) + P(x) i\gamma_5 + \dots$$

while before

$$m(x) = (M - \mathfrak{M}) + S'(x) + P(x) i\gamma_5 + \dots$$
$$= -\mathfrak{M} + S(x) + P(x) i\gamma_5 + \dots$$

Notice now that this coincides exactly with the former calculation which rendered (see (4.38)

$$\left(2M^2 - \tfrac{1}{3}\, m_V^2\, \mathfrak{m}/M\right)(\sigma^2+\pi^2)$$

Inserting here $M = M_0 + \delta M$ and (4.44) gives

$$\begin{aligned} 2M^2 - \tfrac{1}{3}\, m_V^2\, \mathfrak{m}/M &= 2M_0^2 + 4M_0\delta M + 2(\delta M)^2 - 4M_0\left(1+\frac{\delta M}{M_0}\right)\delta M \\ &= 2M_0^2 \end{aligned} \qquad (4.48)$$

Hence the SU(3) symmetric mass M_0 defined by the gap equations (4.41) coincides with the mass M_0 introduced as an abbreviation to the mass combination (4.38).

The Lagrangian (4.34) is recognized as the standard chirally invariant σ model. Its symmetry transformations are for isospin

$$\begin{aligned} &\delta\sigma = i[\alpha,\sigma] \quad ; \quad \delta\pi = i[\alpha,\pi] \\ &\delta V^\mu = i[\alpha, V^\mu] + \tfrac{1}{\gamma}\partial^\mu\alpha \quad ; \quad \delta A^\mu = i[\alpha, A^\mu] \end{aligned} \qquad (4.49)$$

For axial transformations the fields change according to

$$\begin{aligned} &\bar\delta\sigma = -\{\bar\alpha,\pi\} \quad ; \quad \bar\delta\pi = \{\bar\alpha,\sigma\} \\ &\bar\delta V^\mu = i[\bar\alpha, A^\mu] \quad ; \quad \bar\delta A^\mu = i[\bar\alpha, V^\mu] + \tfrac{1}{\gamma}\partial^\mu\bar\alpha \end{aligned} \qquad (4.50)$$

The only term in the Lagrangian which is not invariant is the last linear term. In fact from

$$\bar\delta\mathcal{L} = i\tfrac{2}{3}\, m_V^2\, \mathfrak{m}\, \square\, \{\bar\alpha\pi\} \equiv i\{\bar\alpha, \partial A\} \qquad (4.51)$$

one finds

$$\partial A(x) = f_\pi m_\pi^2 Z_\pi^{-1/2} \pi(x) \tag{4.52}$$

Introducing the conventional pion decay constant via

$$\partial A(x) \equiv f_\pi m_\pi^2 Z_\pi^{-1/2} \pi(x) \tag{4.53}$$

one can read off

$$f_\pi m_\pi^2 = Z_\pi^{1/2} \frac{2}{3} m_V^2 \sqrt{L}\, \delta\sigma \tag{4.54}$$

Inserting m_π^2 from (4.22) this gives

$$f_\pi = Z_\pi^{-1/2} 2M\sqrt{L} \tag{4.55}$$

By squaring this and using $\gamma = \sqrt{\frac{3}{2L}}$, one obtains

$$f_\pi^2 = \frac{6M^2}{Z_\pi}\frac{1}{\gamma^2} = \frac{m_\rho^2}{\gamma^2}\frac{m_A^2 - m_\rho^2}{m_A^2} \tag{4.56}$$

which for $m_A^2 \approx 2m_\rho^2$ renders the well known KSFR relation. The model has the usual predictions

$$g_{\rho\pi\pi} = \gamma_\rho \left(1 - \frac{m_A^2 - m_\rho^2}{2m_A^2}\right) \approx \frac{3}{4}\gamma_\rho \tag{4.57}$$

and

$$g_{A_1\rho\pi} = \frac{1}{2m_\rho} \gamma^2 Z_\pi f_\pi \approx \frac{m_\rho}{2f_\pi} \approx 4 \quad ; h_{A_1\rho\pi} = 0$$

$$g_{A_1\sigma\pi} = \gamma Z_\pi^{1/2} \approx \frac{m_\rho}{f_\pi} \approx 8 \tag{4.58}$$

$$g_{\sigma\pi\pi} = \frac{2}{m_\sigma}\frac{1}{3}\gamma^2 f_\pi Z_\pi^{3/2} \approx \frac{m_\sigma}{f_\pi}\sqrt{2} \approx 9$$

When compared with experiment, the only real defect consists in the d-wave $A_1\rho\pi$ coupling, $h_{A_1\rho\pi}$, being absent. additional chirally invariant terms are needed in the Lagrangian, for example, the so called δ-term:

$$\delta \operatorname{tr}\left[\left(F^{A}_{\mu\nu}+F^{V}_{\mu\nu}\right)D^{\mu}(\sigma+i\pi)D^{\nu}(\sigma-i\pi)-(V\to -V,\ \pi\to -\pi)\right] \tag{4.59}$$

Such terms appear in our derivation if the approximation of large μ^2 is improved by terms which do not grow lagarithmically in μ .

Let us now determine the couplings of π , σ , ρ , A_1 A_1 to external quark fields. The external propagation proceeds via

$$i\bar{\eta}G\eta = i\bar{\eta}G_{M_0}\eta + \bar{\eta}G_{M_0}M'' G_{M_0}\eta - i\ldots \tag{4.60}$$

If one defines the couplings by

$$\begin{aligned}\mathcal{L} \propto\ & g_{\pi QQ}\bar{\psi}i\gamma_5\tau_a\psi\,\pi^a + g_{\sigma QQ}\bar{\psi}\tau_a\psi\,\sigma^a \\ & + g_{VQQ}\bar{\psi}\gamma^\mu\frac{\tau_a}{2}\psi\,V^a_\mu + g_{AQQ}\bar{\psi}\gamma^\mu\gamma_5\frac{\tau_a}{2}\psi\,A^a_\mu\end{aligned} \tag{4.61}$$

and can read off

$$g_{\pi QQ} = \frac{1}{2\sqrt{L}}Z_\pi^{\frac{1}{2}} = \frac{M}{f_\pi}\ ;\ g_{\sigma QQ} = \frac{1}{2\sqrt{L}} = \frac{M}{f_\pi Z_\pi^{1/2}} \tag{4.62}$$

$$g_{\rho QQ} = g_{A_1 QQ} = \gamma$$

We see the vector coupling to quarks agree with vector-meson dominance. Due to PCAC also the Goldberger Treiman relation is respected

$$g_{\pi QQ} = g_A \frac{M}{f_\pi} \tag{4.63}$$

since the axial charge of the quark is $g_A = 1$. Since the quark mass is $M \sim m_N/3$, the pionic coupling to quarks is considerably smaller than to nucleons. Numerically

$$\frac{g_{\pi QQ}^2}{4\pi} \approx \frac{1}{17} \frac{g_{\pi NN}^2}{4\pi} \approx .86 \tag{4.64}$$

The σ meson couples even weaker

$$g_{\sigma QQ}^2/4\pi \approx .43$$

Vector- and axial-vector mesons, on the other hand, couple as strongly as to nucleons which is an expression of universality:

$$\frac{g_{\rho QQ}^2}{4\pi} = \frac{g_{AQQ}^2}{4\pi} = \frac{\gamma^2}{4\pi} \approx 2.6 \left(= \frac{g_{\rho NN}^2}{4\pi} \right) \tag{4.65}$$

We are now ready to extend our consideration to SU(3) (and higher groups). In this case the explicit symmetry breaking in the Lagrangian is too large to be neglected. Thus the bare masses $\mathfrak{M}$ of the quarks have to be considered as a matrix

$$\mathfrak{M} \approx \begin{pmatrix} \mathfrak{M}^u & & \\ & \mathfrak{M}^d & \\ & & \mathfrak{M}^s \end{pmatrix} \tag{4.66}$$

The derivation of the Lagrangian presented above (via the gap equation (4.41)) has shown the complete SU(3) symmetry of M_0^2 . Hence when extending from SU(2) to SU(3), no change occurs except in the last symmetry breaking term of (4.34). As a consequence, the mass expressions for m_p^2 and m_s^2 remain as they are only that the renormalization constants Z_p become more complicated SU(3) dependent quantities due to the involved mixing of pseudoscalar and axial-vector mesons. For a complete discussion of this SU(3) x SU(3) invariant chiral Lagrangian the reader is referred to the review articles[23).] Here we only give a few results:
A best fit to π and K meson masses requires[+)]

$$\mathcal{M} \approx \begin{pmatrix} 15 & & \\ & 15 & \\ & & 435 \end{pmatrix} \mathrm{MeV} \tag{4.67}$$

Thus the explicit symmetry breakdown of SU(3) caused by the bare masses is quite large. The standard parameter[28)] C characterizes this:

$$C \equiv \frac{\mathcal{M}^8}{\mathcal{M}^0} = \frac{\frac{1}{\sqrt{3}}\left(\mathcal{M}^u + \mathcal{M}^d - 2\mathcal{M}^s\right)}{\sqrt{\frac{2}{3}}\left(\mathcal{M}^u + \mathcal{M}^d + \mathcal{M}^s\right)} \approx -1.28 \quad (\approx -\sqrt{2}) \tag{4.68}$$

Inserting into (4.44) we find the shifts in the quark masses caused by dynamics

$$\delta M = \begin{pmatrix} 7 & & \\ & 7 & \\ & & 127 \end{pmatrix} \mathrm{MeV} \tag{4.69}$$

+) For other determinations of   see Ref. 26).

and hence for the "physical" quark masses

$$M \approx \begin{pmatrix} 312 & & \\ & 312 & \\ & & 432 \end{pmatrix} \text{MeV} \tag{4.70}$$

Thus contrary to the large explicit SU(3) violation the bare masses $\mathfrak{M}$, the physical quark masses M show only the moderate violation

$$c' \equiv \frac{M^8}{M^0} = \frac{\frac{1}{\sqrt{3}}(M^u + M^d - 2M^s)}{\sqrt{\frac{2}{3}}(M^u + M^d + M^s)} \approx -16\% \tag{4.71}$$

Since the quark masses M are produced almost completely by dynamical effects we expect some symmetry breakdown to appear also in the vacuum. A measure of this is provided by the expectation values of the scalar quark densities

$$\langle 0|\tilde{u}^a|0\rangle = \langle 0|\bar{\psi}(x)\frac{\lambda^a}{2}\psi(x)|0\rangle \tag{4.72}$$

In the hadronized theory, the scalar densities are identical with the scalar fields , up to a factor:

$$S^i_j(x) = -\frac{g^2}{\mu^2}\bar{\psi}^i(x)\psi_j(x) \tag{4.73}$$

as can be seen most easily by considering the equations of constraint (3.14) following from the Lagrangian (3.11) in the large-μ limit. Hence

$$\langle 0|\tilde{u}^a|0\rangle = -\frac{\mu^2}{g^2}\sum_{i,j}\lambda^a{}_i{}^j\langle 0|S^i{}_j|0\rangle \tag{4.74}$$

$$= -\frac{\mu^2}{2g^2}\operatorname{tr}(M\lambda^a) = -\frac{1}{2}\frac{\mu^2}{g^2}M^a$$

Inserting (4.17) and (4.55) the factor becomes simply

$$\frac{1}{2}\frac{\mu^2}{g^2} = \frac{1}{3}\frac{1}{\frac{2}{3}\frac{g^2}{\mu^2}L}L = \frac{1}{3}m_V^2 Z_\pi \frac{f_\pi^2}{4M^2} \approx f_\pi^2$$

such that

$$\langle 0|\tilde{u}^0|0\rangle \approx -f_\pi^2 M^0 = -f_\pi^2\sqrt{\tfrac{2}{3}}(M^u+M^d+M^s) \approx -8\times 10^{-3}\,\mathrm{GeV}^3$$

$$\langle 0|\tilde{u}^8|0\rangle \approx -f_\pi^2 M^8 = c'\langle 0|\tilde{u}^0|0\rangle \tag{4.75}$$

This shows that the SU(3) violation in the vacuum equals that in the quark masses $\approx -16\%$ [+)]. Notice that the three results (4.22) (4.44) and (4.73) are in complete agreement with what one obtains by very general considerations using only chiral symmetry and PCAC (see App. C).

The extension of the Lagrangian to SU(3) produces additional defects which are well known from general discussions of chiral SU(3) x SU(3) symmetry[23)]. For example the vector mesons ω_1, φ are not mixed (almost) ideally as they should but φ remains close to an SU(3) singlet. In general discussions, additional terms have been added to chiral Lagrangian in order to account for this. There are the so called "current mixing terms":

$$\mathrm{tr}\left\{\left(F_{\mu\nu}^V + F_{\mu\nu}^A\right)^2(\sigma+i\pi)(\sigma-i\pi) + (A\to -A,\ \pi\to-\pi)\right\} \tag{4.76}$$

as well as "mass mixing terms"

+) In Ref. 30), SU(3) breaking in the vacuum was neglected. For a more general discussion and earlier references see Ref. 29).

$$\mathrm{tr}\left\{ (V^\mu + A^\mu)^2 (\sigma + i\pi)(\sigma - i\pi) + (V^\mu - A^\mu)^2 (\sigma - i\pi)(\sigma + i\pi) \right\} \tag{4.77}$$

In our derivation these arise as a next correction to the $\mu^2 \to \infty$ limit. Another problem is the degeneracy of the ideally mixed isosinglet pseudoscalar meson η'_{ideal}, with the pion. In order to account for the fact that the η' (X_0) meson is almost a pure SU(3) singlet and much heavier than the other pseudoscalar mesons one needs some chirally symmetric term

$$\det(\sigma + i\pi) + \det(\sigma - i\pi). \tag{4.78}$$

Such a term breaks PCAC for the ninth axial current. It is well known[30] that the quark gluon triangle anomaly operates in the singlet channel and might be capable of producing such a PCAC violation. In fact, if this was not true, quantum electrodynamics would possess an exactly massless Goldstone boson [31] with η quantum numbers. Also the term (4.78) will appear when μ is not any more very large.

It is obvious that corrections to the $\mu^2 \to \infty$ approximation will become even more important if one tries to extend the consideration to SU(4) since then vector and pseudoscalar masses are quite heavy. In addition, the narrow width of the SU(4) vector meson ψ/J seems to indicate that short-distance parts of the gluon propagator are being probed. Thus the colorless quark gluon theory itself cannot be considered any more a realistic approximation to the colored theory.

At this place we should remark that present explanations of electromagnetic mass differences require also an breakdown of SU(2) symmetry in $\mathfrak{M}$ [32]. This is conventionally parametrized by

$$\alpha \equiv \frac{\mathfrak{M}^3}{\mathfrak{M}^0} = \frac{\mathfrak{M}^u - \mathfrak{M}^d}{\sqrt{\frac{2}{3}}\,(\mathfrak{M}^u + \mathfrak{M}^d + \mathfrak{M}^s)} \tag{4.79}$$

From meson masses (as well as from the electromagnetic $\eta \to 3\pi$ decay) one finds[33]

$$\alpha \approx -3\% \tag{4.80}$$

This amounts to the bare quark masses

$$\mathfrak{M} \approx \begin{pmatrix} 10 & & \\ & 20 & \\ & & 435 \end{pmatrix} \text{MeV} \tag{4.81}$$

giving the "true" masses

$$M \approx \begin{pmatrix} 310 & & \\ & 315 & \\ & & 432 \end{pmatrix} \text{MeV} \tag{4.82}$$

Thus the SU(2) breaking of the vacuum is very small

$$\alpha' \equiv \frac{\langle 0|\tilde{u}^3|0\rangle}{\langle 0|\tilde{u}^0|0\rangle} = \frac{M^3}{M^0} \approx -.6\% \tag{4.83}$$

With all parameters fixed numerically we should finally check whether the approximation of a large gluon mass is self consistent. From (4.55) we have

$$L \approx \frac{1}{2} \frac{f_\pi^2}{M^2} \approx .046. \tag{4.84}$$

Inserting this into (4.29) we calculate

$$\log \mu^2/M^2 \sim (4\pi)^2 \times 0.046 \approx 7.3 \tag{4.85}$$

and hence

$$\mu^2 \approx 1500\, M^2 \quad \gg M^2 \tag{4.86}$$

or

$$\mu \approx 12\ \mathrm{BeV} \tag{4.87}$$

It is gratifying to note that this value is much larger than the mass of the vector mesons. In this way it is assured that higher powers of $q^2/(P_E^2+M^2)$ which were neglected in the derivation of the Lagrangian remain really small as compared to unity for <u>all</u> mesons of the theory (see app. B).

We should point out that the quark gluon theory in the limit $\mu^2 \to \infty$ coincides with the well-known Nambu-Jona Lasinio[24)] model which has proven in the past to be a convenient tool of studying the spontaneous breakdown of chiral symmetry and the dynamical generation of PCAC. Those authors have demonstrated the close analogy of the dynamic structure of this model with that of superconductivity. As we have mentioned before the equation (4.40) removing the tadpoles in the action is analogous to the gap equation for superconductors.

A similar analogy to super-conductors exists also for the hadronized theory. The classical version of it corresponds exactly to the classical Ginzburg-Landau equation for type II super-conductors in which the gap is allowed to be space time dependent. In fact, the classical hadronized theory can be derived alternatively by assuming such a dependence in the gap equation[34,18]. The advantage of our functional derivation is that the hadronized theory is not merely some classical approximation but becomes upon quantization completely equivalent to the original quark gluon theory.

A final comment concerns the Okubo-Zweig-Iizuka rule. As argued in the general section, the meson Lagrangian exactly respects this rule. This can be checked directly for all interaction terms in (4.34). Violations of this rule are all coming from meson loops. The calculation of some important loop diagrams leads to straight-forward estimates for the size of such violations.

V OUTLOOK

We have shown that in the absence of color, quark gluon theories can successfully be hadronized. The resulting quantum field theory incorporates correctly many features of strong interactions. It's basic fields are bilocal and the Feynman rules are topologically similar to dual diagrams. Our considerations have taken place at a rather formal level. Certainly, there are many problems which have been left open. For example, there is need for an understanding of the non-trivial gauge properties of the bilocal theory. Also, a consistent renormalization procedure will have to be developed in future investigations.

The inclusion of color is the challenging problem left open by this investigation. If color quark gluon theory is really equivalent to some kind of dual model the corresponding hadronization program should not produce bilocal but multilocal fields which are characterized by the position of a whole string rather that just its end points. A field theory should by constructed for gauge invariant objects like

$$\bar{\psi}(x) \exp\left(ig \int_x^y G^\nu(z) dz_\nu\right) \psi(y)$$

which depend on the whole path from x to y.

The difficulty in a direct generalization of the previous procedure is the self-interaction of the gluons. Only after the infrared behaviour of gluon propagators will be known, bare hadrons can be constructed inside the corresponding potential well and the "hadronization" methods can serve for the determination of the complete residual interactions.

It is hoped that hadronic Feynman rules in the presence of color will follow a pattern similar to that found here for the non-abelian theory.

Let us finally mention that an interesting field of applications of our methods lies in solid-state physics. Semi-conductors in which conduction and valence band have only small separations may show a phasetransition to what is called excitonic insulator. The critical phenomena taking place inside such an exciton system will find their most appropriate description by studying the scaling properties of the bilocal field theory.

APPENDIX A: Remarks on the Fermion Bose-Salpeter Equation

Consider the four Fermion Green's function

$$G^{(4)}_{\alpha\beta,\alpha'\beta'}(x,y;x'y') \equiv \langle 0|T(\psi_\alpha(x)\bar\psi_\beta(y)\bar\psi_{\alpha'}(x')\psi_{\beta'}(y'))|0\rangle \quad (A.1)$$

which becomes in the interaction picture

$$G^{(4)}_{\alpha\beta,\alpha'\beta'}(x,y;x'y') = \left[\langle 0|T e^{ig\int d^4z\, \bar\psi(z)\gamma^\mu\psi(z)G_\mu(z)}|0\rangle\right]^{-1} \quad (A.2)$$

$$\times \langle 0|T(e^{ig\int d^4z\, \bar\psi(z)\gamma^\mu\psi(z)G_\mu(z)}\psi_\alpha(x)\bar\psi_\beta(y)\psi_{\alpha'}(x')\psi_{\beta'}(y'))|0\rangle$$

Expanding the exponential and keeping only the ladder exchanges corresponding to the Feynman graph in Fig 13,

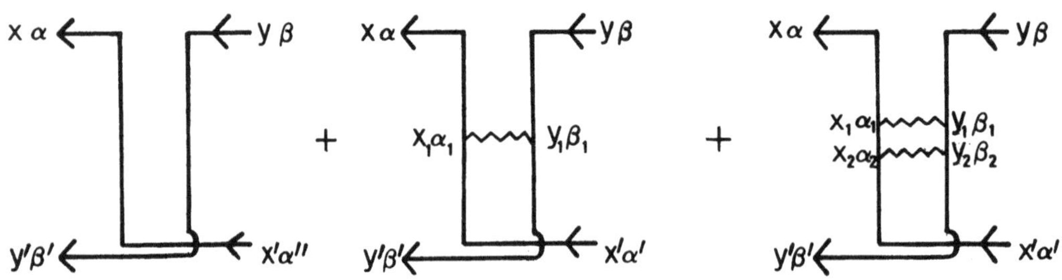

Figure 13

we obtain

$$G^{(4)}_{\alpha\beta,\alpha'\beta'}(xy,x'y') = G_{\alpha\alpha'}(x-x')G_{\beta'\beta}(y'-y) \tag{A.3}$$

$$+g^2\int dx_1 dy_1\, G_{\alpha\alpha_1}(x-x_1)\gamma^{\mu}_{\alpha_1\alpha_1'}G_{\alpha_1'\alpha'}(x_1-x')G_{\beta'\beta_1'}(y'-y_1)\gamma_{\mu\,\beta_1'\beta_1}G_{\beta_1\beta}(y_1-y)D(x-y)$$

$$+g^4\int dx_1 dy_1 dx_2 dy_2\, G_{\alpha\alpha_1}(x-x_1)\gamma^{\mu_1}_{\alpha_1\alpha_1'}G_{\alpha_1'\alpha_2}(x_1-x_2)\gamma^{\mu_2}_{\alpha_2\alpha_2'}G_{\alpha_2'\alpha_2}(x_2-x')$$

$$D(x_1-y_1)\,D(x_2-y_2)\,G_{\beta'\beta_2'}(y'-y_2)\gamma_{\mu_2\,\beta_2'\beta_2}G_{\beta_2\beta_1'}(y_2-y_1)\gamma_{\mu_1\,\beta_1'\beta_1}G_{\beta_1'\beta}(y_1-y)$$

$$+\ \ldots$$

The series can be summed to the integral equation

$$G^{(4)}_{\alpha\beta,\alpha'\beta'}(xy,x'y') = G_{\alpha\alpha'}(x-x')G_{\beta'\beta}(y'-y)$$

$$+g^2\int dx_1 dy_1\, G_{\alpha\alpha_1}(x-x_1)\gamma^{\mu}_{\alpha_1\alpha_1'}G^{(4)}_{\alpha_1'\beta_1',\alpha'\beta'}(x_1y_1,x'y') \tag{A.4}$$

$$\gamma_{\mu\,\beta_1'\beta_1}G_{\beta_1\beta}(y_1-y)\,D(x_1-y_1)$$

With the abbreviation

$$\xi_{\alpha_1\beta_1,\beta_1'\alpha_1'} = \gamma^{\mu}_{\alpha_1\alpha_1'}\gamma_{\mu\,\beta_1'\beta_1} = 1_{\alpha_1\beta_1}1_{\beta_1'\alpha_1'} + (i\gamma_5)_{\alpha_1\beta_1}(i\gamma_5)_{\beta_1'\alpha_1'} \tag{A.5}$$

$$-\tfrac{1}{2}\gamma^{\mu}_{\alpha_1\beta_1}\gamma_{\mu\,\beta_1'\alpha_1'} - \tfrac{1}{2}(\gamma^{\mu}\gamma_5)_{\alpha_1\beta_1}(\gamma_{\mu}\gamma_5)_{\beta_1'\alpha_1'}$$

This can be written as

$$G^{(4)}_{\alpha\beta,\alpha'\beta'}(xy,x'y') = G_{\alpha\alpha'}(x-x')G_{\beta'\beta}(y'-y)$$

$$+g^2\int dx_1 dy_1\, G_{\alpha\alpha_1}(x-x_1)\xi_{\alpha_1\beta_1,\beta_1'\alpha_1'}D(x_1-y_1)G^{(4)}_{\alpha_1'\beta_1',\alpha'\beta'}(x_1y_1,x'y')G_{\beta_1\beta}(y_1-y) \tag{A.6}$$

or , symbolically:

$$G^{(4)} = GG^T + GG^T \xi g^2 D\, G^{(4)} \tag{A.7}$$

The transition matrix T is defined by removing the external particle poles in the connected part of $G^{(4)}$

$$G^{(4)}_{\alpha\beta,\alpha'\beta'}(xy,x'y') \equiv G_{\alpha\alpha'}(x-x')G_{\beta'\beta}(y'-y)$$
$$+ \int dx_1 dy_1 dx_2 dy_2\, G_{\alpha\alpha_1}(x-x_1)G_{\beta'\beta_2}(y'-y_2)\, T_{\alpha_1\beta_1,\alpha_2\beta_2}(x_1y_1,x_2y_2) \tag{A.8}$$
$$G_{\alpha_2\alpha'}(x_2-x')G_{\beta_1\beta}(y_1-y)$$

which may be abbreviated by

$$G^{(4)} = GG^T + GG^T\, T\, GG^T \tag{A.9}$$

From (A.6) and (A.8) the transition matrix satisfies the integral equation

$$T_{\alpha_1\beta_1,\alpha_2\beta_2}(x_1y_1,x_2y_2) = \xi_{\alpha_1\beta_1,\beta_2\alpha_2}\, g^2 D(x_1-y_1)\delta(x_1-x_2)\delta(y_1-y_2)$$
$$(A.10)$$
$$+ \xi_{\alpha_1\beta_1,\bar\beta_1\bar\alpha_1}\, g^2 D(x_1-y_1)\int dx_1' dy_1'\, G_{\bar\alpha_1\alpha_1'}(x_1-x_1')\, T_{\alpha_1'\beta_1',\alpha_2\beta_2}(x_1'y_1',x_2y_2)\, G_{\beta_1'\bar\beta_1}(y_1'-y_1)$$

which is seen to coincide with the equation (3.37) for the propagator of the bilocal field. In a short notation, this equation can be written as

$$T = \xi g^2 D + \xi g^2 D G G^T T \tag{A.11}$$

The perturbation expansion

$$T = \xi g^2 D + \xi g^2 D G G^T \xi g^2 D + \ldots \tag{A.12}$$

reveals the one, two etc. photon exchanges of the ladder diagrams. In momentum space the four particle Green's function is defined by

$$(2\pi)^4 \delta^4(q'-q)\, G^{(4)}_{\alpha\beta,\alpha'\beta'}(P,P'|q)$$
$$= \int dx\,dy\,dx'\,dy'\, e^{i\left[(P+\frac{q}{2})x + (P'-\frac{q'}{2})y' - (P-\frac{q}{2})y - (P'+\frac{q'}{2})x'\right]}\, G^{(4)}_{\alpha\beta\alpha'\beta'}(xy, x'y')$$

(A.13)

where the momenta are indicated in Fig. 14.

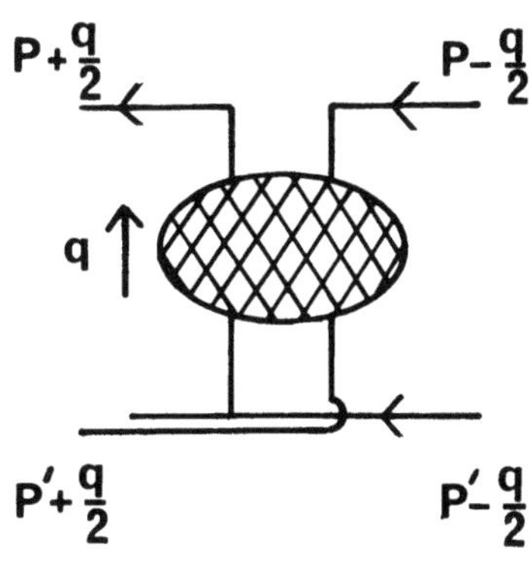

Figure 14

The corresponding scattering matrix $T(P,P'|q)$ satisfies the integral equation

$$T(P,P'|q) = \xi g^2 D(P-P') + \xi g^2 \int \frac{dP''}{(2\pi)^4} D(P-P'') G(P''+\tfrac{q}{2}) \; T(P'',P'|q) G(P''-\tfrac{q}{2}) \tag{A.14}$$

The ladder exchange is in general expected to produce quark anti-quark bound states. Suppose $|H(q)\rangle$ is one of them. Inserting it into (A.1) as an intermediate state gives for $x_0, y_0 > x_0', y_0'$ a contribution

$$G^{(4)}_{\alpha\beta,\alpha'\beta'}(x,y;x'y') = \int \frac{d^3q}{2E_q(2\pi)^3} \,\Theta\!\left(R_0 - R_0' - \tfrac{1}{2}(|z_0| + |z_0'|)\right)$$
$$\times \langle 0 | T(\psi_\alpha(x)\bar\psi_\beta(y)) | H(q)\rangle\langle H(q) | T(\bar\psi_{\alpha'}(x')\psi_{\beta'}(y')) | 0\rangle \tag{A.15}$$

where $R \equiv (x+y)/2$ and $z = x-y$. The Θ function is non-zero if

$$\min(x_0, y_0) > \max(x_0' y_0')$$

Using the integral representation

$$\Theta(x_0) = \frac{i}{2\pi} \int da \; e^{-a x_0} \frac{1}{a + i\varepsilon}$$

we have

$$\Theta(R_0 - R_0' - \tfrac{1}{2}(|z_0| + |z_0'|)) = \frac{i}{2\pi}\int dq_0\, e^{-i(q_0 - E_q)(R_0 - R_0')} \tag{A.16}$$
$$e^{i(q_0 - E_q)\frac{1}{2}(|z_0| + |z_0'|)} \frac{1}{q_0 - E_q + i\varepsilon}$$

Introducing Bethe-Salpeter wave functions

$$\phi_{\alpha\beta}(x,y|q) \equiv \langle 0|T(\psi_\alpha(x)\bar{\psi}_\beta(y))|H(\underset{\sim}{q})\rangle$$
$$\bar{\Phi}_{\beta'\alpha'}(y'x'|q) = \langle H(\underset{\sim}{q})|T(\psi_{\beta'}(y')\bar{\psi}_{\alpha'}(x'))|0\rangle \tag{A.17}$$

and their momentum space versions

$$\phi_{\alpha\beta}(x,y|q) = e^{-i(E_q R_0 - \underset{\sim}{q}\underset{\sim}{R})}\int \frac{d^4P}{(2\pi)^4} e^{-iP(x-y)} \phi_{\alpha\beta}(P|q)$$
$$\bar{\Phi}_{\beta'\alpha'}(y'x'|-q) = e^{i(E_q R_0' - \underset{\sim}{q}\underset{\sim}{R}')}\int \frac{d^4P}{(2\pi)^4} e^{iP'(x'-y')} \bar{\Phi}_{\beta'\alpha'}(P'|-q) \tag{A.18}$$

the four-particle Green's function in momentum space is seen to exhibits a pole at $q_0 \approx E_q$

$$G^{(4)}_{\alpha\beta\alpha'\beta'}(P,P'|q) \approx \frac{-i}{2E_q(q_0 - E_q + i\varepsilon)} \phi_{\alpha\beta}(P|q)\bar{\Phi}_{\beta'\alpha'}(P'|q) \tag{A.19}$$

The opposite time ordering $x_0, y_0 < x_0', y_0'$ contributes a pole at $q_0 = -E_q$. Both poles can be collected by writing in (A.19) the factor

$$-\frac{i}{q^2 - M_H^2 + i\epsilon} .$$

This factorization is consistent with the integral equation only for a specific normalization of the Bethe-Salpeter wave functions. In order to see this write (A.7) in the form

$$\begin{aligned} G^{(4)} &= GG^T + GG^T \xi g^2 D G^{(4)} \\ &= \left(1 - GG^T \xi g^2 D\right)^{-1} GG^T \\ &= GG^T \left(1 - \xi g^2 D GG^T\right)^{-1} \end{aligned} \tag{A.20}$$

Suppose now that a solution is found for different values of the coupling constant g^2. Then the variation of $G^{(4)}$ for small changes of g^2 is

$$\begin{aligned} \frac{\partial G^{(4)}}{\partial g^2} &= \left(1 - GG^T \xi g^2 D\right)^{-1} GG^T \xi D \\ &\quad \times GG^T \left(1 - \xi g^2 D\right)^{-1} \\ &= G^{(4)} \xi D G^{(4)} \end{aligned} \tag{A.21}$$

If one goes in the vicinity of the pole $q^2 \approx M_H^2(g^2)$ this becomes

$$-\frac{\partial}{\partial g^2}\frac{i}{s-M_H^2(g^2)}\Phi_H^q(P|q)\bar{\Phi}_H^q(P|-q)$$

$$= \frac{i}{s-M_H^2(g^2)}\Phi_H^q(P|q)\int\frac{d\bar{P}d\bar{P}'}{(2\pi)^8}\bar{\Phi}_H^q(\bar{P}|-q)\,\xi D(\bar{P}-\bar{P}')\,\Phi_H^q(\bar{P}'|q)\,\bar{\Phi}_H^q(P'|-q)\frac{i}{s-M_H^2(g^2)} \tag{A.22}$$

This can be true at the double pole at $q^2 = M_H^2(g^2)$ only if

$$\frac{\partial M_H^2(g^2)}{\partial g^2} = -i\int\frac{dP\,dP'}{(2\pi)^4(2\pi)^4}\Phi_H(P|q)\,\xi D(P-P')\,\bar{\Phi}_H(P'|-q) \tag{A.23}$$

If we go over to the Bethe Salpeter vertex function (3.29)

$$\Gamma^H(P|q) = N_H\, G_M^{-1}(P+\tfrac{q}{2})\,\Phi^H(P|q)\,G_M^{-1}(P-\tfrac{q}{2}) \tag{A.24}$$

$$\bar{\Gamma}^H(P|-q) = N_H^*\, G_M^{-1}(P-\tfrac{q}{2})\,\bar{\Phi}^H(P|-q)\,G_M^{-1}(P+\tfrac{q}{2})$$

this amounts to

$$g_H^2(q^2)\frac{\partial M_H^2(q^2)}{\partial q^2}=i|N_H|^2\int\frac{dP\,dP'}{(2\pi)^4(2\pi)^4}\,\mathrm{tr}\Big\{G_M(P+\tfrac{q}{2})\Gamma^H(P|q)$$

$$G_M(P-\tfrac{q}{2})\,g_H^2(q^2)\,D(P-P')\,G_M(P'-\tfrac{q}{2})\,\bar{\Gamma}^H(P'|-q)\,G_M(P'+\tfrac{q}{2})\Big\} \tag{A.25}$$

Using the integral equation (3.30) this reduces to the normalization

$$\frac{g_H^2(q^2)}{\dfrac{\partial g_H^2(q^2)}{\partial q^2}}=-i|N_H|^2\int\frac{d^4P}{(2\pi)^4}$$

$$\times\,\mathrm{tr}\left[G_M(P+\tfrac{q}{2})\Gamma^H(P|q)G_M(P-\tfrac{q}{2})\bar{\Gamma}^H(P|-q)\right] \tag{A.26}$$

This determines $|N_H|^2$ as

$$|N_H|^2=g_H^2(q^2)\Big/\frac{\partial g_H^2(q^2)}{\partial q^2} \tag{A.27}$$

Notice that this normalization is defined for all q^2 with some $N_H(q^2)$. For real $\Gamma(P|q)$ one may choose $N_H(q^2)$ real such that

$$\bar{\Gamma}(P|-q)=\Gamma(P|-q)$$

(Both satisfy the same integral equation).
The orthogonality of $\Gamma^H(P|q)$ and $\bar{\Gamma}^{H'}(P|-q)$ for different hadrons is proved as usual by considering (3.30) once for $(\xi g^2 D)^{-1}\Gamma^H$ and once for $(\xi g^2 D)^{-1}\bar{\Gamma}^{H'}$,

multiplying the first by $\bar{\Gamma}^{H'}$ and the second by Γ^{H}, taking the trace and subtracting the results from each other (assuming no degeneracy of $g_H(q^2)$ and $g_{H'}(q^2)$). The normalization (A.27) is seen to be consistent with the expansion of the T matrix given in (3.39)

$$T_{\alpha\beta,\alpha'\beta'}(P,P'|q) = -ig^2 \sum_H \frac{\Gamma^H_{\alpha\beta}(P|q)\bar{\Gamma}^H_{\beta'\alpha'}(P'|-q)}{g_H^2(q^2) - q^2} \tag{A.28}$$

If q^2 runs into a pole M_H^2 this expression is singular as

$$T_{\alpha\beta,\alpha'\beta'}(P,P'|q) \approx -ig^2 \frac{1}{(q^2-M_H^2)\frac{\partial g_H^2(q^2)}{\partial q^2}} \Gamma^H_{\alpha\beta}(P|q)\bar{\Gamma}^H_{\beta'\alpha'}(P'|-q) \tag{A.29}$$

According to (A.9) this produces a singularity in $G^{(4)}$ (in short notation)

$$\begin{aligned} G^{(4)} &\approx -i \frac{g^2}{\frac{\partial g_H^2(q^2)}{\partial q^2}} \frac{1}{q^2-M_H^2} (G\Gamma^H G)(G\bar{\Gamma}^H G) \\ &= -i \frac{g^2}{\frac{\partial g_H^2(q^2)}{\partial q^2}} \frac{1}{q^2-M_H^2} |N_H|^2 \Phi^H \bar{\Phi}^H \end{aligned} \tag{A.30}$$

which coincides with (A.19) by virtue of (A.27).

For completeness we now give the Bethe-Salpeter equation (3.27) the form projected into the different covariants:

$$m'(P|q) = S(P|q) + P(P|q)\, i\gamma_5 + V^\mu(P|q)\,\gamma_\mu + A^\mu(P|q)\,\gamma_\mu\gamma_5 \tag{A.31}$$

If $m_i(P|q)$ (i=1,2,3,4) abbreviates S,P,V,A, one has

$$m_i(P|q) = -4\xi_i\, g^2 \sum_{j=1}^{4} \frac{d^4P'}{(2\pi)^4} \frac{1}{(P-P')_E^2+\mu^2}\, \frac{1}{(P'+\frac{q}{2})_E^2+M^2}\, \frac{1}{(P'-\frac{q}{2})_E^2+M^2} \times t_{ij}(P'|q)\, m_j(P'|q) \tag{A.32}$$

with $t_{ij}(P|q)$ being the traces defined in (4.10) and ξ_i = (4, +4, -2, -2). Explicitly one finds

$$t_{SS}(P|q) = P^2 - \frac{q^2}{4} + M^2 \quad , t_{SV}(P|q) = t_{VS} = 2MP^\mu$$

$$t_{PP}(P|q) = P^2 - \frac{q^2}{4} - M^2 \quad , t_{PA^\mu}(P|q) = -t_{AP} = iMq^\mu$$

$$t_{V^\mu V^\nu}(P|q) = -\left(P^2 - \frac{q^2}{4} - M^2\right)g^{\mu\nu} + 2P^\mu P^\nu - \frac{1}{2}q^\mu q^\nu$$

$$t_{A^\mu A^\nu}(P|q) = -\left(P^2 - \frac{q^2}{4} - M^2\right)g^{\mu\nu} + 2P^\mu P^\nu - \frac{1}{2}q^\mu q^\nu - 2M^2 g^{\mu\nu}$$

with all other traces vanishing. Notice that in the Bethe-Salpeter equation for m' there is no tensor contribution due to the absence of such a term in the Fierz transform of $\gamma^\mu \otimes \gamma_\mu$. The integrals in (A.31) go directly over into (4.9) for large gluon mass μ .

APPENDIX B: The vertices for Heavy Gluons

Here we present the calculation of the vertices $\mathcal{A}_2$, $\mathcal{A}_3$, $\mathcal{A}_4$ for large μ. As discussed in the text, all higher vertices remain finite when the cutoff and μ go to infinity in the order $\Lambda^2 \gg \mu^2 \gg M^2$ and will consequently be neglected.

Consider first $\mathcal{A}_2$ as described in (4.10) and (A.33). The integrals $\mathcal{J}_{ij}(q)$ are evaluated by expanding

$$\left[\left(P+\frac{q}{2}\right)_E^2+M^2\right]^{-1}\left[\left(P-\frac{q}{2}\right)_E^2+M^2\right]^{-1}$$

$$= \left[P_E^2+M^2\right]^{-2}\left\{1+\frac{q^2/2}{P_E^2+M^2}+\frac{(Pq)^2}{(P_E^2+M^2)^2}+O\left(\frac{M^4}{(P_E^2+M^2)^2},\frac{q^4}{(P_E^2+M^2)^2}\right)\right\} \tag{B.1}$$

Since $t_{ij}(P|q)$ grow at most like P_E^2 (see (A.33)) the terms $O(M^4/_E{}^4, q^4/P_E^4)$ contribute finite amounts upon integration and will be neglected. At this place we have assumed q^2 to remain of the same order of M^2. Actually this is not true for vector and axial-vector meson fields$^{+)}$ but since numerically $m_\rho^2, m_{A_1}^2 \lesssim \frac{1}{100}\mu^2$, the neglected terms are indeed very small.

The following integrals are needed in addition to (4.5), (4.11) (neglecting finite amounts)

+) $m_\rho^2 = 6M^2$; $m_{A_1}^2 \approx 12M^2$; $\mu^2 \approx 144\,\mathrm{GeV}^2$

$$\int \frac{d^4P_E}{(2\pi)^4} \frac{1}{(P_E^2+M^2)^2} P^2 = -(Q-M^2L)$$

$$\int \frac{d^4P_E}{(2\pi)^4} \frac{1}{(P_E^2+M^2)^2} P_\mu P_\nu = -(Q-M^2L) \frac{g_{\mu\nu}}{4}$$

$$\int \frac{d^4P_E}{(2\pi)^4} \frac{1}{(P_E^2+M^2)^3} P_\mu P_\nu = -L \frac{g_{\mu\nu}}{4} \tag{B.2}$$

$$\int \frac{d^4P_E}{(2\pi)^4} \frac{1}{(P_E^2+M^2)^4} P_\mu P_\nu P_\lambda P_\varkappa = \frac{L}{24}(g_{\mu\nu}g_{\lambda\varkappa}+g_{\mu\lambda}g_{\nu\varkappa}+g_{\mu\varkappa}g_{\nu\lambda})$$

The results are displayed in equ. (4.12).

There is one subtlety connected with gauge invariance when evaluating the integrals $\mathcal{J}_{VV}(q)$ and $\mathcal{J}_{AA}(q)$. In fact, the first of these integrals coincides with the standard photon self-energy graph in quantum electro-dynamics. There the cutoff procedure is known to produce a non-gauge invariant result. The cutoff calculation yields:

$$\mathcal{J}_{V^\mu V^\nu}(q) = -\frac{1}{3}(q^2 g_{\mu\nu} - q_\mu q_\nu)L - \frac{1}{2}(Q+M^2L)g_{\mu\nu}$$

$$\tag{B.3}$$

$$\mathcal{J}_{A^\mu A^\nu}(q) = -\frac{1}{3}((q^2-6M^2)g_{\mu\nu} - q_\mu q_\nu)L - \frac{1}{2}(Q+M^2L)g_{\mu\nu}$$

There are many equivalent ways to enforce gauge invariance. The simplest one proceeds via dimensional regularization If one evaluates the integrals Q and L in D = 4 − ε dimensions with a small $\varepsilon > 0$, then

$$L = \int \frac{d^D p}{(2\pi)^4} \frac{1}{(P_E^2+M^2)^2} = \frac{\pi^{D/2}}{(2\pi)^4} \frac{1}{(M^2)^{2-\frac{D}{2}}} \frac{\Gamma(2-\frac{D}{2})}{\Gamma(2)}$$

$$= \frac{\pi^2}{(2\pi)^4} \frac{2}{\varepsilon} + O(\varepsilon)$$

$$Q = \int \frac{d^D p_E}{(2\pi)^4} \frac{1}{P_E^2+M^2} = \frac{\pi^{D/2}}{(2\pi)^4} \frac{1}{(M^2)^{1-\frac{D}{2}}} \frac{\Gamma(1-\frac{D}{2})}{\Gamma(2)}$$

$$= \frac{\pi^2}{(2\pi)^4} M^2 \left(-\frac{2}{\varepsilon}\right) + O(\varepsilon)$$

Hence at the pole $\varepsilon = 0$, Q and L become related such that $Q + M^2 L = 0$, cancelling the last terms in (B.3). Notice that when dealing with the renormalizable theory with large gluon mass $\mu^2 \ll \Lambda^2$, this cancellation is still present while the other Q integrals in $J_{ij}(q)$ become unrelated with the L integrals, the first being essentially $\mu^2 \log \Lambda^2/\mu^2$, the other $\log \mu^2/M^2$.

Consider now the interaction terms $\mathcal{A}_3$. Here the traces grow at most as P_E^3. Thus as far as the divergent contributions are concerned, the denominators in the integrals (4.27) can be approximated as

$$\frac{1}{(P+q_2+q_1)_E^2+M^2} \frac{1}{(P+q_1)_E^2+M^2} \frac{1}{P_E^2+M^2} \tag{B.4}$$

$$= \frac{1}{(P_E^2+M^2)^3} \left\{ 1 + \frac{P(2q_1+q_2)}{P_E^2+M^2} + O\left(\frac{M^2}{P_E^2+M^2}, \frac{q_i^2}{P_E^2+M^2}\right) \right\}$$

Since this expression decreases at least as $1/p_E^6$ the traces have to be known only with respect to their leading p_E^3 and p_E^2 behaviours. These are

$$
\begin{aligned}
&t_{SSS}(p|q_2 q_1) \hat{=} 3p^2 M \,, \quad t_{SPP} \\
&t_{SPA^\mu}(p|q_2 q_1) \hat{=} i p^2 p^\mu + 2 i p^\mu [p(q_1+q_2)] - i p^2 q_2^{\ \mu} \\
&t_{SA^\mu P}(p|q_2 q_1) \hat{=} -i p^2 p^\mu - i p^2 (2q_1+q_2)^\mu \\
&t_{SSV^\mu}(p|q_2 q_1) \hat{=} p^2 p^\mu - p^2 q_2^{\ \mu} + 2p^\mu [p(q_1+q_2)] = t_{PPV^\mu}(p|q_2 q_1) \\
&t_{SV^\mu S}(p|q_2 q_1) \hat{=} p^2 p^\mu + p^2 (2q_1+q_2)^\mu = t_{PV^\mu P}(p|q_2 q_1) \\
&t_{SV^\mu V^\nu}(p|q_2 q_1) \hat{=} 4M p^\mu p^\nu - M p^2 g^{\mu\nu} \\
&t_{SA^\mu A^\nu}(p|q_2 q_1) \hat{=} 4M p^\mu p^\nu - 3M p^2 g^{\mu\nu} \\
&t_{V^\mu V^\lambda V^\kappa}(p|q_1 q_2) \hat{=} 4 p^\mu p^\lambda p^\kappa - p^2 (p^\mu g^{\lambda\kappa} + p^\lambda g^{\mu\kappa} + p^\kappa g^{\mu\lambda}) \\
&\qquad + 2 p^\mu p^\lambda q_1^{\ \kappa} + 2 p^\lambda p^\kappa (q_1+q_2)^\mu + 2 p^\mu p^\kappa (2q_1+q_2)^\lambda \\
&\quad - 2 p^\kappa p(q_1+q_2) g^{\mu\lambda} - 2 p^\mu p q_1 g^{\lambda\kappa} - p^2 (-g^{\mu\lambda} q_2 + g^{\lambda\kappa} q_2 + g^{\mu\kappa} (2q_1+q_2 \ldots
\end{aligned}
\tag{B.5}
$$

Using (B.2) one obtains exactly the third order terms in the Lagrangian equ. (4.29) (if this is written in the σ 'form).

The fourth order couplings in $\mathcal{L}_4$ are the simplest to evaluate. Here only the leading p^4 behaviour of $t_{ij}(p|q_3, q_2, q_1)$ contributes proportional to L and the propagator can diractly be used in the form

$$[p_E^2 + M^2]^{-4} .$$

$$t_{SSSS}(P|q_3q_2q_1) \cong t_{SSPP} \cong t_{PPPP} \cong P^4$$

$$t_{SSV^\mu V^\nu}(P|q_3q_2q_1) \cong t_{PPV^\mu V^\nu} \cong t_{SSA^\mu A^\nu} \cong$$

$$\cong t_{PPA^\mu A^\nu} \cong -it_{SPA^\mu V^\nu} \cong it_{PSA^\mu V^\nu}$$

$$\cong 2P^2P^\mu P^\nu - P^4 g^{\mu\nu} \qquad (B.6)$$

etc.

APPENDIX C: Some Algebraic Derivations

Here we want to compare some of our results with traditional derivations[28,29] obtained by purely algebraic considerations together with PCAC.
The vector and axial-vector currents

$$V_\mu^a(x) = \bar{\psi}(x)\gamma_\mu \frac{\lambda^a}{2}\psi(x) \; ; \; A_\mu^a(x) = \bar{\psi}(x)\gamma_\mu\gamma_5\frac{\lambda^a}{2}\psi(x) \quad (C.1)$$

generate chiral SU(3) x SU(3) under which the quark gluon Lagrangian transforms as

$$\mathcal{L} = \mathcal{L}_{\text{chiral invariant}} - u^0 - cu^8 - du^3$$

where

$$u^0 + cu^8 + du^3 = \bar{\psi}\, m\, \psi \equiv \sum_a m^a \bar{\psi}\frac{\lambda^a}{2}\psi$$

$$= \sqrt{\tfrac{2}{3}}(m^u + m^d + m^s)\bar{\psi}\frac{\lambda^0}{2}\psi \quad (C.2)$$

$$+ \frac{1}{\sqrt{3}}(m^u + m^d - 2m^s)\bar{\psi}\frac{\lambda^8}{2}\psi$$

$$+ (m^u - m^d)\bar{\psi}\frac{\lambda^3}{2}\psi$$

Hence

$$u^a \equiv m^0 \bar{\psi}\frac{\lambda^a}{2}\psi \quad (= m^0 \tilde{u}^a) \quad (C.3)$$

$$c \equiv \frac{m^8}{m^0}, \quad d = \frac{m^3}{m^0} \quad (C.4)$$

Defining also the pseudoscalar densities

$$v^a \equiv m^0 \bar{\psi}\, i\gamma_5 \frac{\lambda^a}{2}\psi \quad (C.5)$$

then u^a and v^a form the $(3\bar{3})\pm(\bar{3}3)$ representation of SU(3) x SU(3):

$$[Q_5^a, u^a] = i d^{abc} v^c \quad , [Q_5^a, v^b] = -i d^{abc} u^c \tag{C.6}$$

From the equation of motion one finds the conservation law[+)]

$$\partial^\mu V_\mu^a(x) = \partial^\mu \bar{\Psi} \gamma_\mu \frac{\lambda^a}{2} \Psi = \bar{\Psi} \left[\frac{\lambda^a}{2} \mathfrak{M} \right] \Psi \tag{C.7}$$

$$= i f^{abc} \mathfrak{M}^b \, \bar{\Psi} \frac{\lambda^c}{2} \Psi \qquad a = 0,1,\ldots,8$$

$$\partial^\mu A_\mu^a(x) = \partial^\mu \bar{\Psi} \gamma_\mu \gamma_5 \frac{\lambda^a}{2} \Psi = -\bar{\Psi} \left\{ \frac{\lambda^a}{2} \mathfrak{M} \right\} \gamma_5 \Psi$$

$$= i d^{abc} \mathfrak{M}^b \, \bar{\Psi} i \gamma_5 \frac{\lambda^c}{2} \Psi \qquad a = 1,\ldots,8 \tag{C.8}$$

Let us neglect SU(2) breaking in $\mathfrak{M}$. By taking (C.8) between vacuum and pseudoscalar meson states one finds

$$f_\pi m_\pi^2 = \frac{1}{\sqrt{3}} \left(\sqrt{2}\, \mathfrak{M}^0 + \mathfrak{M}^8 \right) \frac{m_V^2}{3} \sqrt{L} \, Z_\pi^{1/2}$$

$$f_K m_K^2 = \frac{1}{\sqrt{3}} \left(\sqrt{2}\, \mathfrak{M}^0 - \frac{1}{2} \mathfrak{M}^8 \right) \frac{m_V^2}{3} \sqrt{L} \, Z_K^{1/2} \tag{C.9}$$

etc. for the other members of the multiplet, where one has used (see the pseudoscalar version of (4.73)):

$$\langle 0 | \partial^\mu A_\mu^a | \pi \rangle \equiv f_\pi m_\pi^2 \; ; \; \langle 0 | \bar{\Psi} i \gamma_5 \frac{\lambda^\pi}{2} \Psi | \pi \rangle = \frac{\mu^2}{2g^2} \frac{1}{\sqrt{L}} \langle 0 | \pi | \pi \rangle$$

etc.

$$= \frac{\mu^2}{2g^2} \frac{Z_\pi^{1/2}}{\sqrt{L}} = \frac{m_V^2}{3} \sqrt{L} \, Z_\pi^{1/2}$$

By writing $\mathfrak{M}$ as

$$\mathfrak{M} = \begin{pmatrix} \mathfrak{M}^u & & \\ & \mathfrak{M}^d & \\ & & \mathfrak{M}^s \end{pmatrix} = \frac{1}{2\sqrt{3}} \begin{pmatrix} \sqrt{2}\,\mathfrak{M}^0 + \mathfrak{M}^8 & & \\ & \sqrt{2}\,\mathfrak{M}^0 + \mathfrak{M}^8 & \\ & & \sqrt{2}\,\mathfrak{M}^0 - 2\mathfrak{M}^8 \end{pmatrix} \tag{C.10}$$

equ. (C.9) takes the form

$$\begin{aligned} f_\pi^2 m_\pi^2 &= (\mathfrak{M}^u + \mathfrak{M}^d)/2 \;\; Z_\pi^{1/2} \frac{2}{3} m_V^2 \sqrt{L} \\ f_K^2 m_K^2 &= (\mathfrak{M}^u + \mathfrak{M}^s)/2 \;\; Z^{1/2} \frac{2}{3} m_V^2 \sqrt{L} \end{aligned} \tag{C.11}$$

which agrees with (4.54)

By evaluating (C.6) between vacuum states and saturating the commutator with pseudoscalar intermediate state one finds

$$\begin{aligned} f_\pi \mathfrak{M}_0 \frac{\mu^2}{2g^2} \frac{Z_\pi^{1/2}}{\sqrt{L}} &= \frac{1}{\sqrt{3}} \left(\sqrt{2} \langle 0 | u^0 | 0 \rangle + \langle 0 | u^8 | 0 \rangle \right) \\ f_K \mathfrak{M}_0 \frac{\mu^2}{2g^2} \frac{Z_K^{1/2}}{\sqrt{L}} &= \frac{1}{\sqrt{3}} \left(\sqrt{2} \langle 0 | u^0 | 0 \rangle - \frac{1}{2} \langle 0 | u^8 | 0 \rangle \right) \end{aligned} \tag{C.12}$$

and similar for the other partners of the multiplet.

Inserting the result of equ.(4.73)

$$\langle 0 | u^a | 0 \rangle = \mathfrak{M}^0 \langle 0 | \tilde{u}^a | 0 \rangle = -\frac{1}{2} \frac{\mu^2}{g^2} \mathfrak{M}^0 M^a \tag{C.13}$$

and writing M in the same way as $\mathfrak{M}$ in (C.10) brings (C.12) to the form

$$f_\pi Z_\pi^{1/2} = \sqrt{L}\,(M^u + M^d)$$
(C.14)
$$f_K Z_K^{1/2} = \sqrt{L}\,(M^u + M^s)$$

which agrees exactly with (4.54) (written there in SU(3) matrix form). Considerations of this type have led to the determination[28,29,33)]

$$c \approx -1.28$$

or (C.15)

$$\frac{(m^u + m^d)/2}{m^s} \approx \frac{1}{29}$$

Including also SU(2) violation in such a consideration gives[33)]

$$d \approx -.03$$

or (C.16)

$$\frac{m_u - m_d}{m_u + m_d} \approx -\frac{1}{4}$$

There are numerous extensions to SU(4)[35)] but they have to be viewed with great caution since it is hard to see how the large pseudoscalar and vector masses occuring there can dominate the divergence of the axial current and the vector current, respectively.

REFERENCES

1. G. VENEZIANO; Nuovo Cimento 57 A, 190 (1968)
K. BARDAKCI and H. RUEGG; Phys. Letters, 28 B, 342 (1968)
M.A. VIRASORO; Phys. Rev. Letters 22, 37 (1969)
C.J. GOEBEL and B. SAKITA; ibid. 22, 257 (1969)
H.M. CHAN; Phys. Letters 28 B, 485 (1969)
Z. KOBA and H.B. NIELSEN; Nucl. Phys. B 12, 512 (1969)
Y. NAMBU; Proc. Int. Conf. on Symmetries and Quark Models, Wayne State University 1969
H.NIELSEN; 15th Int. Conf. on High Energy Physics, Kiev 1970
L. SUSSKIND; Nuovo Cimento 69, 457 (1970)

2. Y. NAMBU; in Preludes in Theoretical Physics, North Holland (1966)
H. FRITZSCH and M. GELL-MANN; Proc. XVI Intern Conf. on High Energy Physics, Chicago 1972, Vol. 2, p. 135
W. BARDEEN, H. FRITZSCH, M. GELL-MANN in Scale and Conformal Invariance in Hadron Physics, Wiley, New York (1973)
D.J. GROSS and F. WILCZEK; Phys. Rev. Letters 30, 1343 (1973), Phys. Rev. D 8, 3633 (1973)
H. FRITZSCH, M. GELL-MANN and H. LEUTWYLER; Phys. Lett. 47 B, 365 (1973)
G. 't HOOFT; Erice Lectures 1975 (to be published)

3. J.M. CORNWALL and R. JACKIW; Phys. Rev. D 4, 367 (1971)
H. FRITZSCH and M. GELL-MANN; Proc. Intern. Conference on Duality and Symmetry in Hadron Physics (Weizman

ACKNOWLEDGEMENT

The author is grateful to Dr.Annemarie Kleinert for many inspiring discussions.

Science PRSS, 1971) and Proc. XVI Conf. on High Energy Physics, Chicago, 1972, Vol. 2, p. 135
See also: R.A. BRANDT and G. PREPARATA, Nucl. Phys. B 27, 541 (1971) and the review by R.A. BRANDT, Erice Lectures 1972 in Highlights in Particle Physics, ed. by A. ZICHICHI

4. G. 't HOOFT; Nucl. Physics B 75, 461 (1974)
C.R. HAGEN; Nucl. Phys. B 95, 477 (1975)
C.G. CALLAN, N. COOTE, D.J. GROSS; Phys. Rev. D 13, 1649 (1976)
T. APPELQUIST and H.D. POLITZER; Phys. Rev. Letters 34, 43 (1974)

5. G. MORPURGO; Physics 2, 95 (1975) and Erice Lectures 1968, 1971, 1974 ed. by A. ZICHICHI
R.H. DALITZ; Proc. XIIIth International Conference on High Energy Physics (Univ. of Calif. Press) Berkeley, (1967) p. 215
See also the book by J.J.J. KOKKEDEE, The Quark Model, Benjamin, New York 1969
H. LIPKIN; Phys. Rep. 8 C, 173 (1973)
M. BÖHM, H. JOOS and M. KRAMMER; CERN Preprint TH 1949 (1974)

6. H. KLEINERT; Lettere Nuovo Cimento, 4, 285 (1970)
M. KAKU and K.K. KIKKAWA; Phys. Rev. D 10, 1110 (1974), D 10, 1823 (1974), D 10, 3943 (1974)
E. CREMMER and J.L. SHERK; Nucl. Phys. B 90, 410 (1975)

7. For a very detailed introduction see:
J. RZEWUSKI; Quantum Field Theory II, Hefner, New York, 1968
S. COLEMAN; Erice Lectures 1974, in Laws of Hadronic Matter, ed. by A. ZICHICHI, p. 172

8. S. HORI, Nuclear Physics 30, 644 (1962)

9. This is a generalization to bilocal auxiliary fields of an old method of P.T.Mathews and A.Salam; Nuovo Cimento, 12,563 (1954); 2,120 (1955), reviewed by D.J. Gross and A.Neveu, Phys.Rev. D10, 3235 (1974).

10. H. KLEINERT; Phys. Letters 62B, 429 (1976)

11. For a thorough review to
N. NAKANISHI; Progr. Theor. Phys. Suppl. 43, 1 (1969)
The transition matrix T is discussed for scalar particles by
H. ZUR LINDEN; Nuovo Cimento 65 A, 197 (1970),
Phys. Rev. D 3, 1335 (1971)
In the non-relativistic limit:
J. SCHWINGER; Journ. Math. Phys. 5, 1606 (1964)
Recent discussions on fermion-fermion Bethe-Salpeter equation:
W. KUMMER; Nuovo Cimento 31, 219 (1964), 34, 1840 (1964)
K. SETO; Progr. Theor. Phys. 42, 1394, (1969)
H. ITO; Progr. Theor. Phys. 43, 1035 (1970)
N. NAKANISHI; J. Math. Phys. 12, 1578 (1971)

12. K. KIKKAWA et al.; Phys. Rev. 184, 1701 (1969), 187, 2249 (1970), Phys. Rev. D 1, 3258 (1970)
C. LOVELACE; Phys. Lett. 32 B, 703 (1970)
V.A. LESSANDRINI; Nuovo Cimento 2 A, 321 (1971)
G. VENEZIANO; Nucl. Phys. B 74, 365 (1974),
Phys. Letters 52 B, 220 (1974)

13. See Ref. 2., the review articles by
H.D. POLITZER, Phys. Reports 14 C, 129 (1974) and the one dimensional colored quark gluon model in Ref. 4.

14. S. OKUBO; Phys. Letters 5, 165 (1963)
G. ZWEIG; (unpublished)
J. IIZUKA; Supplement to Progress of Theor. Phys. 37, 21 (1966)
P.G.O. FREUND and Y. NAMBU; Phys. Rev. Letters 34, 1645 (1975)
G.F. CHEW and C. ROSENZWEIG; Phys. Letters 58 B, 93 (1975), Phys. Rev. D 12, 3907 (1975), Nucl. Phys. B 104, 290 (1976)
C. ROSENZWEIG; Phys. Rev. D 13, 3080 (1976)

15. H. KLEINERT; Nucl. Phys. B 65, 77 (1973), Erice Lectures 1974, in Lepton and Hadron Structure, ed. by A. ZICHICHI, P. 681, B 79, 526 (1974)

16. R.A. BRANDT; Nucl. Phys. B 83, 60 (1974), Phys. Rev. D 10, 3509 (1974)

17. V.N. GRIBOV, I. Ya. POMERANCHUK and K.A. MARTIROSYAN; Yad. Fiz. 2, 361 (1975), Sov. J. Nucl. Phys. 2, 258 (1966)

18. H. KLEINERT; Phys. Letters 62 B, 77 (1976)

19. O.W. GREENBERG; Phys. Rev. Letters 13, 598 (1964)
M.Y. HAN and Y. NAMBU; Phys. Rev. 139, B 1006 (1965)
M. GELL-MANN; Acta Physica Austriaca Suppl. 9, 733 (1972)

20. See any book on current algebra or
H. KLEINERT; Fortschr. Physik 21, 8 (1973)

21. W.A. BARDEEN, H. FRITZSCH and M. GELL-MANN;in Scale Conformal Symmetry in Hadron Physics, ed. by R. Gatto (John Wiley and Sons, 1973) p. 139

22. G. PREPARATA; Erice Lectures 1972, Highlights in Particle Physics, Editrice Compositori (Bologna), ed. by A. Zichichi, p. 247

23. For a detailed review see
S. GASIOROWICZ and D.A. GEFFEN; Rev. Mod. Phys. 41, 531 (1969)
Extension to see (4):
M. SINGER; Wisconsin Preprint COO-521 (1976)

24. Y.NAMBU and G.JONA LASINIO; Phys.Rev. 122,345 (1961) 124, 246 (1961)
V.G.VAKS and A.I. LARKIN: JETP 40,282 (13,192) (1961)
Further developments:
J.D.BVJORKEN: Ann.Phys. 24, 174 (1963)
I.BIALYNICKE-BIRULA, Phys.Rev. 130,465 (1963)
G.S.GURALNIK; Phys.Rev. 136B,1404,1417 (1963)
H.UMEZAWA; Nuovo Cimento XL, 450 (1965)
Y.FREUNDLICH and D.LURIE; Phys.Rev. D8,2386 (1974)
D.J.GROSS and A.NEVEU; Phys.Rev. D10, 3235 (1974)
H.PAGELS; Phys.Rev. D7, 3689 (1973).
K.LANE; Phys.Rev. D10,2605 (1974)
R.JACKIW and K.JOHNSON; Phys.Rev. D8, 2386 (1974)
P.LANGACKER and H.PAGELS; Phys.Rev.D9, 3413 (1974)
P.LANGACKER; Phys.Rev.Letters 34, 1592 (1975)
H.PAGELS; Rockefeller University preprint COO-2232B-102
See also:
H.MATSUMOTO, H.UMEZAWA, N.J.PAPASTAMATION; Nucl.Phys. B68,236 (1974), B82,45 (1974)
L.LEPLAE, H.UMEZAWA, F.MANCINI; Phys.Rev.10C (1974)

25. For different estimates see Ref.5 and
M.K.GAILLARD, B.W.LEE, J.L.ROSNER; Rev.Mod.Physics 47,277 (1975)
G.PREPARATA; Erice Lectures 1974, in Lepton and Hadron Structure, ed. by A.Zichichi
and references in
K.S.SOH; Phys.Rev. D13,2954 (1967)

26. H.LEUTWYLER; Phys.Letters 48B, 45, 431, NuclPhys. B16, 413 (1974)
See also: R.L.HEIMANN; NuclPhys. B78, 525 (1974)

27. Compilation of Coupling Constants; Springer Tracts in Modern Physics, Vol.55, ed. by G.Höhler

28. M.GELL-MANN, R.OAKES and B.RENNER; Phys.Rev. 175, (1968)

29. P.R.AUVIL and N.G.DESHPANDE; Phys.Rev. 183, 1463 (1969)

30. S.L.ADLER; Phys.Rev. 177,2426 (1969)
P.LANGACKER and H.PAGELS; Phys.Rev. D9,3413 (1974)

31. Y.NAMBU; Phys.Letters 9,214 (1964)
See also:
A.D.MARIS, V.E.HERSCOVITZ and G.JACOB; Phys.Rev. Letters 12,313 (1964)
V.G.VAKS and A.I.LARKIN, JETP 40,792 (1961)(13,556(1961))

32. J.GASSER and H.LEUTWYLER; Nucl.Phys. B94, 269 (1975)

33. G.CICOGNA, F.STROCCHI and R.CAFFARELLI; Phys.Rev. D6,301 (1972)

34. T.EGUCHI and H.SUGAWARA; Phys.Rev. D10, 4257 (1974)
H.SUGAWARA; Phys.Rev. D12, 3212 (1975)
H.KLEINERT; Phys.Letters 59B, 163 (1975)
A.CHAKRABARTI And B.HU; Phys.Rev. D13, 2347 (1976)
G.KONISI, T.SAITO, and K.SHIGEMOTO, Phys. Rev. D15 (1976)
These authors prove the operatorial validity of the "gap wave equation" by using equs. of motion rather than the functional method employed in Ref.18 and

extended to bilocal form in Ref.10.
T.EGUCHI, Chicago Preprint, 1976.

This author studies in detail the local version of hadronization via functional methods which was outlined in Ref.18 on p.78.

H.PAGELS, Rockefeller Preprints, COO-2232 B-102
Here a bilocal generalization of the classical gap wave equation for colored gluon theories is proposed. However, no bilocal field theory equivalent to the quark-gluon theory is constructed.

35. For a typical calculation and earlier references see Y.UEDA; ICTP Preprint, 1976.

D I S C U S S I O N S

CHAIRMAN: Prof. H. Kleinert

Scientific Secretaries: B. Freedman and B. Mathis

DISCUSSION 1

- *GOURDIN:*

Why is colour so difficult to introduce into your formalism?

- *KLEINERT:*

When you have colour, a gluon self-interaction coming from $G_{\mu\nu}^2 = [\partial_\mu G_\nu - \partial_\nu G_\mu - g(G_\mu, G_\nu)]^2$ will be present which cannot be directly integrated out in the generating functional.

- *GOURDIN:*

Is what you have done just a formal transformation?

- *KLEINERT:*

No, this is a re-summation of the perturbation series that yields results which cannot be obtained by a perturbation expansion in the coupling. This we know from solid-state physics where one reaches a new phase by such a technique; to describe superconductivity, one sums bubblewise, while for a collective effect, such as plasmons, one sums ringwise. Here, in hadron physics, we sum ladder by ladder.

It is important to note that one does not expand in powers of the quark-gluon coupling constant. Let me remind you that in electron-positron scattering with two-photon exchange, the parameter in the expansion is not g, but becomes something like $g^4 \log (s - 4m^2)$. When s is near threshold, the effective expansion parameter becomes large. Thus, we have to sum the whole ladder in order to get finite results, and the ladder can be a good approximation even close to the threshold.

- *ROSSI:*

Turning the heavy lines in your diagrams into Preparata's double lines, one obtains the same graphs describing the same phenomena. Did you find a theoretical framework into which Preparata's point of view can be inserted and justified, at least as far as quark-antiquark bound states are concerned? Did you find anything like vector-meson dominance and the direct-coupling term? What about multihadron production in fire sausages?

- *KLEINERT:*

My rules contain all of Preparata's rules but the reverse is not true. For simplicity, Preparata only keeps three- and four-point couplings among hadrons. However, in order to have a hadronic theory that is equivalent to the original quark-gluon theory, one must must keep *all* n-point couplings. The photon couples via vector mesons, and there is an additional direct-coupling term as Preparata wants it. The direct coupling is taken care of by slashing the propagator as I have explained in the lecture.

- *BUCCELLA:*

Preparata's unconventional model incorporates many hadronic features, such as confinement; these are put in so as to agree with known experimental facts. Your formulation provides a natural framework for explaining current algebra, PCAC, etc. How is confinement and the colour-singlet nature of hadrons introduced into your functional integration method?

- *KLEINERT:*

The quark theory that I have hadronized is not yet unconventional because there is no confinement in it. The whole structure that I have shown works for QED for electrons and positrons. The success of Preparata's model indicates that this new structure can be generalized to the confined situation more easily than the original quark-gluon theory. The graphical rules for hadrons may be independent of whether or not there is confinement in the original quark theory.

- *GARCIA:*

In connection with the introduction of colour, you would like to have infrared slavery? What happens to the gluon mass at long distances?

- *KLEINERT:*

This is difficult to answer within my framework because the gluon field has disappeared from the theory upon hadronization. In the colour quark-gluon version, there will be a massless gluon providing the long-range force to confine the quarks. However, this will be true only within the confinement region, the hadron. Outside, the gluon should acquire a dynamically generated effective mass, $\mu \to \to \infty$, to prevent the quark-gluon system from escaping.

- *FREEDMAN:*

Ultimately, when you look for the bound states in the SU(3) × × SU(3) theory, you want to ensure that the poles correspond to hadrons and not to quarks since that is what one means by confinement. Would you care to speculate about how you want to achieve this in the context of your program?

- *KLEINERT:*

There has been a very exciting talk by Migdal at the Tbilisi conference which will be very helpful here. Migdal has *assumed* confinement and has calculated the hadronic spectrum following from quantum chromodynamics. He looked at the vacuum expectation value of operators like $T[\bar{\psi}(x)\psi(x)\bar{\psi}(0)\psi(0)]$, etc., Fourier-transformed to momentum space, and used the behaviour in the far space-like region known from asymptotic freedom. He then assumed a simple pole structure for q^2 positive and found a unique meromorphic function that fits the required asymptotic behaviour: $[\log(-q^2)]^{\gamma}$, with exponential accuracy.

- *FREEDMAN:*

This new development seems very interesting; however, the absence of coloured states from the pole structure as used by Migdal is still an input to the calculation.

- *KLEINERT:*

He never looked at coloured currents.

- *JONES:*

Could you amplify your claim that a colourless theory forms a reasonable description of meson interactions at intermediate energies?

- *KLEINERT:*

I would claim that the low-mass mesons, the π, ρ, σ, and A_1, and their low-energy, $\lesssim$ 1 GeV, scattering amplitudes form a closed set of physical phenomena. This approximation may work to within 20% accuracy. The fact that higher mass resonances do not interfere here shows that colour, which is responsible for the existence of these higher-mass states, can play no role at this level.

- *ORZALESI:*

Your hadronization corresponds roughly to treating hadrons as bound states in a ladder approximation. My first question is whether your approximation is a first step in an iterative approximation scheme and, if so, what is the prescription for calculating hadronic amplitudes to arbitrary order? My second question has to do with gauge invariance: in QED, the ladder approximation is not gauge invariant, and the position of bound-state poles depends on the gauge chosen; furthermore, if one only keeps ladders and rainbow type graphs, the theory is not even renormalizable. How do you deal with such difficulties?

- *KLEINERT:*

My *bare* hadrons are certainly a first approximation, as is any bare field in an interacting field theory. However, after taking into account all hadronic interactions specified in my Lagrangian, there is no more approximation but a complete equivalence to the original quark theory. This also answers your second question: gauge invariance in the original theory has a counterpart in the bilocal theory causing relations among hadron graphs.

- *FREEDMAN:*

I propose that, for gauge theories, it would be useful to work in an arbitrary covariant gauge and let the gauge parameter keep track of the cancellation between gauge-dependent term when calculating physical quantities. I think this is especially useful here where one does not have an intuitive picture of the cancellations occurring in the perturbation series in the effective hadronic coupling.

- *KLEINERT:*

Yes, you will always find the correct family of hadronic diagrams which throws out the gauge dependence.

- *PAULI:*

Today, you wrote down an equation: $m' = \xi g^2 D G_0 m' G_0$, which is essentially a Bethe-Salpeter equation for quark-hadronic vertices. Can this, or should this, give results similar to the constitutent-interchange model of Blancenbecler, Brodsky, Farrar, and Gunion for hadron-hadron scattering?

- *KLEINERT:*

I think you are referring to those rules that apply to rearrangement collisions. An example of this is electron exchange in molecular collisions which give rise to the Van der Waals forces. Certainly the rules given by Brodsky et al. can be rephrased in terms of my ladder re-summation of scattering graphs.

- *POSNER:*

Could you please reassure us that the expansion of $\log(1 + iG_0 m')$ converges?

- *KLEINERT:*

The convergence has not been studied on rigorous grounds. I can only assure you that my re-summation will converge better near thresholds and bound states where normal perturbation theory certainly fails.

- *POSNER:*

In your process of functional integration, you first eliminated the gluons and then the quarks. If you integrate out the fields in the reverse order, another equivalent theory is obtained, which is very different in appearance. What is the theory like? Is it worth studying?

- *KLEINERT:*

If you do it in the other order, you will get what may be called "plasmonization". By first integrating out the fermion fields, you will leave only a dressed photon field, which is now a very complicated object. Plasmonization occurs by eating up all the fermion degrees of freedom bubblewise. Although this theory plays an important role in solid-state physics and in some two-dimensional field theories, e.g. the Schwinger model, I have not explored it further because it seems uninteresting for hadron physics.

DISCUSSION 2

- *MARCIANO:*

Can we understand how infrared divergences emerge from your hadronized version of QED? Their role will be very important in providing confinement in QCD.

- *KLEINERT:*

I have not examined the infrared properties of my formalism.

- *PHAM QUANG HUNG:*

This morning you mentioned the $\eta \to 3\pi$ problem and the neutron-proton mass difference. Do you have any idea how to solve these problems?

- *KLEINERT:*

For these two problems one usually introduced an explicit SU(2) breaking term into the quark mass matrix. Taking the d quark to be 10 MeV heavier than the u-quark yields the correct $(m_n - m_p)$ value. This choice also gives the correct $\eta \to 3\pi$ decay rate. I do find this an unsatisfactory procedure, and I hope a more natural explanation will be found.

- *ALVAREZ:*

How do you propose to recover the sigma model in the colour non-Abelian theory?

- *KLEINERT:*

The sigma model can be recovered by neglecting the effect of higher resonances. Within quantum flavour dynamics, the same thing was achieved by sending μ^2 to infinity. By this trick, quantum flavour dynamics reduces exactly to the sigma model.

- *FREEDMAN:*

One cannot introduce a gluon mass μ due to renormalization requirements. What mass parameter do you plan to use to implement confinement?

- *KLEINERT:*

The only mass parameter available in the colour gauge gluon-quark model, other than the quark mass, is the infrared cut-off that one needs to define the Green's functions. Many people believe that this parameter is related to the momentum cut-off of the quarks inside the confinement region.

- *FREEDMAN:*

But this parameter is arbitrary. How can it determine the hadron masses?

- *KLEINERT:*

I think that eventually one will get an infinite family of solutions, and that the infrared cut-off will be fixed by the lowest hadron mass in the spectrum. All other masses are then determined.

- *WILKIE:*

Are you going to have difficulties forming Regge poles, which are long-distance effects in a $\mu \to \infty$ limit?

- *KLEINERT:*

The $\mu \to \infty$ limit of quantum flavour dynamics gives the sigma model which corresponds to the zero-slope approximation in Regge theory where all higher-mass states have moved to infinity. I assume that when colour is introduced and we make contact with the dual aspect of hadron dynamics, the correct Regge behaviour will be regained.

- *WILKIE:*

Is it obvious in your treatment of the sigma model that the ρ is a $\pi\pi$ resonance?

- *KLEINERT:*

The ρ appears here as a fundamental field as well as a resonance. This is similar to the situation encountered in N/D calculations; one gets only a narrow contribution to the ρ when treating it as a pure $\pi\pi$ system. One has to put in an elementary ρ as well.

- *BERLAD:*

In performing the functional integration over the gluon field, there should appear not only ladder exchanges but also crossed ladders. Where do these appear?

- *KLEINERT:*

This is a point I was trying to explain by showing some examples. The cross-ladder diagrams can be obtained by topological rearrangements of ladder diagrams.

- *PAULI:*

In order to derive the sigma model, the mass of the gluon must be ≥ 12 GeV. This would result in point-like hadrons, yet we know that the hadronic size is roughly the Compton wavelength of the pion. Could you elaborate?

- *KLEINERT:*

In this model, mesons are fundamental fields after the large μ-mass limit has been taken. Obviously, the form factor for the mesons will be point-like. Radiative corrections due to heavy gluons should broaden the form factor by a small amount only. However, for this purpose, this is not a realistic model of hadrons.

- *VON DARDEL:*

What does your model give the for magnetic moments of the baryons? Do small bare quark masses make it difficult to obtain small enough magnetic moments?

- *KLEINERT:*

No. The magnetic moments are determined by the dynamical quark masses $M \approx 312$ MeV. Hence

$$\mu_Q \approx \frac{1}{2M} \approx 3\frac{1}{2M_N}$$

becomes thrice the nucleon Bohr magnetic moment, thus coinciding with the magnetic moment of the proton, as it should.

- *POSNER:*

When you consider the three-point hadronic interaction, a mass relationship involving m' is obtained. Would you give a physical explanation of why the four-point function depends only on m', m'^2, and m'^3?

- *KLEINERT:*

I only took leading order in $\mu \to \infty$ into account. The higher powers of m'^n $(n \geq 4)$ are lower order corrections. They certainly exist, but remember: M^2/μ^2 is very small (< 1%).

- *POSNER:*

The quantity $C' = M_8/M_0$ ($\approx -16\%$) plays a role in πN and KN scattering. Please elaborate on its significance in these interactions.

- *KLEINERT:*

In the exact SU(3) × SU(3) symmetry limit, the πN and KN cross-sections would be equal, but experiment gives $(\sigma_{\pi N} - \sigma_{KN}) \approx 6$ mb. One would expect this difference in the elastic scattering amplitude, governed by the Pomeron, to be caused by the breakdown of the symmetry in the vacuum of the underlying quark dynamics, specifically in the mass matrix. This is only a rough argument, not a quantitative one.

PHENOMENOLOGY OF NEUTRAL CURRENT INTERACTIONS†

J. J. Sakurai

Department of Physics, University of California, Los Angeles,* and CERN, Geneva

1. INTRODUCTION

It was three years ago this summer that the Gargamelle Collaboration discovered neutral-current phenomena[1]--perhaps the most important high-energy physics discovery made on this side of the Atlantic since the discoveries of pions and strange particles in the late 1940s. The initial phase of this three-year period may be called the "discovery era," during which we witnessed breathtaking "alternating neutral currents." It appears that this first era, which may be characterized as allegro con brio,[2] is now over, and we have entered a new era during which we are confronted with the extraordinarily difficult task of determining the detailed space-time and internal properties of the hadronic and leptonic neutral currents. The tempo of the present era may perhaps be described as lento.

Going through the current literature, one finds that most discussions on neutral currents are motivated by gauge-theory considerations. Typical theoretical review papers on this subject start by showing that neutral currents are needed because, otherwise, the cross section for

$$\nu + \bar{\nu} \to W^+ + W^- \tag{1.1}$$

would grow as s. Experimental papers on neutral-current reactions,

† Supported in part by the National Science Foundation and the John Simon Guggenheim Memorial Foundation.

* Permanent address.

after lengthy discussions of background subtractions, etc., all end up with the conclusion

$$\sin^2\theta_W = 0.35 \pm ? \tag{1.2}$$

It appears that theorists and experimentalists are interested only in comparing the available data with gauge models, with the one-parameter Salam-Weinberg model,[3] in particular.

I believe there is some danger in taking one-parameter models too seriously. As I'll show in Section 5, the inclusive cross sections for

$$\overset{(-)}{\nu} + p \to \overset{(-)}{\nu} + \text{any} ,$$

$$\overset{(-)}{\nu} + n \to \overset{(-)}{\nu} + \text{any} \tag{1.3}$$

are potentially sensitive to as many as four coupling parameters. When we analyze the data using a one-parameter model, e.g., the Salam-Weinberg model, we are implicitly assuming that the remaining three parameters either have pre-assigned values or are related in a definite manner to the single parameter we are trying to determine. I hope this point will become transparent as we proceed.

My approach in this set of lectures is somewhat different. I would like to propose a rather general phenomenological framework for discussing the data without commitment to any particular theoretical model. The main objective of my lectures may be summarized by asking three questions as follows:

(i) What do various experiments really measure?
(ii) How can we perform "complete" experiments to determine the neutral-current couplings?
(iii) How can we test <u>any</u> model in an objective, emotionally uninvolved manner?

There are two reasons for following this rather general approach. First, let us note that the observed neutral-current phenomena are genuinely new and therefore interesting regardless of whether gauge-theory considerations are on the right track. Until 1973 essentially the whole of weak interaction physics could be visualized as being based on Fermi's 1934 interaction with only minor modifications:

(i) Change V to V-A.
(ii) Change p and n to u and d_C ($\equiv d\cos\theta_C + s\sin\theta_C$).
(iii) Double the number of leptons:

$$e \to e, \mu, \qquad \nu \to \nu_e, \nu_\mu \quad .$$

With the advent of neutral-current interactions some major modifications are needed! After 40 years of physics with neutrinos, something qualitatively new is finally happening. Whenever we have a genuinely new phenomenon, it is profitable to study it in its own right, without recourse to any particular theoretical framework. We may also keep in mind the possibility that renormalizable gauge theories of weak interactions look attractive now only because theorists don't know anything better at this present moment.

Second, even if the basic philosophy of gauge models is to triumph ultimately, there are now so many models, all within the general gauge-theory framework, that make many different predictions on the neutral-current interactions. If we have learned anything in the past year and a half, it is that the range of model builders' imaginations is unbounded! Once we have a general phenomenological framework, it becomes easier to compare your favorite model with your competitors' in an objective manner.

2. UNORTHODOX POSSIBILITIES

The original Gargamelle paper[1] on the neutral-current discovery is entitled "Observation of Neutrino-like Interactions without Muon or Electron in the Gargamelle Neutrino Experiment." This title is appropriate because the reactions they observed actually looked like

$$\text{"invisible particle"} + N \to \text{possible "invisible particle(s)"} + \text{hadrons} . \qquad (2.1)$$

Most physicists assume from the beginning that the invisible particles in the initial and final states are ordinary neutrinos arising from π and K decays and that the basic couplings involve the neutrino current $i\bar{\nu}\gamma_\lambda(1+\gamma_5)\nu$ interacting with hadronic or leptonic densities of the V and/or A type. It is worth spending some time checking these assumptions. More complete discussions of the various topics covered in this section can be found in my DESY lectures.[4]

There is now good experimental evidence to believe that the initial invisible particle in (2.1) is indeed the neutrino that appears in π or K decay. First of all, all experimental papers on neutral currents show in some detail why observed events of the type (2.1) cannot all be interpreted as being induced by known hadrons such as the neutron and the K_L meson. Furthermore, the observed neutral-to-charged-current ratio

$$\frac{\sigma(\overset{(-)}{\nu} + N \to \overset{(-)}{\nu} + \text{any})}{\sigma(\overset{(-)}{\nu} + N \to \mu^{-(+)} + \text{any})}$$

is independent of the very different manners in which the neutrino beam is prepared from experiment to experiment--wide-band, narrow-band, horn-focused, unfocused--a feature not expected if the neutral-current events were due to some yet unknown background particles that survive a tremendous amount of shielding material. The assumption that the outgoing invisible particle is also an ordinary neutrino is more difficult to prove experimentally. The fact that the neutral-to-charged-current ratio appears to be constant between CERN PS energies (Gargamelle) and Fermilab energies (Harvard-Pennsylvania-Wisconsin-Fermilab[5]), as shown in Fig. 2.1, suggests that there is no obvious threshold effect in the neutral-current interactions, so the outgoing invisible particle in (2.1) is unlikely to be a neutral heavy lepton of appreciable mass. It is, of course, difficult to rule out the possibility that a very light neutral heavy lepton is produced in (2.1) with $\nu + \bar{\nu} + \bar{\nu}$ as its major decay mode.

Even if the final invisible particle in (2.1) is a massless (or nearly massless) "neutrino," it is still meaningful to ask whether the neutrino in the final state is the same as the initial neutrino. This can be checked in $\bar{\nu}_e e$ scattering by examining whether the s channel diagram due to the charged-current interaction can interfere with the t channel diagram due to the neutral-current interaction, as will be discussed in Section 3.[6] In neutrino-induced hadronic reactions we can test the question of neutrino identity in a more indirect manner as follows. First, it can be shown that the equality

$$\lim_{y\to 0} \frac{d\sigma}{dy} (\nu + N \to \nu + \text{any}) = \lim_{y\to 0} \frac{d\sigma}{dy} (\bar{\nu} + N \to \bar{\nu} + \text{any}) \ , \qquad (2.2)$$

where y stands for ν/E (the energy transfer divided by the incident energy), must hold as long as the hadronic current is Hermitian. [At $y \neq 0$ the two sides of (2.2) are in general different because of VA interference, but VA interference vanishes as $y \to 0$.] Now the Hermiticity of the hadronic current is required by the CPT theorem in all models where the incident and outgoing neutrinos are the same. The equality (2.2) can therefore be regarded as a test of ν identity.[7]

Some theorists[8] have speculated that the so-called neutral-current reactions are actually due to the electromagnetic interactions. Suppose the neutrino had, for some reason, an anomalously large charge radius. It can be scattered elastically or inelastically by a nucleon via single-photon exchange. Since the neutrino is strictly neutral, the Dirac form factor of the neutrino must start as

$$F(q^2) = \frac{1}{6} \langle r^2 \rangle q^2 + \cdots \ . \qquad (2.3)$$

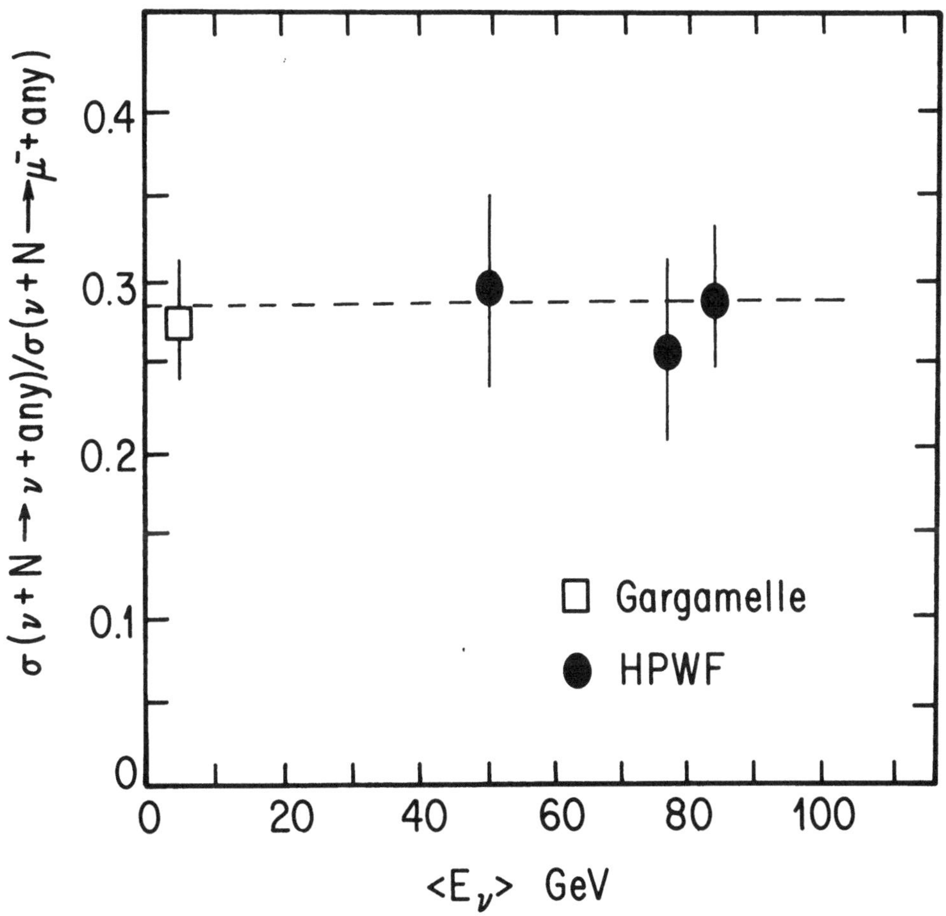

Fig. 2.1. Neutral-to-charged current ratio as a function of the mean neutrino energy.

Now, in scattering processes the q^2 dependence of the leading term in (2.3) just cancels the $1/q^2$ coming from the propagator of the exchanged photon. The net result is exactly what you would expect from a current-current interaction of the standard form. There are, however, two difficulties with this interesting proposal. First, with the charge radius we may compute on the basis of conventional mechanisms, the resulting cross section is a factor of $\sim \alpha^2$ too small to account for the observed cross section.[9] What is even more serious, the form factor $F(q^2)$ of ν_μ (ν_e) must be the same by the CPT theorem as that of $\bar{\nu}_\mu$ ($\bar{\nu}_e$) except in sign. So the cross section for any ν induced reaction, exclusive or inclusive, mediated by photon exchange is predicted to be equal to the cross section for the corresponding $\bar{\nu}$ induced reaction:

$$\sigma(\nu + A \to \nu + B) = \sigma(\bar{\nu} + A \to \bar{\nu} + B) \quad . \tag{2.4}$$

In Sections 5 and 6 I'll present a substantial amount of evidence against (2.4) in both inclusive and exclusive reactions.

In review talks on neutral currents it has become almost traditional to report on the status of "SPT heresy." A number of authors[10] speculated on the possibility that covariants other than V and A are involved in the neutral-current (or better "neutral-density") interactions. It is relatively easy to test pure S and/or P by studying the y distributions in the inclusive reactions

$$\overset{(-)}{\nu} + N \to \overset{(-)}{\nu} + \text{any} \ . \tag{2.5}$$

If the total cross section for (2.5) scales, i.e., if it rises linearly with the incident neutrino energy, as seems to be the case experimentally, the hypothesis of pure S and/or P predicts a y distribution that goes as y^2; in contrast, with a linear combination of V and A, the y distribution is <u>not</u> predicted to rise with increasing y. The data of the Caltech[11] and HPWF[5] groups who studied the hadron energy distributions in (2.5) are in serious disagreement with the pure S and/or P hypothesis. It can be shown, however, that any inclusive distribution obtained by V and A can also be simulated by suitable linear combinations of SPT, which is known as the "Confusion Theorem."[10] As a result, it is impossible to rule out SPT by studying the inclusive distributions.

Recently an astrophysical argument has been advanced against a sizable tensor component in the neutral-current interactions.[12] A tensor coupling between neutrinos and charged fermions of the form $\bar{\nu}\sigma_{\lambda\tau}\nu\,\bar{F}\sigma_{\lambda\tau}F$ where F may stand for a charged lepton or a quark would lead to a finite magnetic moment for the neutrino via charged fermion pairs which couple to the photon. The resulting magnetic moment interaction leads to copious production of $\nu\bar{\nu}$ pairs by plasma photons in stars:

$$\gamma_{plasma} \rightarrow \nu + \bar{\nu} \quad . \tag{2.6}$$

This has interesting implications on stellar evolution. Too high a production rate for neutrino pair creation implies that the cooling of a white dwarf would be too rapid to be compatible with the observed distribution of white dwarfs. To calculate the neutrino magnetic moment we must know what kind of charged fermion pairs couple to neutrinos but with a reasonable estimate based on the known (?) spectrum of leptons and quarks we may deduce from the observed astrophysical data a limit on the tensor coupling constant at least an order of magnitude lower than the Fermi's coupling constant G, hence insufficient to explain the observed neutral-current phenomena.

I may also mention that, if atomic physics parity-violation experiments, to be discussed in Section 7, are to give a positive result, models with SPT only will be conclusively ruled out. This is because parity violation with SPT implies that time reversal invariance must also be violated, but there is already a very stringent limit on the validity of time reversal invariance from the apparent absence of diagonal electric dipoles in atoms.[13] Furthermore parity violation with SPT does not give rise to parity-violating effects in atomic physics experiments of the kind to be discussed in Section 7 because of the pseudoscalar term measured in such experiments is consistent with time reversal invariance.

3. NEUTRINO-ELECTRON SCATTERING

Being purely leptonic, neutrino-electron scattering is the simplest neutrino-induced reaction. It serves as a pedagogical laboratory for exhibiting many of our theoretical concepts--the helicity rule, VA interference, scaling, etc.--in a clean manner.

There are four reactions that can be studied:

$$\nu_e + e^- \rightarrow \nu_e + e^- \ , \tag{3.1a}$$

$$\bar{\nu}_e + e^- \rightarrow \bar{\nu}_e + e^- \ , \tag{3.1b}$$

$$\nu_\mu + e^- \rightarrow \nu_\mu + e^- \ , \tag{3.1c}$$

$$\bar{\nu}_\mu + e^- \rightarrow \bar{\nu}_\mu + e^- \quad . \tag{3.1d}$$

The last two reactions (3.1c) and (3.1d) induced by ν_μ and $\bar{\nu}_\mu$ are of particular interest because they are forbidden by the old charged-current V-A theory; they proceed only via the neutral-current interactions. In contrast, the first two reactions induced by ν_e and $\bar{\nu}_e$ are actually allowed also by the old V-A charged-current coupling because the charged current can appear in the u and s channels, respectively, without violating muon number conservation. All this should be clear from Fig. 3.1.

Since this is a summer school, not a conference for experts, let me begin by dwelling on extremely simple considerations based on angular momentum conservation. First, I define a scaling variable y as follows:

$$y \equiv [(p_\nu - p'_\nu)\cdot p_e]/p_\nu\cdot p_e \tag{3.2}$$

where p_ν, p'_ν and p_e are the four momenta of the incident ν, the outgoing ν and the initial electron, respectively. In the laboratory system in which the target electron is at rest, we clearly have

$$y = \nu/E \tag{3.3}$$

where ν is the energy transfer, and E, the incident neutrino energy. If we go to the center-of-mass system, we get, in the approximation $m_e = 0$,

$$y = \frac{1}{2}(1 - \cos\theta_{cm}) = \sin^2(\theta_{cm}/2) \quad . \tag{3.4}$$

Note that the range of y is from 0 to 1.

For definiteness let us consider $\nu_\mu e^-$ scattering (3.1c) and assume that the basic neutral-current coupling is V-A, which is probably not the case experimentally. The usual V-A $(1 + \gamma_5)$ rule says that at relativistic energies the interaction takes place only when fermions are left-handed (negative helicity) and antifermions, right-handed (positive helicity). It is easily seen that the rule forbids neither forward scattering (y = 0) nor backward scattering (y = 1). See Fig. 3.2. One can, in fact, show that the reaction proceeds only for a J = 0 spin-singlet state; the resulting angular distribution is therefore isotropic. The situation is very different for $\bar{\nu}_\mu e^-$ scattering. This time forward scattering (y = 0) is fully allowed as in the $\nu_\mu e^-$ case but backward scattering (y = 1) is impossible because of angular momentum conservation. So the configuration y = 1 is forbidden for $\bar{\nu}_\mu e^-$ scattering with V-A. Furthermore, if you are familiar with the Jacob-Wick formalism, you'll readily see that backward scattering is "doubly forbidden" at $\theta_{cm} = \pi$ with an angular distribution going like

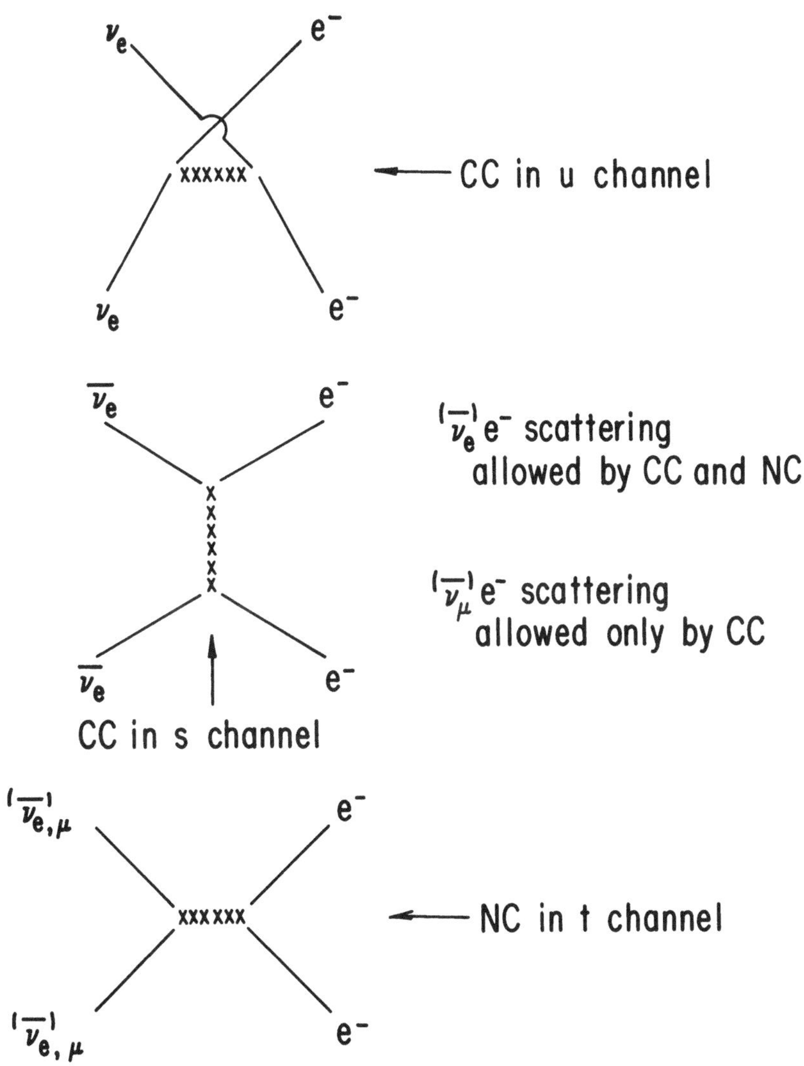

Fig. 3.1. $\overset{(-)}{\nu} e^-$ scattering.

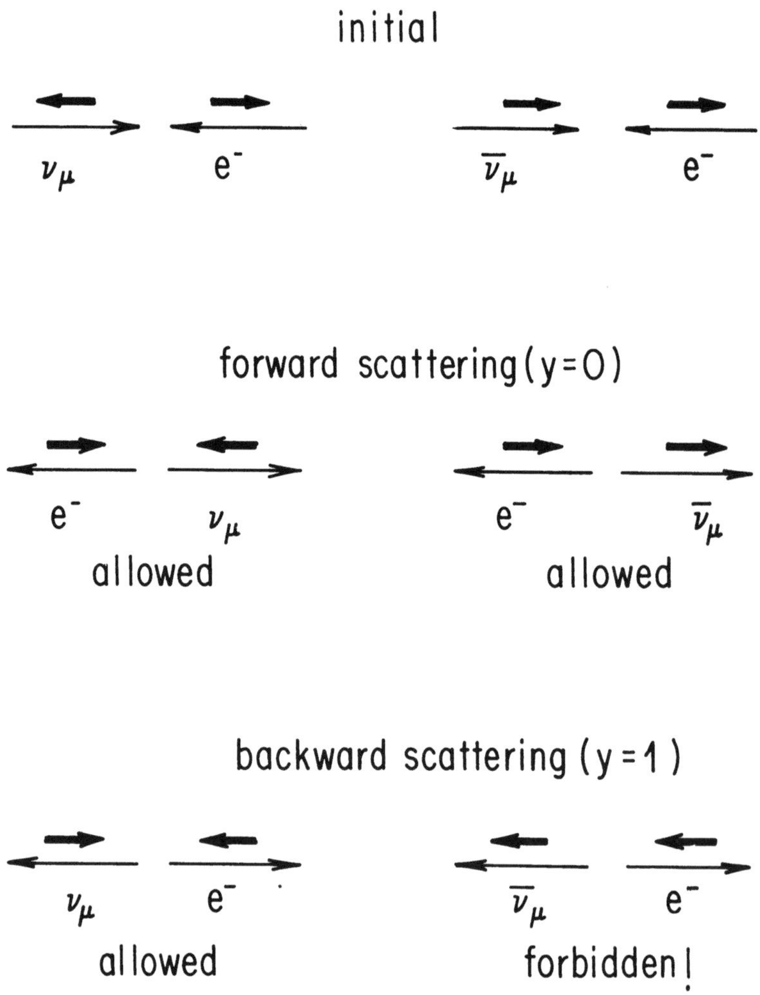

Fig. 3.2. Helicity rule in $\overset{(-)}{\nu}_\mu e^-$ scattering.

$[d_{11}^{(1)}(\theta)]^2 = \cos^4(\theta_{cm}/2)$ or, equivalently, $(1-y)^2$. A similar exercise may be worked out for V+A, and we can construct a simple table for the y distributions as follows:

	V-A	V+A	
$\nu_\mu e^-$ scattering	1	$(1-y)^2$	(3.5)
$\bar{\nu}_\mu e^-$ scattering	$(1-y)^2$	1	

The most general coupling with V and A can be regarded as a linear combination of V-A and V+A. For the cross section there is no interference between V-A and V+A--states of opposite helicities do not interfere--and we can just add the two contributions incoherently to obtain

$$\frac{d\sigma}{dy} = |C_\mp|^2 + |C_\pm|^2 (1-y)^2 \text{ for } \left\{ \begin{matrix} \nu_\mu e^- \\ \bar{\nu}_\mu e^- \end{matrix} \right\} \text{ scattering} \tag{3.6}$$

where $C_\mp$ are the amplitudes for V $\mp$ A. Note in particular that for pure V or pure A we have

$$\frac{d\sigma}{dy} = |C|^2 \{ 1 + (1-y)^2 \} \tag{3.7}$$

for both $\nu_\mu e^-$ and $\bar{\nu}_\mu e^-$. Quite generally, any difference between the ν and $\bar{\nu}$ cross section is due to VA interference; this can be traced to the fact that the V and the A current behave oppositely under charge conjugation.

All these results, which we have obtained using pictures, can, of course, be more rigorously derived by explicit calculations. We start with an effective Lagrangian

$$\mathcal{L}_{NC} = - \frac{G}{\sqrt{2}} [\bar{\nu}_\mu \gamma_\lambda (1+\gamma_5) \nu_\mu][\bar{e}\gamma_\lambda (g_V + g_A \gamma_5) e] \quad , \tag{3.8}$$

which is the most general current-current interaction between ν and e involving V and A as the incident ν ($\bar{\nu}$) from π and K decays is left-handed (right-handed). Throughout these lectures G stands for Fermi's coupling constant $\simeq 10^{-5}/m_p^2$. My metric is such that $1+\gamma_5$ appears with a left-handed coupling, hence $g_V = +g_A$ for V <u>minus</u> A. Using (3.8), we can derive[14]

$$\frac{d\sigma}{dy} = (2G^2 m_e E/\pi) \left[\frac{1}{4} (g_V \pm g_A)^2 + \frac{1}{4} (g_V \mp g_A)^2 (1-y)^2 \right] \text{ for } \left\{ \begin{matrix} \nu e \\ \bar{\nu} e \end{matrix} \right\} \tag{3.9}$$

where E is the incident neutrino energy in the laboratory system, and $E \gg m_e$ has been assumed.

There are two points worth mentioning in connection with (3.9). First, both the differential cross section at fixed y and the total cross section (integrated over y) rises linearly with E. This is the simplest form of "scale invariance" à la Bjorken.[15] If no masses appear explicitly at high energies, the energy dependence of the cross section (assumed to be nontrivial) must be given by dimensional considerations. Now $d\sigma/dy$ or σ has the dimension of $1/M^2$ in natural units; Fermi's constant G also has the dimension of $1/M^2$. Since the cross section is proportional to G^2, the only way to match the dimensions is to have σ proportional to s, the center-of-mass energy squared. Note, in this connection,

$$2G^2 m_e E/\pi = G^2 s/\pi \tag{3.10}$$

for $E \gg m_e$.

Second, the $\bar{\nu}e$ to νe cross section ratio is bounded by

$$\frac{1}{3} \leq \frac{\sigma(\bar{\nu}e)}{\sigma(\nu e)} \leq 3 \quad , \tag{3.11}$$

as can be seen by using

$$\frac{\int_0^1 (1-y)^2\, dy}{\int_0^1 1\, dy} = \frac{1}{3} \quad . \tag{3.12}$$

The lower bound, 1/3, is obtained for pure V-A while the upper bound, 3, is obtained for pure V+A. It may be mentioned that in the SPT case the allowed range for the $\bar{\nu}e$ to νe ratio is considerably wider, between 0.139 and 7.20.[16]

The scattering of ν_μ and $\bar{\nu}_\mu$ by electrons was first studied in a heavy liquid bubble chamber by the Gargamlle Collaboration using the CERN neutrino and antineutrino beams.[17] To date there are three likely events for $\bar{\nu}_\mu e^-$ and no candidate for $\nu_\mu e^-$. In terms of the cross sections we have

$$\sigma(\nu_\mu e^-) < 2.6 \times 10^{-42}\ \text{cm}^2 \times E/\text{GeV} \ , \tag{3.13a}$$

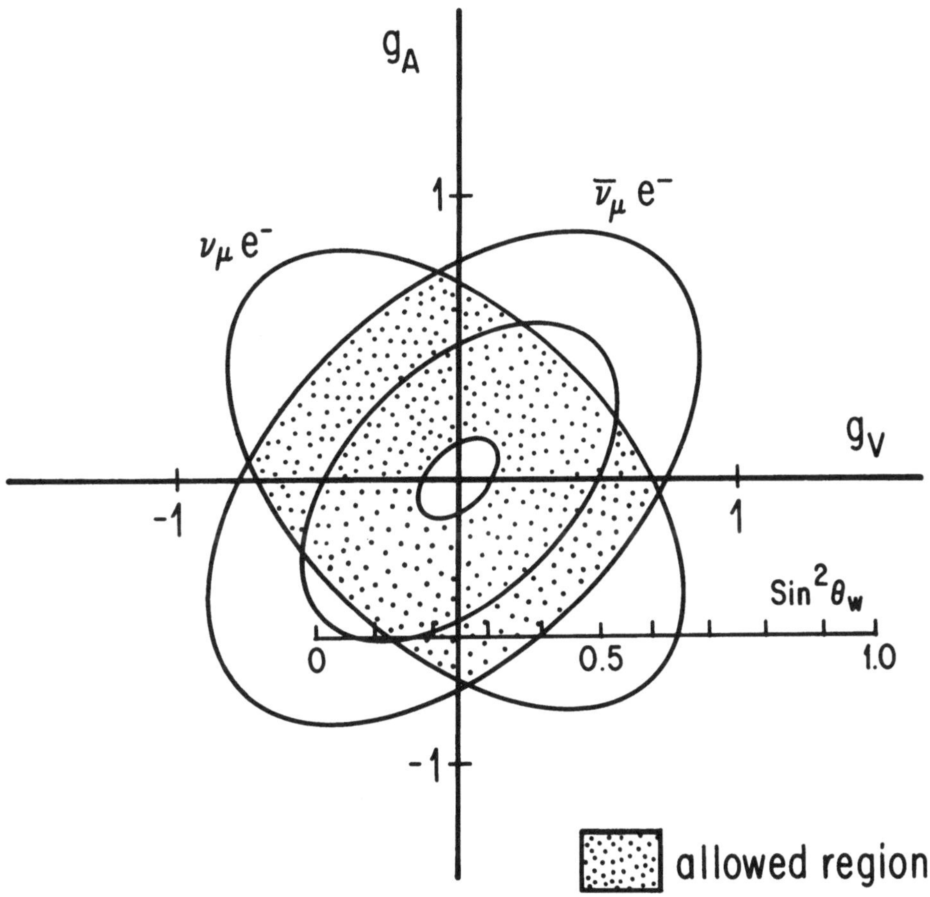

Fig.3.3. Coupling constant plane for $\overset{(-)}{\nu}_\mu e^-$ scattering.

$$\sigma(\bar{\nu}_\mu e^-) = 1.0 \left\{ {+2.1 \atop -0.9} \right\} \times 10^{-42}\ \text{cm}^2 \times E/\text{GeV} \quad . \tag{3.13b}$$

More recently, there is a counter-spark chamber experiment by an Aachen-Padua Collaboration[18] which has reported a somewhat large value of $\sigma(\bar{\nu}_\mu e^-)$ but I understand that the value depends sensitively on the energy cuts one imposes to reduce π^0 background. (See Signora Baldo-Ceolin's talk and also discussion remarks at the end of this session.) The Aachen-Padua Collaboration also has positive evidence for $\nu_\mu e^-$ scattering.

The most objective way to compare experimental data with the Lagrangian (3.9) is to display the range of g_V and g_A compatible with the data in a two-dimensional g_V-g_A plane.[19] See Fig. 3.3. The upper limit (3.13a) for $\nu_\mu e^-$ scattering obtained by the Gargamelle Collaboration implies that the allowed values of g_V and g_A must lie inside the ellipse denoted by $\nu_\mu e^-$. A similar remark holds for $\bar{\nu}_\mu e^-$ scattering except that here the middle region ($g_V \simeq 0$, $g_A \simeq 0$) is excluded because we have positive evidence for this process.

So far I have tried to present a general phenomenological framework for discussing $\overset{(-)}{\nu}_\mu e^-$ scattering. If you like, you can always specialize to some particular model. For example, let us look at the one-parameter Salam-Weinberg model[3] which predicts

$$g_A = -\frac{1}{2}$$

$$g_V = -\frac{1}{2} + 2\sin^2\theta_W \quad . \tag{3.14}$$

We see from Fig. 3.3 that the Gargamelle data are compatible with $\sin^2\theta_W$ between 0.1 and 0.4. As will be shown in Section 5, this range of $\sin^2\theta_W$ is just what we deduce by analyzing the hadronic data. It is important to keep in mind, however, that when one expresses the experimental result in terms of the Weinberg angle, one is implicitly assuming that g_A is equal to -1/2. It is just as important to check this assumption as to compare the various ways of obtaining the Weinberg angle.

Let us now consider $\nu_e e^-$ and $\bar{\nu}_e e^-$ scattering. The effective Lagrangian is taken to be

$$L_{CC+NC} = -\frac{G}{\sqrt{2}}\,[\bar{\nu}_e\gamma_\lambda(1+\gamma_5)\nu_e][\bar{e}\gamma_\lambda(G_V+G_A\gamma_5)e] \quad . \tag{3.15}$$

This Lagrangian contains the charged-current contribution as well; even though the charged-current coupling is usually written as $(\bar{\nu}e)(\bar{e}\nu)$, it can be rearranged into the $(\bar{\nu}\nu)(\bar{e}e)$ form using a Fierz transformation. If the charged-current interaction is correctly given by the usual V-A coupling, G_V and G_A can be written as

$$G_V = 1 + g_V , \qquad G_A = 1 + g_A \tag{3.16}$$

where the "1" stands for the charged-current contribution, and g_V and g_A arise from the neutral-current contribution. Furthermore, if μe universality holds, the g_V and g_A appearing in (3.16) must be the same as the g_V and g_A appearing in the $\overset{(-)}{\nu}_\mu e^-$ scattering Lagrangian (3.8). The expression for $d\sigma/dy$ is the same as (3.9) apart from the substitutions $g_V \to G_V$, $g_A \to G_A$.

Experimentally studies of $\overset{(-)}{\nu}_e e^-$ scattering are difficult at accelerator energies because the $\overset{(-)}{\nu}_e$ flux due to K_{e3} decay is lower than the $\overset{(-)}{\nu}_\mu$ flux due to π and $K_{\mu 2}$ decay by about two orders of magnitude. However, $\bar{\nu}_e e^-$ scattering can be detected using antineutrinos from fission reactors where the antineutrino energy is typically a few MeV. This scattering process has been, and is still being, studied by Reines and collaborators for the past several years at the Savannah River reactor, which is a very copious source of $\bar{\nu}_e$'s. Their most recent results are shown in Fig. 3.4.[20] Notice that to compare Fig. 3.3 and Fig. 3.4, the $\bar{\nu}_e e^-$ ellipse must be shifted in accordance with (3.16).

It is clear from (3.9) and its $\overset{(-)}{\nu}_e e^-$ analog that the observable cross sections are invariant under $g_V \leftrightarrow g_A$, $G_V \leftrightarrow G_A$. In terms of the ellipses of Fig. 3.3 and 3.4, the axes of each ellipse make angles of 45° and 135° with the g_V (G_V) and the g_A (G_A) axis. More physically, given the observable cross section, there is still a two-fold ambiguity; we cannot tell an axial-vector dominant solution from a vector dominant solution. Actually, in writing down (3.9), we have ignored nonasymptotic terms like m_e/E. The coefficient of $m_e y/E$ to be added to (3.9) is given by $\frac{1}{4}(g_V{}^2 - g_A{}^2)$ or $\frac{1}{4}(G_V{}^2 - G_A{}^2)$ in the $\overset{(-)}{\nu}_e e$ case.[21] So, by measuring this term, it is, in principle, possible to resolve the ambiguity. In addition, the magnitude of this term in the case of $\bar{\nu}_e e^-$ scattering can settle the question of whether the s channel diagram indeed interferes with the t channel diagram (see Fig. 3.1), as required by models in which the final neutral particle in the neutral current interaction is the

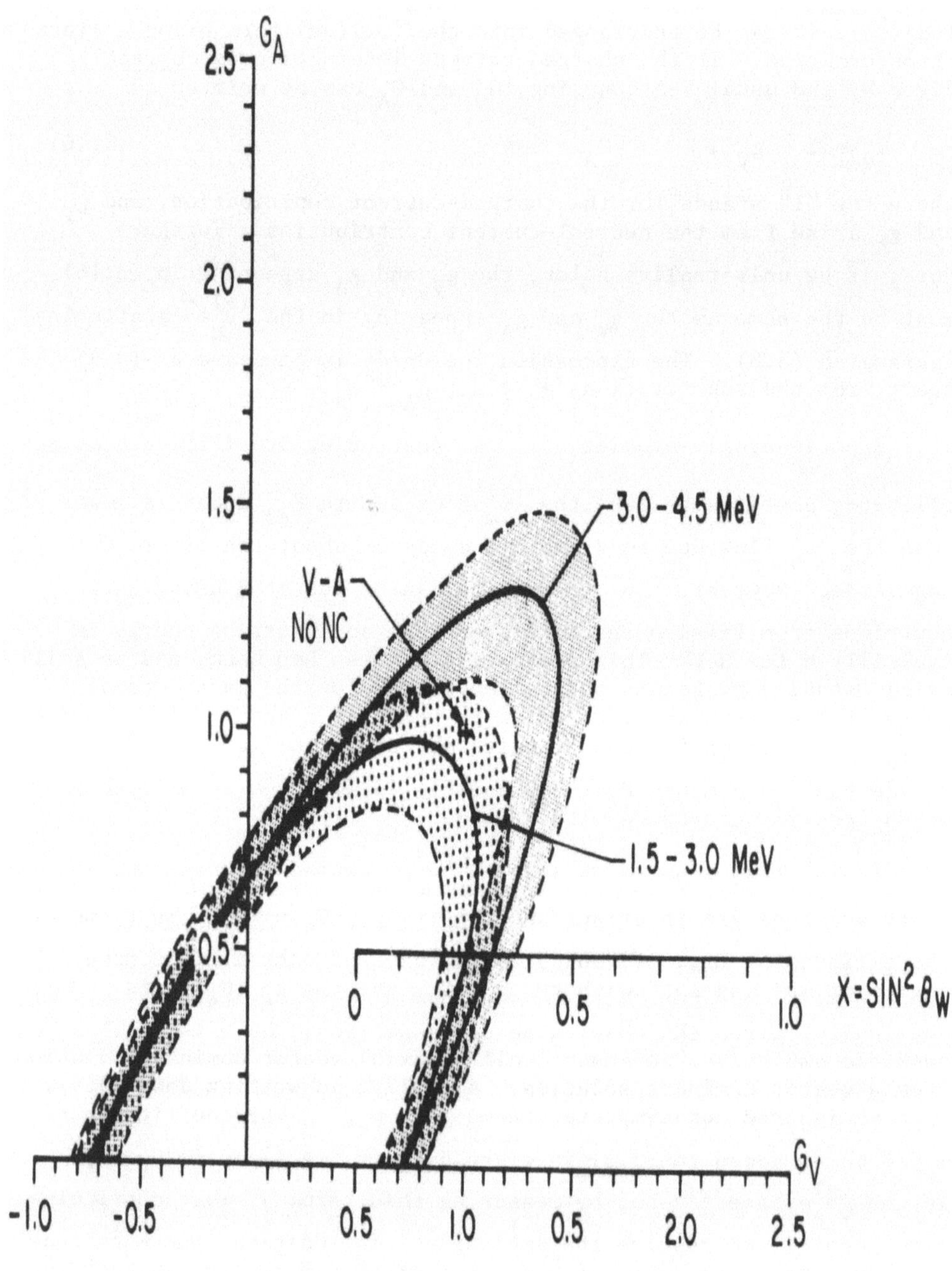

Fig. 3.4. Coupling constant plane for $\bar{\nu}_e e^-$ scattering.

same as the ordinary $\bar{\nu}_e$.[6] In practice, however, detection of this nonasymptotic term appears difficult.

4. HADRONIC CURRENTS AND MODELS

Leaving the purely leptonic world, I now start discussing the space-time and internal properties of the hadronic part of the weak neutral currents.[22] The first question we may ask is: How many independent coupling constants are needed to specify completely the interaction Lagrangian responsible for neutrino-induced hadronic reactions? If SU(2) were the ultimate symmetry of the hadronic world, we would naturally consider four constants--isovector vector, isovector axial-vector, isoscalar vector, and isoscalar axial-vector, which, for later purposes, we may call α, β, γ, and δ respectively. We are prejudiced against currents with isospin equal to two units or higher; such objects cannot be readily constructed out of quark fields. Besides, there is no evidence for isotensor in the electromagnetic or in the charged-current neutrino interactions.

With SU(3) we must distinguish two kinds of isoscalars--the 8th component of an octet (hypercharge-like) and a unitary singlet (baryon-like). It is actually simpler to use the quark model notation and decompose the isoscalar current into two parts--ω like ($\bar{u}u + \bar{d}d$) and ϕ like ($\bar{s}s$). In either way of counting there are now 6 independent constants altogether. If you like charm and other flavors, you may speculate on how $\bar{c}c$, $\bar{t}t$, $\bar{b}b$ like components enter. So there are as many as 8, 10, 12.... constants to be determined--even more, if you start considering flavor-changing neutral currents.

To set up the normalization convention, I write down the effective Lagrangian appropriate for reactions induced by left-handed (right-handed) neutrinos (antineutrinos) using the quark model notation:

$$L = -\frac{G}{\sqrt{2}}\bar{\nu}\gamma_\lambda(1+\gamma_5)\nu\ \Big\{\frac{1}{2}[\bar{u}\gamma_\lambda(\alpha+\beta\gamma_5)u - \bar{d}\gamma_\lambda(\alpha+\beta\gamma_5)d]$$
$$+\frac{1}{2}[\bar{u}\gamma_\lambda(\gamma+\delta\gamma_5)u + \bar{d}\gamma_\lambda(\gamma+\delta\gamma_5)d] + \bar{s}(\gamma'+\delta'\gamma_5)s$$
$$+\text{ possible } \bar{c}c,\ \bar{t}t,\ \bar{b}b \text{ terms}\Big\} \quad . \qquad (4.1)$$

As in the νe Lagrangian, $\alpha = +\beta$, $\gamma = +\delta$ for V minus A. I may mention that every theorist--or even every paper written by the same theorist--uses a different normalization convention for isoscalar currents.[23] I would like to urge every author to specify how his or her currents are normalized. Don't just write down $J_\lambda^{(0)}$ without explaining what it means, say, in terms of quark fields. The only practical way of punishing violators of this rule is, in my opinion, not to quote those papers which do not explain clearly how the currents are normalized!

Various models can be classified by specifying α, β, γ, ... For example, in the one-parameter Salam-Weinberg model[3] we have

$$\alpha = 1 - 2\sin^2\theta_W , \qquad \beta = 1 ,$$

$$\gamma = -\frac{2}{3}\sin^2\theta_W , \qquad \delta = 0 . \tag{4.2}$$

When this model is supplemented by the GIM[24] mechanism for eliminating the stangeness-changing neutral currents, we get predictions for γ' and δ' as follows:

$$\gamma' = -\frac{1}{2} + \frac{2}{3}\sin^2\theta_W , \qquad \delta' = -\frac{1}{2} . \tag{4.3}$$

When we compare experimental data with the Salam-Weinberg model, we are implicitly assuming that β is unity, δ is zero, and α and γ are related by

$$\gamma = \frac{\alpha-1}{3} . \tag{4.4}$$

Perhaps some experimentalists or theorists prefer to analyze the data without such constraints. It is to such physicists that my approach to the neutral currents is primarily addressed.

If you are interested in other models, you may look at Table 4.1. I don't expect you to digest this Table in detail; the important point is that various theoretical models predict very different sets of coupling constants. In gauge models there is a strength parameter λ that depends on the details of the Higgs mechanism; it is fixed once we specify how m_Z and m_W are related. For historical reasons, in the Salam-Weinberg model that strength parameter is usually set equal to unity so that we have a simple relation

$$m_W/m_Z = \cos\theta_W . \tag{4.5}$$

In addition, when the data are analyzed using the two-parameter version of the Salam-Weinberg model, the value of the strength parameter indeed appears to be roughly unity after all.

In Table 4.1 I did not list the predictions for γ' and δ'. With the exception of the isovector V-A model of Adler and Tuan,[29] which gives $\gamma' = \delta' = 0$, all the models in the Table predict

$$\gamma' = \frac{\gamma-\alpha}{2} , \qquad \delta' = \frac{\delta-\beta}{2} . \tag{4.6}$$

Table 4.1 Classification of models.

Model	α	β	γ	δ
S-W[3]	$1 - 2\sin^2\theta_W$	1	$-\frac{2}{3}\sin^2\theta_W$	0
EM[25]	λ	0	$\frac{1}{3}\lambda$	0
Vectorlike[26]	$\lambda(1 - \sin^2\theta_V)$	0	$-\frac{1}{3}\lambda\sin^2\theta_V$	0
AKW[27]	$\lambda(1 + \frac{1}{2}\cos^2\phi - 2\sin^2\theta_W)$	$\lambda(1 - \frac{1}{2}\cos^2\phi)$	$\lambda(-\frac{2}{3}\sin^2\theta_W + \frac{1}{2}\cos^2\phi)$	$-\frac{1}{2}\lambda\cos^2\phi$
G-S(B)[28]	$\lambda\left(\frac{3}{2} - 2\sin^2\theta_W\right)$	$\frac{1}{2}\cdot\lambda$	$\lambda\left(\frac{1}{2} - \frac{2}{3}\sin^2\theta_W\right)$	$-\frac{1}{2}\lambda$
G-S(C)[28]	$\lambda(1 + \frac{1}{2}\cos^2\frac{\alpha}{2} - 2\sin^2\theta_W)$	$\frac{1}{2}\lambda\cos^2\frac{\alpha}{2}$	$\lambda(1 - \frac{1}{2}\cos^2\frac{\alpha}{2} - \frac{2}{3}\sin^2\theta_W)$	$-\frac{1}{2}\lambda\cos^2\frac{\alpha}{2}$
Isov. V-A[29]	λ	λ	0	0

This is a simple consequence of the fact that in all these models $\bar{d}d$ and $\bar{s}s$ appear in a definite combination:

$$\bar{d}d + \bar{s}s = \bar{d}_C d_C + \bar{s}_C s_C \quad , \tag{4.7}$$

which is invariant under a Cabibbo rotation (a particular form of U spin rotations)

$$d \to d_C = d \cos\theta_C + s \sin\theta_C \ ,$$
$$s \to s_C = - d \sin\theta_C + s \cos\theta_C \quad . \tag{4.8}$$

The requirement (4.7) is automatically satisfied in gauge-type models which eliminate the strangeness-changing neutral currents via the GIM trick.[30]

5. NEUTRINO-INDUCED INCLUSIVE HADRONIC REACTIONS

We are now in a position to discuss various hadronic reactions. Let us begin with the inclusive reactions

$$\overset{(-)}{\nu} + p \to \overset{(-)}{\nu} + \text{any} \ ,$$
$$\overset{(-)}{\nu} + n \to \overset{(-)}{\nu} + \text{any} \quad . \tag{5.1}$$

The basic kinematical variables are: (i) E, the incident neutrino energy, (ii) q^2, the invariant momentum transfer squared, and (iii) ν, the energy transfer. Alternatively, we may use the two scaling variables

$$x \equiv q^2/2m_N\nu \quad ,$$
$$y \equiv \nu/E \quad , \tag{5.2}$$

and q^2. If scaling holds, $(1/E)d\sigma/dxdy$ is a universal function of x and y, independent of q^2.

Can we completely determine α, β, γ, and δ by studying these inclusive reactions? The answer turns out to be: almost yes, but we need data on protons, not just on isoscalar targets. I say "almost" because there are quadratic ambiguities that can be settled only by studying exclusive reactions, as will be discussed shortly.

Provided that we are below "new thresholds" in the charged as well as in the neutral current reactions, the constants α, β, γ, and δ can be extracted from the following "master equations"[31,32]:

$$\frac{(d\sigma_{NC}^{\nu p} + d\sigma_{NC}^{\nu n}) + (d\sigma_{NC}^{\bar{\nu} p} + d\sigma_{NC}^{\bar{\nu} n})}{(d\sigma_{CC}^{\nu p} + d\sigma_{CC}^{\nu n}) + (d\sigma_{CC}^{\bar{\nu} p} + d\sigma_{CC}^{\bar{\nu} n})} = \frac{1}{4}(\alpha^2+\beta^2+\gamma^2+\delta^2) + \bar{s}s \text{ corrections} \tag{5.3a}$$

$$\frac{(d\sigma_{NC}^{\nu p} + d\sigma_{NC}^{\nu n}) - (d\sigma_{NC}^{\bar{\nu} p} + d\sigma_{NC}^{\bar{\nu} n})}{(d\sigma_{CC}^{\nu p} + d\sigma_{CC}^{\nu n}) - (d\sigma_{CC}^{\bar{\nu} p} + d\sigma_{CC}^{\bar{\nu} n})} = \frac{1}{2}(\alpha\beta+\gamma\delta) , \tag{5.3b}$$

$$\frac{(d\sigma_{NC}^{\nu p} + d\sigma_{NC}^{\bar{\nu} p}) - (d\sigma_{NC}^{\nu n} + d\sigma_{NC}^{\bar{\nu} n})}{(d\sigma_{CC}^{\nu p} + d\sigma_{CC}^{\bar{\nu} n}) - (d\sigma_{CC}^{\nu n} + d\sigma_{CC}^{\bar{\nu} p})} = -\frac{1}{2}(\alpha\gamma+\beta\delta) , \tag{5.3c}$$

$$\frac{(d\sigma_{NC}^{\nu p} - d\sigma_{NC}^{\bar{\nu} p}) - (d\sigma_{NC}^{\nu n} - d\sigma_{NC}^{\bar{\nu} n})}{(d\sigma_{CC}^{\nu p} - d\sigma_{CC}^{\bar{\nu} n}) - (d\sigma_{CC}^{\nu n} - d\sigma_{CC}^{\bar{\nu} p})} = -\frac{1}{2}(\alpha\delta+\beta\delta) , \tag{5.3d}$$

In (5.3a)-(5.3d) NC and CC refer to the neutral-current and charged-current inclusive reactions, respectively, and $d\sigma$ may stand for the differential cross sections $d\sigma/dxdy$, $d\sigma/dx$ (y integrated out) or even the total cross section; in the last case we must be a little careful with the "$\bar{s}s$ corrections," which are x dependent, as will be seen later.

The proof of these relations is extremely simple within the framework of the valence quark model[33] that ignores the sea of quark-antiquark pairs as well as strange quarks and antiquarks. It is sketched in Appendix. Even with the sea, as long as the sea is SU(2) symmetric, there is no correction due to the ordinary (non-strange) $q\bar{q}$ pairs which affect the neutral- and charged-current processes in the same way. The only corrections we get are due to $\bar{s}s$ pairs which affect just (5.3a); when explicitly written the correction term reads

$$\bar{s}s \text{ corrections} = \frac{(\gamma'^2 + \delta'^2)\, S(x)}{Q(x) + \bar{Q}(x)} \tag{5.4}$$

where $S(x)$, $Q(x)$, and $\bar{Q}(x)$ are, respectively, the distribution functions for the strange quark (or antiquark), the quark distribution function for u and d averaged, and the antiquark distribution function for $\bar{u}$ and $\bar{d}$ averaged. There is reason to believe that the $\bar{s}s$ corrections are important only for $x \lesssim 0.15$ and, in any case, affects the total cross section by at most $\sim 3\%$ in most models. The actual contribution can even be smaller particularly because a comparison between electroproduction and the charged-current reactions reveals that the strange quark distribution function S(x) is smaller than is predicted by the hypothesis of an SU(3) symmetric sea.

The original quark parton model is constructed to accommodate (or explain?) scaling. There is now some experimental evidence against exact scaling in both inelastic electron-nucleon scattering[34] and the charged-current inclusive reactions.[35] So we may naturally ask: How would (5.3a)-(5.3d) be affected? If scaling is violated in such a way that the relative importance of the sea and the valence quarks varies at high q^2 or high E, then (5.3a)-(5.3d) may still hold provided, of course, that the differential cross sections for the charged- and neutral-current reactions are compared at the same values of q^2 or E. On the other hand, if scaling violation is due to new threshold channels opening up, then (5.3a)-(5.3d) must clearly be modified; so it is safer to work at energies not too high for the purpose of extracting α, β, γ, and δ. It is also worth mentioning that the isovector part of (5.3a) and (5.3b) follows from charge independence and chiral symmetry alone, and to fix the scale of the isoscalar contributions we do not really need the full machinery of the quark parton model even though some quark model arguments are needed--of the kind that leads to $\sigma(\rho N) = \sigma(\omega N)$.[36] In summary I believe that the results (5.3a)-(5.3d) are more general than the derivations based on the quark parton model.

Physically the four ratios (5.3a)-(5.3d) measure. respectively, the overall strength, VA (but not isoscalar-isovector) interference, isoscalar-isovector (but not VA) interference, and VA and isoscalar-isovector interference. These relations are intuitively reasonable if we recall that the difference between the neutrino and antineutrino cross sections is a measure of VA interference while the difference between the proton and neutron target cross sections isolates isoscalar-isovector interference.

Our information is quite incomplete when only data with isoscalar targets (targets with equal numbers of protons and neutrons) are available; we can then determine just the first two of the four ratios. Clearly no matter how skillful an experimentalist you may be, you cannot extract 4 constants from 2 ratios! At the end of this section I'll mention what we currently know experimentally about the two ratios.

When data on proton targets are available, we'll have four measured ratios and four coupling parameters to be determined. So, apart from the quadratic ambiguities to be discussed in a moment, it is, in principle, possible to determine the four constants. It is very important to note that each of the four relations is supposed to hold at every value of x and y; we must therefore have four "universal" ratios independent of x and y everywhere except for a small correction in the region $x \lesssim 0.15$ for (5.3a). This is actually a very powerful consistency check of our basic assumption that the neutral and charged currents are related in the manner indicated by the usual quark field currents. If the neutral current phenomena are due to SPT or second class V, A, or if there are "new thresholds" opening up, universality of this kind is not expected.

I now discuss the "ambiguities." Apart from the overall sign ambiguity, the right-hand sides of (5.3a)-(5.3d) are invariant under

(i) VA ambiguity: $\alpha \leftrightarrow \beta,\ \gamma \leftrightarrow \delta,$ (5.5)

(ii) isoscalar-isovector ambiguity: $\alpha \leftrightarrow \gamma,\ \beta \leftrightarrow \delta.$ (5.6)

From the inclusive distributions alone it is impossible to resolve these two ambiguities. When one compares data with the Salam-Weinberg model, one is implicitly assuming that the $\beta > \alpha$, $\delta = 0$ solution is the right one.

Equation (5.3a) defines a sphere in a four-dimensional α, β, γ, δ space while (5.3b)-(5.3d) define three hypersurfaces. Our objective then is to obtain the common intersection points of the sphere and the three hypersurfaces. Unfortunately our human mind is not accustomed to visualizing hypersurfaces in four dimensions. For pedagogical purposes let us look at a more simplified (if unrealistic) situation where the isoscalar constants γ and δ are set equal to zero by assumption--"pure isovector models." Equation (5.3a) then defines a circle in a two-dimensional α-β plane while (5.3b) defines hyperboles. We look for the intersection points of the circle ($\alpha^2+\beta^2$ = constant) and the hyperbolas ($\alpha\beta$ = constant). This is shown in Fig. 5.1. The existence of the VA ambiguity is evident from the figure; it amounts to the question of choosing between the intersection point near the α axis (vector dominant solution) and the one near the β axis (axial-vector dominant solution). The ambiguity disappears when the hyperbolas just touch the circle; VA interference then becomes maximal, and we have pure V-A (or pure V+A if the hyperbolas turn out to lie in the 2nd and 4th quadrants).

One of the most important questions in this field is: How large is isoscalar-isovector interference or, equivalently, the p-n difference in the ν and $\bar{\nu}$ induced neutral-current reactions? Within the framework of the valence quark model it is easy to relate the p-n difference in the neutral-current reactions to the p-n difference in electroproduction. The reason is that in both cases the difference is proportional to U(x)-D(x) where U and D stand for the quark distribution functions for the u type and the d type quark, respectively, inside the proton. We obtain[31]

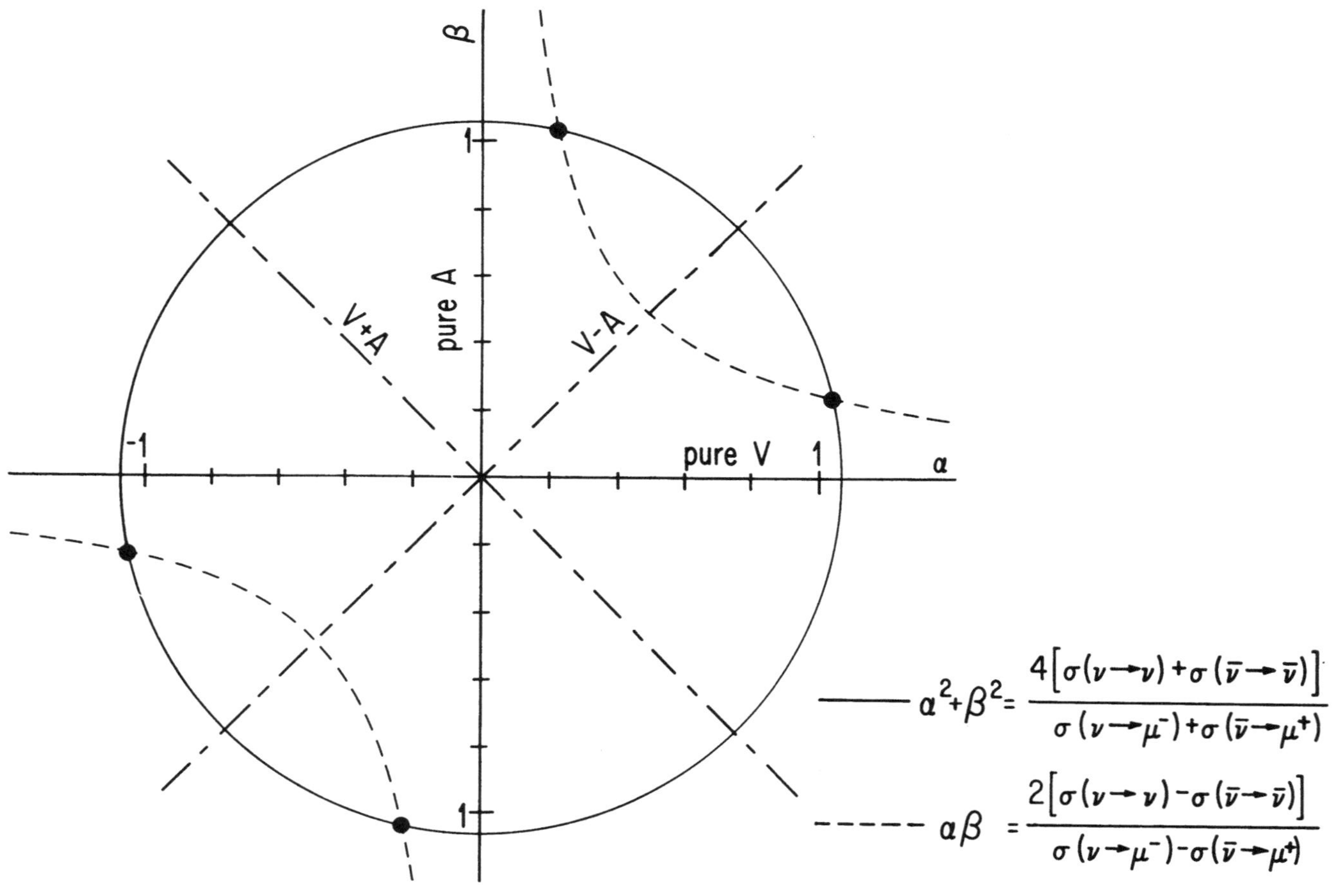

Fig. 5.1. Coupling constant determination in pure isovector models.

$$\frac{\dfrac{d\sigma^{\nu p}_{NC}}{dx} - \dfrac{d\sigma^{\nu n}_{NC}}{dx}}{\dfrac{d\sigma^{\nu p}_{NC}}{dx} + \dfrac{d\sigma^{\nu n}_{NC}}{dx}} = \frac{5}{3}\,\frac{(2\alpha\gamma+2\beta\delta+\alpha\delta+\beta\gamma)}{(\alpha^2+\beta^2+\gamma^2+\delta^2+\alpha\beta+\gamma\delta)}\left[\frac{F_2^{ep}(x) - F_2^{en}(x)}{F_2^{ep}(x) + F_2^{en}(x)}\right],$$

$$\frac{\dfrac{d\sigma^{\bar{\nu} p}_{NC}}{dx} - \dfrac{d\sigma^{\bar{\nu} n}_{NC}}{dx}}{\dfrac{d\sigma^{\bar{\nu} p}_{NC}}{dx} + \dfrac{d\sigma^{\bar{\nu} n}_{NC}}{dx}} = \frac{5}{3}\,\frac{(2\alpha\gamma+2\beta\delta-\alpha\delta-\beta\gamma)}{(\alpha^2+\beta^2+\gamma^2+\delta^2-\alpha\beta-\gamma\delta)}\left[\frac{F_2^{ep}(x) - F_2^{en}(x)}{F_2^{ep}(x) + F_2^{en}(x)}\right]. \tag{5.7}$$

Even though these results can be most straightforwardly derived using the valence quark model, much of the above relations can be shown to follow also from the assumptions of generalized CVC and chiral symmetry.

Armed with these relations, we can solve for the n-to-p ratio as a function of x. Model predictions are shown in Fig. 5.2 where the electroproduction n-to-p ratio needed was taken from the SLAC-MIT data.[37,38] In the electromagnetic current (EM) model[25] the n-to-p ratio is, of course, the same as in electroproduction. In models with pure isovector currents the ratio is obviously equal to unity at all x. Isoscalar-isovector interference in the Salam-Weinberg model[3] is not very large if $\sin^2\theta_W \lesssim 0.35$. It arises primarily from the dominant isovector axial amplitude (β term) interfering with the small isoscalar vector amplitude (γ term); it therefore changes sign as we go from ν to $\bar{\nu}$. If the neutral current is of the pure $\bar{d}d$ form, as in vectorlike models[26] with $\sin^2\theta_V = 3/4$, then the n-to-p ratio is predicted to become very large, as indicated by the top curve of Fig. 5.2. To see the physical origin of this spectacular behavior, let us recall that experimentally the n-to-p ratio in electroproduction appears to approach 1/4 as x approaches 1, which, in the quark model, is possible only if $D(x) \to 0$. So, in a model in which the neutral current interaction explores only the d type quark distribution inside the proton, the n-to-p ratio goes to ∞ as $x \to 1$.

One can also compute the total cross section ratios, as shown in Table 5.1. It is seen that, when we integrate over x, the isoscalar-isovector interference effect is less spectacular, which illustrates the importance of studying the x distribution.

Neutrino physicists often talk of comparing $R_{\nu p} \equiv \sigma^{\nu p}_{NC}/\sigma^{\nu p}_{CC}$ with $R_{\nu N} \equiv \sigma^{\nu N}_{NC}/\sigma^{\nu N}_{CC}$ where N stands for the average of proton and neutron.

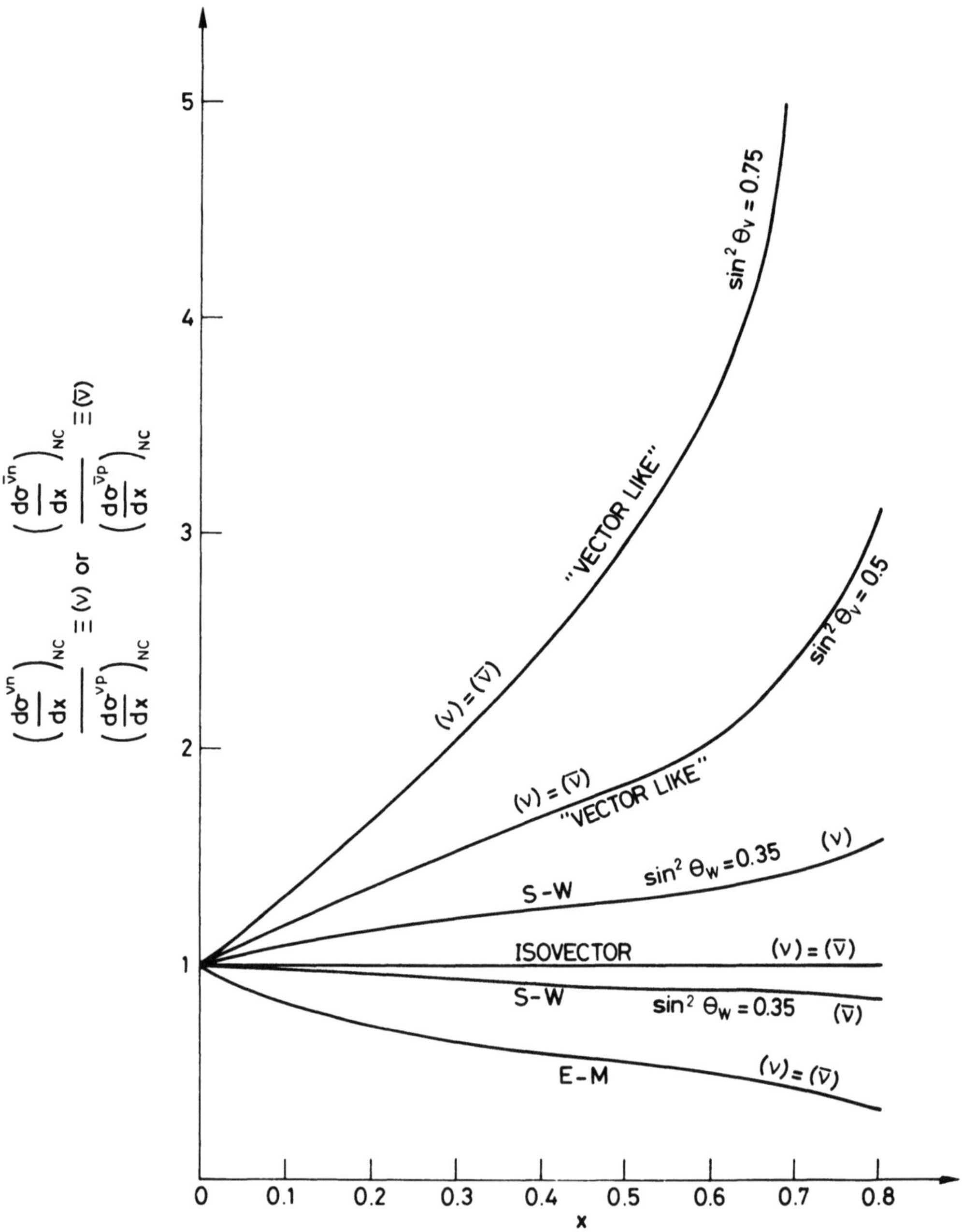

Fig. 5.2. Cross section ratios in the inclusive reactions (5.1).

Table 5.1.
Cross section ratios in the inclusive reactions (5.1).

	$\sigma_{NC}^{\nu n}/\sigma_{NC}^{\nu p}$	$\sigma_{NC}^{\bar{\nu} n}/\sigma_{NC}^{\bar{\nu} p}$
Pure isovector (pure isoscalar)[29]	1	1
Salam-Weinberg ($\sin^2\theta_W = 0.35$)[3]	1.13	0.955
Vectorlike ($\sin^2\theta_V = 0.5$)[26]	1.30	1.30
Vectorlike ($\sin^2\theta_V = 0.75$)[26]	1.56	1.56
Electromagnetic current (EM)[25]	0.75	0.75

For this comparison, the charged-current n-to-p ratio also enters. Even if there is no isoscalar-isovector interference in the neutral-current interactions, we expect

$$R_{\nu p} \simeq 1.3\, R_{\nu N} \tag{5.8}$$

where the factor 1.3 arises from the fact that the n-to-p ratio for the charged-current interaction is about 1.6. We have, quite generally,

$$R_{\nu p} \geq R_{\nu N} \quad . \tag{5.9}$$

Only in models where the neutral current is of the pure $\bar{d}d$ form does $R_{\nu p}$ become equal to $R_{\nu N}$; this readily follows because in such models the neutral- and charged-current interactions both explore the same quark distributions, viz. D(x) on proton targets, U(x) on neutron targets.

I have discussed at some length how we might extract the coupling constants α, β, γ, and δ given good data on the inclusive distributions. Unfortunately with the data currently available to us, what we can do is rather limited. First of all, most of our knowledge on the inclusive distributions comes from experiments in which the target material is made up of complex nuclei, nearly "isoscalar targets." This means that we have no information on isovector-isoscalar interference from the inclusive reactions. Second, we have no experimental information on the x distributions because to determine q^2 it is necessary to study the direction of the hadronic showers, which has not been done by any of the experiments performed so far. In both the Gargamelle Collaboration experiment[1,39] and the two counter experiments at Fermilab (Caltech[11] and HPWF[5]) what has been measured is the hadronic energy distribution or the ν distribution. So knowing the incident neutrino spectrum, we have some information on the y distribution even though the variable y is not directly determined for each event. What is usually done is to assume scaling, spin 1/2 constituents, and Hermiticity to write the cross sections as

$$(\pi/2G^2 m_N E)\,\frac{d\sigma}{dy} = A_{\mp} + A_{\pm}\,(1-y)^2\ , \qquad \left\{ \begin{matrix} \text{for } \nu N \\ \text{for } \bar{\nu} N \end{matrix} \right\} \tag{5.10}$$

and determine A_+ and A_-, taking into account the effect of the energy cut. Now, if we had no antiquarks, the coefficients A_- and A_+ would directly correspond to V-A and V+A, respectively. Because the antiquark content is known to be nonvanishing, there are corrections to this naive valence quark approach, which, however, can be estimated by studying the analogous coefficients for the charged-current reactions. (We, of course, assume that the charged-current interaction is pure V-A.) In any case, knowing $A_\pm$ and their charged current analogs, it is possible to compute the left-hand sides of (5.3a) and (5.3b) so that we can determine $\alpha^2 + \beta^2 + \gamma^2 + \delta^2$ and $\alpha\beta + \gamma\delta$.

Using the experimental data reported at the Aachen Conference (June 1976), I have deduced the following quantities[22]:

$$\alpha^2 + \beta^2 + \gamma^2 + \delta^2 = \left\{ \begin{array}{ll} 1.15 \pm 0.16 & (\text{Gargamelle}^{39}) \\ 1.18 \pm 0.19 & (\text{HPWF}^{5}) \\ 1.24 \pm 0.17 & (\text{Caltech}^{11})\ , \end{array} \right. \tag{5.11a}$$

$$\frac{2(\alpha\beta + \gamma\delta)}{\alpha^2 + \beta^2 + \gamma^2 + \delta^2} = \left\{ \begin{array}{ll} 0.57 \pm 0.18 & (\text{Gargamelle}^{39}) \\ 0.93 \pm 0.38 & (\text{HPWF}^{5}) \\ 0.44 \pm 0.25 & (\text{Caltech}^{11})\ . \end{array} \right. \tag{5.11b}$$

In obtaining these numbers I assumed: (i) the strange quark-antiquark pairs (γ' and δ' terms) contribute negligibly to the total cross sections, an assumption justifiable to an accuracy of a few % in most models, and (ii) the observed deviations from the standard-one-to-three ratio in the charged-current reactions are due to the presence of quark-antiquark pairs rather than new particle production with right-handed currents.

There are a few remarks to be made in connection with (5.11a) and (5.11b). First, from (5.11a) we can conclude that we now know the overall strength of the neutral-current interactions to an accuracy of about 20%. The three different groups with very different tastes in physics, using very different experimental techniques and neutrino beams (different in both the mean energy and the spectrum shape) came up with the same number within errors. This, in my opinion, is a remarkable achievement.

Let us now look at (5.11b). This quantity, which measures the amount of VA interference, is +1 for V-A, 0 for pure V/pure A, and -1 for V+A. Experimentally, even though the three groups do not completely agree on the central values, it appears that this quantity is nonvanishing and positive. So we are led to conclude that both V and A coexist in the hadronic part of the weak neutral currents. Models with pure V (or pure A) are ruled out to the extent that the quantity $\alpha\beta+\gamma\delta$ does not seem to vanish.

Given (5.11a) and (5.11b) we can determine the parameter(s) of any of the models listed in Table 4.1. For example, if we specialize to the one-parameter Salam-Weinberg model,[3] we see

$$\alpha^2 + \beta^2 + \gamma^2 + \delta^2 = 1.2 \Rightarrow \sin^2\theta_W = 0.30 \tag{5.12a}$$

$$\frac{2(\alpha\beta + \gamma\delta)}{\alpha^2 + \beta^2 + \gamma^2 + \delta^2} = \left\{ \begin{matrix} 0.5 \\ 0.9 \end{matrix} \right\} \Rightarrow \sin^2\theta_W = \left\{ \begin{matrix} 0.35 \\ 0.21 \end{matrix} \right\} . \tag{5.12b}$$

We note that the two ways of determining θ_W yields consistent results within errors. A similar exercise can be done with any of your favorite models.

6. NEUTRINO-INDUCED EXCLUSIVE HADRONIC REACTIONS

We have seen in the previous section that the inclusive distributions do not completely determine the coupling constants because of the VA and isoscalar-isovector ambiguities. To settle this question let us now turn to exclusive reactions.

As in beta decay physics, low energy nuclear transitions between states of definite isospin and spin-parity are sensitive to specific pieces of the neutral currents. A classical example along this line is[40]

$$\bar{\nu}_e + {}^7\mathrm{Li}\ (3/2^-) \rightarrow \bar{\nu}_e + {}^7\mathrm{Li}^*\ (1/2^-) \tag{6.1}$$

which may be studied using reactor antineutrinos with ${}^7\mathrm{Li}^*$ identified by its subsequent γ transition. This is a typical Gamow-Teller transition which isolates the isovector axial-vector part (β term) of the neutral current interactions. Another reaction of interest is

$$\bar{\nu}_e + D \rightarrow \bar{\nu}_e + n + p \tag{6.2}$$

which is the weak analog of the photodisintegration of the deuteron.

At reactor energies the vector matrix element is just the overlap integral between the initial and final wave functions, which vanishes by orthogonality. The reaction is allowed with isovector axial-vector (β term again) because the deuteron ($^3S_1$, I = 0) can disintegrate into a $^1S_0$, I = 1 state of the n-p system. A preliminary feasibility study by Reines and co-workers[41] shows that β^2, which should be 1 in the Salam-Weinberg model, is given by

$$\beta^2 = -\ 0.9 \pm 2.2 \quad . \tag{6.3}$$

It is expected, however, that eventually the errors will shrink to ±0.2.

Coherent elastic scattering of complex nuclei

$$\nu + A \to \nu + A \quad , \tag{6.4}$$

first considered by Freedman,[42,43] is another exclusive process of considerable importance. This scattering reaction received much attention because of its possible relevance to supernova explosion. On isoscalar targets it is sensitive to only the isoscalar vector part of the current (γ term). Unfortunately astrophysical calculations on this subject appear to be rather uncertain at this moment, and the experts in this field do not seem to agree on whether supernova explosion is possible, for example, with $\gamma \simeq -0.25$, a value predicted by Salam-Weinberg model with the currently accepted Weinberg angle.[44]

I now turn to diffractive production of vector and axial-vector mesons. We first recall that in photoproduction and also in low q^2 electroproduction, the reactions

$$\gamma(\text{real or virtual}) + N \to (\rho^o,\ \omega,\ \phi) + N \tag{6.5}$$

show up conspicuously. Furthermore, these vector meson states, having the same quantum numbers as the photon ($J^{PC} = 1^{--}$), are produced with characteristic features we expect from "diffraction"--energy-independent cross sections, diffractive slopes typical of elastic scattering, sharp coherent peaks when nuclear targets are used, etc. Likewise we expect that in neutrino reactions

$$\nu + N \to \nu + (\rho^o,\ \omega,\ \phi,\ A_1^{\ o} \cdots) + N \tag{6.6}$$

the meason states with the same quantum numbers as the current are copiously produced via diffractive mechanism:

ρ^o for isovector vector (α term)

$A_1^{\ o}$ for isovector axial-vector (β term)

ω and ϕ for isoscalar vector (γ and γ' terms) .

There have been many calculations on these diffractive processes based on vector (and axial-vector) meson dominance.[45] Such estimates are somewhat model dependent and may not be too reliable in quantitative details. Fortunately, for the purpose of estimating the coupling constants α and β, it is sufficient to compare the neutral-current diffractive processes with the changed-current diffractive processes as follows:

$$\left.\frac{\frac{d\sigma}{dq^2}(\nu N \to \nu \rho^o N)}{\frac{d\sigma}{dq^2}(\nu N \to \mu^- \rho^+ N)}\right|_{\text{diffractive}} = \frac{1}{2}\alpha^2 \tag{6.7a}$$

$$\left.\frac{\frac{d\sigma}{dq^2}(\nu N \to \nu A_1{}^o N)}{\frac{d\sigma}{dq^2}(\nu N \to \mu^- A_1{}^+ N)}\right|_{\text{diffractive}} = \frac{1}{2}\beta^2 \quad . \tag{6.7b}$$

In this way much of the model uncertainties cancels out. To resolve the ambiguities in inclusive data, only crude knowledge of diffractive reactions is needed, e.g., whether $|\alpha| > |\beta|$ or $|\beta| > |\alpha|$; whether $\alpha^2+\beta^2$ is a sizable fraction of, or much smaller than, $\alpha^2+\beta^2+\gamma^2+\delta^2$. It is also interesting to note that the elusive coupling constant γ', which measures the strength of the vector part of $\bar{s}s$ (ϕ like), can be estimated by comparing the ϕ-to-ω ratio in the neutral-current reactions to the corresponding ratio in the electroproduction reactions:

$$\left.\frac{\frac{d\sigma}{dq^2}(\nu N \to \nu \phi N)}{\frac{d\sigma}{dq^2}(\nu N \to \nu \omega N)}\right|_{\text{diffractive}} = \frac{\gamma'^2}{\gamma^2}\left.\frac{\frac{d\sigma}{dq^2}(eN \to e\phi N)}{\frac{d\sigma}{dq^2}(eN \to e\omega N)}\right|_{\text{diffractive}} \tag{6.8}$$

at the same value of q^2.

Elastic scattering

$$\nu + p \to \nu + p \tag{6.9}$$

is, without doubt, one of the most important reactions in high-energy physics. It is, in a certain sense, as fundamental as neutron beta decay. At $q^2 = 0$ we have a clean prediction[46]

$$R_{el}^{(q^2=0)} \equiv \left. \frac{\frac{d\sigma}{dq^2}(\nu p \to \nu p)}{\frac{d\sigma}{dq^2}(\nu n \to \mu^- p)} \right|_{q^2=0} = \frac{(\alpha+3\gamma) \quad + (1.25\ \beta + g_A^{(s)}\delta)^2}{4[1+(1.25)^2]} . \tag{6.10}$$

Possibly the only uncertainty in the $q^2 = 0$ prediction lies in the isoscalar axial-vector renormalization factor $g_A^{(s)}$, which is the analog of the g_A/g_V ratio in neutron beta decay. According to the way I normalized the coupling constants, it assumes the value 1 in the naive SU(6) limit; however, the SU(6) prediction for g_A/g_V in neutron decay is wrong by a factor of 1.25/(5/3), so a reasonable attitude to take may be to assume the same correction factor to obtain

$$g_A^{(s)} = 1.25/(5/3) = 0.75 , \tag{6.11}$$

a result derived by Adler and collaborators[47] using a more involved argument.[48]

At $q^2 = 0$ this elastic reaction is unfortunately unobservable; recall

$$q^2 = 2m_p T_p \tag{6.12}$$

where T_p is the kinetic energy of the recoil proton, so both particles in the final state are invisible as is the incoming particle! There are two groups which have reported experimental studies of this elastic reaction, a Harvard-Pennsylvania-Wisconsin Collaboration[49] and a Columbia-Illinois-Rockefeller Collaboration,[50] both working at Brookhaven where the mean neutrino energy is about 1 GeV. Both groups study elastic events with $q^2 > 0.3$ (GeV)^2;

$$R_{q^2>0.3}^{(el)} = \begin{cases} 0.17 \pm 0.05 & \text{HPW} \\ 0.23 \pm 0.09 & \text{CIR} \end{cases} . \tag{6.13}$$

Unfortunately it is not straightforward to compare (6.13) with the $q^2 = 0$ prediction (6.10). As we go away from $q^2 = 0$, there are uncertainties with the axial-vector form factors, and furthermore, some form factors which are dormant at $q^2 = 0$ could contribute. In any case, the correction factor needed to get $R_{q^2=0}^{(el)}$ from $R_{q^2>0.3}^{(el)}$ depends on the relative ratios of α, β, γ, and δ, the very quantities we are trying to determine! In typical models $R_{q^2=0}^{(el)}$ at

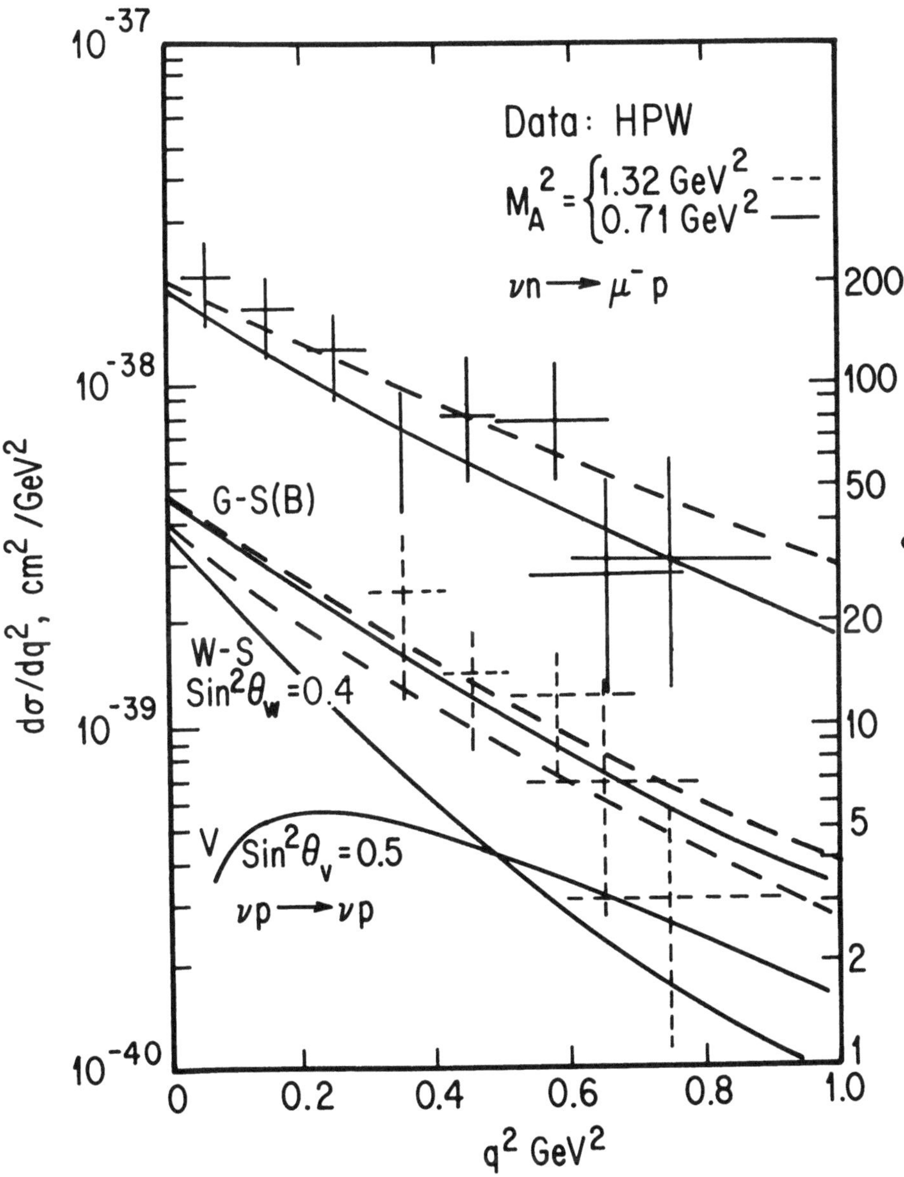

Fig. 6.1. νp elastic scattering

Brookhaven energies is larger than $R^{(el)}_{q^2>0.3}$ by about a factor of 1.5 to 2. Ideally it would be nice if we could extrapolate $d\sigma/dq^2$ observed in the $q^2 > 0.3$ GeV^2 range to $q^2 = 0$ in a relatively model independent manner. But this appears to be difficult, as you can see from some typical model calculations shown in Fig. 6.1, taken from a recent paper of Albright et al.[51] Notice, in particular, how sensitive the predicted curves are to the mass of the axial-vector form factor assumed to be of the dipole form. The situation is presumably more favorable with accurate data down to $q^2 \simeq 0.2$ GeV^2.

The HPW Collaboration[52] has also studied $\bar{\nu}p$ scattering. Under nearly the same kinematical conditions they report

$$\left.\frac{\sigma(\bar{\nu}p \to \bar{\nu}p)}{\sigma(\nu p \to \nu p)}\right|_{q^2>0.3} = 0.4 \pm 0.2 \quad . \tag{6.14}$$

This provides further evidence against models based on pure V or pure A.

I now turn to single pion production--Δ and $N^*_{1/2}$'s. There are many papers on this topic, using static model calculations with $O(q)$ corrections, taking into account nuclear charge-exchange effects, etc. Here, however, I wish to concentrate on results that follow from simple isospin considerations. There are clean predictions we can make in models with pure isovector--or almost pure isovector as in SW with $\sin^2\theta_W \simeq 0.35$--models[53]:

$$R(\Delta^{+,0}) \equiv \frac{\sigma(\nu p\to\nu\Delta^+) + \sigma(\nu n\to\nu\Delta^o)}{2\sigma(\nu n \to \mu^-\Delta^+)} = \frac{\alpha^2\sigma_V + \beta^2\sigma_A + \alpha\beta\sigma_I}{\sigma_V + \sigma_A + \sigma_I} \tag{6.15a}$$

$$R(N^*_{1/2}) \equiv \frac{\sigma(\nu p\to\nu N^{*+}) + \sigma(\nu n\to\nu N^{*o})}{2\sigma(\nu n \to \mu^- N^{*+})} = \frac{1}{4}\; \frac{\alpha^2\sigma_V + \beta^2\sigma_A + \alpha\beta\sigma_I}{\sigma_V + \sigma_A + \sigma_I} \tag{6.15b}$$

where σ_V, σ_A, and σ_I are the vector, the axial-vector, and the interference contributions to the cross section for the charged-current reaction appearing in the denominator, and Δ and $N^*_{1/2}$ stand, respectively, for any $I = 3/2$ and $I = 1/2$ πN (or any $B = 1$, $S = 0$) system, not necessarily nucleon isobars of definite spin-parity. To make use of these equations we must know the relative size of σ_V and σ_A; it turns out that most model calculations indicate $\sigma_V \simeq \sigma_A$ for almost any nucleon isobar, and this can simply be understood by applying Bloom-Gilman duality[54] to the neutrino and antineutrino cross sections at resonance energies. The relations (6.15a) and

(6.15b) mean that when the same charge states are selected, there is a factor of ~ 4 decrease in the neutral-to-charged current ratio as we go from the Δ dominant region to the $N^*_{1/2}$ dominant region. In reality the situation is not so simple because of nonresonant backgrounds, but we can conclude that: (i) a possible $I = 1/2$ contamination in the Δ region should be less serious in the neutral-current than in the charged-current data, and (ii) a possible distortion in the shape of the Δ peak is also expected to present a less serious problem in the neutral-current data. All these simple conclusions based just on isospin invariance are borne out by much more elaborate calculations performed by Adler and collaborators[55] and by Paschos.[56]

Even though data available so far on the πN mass distribution in the Δ region are still inconclusive as to the presence or absence of Δ, there is one piece of information relevant to the isospin properties of the neutral currents. Recently the Gargamelle Collaboration[57] has examined the pion charge ratio in

$$\overset{(-)}{\nu} + N \rightarrow \overset{(-)}{\nu} + \pi^{\pm,o} + N' \quad . \tag{6.16}$$

If the current is pure isoscalar and if the target is also isoscalar, we must have

$$\sigma(\pi^+) : \sigma(\pi^o) : \sigma(\pi^-) = 1 : 1 : 1 \quad , \tag{6.17}$$

which is clear because there is no preferred direction in isospin space in the initial state. Work of the Gargamelle Collaboration shows that the π^o-to-π^- ratio is more like 1.4 ± 0.4 for the neutrino reaction and 2.1 ± 0.4 for the antineutrino reaction. These numbers should be compared to the isoscalar-current prediction 0.9, a value expected when we take into account the fact that freon is not quite pure isoscalar. Provided the experimental group knows how to estimate the relative efficiencies of π^o and π^-, we must conclude that pure isoscalar models are ruled out, i.e., α and/or β must be nonvanishing.

Instead of single pion production we may also examine pion semi-inclusive reactions

$$\overset{(-)}{\nu} + N \rightarrow \overset{(-)}{\nu} + \pi^{\pm,o} + \text{any} \quad . \tag{6.18}$$

Any difference between π^+ and π^- in this set of reactions is a measure of isoscalar-isovector interference when isoscalar targets are used. However, to extract the relevant coupling constants we must rely on the details of parton fragmentation models[58] which are only now being critically tested in electroproduction.

7. NEUTRAL-CURRENT PHENOMENA WITHOUT NEUTRINOS

The total neutral currents to which the Z boson may be coupled are made up of three pieces, $(\bar{\nu}\nu)$, $(\bar{\ell}\ell)$, and $(\bar{q}q)$, where ℓ stands for e or μ. The interaction of $(\bar{\nu}\nu)$ with $(\bar{\ell}\ell)$ gives rise to $\overset{(-)}{\nu}e$ scattering discussed in Section 3 while the neutrino-induced hadronic reactions treated in Sections 5 and 6 take place via $(\bar{\nu}\nu)(\bar{q}q)$. As examples of $(\bar{\ell}\ell)$ interacting with itself, I may mention weak interaction effects in Bhabha scattering

$$e^+ + e^- \to e^+ + e^- \tag{7.1}$$

and muon pair production in electron-positron collisions

$$e^+ + e^- \to \mu^+ + \mu^- \quad . \tag{7.2}$$

There are also phenomena due to $(\bar{\ell}\ell)(\bar{q}q)$: parity violation effects in inelastic muon-nucleon scattering, electron-positron annihilations into hadrons, and atomic radiative transitions.

I would like to emphasize that it is of vital importance to study at least one neutral-current reaction not involving neutrinos. We recall that $\bar{\nu}_\mu e$ scattering and $\overset{(-)}{\nu}$ induced inclusive hadronic reactions are both characterized by cross sections of order G^2s/π. This means that the strength of $(\bar{\nu}\nu)(\bar{\ell}\ell)$ is similar to that of $(\bar{\nu}\nu)(\bar{q}q)$ so that we have rough "universality." However, we can still conceive of an unconventional model in which the coefficients in front of $\bar{\ell}\ell$ and $\bar{q}q$ are both enhanced by a factor r compared to normal models with universality while the coefficient in front of $\bar{\nu}\nu$ is down by 1/2. Such a model, of course, would give large neutral-current effects in processes not involving neutrinos.[59]

Among the various neutrinoless neutral-current reactions mentioned earlier, I would like to concentrate on just two topics: electron-positron annihilation into muon pairs and parity violation in atomic radiative transitions. As you'll see, both topics are of high current interest.

We begin with muon pair production (7.2). First, before starting to perform detailed computations, let us look at the order of magnitude of weak neutral-current effects. At low (e.g., SPEAR or DORIS) energies muon production is dominated by QED one-photon exchange. Its amplitude (the Dirac spinors omitted) goes like e^2/s where 1/s comes from the photon propagator in the s channel. The process can also proceed via the neutral-current interactions where the current is again in the s channel. In any model with rough universality, the weak amplitude must be of order G [or $Gm_Z{}^2/(m_Z{}^2-s)$ if the center-of-mass energy $\sqrt{s}$ is not so small compared to the boson mass]. So the ratio of the weak to electromagnetic amplitude goes as

$$G/(e^2/s) \sim 10^{-4}\ s \tag{7.3}$$

where s is in GeV^2. If the weak interactions continue to be pointlike (i.e., of the Fermi-type) at very high energies, the weak amplitude is going to dominate the electromagnetic amplitude at center-of-mass energies of several hundred GeV. Actually most models which attempt to unify the electromagnetic and weak interactions promise us a spectacular resonance peak due to the Z boson before reaching such an energy region.

Even at $\sqrt{s}$ as low as tens of GeV, the weak amplitude could be several per cent of the electromagnetic amplitude, and weak interaction effects may well be detectable via weak-electromagnetic interference. This is of immediate interest because there are now colliding beam machines--PETRA and PEP--under construction precisely in this energy range.

To be quantitative let us start with the most general V, A couples compatible with μe universality[60]:

$$\begin{aligned} L = -\frac{G}{\sqrt{2}}\,[&h_{VV}(\bar{e}\gamma_\lambda e + \bar{\mu}\gamma_\lambda\mu)(\bar{e}\gamma_\lambda e + \bar{\mu}\gamma_\lambda\mu) \\ &+ 2h_{VA}(\bar{e}\gamma_\lambda e + \bar{\mu}\gamma_\lambda\mu)(\bar{e}\gamma_\lambda\gamma_5 e + \bar{\mu}\gamma_\lambda\gamma_5\mu) \\ &+ h_{AA}(\bar{e}\gamma_\lambda\gamma_5 e + \bar{\mu}\gamma_\lambda\gamma_5\mu)(\bar{e}\gamma_\lambda\gamma_5 e + \bar{\mu}\gamma_\lambda\gamma_5\mu)] \quad . \end{aligned} \tag{7.4}$$

In writing (7.4) I have assumed that the center-of-mass energy $\sqrt{s}$ is much lower than the Z boson mass; otherwise just multiply the whole expression by $m_Z{}^2/(m_Z{}^2 + s - im_Z\Gamma_Z)$. Even though models with a single Z boson predict

$$h_{VA}^2 = h_{VV}h_{AA} \quad , \tag{7.5}$$

from a general phenomenological point of view the three constants h_{VV}, h_{AA}, and h_{VA} must be treated as independent parameters. It is therefore desirable to plan experiments that measure the three terms separately.

In 1961 Cabibbo and Gatto,[61] in one of the earliest papers discussing the raison-d'être of e^+e^- colliding beam machines, pointed out that if there are weak neutral currents, weak-electromagnetic interference may be studied in electron-positron annihilation into muons. In particular they proposed to look for neutral-current effects in the following three places:

(i) the magnitude and s dependence of the cross section,

(ii) forward-backward asymmetry,

(iii) the longitudinal muon polarization.

Just to appreciate the time scale involved in weak interaction physics, I may mention that the first positive results along this line will be obtained probably in the early 1980s when PETRA and PEP come into operation--twenty years after the Cabibbo-Gatto paper!

According to QED, the cross section for (7.2) to order α^2 is given by ($m_\mu^2 \ll s$)

$$\sigma_{QED} = \frac{4}{3}\pi\alpha^2/s \quad . \qquad (7.6)$$

With neutral currents added, the muon pair cross section is predicted to deviate from the QED prediction (7.6) both in the magnitude and in the s dependence. For s not too large the deviation is sensitive only to the h_{VV} term[61]

$$\Delta\sigma/\sigma_{QED} = (G/\sqrt{2}\,\pi\alpha)h_{VV}s \quad . \qquad (7.7)$$

Let us now turn to the angular distribution for which the QED prediction is

$$d\sigma/d\Omega \propto 1 + \cos^2\theta \quad , \qquad (7.8)$$

forward-backward symmetric in the one-photon exchange approximation. With an axial-vector piece in the leptonic current (h_{AA} term), forward-backward asymmetry is possible via weak-electromagnetic interference. The relevant formula is[61]

$$A(\theta) \equiv \frac{\sigma(\theta) - \sigma(\pi-\theta)}{\sigma(\theta) + \sigma(\pi-\theta)} = (G/\sqrt{2}\,\pi\alpha)h_{AA}s\,\frac{2\cos\theta}{1+\cos^2\theta} \quad . \qquad (7.9)$$

Finally, there is a genuinely parity-violating effect, muon polarization, due to the h_{VA} term. The formula for the longitudinal polarization with unpolarized initial beam is[61]

$$P(\mu^+) = -P(\mu^-) = (G/\sqrt{2}\,\pi\alpha)h_{VV}s\left[1 + \frac{2\cos\theta}{1+\cos^2\theta}\right] \quad . \qquad (7.10)$$

Notice that the three different experiments are sensitive to the three different terms of the Lagrangian (7.4). It is also worth keeping in mind that the formulas (7.7), (7.9), and (7.10) are valid only if the weak amplitude is still small compared to the electromagnetic amplitude; in particular, they are not valid if there is a Z boson with mass close to $\sqrt{s}$.

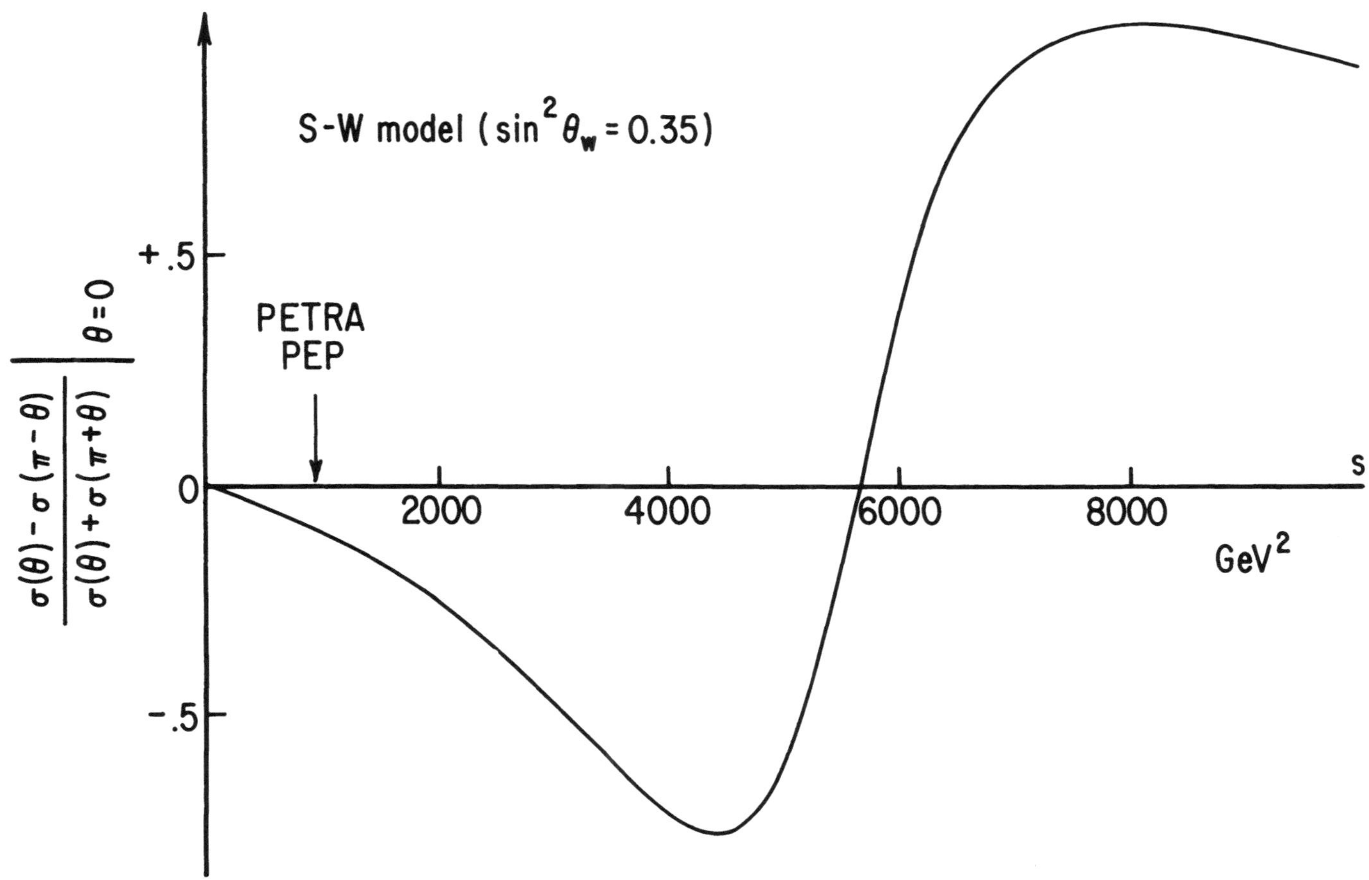

Fig. 7.1. Forward-backward asymmetry in $e^+ + e^- \rightarrow \mu^+ + \mu^-$.

To obtain numerical results expected in experiments of this type, let us now specialize to the Salam-Weinberg model.[3] The coupling constants in this model are given by[62]

$$h_{VV} = \frac{1}{4}(1 - 4\sin^2\theta)^2 \quad ,$$

$$h_{VA} = \frac{1}{4}(1 - 4\sin^2\theta) \quad , \qquad (7.11)$$

$$h_{AA} = \frac{1}{4} \quad .$$

In Fig. 7.1 I show forward-backward asymmetry expected for $\sin^2\theta_W = 0.35$, taken from a recent review talk by Gourdin.[63] At low values of s the asymmetry is linear in s in agreement with (7.9) but there is a spectacular deviation from the linear extrapolation as s approaches the squared boson mass, at which the asymmetry changes sign. At a typical PETRA or PEP energy of $s = 900\ \text{GeV}^2$ (15 GeV beam energy) the asymmetry $A(\theta)$ evaluated at $\theta = 0°$ is predicted to be as large as 8%, which should be comfortably measurable.[64]

In models in which the charged-leptonic part of the neutral currents is purely or predominantly vector, a measurable angular asymmetry is not expected. However, in such models the deviation from the QED prediction given by (7.7) may be sizable. The coefficient $(G/\sqrt{2}\,\pi\alpha)$ is $3.5 \times 10^{-4}/\text{GeV}^2$, so in any model with $h_{VV} \simeq 1/4$, the muon pair cross section may deviate from the QED prediction by as much as 8% at $s = 900\ \text{GeV}^2$. So regardless of whether nature chooses h_{AA} to be sizable (as in the Salam-Weinberg model) or h_{VV} to be sizable, something interesting is bound to happen at PETRA or PEP.

We now turn to the subject of parity violation in atomic physics. Consider the axial part of the electronic current interacting with the vector part of the hadronic current. The interaction is clearly parity-violating, and, moreover, its effect in complex atoms is coherent in the sense that the contributions from the various nucleons in the nucleus are additive as the charges add up. In principle, we can also consider a parity-violating effect due to the vector part of the electronic current interacting with the axial-vector part of the hadronic current but this latter effect is not so easy to detect because the spin orientations inside the nucleus don't add up to give rise to a huge coherence effect.

Let us look more closely at the $V_{had}A_{lept}$ interaction. For the hadronic side we must consider the time component of the vector current because for nonrelativistic nucleons γ_λ is large only for the time component. The leptonic side is then given by $\bar{e}\gamma_4\gamma_5 e = e^\dagger\gamma_5 e$, which reduces in the plane-wave representation to $-\vec{\sigma}\cdot(\vec{p}_i+\vec{p}_f)/2m_e$.

The resulting interaction can be represented by an effective parity violation potential between the electron and the nucleus (treated to be pointlike) as follows[65]

$$H_{p.v.} = (G/4\sqrt{2}\, m_e)\, Q_W\, \vec{\sigma}\cdot\{\vec{p},\delta^{(3)}(\vec{x})\} \tag{7.12}$$

with Q_W given by

$$Q_W = -(a_e/a_\nu)[\alpha(Z-N) + 3\gamma(Z+N)] \tag{7.13}$$

In (7.13) Z and N are the number of protons and the number of neutrons, respectively, and a_e and a_ν are the coefficients in front of $i\bar{e}\gamma_\lambda\gamma_5 e$ and $i\bar{\nu}\gamma_\lambda(1+\gamma_5)\nu$ that enter in the leptonic neutral current to which the Z boson is coupled. Notice, how the contributions from the individual nucleons add up coherently for the isoscalar piece. Just as an example, I may mention that a_e/a_ν is -1 in all $SU(2) \otimes U(1)$ models where the 3rd components of left-handed weak isospin have opposite signs between ν and e, and that in the Salam-Weinberg model with $\sin^2\theta_W \simeq 0.35$, Q_W is -160 for bismuth atoms.

Because of the potential (7.12) a given atomic state $|a\rangle$ is no longer an exact parity eigenstate but contains an opposite parity admixture with amplitude going like $\langle b|H_{p.v.}|a\rangle/(E_b-E_a)$ where $|b\rangle$ is a nearby state of opposite parity. As a result, a radiative transition between a pair of atomic levels acquires a parity-violating component; a dominantly M1 transition, for instance, contains a very small amount of E1 as well. This means that photons emitted by excited atomic states are, in general, circularly polarized.

There are now two experimental groups--Fortson and collaborators[66] at Seattle (the University of Washington) and Sandars and collaborators[67] at Oxford--who are carrying out experiments of sufficient sensitivity to detect parity violation in atoms. In both experiments ^{209}Bi (Z = 83) atoms are used. A linearly polarized laser beam whose wavelength corresponds to the level difference in question--between $^4S_{3/2}$ (the ground state) and $^2D_{3/2}$ (the first excited state) in the Seattle experiment[66] and between $^2D_{3/2}$ and $^2D_{5/2}$ (the second excited state) in the Oxford experiment[67]--is allowed to pass through an oven containing bismuth vapor. The basic principle of the experiments is to detect parity violation by observing an optical rotation of the linear polarization expected when right- and left-circularly polarized beams have different indices of refraction. Theoretical calculations which, incidentally, require good knowledge of the atomic wave functions, show that with a realistic experimental arrangement, the expected rotation is in the neighborhood of

10^{-7} - 10^{-6} radians with Q_W given by the Salam-Weinberg model.[68] At the time of this summer school, despite some earlier rumors, there is no firm number that can be announced publicly by either group.

It is worth remarking that these atomic physics experiments can be very powerful in discriminating various theoretical models. It is clear from (7.13) that by varying the neutron-to-proton ratio, one can, in principle, measure different linear combinations of the isovector and isoscalar coupling constants. Furthermore, models that give very similar predictions on neutrino-induced hadronic reactions can be shown to give very different values of Q_W in both sign and magnitude.[69]

8. CONCLUSION

Three years have passed since the Gargamelle Collaboration announced the great discovery of neutral currents. Having spent tens of millions of Swiss francs (or U.S. dollars) on both sides of the Atlantic, what have we actually learned since then? First, the existence of neutral currents was confirmed by several independent experiments. It has been demonstrated to everybody's satisfaction that neutral currents indeed exist and are here to stay. The currents are no longer alternating.

Second, some extreme models have been ruled out; the casualty list goes as follows:

(i) models with pure S and/or P hadronic densities,
(ii) models with pure V or pure A hadronic currents,
(iii) models with pure isoscalar ($I = 0$) hadronic currents,
(iv) models with very different strengths between $(\bar{\nu}\nu)(\bar{q}q)$ and $(\bar{\nu}\nu)(\bar{e}e)$,
(v) models that violate scaling violently.

Considering, however, that the history of weak-interaction physics is marred by wrong experiments and bad theoretical models inspired by the wrong experiments, we should not regard this casualty list as absolute final.

In terms of the coupling constants α, β, γ, and δ defined by (4.1) we know that the overall strength $\alpha^2 + \beta^2 + \gamma^2 + \delta^2$ is equal to 1.2 to an accuracy of about 20% and that the quantity $\alpha\beta + \gamma\delta$ is positive and nonvanishing, indicating that the hadronic current is somewhere between pure V-A and pure V/pure A. In addition, from single pion reactions we must have

$$\alpha \neq 0 \quad \text{and/or} \quad \beta \neq 0 \quad . \tag{8.1}$$

Assiduous model builders can still construct an infinite variety of models subject to the constraints above!

What further experiments are needed to completely determine the properties of the neutral-current interactions?

As far as the inclusive hadronic reactions are concerned, clearly the next major progress will be made when we obtain accurate data on proton targets, not just on isoscalar targets. Furthermore, it is desirable to measure the neutron-to-proton-target ratio as a function of x because isoscalar-isovector interference, if present, is predicted to show up more conspicuously for $x \gtrsim 0.5$; the effect tends to get diluted when we integrate over x. In new generation bubble chambers with track-sensitive targets, it is not beyond the realm of feasibility to study the x distributions of the neutral-current reactions separately for protons and neutrons.

I have already emphasized in Section 6 the importance of studying diffractive vector and axial-vector production in resolving the two "ambiguities." A careful study of ϕ and ω diffractive production will determine the elusive constant γ' and test the details of the GIM mechanism or U spin invariance. As for elastic νp scattering, it is desirable to design experiments to study $d\sigma/dq^2$ down to q^2 values as low as possible.

Turning now to neutrino-electron scattering, we would like to see how the three ellipses for $\nu_\mu e$, $\bar{\nu}_\mu e$, and $\bar{\nu}_e e$ with negligible error bands intersect in a g_V-g_A plane. Very accurate data for $\bar{\nu}_e e$ scattering at reactor energies are needed to see whether the vector dominant or axial-vector dominant solution is the right one.

The prospects for detecting weak-electromagnetic interference in electron-positron collisions at PETRA and PEP energies are extremely promising. We look forward to hearing about the first experimental report on weak neutral-current effects in electron-positron annihilations into muon pairs at the 1981 Erice School.

Of more immediate interest is the status of parity violation experiments with atoms. Even the sign and crude order of magnitude of Q_W are of enormous value. Measurements of Q_W with different atoms should be very powerful in eliminating various theoretical models (the experimentalists' delight!).

Some of us theorists are often discouraged by the slow rate of progress made in neutral-current physics--slow, when compared to the fantastic progress made in the field of new-particle spectroscopy since that fateful Sunday, November 11, 1974. But we have to be patient. It took 25 years to demonstrate V, A structure in beta decays where the number of invisible particles is only one. You may speculate how long it will take to determine completely the structure of the leptonic and hadronic parts of weak neutral currents by studying reactions with <u>two</u> invisible particles.

ACKNOWLEDGMENT

Much of the lecture material was prepared while I was a scientific associate at CERN holding a John Simon Guggenheim Memorial Foundation Fellowship. I wish to take this opportunity to acknowledge the hospitality of the Theory Division at CERN.

APPENDIX

In this appendix I present a simple derivation of the master equations (5.3a) - (5.3d) within the framework of the valence quark model.

In the quark language the charged-current reactions are visualized to be due to the "fundamental processes"

$$\nu + d \rightarrow \mu^- + u \ , \qquad \bar{\nu} + u \rightarrow \mu^+ + d \ . \tag{A1}$$

To the extent that the antiquark content within the nucleon can be ignored, the ν induced inclusive charged-current reaction on protons explores $D(x)$, the momentum distribution of the d type quark within the proton. Likewise the $\bar{\nu}$ induced reaction on protons explores $U(x)$, the distribution function for the u type quark within the proton. Now, because of the charge symmetry of the strong interactions, $D(x)$ is also the momentum distribution of the u type quark within the neutron; so the $\bar{\nu}$ induced reaction on neutrons measures $D(x)$. Similarly the ν induced reaction on neutrons is sensitive to $U(x)$. Recalling now the helicity rule for V-A--a flat y distribution for fermion-fermion scattering, $(1-y)^2$ for antifermion-fermion scattering--discussed in Section 3, we obtain

$$\frac{d\sigma^{\nu p}_{CC}}{dxdy} = (2G^2 m_N E/\pi) \times D(x) \ , \qquad \frac{d\sigma^{\bar{\nu} p}_{CC}}{dxdy} = (2G^2 m_N E/\pi) \times U(x)\ (1-y)^2 \tag{A2}$$

and similar expressions for the neutron target with U and D interchanged.

If we now go to the neutral-current case, we must keep in mind that both V-A and V+A are in general present. Now the coupling constant combinations $\alpha \pm \beta$ and $\gamma \pm \delta$ go with $V \mp A$, and $\gamma \pm \alpha$ and $\delta \pm \beta$ go with the $\{{u \atop d}\}$ type quark [see (4.1)]. Combining all this information with the helicity rule, we obtain

$$\frac{d\sigma_{NC}^{\nu p}}{dxdy} = (2G^2 m_N E/\pi) \times [U(x)\{a_L{}^2 + a_R{}^2(1-y)^2\} + D(x)\{b_L{}^2 + b_R{}^2(1-y)^2\}]$$

(A3)

$$\frac{d\sigma_{NC}^{\bar{\nu} p}}{dxdy} = (2G^2 m_N E/\pi) \times [U(x)\{a_R{}^2 + a_L{}^2(1-y)^2\} + D(x)\{b_R{}^2 + b_L{}^2(1-y)^2\}]$$

where $a_{L,R}$ and $b_{L,R}$ are given by

$$a_L \equiv \frac{1}{4}(\alpha+\beta+\gamma+\delta) , \qquad a_R \equiv \frac{1}{4}(\alpha-\beta+\gamma-\delta) ,$$

(A4)

$$b_L \equiv \frac{1}{4}(-\alpha-\beta+\gamma+\delta) , \qquad b_R \equiv \frac{1}{4}(-\alpha+\beta+\gamma-\delta) .$$

The expressions for the neutron target reactions can again be obtained by interchanging U and D.

Notice that in this naive valence quark approximation, all the differential distributions are completely determined once U(x) and D(x) are given together with the coupling constants. Conversely, by taking the ratios of suitable linear combinations of (A2) and (A3) and their neutron analogs, it is possible to eliminate U(x) and D(x). In this manner we readily derive the master equations (5.3a) - (5.3d) which do not involve U(x) and D(x).

REFERENCES AND FOOTNOTES

1) F. J. Hasert et al., Phys. Lett. 46B, 138 (1973).
2) I am following Professor H. Faissner who compared the Neutrino Conference (Aachen, June 1976) he organized to a well-structured Viennese symphony.
3) A. Salam, Proceedings 8th Nobel Symposium, ed. N. Svartholm (Almquist and Wiksells, Stockholm, 1968), p. 367; S. Weinberg, Phys. Rev. Lett. 19, 1264 (1967).
4) J. J. Sakurai, CERN TH-2099 (to be published in Proceedings International Summer Institute of Theoretical Physics, DESY, Hamburg, September 1975).
5) T. Y. Ling, Proceedings Neutrino Conference (Aachen, June 1976).
6) L. M. Sehgal, Phys. Lett. 55B, 205 (1975).
7) L. Wolfenstein, Nucl. Phys. B91, 95 (1975).
8) J. E. Kim, V. S. Mathur, and S. Okubo, Phys. Rev. D9, 3050 (1974).
9) J. Bernstein and T. D. Lee, Phys. Rev. Lett. 11, 512 (1963); C. Bouchiat, J. Iliopoulos, and Ph. Meyer, Phys. Lett. 42B, 91 (1972).

10) B. Kayser et al., Phys. Lett. 52B, 385 (1974); R. L. Kingsley, F. Wilczek, and A. Zee, Phys. Rev. D10, 2216 (1974).
11) D. Buchholz, Proceedings Neutrino Conference (Aachen, June 1976).
12) P. Sutherland et al., Phys. Rev. D13, 2700 (1976).
13) C. Bouchiat, Phys. Lett. 57B, 284 (1975); E. A. Hinds et al., Phys. Lett. 62B, 97 (1976).
14) G. t'Hooft, Phys. Lett. 37B, 195 (1971).
15) J. D. Bjorken, Phys. Rev. 179, 1547 (1969).
16) S. Pakvasa and G. Rajasekaran, Phys. Rev. D12, 113 (1975); M. Gronau, Nuovo Cimento Lett. 14, 204 (1975).
17) F. J. Hasert et al., Phys. Rev. Lett. 46B, 121 (1973); J. Blietschau et al., CERN/EP/PHYS 76-42 (submitted to Nucl. Phys. B).
18) F. Bobisut, Proceedings Neutrino Conference (Aachen, June 1976).
19) A plot of this kind was first proposed in H. H. Chen and G. W. Lee, Phys. Rev. D5, 1874 (1974).
20) F. Reines, H. S. Gurr, and H. W. Sobel, Phys. Rev. Lett. 37, 315 (1976).
21) When the electron mass is not ignored, the variable y in the formula for $d\sigma/dy$ should be understood as E_e/E where E_e is the final electron energy in the laboratory system.
22) Sections 4 - 6 are largely based on an invited talk I gave at the Aachen Conference; J. J. Sakurai, Proceedings Neutrino Conference (Aachen, June 1976).
23) For example, the constant γ in this paper is related to the constant v_s of Reference 4) by $\gamma = \frac{2}{3} v_s$.
24. S. L. Glashow, J. Iliopoulos, and L. Maiani, Phys. Rev. D2, 1285 (1970).
25) M. A. Bég and A. Zee, Phys. Rev. Lett. 30, 675 (1973); V. S. Mathur, S. Okubo, and J. E. Kim, Phys. Rev. D11, 1059 (1975).
26) A. De Rújula, H. Georgi, and S. L. Glashow, Phys. Rev. Lett. 35, 65 (1975); H. Fritzsch, M. Gell-Mann, and P. Minkowski, Phys. Lett. 59B, 256 (1975).
27) Y. Achiman, K. Koller, and T. F. Walsh, Phys. Lett. 59B, 261 (1975).
28) F. Gürsey and P. Sikivie, Phys. Rev. Lett. 36, 775 (1976).
29) S. L. Adler and S. F. Tuan, Phys. Rev. D11, 129 (1975).
30) The importance of U spin invariance in understanding the absence of the strangeness-changing neutral currents was emphasized in: J. J. Sakurai, Phys. Rev. D9, 250 (1974); V. S. Mathur, S. Okubo, and J. E. Kim, Phys. Rev. D10, 3648 (1974).
31) Pham Quang Hung and J. J. Sakurai, Phys. Lett. 63B, 295 (1976).
32) Special cases of these equations have been derived by a number of authors: E. A. Paschos and L. Wolfenstein, Phys. Rev. D7, 91 (1973); G. Rajasekaran and K. V. L. Sarma, Pramana 2, 62 (1974); A. De Rújula et al., Rev. Mod. Phys. 46, 391 (1974).
33) R. P. Feynman, "Photon Hadron Interactions" (W. A. Benjamin, New York, 1972), p. 132; J. D. Bjorken and E. A. Paschos, Phys. Rev. D1, 3151 (1970).

34) C. Chang et al., Phys. Rev. Lett. 35, 901 (1975);
H. L. Anderson et al., Phys. Rev. Lett. 37, 4 (1976).
35) A. Benvenuti et al., Phys. Rev. Lett. 36, 1478 (1976);
A. Benvenuti et al., Phys. Rev. Lett. 37, 189 (1976).
36) H. J. Lipkin, Phys. Reports 8, 175 (1973).
37) W. B. Atwood, SLAC Report 185 (1975).
38) After Fig. 5.2 [taken from Reference 31)] was prepared, we have learned that the EM and vectorlike models considered in the figure are ruled out by recent data. However, variants of the models where we add axial-vector currents with the same isovector-isoscalar ratios ($\beta/\delta = \alpha/\gamma$) give the same predictions for neutron-to-proton ratios, and they are not yet ruled out by the data.
39) W. Van Doninck, Proceedings Neutrino Conference (Aachen, June 1976).
40) T. W. Donnelly et al., Phys. Lett. 49B, 8 (1974).
41) H. S. Gurr, F. Reines, and H. W. Sobel, Phys. Rev. Lett. 33, 179 (1974).
42) D. Z. Freedman, Phys. Rev. D9, 1389 (1974).
43) The constant a_o in Reference 42) is related to our γ by $a_o = \frac{3}{2}\gamma$.
44) J. R. Wilson, Phys. Rev. Lett. 32, 849 (1974); D. N. Schramm and W. D. Arnett, Phys. Rev. Lett. 34, 113 (1975).
45) M. K. Gaillard, S. Jackson, and D. Nanopoulos, Nucl. Phys. B102, 326 (1976).
46) Equation (6.10) is a generalization of a relation first obtained in: S. Weinberg, Phys. Rev. 5, 1412 (1972).
47) S. L. Adler et al., Phys. Rev. D11, 3309 (1975);
S. L. Adler, Phys. Rev. D12, 2644 (1975).
48) A somewhat smaller value of $g_A^{(s)}$ is obtained if SU(3) is used together with the D/F ratio inferred from hyperon decay.
49) D. Cline et al., Phys. Rev. Lett. 37, 252 (1976).
50) W. Lee et al., Phys. Rev. Lett. 37, 186 (1976).
51) C. H. Albright et al., FERMILAB-PUB-76/45-TH.
52) D. Cline et al., Phys. Rev. Lett. 37, 648 (1976).
53) J. J. Sakurai, Phys. Rev. Lett. 35, 1037 (1975).
54) E. D. Bloom and F. J. Gilman, Phys. Rev. Lett. 25, 1140 (1970).
55) S. L. Adler et al., Phys. Rev. D13, 1216 (1976).
56) E. A. Paschos, Proceedings 10th Rencontre de Moriond, ed. Tran Thanh Van (CNRS, Paris), p. 341.
57) C. H. Bertrand Coremans et al., Phys. Lett. 61B, 207 (1976).
58) C. H. Albright and J. Cleymans, Nucl. Phys. B76, 48 (1974);
L. M. Sehgal, Nucl. Phys. B90, 471 (1975).
59) C. H. Llewellyn Smith and D. V. Nanopoulos, Nucl. Phys. B78, 205 (1974).
60) The spirit of my approach to this subject closely parallels that of L. Wolfenstein [AIP Proceedings No. 23, Division of Particles and Fields, Williamsburg, Sept. 1974 (ed. C. E. Carlson), p. 84] who discussed the reaction (7.2) without reference to particular models.

61) N. Cabibbo and R. Gatto, Phys. Rev. 124, 1577 (1961); A. Love, Nuovo Cimento Lett. 5, 113 (1972); J. Godine and A. Hankey, Phys. Rev. D6, 3301 (1972); V. K. Cung, A. K. Mann, and E. A. Paschos, Phys. Lett. 41B, 355 (1972); R. Bundy, Phys. Lett. 45B, 340 (1973).

62) Equation (8.12) of my DESY paper [Reference 4)] is wrong.

63) M. Gourdin, Proceedings Neutrino Conference (Aachen, June 1976).

64) Note from (7.9) and (7.11) that at low values of s the asymmetry prediction in the Salam-Weinberg model is independent of θ_W.

65) M. A. Bouchiat and C. C. Bouchiat, Phys. Lett. 46B, 111 (1974).

66) P. C. Soreide et al., Phys. Rev. Lett. 36, 352 (1976).

67) P. G. H. Sandars, "Atomic Physics" 4, ed. G. Zu Putlitz, E. W. Weber, and A. Winnacker (Plenum, New York, 1975).

68) E. M. Henley, Proceedings 9th International Conference on the Physics of Electronic and Atomic Collisions, Seattle, Washington (1975); E. M. Henley and L. Wilets, Phys. Rev. A (to be published).

69) Some examples of this can be found in a recent paper by J. Bernabéu and C. Jarlskog, CERN TH-2206.

DISCUSSIONS

CHAIRMAN: Prof. J.J. Sakurai

Scientific Secretaries: H. Pham Quang and M.R. Pauli

DISCUSSION 1

- *MARCIANO:*

In your local leptonic Lagrangian, you assumed μe universality for the neutral currents. Is there any experimental evidence that this is indeed the case?

- *SAKURAI:*

At present, there is no experimental evidence for μe universality in the neutral currents. Even in the charged current inclusive reactions, μe universality is only crudely tested.

- *MARCIANO:*

Will the neutrino beam experiments allow us to examine interesting questions about the nature of the neutrinos, i.e. $\nu_e \nu_\mu$ mixing, whether they are massless, and, if not, what is the decay rate for, say, $\nu_\mu \to \nu_e + \gamma$?

- *SAKURAI:*

Starting with a beam of neutrinos, from a reactor for example, we can examine the distance dependence of the cross-section and infer various properties from the data. The neutrino from a reactor is $\bar{\nu}_e$ to start with, but if there is a finite mass difference between the muon neutrino and the electron neutrino, the beam may acquire a $\bar{\nu}_\mu$ component. Thus, there might be a reduction in the cross-section for inverse beta decay.

- *ZICHICHI:*

In this context, I must tell you something which has already been brought up on several occasions concerning the neutral currents generated by the muon neutrino. One has an incoming muon neutrino and no muon in the final state. The question is what would happen if an electron were found in the final state? In this case, it would not be the neutral currents, but it would be a violation of the leptonic selection rule. How can one exclude this? It would be very poor to use the energy distribution of the calorimater.

- *SAKURAI:*

Comparison of the visible energies in the final state shows that in the neutral current case, there is a sizeable fraction of the total energy missing.

- *BUCCELLA:*

What is the evidence of the neutral leptonic currents?

- *SAKURAI:*

The Gargamelle evidence is based on three events for $\bar{\nu}_\mu e \to \bar{\nu}_\mu e$; however, each of the events has some small (0.5%) probability for being due to $\nu_e + n \to e^- + p$ where the proton is not seen. The Aachen-Padua evidence, based on a much larger sample of candidates, is considered to be a 2.5 standard deviation effect for both $\nu_\mu e$ and $\bar{\nu}_\mu e$. However, since this is a spark chamber experiment, the background due to single π^0 production is serious.

- *KLEINERT:*

Can the experts explain what made the elastic νe and $\bar{\nu} e$ cross-sections drop by a factor of two?

- *SAKURAI:*

I believe you are referring to the data of the Aachen-Padua Collaboration. As I understand the situation, the data itself has not changed since the Aachen Conference, but if we use a subsample of data with a large E_e to reduce the neutral pion background, the cross-section becomes smaller, perhaps more in agreement with the Gargamelle data.

- *JONES:*

Could you explain the significance of the various ellipses on the g_V/g_A allowed region plot?

- *SAKURAI:*

(See lecture notes for the ellipse diagram.)

The outer ellipse denoted by $\bar{\nu}_\mu e$ is the 90% confidence level upper limit for $\nu_\mu e \rightarrow \nu_\mu e$; the inner ellipse is the 90% confidence level lower limit. The ellipse denoted by $\nu_\mu e$ is the 90% confidence level upper limit for $\nu_\mu e \rightarrow \nu_\mu e$.

- *SOHNIUS:*

How good is the evidence that no S or P contribution is present in neutral currents?

- *SAKURAI:*

It is very good for the inclusive hadron reactions, provided that T is not simultaneously present, but it is very poor for neutrino-electron scattering.

- *SOHNIUS:*

Are possible contributions included in your ellipse graph?

- *SAKURAI:*

No.

- *SOHNIUS:*

How would the ellipses change using the new data presented here earlier by Baldo-Ceolin?

- *SAKURAI:*

The ellipse of $\bar{\nu}_\mu e$ would become larger. In terms of the Weinberg angle, the Gargamelle ellipse corresponds to $0.1 < \sin^2 \theta_W < 0.4$ -- see the diagram in the lecture notes -- while the Aachen-Padua ellipse, as presented at the time of the Aachen Conference, corresponds to $0.4 < \sin^2 \theta_W < 0.7$. However, I understand from Baldo-Ceolin that the Aachen ellipse becomes smaller when they select events with high electron energies.

- *GOURDIN:*

Let me make a general comment concerning the SPT situation. As it is well known, both in the leptonic and hadronic cases, we have the so-called confusion theorem which tells us that to a (V,A) solution corresponds an (S,P,T) solution and an infinity of (S,P,T,V,A) solutions.

Therefore, we must add some information to eliminate (S,P,T) components. In the leptonic case, we need to study the low-energy correction and to study the charged current-neutral current interference in electron neutrino-electron scattering by comparing with muon neutrino-electron scattering. In the hadronic case, we need to study a very complicated transfer of polarization measurements at the hadronic vertex, for instance, in neutrino-proton scattering.

- *SAKURAI:*

That is quite correct.

DISCUSSION 2

- *GOURDIN:*

You showed a graph for $R^{(\nu)}$ as a function of energy where Gargamelle data and HPWF results are in agreement. Do you have an analogous graph for $R^{(\bar{\nu})}$, because if charged current data coming from HPWF are right, we must have an increase of charged current antineutrino cross-sections, and therefore a decrease of $R^{(\bar{\nu})}$ with energy?

- *SAKURAI:*

The mean energy in the HPWF experiment is 40 GeV for the antineutrino. This is in the energy range where the famous "anomalies" in the charged current antineutrino data start showing up. I believe that the ratio $\sigma(\bar{\nu})/\sigma(\nu)$ for the charged current reactions is about 0.45. You are, of course, quite right in saying that $R^{(\bar{\nu})}$ should decrease if the neutral current data does not exhibit anomalies.

- *GOURDIN:*

Your plot in the $\alpha\beta$ plane with a pure isovector neutral current can be used in a more complex way, independently of any valence quark model. This was dine in my 1975 Schladming lectures and it is possible to define, by positivity, a domain of allowed values for α and β. The limit of that domain obviously corresponds to a pure isovector neutral current.

- *SAKURAI:*

Yes, other plots are possible to exhibit the constraints imposed by the data on the various coupling constants. For pedagogical reasons, I have shown the simplest plot.

- *MARCIANO:*

Concerning the mass of the neutral Z^0 boson, we are all influenced by the Salam-Weinberg model into thinking that the mass of the Z^0 must be very large. However, merely from experiment, what bound can we put on the mass?

- *SAKURAI:*

One can probably infer from the linear rise of the cross-section as a function of E -- assuming scaling -- that the mass of the Z^0 is rather large. The data still has large errors but we can say that at least the mass of the Z^0 is greater than 10 GeV. We can also rule out a low mass Z^0 boson using the fact that the QED calculations for $e^+ + e^- \to \mu^+ + \mu^-$ fit the data at SPEAR energies.

- *MARCIANO:*

Assuming a coupling constant of the order of the electric charge, what does the mass of the Z^0 seem to be?

- *SAKURAI:*

From purely dimensional arguments, the assumption that the dimensionless coupling constant is of the order of e would infer a very large mass for the Z^0. Remember that Fermi's G has dimensions of M^{-2}; the squared dimensionless coupling constant is of the order of 10^{-5} -- where the reference mass is identified with the proton mass. In any model where the dimensionless squared coupling constant is as large as e^2, the reference mass must be as large as the mass of a heavy nucleus. All this is completely independent of whether the $SU(2) \times U(1)$ gauge model is right.

- *PAULI:*

You had written down data for neutrino and antineutrino in inclusive reactions and obtained 90% V-A and 10% V+A. How does this agree with the data from HPWF $\sigma(\bar{\nu}p \to \bar{\nu}p)/\sigma(\nu p \to \nu p) = 0.4 \pm 0.2$ and the data from CERN $\sigma(\nu p \to \nu\pi p)/\sigma(\nu p \to \nu\pi p) = 0.49 \pm 0.12$ for exclusive reactions? Are the experiments that are being done to determine pure V versus pure A versus V+A versus V-A consistent with each other?

- *SAKURAI:*

The inclusive reactions and the exclusive reactions agree to the extent that they both indicate the presence of sizable V,A interference. However, to extract quantitatively the relative amounts of V and A from the exclusive reactions, more work is needed. It is very important to note, in this connection, that the ratio

$\sigma(\bar{\nu}p \to \bar{\nu}p)/\sigma(\nu p \to \nu p)$ is dependent on the beam energy. On general considerations, it is possible to show that this ratio must approach unity for $E_\nu >> m_p$, even if both V and A are present. A similar remark applies to single pion production.

- *PUGLIERIN:*

For $(d\sigma/dy) = 0.9 \pm 0.1\ (1-y)^2$, the values of 0.9 and 0.1 are derived from the R value in the HPWF experiment. For a direct measurement, it is probably better to refer to the Caltech data since they have a narrow-band beam. Do you know if there are some new data?

- *SAKURAI:*

It is true that there is a considerable advantage of using a narrow-band beam if we are to measure the y distribution. In fact, for a strictly monoenergetic beam, the y distribution is essentially the same as the E_{had} distribution, where E_{had} is the total hadronic energy deposited in the calorimeter. There were some new E_{had} plots by the Caltech Group presented at the Aachen-Conference. The coupling constants deduced implicitly depend on the E_{had} distributions as well as the neutral current/charged current ratios with certain energy cuts.

WEAK CURRENTS AND NEW QUARKS

Michel GOURDIN

Laboratoire de Physique Théorique et Hautes Energies
Tour 16 - 1er étage - 4, place Jussieu
75230 PARIS cedex 05

The aim of these lectures is to give a review of the situation concerning neutrino and antineutrino inelastic scattering and the implication of recent high energy data concerning the number and the properties of quarks.

The obvious reference for a comparison is the set of experiments performed at Gargamelle which together with SLAC electroproduction experiments can receive a simple interpretation in terms of a quark parton model. We therefore first study low energy data using the two component quark parton model and we show how well it fits the data by making predictions which are in agreement with experiments.

Anomalies with respect to the low energy description have been claimed to appear at Fermilab energies but the experimental situation is still unclear. Nevertheless, we discuss two possible schemes for weak currents and we show how various data can be explained. The first class of models use only left handed charged current and it is the natural extension of the low energy situation where the Cabibbo current describes nicely experimental data. Variation between low energy and high energy of the quark parton model parameters are discussed. The second class of models involves new types of currents with right handed helicity and consequently new quarks with high masses. It follows that the quark parton model formalism has to be modified in order to take into account threshold effects for heavy quark production. We discuss this point and the slow rescaling in the Bjorken variable x implies a continuous variation with energy of these new contributions.

Finally in part III, we look at the implications for neutral currents of the previous considerations using the simplest $SU(2) \otimes U(1)$ gauge group. The mixing angle θ_W is then determined after comparison with low and high energy data. Again consistency with experiment is found.

PART I

GARGAMELLE DATA

In this part we briefly review the main features of neutrino and antineutrino inclusive reactions as observed in the Gargamelle experiment made at CERN. The electromagnetic counterpart is the set of inclusive electroproduction experiments performed at SLAC and DESY. They will not be discussed here and we only assume they are well known.

I - BJORKEN CANONICAL SCALING

1°) High Energy Cross-Sections

High energy incident beams in the laboratory frame mean

$$E \gg M$$

In that case the fixed x, y double differential cross-sections have a very simple form. As a consequence of spin one exchange in the t channel they are described by three Lorentz invariant structure functions $F_\lambda(x, q^2)$ where $\lambda = \pm 1, 0$ is the helicity of the weak current. We simply obtain *

$$\frac{d^2\sigma^{\nu}}{dx\,dy} = \frac{G^2 M E}{\pi}\, x \left\{ F_-^{\nu}(x,q^2) + (1-y)^2 F_+^{\nu}(x,q^2) + 2(1-y) F_0^{\nu}(x,q^2) \right\}$$
$$\frac{d^2\sigma^{\bar\nu}}{dx\,dy} = \frac{G^2 M E}{\pi}\, x \left\{ F_+^{\bar\nu}(x,q^2) + (1-y)^2 F_-^{\bar\nu}(x,q^2) + 2(1-y) F_0^{\bar\nu}(x,q^2) \right\} \quad (1)$$

As usual G is the Fermi coupling constant as measured in μ decay. Experimentally, effects due to an intermediate boson propagator have not been observed at present energies and we shall work in the Fermi local approximation.

2°) Bjorken Canonical Scaling

With this choice of structure functions, Bjorken scaling is a very simple limiting statement at large q^2

---

* The structure functions usually used, F_1, F_2, F_3 are related to F_λ's by

$$F_1 = \frac{1}{2}(F_+ + F_-) \qquad F_2 = x\left[F_+ + F_- + 2F_0\right] \qquad F_3 = F_+ - F_-$$

$$\lim_{\substack{q^2 \Rightarrow \infty \\ x \text{ fixed}}} \overline{F}_\lambda(x, q^2) = \overline{F}_\lambda(x) \qquad (2)$$

Such a property has been observed at SLAC for electroproduction and it is expected to hold for weak processes in the same range of values for q^2 and W : $q^2 > 1$ GeV^2 $W > 2.6$ GeV.

Moreover small breakdown of exact scaling seem to be present both at SLAC * and Fermilab energies. They have a natural explanation in the framework of asymptotically free gauge theories. We shall not discuss this problem in that part, being essentially interested to the dominant features of the data.

We then assume exact scaling in what follows and the structure functions F_λ being q^2 independent, the y and E dependence of the double differential cross-sections (1) is entirely known and the two important features are

i) the cross sections rise linearly with E ;

ii) the cross sections are second order polynomials in y .

Therefore all the dynamics is concentrated in the x dependence of the scaling functions F_λ (x).

3°) Energy Distributions

The fixed y distributions are obtained integrating over x the double differential cross-sections. Such an integration defines the first moments I_λ of the scaling functions

$$I_\lambda = \int_0^1 x \overline{F}_\lambda(x)\, dx$$

and we get

$$\frac{d\sigma^\nu}{dy} = \frac{G^2 ME}{\pi} \left\{ I_-^\nu + (1-y)^2 I_+^\nu + 2(1-y) I_0^\nu \right\} \qquad (3)$$

$$\frac{d\sigma^{\bar\nu}}{dy} = \frac{G^2 ME}{\pi} \left\{ I_+^{\bar\nu} + (1-y)^2 I_-^{\bar\nu} + 2(1-y) I_0^{\bar\nu} \right\}$$

Properties i) and ii) obviously hold for the energy distributions (3)

4°) Total Cross-Sections

Integrating now over y the energy distributions, we obtain the

---

* At SLAC energies, it is not completely clear what type of scaling variable : x,x', etc and of scaling function we must use and quantitative statements depend crucially of these choices.

total cross-sections

$$\sigma = \frac{G^2 M E}{\pi} A \qquad (4)$$

where the slope parameters A^{ν} and $A^{\bar\nu}$ are linear combinations of the first moments I_λ given by

$$A^{\nu} = I_-^{\nu} + \frac{1}{3} I_+^{\nu} + I_0^{\nu}$$
$$A^{\bar\nu} = I_+^{\bar\nu} + \frac{1}{3} I_-^{\bar\nu} + I_0^{\bar\nu} \qquad (5)$$

Again, property i) of linear rising with incident energy emerges as a consequence of scaling.

5°) Distributions in x

The x distribution are obtained integrating the double differential cross-sections over y. They contain the complete information over the dynamics but they are relatively difficult to measure with accuracy at low energy ($E \simeq 1 - 10$ GeV) because of poor statistics.

6°) Constituents

It is convenient to introduce two constant parameters A and B measuring the relative importance of the three helicity contributions. The usual definition is

$$A = \frac{I_- + I_+}{I_- + I_+ + 2I_0} \qquad B = \frac{I_- - I_+}{I_- + I_+ + 2I_0} \qquad (6)$$

and we immediately deduce the positivity constraints

$$|B| \leq A < 1$$

The energy distributions (3) can be equivalently written in the form *

$$\frac{d\sigma}{dy} = \frac{G^2 M E}{\pi} \frac{I_- + I_+ + 2I_0}{2} \left\{ (1 \pm B) + (1-y)^2 (1 \mp B) - y^2 (1-A) \right\}$$

From electroproduction experiments at SLAC we know that the longitudinal contribution is small as compared to the transverse one. It means, in this language, that A is close to unity. In fact, A = 1 is the integrated form of the Callan-Gross relation.

---

* The upper sign refers to ν reactions and the lower sign to $\bar\nu$ reactions.

In what follows, we shall neglect the longitudinal part for weak processes replacing in equation (7) A by unity. The only remaining parameter B describes the shape of the y distribution.

II - GARGAMELLE DATA

The experiments were made using neutrino and antineutrino beams with energy E in the range 1 - 10 GeV. The Gargamelle bubble chamber was filled with freon and the main results are the following.

1°) Total Cross-Sections

They are compatible with a linear rising with energy

$$\sigma^{\nu} = (0,81 \pm 0,03)\ E\ 10^{-38}\ cm^2/GeV$$
$$\sigma^{\bar{\nu}} = (0,28 \pm 0,01)\ E\ 10^{-38}\ cm^2/GeV$$

and their ratio $r_c = \sigma^{\bar{\nu}}/\sigma^{\nu}$ remains constant with energy

$$r_c = 0,35 \pm 0,03$$

2°) Energy Distributions

For neutrino the y distribution is essentially constant and for antineutrino it behaves like $(1 - y)^2$ to a good approximation. Using equation (7) with A = 1 we immediately conclude that the two parameters B and $\bar{B}$ must be close to unity. To a first approximation we have the simple relation

$$B \simeq \bar{B} \simeq 2\ \frac{1-r_c}{1+r_c} \simeq 0,96 \pm 0,06 \qquad (8)$$

From equations (6) it follows that the contribution due to helicity $\lambda = -1$ strongly dominate over the contribution to helicity $\lambda = +1$.

3°) Conclusion

The data are consistent with the consequences of Bjorken scaling but we do not have a direct and unambiguous proof of scaling from these experiments because of the poor statistics at large values of q^2 and W^2.

III - TWO COMPONENT QUARK PARTON MODEL

How electroproduction, neutrino and antineutrino data at these energies can be interpreted with a parton language is now folklore. We only wish to show in this section how a particle quark parton model can reproduce the existing data and make predictions.

1°) Basic Ingredients

For a quark parton model we need two types of ingredients. Firstly, the distribution functions for each type of quark and antiquark in the hadron. We call it D_j (x) taking in mind that these quantities are probability distributions and therefore they are positive. Secondly, we have point-like coupling constants between the current and the quarks or antiquarks. These coupling constants depend on the internal symmetry group used for strong interaction via the definition of currents.

2°) Two-Component Model

In the so-called two component model, we have two types of quarks.

a) the valence quarks, u, d, giving rise to non-diffractive distribution V_u (x) and V_d (x). The integral over x of these distributions is just the number of valence quark in the nucleon and it is therefore finite.

b) the quark antiquark pairs building an isoscalar sea, $u\bar{u}$ + $d\bar{d}$, $s\bar{s}$, $c\bar{c}$, etc. The integral of these diffractive distributions $S_q(x)$ can be infinite if the number of pairs in the sea is infinite. Such a quantity is related in the quark parton model to the limit at x = 0 of the scaling function $F_2(x)$. Recent electroproduction data at Fermilab indicate a continuous decreasing of F_2 (x) when $\omega = 1/x$ increases. The limit at x = 0 requires at infinite incident energy but in the present situation the extrapolation of existing data suggests the possibility for F_2 (0) to be zero or something finite. Therefore such a question remains open.

3°) SU(3) Quark Parton Model

In this range of energies, it is reasonable to work with a parton model based on SU(3) symmetry. We have three basic quarks, u, d, s and the weak charged current is the usual Cabibbo left-handed current.

The first moments d_j of the quark distributions $D_j(x)$ being defined by

$$d_j = \int_0^1 x\, D_j(x)\, dx$$

We have only four independent parameters in the two component quark parton model.

a) Valence parameters V_u, V_d

b) Sea parameters $S_u = S_d$, S_s

For simplicity we shall choose a SU(3) symmetric sea $S_u = S_d = S_s = S$ and we get the following expressions for the quantities I_λ which govern the energy distributions and the total cross-sections

Proton target

$$I_+^{\nu p} = S \qquad I_-^{\nu p} = S + 2 V_d \cos^2\theta_c$$

$$I_+^{\bar{\nu} p} = S \qquad I_-^{\bar{\nu} p} = S + 2 V_u$$

Neutron target

$$I_+^{\nu n} = S \qquad I_-^{\nu n} = S + 2 V_u \cos^2\theta_c$$

$$I_+^{\bar{\nu} n} = S \qquad I_-^{\bar{\nu} n} = S + 2 V_d$$

IV - ANALYSIS OF GARGAMELLE DATA

The small number of free parameter of the simple two-component quark parton model allows us to take into account in a consistent way two small corrections.

i) the difference between the number of protons and neutrons in the freon $N_n / N_p = 1,19$;

ii) The $|\Delta S| = 1$ contributions of the Cabibbo current .

The input quantities are

i) the neutrino and antineutrino total cross-sections in freon σ^{ν} and $\sigma^{\bar{\nu}}$

ii) the difference between proton and neutron electroproduction scaling functions integrated over x

$$I^{ep} - I^{en} = \int_0^1 \left[F_2^{ep}(x) - F_2^{en}(x)\right] dx = 0,04 \pm 0,02$$

which fixes the isovector component of the valence parameters

$$V_u - V_d = 0,12 \pm 0,06$$

The output of the calculations are the value of the isoscalar component $V = V_u + V_d$ of the valence parameter

$$V = 0,53 \pm 0,05 \qquad (10)$$

and an inaccurate estimate of the sea parameter

$$S = 0,004 \pm 0,019 \tag{11}$$

Let us notice that the origin of the large error for S is closely -if not entirely- related to the large uncertainty entering in the electroproduction difference .

1°) Main Features

Let us first remark that the model is consistent with neutrino and antineutrino experiments, the non trivial positivity constraint being satisfied $S \geqslant 0$. Moreover in this framework the dominance of the helicity $\lambda = -1$ contribution implies the dominance of the valence quark contribution and in the one standard deviation limit the sea amount S/V is less than 4,4%.

The shape parameter for the y distributions is easily computed and its value

$$B = \bar{B} = 0,98 \pm 0,07 \tag{12}$$

is close to the estimate made directly from the data neglecting small corrections.

Finally, it is well known that non interacting partons called gluons are needed in this class of model in order to insure the conservation of energy momentum *. From

$$\varepsilon = 1 - \sum_j' d_j$$

a large amount of gluons is needed

$$\varepsilon = 0,46 \pm 0,02 \tag{13}$$

2°) Predictions

We are now able to make predictions for other quantities a part of them being already measured.

i) Isoscalar electroproduction integral

$$\underline{I}^{ep} + \underline{I}^{en} = \int_0^1 \left[\bar{F}_2^{ep}(x) + \bar{F}_2^{en}(x) \right] dx = 0,298 \pm 0,011 \tag{14}$$

The value measured at SLAC : 0,30 ± 0,02 is remarkably close to the computed value and we see that in addition electroproduction data are compatible with this simple model.

---

* We do not discuss the dynamical role of gluons in this paper.

ii) Neutrino and antineutrino total cross-sections on proton and neutron separately. The predictions are

$$\frac{\sigma^{\nu n}}{\sigma^{\nu p}} \simeq \frac{\sigma^{\bar{\nu} p}}{\sigma^{\bar{\nu} n}} \simeq 1.6 \pm 0.4 \qquad (15)$$

Again the large error is to be attributed to the large uncertainty on proton-neutron difference in electroproduction.

A preliminary result from Argonne

$$\sigma^{\nu n}/\sigma^{\nu p} = 2{,}08 \pm 0{;}23$$

is in good agreement with the model prediction

iii) Strangeness violating cross-sections can also be evaluated and the result is

$$\frac{\sigma^{\nu}(\Delta S=1)}{\sigma^{\nu}_{tot}} < 0{,}4\% \qquad (16)$$

$$\frac{\sigma^{\bar{\nu}}(\Delta S=-1)}{\sigma^{\bar{\nu}}_{tot}} \simeq 5{,}3\%$$

No reliable measurement of these quantities has been obtained because of the difficulty of separating $|\Delta S| = 1$ events from associated production of strange particles. Nevertheless they have been observed in both neutrino and antineutrino scattering showing, in particular, that the sea amount is not strictly zero.

PART II

FERMILAB DATA

We now study the data obtained at Fermilab by various groups for charged current reactions. We can classify the experiments into two categories.

1 - Counter experiments;

2 - Bubble chamber experiments.

It is not yet absolutely clear what is happening at these energies. Anomalies in the y distribution for antineutrino scattering have been observed from a long time by the HPWF collaboration. On the other hand, all groups agree about the existence of dimuon events not coming from the decay of a heavy neutral lepton. But the precise magnitude of these new effects is not clear for a theorist and it is always very delicate to choose between various experiments.

On the theoretical side we can classify the various types of models into two categories

1 - without right handed currents: standard models;

2 - with right handed currents and new quarks.

We discuss successively the two possibilities using a simple two component quark parton model and we compare the result of the analysis with what we have obtained at Gargamelle energies in Part I. A lot of interesting experimental informations are missing, in particular absolute differential and total cross-sections. Nevertheless some conclusions can be reached but they strongly depend on what data we use.

I - LEFT HANDED CURRENT MODELS

1°) General Structure

We assume, in this section, that only left handed quark doublets are present and the charged current coupling constants are chosen in such a way that the associated neutral current is diagonal in the quark basis. (see Appendix).

It then follows for the parameters I_+ and I_-, a general structure independently of the precise number of quarks doublets. For an isoscalar target we get

$$I_+^{\nu N} = S_1 \qquad I_-^{\nu N} = V + S_2$$
$$I_+^{\bar{\nu} N} = S_2 \qquad I_-^{\bar{\nu} N} = V + S_1 \qquad (17)$$

where V is the isoscalar valence first moment and S_1 and S_2 two linear combinations of sea first moments given by

$$S_1 = \sum_{d \in Q^+} S_d \qquad S_2 = \sum_{k \in Q^-} S_k \qquad (18)$$

In these models, we implicitely assume that only quarks with electric charge +2/3 (set Q^+) and -1/3 (set Q^-) are present.

Let us consider as the simplest example the SU(4) case with the Glashow-Iliopoulos-Maiani mechanism. We have two doublets

$$\begin{vmatrix} u \\ d_c \end{vmatrix}_L \qquad \begin{vmatrix} c \\ S_c \end{vmatrix}_L$$

where the index c means here the usual Cabibbo rotation for d and S quarks. The expressions of S_1 and S_2 are

$$S_1 = S_u + S_c \qquad S_2 = S_d + S_c$$

with the relation $S_u = S_d$ implied by the isoscalar character of the sea.

In general, the above current is not charge symmetric. We only have the weaker relation

$$\bar{I}_+^{\nu N} + \bar{I}_-^{\nu N} = \bar{I}_+^{\bar{\nu} N} + \bar{I}_-^{\bar{\nu} N} = V + S_1 + S_2 \qquad (19)$$

and violations of charge symmetry occur via the sea excepted if the sea is SU(4) symmetric.

2°) Energy Distributions and Total Cross-Sections

The shape parameters B and $\bar{B}$ and the ratio of total cross-sections r_c are easily computed from the previous expressions of I_+ and I_-

$$B = \frac{V - S_1 + S_2}{V + S_1 + S_2} \qquad \bar{B} = \frac{V + S_1 - S_2}{V + S_1 + S_2} \qquad (20)$$

$$r_c = \frac{V + S_1 + 3S_2}{3V + S_1 + 3S_2}$$

It the amount of sea is small as it is at Gargamelle energies $S_j \ll V$ we get

$$B \simeq 1 - \frac{2S_1}{V} \qquad \bar{B} \simeq 1 - \frac{2S_2}{V}$$

$$r_c = \frac{1}{3} + \frac{2}{9}\,\frac{S_1 + 3S_2}{V}$$

and we obtain the following qualitative features

i) B and $\bar{B}$ are both close to unity;
ii) Charge symmetry violation is small;
iii) The ratio r_c is close to 1/3 .

Conversely, a dramatic change between low and high energy for one of the shape parameter B and $\bar{B}$ would imply a dramatic change for r_c and it can only be explained, in this framework, by a considerable increasing of the sea amount.

In fact in this class of models, where the first equality(19) is satisfied, the ratio r_c is related to the shape parameters B and B by *

---

* Such a relation has already been used in Part I to obtain a first estimate of the shape parameter B using only the ratio of total cross-section given by the Gargamelle experiment.

$$r_c = \frac{2-\bar{B}}{2+B} \tag{21}$$

3°) Harvard-Pennsylvania-Wisconsin-Fermilab Collaboration Results.

When the data for neutrino and antineutrino y distributions- integrated over x for $x < 0.6$- are represented with the shape parameters B and $\bar{B}$, they exhibit noticeable differences with the Gargamelle resutls.

$$\begin{aligned} B\ \text{exp} &= 0.75 \pm 0.15 \\ \bar{B}\ \text{exp} &= 0.55 \pm 0.13 \end{aligned} \tag{22}$$

the incident beam energy being larger than 50 GeV.

From the previous considerations, the only possible fit of these data, in the framework of left handed current models, implies important sea contributions.

Absolute cross-sections being not available, we can estimate only the ratio S/V. Neglecting new quarks contributions and using as previously an SU(3) symmetric sea

$$S_1 = S \qquad\qquad S_2 = 2S$$

a typical value for the ratio S/V is

$$S/V = 0.17$$

within 40%. The corresponding values of B and B

$$B = 0.77 \qquad\qquad \bar{B} = 0.56$$

fit very well the data and the corresponding ratio of total cross-section

$$r_c = 0.52$$

is also in agreement with HPWF results.

Fits of the antineutrino y distributions integrated over x in the two ranges $x < 0.15$ and $0.15 < x < 0.60$ are shown on Fig. 2 for three values of the ratio S/V : 0.10 - 0.17 - 0.24. The y distributions integrated over x with $x < 0.6$ are less sensitive to the value of this ratio as it is clear from Fig. 3 for antineutrino data and Fig 4 for neutrino data.

4°) The main feature of this analysis of HPWF data with left handed

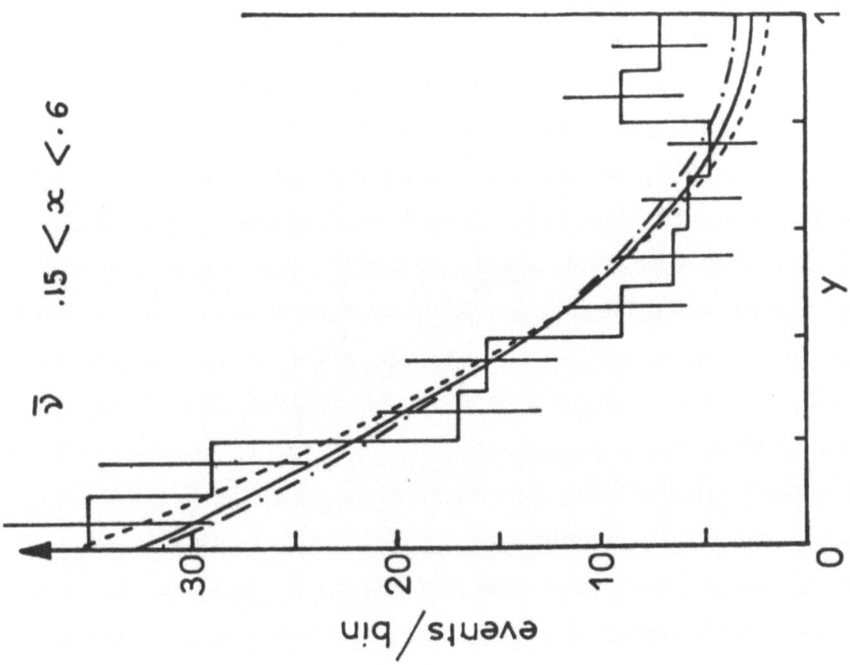

Figure 2b

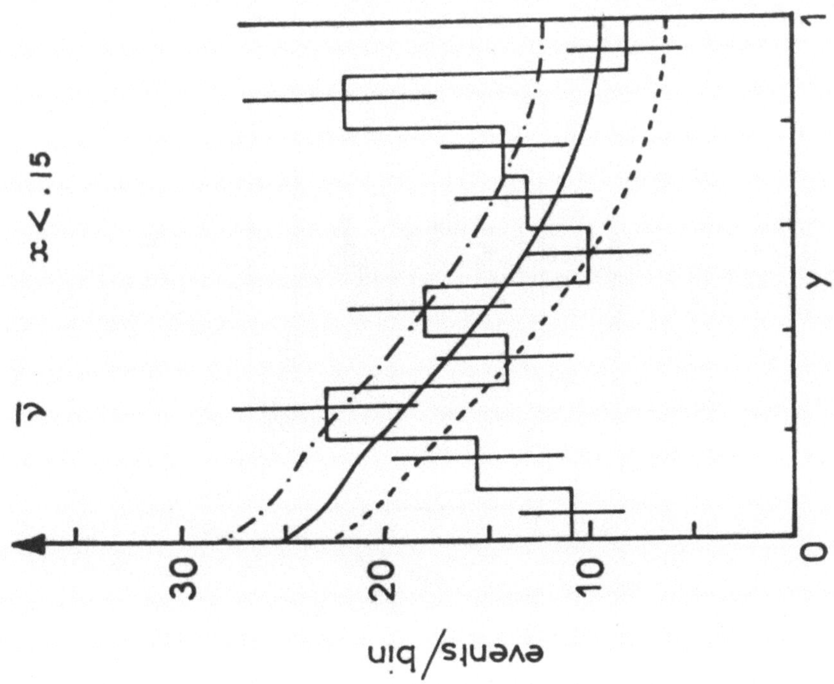

Figure 2a

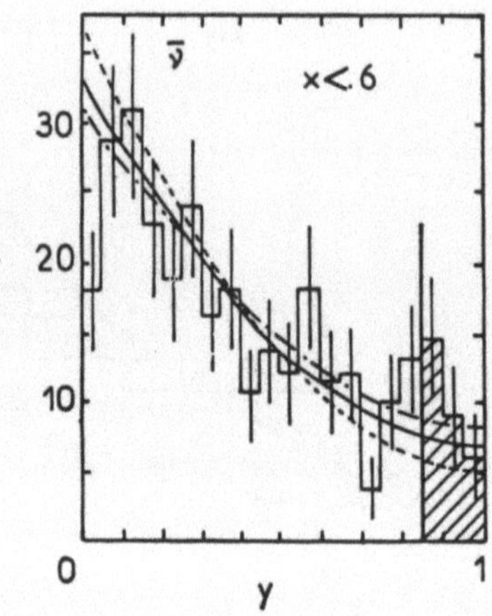

Figure 3

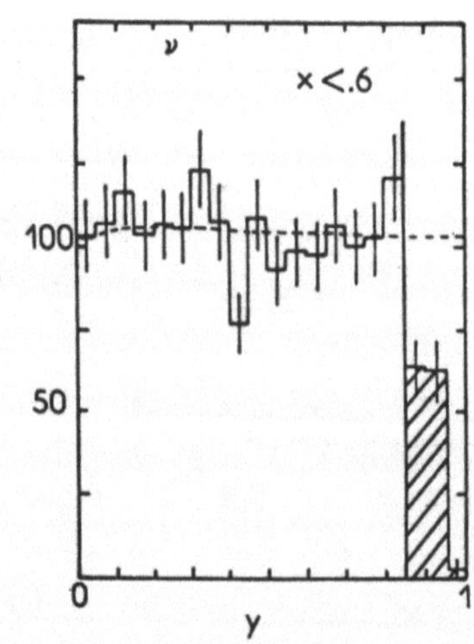

Figure 4

current model is a considerable increasing of the sea amount by at least a factor of 4 with respect to its value at Gargamelle energies.

It is obvious that such an important change associated to the nucleon structure will have observable consequences in electroproduction. In the same two component quark parton model, the isoscalar electroproduction integral is given by

$$I^{ep} + I^{en} = \frac{5}{9} V + \frac{8 S_1 + 2 S_2}{9} \tag{23}$$

or using the same sea as previously

$$I^{ep} + I^{en} = \frac{5}{9} V \left[1 + \frac{12}{5} \frac{S}{V} \right]$$

Using our previous estimate of S/V at Fermilab energies the bracket becomes 1.41 and it looks very unlikely to compensate such a large increasing by an equivalent decreasing of V. Therefore, a large effects would appear in μ inelastic scattering experiments performed at Fermilab. Up to now, there is no indication of such an effect.

On the theoretical side, the only consistent explanation of such a modification of the sea amount is a breakdown of canonical Bjorken scaling. In asymptotically free gauge theories such a breakdown exists with logarithmic factors of the form Log q^2/μ^2 where μ is the group renormalization unknown mass parameter. In this framework, the valence moment V decreases and the sea moment S increases when q^2 becomes larger and larger. However the magnitude of this effect is expected to be considerably smaller than what seems to be needed to accomodate HPWF data in their present form with left handed current model.

5°) Conclusions

i) The standard models with only left handed charged currents are not excluded by existing experiments.

ii) The counter experiments performed by the HPWF and CTF groups require a very large value of the sea over valence ratio S/V.

iii) The bubble chamber experiment E 180 does not need a large value of S/V and

$$\text{with } \frac{S}{V} = 0.06 \qquad \text{we get } \bar{B} = 0.8$$

which is the best fit of the data at all energies.

iv) The production of new quarks is always very small in these models. Consider, as an example, the production of charmed quarks. It can occur either on the sea quarks or antiquarks - for ν and $\bar{\nu}$ scattering - or on the d valence quark with a damping due to the Cabibbo factor $tg^2\theta_c$ - for ν scattering - and we have the constraint

$$\sigma^{\bar{\nu}}_{CH} / \sigma^{\nu}_{CH} < 1$$

where **CH** means charmed hadrons.

Therefore in order to account for dimuon experiments we need a large leptonic branching ratio for the decay of charmed particles which has to be between 10% and 15%.

II - MODELS WITH RIGHT HANDED CURRENTS

The terminology means that in addition to the previously studied left handed currents, we allow the possibility of having right handed currents.

If one believes the HPWF and CTF data, and if we refuse the considerable change in the sea amount found in the previous section as unphysical, we are obliged to introduce new currents with right handed helicity as an alternative way for analyzing the data. The sea amount will remain small if these new contributions involve valence quarks. We then can have two doublets of this type

$$\begin{vmatrix} t \\ d \end{vmatrix}_R \qquad \begin{vmatrix} u \\ b \end{vmatrix}_R$$

where t and b are new heavy quarks of respective charges +2/3 and -1/3. The first doublet will give valence contribution to neutrino scattering and the second doublet valences contributions to antineutrino scattering. A comparison of Gargamelle and HPWF results shows clearly that the sensible point is antineutrino scattering. Therefore, we begin the discussion by considering the second handed doublet.

1°) General Structure

The two parameters I_+ and I_- for antineutrino scattering off an isoscalar nucleon taget are now given by

$$I_+^{\bar{\nu}N} = W + S_2 \qquad\qquad I_-^{\bar{\nu}N} = V + S_1 \qquad (24)$$

where the u-quark valence moment W has to be equal to V if universality holds and if x scaling is valid.

The shape parameter $\bar{B}$ of the y distribution becomes

$$\bar{B} = \frac{V + S_1 - W - S_2}{V + S_1 + W + S_2}$$

and using the HPWF value $\bar{B} = 0.45 \pm 0.13$ we get

$$\frac{W + S_2}{V + S_1} = 0.38 \pm 0.14$$

We then get, in that way, an ad hoc explanation for the large violation of charge symmetry without the necessity to modify the sea amount by a large factor.

Clearly the W/V is far to unity and one can think to a breaking of universality. That is deeply unaesthetic and we must keep in mind that the new b quark we introduce in our doublet cannot be a light quark and its mass is expected larger than the charmed quark one. Therefore, thresholds effects are very important and they give the explanation why the ratio W/V obtained assuming x scaling is not unity even if all doublets enter in the current with the same coupling constant.

2°) Modified Quark Parton Model

In their analysis of operator product expansion near the light-cone for color gauge theories, Georgi and Politzer have shown that the correct scaling variable to use for the transition of a light quark to a heavy quark is not Bjorken variable x but the quantity ξ defined by

$$\xi = \frac{q^2 + m_f^2}{2M\nu} = x + \frac{m_f^2}{2MEy} \tag{25}$$

where m_f is the heavy quark mass. Of course, for light quark light quark transitions we put $m_f = 0$ and we recover the x variable.

The consequences of this observation have been studied by Barnett and many authors where we refer for details. We only give here the rule of the game. In units of G^2ME/π the quark model expressions for the contribution to the double differential cross-section of the transition

light quark j $\Rightarrow$ heavy quark k

is given by

ν reactions with V−A weak current, $\bar{\nu}$ reactions with V+A weak current

$$\xi_k\, q_j(\xi_k)\left[1 - \frac{m_k^2}{2ME\xi_k}\right] \tag{26a}$$

ν reactions with V+A weak current, $\bar{\nu}$ reactions with V−A weak current

$$\xi_k\, q_j(\xi_k)\left[(1-y)^2 + (1-y)\frac{m_k^2}{2ME\xi_k}\right] \tag{26b}$$

For antiquark - antiquark transitions the two expressions are exchanged.

The main effect of the change of scaling variable from x to ξ is a suppression of the production of heavy quarks off light quarks as compared to what is expected in the x scaling region. Kinematics is responsible of this fact

$$x \geq 0 \quad \Rightarrow \quad \xi_R \geq \frac{m_R^2}{2ME y} \qquad (27)$$

by the suppression of the low ξ region. Explicit calculations using realistic valence and sea quark distributions show that x rescaling is very slow and it is far to be achieved at Fermilab energies even for the production of charmed quark. We then understand better why the effective valence quark ratio W/V is only 0.3 at 100 GeV.

3°) Analysis of experiments

It is clear that the magnitude of the threshold effects depend crucially on the heavy quark masses. Fixing the charmed quark at m_c = 1.65 GeV, the b quark mass is considered as a free parameter one has to determine from experiment.

We show, in Fig. 5, a fit of HPWF y distributions integrated over x in the two ranges x < 0.15 and 0.15 <a< 0.60. The sea over valence ratio is fixed to 10% and the curves correspond to quark parton model predictions with m_b = 3 GeV, 5 GeV and 6 GeV. The lowest value m_b = 3 GeV seems to be excluded by the present data and acceptable fits of the data are obtained with m_b in the range 5 - 6 GeV.

An analysis of CTF data gives essentially the same result for m_b.

4°) Other doublets

The second right handed doublet involving the valence quark u and a Q = 2/3 charged quark t will give analogous effects in neutrino scattering. But in that case experimental data do not seem to require such a doublet and standard models with only left handed currents fit easily HPWF neutrino y distributions with S/V $\simeq$ 10%.

However, if one insits for the presence of such a doublet, the data imply a t quark mass as large as 5 or 6 GeV. This means that the hypothetical t quark cannot be identified with the charmed quark.

Ohter possible right handed doublets not involving valence quarks can be introduced. They will not destroy the agreement

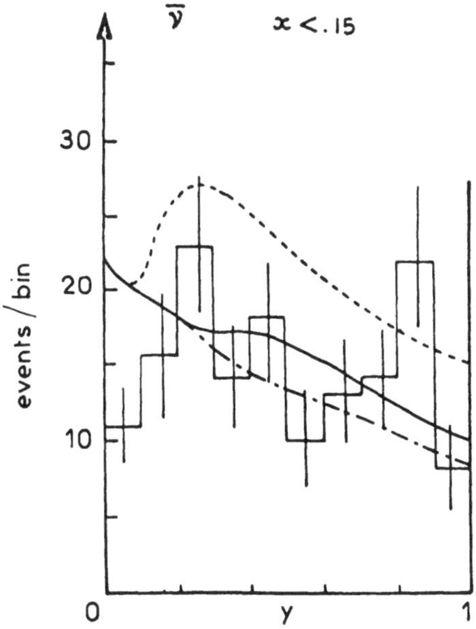

Figure 5b

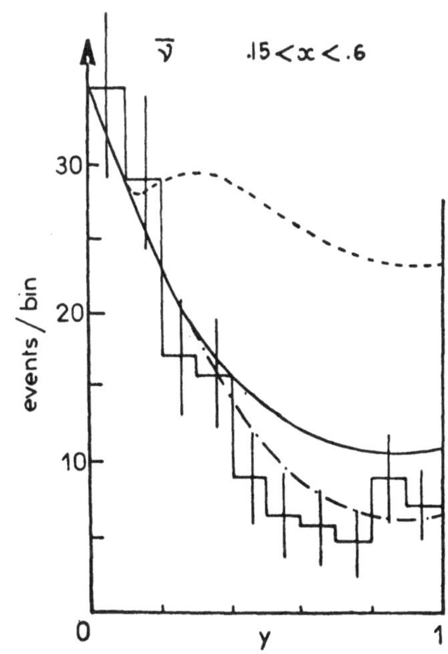

Figure 5a

between theory and experiment as long as the sea over valence ratio remains small. The simplest example of such a possibility is the $\begin{vmatrix} c \\ s \end{vmatrix}_R$ doublet.

From the point of view of anomalies we expect each new doublet of quark to be accompanied by a doublet of leptons of same helicity. There is some evidence for a new heavy lepton at SPEAR of mass around 2 GeV. It can be the leptonic mirror of the new b quark considered here.

In that case the observation of the charged lepton spectrum in the decay mode

$$L^- \Rightarrow \nu_L + \ell^- + \bar{\nu}_\ell$$

must exhibit the characteristics of a V + A interaction at the heavy lepton vertex.

5°) Conclusion

i) Reasonable fits of HPWF and CTF data are obtained using only one right handed doublet $\begin{vmatrix} u \\ b \end{vmatrix}_R$ in addition to the usual left handed one of the standard model. The mass of the new heavy quark b has to be as large as 5 - 6 GeV.

ii) Such fits use a sea over valence ratio of 10%, e.g. the variation of S/V between Gargamelle and Fermilab energies is considerably smaller than what is needed in the standard model. Moreover quantities like the ratio r_c of total cross-sections and the averaged value of y for antineutrino energy distributions exhibit in the 30 - 50 GeV region a variation with energy which cannot be reproduce by threshold effects only. If correct, this feature of the data implies an increasing with energy of the ratio S/V and now asymptotic freedom can account for.

iii) The introduction of the right handed doublet $\begin{vmatrix} u \\ b \end{vmatrix}_R$ provides a new source of dimuon events in antineutrino scattering. Such a contribution turns out to be important because the transition u $\Rightarrow$ b occurs off a valence quark without Cabibbo suppression factor. But there is a damping due to heavy quark mass effect which decreases when energy increases. We therefore expect more dimuon events induced by antineutrinos than by neutrino and the associated ratio will increase with energy.

iv) We now have two types of new hadrons, those containing c or $\bar{c}$ quarks and those containing b or $\bar{b}$ quarks. The branching ratio for the leptonic decay of these new hadrons is now estimated between 5% and 10% from the dimuon rates and it is compatible with the same value for both types of new hadrons.

v) The rescaling in the x variable is very slow and it is far to be reached at Fermilab energies even for c- quark production. Therefore, we expect the shape parameter $\bar{B}$ to continue to decrease and the averaged value $\bar{y}_{\bar{\nu}}$ to continue to increase when the incident energy increases. In a minimal model with two left handed and one right handed quark doublets, we get

$$\bar{B} \Rightarrow \frac{S_1 - S_2}{2V + S_1 + S_2} \qquad \bar{y}_{\bar{\nu}} = \frac{7 - 5\bar{B}}{8(2 - \bar{B})} \Rightarrow \frac{7}{16} - \frac{3}{64}\,\frac{S_1 - S_2}{V}$$

PART III

IMPLICATIONS FOR HADRONIC NEUTRAL CURRENTS

It seems reasonable to relate weak charged, weak neutral and electromagnetic currents in a unified scheme and gauge theories provide a nice framework to that purpose. The aim of this part is to make predictions for neutral current cross-sections using the results of the analysis made in Parts I and II.

We make the usual choice for the gauge group $SU(2) \otimes U(1)$ and we have the mixing angle θ_W as a free parameter. Therefore the neutral current cross-sections are functions of θ_W. A comparison with experiment determines an allowed range of values for θ_W and the consistency of the scheme is obtained from positivity constraints.

I - LEFT HANDED CURRENT MODELS

1°) In the quark representation, the neutral current is assumed to be diagonal. When the charged current is purely left handed we have the following structure

$$J^{NC}_{\mu} = i \sum_{j \in Q^+}' b^L_j \bar{q}_j \gamma_\mu \frac{1+\gamma_5}{2} q_j - i \sum_{k \in Q^-}' b^L_k \bar{q}_k \gamma_\mu \frac{1+\gamma_5}{2} q_k - 2x J^{EM}_{\mu}$$

wher $x = \sin^2\theta_W$. As previously Q^+ (Q^-) is the set of quarks with electric charge + 2/3 (- 1/3). The coupling constants b_j and b_k are 1 or 0 depending of the details of the model.

2°) It is now straightforward to compute the scaling moments I_+ and I_- entering in the y distributions and in the total cross-sections. Because of the hermeticity of the neutral current, neutrino and antineutrino parameters are simply related

$$\underline{I}_{\pm}^{\bar{\nu}} = \underline{I}_{\pm}^{\nu}$$

and we get the following results

<u>Proton target</u>

$$\underline{I}_{+}^{p} = \frac{2x^2}{9}(4V_u + V_d) + S_{NC}$$

$$\underline{I}_{-}^{p} = 2\left(\frac{1}{2} - \frac{2x}{3}\right)^2 V_u + 2\left(\frac{1}{2} - \frac{x}{3}\right)^2 V_d + S_{NC}$$

<u>Neutron target</u>

$$\underline{I}_{+}^{n} = \frac{2x^2}{9}(V_u + 4V_d) + S_{NC}$$

$$\underline{I}_{-}^{n} = 2\left(\frac{1}{2} - \frac{x}{3}\right)^2 V_u + 2\left(\frac{1}{2} - \frac{2x}{3}\right)^2 V_d + S_{NC}$$

where the sea contribution S_{NC} is given by

$$S_{NC} = \sum_{j \in Q^+} \left\{ \left(\frac{b_j^L}{2} - \frac{2x}{3}\right)^2 + \frac{4x^2}{9} \right\} S_j + \sum_{k \in Q^-} \left\{ \left(\frac{b_k^L}{2} - \frac{x}{3}\right)^2 + \frac{x^2}{9} \right\} S_k$$

3°) In order to analyze the Gargamelle data, we use the SU(4) symmetry group. In that case

$$j = u, c \qquad b_u^L = b_c^L = 1$$

$$k = d, s \qquad b_d^L = b_s^L = 1$$

The neutral current sea is simply a linear combination of the parameters S_1 and S_2 already introduced for charged currents.

$$S_{NC} = \left(\frac{1}{4} - \frac{2x}{3} + \frac{8x^2}{9}\right) S_1 + \left(\frac{1}{4} - \frac{x}{3} + \frac{2x^2}{9}\right) S_2$$

with

$$S_1 = S_u + S_c \qquad S_2 = S_d + S_s$$

Neglecting charmed quark and antiquark distributions in the nucleon and using as previously an SU(3) symmetric sea

$$S_u = S_d = S_s = S \qquad S_c = 0$$

we get

$$S_{NC} = \left(\frac{3}{4} - \frac{4x}{3} + \frac{4x^2}{3}\right) S$$

It is now very easy to compute neutral current total cross-sections in Freon and we obtain quadratic functions of x. But the usual procedure for presenting the data is to compare inclusive neutral current and charged current cross-sections.

$$R^{\nu} = \frac{\sigma^{\nu}_{NC}}{\sigma^{\nu}_{CC}} \qquad R^{\bar{\nu}} = \frac{\sigma^{\bar{\nu}}_{NC}}{\sigma^{\bar{\nu}}_{CC}}$$

Again these ratios R^{ν} and $R^{\bar{\nu}}$ are quadratic functions of x. They have been represented in Fig 6 and 7 using for the parameters V_u, V_d and S the values obtained in Part I from the quark parton model analysis of electroproduction and charged current data.

The most recent Gargamelle data are

$$R^{\nu} = 0.25 \pm 0.04$$

$$R^{\bar{\nu}} = 0.39 \pm 0.06$$

and we can easily deduce the values of x allowed by experiment.

for ν scattering $x = 0.32 \pm 0.12$
for $\bar{\nu}$ scattering $x = 0.08 - 0.66$

From these results we see that the quark parton model with SU(4) symmetry is compatible with neutral current experiments when the parameter x is chosen in the range

$$0.2 < \sin^2\theta_W < 0.4$$

4°) For high energy experiments we use a model of the type previously described and we restrict the calculations to an isoscalar target N. The slope parameters for the total cross-sections are of the form

$$A^{\nu N}_{NC} = \left(\frac{1}{2} - x + \frac{20}{27}x^2\right)V + \left(1 - \frac{16x}{9} + \frac{16x^2}{9}\right)S$$
$$A^{\bar{\nu} N}_{NC} = \left(\frac{1}{6} - \frac{x}{3} + \frac{20}{27}x^2\right)V + \left(1 - \frac{16x}{9} + \frac{16x^2}{9}\right)S \qquad (28)$$

We have computed the three ratios R^{ν}, $R^{\bar{\nu}}$ and r_N

$$r_N = \sigma^{\bar{\nu}}_{NC} / \sigma^{\nu}_{NC}$$

for x = 0.3 x = 0.35 and x = 0.40 which are values in the range suggested by low energy data. The three ratios depend only on the sea amount S/V and we consider the high value S/V = 0.17 implied by HPWF and CTF experiments and the value S/V = 0.06 which fits the E 180 bubble chamber experiment. The results are given in Table 1

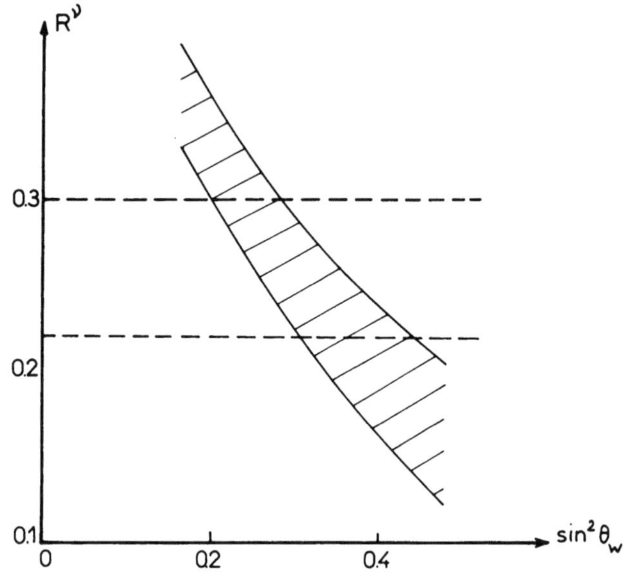

Figure 6

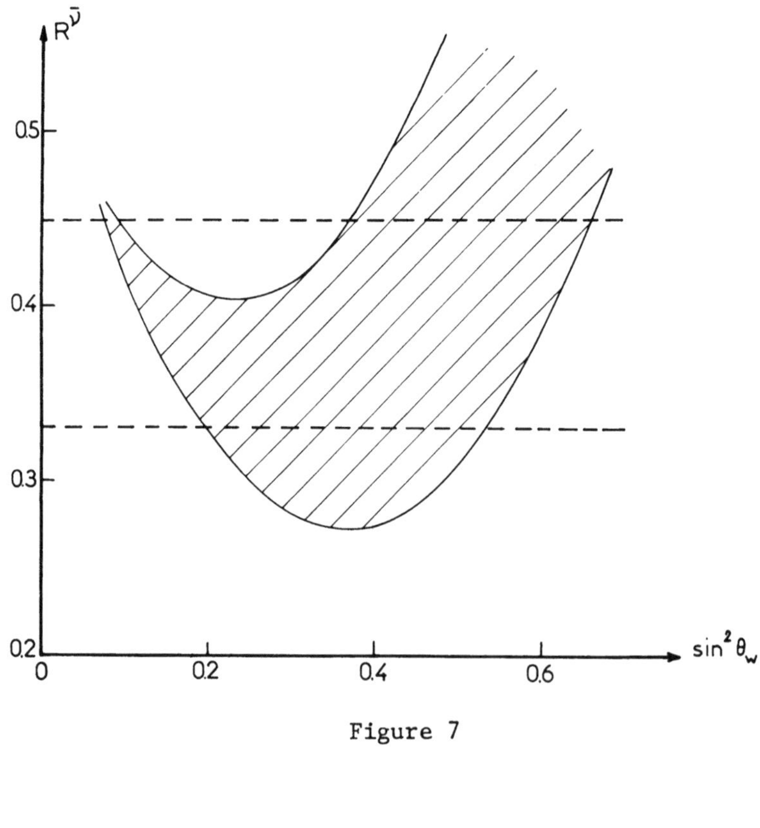

Figure 7

x		0.30	0.35	0.40
R^{ν}	S/V = 0.17	0.27	0.24	0.23
	S/V = 0.06	0.27	0.24	0.22
$R^{\bar{\nu}}$	S/V = 0.17	0.33	0.33	0.34
	S/V = 0.06	0.36	0.37	0.39
r_N	S/V = 0.17	0.64	0.71	0.88
	S/V = 0.06	0.56	0.64	0.74

Table 1

Interesting qualitative features can be read on Table 1

i) the ratio R^{ν} is rather insensitive to the parameter S/V. We then expect R^{ν} to be energy independent.

ii) the ratio $R^{\bar{\nu}}$ decreases when the sea amount S/V increases. On the other hand, it varies slowly with x.

iii) the ratio r_N increases when S/V increases and it varies rapidly with x.

The presently available experimental data at Fermilab energies come from counter experiments.

HPWF collaboration

$$R^{\nu} = 0.29 \pm 0.04 \qquad R^{\bar{\nu}} = 0.39 \pm 0.10$$

$$r_N = 0.48 \pm 0.20$$

CTF collaboration

$$R^{\nu} = 0.25 \pm 0.04 \qquad R^{\bar{\nu}} = 0.36 \pm 0.11$$

$$r_N = 0.75 \pm 0.14$$

The results of the three experiments for R^{ν} and $R^{\bar{\nu}}$ do not exhibit any dependence with energy but they do not exclude such a dependence as that shown on Table 1.

Concerning the quantity r_N, we must keep in mind that it has not been directly measured and the quoted numbers are the results of computations involving R^{ν}, $R^{\bar{\nu}}$ and r_c the ratio of charged current total cross-sections.

$$r_N = \frac{R^{\bar{\nu}}}{R^{\nu}} \times r_c$$

In particular, the HPWF and CTF groups take different values for r_c.

II - RIGHT HANDED CURRENT CONTRIBUTION

1°) Let us discuss only for simplicity the minimal case of two left handed doublets as in SU(4) and one right handed doublet $\begin{vmatrix} u \\ d \end{vmatrix}_R$. The moment integrals I_+ and I_- become for an isoscalar target

$$I_+^N = \left(\frac{1}{4} - \frac{2x}{3} + \frac{5x^2}{9}\right) V + S_{NC}$$

$$I_-^N = \left(\frac{1}{2} - x + \frac{5x^2}{9}\right) V + S_{NC}$$

As previously, we retain only, in the sea, contributions coming from $u\bar{u}$, $d\bar{d}$, $s\bar{s}$ pairs in an SU(3) symmetric way

$$S_{NC} = \left(1 - \frac{5x}{3} + \frac{10x^2}{9}\right) S$$

The slope parameters for the total cross-sections follow immediately from these expressions

$$A_{NC}^{\nu N} = \left(\frac{7}{12} - \frac{11x}{9} + \frac{20x^2}{27}\right) V + \left(\frac{4}{3} - \frac{20x}{9} + \frac{16x^2}{9}\right) S \qquad (29)$$

$$A_{NC}^{\bar{\nu} N} = \left(\frac{5}{12} - x + \frac{20x^2}{27}\right) V + \left(\frac{4}{3} - \frac{20x}{9} + \frac{16x^2}{9}\right) S$$

2°) By comparing equations (28) and (29), we see that the variation with x of the slope parameters $A^{\nu N}$ and $A^{\bar{\nu} N}$ are very different in the two models. In the range $0.3 < x < 0.4$ previously considered the ratio r_N of neutral current total cross-sections is essentially independent of x and it increases substancially with the sea amount S/V as shown on table 2.

S/V \ x	0.30	0.35	0.40
0	0.65	0.64	0.63
1%	0.66	0.65	0.65
10%	0.73	0.73	0.73

Table 2

Because of the large experimental uncertainties these numbers agree with low and high energy data but the comparison is unconclusive.

As noticed by Barger and Nanopoulos the valence contributions are the same in the two models for the particular value x = 0.375 and this fact makes more difficult a clear choice between the two types of models if nature choose a value of x close to that value.

3°) Let us consider the ratios R^{ν} and $R^{\bar{\nu}}$ which compare charged and neutral current total cross-sections. At fixed values of x threshold effects are present via charged current cross-sections and we expect a variation with energy of these quantities. We give numerical estimates in Table 3 using an effective quark model where right handed current effects are effective only at 30% at Fermilab energies

x		0.30	0.35	0.40
R^{ν}	LE	0.28	0.25	0.22
	HE	0.29	0.26	0.23
$R^{\bar{\nu}}$	LE	0.54	0.46	0.40
	HE	0.29	0.25	0.22

Table 3

L.E means low energy with $S/V \simeq 1\%$

H.E means high energy with $S/V \simeq 10\%$

Again we observe that the neutrino ratio R^{ν} remains essentially independent of the incident energy E whereas the antineutrino ratio $R^{\bar{\nu}}$ decreases rapidly with E. The effect is clearly more important than with left handed current models but the experimental data are not accurate enough to allow any definite conclusion.

APPENDIX A

KINEMATICS

The relevant diagram for inelastic lepton or antilepton scattering is shown on Fig 1

Three Lorentz scalar variables can be defined

$$q^2 = -(k - k')^2 \qquad W^2 = -(p + q)^2 \qquad \nu M = -p.q$$

and they are not independent

$$W^2 = M^2 + 2M\nu - q^2$$

AS usual M is the nucleon mass.

In the laboratory frame, we call as E the incident beam energy $-p.k = M\,E$ and in this frame ν represents the transfert of energy between lepton e.g. the energy carried out by the current.

Finally, we introduce two dimensionless scalar variables

$$x = \frac{q^2}{2M\nu} \qquad y = \frac{p.q}{p.k} = \frac{\nu}{E}$$

and we get the useful relations

$$q^2 = 2\,M\,E\,xy \qquad W^2 = M^2 + 2M\,E\,y\,(1 - x)$$

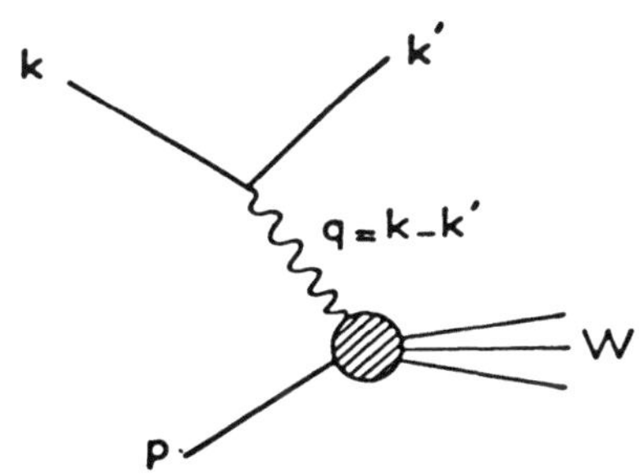

Figure 1

APPENDIX B

ELEMENTARY QUARK TRANSITIONS

Neutrino Scattering with SU(4)

$\Delta C = 0$ $\Delta S = 0$

d	$\Rightarrow$	u	$\cos\theta_c$	VALENCE
$d\bar{d}$	$\Rightarrow$	$u\bar{d}$	$\sin\theta_c$	SEA
$\bar{u}u$	$\Rightarrow$	$\bar{d}u$	$\sin\theta_c$	SEA

$\Delta C = 0$ $\Delta S = 1$

$s\bar{s}$	$\Rightarrow$	$u\bar{s}$	$\sin\theta_c$	SEA
$\bar{u}u$	$\Rightarrow$	$\bar{s}u$	$\sin\theta_c$	SEA

$\Delta C = 1$ $\Delta S = 0$

d	$\Rightarrow$	c	$\sin\theta_c$	VALENCE
$d\bar{d}$	$\Rightarrow$	$c\bar{d}$	$\sin\theta_c$	SEA
$\bar{c}c$	$\Rightarrow$	$\bar{d}c$	$\sin\theta_c$	SEA

$\Delta C = 1$ $\Delta S = 1$

$s\bar{s}$	$\Rightarrow$	$c\bar{s}$	$\cos\theta_c$	SEA
$\bar{c}c$	$\Rightarrow$	$\bar{s}c$	$\cos\theta_c$	SEA

Antineutrino Scattering with SU(4)

$\Delta C = 0$ $\Delta S = 0$

u	$\Rightarrow$	d	$\cos\theta_c$	VALENCE
$u\bar{u}$	$\Rightarrow$	$d\bar{u}$	$\cos\theta_c$	SEA
$\bar{d}d$	$\Rightarrow$	$\bar{u}d$	$\cos\ _c$	SEA

$\Delta C = 0 \quad \Delta S = -1$

$u \Rightarrow s$		$\sin\theta_c$	VALENCE
$u\bar{u} \Rightarrow s\bar{u}$		$\sin\theta_c$	SEA
$\bar{s}s \Rightarrow \bar{u}s$		$\sin\theta_c$	SEA

$\Delta C = -1 \quad \Delta S = 0$

$c\bar{c} \Rightarrow d\bar{c}$		$\sin\theta_c$	VALENCE
$\bar{d}d \Rightarrow \bar{c}d$		$\sin\theta_c$	SEA

$\Delta C = -1 \quad \Delta S = -1$

$s\bar{s} \Rightarrow c\bar{s}$		$\cos\theta_c$	SEA
$\bar{c}c \Rightarrow \bar{s}c$		$\cos\theta_c$	SEA

$\begin{vmatrix} u \\ b \end{vmatrix}_R$ Doublet added

New quantum number T associated to the b quark. All transitions $\Delta C = 0 \quad \Delta S = 0$

Neutrino Scattering $\Delta T = +1$

$b\bar{b} \Rightarrow u\bar{b}$		1	SEA
$\bar{u}u \Rightarrow \bar{b}u$		1	SEA

Antineutrino Scattering $\Delta T = -1$

$u \rightarrow b$		1	VALENCE
$u\bar{u} \Rightarrow b\bar{u}$		1	SEA
$\bar{b}b \Rightarrow \bar{u}b$		1	SEA

$\begin{vmatrix} c \\ s \end{vmatrix}_R$ Doublet added All transitions $\Delta S = \Delta C$

Neutrino Scattering $\Delta C = +1$ $\Delta S = 1$

$s\bar{s} \Rightarrow c\bar{s}$	1	SEA
$\bar{c}c \Rightarrow \bar{s}c$	1	SEA

Antineutrino Scattering $\Delta C = -1$ $\Delta S = -1$

$c\bar{c} \Rightarrow s\bar{c}$	1	SEA
$\bar{s}s \Rightarrow \bar{c}s$	1	SEA

APPENDIX C

NEUTRAL CURRENTS

We study models with only two types of quarks with electric charge + 2/3 (Q^+) and - 1/3 (Q^-).

m quarks $q_j \in Q^+$

n quarks $q_k \in Q^-$

The electromagnetic current has then the following quark representation

$$J^{EM}_\mu = \frac{2}{3}\, i \sum_{j \in Q^+} \bar{q}_j \gamma_\mu q_j - \frac{1}{3}\, i \sum_{k \in Q^-} \bar{q}_k \gamma_\mu q_k$$

For the weak charged current, only quarks of the two sets are related and we introduce two m x n matrices respectively associated to left handed and right handed couplings.

$$J^{cc}_\mu = i \bar{q} \gamma_\mu (1+\gamma_5) \begin{vmatrix} 0 & A_L \\ 0 & 0 \end{vmatrix} q + i \bar{q} \gamma_\mu (1-\gamma_5) \begin{vmatrix} 0 & A_R \\ 0 & 0 \end{vmatrix} q$$

1°) Neutral currents in SU(2)⊗ U (1) models

In this class of models, the weak charged current is simply related to the weak isotopic spin current.

$$J^{cc}_\mu = 2(J^1_\mu + i J^2_\mu) \qquad J^{cc*}_\mu = 2(J^1_\mu - i J^2_\mu)$$

The third component of the weak isotopic spin current is obtained from the commutation relations

$$[J^1 + i J^2, J^1 - i J^2] = 2 J^3$$

All these expressions are valid for both the left handed and right handed parts using the projection operators $\frac{1 \pm \gamma_5}{2}$

In the class of models with a SU(2)⊗ U(1) structure the neutral current is given by

$$J^{NC}_\mu = 2 J^3_\mu - 2 \sin^2\theta_W J^{EM}_\mu$$

Using the previous notations, we obtain

$$2 J^3_\mu = \frac{i}{2} \bar{q} \gamma_\mu (1+\gamma_5) \begin{vmatrix} A_L A_L^* & 0 \\ 0 & -A_L^* A_L \end{vmatrix} q + \frac{i}{2} \bar{q} \gamma_\mu (1-\gamma_5) \begin{vmatrix} A_R A_R^* & 0 \\ 0 & -A_R^* A_R \end{vmatrix} q$$

2°) Diagonal neutral current

Starting with the Cabibbo current constructed from only three quarks u, d, s, we obtain a neutral current having a strangness violating component. From Experiment such a component is not present at the same level as the strangness conserving one. We have to cancel the unwanted part and the simplest way is to introduce a new quark c as proposed by Glashow, Iliopoulos and Maiani. In this model the neutral current is diagonal in the quark representation and the matrix A_L is orthogonal

$$A_L = \begin{vmatrix} \cos\theta_c & \sin\theta_c \\ -\sin\theta_c & \cos\theta_c \end{vmatrix}$$

θ_c being the Cabibbo angle as usual.

We now restrict ourselves to models where the neutral current is diagonal in the quark representation. From the previous calculation, the constraints on the matrices A are

$$AA^{*} \text{ and } A^{*}A \text{ are diagonal matrices}$$

for both the left handed and the right handed parts.

In the particular case where A is a squared matrix $m = n$, we get the following properties

i) $AA^{*} = A^{*}A$

ii) The irreducible parts of A are proportional to unitary matrices

$$A = \begin{vmatrix} \lambda_1 U_1 & & \\ & \lambda_2 U_2 & \\ & & \ddots & \\ & & & \lambda_p U_p \end{vmatrix}$$

If time reversal holds, A can be chosen as real and the hermitian conjugate matrix A^{*} reduces to the transposed matrix A^T.

Consider now a $m \times m$ matrix A which satisfies the constraints

$$AA^{*} = A^{*}A = \text{Diagonal}$$

Let us add a new quark of electric charge + 2/3

$$B = \begin{vmatrix} & A & \\ b_1 \; b_2 & \cdots & b_m \end{vmatrix}$$

It is straightforward to prove that the constraints BB^{*} and

$B^{*}B$ diagonal matrices imply $b_1 = b_2 = \ldots = b_m = 0$

The physical consequence of this result is the following : a theory with (m + 1, m) quarks or (m, m + 1) quarks is equivalent to a theory with (m, m) quarks. Obviously the equivalence can be different for left handed and right handed current and we finally get a new model.

Let us give an example with five quarks u, d, s, c, b. For the left handed current in the usual SU(4) model, the matrix A_L is that previously given. The introduction of the quark b does not change this part of the current.

$$A_L = \begin{vmatrix} \cos\theta_c & \sin\theta_c & 0 \\ -\sin_c & \cos\theta_c & 0 \end{vmatrix}$$

But the new quark can be used for a right handed current. If it is coupled to the u quark the two possible solutions, assuming universality, are

$$A_R = \begin{vmatrix} 0 & 0 & 1 \\ 1 & 0 & 0 \end{vmatrix} \qquad \text{or} \qquad A_R = \begin{vmatrix} 0 & 0 & 1 \\ 0 & 1 & 0 \end{vmatrix}$$

corresponding to right handed doublets $\begin{vmatrix} u \\ b \end{vmatrix}_R$ and

$$\begin{vmatrix} c \\ d \end{vmatrix}_R \qquad \text{or} \qquad \begin{vmatrix} c \\ s \end{vmatrix}_R$$

With six quarks, u, d, s, c, b, t, three of charge + 2/3 and three of charge -1/3 the situation is entirely different. The matrix A_L is chosen as unitary and we allow in that way a small violation of time reversal invariance constrained by the structure of the Cabibbo current. The matrix A_R is free and only experiment will determine its precise structure. From parity violation in neutral current reactions we know that it cannot be an unitary matrix but only a part of it, the matrices $A_R A_R^{*}$ and $A_R^{*} A_R$ being only diagonal matrices with some eigenvalues zero. The other eigenvalues can be unity if universality holds.

REFERENCES

A - EXPERIMENTS

1°) Gargamelle Collaboration

V. BRISSON Talk at the Moriond Meeting Flaine (march 1976)

W. KRENZ Talk at the Aachen Conference (June 1976)

W. Van DONNINCK Talk at the Aachen Conference (June 1976)

2°) Harvard Pennsylvania-Wisconsin-Fermilab Collaboration

A. BENVENUTI et al HPWF Reports 76/1 76/4

C. RUBBIA Talk at the ENS meeting Paris (January 1976)

A. BENVENUTI Talk at the Vanderbilt Conference

D.K. MANN Talk at the Storage Ring Meeting Flaine (Feb. 1976)

T.Y. LING Talk at the Madison Conference (1976)

P. WANDERER Talk at the Madison Conference (1976)

A. BENVENUTI Talk at the Aachen Conference (June 1976)

T.Y. LING Talk at the Aachen Conference (June 1976)

3°) Caltech-Fermilab Collaboration

B.C. BARISH et al. Phys. Rev. Lett. 36 939 (1976)

A. BODEK Talk at the Moriond Meeting Flaine (March 1976)

L. STUTTE Talk at the Madison Conference (1976)

B.C. BARISH Talk at the Aachen Conference (June 1976)

D. BUCKHOLZ Talk at the Aachen Conference (June 1976)

4°) ANL - CMU - Purdue collaboration

M. DERRICK Talk at the Aachen Conference (June 1976)

5°) American Russian Collaboration Hydrogen-neon Bubble Chamber E 180

F.A NEZRICK Talk at the Aachen Conference (June 1976)

$\bar{B}$ Parameter at Fermi Lab Energies

	E	B
HPWF	25	0,95 ± 0,10
	78	0,45 $^{+0,15}_{-0,10}$
CTF	50	0,64 $^{+0,22}_{-0,26}$
	150	0,36 $^{+0,30}_{-0,36}$
E 180	20	0,76 $^{+0,08}_{-0,10}$
	40	0,90 $^{+0,08}_{-0,10}$
	62	0,73 $^{+0,12}_{-0,18}$

Fitted $\bar{B}$ values for antineutrino Scattering

B - THEORY

1°) Models for weak currents

S. WEINBERG Phys. Rev. Lett 19 1364 (1967)

Phys. Rev. Lett 27 1688 (1971)

A. SALAMI Proceedings of the XVIIIth Nobel Symposium
Aspenäsgarden (1968)

S.L. GLASHOW, J. ILIOPOULOS and L. MAIANI
Phys. Rev. D 2 1285 (1970)

S. PAKVASA, W.A. SIMMONS and S.F. JUAN
Phys. Rev. Lett. 35 70(1975)

H. FRITZSCH, M. GELL-MANN and P. MINKOWSKI
Phys. Lett. 59 B 256 (1975)

F. WILCZEK , A. ZEE, R.L. KINGSLEY and S.B. FREIMAN
Phys. Rev. D 12 2768 (1975)

A. de RUJULA, H. GEORGI and S.L. GLASHOW
Phys. Rev. D 12 3589 (1975)

H. HARARI Phys. Lett. 57 B 265 (1975)

F.E. CLOSE, R.L. KINGSLEY, G.A. RINGLAND, D. SIVERS
Preprint Rutherford (March 1976)

2°) Quark Positon model

M. GOURDIN Lectures given at the Bonn Summer Institute
(August 1974) and references

V. BARGER, T. WEILER and R.J.N. PHILLIPS
Nuclear Physics B 102 433 (1976)
University of Wisconsin Preprint (1976)

3°) Asymptotic Freedom in non abelian Gauge Theories

H. GEORGI and H.D. POLITZER Phys. Rev. D 9 416 (1974)

D. GROSS and F. WILCZEK Phys. Rev. D 10 2141 (1974)

D. GROSS Phys. Rev. Lett. 32 1071 (1974)
Lectures given at the Ettore Majorana
Center (July 1975)

G. ALTARELLI, G. PARISI and R. PETRONZIO
University of Rome preprint (Feb.1976)

P.W. JOHNSON and WU-KI TUNG
Illinois Institute of Technology
preprint (1976)

4°) Threshold Effects

H. GEORGI and H.D. POLITZER	Harvard Preprints (1975)
H.D. POLITZER	Talk at the Coral Gable Conference (Jan. 1976)
R. BARNETT	Phys. Rev. Lett. 36 1163 (1976)
A. de RUJULA	Talk at the Coral Gable Conference (Jan. 1976)
E. DERMAN	Nucl. Phys. B 110 40 (1976)
J. KAPLAN and F. MARTIN	University of Paris Preprint (May 1976)
C. ALBRIGHT and R. SHROCK	Fermilab-Conference 76/50 THY

REVIEW OF MASSIVE DILEPTON PRODUCTION IN PROTON NUCLEUS COLLISIONS

J. W. Cronin

The Enrico Fermi Institute

University of Chicago, Chicago, Il 60637 U.S.A.

In this lecture we will review recent experiments on production of lepton pairs in proton-nucleus collisions at large effective dilepton mass. For the purpose of this talk we will consider the production of lepton pairs with masses above the J/ψ and ψ' region ($m \geq 4$ GeV/c^2). This range appears to be above the region dominated by resonant production[1] and we will discuss it in terms of continuum dilepton production. Dilepton production for $m \leq 4$ GeV/c^2 will be discussed by Prof. A.J.S. Smith of Princeton University.

It should be noted that the study of dilepton production in hadronic collisions was begun about eight years ago with the pioneering experiment of Christenson et al. at Brookhaven National Laboratory.[2] In this experiment which was characterized by high sensitivity, but poor resolution, a significant dimuon continuum was established. In retrospect a major part of the signal in this experiment was due to the J/ψ which was subsequently discovered at BNL and SLAC.[3,4] Nevertheless, when these resonances were subtracted, a substantial dimuon continuum remained with dimuon masses which were a significant fraction of the total available center of mass energy.[5]

Within the framework of our current understanding, dilepton production experiments are important for two reasons. First, Yamaguchi[6] pointed out that the production cross section in nucleon-nucleon collisions for the charged intermediate vector boson which mediates the weak interaction can be related by the conserved vector current hypothesis (CVC) to the production cross section of virtual photons and hence to the production of dilepton pairs. It is common to make the assumption[7] that the dilepton production cross section has the scaling form

$$\frac{d\sigma}{dm} = m^{-3} f(\frac{m^2}{s})$$

where s is the square of the center of mass energy and f is a universal function. Then one can use measurements of dilepton production at the existing low energy machines to predict the production of intermediate vector bosons in colliding beam machines of the future.

A second reason for the importance of these experiments is the connection pointed out by Drell and Yan[8] of the dimuon production process with the constituent structure of the nucleon which has emerged during the past eight years. In this picture the production of the dileptons is due to the annihilation of a constituent (quark) in one nucleon with its corresponding anti-constituent in the other nucleon. Recently a number of authors[9,10,11,12] have made quantitative predictions of the cross section for dilepton production. They use as input to their calculations the distributions of fractional momentum of the nucleon carried by the quarks and antiquarks as deduced from deep inelastic neutrino and electron scattering. As we shall see later, the predicted cross sections depend sensitively on the assumed antiquark distributions and exact quantitative agreement between the predictions and experiment cannot be expected. Another important consequence of the Drell-Yan process is the scaling prediction referred to above. This latter prediction is a necessary condition for the validity of quark-antiquark annihilation as the dilepton source.

Recently two experiments have been completed at Fermilab which have measured dilepton production in proton-nucleus collisions. The first of these has been carried out by a Columbia-Fermilab-Stony Brook (CFS) group. Part of the results on dielectron production have been published.[1] Preliminary results on dimuon production are also available.[13]

The CFS experiment has observed both electron pairs and muon pairs produced in collisions of 400 GeV protons with beryllium and copper targets respectively. Figure 1 shows a sketch of their apparatus which is a symmetric double arm magnetic spectrometer. Each arm views the production target at an angle of ~90° in the c.m. system of the nucleon-nucleon system. The electrons are identified by lead glass detectors. Discrimination against hadrons is achieved by the requirement that the momentum measured in the magnet agree with the energy measured in the lead glass. Background consists of accidental coincidences between the electrons detected in each arm and real coincidences between an electron and a hadron misidentified as an electron. The backgrounds become small for $m > 5.5$ GeV/c^2.

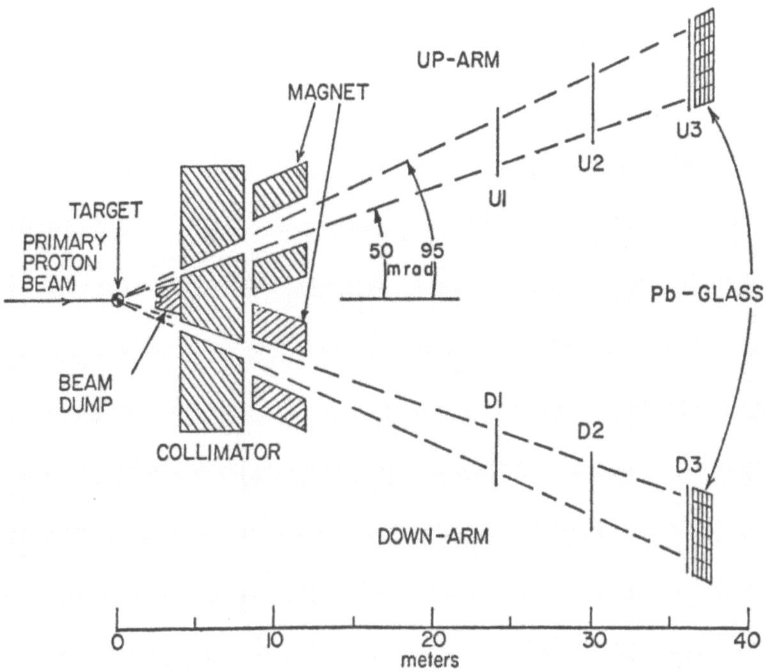

Fig. 1. Apparatus of the Columbia-Fermilab-Stony Brook group.

Data on dimuon production was also taken when beryllium hadron absorbers were inserted in each arm near the target, and muon identifiers were installed behind the lead glass.

Figures 2 and 3 show the measured cross section $d\sigma/dmdy|_{y=0}$ for dielectrons and dimuons respectively. Here y is the c.m. rapidity of the dilepton system. This cross section is quoted per nucleon assuming a linear A dependence of the cross section. The authors quote this cross section rather than $d\sigma/dm$ because the apparatus is sensitive to dileptons produced nearly at rest in the c.m. system (i.e., $y \sim 0$).

Another important characteristic of their observations is that the vector sum of the two muon momenta has a net transverse momentum. The mean transverse momentum of the dilepton system is observed to be ~ 1.5 GeV/c in the mass range 4.5 to 8 GeV/c^2.

A second experiment[14] by a Chicago-Princeton (CP) group has recently been completed. They have observed the production of dimuons in collisions of 400 GeV protons with a copper target. Figure 4 shows a schematic view of the apparatus, which consists of a highly asymmetric double arm spectrometer. One arm is a

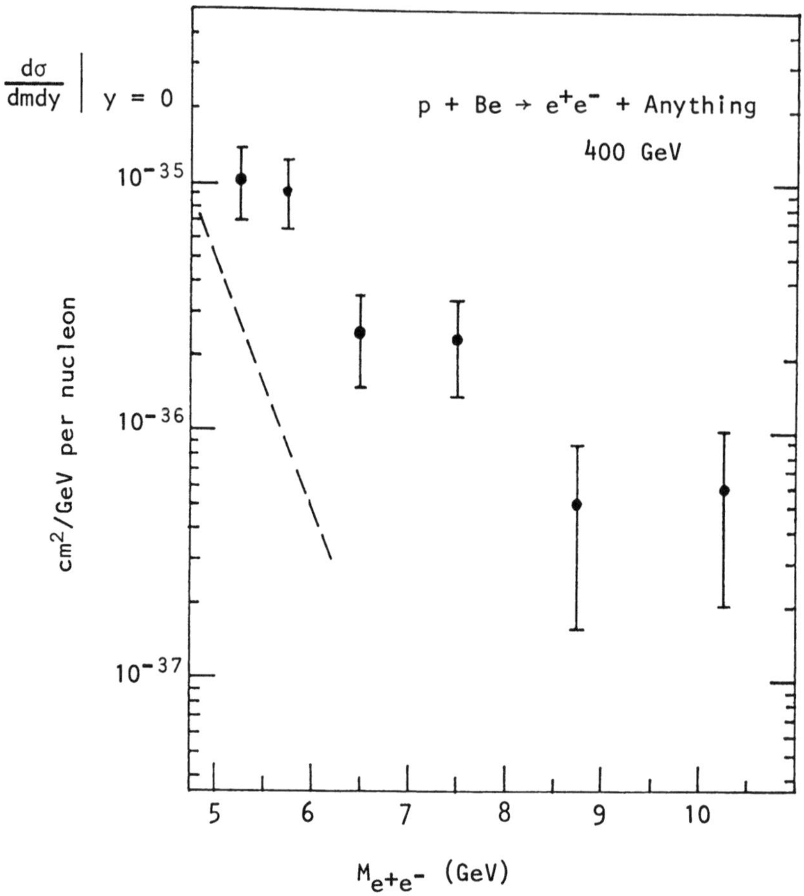

Fig. 2. Dielectron production cross section measured by CFS group.

magnetic focusing spectrometer of small solid angle which can be adjusted to select, with high purity, direct single muons produced at ~90° in the c.m. system. The other arm consists of an array of 10 large scintillation detectors shielded from the copper target by earth and steel. The shielding thickness and configuration is such that a muon must have a transverse momentum, $(p_\perp) \gtrsim 3.2$ GeV in order to register in the detector. The large arm covers in the c.m. system a polar angle of $60° < \theta < 126°$, and an azimuthal angle of $-8° < \phi < 25°$. The angle ϕ is defined with respect to the plane containing the beam and the magnetic spectrometer.

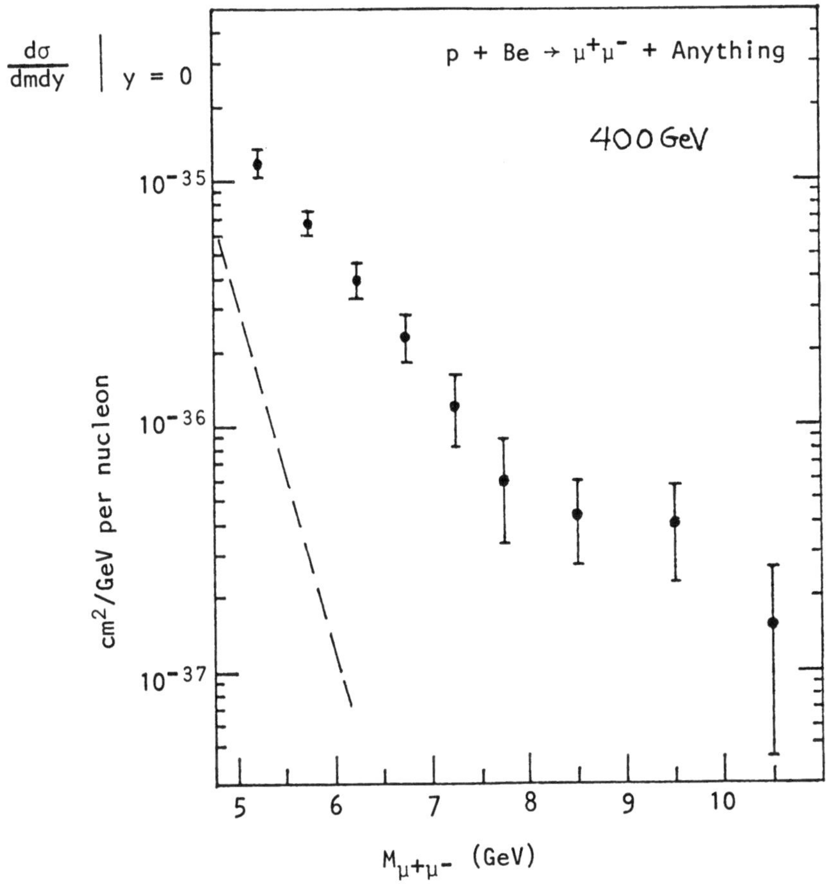

Fig. 3. Dimuon production cross section measured by CFS group.

Data is taken as a function of $p_\perp^S$, the transverse momentum selected for the muon observed in the magnetic spectrometer. The rate of events in which a second coincident muon is detected in the large arm is recorded. In the circumstance that the dimuon system has very small transverse momentum, then the $p_\perp$ of the other muon which is detected in the large arm must balance the $p_\perp^S$ of the selected muon and the dimuon mass is approximately $m \simeq 2\, p_\perp^S$. If the parent of the dimuon is produced with significant transverse momentum then the mass acceptance of the apparatus is broadened but is still centered about $2\, p_\perp^S$.

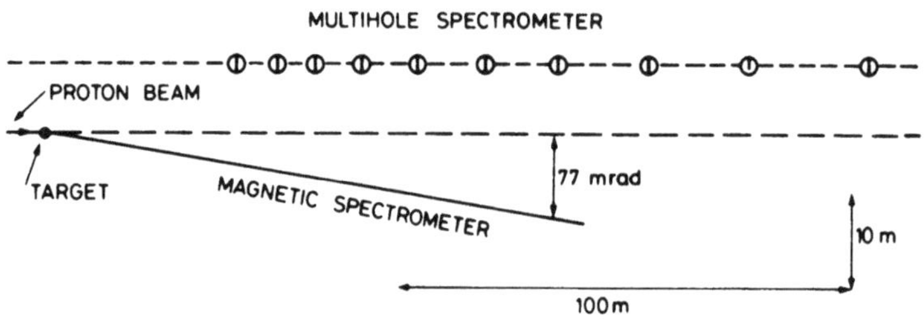

Fig. 4. Schematic view of dimuon apparatus of CP experiment.

Knowledge of the mean transverse momentum of the dimuon parent is necessary to relate the coincidence rate at a given $p_\perp^s$ to the average mass being observed and the cross section. In this experiment the mean transverse momentum can be inferred from the azimuthal angular distribution of the muon detected in the large arm with respect to the plane of the magnetic spectrometer and the incident beam. For $m \approx 9$ GeV where an adequate number of events was available the mean transverse momentum of the dimuon parent was found to be ~ 1.25 GeV/c. This is about the same value observed by the CFS experiment. In order to evaluate the mean masses and the cross sections observed in the CP experiment a mean transverse momentum equal to 1.25 GeV/c independent of m was assumed. For a given $p_\perp^s$, the mass acceptance of the spectrometer was about 2 GeV/c^2 FWHM. Thus the CP experiment cannot resolve narrow resonances whose total yield is less than or equal to the continuum yield in a 2 GeV/c^2 wide mass band. For what follows we assume we are dealing with a continuum.

Figure 5 shows the results quoted as $d\sigma/dmdy|_{y=0}$. Like the CFS experiment, the CP apparatus was only sensitive to dimuons produced in the neighborhood of $y = 0$ (at rest) in the c.m. system. Also the cross sections are quoted per nucleon assuming that the A dependence is linear. Measurements of dimuon production at lower m give an A dependence which is consistent with a linear A dependence.[15]

In the CP experiment measurements were also made for 300 GeV incident protons. These results will be discussed at a later point

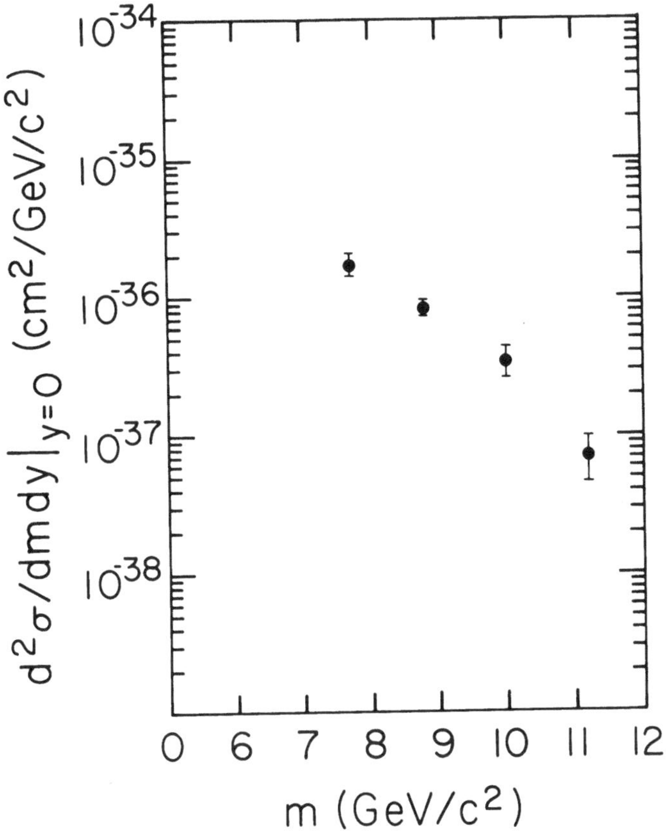

Fig. 5. Dimuon production cross section measured by CP group.

in the lecture when we will discuss the evidence relevant to scaling.

Another observation of the CP experiment bears directly on the question of the source of direct single lepton production.[16] For single direct muons selected with $p_\perp^S \gtrsim 4.5$ GeV/c, the fraction that is accompanied by a second muon in the large solid angle arm is significant. For example, for $p_\perp^S = 5.25$ GeV/c, (13±2)% of the single muons are accompanied by a second muon. Using only up-down symmetry with respect to the production plane and a conservative extrapolation of the observed distribution beyond the solid angle subtended, one can estimate in a model independent way that $\gtrsim 30\%$ of the single direct muons have their origin as one member of a high mass muon pair.

In fact if one parametrizes the dimuon production cross section as $Ed\sigma/dmd^3p \sim e^{-m}\, e^{-1.6p_\perp}(1-x_F)^4$ one predicts a ratio of double muons to single direct muons of ~ 10%. Thus it is possible to account for all of the single direct leptons at large $p_\perp$ ($p_\perp \gtrsim 5$ GeV/c) by the production of massive dilepton pairs. In the conventional Drell-Yan model calculations it is not possible to account for the observed yield of single direct leptons. The reason for this result is that the Drell-Yan model in its simplest form does not provide any transverse momentum for the dilepton system. If a substantial transverse momentum (~1 GeV/c) is arbitrarily added to the model, then the single lepton production can be explained. This fact was pointed out by M. Duong-Van[17] before the current observations were made. It is important to stress, however, that such a large transverse momentum does not fit into the Drell-Yan model in any natural way.

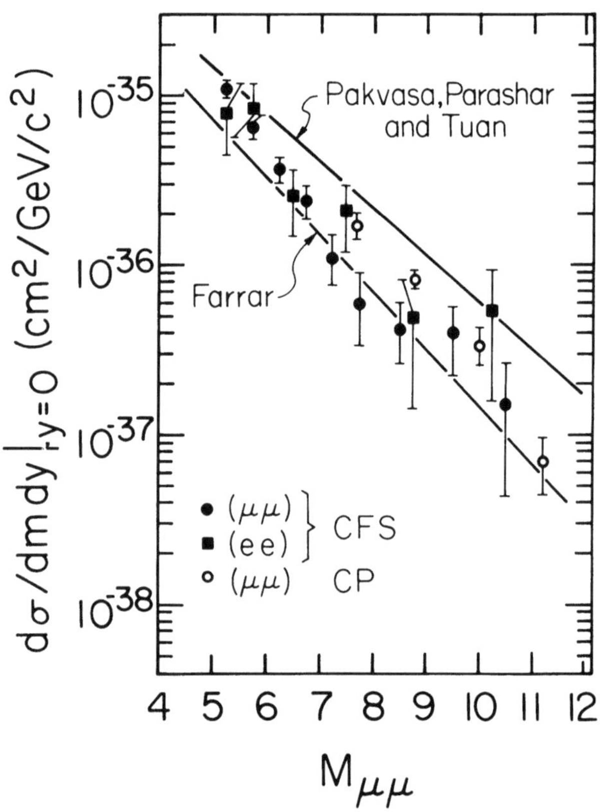

Fig. 6. Plot of all dilepton cross sections measured in the CP and CFS experiments. For comparison two Drell-Yan calculations for the cross section are also plotted.

In Fig. 6 we have plotted the results of both experiments on the same graph. As one can see the agreement is acceptable given the statistical quality of the data. In the range $5 \leq m \leq 11$ the cross section $d\sigma/dmdy|_{y=0}$ can be fit by an expression $\sim e^{-.72m}$. Also plotted are the predictions of the cross section predicted by the Drell-Yan process as calculated by two different authors.

The two models differ principally in the parametrization of the antiquark distributions. The model of Pakvasa, Parashar, and Tuan[10] parametrizes the antiquark distribution as $(1-x)^{7/2}$ while the parametrization of Farrar[9] is $(1-x)^7$. Here x is the fraction of the nucleon momentum carried by the quark. Since the latter model is less rich in antiquarks at large x, the predicted cross section for massive dimuon production is smaller and falls more rapidly with increasing mass than the former model. Both models assume that the quarks possess a color variable.

The data fall in between the two calculations which indicate that the Drell-Yan process can be easily adjusted to fit the observations. It is interesting to compare the dimuon cross section predicted by a two photon process. This two photon process has been calculated in detail by Chen, Muzinich, Terazawa and Cheng[18] and must be observable if no stronger process intervenes.

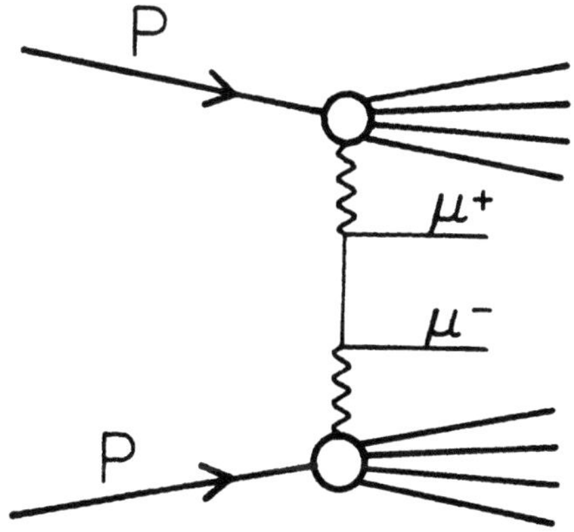

Fig. 7. Diagram of dilepton production by the two photon process.

Figure 7 shows the diagram of the two photon process which can be related to measured elastic and inelastic electron-nucleon scattering. The kinematics of the process are such that the dilepton system produced by the two photons has a large rapidity in the nucleon-nucleon c.m. system. The two experiments described here are insensitive to a dilepton system produced with large rapidity. Nevertheless it is interesting to compare the total cross section $d\sigma/dm$ measured in the experiments with the predictions of the two photon process.

To calculate the total dimuon cross section at a given mass we assume that the dependence of the cross section on x_F is $(1-x_F)^4$. (x_F is the Feynman x of the dilepton system.) This dependence was observed for J/ψ production[19] and is assumed somewhat arbitrarily to hold for all dilepton production. In Fig. 8 we plot the total

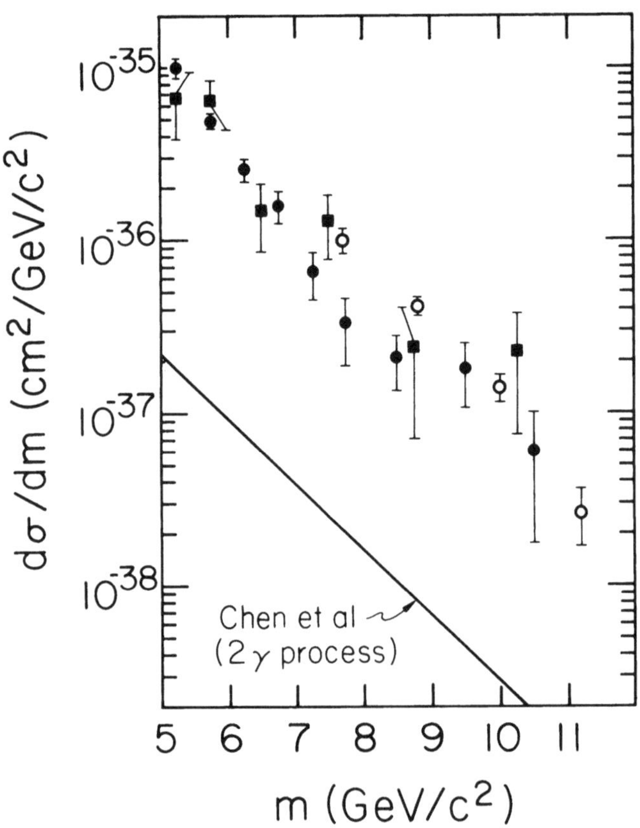

Fig. 8. Plot of total dilepton production cross section per unit mass. Expected cross section for two photon process is also plotted.

cross section so calculated and compare with the predicted dilepton yield for the two photon process. What is clearly demonstrated in Fig. 8 is that the yield of dimuons is at least 20 times larger than the yield predicted on the basis of the two photon process, and hence must come from some other process which is much stronger. The most attractive interpretation is the Drell-Yan process since the observed rate is consistent with the predictions of that process.

There is, however, the one major reservation to the above conclusion and that is the fact that the average transverse momentum of the parent of the dimuon is 1.2 - 1.5 GeV/c. This is a large amount since one conventionally believes that the mean transverse momentum of the quarks should be comparable to the typical average tranverse momentum in hadronic processes, 300 MeV/c. We regard this observation of large transverse momentum as an important fact which must be understood before one can accept the Drell-Yan process as being demonstrated.

Regardless of the source of the dimuon continuum, one can apply the CVC argument to obtain lower limits on the production of the charged vector bosons ($W^{\pm}$) as a function of their mass. Since the recent gauge theories which unite the weak and electromagnetic interactions predict the mass of the $W^{\pm}$ and heavy neutral bosons to be in the range from 50 to 100 GeV/c^2, it is necessary to establish the energy dependence of the dimuon production. If the dimuon production process is indeed Drell-Yan, then the predicted scaling behavior gives a precise definition of the energy dependence. Thus it is of great interest to investigate the scaling behavior of the dimuon process. The most comprehensive and recent review of scaling has been given by Palmer, Paschos, Samios, and Wang in the 1975 Isabelle Summer Study.[20]

At present the data on the scaling question are rather sparse. In order to give as complete a view as possible we will include continuum muon data at lower masses. These data come from the Columbia-Hawaii-Illinois-Fermilab group (CHIF)[21] and a second Chicago-Princeton Group[22,23] (CP-II) using the Chicago Cyclotron Facility at the Fermilab.

The former group has measured continuum dimuon production by ~320 GeV neutrons in the mass region from 1.2 to 2.6 GeV/c^2, while the latter has measured the production by 150 and 225 GeV protons of dimuons in the mass region from 1.4 to 2.6 GeV/c^2. All the measurements were made on nuclei. The cross sections per nucleon were extracted using the measured A dependence in the case of the CHIF group. The cross sections/nucleon for the CP-II group were deduced assuming an A^1 dependence. Finally we have also used the original data of the Columbia group at BNL for 29 GeV protons incident on uranium.[24]

The Chicago-Princeton experiment has measured two points at m = 7.6 and m = 8.6 GeV/c^2 for 300 GeV incident protons. This is to be compared with the measurements at m = 7.7, 8.8, 10.0 and 11.2 GeV previously discussed. These measurements are the only measurements of the energy dependence of massive dimuon production with the same apparatus. These measurements unfortunately lack the statistical precision required to make a good check of scaling.

With all the qualifications discussed above, we present the results in Fig. 9. Here we plot $m^3 d\sigma/dm$ vs m^2/s. If scaling holds all these points should lie on a universal curve. The solid diamond points from the BNL experiment are assigned no errors because of the many uncertain operations performed on the original data. One sees that despite all the corrections the 29 GeV data lies about a factor 3 above the mean of the points measured for 300 and 400 GeV incident protons.

The most reliable comparison of two energies are the open and

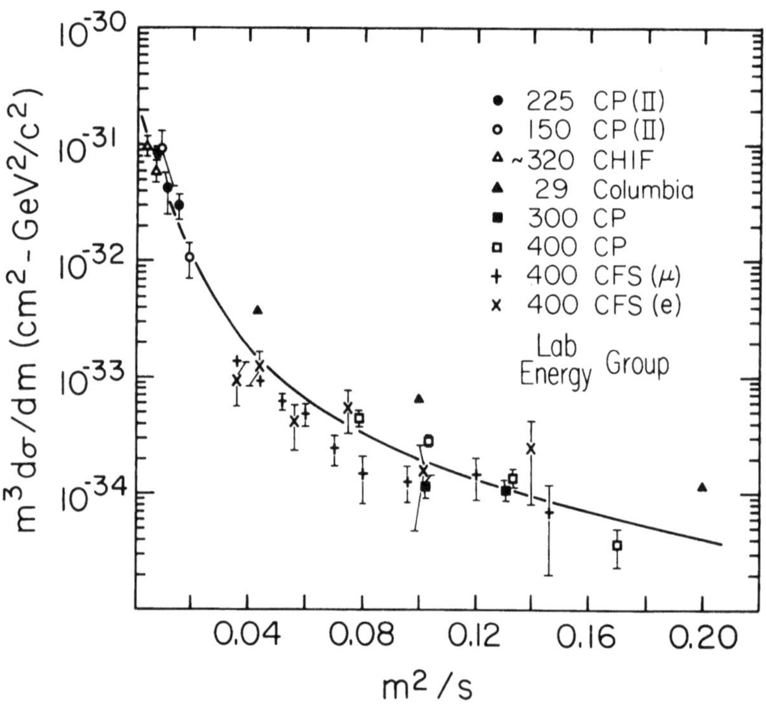

Fig. 9. Plot of $m^3 d\sigma/dm$ as a function of m^2/s for available dilepton production data.

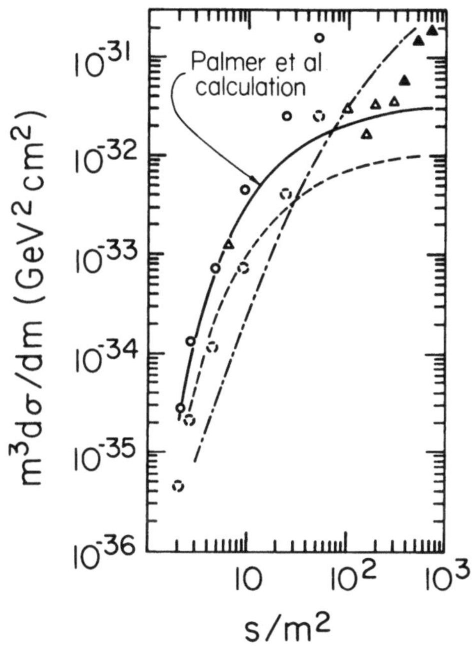

Fig. 10. Comparison of scaling curve obtained from recent data with scaling curve of 1975 Isabelle Summer Study (Ref. 20).

closed squares at 300 and 400 GeV incident protons. In one case there is good agreement with scaling, while in the other there is a three standard deviation discrepancy. Clearly the evidence for or against scaling is not established. Nevertheless, if we assume that scaling will eventually be established, then the universal curve should lie close to the solid curve drawn by eye on Fig. 9. We will use this curve to make some predictions for rates of dilepton production and $W^{\pm}$ production as a function of its assumed mass.

In Fig. 10 we plot the solid curve of Fig. 9 (now shown as a dot-dash curve) on a similar curve obtained from the Isabelle study. The solid curve is the result of a Drell-Yan calculation by the Isabelle study group which neglects the color variable for the quarks. The dashed curve is the result of the Isabelle study when divided by three to account for color. The solid circles are the 29 GeV data from BNL, while the dashed circles are the BNL data assuming an A^1 dependence. One sees that the scaling curve obtained from the data is qualitatively different from that of the Isabelle study. For low m^2/s the data is higher, while for high m^2/s the data is lower. As one can see below, studies of dilepton production at the ISR will be most valuable in verifying the scaling curve at low m^2/s.

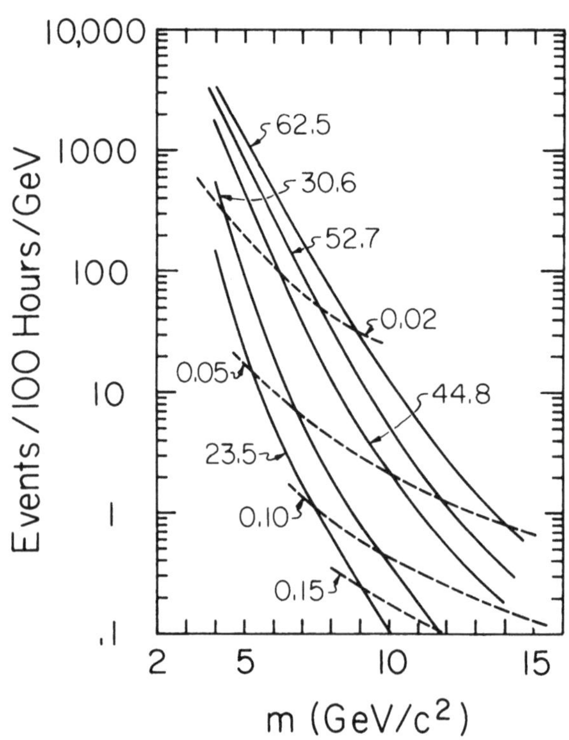

Fig. 11. Expected dilepton yields at ISR assuming the scaling curve of Fig. 9. Dashed lines are lines of constant m^2/s. Solid lines correspond to the indicated energy in GeV/c.

We now assume that the solid curve of Fig. 9 is indeed a scaling curve and calculate the dilepton production rates that can be expected for the CERN ISR. We assume a luminosity of 2.5×10^{31} sec^{-1} cm^{-2} which corresponds to an interaction rate of 10^6/sec. We neglect the small difference that may arise because the data here is derived from nuclei with roughly equal mixtures of neutrons and protons extrapolated to A = 1, with a case of pure proton-proton collisions. We assume an ISR detector that has a product of detection efficiency times solid angle which is 25% of 4π. In Fig. 11 we plot the expected yields for 100 hours of operation at each of the standard ISR operating energies as a function of the dimuon mass. We also show lines of constant m^2/s. One can see that the ISR should be a good device to investigate scaling for $m^2/s \leq 0.05$. For a dilepton mass of 6 GeV/c^2 the yield is predicted to rise by two orders of magnitude between the lowest and highest ISR energies.

We can now discuss the CVC relation between the dilepton production and the production cross section for charged intermediate vector bosons. If one uses an isoscalar target with an incident proton beam it can be shown[7] that

$$\sigma_W^+ + \sigma_W^- \geq \frac{3}{2\sqrt{2}} \frac{G\cos^2\theta\, m_W^3}{\alpha^2} \frac{d\sigma}{dm}(m_W)\Big|_{\text{isovector}}$$

where $G = 1.16 \times 10^{-5}$ GeV^{-2} is the Fermi coupling constant, m_W is the mass of the intermediate boson, $\cos\theta$ is the Cabbibo angle, $\sigma_W^+ + \sigma_W^-$ is the sum of the cross section for the production of W bosons of <u>both charges</u>, α is the fine structure constant and $\frac{d\sigma}{dm}(m_W)\Big|_{\text{isovector}}$ is the continuum isovector dilepton production cross section evaluated where the effective mass of the dilepton system is equal to the vector boson mass. If we assume that the isoscalar part is small, then the above relation becomes numerically:

$$\sigma_W^+ + \sigma_W^- \geq 0.23\, m_W^3 \frac{d\sigma}{dm}(m_W)$$

where all masses are given in GeV/c^2.

We can use the smooth curve of Fig. 9 to estimate the cross sections for W production as a function of mass. The expression above gives lower limits on the production of W bosons on an isoscalar target. The smooth curve of Fig. 9 is the result of measurements of incident protons on nearly isoscalar targets and is reduced to a cross section per nucleon. We use these cross sections, the scaling assumption, and the inequality above to find the lower limits for W boson production for collisions at 200, 400

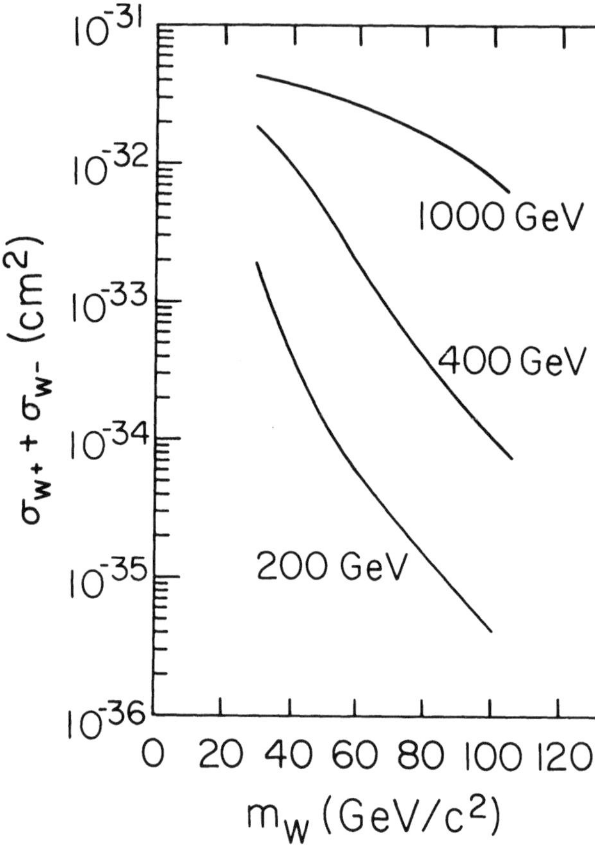

Fig. 12. Production cross sections for W bosons as a function of mass for various total c.m. energies.

and 1000 GeV c.m. energy. While these are lower limits for incident protons on an isoscalar target, they should be very close to the values expected for proton proton collisions which would be obtained with colliding beams. These results are plotted in Fig. 12. The striking feature of these curves is the increase with c.m. energy in cross section for large W mass. For a 100 GeV/c^2 mass, the increase between 200 GeV and 1000 GeV is greater than three orders of magnitude. In Fig. 13 we plot the estimated rates for $W^\pm$ detection assuming a luminosity of 2.5×10^{31} cm^{-2} sec^{-1} (10^6 interactions/sec) and a detector with a solid angle efficiency product of 0.25 of 4π. The branching ratio for $W^\pm \rightarrow \mu^\pm + \nu$ was assumed to be 0.10. These curves give one some idea of what can be expected from various colliding beam devices that one might contemplate.

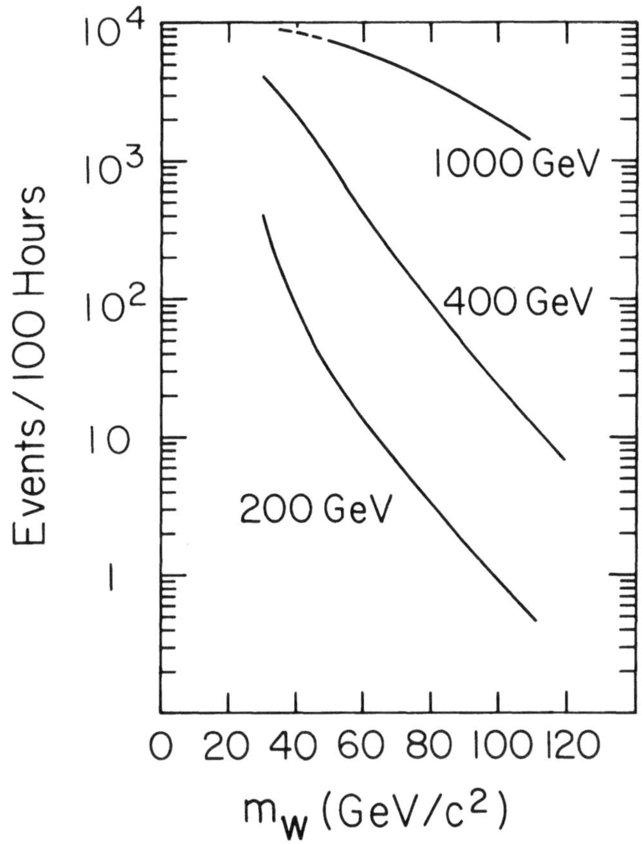

Fig. 13. Yield per hour of $W^{\pm} \to \mu^{\pm} + \nu$ events detected in a colliding beam experiment. See text for details.

Some cautionary remarks should be made concerning the $W^{\pm}$ cross section estimates used above. The predictions for 1000 GeV c.m. energy come from very low values of m^2/s. The data come in fact from measurements at Fermilab where the actual masses are ~ 2 GeV/c^2 which may not be in the scaling region. Also the evidence for scaling in m^2/s is at present not very strong. The estimates depend totally in the assumption of scaling. Finally all the cross sections per nucleon are deduced assuming an atomic number dependence A^1 which has not been verified for dilepton masses larger than 3 GeV/c^2.

I hope this review has succeeded in demonstrating that a great deal of progress has been made in the study of dilepton production in proton-hadron collisions. These studies aim to verify that the hadron structure, as deduced from the study of a single hadron with a point lepton probe, is sufficient to understand a more complicated process involving the collision of two hadrons. The important fact, that the mean transverse momentum of the dilepton parent is ~1.5 GeV/c, is not understood in terms of the present picture of hadron structure. These studies also serve the practical purpose of defining the cross section for production of a $W^{\pm}$ boson, given knowledge of its mass.

Over the next few years we can expect further progress. The question of how well the dilepton production process scales in m^2/s should be answered. Also one will begin to measure dilepton production by incident pions at larger m^2/s. This process should provide a richer source of antiquarks than one has in proton-nucleon collisions. One can in fact anticipate the penultimate experiment in which colliding beam experiments can be made between protons and antiprotons.[25]

The author would like to acknowledge the hospitality of the Aspen Center for Physics where this report was written.

REFERENCES

1. At the present time the existence of a possible resonance at m = 6 GeV/c^2 remains unconfirmed. See D. C. Hom *et al*., Phys. Rev. Lett. 36, 1236 (1976).
2. J. H. Christenson *et al*., Phys. Rev. Lett. 25, 1523 (1970); ibid, Phys. Rev. D 8, 2016 (1973).
3. J. J. Aubert *et al*., Phys. Rev. Lett. 33, 1404 (1974).
4. J. E. Augustin *et al*., Phys. Rev. Lett. 33, 1406, 1453 (1974).
5. L. M. Lederman, Columbia University preprint (1974).
6. Y. Yamaguchi, Nuovo Cimento 43, 193 (1966).
7. L. M. Lederman and B. G. Pope, Phys. Rev. Lett. 27, 765 (1971).
8. S. D. Drell and T.-M. Yan, Phys. Rev. Lett. 25, 316 (1970), and Ann. Phys. 66, 578 (1971).
9. G. Farrar, Nucl. Phys. B 77, 429 (1974).
10. S. Pakvasa, D. Parashar, and S. F. Tuan, Phys. Rev. D 10, 2124 (1974).
11. H. P. Paar and E. A. Paschos, Phys. Rev. D 10, 1502 (1974).
12. J. Finjord and F. Ravndal, Phys. Lett. 62 B, 438 (1976).

13. The author is most grateful to Prof. Lederman for permission to incorporate these results in this review.

14. L. Kluberg et al., Phys. Rev. Lett. to be published.

15. M. Binkley et al., "Dimuon Production on Nuclear Targets," Fermilab preprint (1976).

16. Recent reviews of this subject have been given by J. W. Cronin, Lectures of the School of Subnuclear Physics (Erice, 1975), and L. Lederman, Proceedings of the International Symposium on Lepton and Photon Interactions at High Energies (SLAC, 1975).

17. M. Duong-Van, SLAC preprint (SLAC-PUB-1604, 1975).

18. M.-S. Chen et al., Phys. Rev. 7 D, 3485 (1973).

19. See for example: H. D. Snyder et al., Phys. Rev. Lett. 36, 1415 (1976).

20. R. B. Palmer et al., B.N.L. preprint (BNL-20634, 1975).

21. M. Binkley et al., Phys. Rev. Lett. 37, 574 (1976).

22. K. J. Anderson et al., "Inclusive μ-Pair Production at 150 GeV by π^+ Mesons and Protons," (submitted to XVIII Conference on High Energy Physics, Tbilisi, USSR, 1976).

23. K. J. Anderson et al., "Production of Continuum Muon Pairs at 225 GeV by Pions and Protons," (submitted to XVIII Conference on High Energy Physics, Tbilisi, USSR, 1976).

24. We use these data in a rather indirect fashion. We take the points from Ref. 20. The data have been corrected for the fact that the laboratory momentum of the dimuon was required to be $\geq$ 12 GeV/c. In Ref. 20 the cross sections per nucleon were extracted assuming an $A^{2/3}$ dependence. We have corrected these cross sections assuming instead an A^1 dependence. This is equivalent to dividing the BNL data obtained from Ref. 20 by a factor 6.2.

25. C. Rubbia, P. McIntyre, and D. Cline, Harvard University preprint (March, 1976).

D I S C U S S I O N

CHAIRMAN: Prof. J.W. Cronin

Scientific Secretaries: Y. Afek and G.J. Tarnopolsky

DISCUSSION

- *YPSILANTIS:*

Except for your two points at 300 and 400 GeV/c, there seems to be no experiment at the same values of M^2/s to check scaling. Would you comment on this.

- *CRONIN:*

In addition to our points, some measurements overlap at the lowest M^2/s values; but the curve is rather steep there. For $W^{\pm}$ boson production you want to be in the range of M^2/s of 0.1 to 0.2.

- *WIGNER:*

Is your relation $M^3_{\mu\mu}\ d\sigma/dM_{\mu\mu} = f(M^2_{\mu\mu}/s)$ compatible with the existence of singularities at all thresholds or is it not to be considered such an exact relation?

- *CRONIN:*

The scaling relationship would not hold if new thresholds open in $M^2_{\mu\mu}$.

- *ETIM:*

If the μ-pairs came from the decay of a series of vector mesons, perhaps infinite in number, could not the large p_T events be easily understood?

- *CRONIN:*

The experiment of CFS has good resolution and extends to ~ 7 GeV/c^2. If the μ pairs are a series of vector mesons then their decay widths and spacing would have to be such that the result is an apparent continuum. Above 7 GeV/c^2 the CP experiment has poor resolution and could not detect a narrow resonance. Hence there is certainly the possibility that the dimuon continuum is a series of vector mesons.

- *CLEYMANS:*

Will you measure this process with pion beams?

- *CRONIN:*

We have an approved experiment at the Fermilab in which we plan to study dimuon production by incident pions. We expect to begin in about two years.

- *KLEINERT:*

In the Chicago-Princeton experiment, what is the background due to secondary muons from weak decays of π's and K's? Or of charmed D mesons?

- *CRONIN:*

π and K meson decays are experimentally removed. We observe highly correlated pairs. It is kinematically unlikely that the μ's come from different particles. We will check this correlation by triggering on electrons in the precision spectrometer and looking for muons in the M.H.S.

- *ZICHICHI:*

In the Chicago-Princeton experiment, what are the solid angles of the spectrometers? The acceptance really matters for uncorrelated pairs.

- *CRONIN:*

The solid angles are

$$\Omega_{MAG} = 2 \times 10^{-3} \text{ sr}, \quad \Omega_{MHS} = \left(\frac{4\pi}{20}\right) \text{ sr} .$$

In addition to the solid angle constraint we also demand at least 3 GeV/c on each side. This requires a rather stringent correlation.

- *YPSILANTIS:*

Assuming perfect energy and angular resolution on the μ's detected in the M.H.S., what would be the μμ mass resolution du only to the multiple scattering in the magnetic spectrometer?

- *CRONIN:*

It becomes clearly much better and perhaps comes to 5-7%.

- *LIPKIN:*

When a muon pair has a high transverse momentum, there must be something recoiling against the pair with opposite transverse momentum. Could this be the result of a higher mass particle decaying into a pion, or something else, and a vector meson with high p_T and the vector vector then decaying into $\mu^+\mu^-$?

- *CRONIN:*

It is possible. It would be a very heavy particle since the $\mu^+\mu^-$ pair consumes around 40% of the centre-of-mass energy, at the highest mass measured.

ARE JETS REALLY THERE ?

Endre Lillethun

Department of Physics, University of Bergen

N-5014 Bergen-Univ., Norway

INTRODUCTION

In introducing the concept of jets I shall go back to some results from high energy proton-proton collisions obtained at the CERN Intersecting Storage Rings (ISR) in 1972 - 1973 [1,2].

The data are shown in Fig. 1. The initial work measured the differential production cross section for pions only in the region of transverse momentum, p_T , up to about 1 GeV/c. In this interval a straight line fit to the logarithm of the invariant cross section $Ed^3\sigma/d^3p$ as function of p_T appeared very good, indicating the functional relationship

$$E\frac{d^3\sigma}{d^3p}\ p_T \quad = \quad A\cdot\exp(-B\ p_T)\ .$$

This function was indeed expected from the ideas of Feynman scaling[3] which also predicted that the constants A and B should be independent of the centre of mass (c.m.) energy, $\sqrt{s}$, of the proton-proton system at very high energies.

A simple picture of the collision process is shown in Fig. 2.a. On colliding, the two protons (Lorentz-contracted) break into two fragments each. The two parts which really hit each other remain almost at rest in the c.m. system, forming a lump of very excited energy, boiling off the energy as particles of relatively low momentum. The other two parts continue almost in their original direction with rather high momenta, "decaying" into a few particles with low transverse momentum.

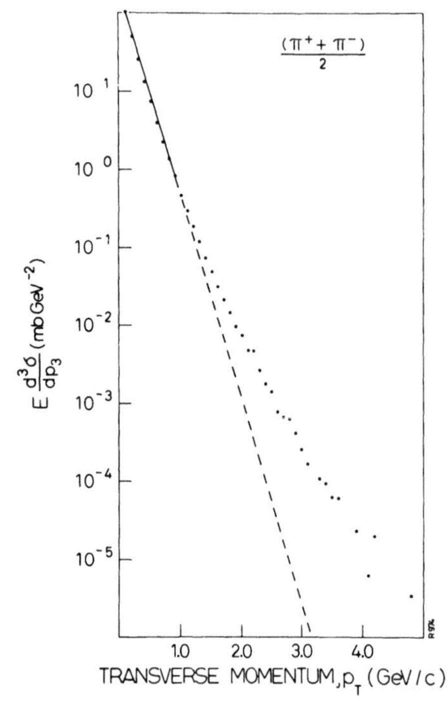

Fig. 1. The mean invariant cross section for production of π^+ and π^- in proton-proton collisions with centre of mass energy 52.8 GeV plotted as function of the transverse momentum of the $\pi^\pm$. The straight line is a fit to early data (see Ref. 1).

When the measurements at the ISR were continued to higher p_T [4-6], it turned out that the cross section did not follow the simple exponential p_T-dependence. The measured values were very much higher, as can be seen from Fig. 1.

The explanation for this behaviour may be that there are hard grains (quarks, partons) within the protons and when the grain from the two protons meet they enter in a hard collision, in a sense like the semi-elastic scattering of protons in nucleus-nucleus collisions. The idea is pictured in Fig. 2b. The grains that undergo the hard collision may leave the interaction region with high transverse momenta, carrying along with them some part of, or the entire central lump of energy, "decaying" as they fly away. It is the "decay" products of such hard collisions, indicated by the double arrows at the bottom of Fig. 2b, which we call jets.

It seems to me that the hard scattering also could occur

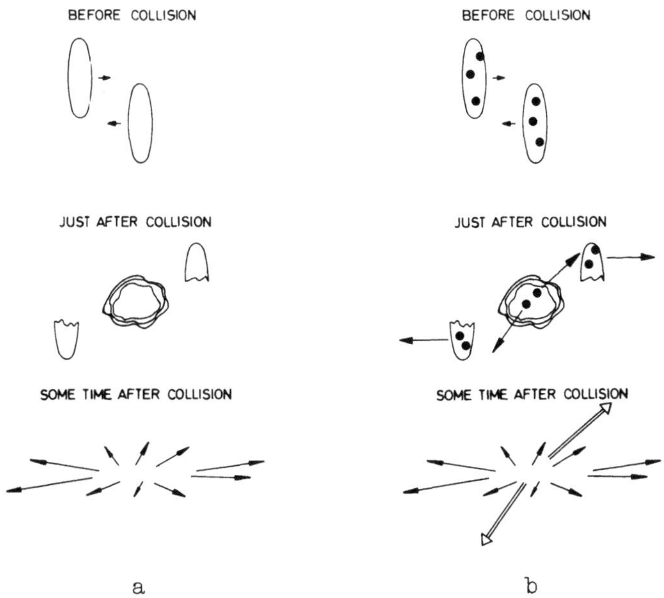

Fig. 2a. Sketch indicating a proton-proton collision resulting in particle production following Feynman scaling.

Fig. 2b. Sketch indicating a proton-proton collision including hard scattering of constituents of the protons.

between two regions of very high energy density (due to statistical fluctuations) in the kinetic part of the energy associated with the protons. Therefore I shall in this talk frequently use the word "flucton" to describe any part of the energy, associated with the protons, that takes part in the hard collision leading to jets, including quarks, partons, gluons and constituents. Although it is not necessary for an energy density fluctuation to have fixed quantum numbers, the jets resulting from the final state interactions ("decays") must have specific quantum numbers, but their invariant mass distribution may be continuous.

The inclusion of hard scattering within the protons in high energy collisions has led to the description of many models and parametrizations. See Ref. 7 for a review of Large Transverse Momentum Processes as of June 1975 and Ref. 8 for parametrization of data on inclusive production in terms of such models.

The properties of jets have been discussed in several papers[9-13] beginning with Bjorken and Brodsky's[9] article in 1970 about hadron

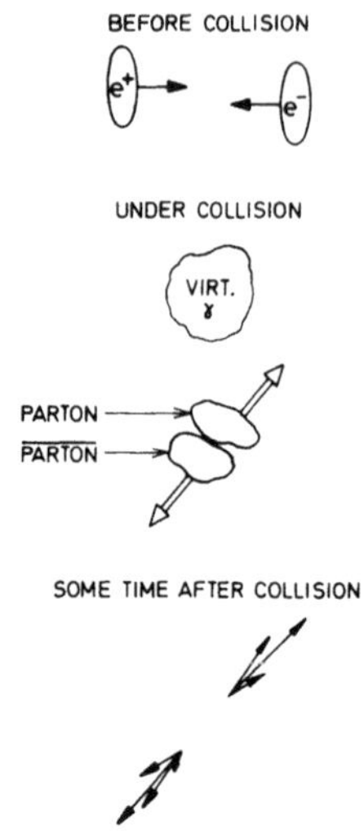

Fig. 3. Sketch indicating an electron-positron collision resulting in the production of hadrons.

production in electron-positron collisions. In the following I shall describe as an experimentalist the different main features of jets, leaving you to study the more mathematical approaches in the above references.

The simplest picture (theoretical and experimental) of jets may actually come from the production of hadrons in electron-positron (e^+e^-) scattering, since when these leptons collide they annihilate, leaving a very highly condensed amount of energy (a virtual photon) which in some cases will appear as a single parton-antiparton pair as shown in Fig. 3. The pair of partons may then fly in any direction, back to back in their c.m. system, "decaying" very quickly into two jets of hadrons with rather low momentum transverse to the direction of the parent partons.

It is reasonable to assume that these jets, produced in e^+e^- collisions are similar to the jets produced in proton-proton collisions. Actually the structure of the jets in their c.m. system may look like the structure of a proton-proton inelastic event without high transverse momentum particles. See bottom part of Fig. 2a. The e^+e^- events frequently have a single set of jets, while the proton-proton, pion-proton or any hadron-hadron event with jets will include also the particles from the forward-backward flying fragments, thereby confusing the structure: which particle belongs to the jets, which belongs to the forward-backward fragments?

An extra complication in proton-proton events comes from the (assumed) fact that the protons consist of at least two types of quarks and a number of quark-antiquark pairs of different kinds in addition to other types of fluctons. The jets can therefore be the result of hard scatterings of several types of pairs of fluctons and/or antifluctons, while in the case of e^+e^- scattering the colliding fluctons must be expected to be pairs of flucton-antiflucton of the same kind (in quark language $(u\bar{u})$, $(d\bar{d})$ or $(s\bar{s})$).

JETS IN e^+e^- COLLISIONS

At SPEAR, the e^+e^- colliding beam facility at Stanford Linear Accelerator Centre, a study has been made of the hadronic jets produced in e^+e^- collisions at several c.m. emergies.[14,15] The property measured is called sphericity, s, and is defined as

$$S = \frac{3\,\lambda_3}{\lambda_1+\lambda_2+\lambda_3} = \frac{3\left(\sum_i p_{T_i}^2\right)_{\min}}{2\,\sum_i \vec{p}_i^{\,2}}$$

where λ_i are the eigenvalues of the tensor

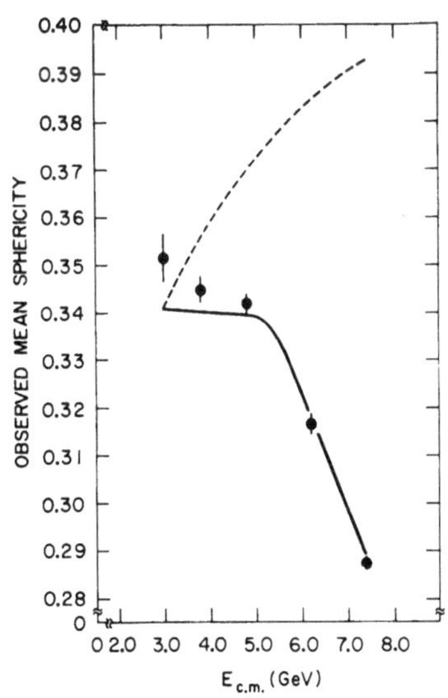

Fig. 4. Observed mean sphericity versus centre of mass energy for events of electron-positron collisions producing hadrons. The solid (dashed) curve shows a mean sphericity obtained from a Monte Carlo computation based on a jet model (phase space model). See Ref. 15.

$$T^{\alpha\beta} = \sum_i \left(\delta^{\alpha\beta} \vec{p}_i^{\,2} - p_i^{\alpha} p_i^{\beta} \right) .$$

The λ_i are the sums of the squares of the transverse momenta with respect to the corresponding eigenvector directions. λ_3 is the minimum eigenvalue and the eigenvector corresponding to λ_3 is defined as the jet axis.

Events with high multiplicity and a uniform distribution are characterized by $S \simeq 1$, for events where all particles follow the jet-axis the value of $S = 0$. The value of S therefore is a measure of the jet structure of the events.

The group has found evidence for jets as we shall see in the following figures. However, because of the non-uniformity of the acceptance of the particles produced, the measured results had to be compared to Monte Carlo simulations of the same distributions, based on either an isotropic phase space model or a jet model with a mean transverse momentum of 315 MeV/c with respect to the jet axis. Note that the computations assume production of pions only (neutral and charged) and that all charged tracks found experimentally are assumed to be pions. (We shall come back to the importance of really identifying the particles later).

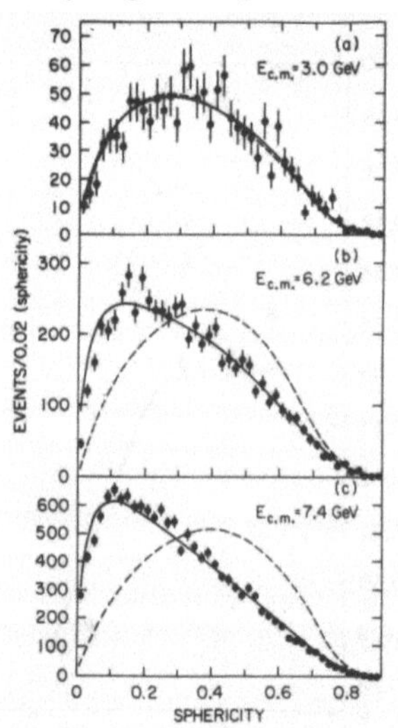

Fig. 5. Observed sphericity distributions for events of electron-positron collisions producing hadrons. The curves were obtained from computations as explained for Fig. 4 and were normalized to the number of events in the data. The data were recorded at centre of mass energies 3.0 , 6.2 and 7.4 GeV for particles of $x = 2p/E_{c.m.} < 0.4$ (p is the particle momentum).

Fig. 4 shows the mean spericity, <S> , versus c.m. energy. The measured values of <S> vary from 0.35 at $\sqrt{s}$ = 3.0 GeV to 0.29 at $\sqrt{s}$ = 7.4 GeV. The energy dependence is in good agreement with the jet model but disagrees with the phase space model.

The distribution of events in the sphericity space is shown in Fig. 5 for c.m. energies 3.0 , 6.2 and 7.4 GeV. At the lowest energy both model calculations agree very well with the measured values. For the higher energies the discrepancy between experimental data and the phase space model increases with energy while there is always good agreement between data and the jet model distributions. From observations it therefore seems clear that hadrons in e^+e^- collisions are produced in jets. However, there could be other effects that might simulate jets, e.g. statistical fluctuations, kinematical correlations and production of resonances. These effects have been studied theoretically by Grassberger and De Groot[16] who introduce a somewhat different definition of sphericity, or rather asphericity, which eliminates the effect of statistical fluctuations. From the above mentioned effects they estimate contributions to the asphericity of less than 20 % while they expect the value of about 40 % at the highest SPEAR energies if the hadrons are produced in jets with transverse momentum distributions similar to that of hadronic reactions.

A definitive measurement and analysis following this method has not yet been carried out, so, although jet-like production of hadrons in e^+e^- collisions seems very likely, it is not quite proved to be so.

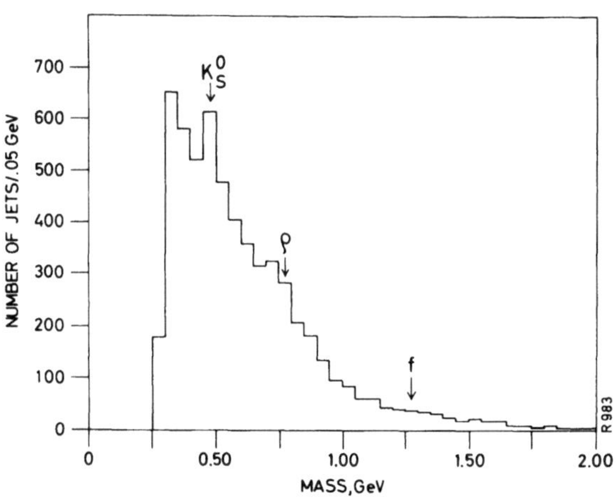

Fig. 6. Invariant mass distribution for neutral two-particle combinations of hadrons produced in electron-positron collisions (all hadrons interpreted as pions).

It is natural to ask whether these jets are something new or just known resonances. The SPEAR-group has studied the invariant mass distribution for two-particle combinations in different categories. One of the distributions is shown in Fig. 6 for events with at least three tracks, c.m. energy 7.4 GeV and for neutral combinations of particles, all assumed to be pions. The masses of some resonances are indicated in the diagram, and it is clear that no large production of resonances is found. However, the non-identification of particles, if there is a fair fraction of particles heavier than the pions, washes out the possible resonance structures both because a too large number of combinations are formed and because possible resonances of pion and a heavy particle spread over relatively wide mass ranges due to the misidentification.

JETS IN PROTON-PROTON COLLISIONS

In dealing with the jet concept in proton-proton collisions it has so far not been found practical to use sphericity as the observable. Instead one has concentrated on the study of other correlation effects which will be discussed in the following. Several experiments [17-23] have been performed in order to study correlation effects in high energy proton-proton collisions. I believe that the clearest information about jets will come from experiments involving particles in the large-angle region and with the identification of as many particles as possible. Therefore I shall in this talk mainly use

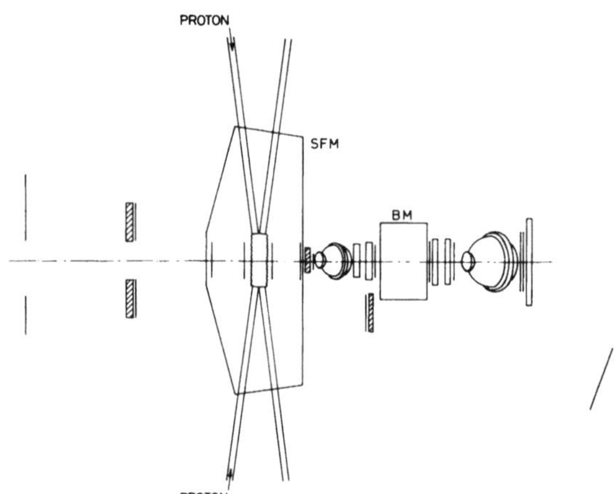

Fig. 7. Sketch of an experimental set-up used at the proton-proton intersecting storage rings (ISR) for the study of events with high transverse momentum. SFM and BM indicate bending magnets, single straight lines show scintillator hodoscopes, shaded rectangles show multiwire proportional chambers and the open rectangles spark chambers. The round structures are Čerenkov counters.

some new preliminary data from the British-Scandinavian-Orsay Group (BSO)[22] and results from a CERN group (Darriulat et al.)[23] which have been published and already reviewed several times.

The sets of data have both been obtained at the ISR, using a general experiment device, the Split Field Magnet (SFM), and spesific trigger systems. The set-up for the BSO group is shown in Fig. 7. The SFM-system contains a large number of multiwire proportional chambers to record the tracks of "all" the particles from an event. The BSO trigger system consists of a magnetic spectrometer placed at 90° to the bisector of the 15° angle between the proton beams and almost in the horizontal plane, capable of identifying charged particles of momenta up to 5 GeV/c. The CERN group used an array of lead glass blocks, also placed at about 90° to the bisector and centered on the horizontal plane, but on the side opposite to the direction of motion of the c.m. system of the colliding protons.

The coordinate system used in the presentation of the data is described in Appendix A and indicated in Fig. 8, where the protons come in from left and right along the y-axis.

In the present description I have neglected small differences between the systems used by the two groups in order to give a coherent, simple picture of the data. The simplifications can cause only very minor changes to the distribution, well within present accuracies.

In Appendix B a description is given of the difficulties in presenting the data because of the non-uniform acceptances. In particular a ratio R = (413/min.bias) is defined, normalizing event rates with a high p_T trigger to those with a socalled minimum bias trigger.

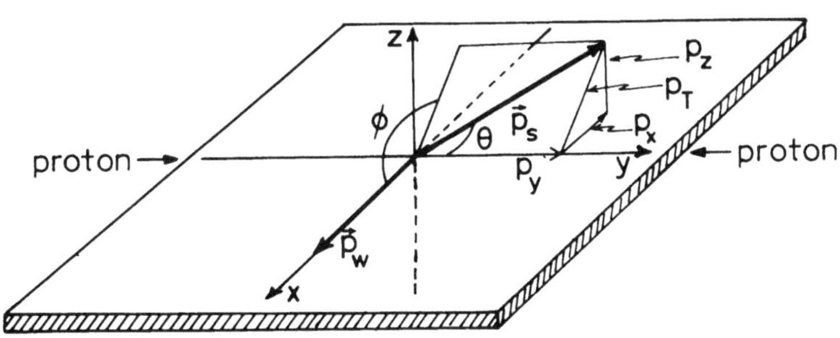

Fig. 8. The coordinate system used to describe the events. The protons are coming in along the y-axis, and the positive x-axis is defined by the momentum of the trigger particle, $\vec{p}_W$. See Appendix A for further explanations.

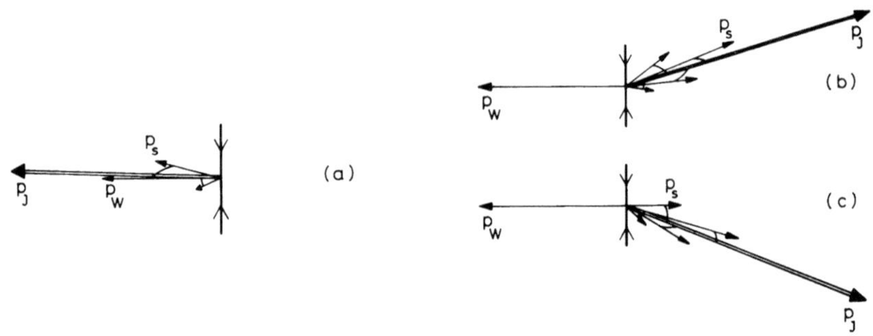

Fig. 9. Sketch indicating production of jets of particles in proton-proton collisions, the protons entering from top and bottom for each figure. p_W , p_J and p_S represent respectively the momenta of the trigger particle, the jet and the secondary particles making up the jet. Figure a may be substituted for the left side of the Figures b and c .

General Jet-Features

We shall assume that a jet is characterized by a set of particles within a cone of relatively small opening angle ($\sim 45^{\circ}$), and that in general two jets are produced through a hard collision between two fluctons, the fluctons flying apart back to back in their own c.m. system.

Fig. 9 shows jet configurations where the total momentum of the jet (or the initial flucton) is drawn as a double lined arrow p_J . The momenta, p_S of the final jet particles (the secondary particles) are drawn as single line arrows. p_W represents the trigger particle, usually of high transverse momentum, p_T . The jet on the trigger side may consist of the trigger particle only or a jet of the type shown in Fig. 9a. As indicated, $p_J \neq p_W$. The full picture may look as Fig. 9b or 9c , or with p_W in those drawings replaced by Fig. 9a. The jet opposite to the trigger may of course also consist of only one particle. The angles between the particles are chosen smaller than expected on the average[24] , in order to bring out the features more clearly.

It should be noted that the polar angle of the jet opposite to p_W depends on the fraction of the proton momenta carried by the initial fluctons. The azimuthal angle between p_W and $p_{J(opp)}$ is expected to depend only on a relatively small internal motion of the fluctons in the protons (similar to Fermi motion in the nuclei) so that the two jet-axes should be approximately coplanar. The experimental study of jets is complicated by the fact that only a fraction of the jet-particles are detected, e.g. only

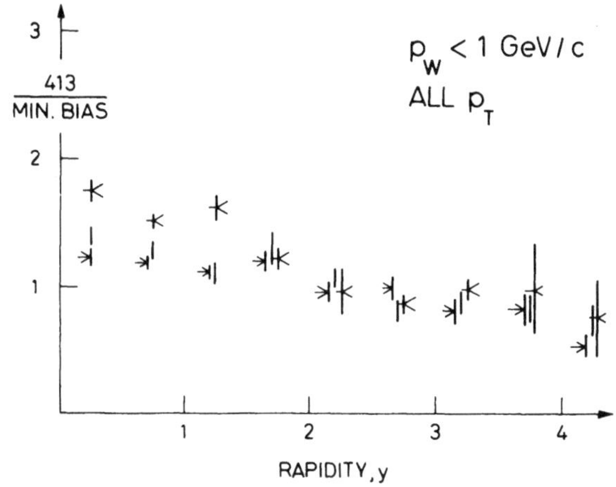

Fig. 10. Rapidity distribution of charged particles produced in proton-proton collisions with centre of mass energy 52.8 GeV, in events with a charged trigger particle of transverse momentum, $p_W < 1$ GeV/c. The vertical scale (for experiment 413) is normalized to a distribution obtained with a minimum bias trigger. Secondary charged particles of any transverse momentum are included. Three distributions are shown for particles in the following ranges of azimuthal angle with respect to the plane defined by the trigger particle and the incoming protons:
$|\phi| < 30^\circ$ (⇥), $|180^\circ - \phi| < 30^\circ$ (⊣K) and $75^\circ < |\phi| < 105^\circ$ (|).

the charged ones and not even all of them.

Turning now to experimental data, Fig. 10, displays R, the ratio of the rate of p_W-trigger events to that of minimum bias events (see Appendix B) as a function of rapidity, y of the particles. Rapidity is defined as

$$y = \frac{1}{2} \ln \frac{E + p_y}{E - p_y}$$

where E is the total energy of the particle and p_y its longitudinal momentum. In this case $p_W < 1$ GeV/c, no choice of trigger particle type has been made, and all charged particle tracks found in the following regions are included:

Region I	("same side")	defined by	$\lvert\phi\rvert < 30^\circ$
Region II	("opposite side")	defined by	$\lvert 180^\circ - \phi\rvert < 30^\circ$
Region III	("up")	defined by	$75^\circ < \phi < 105^\circ$
Region IV	("down")	defined by	$75^\circ < (-\phi) < 105^\circ$

Regions III and IV are added in the plots.

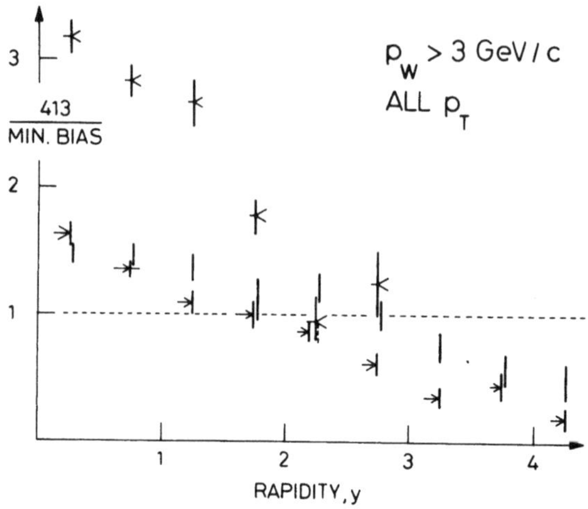

Fig. 11. Same type of distribution as in Fig. 10, but with $p_W > 3$ GeV/c .

Even for these low values of p_W there is a clear trend for the particles to accumulate in the neighbourhood of the trigger particle, R decreasing from about 1.4 at $y = 0$ to about 0.7 at $y = 4$. For low values of y the ratio R is largest for the opposite side.

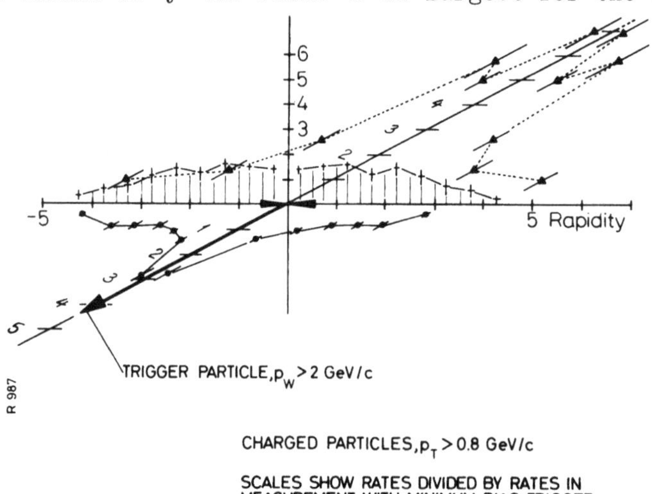

Fig. 12. Same type of distribution as in Fig. 10, but with the different azimuthal ranges represented "three-dimensionally" by the respective coordinate planes. In this case $p_W > 2$ GeV/c and secondary particles are required to have transverse momentum, $p_T > 0.8$ GeV/c .

Fig. 11 shows a similar plot of R, but now $p_W > 3$ GeV/c. The most striking difference between Figures 10 and 11 is the increased number of particles near $y = 0$ for the opposite side as would be expected according to the hard scattering model. Also the multiplicities of regions I, III and IV have increased, but only by about 30 %. In addition one notes that the multiplicity in the forward direction $y \sim 4$ is very low $(R \lesssim 0.3)$ for the same side. (The data for $y \sim 4$ are lacking for the opposite side.)

It is interesting to notice the similarity between these distributions for the up, down and same side regions (the trigger particle is not counted on the same side). This property changes drastically when one requires $p_T > 0.8$ GeV for the particles, as shown in Fig. 12 where $p_W > 2$ GeV/c. The figure is plotted in a 3-dimensional way to bring out the picture of coplanarity. The values of R around $y = 0$ are $R \sim 3$ for the same side, $R \sim 1.5$ for the up-down regions and $R \sim 7$ on the opposite side, in good agreement with a jet-picture.

Note that the multiplicity on the same side is lower than on the opposite side due (at least in part) to the requirement of a large value of p_W.[11,24]

The distribution on the opposite side is much broader than that

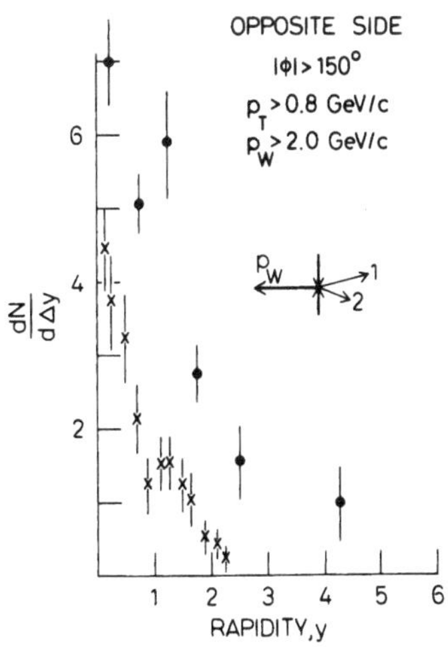

Fig. 13. Distribution of two-particle combinations as function of the rapidity difference Δy between the two particles (✗). The two particles are on the opposite side to the trigger particle ($|180^\circ - \phi| < 30^\circ$) and are each required to have $p_T > 0.8$ GeV/c. $p_W > 2$ GeV/c. For comparison is shown the distribution of Fig. 12 of single particles with $|180^\circ - \phi| < 30^\circ$ (♦).

on the same side. This, as stated above, is expected because of the motion of the c.m. system of the fluctons. A selection of events with two particles on the opposite side, each with $p_T > 0.8$ GeV/c, was made in order to look for the widths of the jets themselves. In Fig. 13 we plot the rates of such events as function of the difference between the rapidities of these two particles, and for comparison the R as function of y as also shown in Fig. 12. (The two rates are not to the same scale.)

The Δy distribution is clearly much narrower, with a width approximately like the one for the same side in Fig. 12. Although the Δy measurement by no means is a correct measure of the Δy distribution of the whole jet, the result at least seems consistent with the expectations outlined above for the jet model.

The last figure concerned with these general features of jets is due to the CERN group. Using a π^0 trigger with $p_W > 2$ GeV/c they select the particle on the opposite side balancing the largest fraction of the p_W, (x_E), with a $p_T > 1.2$ GeV/c and lying in a particular rapidity interval. The rapidity distribution of the remaining particles (without any momentum cut) with respect to the chosen particle is shown in Fig. 14. As the rapidity interval for the chosen particle moves away from $y = 0$ of the proton-proton c.m. system, an enhancement in the distribution follows it, again

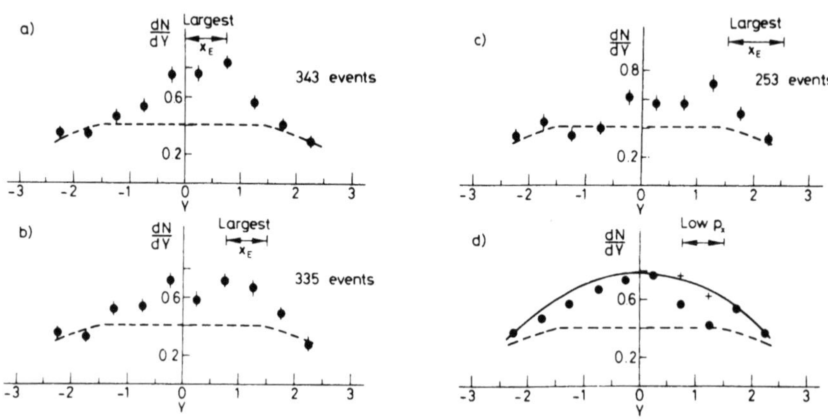

Fig. 14. Rapidity distributions of charged particles of any transverse momentum in the hemisphere opposite to a π^0 trigger particle with $p_W > 2$ GeV/c. In Figures a, b and c events have been chosen with a particle of $p_T > 1.2$ GeV/c (not included in the plot) in the rapidity range indicated by double arrows (<——>). In Fig. d similarly a particle of $p_T < 0.6$ GeV/c has been chosen. The dashed lines show the rapidity distribution observed in minimum bias events and the solid curve is the rapidity distribution observed in average events with $p_W > 2$ GeV/c.

consistent with the picture of a jet of particles following the high p_T particle.

Note that an enhancement of the particle multiplicity with respect to the minimum bias values also remains at $y = 0$, i.e. opposite to the π^0 trigger particle, independent of the y-region of the selected high p_T particle on the opposite side.

So far we have mostly looked at the very general features of the jets, mixing the results from recordings with triggers of all charged particle types. In looking for a dependence on the particle type we show in Fig. 15 the ratio of the multiplicities associated with a $K^\pm$ or $p^\pm$ trigger to that with a π trigger, with $p_W > 2$ GeV/c. The ratio is given as function of rapidity for particles on the same side and on the opposite side. The ratio is consistent with 1 over the entire rapidity interval available, maybe showing a slight trend toward higher multiplicities at large y for a K-trigger with associated particles of $p_T > 0.4$ GeV/c. In the jet interval, $y \lesssim 1$, there is no difference within the present statistics, but it shall be interesting to get more accurate results and separate data for the different charges of the trigger particles.

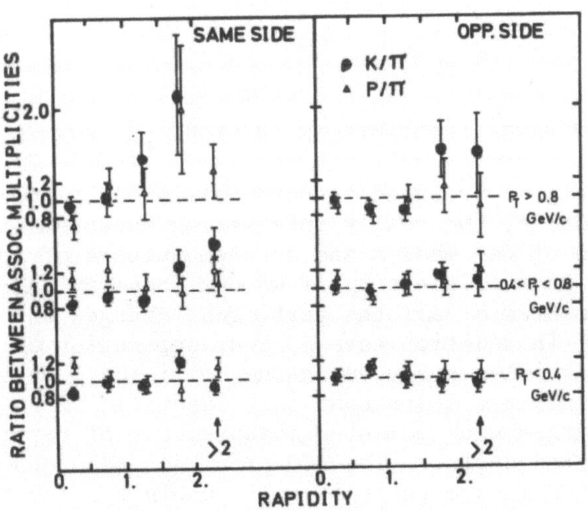

Fig. 15. The multiplicity associated with $K^\pm$ (⧫) or $p^\pm$ (▲) triggers divided by those associated with $\pi^\pm$ as function of rapidity, y. Trigger momentum $p_W > 2$ GeV/c. In the left figure $|\phi| < 30^\circ$. in the right one $|180^\circ - \phi| < 30^\circ$. Three sets of data are shown for each side, viz. for associated particles with transverse momentum $p_T < 0.4$ GeV/c, $0.4 < p_T < 0.8$ GeV/c and $p_T > 0.8$ GeV/c as indicated. The points shown at $y = 2.25$ include all particles with $y > 2$.

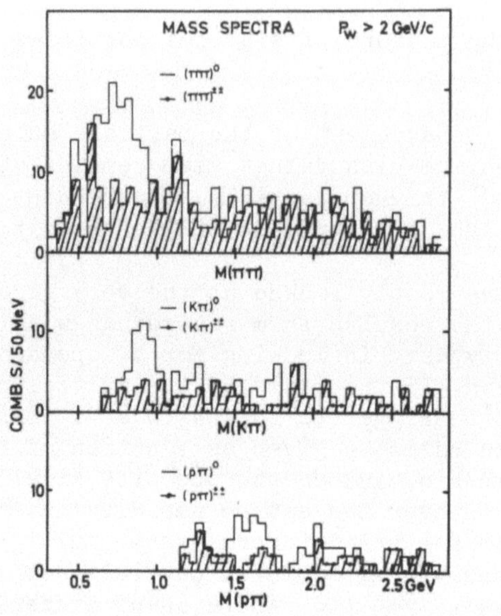

Fig. 16. Invariant mass spectra of two-particle combinations of trigger particles $\pi^{\pm}$, $K^{\pm}$ or $p^{\pm}$ of $p_W > 2$ GeV/c with other particles in the event with $p_x > 0.8$ GeV/c and all assumed to be pions. Both neutral and doubly charged combinations (shaded areas) are shown.

INVARIANT MASS DISTRIBUTIONS IN HIGH p_T EVENTS

If there are jets produced in hard collisions between constituents of the protons, the really interesting measurements would be the determination of the masses and quantum numbers of these constituents. This is possible only if we can identify and measure the energy or momentum of all the particles (charged and neutral) in the events and in addition have a clear separation of the jets from the other particles produced in the collision. We are far from this goal, and have at present only identification of a trigger particle together with momentum measurements of particles over a fairly large solid angle. (The BSO group has recorded information which in their future analysis will enable the identification of a few additional charged particles per event.)

While waiting for better data the BSO group has produced invariant mass (M) plots of two-particle combinations of an identified trigger particle with another particle assumed to be a $\pi^{\pm}$. Fig. 16 shows histograms of the number of such combinations per $\Delta M = 50$ MeV as function of M when $p_W > 2$ GeV/c, the particles appear on the same side with $p_x > 0.8$ GeV/c and their momentum is measured with an accuracy better than 20 %.

Combinations $(\pi\pi)^{\circ}$ and $(\pi\pi)^{\pm\pm}$ are plotted in the same diagram, showing that in the high mass region they are almost equal. In the low mass region the $(\pi\pi)^{\circ}$ distribution shows a strong, broad peak around the mass of the ρ. Similarly the $(K\pi)^{\circ}$ and $(K\pi)^{\pm\pm}$ distributions show a marked difference only in the region of the K* (890) and the $(p\pi)$ distributions show a significant difference only in the region of the N* (1500-1700 MeV) where again the frequency of neutral combinations exceeds that of the charged ones.

The neutral combinations stand out more clearly in Fig. 17 which shows the difference between the number of neutral and the charged combinations.

The finer details of these invariant mass distributions can only be studied with more statistics together with a clear identification also of the particles associated with the trigger. Measurements of inclusive spectra show a decreasing fractional component of pions among the particles produced as p_T increases (Fig. 18). The component of heavier particles increases from about 25 % to

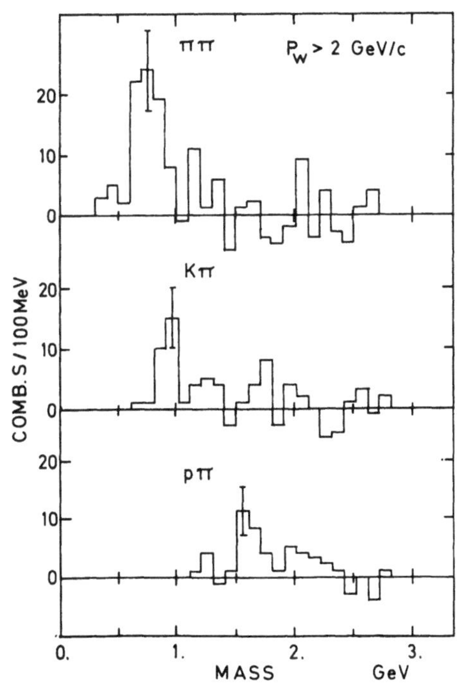

Fig. 17. The invariant mass spectrum obtained as the difference between the spectra of the neutral and the doubly charged combinations shown in Fig. 16.

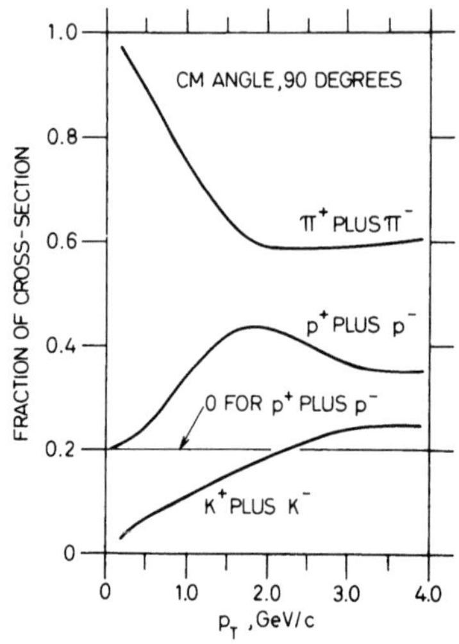

Fig. 18. The fraction of the charged particle production cross section in proton-proton collisions at $\alpha = 90^{o}$ going into pions, kaons and protons (summed over both charge types) as a function of transverse momentum. The curves indicate a smoothed average of data for proton-proton centre of mass energies between 23 GeV and 63 GeV. Note that the curve for protons plus antiprotons is shifted two units upwards.

about 40 % in the range $0.8 < p_T < 1.6$ GeV/c so the contamination of heavy particles in the combinations of Fig. 16 may be expected to be large.

I also want to present some of the invariant mass plots from the CERN group measurements. Fig. 19 shows to the left the number of particles on the same side as the trigger π^o, per unit rapidity for 4 different regions of the momentum component p_x of the particle ($p_{\pi^o} = p_W > 2$ GeV/c). There is a clear enhancement of the distribution around the π^o ($y \sim 0$) which sharpens as the required p_x is increased. For values of $|y| > 1$ the distribution is roughly equal to that obtained with a minimum bias trigger as indicated by the dashed lines (compare also with Figures 10 and 11.)

To the right in Fig. 19 are shown the invariant mass plots for the two-particle combination π^o together with any other particle in the given p_x-range, assuming the latter to be a pion. The distributions are compared to background curves calculated for uncorrelated minimum bias particle distributions and show deviations from the latter only in the low mass region. The deviation increases with increasing p_x and is more or less centered on the mass of the ρ, but with a width too large for a ρ.

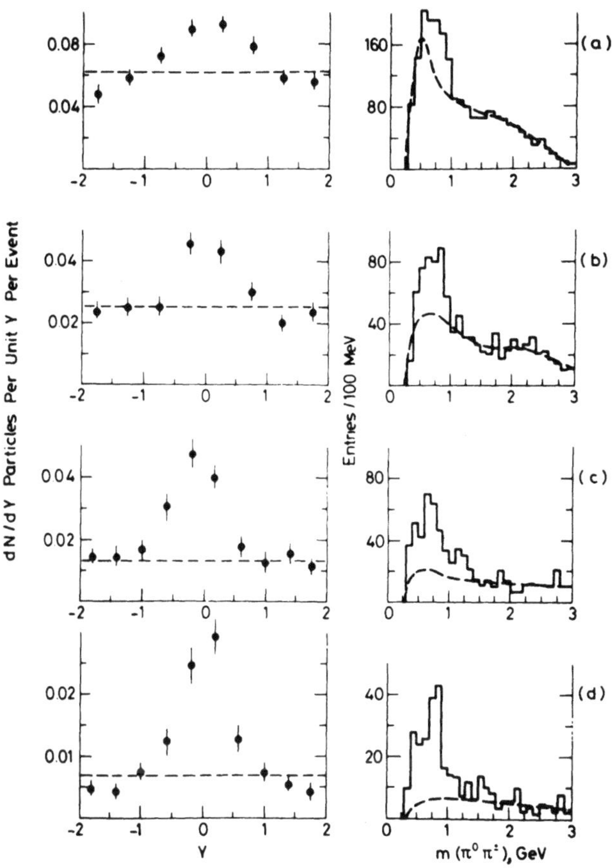

Fig. 19. Rapidity distributions (left) for particles associated with a π^0 of momentum $p_W > 2$ GeV/c and invariant mass distributions (right) of the two-particle combination of the high transverse momentum π^0 with another particle in the event assumed to be a π^+ or π^-. The particles accepted were within $|\phi| < 27^0$ and the data have been grouped in the following intervals of horizontal transverse momentum component, p_x :

a) $0.4 < p_x < 0.6$ GeV/c
b) $0.6 < p_x < 0.8$ GeV/c
c) $0.8 < p_x < 1.1$ GeV/c
d) $1.1 < p_x < 1.7$ GeV/c

The mass distributions include particles with $|y| < 2$ only. The dashed lines represent the minimum bias rapidity distributions and the dashed curves the mass distributions expected for uncorrelated particles.

In order to bring out the mass peak features more clearly the CERN group has added up all the data for $0.7 < p_x < 1.7$ GeV/c which show a mass distribution as seen in Fig. 20. The dashed curve again represents a computed distribution expected for uncorrelated particles.

The mass distribution shows a clear peak centered at the ρ mass, with a width consistent with that of the ρ and the accuracy of the measurements. From the ρ peak and the assumed background rate shown as a dot-dash line, the group estimates the ρ production cross section and finds a value

$$\frac{1}{2}\,\frac{\sigma_{\rho^+} + \sigma_{\rho^-}}{\sigma_{\pi^0}} = 0.9 \pm 0.2$$

for the ratio between the average of ρ^+ and ρ^- production and the direct π^0 production (π^0 from ρ-decay removed).

However, the distribution in Fig. 20 shows an indication of a peak also around $M \sim 400$ MeV. This could come from K* (890) production, which we already have seen is not unimportant (Fig. 17) because if the K in the (Kπ) decay of K* is wrongly labelled π, the invariant mass of the (Kπ) system from K* will appear close to $M = 400$ MeV for the particle momentum intervals used. If this interpretation is correct the cross section for ρ may have been underestimated.

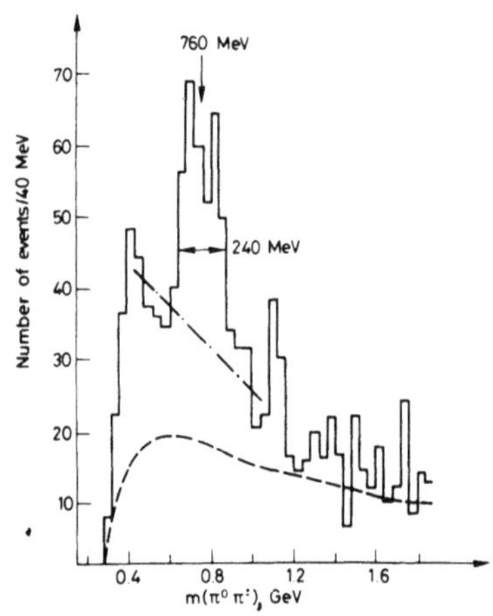

Fig. 20. The invariant mass distribution for two-particle combinations of π^0 and π^+ or π^- as explained for Fig. 19, but with $0.7 < p_x < 1.7$ GeV/c. The dashed curve is the mass distribution expected for uncorrelated particles and the dash-dot-line represents the background used when estimating the ρ production cross section.

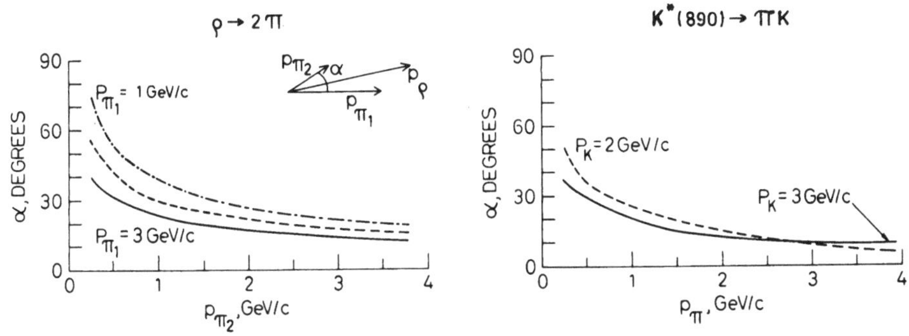

Fig. 21. The opening angle, α , for a ρ decaying into two pions (left) or a K* (890) decaying into a pion and a kaon (right), as function of pion momentum for fixed values of the momentum of the other particle.

Having now found that a very large part of the particle combinations are due to ρ we again turn to the left part of Fig. 19. The distribution was found to become more and more peaked as p_X was increased, with a width of the peak with $1.1 < p_X < 1.7$ GeV/c corresponding to an opening angle of 25°.[23] This is very reasonable if the pions come from the decay of ρ as is shown in Fig. 21, left, where three curves are drawn showing the opening angle between the pions for different pion momenta. It is seen that if $p_{\pi_1} = 2$ GeV/c and $p_{\pi_2} = 1.5$ GeV/c , the opening angle is 25° . Similarly the opening angle for the $K\pi$ system of the K* is shown in Fig. 21, right. Here the angle is found to be even smaller.

The trend towards more peaked distributions of the associated particles when their momenta are required to be higher, could therefore just reflect the production of low mass resonances with high transverse momentum in the proton-proton system.

The same jet-like trend as described above for particles on the same side as the trigger particle, holds also for particles on the opposite side. Fig. 22 shows the number of particle combinations on the opposite side where each particle has $p_T > 0.8$ GeV/c, per unit rapidity difference between the two particles. The trigger particle is required to have $p_W > 2$ GeV/c . For comparison is drawn in the distributions obtained when a particle from one high p_T event is combined with particles from different high p_T events. It is clear that the proper combinations show a distribution more peaked around $\Delta y = 0$ than does the "background" distribution. Within the poor statistics there is also here an indication of a larger number of neutral than doubly charged combinations in this region.

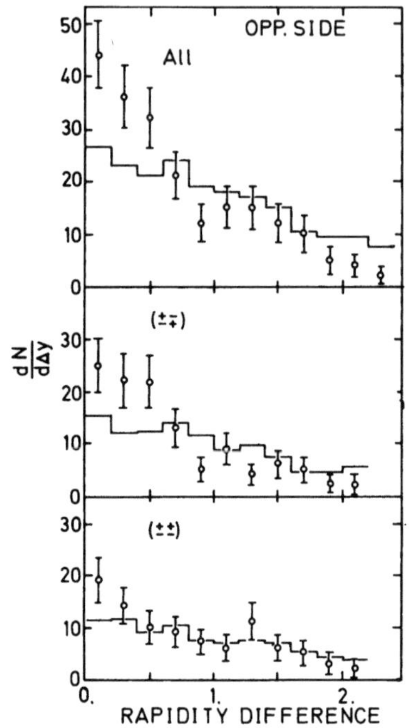

Fig. 22. Distribution of two-particle combinations of particles on the opposite side to a trigger particle with $p_W > 2$ GeV/c, as function of the difference between the rapidities of the two particles. Both particles are required to have $p_T > 0.8$ GeV/c. The histograms represent combinations obtained with one particle taken from one event, the other from another high p_W-event. No distinction between different types of particles. Data are shown for neutral and doubly charged combinations and the sum of these.

In order to get a measure of the mean transverse momentum, q , of a secondary particle in a jet with respect to the jet axis, the BSO group has chosen to look at two-particle combinations once more. The momenta of the trigger particle and the other particle are added vectorially and the components of momentum of each particle perpendicular to this direction is called q (see upper right corner of Fig. 23). p_W is chosen in three different intervals as indicated in Fig. 23 and the other particle on the same side is required to have $p_T > 1$ GeV . The distribution of number of particle combinations as function of q shows a double peaked structure. One peak stays near $q = 0$ GeV/c and shows a tendency of shrinking as p_W is increased. Due to experimental effects the first bin in the diagram is nearly empty so a quantitative value of a mean q cannot be given but a value around $q = 0.4$ GeV/c for $p_W > 3$ GeV/c is a

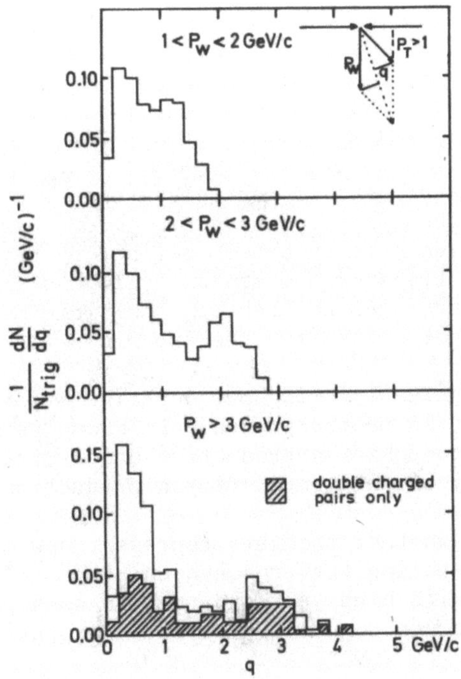

Fig. 23. Distribution of two-particle combinations as function of the particle momentum (q) transverse to the vector sum of the momenta of the two particles. The trigger particle is chosen with p_W in the intervals indicated. The other particle is required to have $p_T > 1$ GeV/c and $|\phi| < 90^o$.

reasonable estimate.

In the figure for $p_W > 3$ GeV/c the shaded areas show the distribution for doubly charged combinations. The peaking is by no means as clear as for the rest, and the peaking may therefore still be due mainly to neutral resonances.

The second peak in Fig. 23, moving towards higher q-values as p_W is increased, reflects only the fact that there are high-momentum particles in the forward direction which, when combined with the trigger particle, whose momentum spectrum is very steeply falling (see Fig. 1), produce the q-values observed.

CONCLUSION

We have used a simple definition of jets as the result of a hard collision between two fluctons (quarks, partons, constituents, gluons or statistical fluctuation giving regions of very high energy density) of the colliding particles. This leads to a picture of coplanar jets of particles (one or more) coming out back to back in the centre of mass system of these constituents. The jets studied here have been defined by having at least one particle of high transverse momentum. If jets frequently break up into many particles no high p_T particle may be present and a different trigger system for the experiment is needed.

The experiments clearly bring out jet-like events both in electron-positron and in proton-proton collisions as measured by sphericity in electron-positron production and by the jet opening angle and coplanarity in the proton-proton production.

It is not clear whether the jets represent new physics or is just another way of stating that resonances (ρ , K^* , Δ , N^* , etc.) are produced with high p_T and that in such productions the high transverse momentum must be balanced essentially locally in the collision.

In order to obtain a clear picture of what happens in high p_T events it is necessary to continue experimentation with large solid angle for recording identified high p_T particles (low cross section) and for identifying as many as possible also of the other particles (both charged and neutral). Only in this way can we hope to separate out jets and study the quantum numbers of the jet-parents available in the process of two particles colliding with high kinetic energi.

REFERENCES

1. B. Alper et al, Proceedings of the 4th Int. Conf. on High Energy Collisions, Oxford, 2 (1972) 197 and Phys.Letters 47B (1973) 75.

2. M. Banner et al, Phys.Letters 41B (1972) 547.

3. R.P. Feynman, Phys.Rev.Letters 23 (1969) 1415.

4. B. Alper et al, Phys.Letters 44B (1973) 521 and 527 and Nuclear Physics B87 (1975) 19.

5. M. Banner et al, Phys.Letters 44B (1973) 537.

6. F.W. Büsser et al, Phys.Letters 46B (1973) 471.

7. D.Sivers, S.J. Brodsky and R. Blankenbecler, Phys.Reports 23C (1976) 1.

8. B. Alper et.al, Nuclear Physics B100 (1975) 237.

9. J.D. Bjorken and S.J. Brodsky, Phys.Rev. D1 (1970) 1416.

10. J.D. Bjorken, Proc. of Summer Institute on Particle Physics, SLAC-191 (1975) 85.

11. S.D. Ellis, M. Jacob and P.V. Landshoff, CERN preprint TH.2109 (1975)

12. P.V. Landshoff, Jets in High p_T Reactions. Talk at the XI Rencontre de Moriond, March 1976.

13. G. Preparata and G. Rossi, CERN preprint TH.2163 (1976).

14. G. Hanson, Proc. of Summer Institute on Particle Physics, SLAC-191 (1975) 237.

15. G. Hanson et al, Phys.Rev.Letters 35 (1975) 1609.

16. P. Grassberger and E.H. de Groot, Nuclear Physics B102 (1976) 297.

17. S.R. Amendolia et al, Phys.Letters 48B (1974) 359.

18. S.R. Amendolia et al, Il Nuovo Cimento 31A (1976) 17.

19. F.W. Büsser et al, Phys.Letters 51B (1974) 306 and 311.

20. F.W. Büsser et al, Nuclear Physics B106 (1976) 1.

21. K. Eggert et al, Nuclear Physics B98 (1975) 49 and 73.

22. British-Scandinavian-Orsay Collaboration, data presented by R. Møller at the 7th Int. Colloqium on Multiparticle Reactions, Tutzing, June 1976, and private communication.

23. P. Darriulat et al, Structure of Final States with a High Transverse Momentum π^0 in Proton-Proton Collisions, to be published in Nuclear Physics.

24. M. Jacob and P.V. Landshoff, CERN preprint TH.2182 (1976).

APPENDIX A.

Coordinate Systems.

The main coordinate system used in the description of the results is shown in Fig. 8. The x-axis is taken to be at 90° to the incoming protons in their centre-of-mass system with positive direction along the trigger particle. The x - y plane is defined by the trigger particle and the incoming protons and does not usually coincide with the horizontal plane in the laboratory system (The trigger particle of the BSO system is at $\sim 4^{\circ}$ and that of the CERN system within $\pm 8^{\circ}$ of the horizontal plane in the laboratory). The z-axis of this righthanded system points upwards as shown in Fig. 8.

The polar angle (θ) in the system is the angle made with the y-axis, the azimuthal angle (ϕ) is measured in the x - z plane, the positive x-axis at $\phi = 0$.

In addition to this system there is a coordinate system attached to the jets, with an axis along the directions of the parents of the jets in the c.m. system of these particles. We have at the moment no possibility to find this system because we have no means of knowing that we have found all the particles belonging to a jet (e.g. the SFM detector records charged particles only). In some cases we approximate this jet system with an axis along the trigger particle on the one side and that of the particle with the highest transverse momentum, p_T , on the other side. In other cases the axis is taken to be along the vector sum of the momenta of the particles included in the set defined as a jet. The opening angle in the jet system is called α, often taken to be the angle between two particles of high p_T. The transverse momentum of a particle with respect to the jet axis is called q.

APPENDIX B.

Acceptance and Normalization.

The acceptances of the equipment and the normalizations should be considered before a discussion of the measured jet-relevant data is started.

So far no other particle than the trigger particle has been identified. In principle one should want a detector with uniform acceptance over the total solid angle of 4π . The non-uniformity of acceptance (for charged particles only) is indicated in Fig. 24

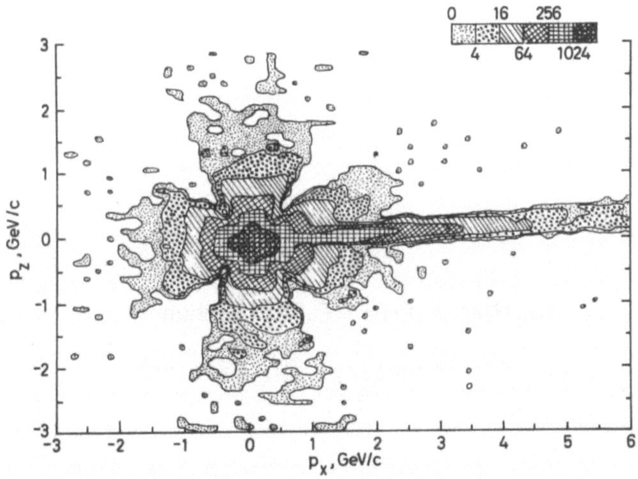

Fig. 24. Diagram representing the particle distribution in the Split Field Magnet detector associated with a trigger particle in the spectrometer of the British-Scandinavian-Orsay group. The horizontal and vertical axes represent the horizontal transverse (p_x) and the vertical (p_z) momentum component respectively (see Figures 7 and 8). The different shadings represent the number of particles recorded as indicated in the upper right corner. The centre of mass energy of the proton-proton system was 52.8 GeV and only particles with longitudinal momentum, $p_y < 1$ GeV/c were included in the plot.

which shows the spatial distribution in terms of p_x and p_z. The "four-leaf clover" has large openings between its leaves where there is either no detector, or the detector is shaded by obstacles. The computing of acceptance for such a system requires a very large amount of Monte Carlo simulations. The data of the BSO group[22] (experiment 413) is instead normalized by dividing the rates observed per event with a high p_T trigger by the rates obtained with a "minimum bias" trigger which requires only that minimum two tracks are found in the SFM detector. A similar normalization has also been used by the CERN group.[23]

From Fig. 23 one can see that the distribution is skew, with more particles in the positive than in the negative x-direction. This is due mainly to the motion of the c.m. system of the two incoming protons, moving with a velocity 0.13 c in the direction of the trigger particle, coupled with the steeply falling production cross section as function of p_T (Fig. 1).

D I S C U S S I O N

CHAIRMAN: Prof. E. Lillethun

Scientific Secretary: W. Marciano

DISCUSSION

- *McPHERSON:*

Do you have a model which explains the relationship between the high mass p_T trigger particle and the extension to higher y of the associated pions?

- *LILLETHUN:*

We have no model for this at the moment.

- *McPHERSON:*

Are the "jets" opposite to the trigger particle better collimated on an event-by-event basis than by the sum of all events which includes an averaging of any angle between the "jets" and $-\vec{p}$ trigger?

- *LILLETHUN:*

In some events we have observed rather well collimated jets; but no real analysis of this has been performed. They could, for example, be pure statistical fluctuations. However, as I showed in one of the transparencies, the y_1-y_2 distribution for two particles on the side opposite to the trigger with high p_T is much narrower than the y distribution of the same particles, indicating a yes to your question.

- *CLEYMANS:*

What is known about the variation of the average multiplicity of particles in either the trigger side jet or the opposite side jet?

- *LILLETHUN:*

We have not tried to look into that as yet.

- *CLEYMANS:*

Do you have any hope of obtaining information on the mass of the jets?

- *LILLETHUN:*

In order to obtain the jet mass, we have to identify, get the masses, of the particles in addition to their momenta. With the present experimental set-up, this information is available for some particles and we hope to study the mass of some particle combinations. We do not, however, detect neutral particles. A new experiment at the ISR with particle identification over much larger solid angles is being planned. Here, jet mass information should become very much better.

- *CLEYMANS:*

Have you corrected your data so as to take into account the centre-of-mass motion at ISR?

- *LILLETHUN:*

Yes. There are, however, some uncertainties since at the present state of analysis, only the trigger particle is identified. All the others are taken to be pions. The data includes information on the identitiy of the particles within some solid angle region; therefore, in the future, better analysis will be possible.

- *FERBEL:*

Are there any new results on the multiplicity of particles associated with any particle as a function of the transverse momentum of that particle? About two years ago, there was evidence for an increase in multiplicity for triggers with $p_T > 1$ GeV/c. Has this been confirmed?

- *LILLETHUN:*

(Added after the discussion session.)

There is new data from the British-Scandinavian-Pisa-Stony Brook Collaboration at the ISR presented at the Tbilisi Conference. The total multiplicity of the event clearly rises with the transverse momentum.

- *SMITH:*

Why do people studying jet structure in pp interactions not use the sphericity parameter that is so famous in the analysis of jet structure at SPEAR?

- *LILLETHUN:*

The sphericity concept is less useful for the following reason: as I indicated under the discussion of Fig. 2b, a proton-proton collision will include a set of particles in the forward-backward direction, i.e. a jet-like structure in a fairly well-defined direction. This comes in addition to a possible jet of the type we look for -- it will both confuse the picture and reduce the sphericity effect. Perhaps, in the future, we may find a way to obtain these two axes and use some more complicated sphericity analysis.

- *BASILE:*

Is the "over-all" charged multiplicity of minimum bias events the same as that of events with jets; high p_T events?

- *LILLETHUN:*

No. The jets have a higher mean charge multiplicity than the minimum bias events. In comparing data with a high p_T trigger with minimum bias trigger data, we shall look into the necessity of choosing minimum bias events with the same multiplicity distribution.

CHARACTERISTICS OF $\mu^- e^+ K^\circ$ EVENTS PRODUCED BY A NEUTRINO BEAM

W. F. Fry

Department of Physics, University of Wisconsin-Madison

1150 University Avenue, Madison, Wisconsin 53706

INTRODUCTION

A search for new phenomena in neutrino interactions has been made in film from the 15 foot bubble chamber exposed to the neutrino horn beam at FNAL. The chamber was filled with 20 percent neon. An external muon identifier (EMI) which surrounded a part of the chamber was very effective in separating muons from strongly interacing particles. About 80,000 pictures were taken which yielded approximately one neutrino interaction in every 10 frames.

One goal of the early analysis was motivated by the observation by the Harvard-Pennsylvania-Wisconsin-FNAL collaboration of neutrino induced events with two muons in the final state.[1] A search for events with a muon and an electron in the final state was made in the bubble chamber film and a sample of events has been isolated. Early results from the study have been previously reported[2] and more detailed results on a significantly larger sample will be given in this publication. These results are to be compared to results from similar studies in the BNL[3] and Gargamelle[4] chambers.

A surprising characteristic of the events is the large number of K mesons produced. The confirmation of dilepton events is these data strengthens the interpretation as production of a new family of hadrons characterized by a new quantum number.

This talk will report in detail the procedures and techniques used to select the events; the characteristics of the events; the associated kaon phenomena, and the systematic errors pertinent to the results.

EXPERIMENTAL DETAILS

The 300 GeV proton beam was extracted from the accelerator and directed on a Berylium target one interaction length long. The resultant spray of positive mesons was focussed by two horns. A typical burst of 6×10^{12} protons produced ν_μ neutrinos. This beam produced about one neutrino event per ten frames. The electron neutrino flux (ν_e) was estimated by Monte Carlo calculation to be about 2% of the neutrinos. This study also found the electron antineutrino ($\bar{\nu}_e$) flux to be about one order of magnitude smaller than ν_e. Although the absolute flux is difficult to determine by calculation, the flux is probably known to about 20%.

For this exposure the chamber was filled with a mixture of neon and hydrogen (20% neon by volume). The density of this mixture is 0.281 gm/cm^3, the radiation length (λ_R) is 116. cm., and the characteristic interaction length is about 3 m.

An important feature of this chamber is the addition of apparatus to distinguish muons from charged hadrons. This External Muon Identifier (EMI) consists of 24 proportional wire chambers each 1 meter square in area located behind and outside the vacuum vessel of the chamber. Metalic zinc has been placed between the coils such that about four interaction lengths is provided by the zinc in the central region. A track is extrapolated from the sensitive volume of the bubble chamber through the absorber and the magnetic field to the position of the wire chambers. The track is called a muon if there is an isolated "hit" in the wire chambers near the position of the extrapolated track.

A hadron will interact in the absorber and the shower of secondary particles will record multiple hits in the wire chambers which have little correlation with the extrapolated parent track. The misidentification of a true muon as a hadron is due mainly to chamber inefficiency and is about 1%. The misidentification of a hadron as a muon is due chiefly to the "punch through" of the energetic shower into the wire chambers and the accidently correspondence of one hit to the parent track. The probability of incorrectly identifying a muon is thus dependent on the number and energy of the hadrons produced in the ν interaction.

For the study of $\mu^- e^+$ events to be discussed here the background of punch through is sufficiently low as to be unimportant.

SCANNING

In the study of rare event type phenomena it is imperative that the scanning and selection process be well understood in order that the results will be quantitatively meaningful. In the experiment reported here, the procedure involved scanning of a segment of the film by physicists and subsequent scanning and rescanning by professional scanners. All candidates for neutrino interactions of all types were studied by physicists, looking for events where an identifiable positron or electron was associated with the neutrino interaction. An electron is said to be identified if it exhibited one or more of the electromagnetic phenomena: (1) visible bremmstrahlung where the change in momentum was clearly discernible by visual inspection of the curvature (abbreviated as V.B.S.), (2) visible bremmstrahlung accompanied by the conversion of the γ ray into an electron pair or compton electron (abbreviated as V.B.S.+e.p.), (3) non-visible bremmstrahlung plus electron pair where the electron pair did not point to the vertex but tangentially to the electron track or (4) the superposition of an electron pair on the track in question (triplet) which comes from the bremmstrahlung and conversion of the γ ray in a short distance so that the electron pair is superimposed on the incident electron.

Because there is a finite probability that the electron pair, for some of the above processes could come directly from the main vertex, it was necessary to apply an origin-electron pair distrance criteria such that this background would be less than 10 percent of the signal.

ELECTRON DETECTION EFFICIENCY

The detection efficiency for electrons and its energy dependence was determined from a study of electron pairs. In this study an electron pair must satisfy the following criteria: (1) zero opening angle and (2) point to a vertex in a chamber. The following quantities were recorded: (1) the type and number of electromagnetic processes which lead to its identification as an electron (if it was identified), (2) its position in the chamber and (3) its momentum. The chamber was divided into five zones along the length of the chamber. The detection efficiency was determined for each zone. To set the average detection efficiency each zone was weighted by its volume and the average taken. The results are tabulated in Table I.

Although a quantitative study of the detection efficiency above 5 GeV/c has not been made, the fact that electron neutrino events have been found which have electrons up to about 100 GeV/c and about at the correct rate is considered to show that this detection efficiency cannot be much lower than at 5 GeV/c.

All events which had an electron of either sign were taken as electron candidates and treated separately from all of the other neutrino interactions. Each of these events were then examined under higher magnification to see if there was any evidence for a

V° or a changed $K^{\pm}$ meson. Since the most striking characteristics of the dilepton events is the number of V° decays, the question arises as to whether the higher percentage of V's comes from a more detailed study under high magnification of the electron events. It was found that an additional V° was found by this more detailed study of the vertex in only about 5-10 percent of the cases and hence their effect is small.

CHARACTERISTICS OF $e^+\mu^-$ EVENTS

Some of the more obvious characteristics of the events are given in the following tables of those events where a e^+ is clearly identified and a μ^- is clearly identified by the EMI or where there is a μ^- candidate.

Table 1
Electron Detection Efficiency

	Detection Efficiency	
$100 < p < 300$ MeV/c	0.29 ± 0.08	
$300 < p < 800$ MeV/c	0.47 ± 0.01	
$800 < p < 3000$ MeV/c	0.52 ± 0.08	} 0.48 ± 0.07
$3000 < p < 5000$ MeV/c	0.41 ± 0.11	

RATE OF μ^-e^+/CC

The production rate of μ^-e^+ events for the cuts $E_{e^+} > .8$ GeV and $E_{vis} \gtrsim 5$ GeV can be estimated by correcting for the electron detection efficiency which has been given previously. The contribution of neutral currents in the denominator has been subtracted if the partial cross section is to be compared to the charged current cross section. To obtain the true production rate, however, two more corrections have to be made.

Loss of Events Due to Simulated Dalitz Pairs

True μ^-e^+ events will be lost in our selection process if the positron, combined with one of the negative non-interacting hadrons in the reaction, gives a mass $m(e^+L^-) < m(\pi^\circ)$. This loss was determined empirically by studying events with Dalitz pairs, where both $e^{\pm}$ were identified. For these events, the positron simulated a Dalitz pair with a leaving negative track in approximately 20% of the times.

Table II
Topology of the Events

ID#	How e^+ Ident.	μ^- Ident.	K°_S	K°_L	$K^\pm$	$\pi^\pm$	$L^\pm I^\pm S^\pm$
1	1NVB+PR 1VB+PR	EMI	+ -				
2	1NVB+PR 2VB+PR	Geom miss of EMI	+ -				
3	2(VB+PR) ANNIH+PR	EMI	+ -				
4	1DELTA 1VB 1TRID	Geom miss $L^+(2.8)$ in EMI	+ -				
5	1(VB+PR) 1NVB+C $\delta(-2)$at VTX	EMI	+-,00	$1d(1I^\circ?,1I^+?)$			
6	1NVB+C 1NVB+PR 1NVB+C?	EMI	+ -	$1d(1I^\circ?,1I^+)$			
7	1NVB+CPR 1VB+CPR		$\Lambda(+-)$			1^+d	
8	1VB+PR	SHWR+NO EMI	$\Lambda(+)$				
9	1VB+PR	Geom miss EMI		$1I\to\Lambda$ $1I\to K^\circ_\pi?$	1^-d		
10	1NVB+PR 1VB KINK or VB	EMI	+ -	1I,1n		$3I^+$ 1^-CH	$2L^+,6\gamma$ $1^-(K)$
11	1VB+PR	EMI	00				
12	1NVB+PR 1VB+PR	EMI	or Λ				
13	2(VB+PR)	EMI			1^+d		
14	1NVB+PR 1VB+PR	EMI	or $\Lambda(+-)$				

NVB Non-visible brems.
VB Visible brems.
TRID $e^\pm$ pair on e^+ track
PR Electron-positron pair
CPR Close electron-positron pair
C Compton electron
n Neutron
γ $e^\pm$ pair from prim. VTX

L Leaving track
I Interacting track
S Stopping track
d Decay
CH Charge exchange

Table III
Lepton and Strange Particle Kinematics

ID#	E_{vis}	Momenta					Visible				
		μ^-	e^+	K°_S	K°_L	$K^\pm$	Q^2	ν	x	y	W^2
1	33	14	1.6	6.1			.13	18	.004	.56	35
2	12	3.4	1.1	3.1			1.6	8.1	.10	.71	15
3	26	21.8	2.2	1.6			6.2	3.1	1.	.12	.4
4	28	9.3	5.2	5.8			7.2	18	.21	.66	28
5	98	58	2.1	+-00 4.4	2.1		.21	39	.003	.41	75
6	24	15	.4	2.2	3.3		5.5	7.8	.37	.34	10
7	15	SS 2.0	5.6	1.2Λ			.16	12	.007	.86	24
8	29	(5.4)	.64	11Λ			1.0	23	.024	.81	43
9	20	(.80)	3.9		3.6 4.6	.6	16	18	.49	.96	18
10	97	±S 9.3	5.7	38	1.0		9.4	87	.058	.90	154
11	21	14.1	1.3	00 1.1			1.2	5.7	.11	.29	10
12	30	26.2	1.4	1.6 orΛ			2.7	2.9	.49	.10	3.6
13	61	SS 17.9	4.0			+ .25	.70	42	.009	.70	78
14	42	33.6	.80	1.5 orΛ			3.8	7.3	.29	.17	10

SS "same side" muon

Table IV
Mass, Momentum Correlations $\perp$ to $\vec{\nu}_{inc}$

ID#	M_{e^+K}	Positron $P^e_{\hat{n}}$	Positron $P^e_{\hat{\mu}}$	Positron $\lvert\vec{P}^e_\perp\rvert$	Strange $P^K_{\hat{n}}$	Strange $P^K_{\hat{\mu}}$	Strange $\lvert\vec{P}^K_\perp\rvert$	Event Total $P^{miss}_{\hat{n}}$	Event Total $P^{miss}_{\hat{\mu}}$	Event Total $\lvert\vec{P}^{miss}_\perp\rvert$
1	.70	-.018	-.006	.019	-.552	.776	.952	1.897	.512	1.964
2	.81	-.051	-.040	.065	.277	-.924	.965	.170	.234	.289
3	1.04	-.321	-.503	.597	.180	-.862	.881	.108	-.951	.957
4	.70	-.187	-.499	.531	-.279	-.550	.617	.260	.102	.279
5	.60 .69	.007	-.256	.257	-.089 .272	-.205 .556	.229 .619	-.369	-.234	.437
6	.66 .67	-.047	.067	.081	.143 .024	-.219 -.573	.261 .573	-.091	-.807	.812
7	2.6(Λ) (1.5(K))	-.390	-.843	.929	.320	-.221	.389	.033	-.581	.581
8	1.3(Λ)	.085	.079	.116	-.490	.592	.769	-.237	1.320	1.341
9	.85 .75 1.4K⁻	-.052	-.213	.219	-.082 -.371 .026	-.629 -.391 -.241	.634 .539 .242	.407	1.052	1.129
10	1.3 2.1	.054	-.577	.580	.249 .049	1.296 -.254	1.320 .259	-.019	-1.251	1.252
11	1.08	.379	-.416	.562	-.002	.327	.327	-.740	-.487	.886
12	1.38(K)	-.397	.288	.491	-.194	-.925	.945	.678	-.990	1.200
13	1.89	-.334	.482	.587	.210	.068	.221	.062	-.593	.596
14	.95K (1.5Λ)	.420	-.164	.451	-.104	-.024	.107	-.240	-.814	.849

Loss of Events at Energies $E_{e^+} < .8$ GeV

Unfortunately, this loss cannot be determined experimentally since it depends on the mechanism by which the positrons are being produced. If the positron originates indeed from the decay of a particle with a mass of several GeV, the momentum spectrum of the positron will extend to quite small values and the effect of the $E_{e^+} > .8$ GeV cut will be substantial. It has been estimated using 3-body decay of a 2 GeV mass particle, that as much as 50% of the events will be lost with our cut.

The best estimate for the true rate is then

$$\mu^- e^+/CC = 1.5\%.$$

ENERGY DEPENDENCE OF THE PRODUCTION RATE

The sample of $(e^+\mu^-)$ events have been divided into two groups, those where the total visible is below 30 GeV and those above. This ratio is then to be compared to the same ratio of number of events in a partial sample of charged current events where in each event a μ^- was identified by a hit in the EMI. These data are summarized in the following table.

Table V

	$e^+\mu^-$ Events	CC Events
Number of events $E_{vis} > 30$ GeV*	6	31.5%
Number of events $E_V < 30$ GeV*	9	68.5%
Number E > 30/Number E < 30 GeV	.66 ± .18	0.46 ± .05

*The average energy for $E_V < 30$ GeV is 21 GeV while the average for $E_{vis} > 30$ GeV is 62 GeV for $e^+\mu^-$ events.

There is no meaningful evidence for an energy dependent in the cross section for $e^+\mu^-$ events from these relatively poor data.

STRANGE PARTICLES

In order that the K^0 multiplicity m_0 that has been found in $e^+\mu^-$ events be meaningful it must be compared to that found in "normal" charged current interactions. A sample of charged current interactions have been studied looking for associated V's which come from decays of Λ°, K_S° and K_L°. Very preliminary results of this study in a partial sample is summarized in the following table (VI). The V° events listed in this tables comes from a sample of about 970 ν interactions which have a visible energy above about 5 GeV. An attempt was made to eliminate neutral currents in this sample (see Table VI).

Since about 85% of the V's in the $\mu^- e^+$ events are K_S° decays it is interesting to compare the ratio of various K_S° decays to charge current interactions in the $e^+\mu^-$ events with that in the charge currents. This comparison is shown in Table VII.

The total V°/(interaction) in $(e^+\mu^-)$ compared to V°/(CC int.) is a ratio of

$$12/14 \times .05 = 17.$$

A more detailed analysis of the strange K particle production in $(e^+\mu^-)$ now follows.

K° PHENOMENA

Although the available data on strange particle production in the $e^+\mu^-$ events is somewhat meager, it is enough to give some important results. The most quantitative results come from the K° phenomena while the evidence for charged K mesons is at best only collaborative. A listing of the events along with the strange particle phenomena is given in Table VIII.

The most definitive mode of identification of the K° is its two body decay; $K_S^\circ \to \pi^+ + \pi^-$. Typically it is a 3C fit to a vertex. The mean error in the invariant mass is about 25 MeV in this bubble chamber so that the chance coincidence of a neutron background event is negligible.

It is important to note that in two cases the V° has been identified as a Λ° and not a K°. In two other cases there is a K°-Λ° ambiguity. One possible source for the Λ°'s is the production of Λ°'s by a $\overline{K^\circ}$ meson which was produced in the initial neutrino interaction. A crude estimate of this probability gives a number which is consistent with the 2 Λ° events observed. (The two Λ°-K° events which are ambiguous are more likely to be K° than Λ° from a statistical consideration of probabilities.)

Perhaps the simplest way to estimate the mean number of neutral K mesons produced per event is to use only the $K_S^\circ \pi^+\pi^-$ mode and correct for the branching ratio. It will be assumed that the numbers of K_S° and K_L° are equal and that the branching ratio of $K_S^\circ(\pi^+\pi^-)$ to K_S° total is 2/3, this gives the efficiency for detecting a K_S° as 1/3. In Table VIII the total number of events with one K_S° or other V's in addition is 10 out of a total of 14 $e^+\mu^-$ events. This gives for the average multiplicity of K° production

$$m(K_{sL}) = (10/14) \times 3 = 2.1 \pm 0.7.$$

The error is just the statistical error in 10 events.

A more sophisticated method to evaluate this same multiplicity is to use all of the K° phenomena (reactions 1 through 5) and to weight the value of the mean given by the various reactions by the statistical weight of each reaction. The reactions considered are:

$K_S^{\circ} \rightarrow \pi^+ + \pi^-$ (1)

$K_S^{\circ} \rightarrow \pi^{\circ} + \pi^{\circ}$ (2)

K_L° Decay $\rightarrow$ charged modes (3)

K° interactions $\rightarrow K_S^{\circ}$ or Λ (4)

K° interactions in the ν target nucleus before escaping (5)

The evaluation of the detection efficiency for each reaction is involved and a breakdown of the probabilities for each reaction is given in Table IX.

Not only is the mean multiplicity m_0 important but also the second moment of the distribution. Clearly the data are insufficient to give a quantitative evaluation of shape of the distribution of the multiplicity. However, in events 4 and 6 there are three charged V decays. Even though one of the V°'s in each of the two events may have come from a charged exchange of a K^+ it does show that at least three or more strange particles were produced. event 4 there is good evidence for four strange particles, three of which were neutral K°'s.

The expected numbers of events using the detection efficiencies given in Table IX for estimating the various types of events are given in Table X. There is reasonable agreement of the number of events of the various types with what is expected.

OTHER ELECTRON PHENOMENA

In the scanning, all tracks were examined to see if there was an electromagnetic process which would identify it as an electron track; either as an electron or a positron. Five processes lead to the production of electrons or positrons; namely:

1) Genuine μ^-e^+ (or μ^+e^- from $\bar{\nu}_\mu$ contamination).

Table VI
Number and Type of V Events in "Normal" Charged Current ν_μ Interactions in 20% Neon

Decay Mode	Number	Number of V's/CC
K_S°	19	19/970 = .02
Λ°	28	28/970 = .03
K_L°	2	2/970 = .002
$K_S^\circ \Lambda^\circ$	2	2/970 = .002
$K_S^\circ K_S^\circ$	0	
Total	51	5%
(Total sample CL = 97)		

Table VII
Comparison of V Event Rates in $e^+\mu^-$ Events with CC Events

Decay Mode in V° Events	V°'s in $e^+\mu^-$ Number	V°'s in $e^+\mu^-$ N/Event	V° in CC
$K_S^\circ(\pi^+\pi^-)$	6	6/14 = 42%	2%
$\Lambda^\circ(p\pi^-)$	2	2/14 = 14%	3%
K_L°	1	1/14 = 7%	0.2%
$K_S^\circ(\pi^+\pi^-)K_L^\circ$	1(K_L° int)	1/14 = 7%	$0 \rightarrow \leq 0.1\%$
*$K_S^\circ(\pi^+\pi^-)$, K_L°, K_L°	2	2/14 = 14%	$0 \rightarrow \leq 0.1\%$
One or more V's/Event		12/14 = 85%	5%

*The K_L° decay is taken in both $(e^+\mu^-)$ events and (CC) events even though it may come from a charge exchange of a $K^\pm$.

Table VIII

Event	Type of V	Event	Type of V
1	$K_S(+-)$	8	$\Lambda°$
2	$K_S(++)$	9	$(K_L?)$
3	$K_S(+-)$	10	K_S $K_L(\text{int}\to\Lambda°)$
4	$K_S(+-)$, $K_S(00?)$, K_L, K_L	11	$K_S(00?)$
5	$K_S(+-)$, K_L, K_L	12	$K_S(+-)$
6	$\Lambda°(p\pi^-)$	13	
7	$K_S(+-)$	14	$K_S(+-)$

2) ν_e interactions and hence an absence of a muon.
3) Dalitz pairs from $\pi°$ or $\eta°$.
4) High energy knock knock-on electrons along a track very close to the neutrino vertex, and
5) Electron pairs where the conversion of the γ occurred so close to the ν interaction that the gap could not be observed. If the pair is very asymmetric the lower energy electron or positron may be unobserved and hence only one electron of either sign seen.

Background classes 3, 4 and 5 can be reduced to a level below the μ^-e^+ signal by a suitable choice of electron energy and asymmetry parameters as was discussed in the section on backgrounds.

The number of electrons and positrons which are consistent with or identified as Dalitz pairs are estimated from the average number of pions in normal neutrino interactions. There are about 4-5 Dalitz pair candidates for each identified $e^+\mu^-$ events.

Table IX
Evaluation of Detection Efficiency for Reactions

Reaction	Overall Det. Efficiency ξ	Comments
$K_S^\circ \to \pi^+ + \pi^-$	$\xi = 0.95$	If the decay occurs within $\sim$1.0 cm of origin it may be missed. This percentage is $\sim$0.5%.
$K_S^\circ \to \pi^\circ + \pi^\circ$	$\xi \cong .38$	Difficult to estimate. The probability of observing 3 out of 4 γ's is about 0.5. The decay must occur at a distance >5 cm $\xi = 0.5 \times 0.75 = .38$
K_L° Decay	$\xi \cong .03$	The mean usuable decay length in the chamber is about 1.8 meters giving a probability of decay into charged modes of 3%.
K_L° Interaction	$\xi \cong .12$	The detectable reactions are $K_L^\circ + \eta \to K_S^\circ + \eta'$ (1) $\overline{K^\circ} + \eta \to \Lambda^\circ + \eta'$ (2) Reaction (1) gives 0.6 for probability of K° interacting times 0.2 for K_S° regeneration making the total of 6%. Reaction (2) gives $\sim$30% for K° interacting in $\overline{K^\circ}$ mode times 20% for Λ production making total of 6%.
K° interaction in target nucleus giving Λ°	$\xi = .07$	$\overline{K^\circ} + \eta \to \Lambda + \eta'$ in the ν target nucleus. (Assume that the number of K°'s available for this reaction is one less than $\langle m\rangle$. Probability of interaction is $\overline{K^\circ}/K^\circ = 1/2$ times probability of interaction giving $\Lambda^\circ \cong 20\%$ from $\overline{K^\circ}$ mode. $\xi = (\langle m\rangle^{-1}/\langle m\rangle) \times 1/3 \times .2 \cong .07$.

Table X
Expected K° Phenomena

Based on m(0) = 2 (for 13 $\mu^{-}e^{+}$ events)

	$K_s(+-)$	$K_s(00)$	K_L Decays	K_L int. in liq.	K_L int. in target nucleus $\rightarrow \Lambda^{\circ}$
observed	8	2	4	1	2
expected	8	1.6	0.4	0.8	1.8

There is a non-negligible contamination of electron neutrinos in the muon neutrino beam. No quantitative measure of the ratio of electron to muon neutrinos has been made, but an estimate can be obtained from the ratio of K/π from the proton horn target. This gives a ratio of about 4 percent. In the sample of filmed scanned for $e^{+}\mu^{-}$ events there are with the number estimated from the ν_e flux, the number of ν_e candidates is consistent. These events which have all the characteristics to be expected from ν_e interactions; namely the x and y distributions are reasonable and they do not have a muon associated with the interaction. Not only are these events of interest in the study of electron neutrinos, but they clearly show that the detection efficiency for high energy electrons does not become small. The electron energies go to as high as about 90 GeV. In summary, these ν_e interactions look like the normal ν_μ interactions expected for the lepton and they both differ in the ways from the $e^{+}\mu^{-}$ events in that the normal ν_e and ν_μ charged current interactions have a low V° multiplicity and the electron energy in the $e^{+}\mu^{-}$ events peaks at a much lower value.

SUMMARY AND CONCLUSIONS

The most important properties of the $\mu^{-}e^{+}$ events are summarized in the following paragraphs.

a) The most striking characteristics of the $\mu^{-}e^{+}$ events is the high K° multiplicity. A search for decays of charged K mesons is clearly less conclusive but the results are consistent with the charged multiplicity being equal to the neutral multiplicity. The best estimate for the average neutral K multiplicity is about 2 $K^{\circ}/\mu^{-}e^{+}$ events. This is to be compared with an observed K° multiplicity of 0.06 for the charged current neutrino interactions.

b) The Λ° production is much lower than the K° production. It is possible that the observed Λ°'s come entirely from secondary $\overline{K^{\circ}}$ interactions in the target nucleus. The observed $K^{\circ}/\Lambda^{\circ}$ ratio associated with the $\mu^{-}e^{+}$ events corrected for the probability of observing each, is found to be 8 while in ν CC interaction this ratio is found to be about 1.3. Clearly the strange production is different for the μ^-e^+ events than in the ν_μ CC events.

c) The rate of production of $e^+\mu^-$/CC is estimated to be 1.5%.

d) There is no meaningful evidence for any energy dependence in the production of the μ^-e^+ events, although the statistics are poor.

e) The values of x and y are not significantly different for the μ^-e^+ events than for CC ν_μ interactions.

f) The ratio of the momenta of the two leptons is a significantly large number $\langle p_\mu \rangle / \langle p_e \rangle = 6.28$ suggesting that the two leptons do not come from the same vertex.

g) There has been observed one event of the type μ^+e^- which is consistent with an interaction of an anti-mu-neutrino giving $\mu^+ + e^- +$ (hadrons).

h) The remaining ν interactions which have an electron and not a muon are consistent with interactions of electron neutrinos.

The high K° multiplicity suggests that there could be the strong decay of a charmed particle into another lower state which subsequently decay leptonically into a positron, a neutral K meson and a neutrino. Several such schemes have been proposed.

REFERENCES AND FOOTNOTES

1. A. Benvenuti, D. Cline, W. T. Ford, R. Imlay, T. Y. Ling, A. K. Mann, R. Orr, D. D. Reeder, C. Rubbia, R. Stefanski, L. Sulak and P. Wanderer, Phys. Rev. Letters 35, 1199 (1975), 35, 1249 (1975).

2. J. von Krogh, W. F. Fry, U. Camerini, D. Cline, R. J. Loveless, J. Mapp, R. H. March, D. D. Reeder, A. Barbaro-Galtieri, P. Bosetti, G. Lynch, J. Marriner, F. Solmitz, M. L. Stevenson, D. Haidt, G. Harigel, H. Wachsmuth, R. Cence, F. Harris, S. I. Parker, M. Peters, V. Peterson and V. Stenger, Phys. Rev. Letters 36, 710 (1976).

3. E. Cazzoli et al., Phys. Rev. Letters 34, 1125 (1975).

4. H. Deden et al., Phys. Letters 58B, 361 (1975).
J. Blietschau et al., Phys. Letters 60B, 207 (1976).

D I S C U S S I O N

CHAIRMAN: Prof. W.F. Fry

Scientific Secretaries: J. Shigemitsu and A. Bisello

DISCUSSION

- *BALDO CEOLIN:*

You said that you can check the hypothesis of two K^0's for each charmed event by comparing the interactions of the naturally produced K^0 in the ν and $\bar{\nu}$ cases, measuring, for example, Λ^0 production, K_1^0-K_2^0 mass difference, etc. Do you not always get ν, $\bar{\nu}$ producing dimuons and dikaons in the two cases with the same couple of K^0 and $\bar{K}^0$, the only difference being the exchange between the naturally produced and the other?

- *FRY:*

There is a large difference between ν_μ and $\bar{\nu}_\mu$. For neutrinos, the D is positively charged and decays into a $\bar{K}^0$; while the K^0 produced in the nucleus is K^0. In antineutrinos, the D is negative, giving a K^0, while the K in the nucleus is a $\bar{K}^0$ which has a high probability of producing Λ^0 hyperons. For this reason, antineutrino production may lead to a high number of Λ^0 hyperons.

- *WIGNER:*

Three questions: What is the order of magnitude of the nuclear cross-section in these energies? What is the chance that a hadron interacts? and What is the neutrino energy involved in these reactions?

- *FRY:*

The answer to your first question is of the order of 10^{-38} cm^2 per nucleon. It is quite likely that a hadron interacts because the geometrical mean free path is about 2.8 m, while the diameter of the chamber is a little more than 4 m. The mean neutrino energy is about 30 GeV, but the spectrum goes beyong 150 GeV.

- *MARCIANO:*

Does the rate $(e + \mu^-)/(\text{charged currents } \nu) = 1.3\text{-}1.5\%$ include the $\mu^+\mu^-$ mode? If not, is this not rather high for the charm picture? And could you elaborate a bit on how the contamination of the beam occurs?

- *FRY:*

The rate 1.3-1.5% did not include the $\mu^+\mu^-$ mode. If one naively doubles this percentage, it is high and could, under certain assumptions, present somewhat of a problem only if the non-leptonic decay is high; however, I will show in a moment that the non-leptonic mode does not contribute a large percentage.

Most of the ν_μ's come from $\pi\mu$ decays and K_e decays, which strongly favour ν_μ and are lightly peaked in energy. The main source of electron neutrinos is K_{e3} decay which produce more energetic ν_e's than the ν_μ from π decays.

- *SMITH:*

Why are there so many K's? Does the charm picture not predict only one kaon?

- *FRY:*

The extra K can result from the neutrino acting on a strange quark from the sea. This would give a K^0 or K^+ from production, along with a $\bar{K}^0$ from the leptonic decay of the charmed particle. For example

$$\nu + N \rightarrow D^+ K^0 N \mu^-$$
$$\quad\quad \hookrightarrow \bar{K}^0 e\nu \; .$$

For antineutrino beams, there will be a K^0 or K^- produced in the Ne nucleus. This should result in greater Λ^0 production from $\bar{\nu}$ beams than from ν beams.

- *FERBEL:*

Would you not expect a lot of charged K production?

- *FRY:*

Yes. This is a harder question to evaluate; we need more data than exists at the present time in order to quantitative evaluate the charged K production.

HADRON PHYSICS AT FERMILAB*

T. Ferbel

Department of Physics and Astronomy

University of Rochester, Rochester, New York, 14627 USA

ABSTRACT

I survey recent experimental results from studies of hadron interactions at Fermilab. Elastic, total and charge-exchange cross section measurements, diffractive phenomena, and inclusive production, using nuclear as well as hydrogen targets, are discussed in these lectures.

INTRODUCTION

The momentous J/ψ discoveries of almost two years ago [1] have put an unusually depressing damper on the entire field of "old-fashioned" hadron physics. Subsequent to the J/ψ announcements, every able-bodied experimentalist who had equipment installed on the experimental floor of Fermilab abandoned all his previous well-planned efforts to join in the mad search for charm, color or what have you. (Unless your experiment involves a prompt lepton trigger or has such key words as "search for narrow.." on the proposal, you are still regarded in some circles of Fermilab as a myopic pariah.) Despite all the frenetic J/ψ related activity, several groups have nevertheless managed to generate some outstanding results on "hard-core" hadron physics. In these lectures I will summarize some recent experimental and phenomenological investigations which I find particularly stimulating. I will try, where possible, to present those results which tend to raise new questions rather than answer old ones.

---

*Research supported by the United States Energy Research and Development Administration.

I. TOTAL, ELASTIC AND CHARGE-EXCHANGE CROSS SECTIONS

Total Cross Sections

Precision measurements of total and of elastic scattering cross sections are now available for $K^{\pm}$, $\pi^{\pm}$, n, p and $\bar{p}$ projectiles. The latest data for total cross sections have recently been published [2]. I reproduce the results of these measurements on hydrogen and deuterium in Fig. 1. It is clear that all hadronic total cross sections appear to fall from values measured at lower energies; after passing broad minima, which occur near $\sim$10 GeV/c for K^+p and perhaps as high as $\sim$300 GeV/c for $\bar{p}p$ (no $\bar{p}p$ measurements are as yet available above 200 GeV/c!), the cross sections start to rise again. The differences between antiparticle-nucleon and particle-nucleon cross sections shown in Fig. 2 can be parameterized as a function of the square of the energy in the center of mass (s) in terms of a power-law dependence of the form $As^{\alpha-1}$. The values of α obtained from a fit to the difference in the $\bar{p}p$ and pp cross-section (α_{pp}=0.39±0.02) and for the K^-p and K^+p difference (α_{Kp} = 0.40 ± 0.04) are the same within error. From Regge phenomenology we expect these differences in the particle-antiparticle cross sections to isolate the dominant contribution of the ω Regge trajectory (although the differences also contain ρ and A_2 terms, these mesons do not couple nearly as strongly to the proton as the ω does). Recent measurements of K_S^0 regeneration on carbon [3] yield a value for the ω trajectory at t=0 of $\alpha_\omega(t=0)=0.43\pm0.01$, a result in good agreement with the above given α_{pp} and α_{Kp}. Similarly, the value of $\alpha_{\pi p}=0.55\pm0.03$, obtained from the data in Fig. 2, is reasonably consistent with that expected from the contribution of the ρ trajectory to the difference between the π^-p and the π^+p cross section. (A value of $\alpha_\rho(0)=0.48\pm0.04$ has recently been extracted from measurements of $\pi^-p\to\pi^0n$ at Fermilab [4].)

The data in Fig. 2 indicate that $\Delta\sigma_T(p^{\pm}p)$ is about 2-3 times larger than $\Delta\sigma_T(K^{\pm}p)$; also, $\Delta\sigma_T(p^{\pm}n)$ is about 4 times larger than $\Delta\sigma_T(K^{\pm}n)$. These kind of differences in the hadron-hadron cross sections can be used to test various symmetry relations. In particular, the ratio $3[\Delta\sigma(K^{\pm}p)+\Delta\sigma(K^{\pm}n)]/[\Delta\sigma(p^{\pm}p)+\Delta\sigma(p^{\pm}n)]$ measures the contribution of ω-exchange to kaon-nucleon relative to baryon-nucleon total cross sections. The constancy observed for this ratio from $\sim$15 GeV/c through 200 GeV/c speaks for an energy-independent universal coupling of the ω meson to kaons and baryons. I will not elaborate further on this subject, I will note only in passing that, besides ω universality, ρ-ω universality and the Johnson-Treiman relations are all in agreement with the data, to $\sim$(10-20)% accuracy, in the 30-250 GeV/c momentum range. (These agreements may be short-lived if the trend observed for the

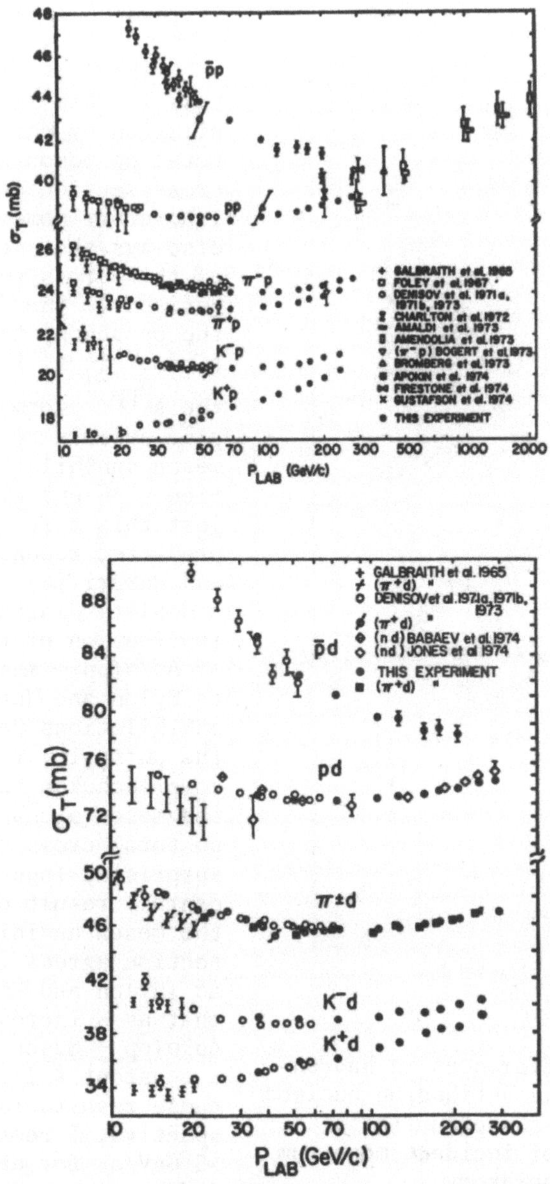

Figure 1: Total cross sections for hadron interactions with hydrogen and deuterium target particles.

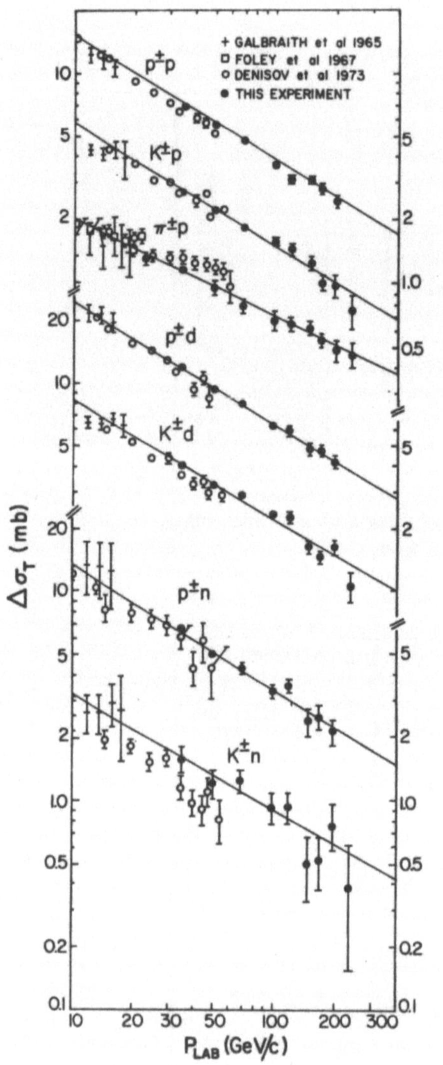

Figure 2: Differences in hadron-nucleon and antihadron-nucleon total cross sections as a function of incident momentum in the laboratory.

K^-p-K^+p difference to decrease more rapidly for momenta above 100 GeV/c persists with further increase in energy or in data!)

The last item I will discuss in this section is a compilation by Whitmore [5] of the total $\bar{p}p$ meson-annihilation cross section as a function of laboratory momentum. Figure 3 displays the energy dependence of this cross section in comparison to the difference of the $\bar{p}p$ and pp total cross sections. It is interesting that the $s^{\alpha_{pp}-1}$ form given in Fig.2 also appears to describe the meson annihilation cross section. An energy variation of just this sort was, in fact, predicted recently by Eylon and Harari [6] on the basis of a duality-diagram model. The implication of the constancy of $\Delta\sigma/\sigma(\bar{p}p \to \text{mesons})$ according to Eylon and Harari, is that $\bar{p}p$ annihilations contribute through the unitarity relation to the Pomeranchukon rather than to the meson exchange terms in the $\bar{p}p$ total cross section. The surprising (and perhaps accidental) result of Fig. 3 is that the meson annihilation cross section agrees in magnitude with $\Delta\sigma$ (Eylon and Harari predict that as s increases $\Delta\sigma/\sigma(\bar{p}p \to \text{mesons})$ approaches a constant $\leqslant 1$). To put these results in some perspective, I remind you that at $\sim$5 GeV/c, for example, the difference in the $\bar{p}p$ and pp elastic cross section is $\sim$5 mb, which means that the $\bar{p}p$ inelastic non-annihilation cross section is smaller than

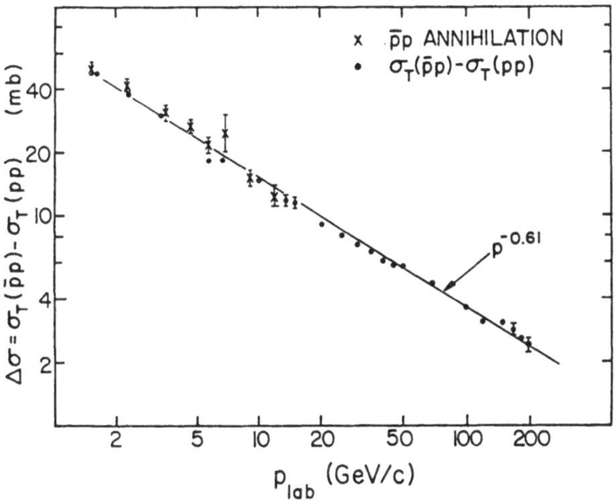

Figure 3: Comparison of the difference in p̄p and pp total cross sections with the cross section for p̄p annihilation into mesons.

the inelastic pp cross section, and so it is clearly not proper to regard the p̄p interaction as simply the super-position of the pp interaction and the additional annihilation channel which is not available to the pp system. Absorption effects in the final states certainly play an important role in the generation of $\Delta\sigma$; this is a theme I will periodically return to in the course of this lecture.

Elastic Scattering

Data on the elastic scattering of $\pi^{\pm}$, $K^{\pm}$ and $\bar{p}$ on hydrogen in the 50-200 GeV/c momentum range have become available during this past year. Figure 4 provides the character of the data obtained at small momentum transfers [7]. The differential cross sections for the scattering of particles on hydrogen are observed to have shapes similar to those found for the scattering of their respective antiparticles on hydrogen. All the differential cross sections appear to have curvature in t, and all except the p̄p channel tend to become steeper with increasing momentum (only the 100 GeV/c data are shown in Fig. 4). The latter result is emphasized in Fig. 5, where the slopes of the elastic spectra $B(t) = d(\ell n\, d\sigma/dt)/dt$ at $t=0.2$ GeV^2 are plotted as a function of s. The difference in the steepness of the K^+p and K^-p diffraction peaks, as well as of the p̄p and pp data, is slowly disappearing with increasing energy. All meson channels appear to be approaching a common value for B, a value which is substantially smaller than that describing the p̄p and pp slopes. At the smallest values

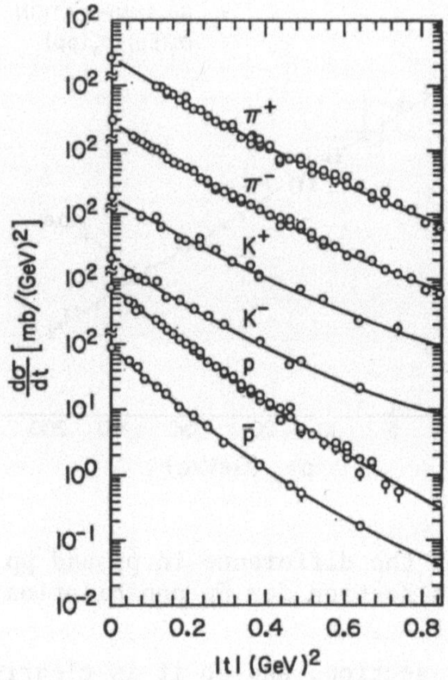

Figure 4: Elastic scattering differential cross section for hadrons incident on protons at 100 GeV/c.

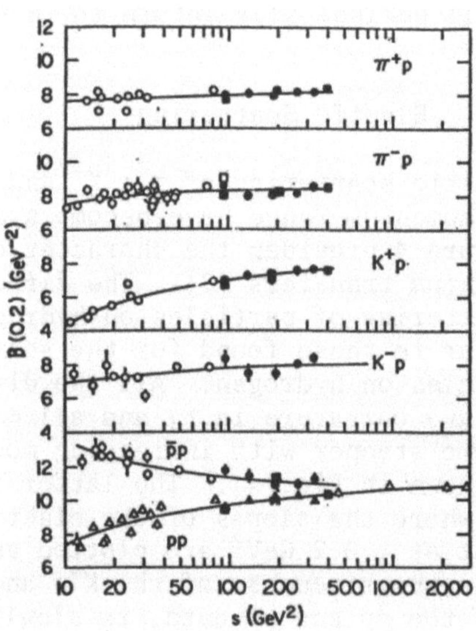

Figure 5: Slopes of the t-distributions at $|t|=0.2$ GeV^2 for hadron-proton and antihadron-proton elastic scattering as a function of s (see ref. 7).

of t, negative-particle cross sections tend to be larger than positive-particle cross sections, and vice versa at larger t. This effect, referred to as the crossover phenomenon, has been examined extensively at low energies [8].

In terms of a Regge-exchange picture, the observed difference between antiparticle and particle elastic scattering can be understood as follows. At small values of t the Pomeranchukon is the dominant exchange trajectory. The Pomeranchukon contribution to the elastic amplitude is mainly imaginary and non-flip. As in the case of total cross sections, the difference between antiparticle and particle elastic scattering can be attributed to the exchange of meson trajectories with negative charge conjugation. Consequently, the difference in elastic scattering must correspond to an interference term between the Pomeranchukon and the non-flip imaginary part of the relevant vector meson trajectories ($\text{Im } V_{\Delta\lambda=0}$). In the case of $K^{\pm}p$ and $p^{\pm}p$ the strongly coupled ω trajectory dominates the C=-1 exchange and the difference in K^-p and K^+p, as well as $\bar{p}p$ and pp, elastic scattering is therefore expected to be large. For the $\pi^{\pm}p$ channels, wherein ω cannot be exchanged because of the constraint of G-parity, only ρ contributes and consequently, because of the weak ρ-nucleon coupling, the $(\pi^-p)-(\pi^+p)$ difference in elastic cross sections is expected to be small. The t-values at the positions of the cross overs are $t_c^K=0.19\pm0.04$ GeV^2 for $K^{\pm}p$ and $t_c^p=0.11\pm0.02$ GeV^2 for $p^{\pm}p$ data, in the momentum range 50-175 GeV/c [9]. There appears to be a weak energy dependence in t_c^p (t_c^p was measured to be 0.162 ± 0.004 GeV^2 near 5 GeV/c [8]). The latest data [9] pertaining to the cross-over phenomenon for $K^{\pm}p$ and $p^{\pm}p$ channels are shown in Fig. 6 in terms of the ratios $[(d\sigma/dt)^-/(d\sigma/dt)_f^+]$ and $[(d\sigma/dt)^+/(d\sigma/dt)_f^+]$, where $(d\sigma/dt)_f^+$ are the fitted (smoothed) data for positively charged projectiles.

The fact that t_c is so small means that $\text{Im}V_{\Delta\lambda=0}$ goes through zero for $t \lesssim 0.2$, which implies that the exchanged C=-1 contribution must be exceedingly peripheral. Davier and Harari [10] have, in fact, shown that within the framework of a dual absorption model the non-Pomeranchukon part of the elastic scattering amplitude is strongly dominated by the largest partial waves. (The Pomeranchuk contribution is central while the other Regge terms peak near an impact parameter of $\sim$1 fermi.) Consequently, these results suggest that a simple Regge-pole description of the crossover phenomenon cannot be adequate - strong absorption, cuts, or the like, are required to parameterize the effect.

For s-values above $\sim$25 GeV^2 the ratios of elastic to total cross sections appear to be constant and same (to within $\sim$5% accuracy) for antiparticles and particles [11]: $\pi^{\pm}p/K^{\pm}p/p^{\pm}p$ = 0.15/0.12/0.18. Similarly, the ratios of forward elastic slopes to total cross sections are energy independent [12]:0.34/0.39/0.26.

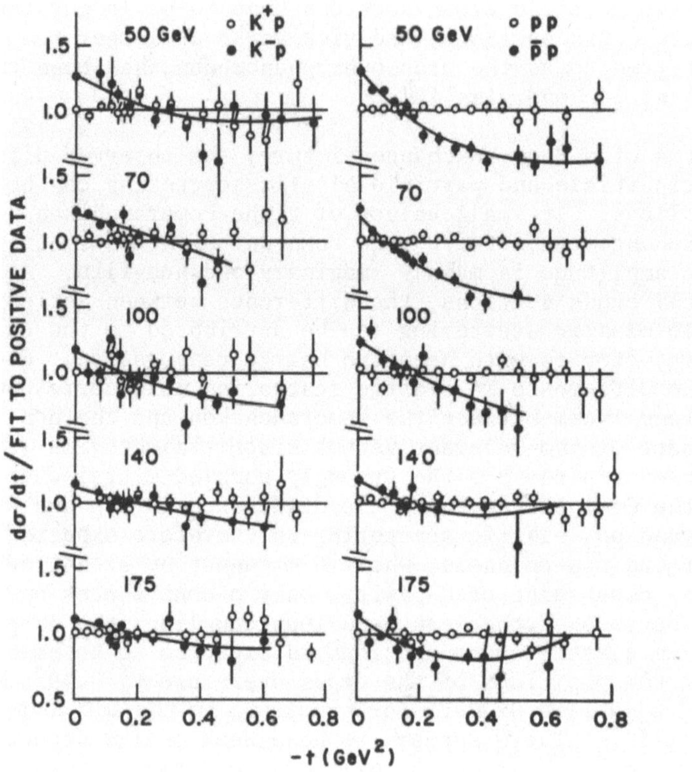

Figure 6: Study of the crossover phenomenon in elastic scattering between 50 GeV/c and 175 GeV/c.

The energy independence of these ratios can be realized in a model invoking geometrical scaling (GS) of the scattering amplitude in impact parameter (A(b,s)) [13]. The s-dependence in this model is contained entirely in the effective radius of the interaction (the scale), and the amplitude is energy independent when expressed in terms of b/R(s).[12,13] A pictorial way of viewing the GS assumption is that the distribution of nuclear matter does not change with s when it is parameterized in terms of the dimensionless quantity $\sqrt{\pi b^2/\sigma_{inelastic}}$. An immediate consequence of GS is that inelastic, elastic and total cross sections are proportional to R^2, as is the slope of the elastic differential cross sections at small t. Furthermore, because dσ/dt can also be equated to $R^4|f_{el}(R^2t)|^2$, the quantity $(1/\sigma_{tot}^2)(d\sigma/dt)$ should be s-independent when plotted as a function of $\sigma_{tot} \cdot t$. [14] All the above features of the GS hypothesis are in remarkable agreement with the data. Parameterizing a logarithmic growth of R^2 with increasing s also provides the observed rise in σ_{el}, σ_{tot} and B for the FNAL-ISR energy regime. Independent of the success of the GS hypothesis, it is clear that, phenomenologically, elastic scattering at small t is simplifying with increasing s-values.

A very surprising result in the pp elastic cross section at large t has been the rapid onset at 200 GeV/c of the Chou-Yang dip, observed previously at the ISR near t = 1.5 GeV^2 [15]. Figure 7 displays 100 GeV/c and 200 GeV/c pp data from Fermilab illustrating this unusual effect. The dip at 200 GeV/c occurs at t_D=1.5 GeV^2, which is to be compared with a value of $t_D \sim 1.3$ GeV^2 at 1500 GeV/c (ISR). An energy dependence of this kind for t_D is, again, consistent with the simple GS hypothesis.

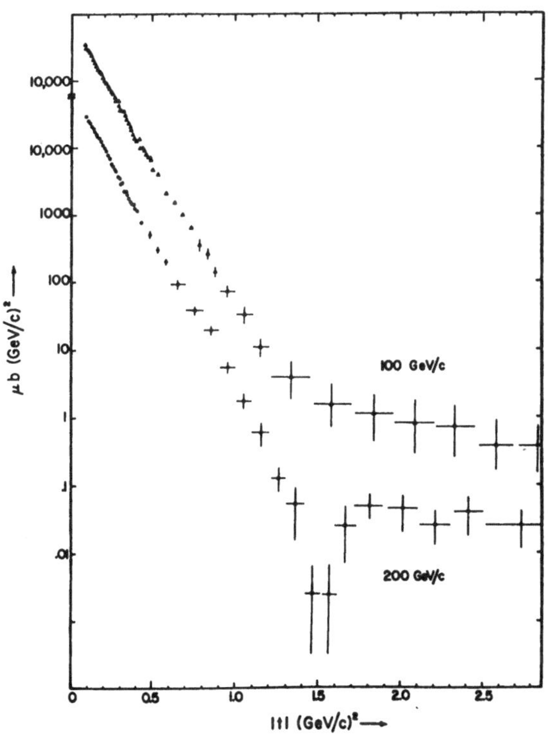

Figure 7: Elastic scattering cross sections for pp collisions at 100 GeV/c and 200 GeV/c.

Finally, preliminary measurements of the ratio of the real to imaginary part of the foward scattering amplitude (ρ), obtained using nuclear-coulomb interference at small t, for $\pi^{\pm}p$, $K^{\pm}p$ and $p^{\pm}p$ in the 70 GeV/c to 150 GeV/c range of incident momenta [16] appear to be in general agreement with the latest calculations [17] using dispersion relations and the new measurements of total cross sections [2]. The values of ρ are within ∿5% of zero near 100 GeV/c momentum, except for ρ_{pp}, which is ≈ -0.1, consistent with previous measurements [18] at Fermilab. The only sizeable discrepancy between the new data and the calculations may be for the K^-p channel; however, the experimenters go out of their way to warn of the preliminary nature of their results.

Two-Body Charge-Exchange Reactions

Pion charge-exchange scattering has been measured at Fermilab in the π^- momentum range of 20 GeV/c to 200 GeV/c [4]. The differential cross sections and phenomenological fits to these spectra are shown in Fig. 8. The data display the well known helicity-flip turn-over at t=0 and the dip near t=0.5 GeV^2; these features have been attributed to the dominance of the ρ trajectory in the production. Although the authors point out that it is likely that the charge exchange amplitude has small contributions from sources other than just the ρ-trajectory (an asymmetry has been observed in the scattering from polarized targets near 5 GeV/c), they have nevertheless attempted to fit their data to the simple Regge-pole form $d\sigma/dt=\beta(t)\nu^{2\alpha(t)-2}$, where $\nu=(s-u)/4M$, u is the square of the four-momentum transfer between the incident π^- and the neutron, and M is the nucleon mass. The authors have performed a 7 parameter fit to $\beta(t)$ and a 3 parameter fit to the effective trajectory $\alpha_\rho=\alpha_0+\alpha_1 t+\alpha_2 t^2$. The result of their fit to $\alpha_\rho(t)$ is displayed in Fig. 9. A straight line through points corresponding to ρ and g mesons falls remarkably close to the extracted $\alpha(t)$ for $|t|<0.3$ GeV^2. The value of $\alpha_\rho(0)$ is significantly lower than the value of 0.58±0.03 obtained at lower energies [19]. The relatively poor agreement of the present fit with data at lower energies is displayed in the comparison with the 5.9 GeV/c data in Fig. 8. (There is, in addition, an apparent systematic discrepancy with the data from Serpukhov.)

The authors have also compared their fits to the charge exchange reaction with the difference in the π^-p and π^+p cross sections. Assuming that the forward charge exchange cross section continues to fall with increasing s in the same manner as observed up to 200 GeV/c, the authors determined the ratio of the real to the imaginary part of the forward amplitude using a dispersion relation. This, along with the recognition that the imaginary part of the forward charge exchange amplitude is related to the difference in the π^-p and π^+p total cross section enabled them to check the consistency of their data with that of reference (2). The successful result of this check (and of dispersion relations, isospin invariance, and the optical theorem) is displayed in Fig. 9.

Just as the pion charge exchange reaction is believed to be dominated through the exchange of the ρ trajectory, the reaction $\pi^-p\to\eta n$ is expected to proceed through A_2 exchange. This reaction has now also been measured at Fermilab [4]. Near 100 GeV/c the η

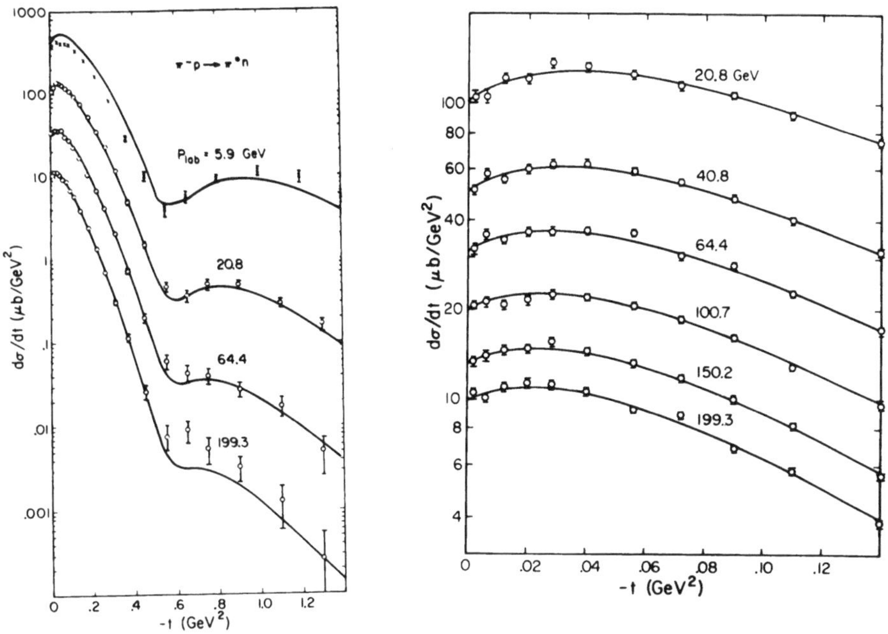

Figure 8: Differential cross sections for $\pi^- p \to \pi^0 n$ between 20.8 GeV/c and 199.3 GeV/c. The data at 5.9 GeV/c are from a previous investigation.

production reaction has a cross section of ∿0.8 μb, compared to the charge exchange cross section of ∿3 μb. The data and the authors' fit to the A_2 trajectory are shown in Fig. 10. The properties of this reaction appear to be quite similar to those noted for the charge exchange process. Again, except in just a qualitative way, the fit to α_{A_2} at high energies is not very consistent with the results at low energy. (In particular, the extracted value of $\alpha_{A_2}(0)=0.37\pm0.01$, is low again.) Because the effective ρ and A_2 trajectories do not have the same slopes or intercepts this obviously means that they are not degenerate for $t<0$. And since there is no established $J^P=4^+$ partner to the A_2, it is not clear what is the situation for $t>0$. In any case, the concept of exchange degeneracy which has often been taken as an article of faith must be considered, at best, a poor approximation to reality.

In contrast to pion charge exchange, neutron-proton charge exchange cross sections exhibit characteristic sharp peaks (with widths of ∿0.02 GeV^2) at $t=0$. Measurements of differential cross

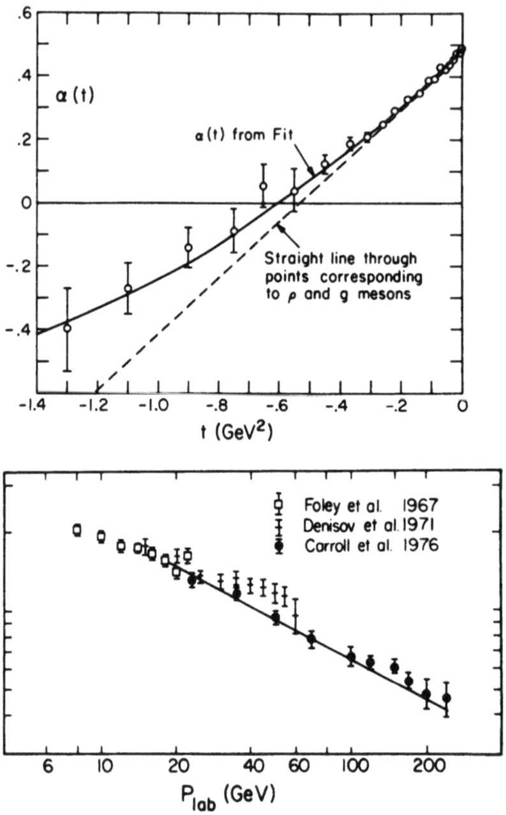

Figure 9: Fit to an effective Regge trajectory $\alpha(t)$ for the reaction $\pi^- p \rightarrow \pi^0 n$. The bottom graph provides a comparison of the measured difference in $\pi^- p$ and $\pi^+ p$ total cross sections with the difference expected from an analysis of the charge exchange process.

sections for neutron-proton charge exhange for incident momenta between 60 GeV/c and 300 GeV/c are shown in Fig. 11. [20] The authors remark that, although the absolute cross section falls by a factor of $\sim 10^5$, the shape of their spectra is essentially the same as observed near 1 GeV/c. The total cross section falls as p_{lab}^{-2} between 2 GeV/c and 12 GeV/c, as $p_{lab}^{-1.5}$ near 50 GeV/c and as p_{lab}^{-1} at FNAL energies. The sharp peak near t=0 is reminiscent of reactions which are dominated by pion exchange. The data at low energies do, in fact, agree with the presence of a large non-flip contribution from pion exchange. The onset of a shoulder in the cross section at $t \approx 0.1$ at high energies may be due to the emergence of contributions from higher lying trajectories

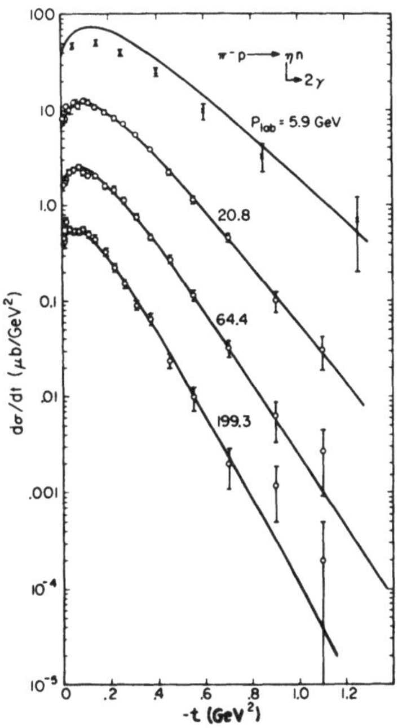

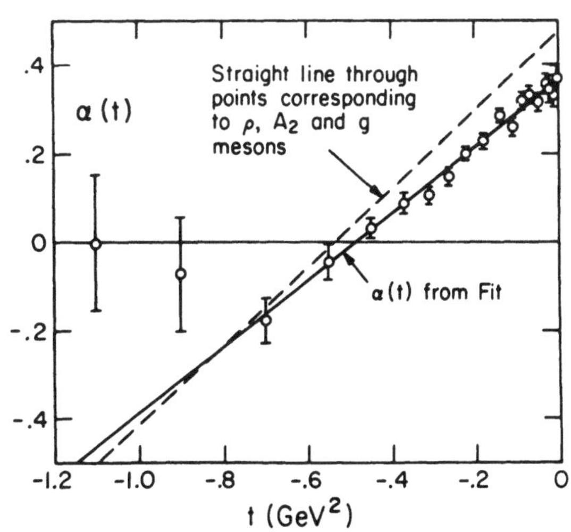

Figure 10: Phenomenological fits to the data for $\pi^- p \to \eta n$. The effective Regge trajectory $\alpha(t)$ should correspond to the A_2 meson.

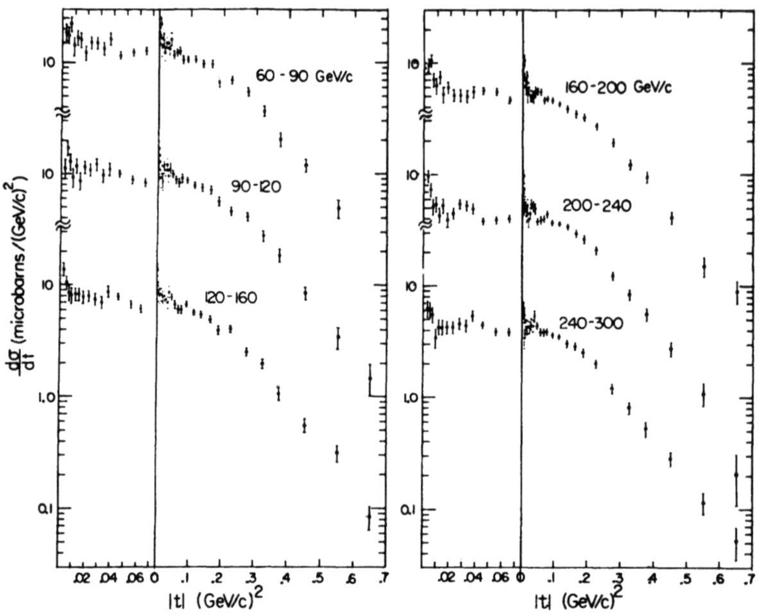

Figure 11: Data for the charge exchange reaction np → pn at Fermilab energies.

which were overwhelmed by the pion term at small s. Figure 12 displays the author's fit to an effective exchange trajectory for the process. Unlike the results at lower energies, the effective $\alpha(t)$ has more resemblance to a vector-meson than to a π trajectory. This is not entirely surprising because the π-exchange contribution falls rapidly with increasing energy. But then arguing this way makes the presence of the steep peak at $|t|\lesssim 0.02$ somewhat enigmatic.

From an extrapolation of the fitted np charge exchange data to t=0, the authors have provided a bound on the absolute difference between the pp and np total cross section as a function of beam momentum. These bounds are far more stringent than those available from direct measurements of the individual total cross sections. At 250 GeV/c, for example, the expected difference can be no greater than 0.30±0.03 mb.

I conclude this section with the note that the simple Regge ideas have fared surprisingly well in their confrontation with the charge-exchange data. Although exchange degeneracy has been dealt a severe blow, the essential features of the Regge-pole

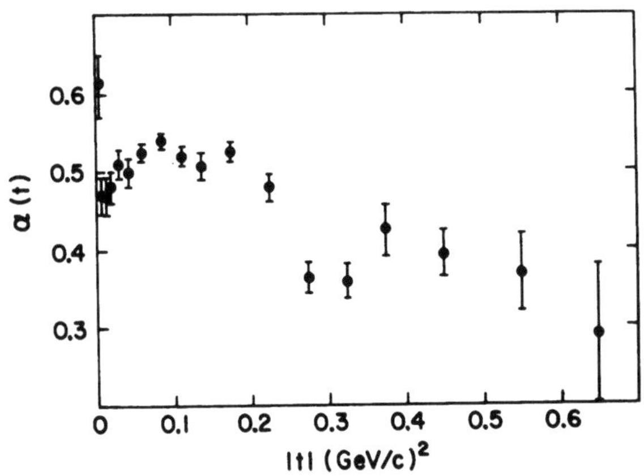

Figure 12: Effective Regge trajectory for the data displayed in Fig. 11.

exchange mechanism have survived the stringent tests of ultra-high energies. The small changes in the effective-trajectory parameters can certainly be attributed to absorption effects or to the presence of contributions from other non-leading exchanges which cannot be ignored at low energies. For the np charge exchange reaction matters are somewhat less clear, but here the low-lying π-trajectory is involved and consequently the test of Regge ideas are more muddled by possible contributions from cuts and strong absorption.

II. INELASTIC DIFFRACTION PRODUCTION

Selected Exclusive Channels

An extensive investigation of the dissociation of neutrons in the reaction $np \to (p\pi^-)p$ is presently in progress [21]. Preliminary results, which have recently been published, [22] indicate that the decay properties of the produced $p\pi^-$ system can be described quite well in terms of a diffractive Deck model involving pion as well as proton exchange graphs. Figure 13 provides the essential features of the data. The $p\pi^-$ mass (M) spectrum has a characteristic accumulation at low mass and shows, in addition, resonance structure near 1500 and 1680 MeV. The t distribution of the $p\pi^-$ system is sharply peaked at small angles, as is the case in all diffractive processes. Although the t spectrum for all M shows substantial curvature near $t \sim 0.1$, it can be approximated rather well by an exponential for $t \lesssim 0.2$ when the data are not integrated over M (i.e., at fixed M). The decay angles of the

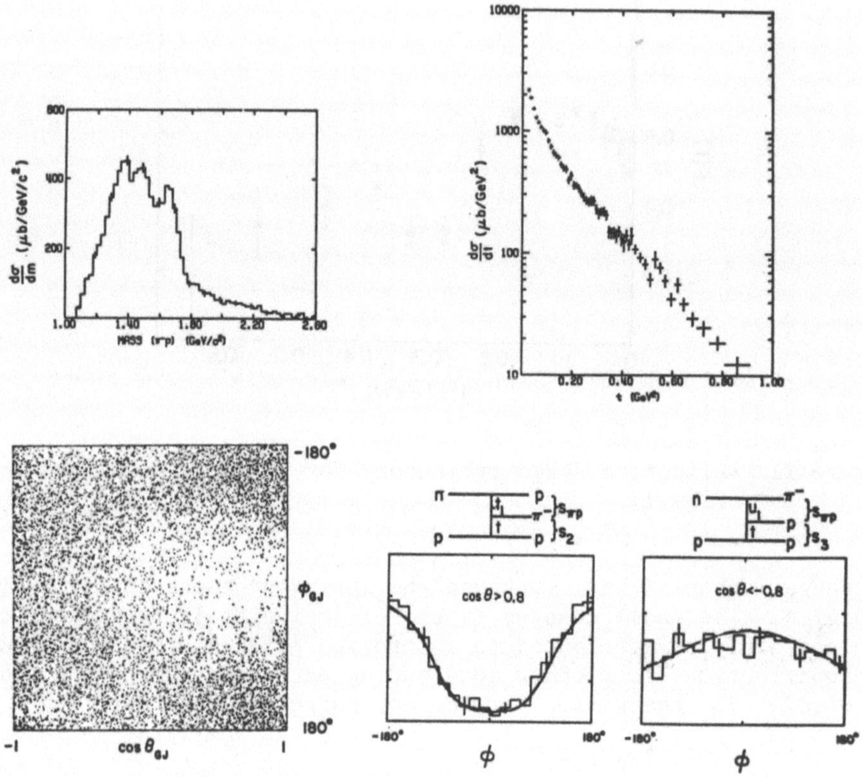

Figure 13: General characteristic of the reaction $np \to p\pi^-p$ at Fermilab energies.

proton as examined in the Gottfried-Jackson frame of the $p\pi^-$ system (for M<1.4 GeV) are also shown in Fig. 13. The scatterplot displays the sort of structure expected from the dominance of Deck processes in the production. The pion-exchange contribution is dominant for $\cos\theta\sim1$ and the proton exchange near $\cos\theta\sim-1$. (Because there are no known $I=\frac{1}{2}$ baryon resonances below M=1.4 GeV, I would regard with suspicion fits to these data which involve only the super-position of the π-exchange diagram and resonance production.)

Figures 14, 15, 16 and 17 display some other striking and rather unexpected features of these same data. Simple exponential fits have been performed to the t spectra as a function of M. Figure 14 displays the slopes in t obtained in these fits. At the threshold in $p\pi^-$ mass the slope is about two times larger than observed in elastic pp scattering at comparable t-values. This immediately suggests that the diffraction process takes place at

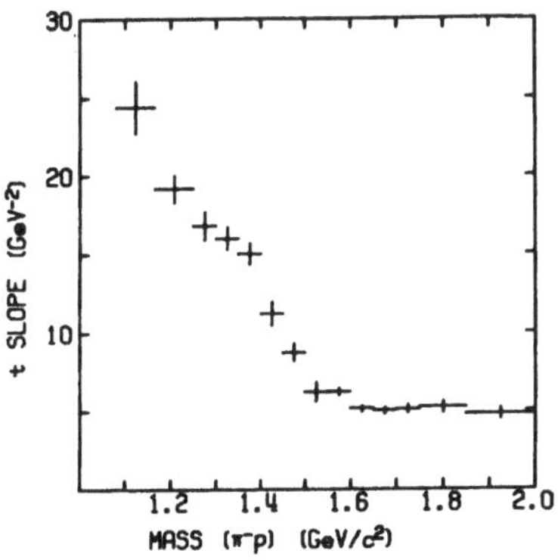

Figure 14: Slope of the t-distribution as a function of the mass of the π^-p system produced in the diffractive reaction $np \to p\pi^-p$.

larger impact parameters than elastic scattering. The slope falls almost linearly to a value of 5 GeV^{-2} at M∿1.55 GeV, above which it is constant. An interesting shoulder is observed in the slope at M∿1.35 GeV.[23] There is essentially no energy dependence observed for the dissociation process. In Figure 15 I show the slope in t for M<1.4 GeV as a function of momentum. There is no apparent variation in this parameter, although a change of the order of ∿5%, as observed in elastic scattering, clearly cannot be excluded by the data. Another characteristic of diffractive processes is the lack of any substantial s-dependence in the cross section. This feature of the data is shown in Fig. 16, where cross sections for specific mass intervals are given as a function of momentum. The absolute values of these cross sections are typically about a factor of two smaller than observed at ∿20 GeV/c.

Last year at my seminar here [22] I showed the t-distribution of the data for two regions of cosθ. E. L. Berger encouraged us to examine the t-dependence on cosθ in more detail and the remarkable result of this effort is shown in Fig. 17 (for M<1.35 GeV). At very forward θ the t spectra display a turn over characteristic of the Deck π-exchange diagram. At large θ the t distribution is rather steep (presumably due to proton exchange). But for θ ~ π/2 there is an unusually steep interference dip near t~0.2 GeV^2 [24]. It is not clear at all what causes this highly peripheral feature of the data. The results for higher mass values are

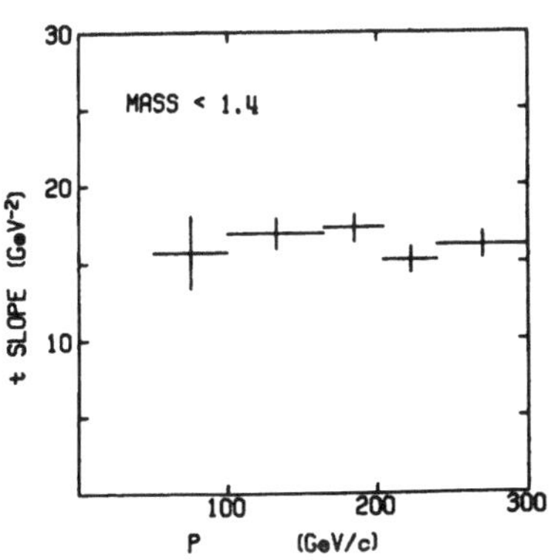

Figure 15: Energy dependence of the t-slope parameter in the reaction $np \to p\pi^-p$.

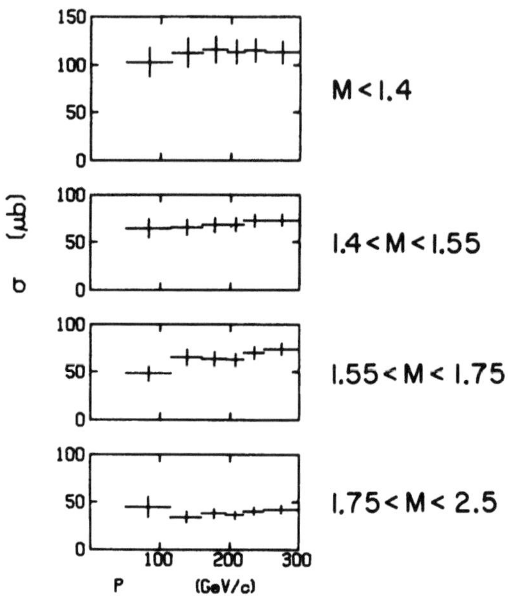

Figure 16: The variation with incident momentum of the cross section for the production of a $p\pi^-$ system of fixed mass M in the reaction $np \to p\pi^- p$.

qualitatively similar, but the sharp dip moves out in t and becomes less pronounced with increasing M (akin to the results in Fig. 14). I would conclude by saying that although the qualitative aspects of these data can be understood on the basis of a Deck model (with absorption), the most striking result, namely the unusual correlation between cosθ and t is still to be explained. In addition, it has been emphasized by several authors [25] that when spin is taken into account in the baryon-exchange Deck diagram (spin was ignored in the work of ref. 22), the predicted ϕ distribution for $\cos\theta \simeq -1$ tends to peak at $|\phi| \sim \pi$, which would be in disagreement with the data. Obviously, more work is required here to establish the nature of the exclusive diffraction production mechanism.

The double diffractive dissociation (DD) of the proton in the reaction $pp \to (p\pi^+\pi^-)+(p\pi^+\pi^-)$ has been studied at 200 GeV/c and 300 GeV/c [5]. This reaction has been compared to the single dissociation (SD) of one proton in the reaction $pp \to (p\pi^+\pi^-)+p$ to see whether there is evidence for factorization in these Pomeranchukon-dominated processes. At fixed t and M, Pomeranchuk factorization

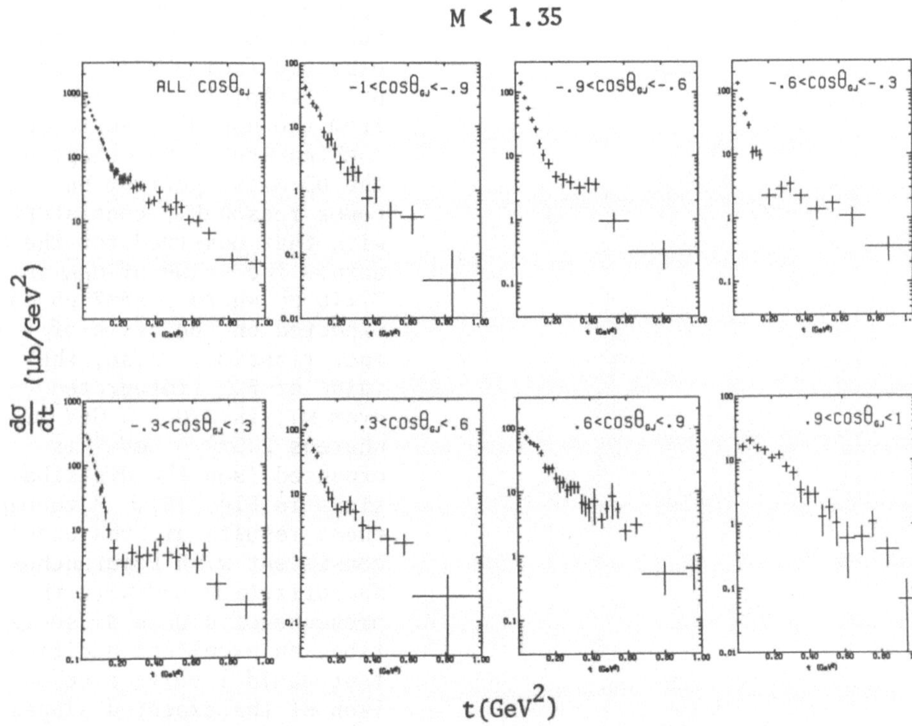

Figure 17: Differential cross sections for the production of a $p\pi^-$ system as a function of the decay angle (θ_{GJ}) of the system in the reaction $np \to p\pi^- p$.

would imply the following relationship between the DD process into masses M_1 and M_2 and the SD processes into M_1 and into M_2:

$$\frac{d\sigma}{dt}(M_1,M_2) = \frac{d\sigma}{dt}(M_1,p)\,\frac{d\sigma}{dt}(M_2,p) \Big/ \frac{d\sigma}{dt}\text{(elastic)}.$$

Because of poor statistics the comparison has been performed integrated over M and t. In this case, assuming an exponential t dependence (e^{Bt}) for all the diffractive processes, the relationship for the integrated cross sections becomes:

$$\sigma_{DD} = \frac{\sigma_{SD}^2}{\sigma_{elastic}} \quad \frac{B_{SD}^2}{B_{DD}B_{el}}$$

where also $B_{DD}=2B_{SD}-B_{el}$ must hold to have factorization in t.

Figure 18 shows the actual comparison. The $p\pi^+\pi^-$ mass for the SD data, again, displays the low-mass enhancement characterizing exclusive dissociation phenomena. The $p\pi^+\pi^-$ mass in the DD

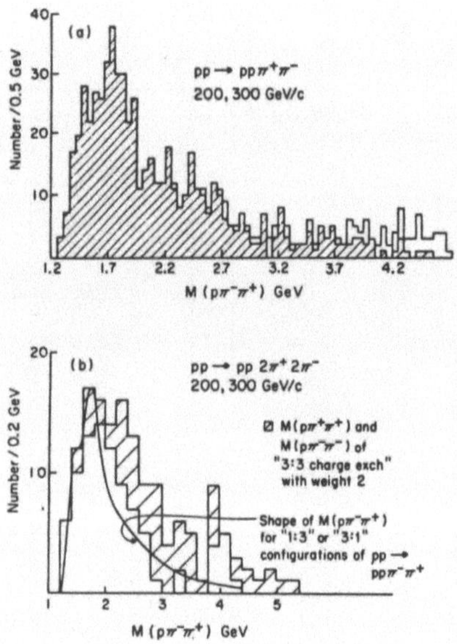

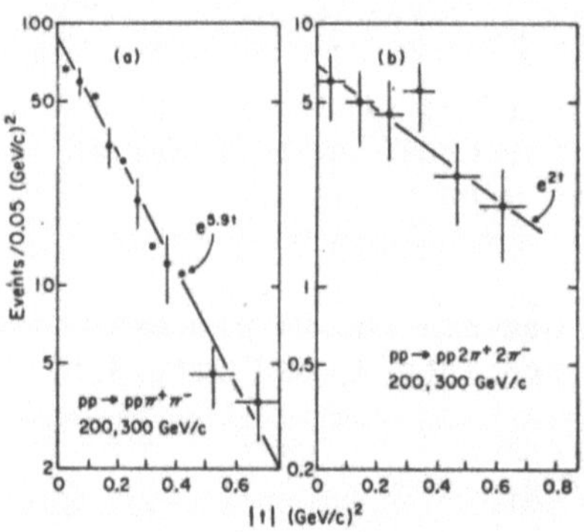

Figure 18: Comparison of single and double dissociation of protons into $p\pi^+\pi^-$ systems.

events appears to be considerably broader; however, upon subtraction of background expected from choosing wrong $p\pi^+\pi^-$ triplets (estimated from the non-diffractive $p\pi^+\pi^+$ combinations), the shape of the DD mass spectrum becomes reasonably consistent with that observed for the SD data. The value of σ_{DD} is 37±10 μb whereas 23±7 μb is expected on the basis of factorization. Also, the value of B_{DD} (integrated over M), is 2.0±0.5 GeV^{-2}, whereas 2.5±0.6 GeV^{-2} is expected (see the distributions in Fig. 18). Although these results are reasonably consistent with Pomeranchukon factorization and with the presence of double dissociation, an excellent additional test would involve a comparison of the expected slopes only for M<1.5 GeV. Here the value of B_{DD} should be larger than B_{el}! (It is, of course, not clear to what accuracy one should expect factorization to hold because of the presence of non-factorizable absorption effects in the final states. In addition, the strong M-t correlation noted in Fig. 14 raises questions concerning the relevance of results on factorization when data are examined integrated over M.)

Inclusive Diffraction Production

As discussed earlier in this lecture, elastic-scattering cross sections are characterized by their very weak energy dependence and their steep angular distributions which peak sharply at t=0. These properties are similar to those expected for the diffractive scattering of light from an absorbing disk. Certain exclusive channels, such as the above described neutron dissociation into low-mass $p\pi^-$ systems, also exhibit characteristic diffractive properties. In addition to the elastic and to the low-mass dissociation processes which occur at low energies, there exists at higher energies a substantial cross section for the diffractive excitation of a particle into what appears to be a large mass ($M \gtrsim 2$ GeV) continuum. The inclusive channels which are relevant here are those which contain leading particles in the final state, namely, reactions of the kind A+B→A+Anything [26]. When $M^2 >> 1$ GeV^2 and $M^2/s << 1$, a regime which exists only at or above Fermilab energies, the triple-Regge formalism becomes appropriate to this inclusive reaction.

Figure 19 displays the M^2 distribution of the Anything system for the inelastic reactions:

$$\pi^- p \rightarrow p + \text{Anything} \quad (1)$$

$$pp \rightarrow p + \text{Anything} \quad (2)$$

The data are from bubble chamber measurements at 200 GeV/c. [27] The dramatic enhancements at $M^2 \lesssim 25$ GeV^2 in Fig. 19 have two contributions: one is from the dissociation of the projectile particle into low-mass resonances or peaks (such as the 3π A_1 enhancement for reaction (1), and the "N*(1400)" $p\pi$ peak for reaction (2)); the other contribution is from the excitation of the projectile into the multiparticle large-M^2 continuum mentioned above. The total cross section for the peak in reaction (1) is ∿2 mb, and in reaction (2) it is ∿3 mb (the latter is for the excitation of only one proton). An analogous peak is observed in the reaction $\pi p \rightarrow \pi$+Anything, also at the 2 mb level. Hence, the sum of the cross sections for the excitation of either a

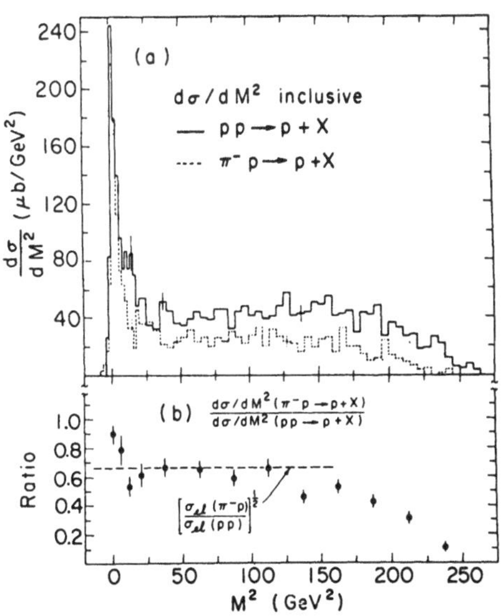

Figure 19: Inclusive cross sections (integrated over t) for the production of protons in the regime of target fragmentation at 200 GeV/c.

target or a beam particle are comparable to elastic scattering cross sections for the same incident channels (3 mb and 7 mb respectively).

Other properties of the peaks at $M^2<25$ GeV^2 can be summarized as follows [26]: (1) The average multiplicity of the events in the enhancement is considerably lower than for events outside of the peak; the multiplicity of the Anything system grows approximately as $\ell n\, M^2$. (2) The cross section for the entire low mass peak is almost energy independent. (3) The t distribution of the M^2 system is sharply peaked at small angles, particularly for the very lowest M^2 (recall the n-dissociation data). (4) These peaks occur only for systems which have the same charge, strangeness, and baryon number as the incident particle. From these characteristics it is reasonable to presume that the low mass enhancement is mainly a diffractive phenomenon involving Pomeranchukon exchange.

Figure 20 illustrates the triple-Regge pole graph which is used to describe the break up of a particle B into a massive system (M). R_i are any Regge poles which can be exchanged. The sum over M is to be interpreted as a sum over all states contributing to the reaction at fixed M^2. The large-M^2 requirement of the model has to do with the substitution of the R_j' Regge pole for the elastic scattering of R on B at $t_j=0$; this is only valid at large M^2, above the resonance region. The triple-Regge formula can be written as

$$E\frac{d\sigma}{d^3p} \simeq \frac{s}{\pi}\frac{d\sigma}{dtdM^2} \simeq \sum_{ij} \beta_{ij}(t)\left(\frac{s}{M^2}\right)^{2[\alpha_i(t)-\frac{1}{2}\alpha_j(o)]} s^{[\alpha_j(o)-1]}$$

where β is an unknown function of the vertex couplings. When i=j=P (Pomeranchukon), then the invariant cross section at fixed $M^2/s \simeq 1-x$ becomes essentially proportional to s/M^2 and s-independent. Also, the cross section $d^2\sigma/dtdM^2$ has a M^{-2} form, independent of energy. When j=V (vector meson) and i=P we obtain another diffractive term which has the approximate form $(s/M^2)^{3/2}\, s^{-\frac{1}{2}}$ for the invariant cross section. At fixed M^2/s this term falls with

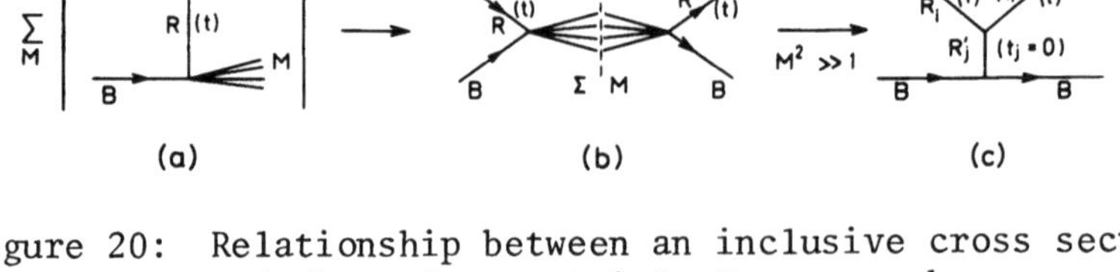

Figure 20: Relationship between an inclusive cross section and the relevant triple-Regge graph.

energy as $s^{-\frac{1}{2}}$. Another interesting situation occurs when i=V and j=P. This non-diffractive R-P-R term scales with energy (i.e., is independent of s) and consequently $d^2\sigma/dtdM^2$ at fixed M^2 falls approximately as s^{-1}. (The above statements are only roughly correct because we have ignored Regge interference [28] terms as well as the t-dependence of the α_j, and have set $\alpha_V(0)=0.5$ and $\alpha_P(0)=1.0$.) The P-P-P term is expected to dominate for small M^2/s. The P-R-P term should also be important at low M^2 and become negligible at very high energy. The R-P-R term should dominate for M^2/s beyond the diffractive peak. Data appear to exhibit the gross features of the triple-Regge parameterization.

Figure 21 displays the M^{-2} fall off (beyond the resonance peak which is observed near $M^2 \sim 3$) expected from the P-P-P term for the large-mass diffractive excitation of target protons in π^+p and pp reactions between 140 GeV/c and 170 GeV/c [29]. The authors have extracted the P-p total cross section and the P-P-P coupling constant from their $\pi^{\pm}p$, $K^{\pm}p$ and $p^{\pm}p$ proton-fragmentation data. The results from all channels agree; i.e., cross sections factorize to ∿10% accuracy and the P-p cross section can be written in the factorized form [29]:

$$\sigma^{Pp}_{tot}(M^2,t) = [M^2 \frac{d^2\sigma}{dtdM^2} \Big/ \frac{d\sigma}{dt}_{elastic}] \cdot \sqrt{16\pi(d\sigma/dt)_{pp}}$$

the expression in the bracket refers to the ratio of inclusive and elastic scattering for the same incident channel. The value of $\sigma_{Pp}(t)$ is independent of M^2 (Fig. 21) and equals $2.9\ \exp(-1.04t+0.35t^2)$mb; at t=0 the P-P-P coupling constant r_o is $0.80\pm0.03\ \mathrm{GeV}^{-1}$. (It is interesting that at larger values of M^2, where the PPP term is no longer expected to dominate, factorization is grossly violated in pp data relative to πp and Kp channels [30]. This can be attributed, for example, to the presence of different RPR terms, such as πPπ, in the pp as opposed to the meson-proton reactions.)

The energy dependence of the πp and Kp data at low M^2/s has been checked and appears to exhibit an $s^{-\frac{1}{2}}$ component, particularly at small t, similar to that observed for older pp results [26].

In summary, there exist at high energy large cross sections for the diffractive excitation of hadrons into multiparticle systems. The cross section for the inelastic excitation of a proton in a pp collision is ∿3 mb, and that of a pion in a πp collision is ∿2 mb. Consequently, the total diffractive, essentially energy independent cross section (elastic, inelastic, and estimated double diffractive) is $\sim 0.4\ \sigma_{TOT}$, for both πp and pp data. The inclusive leading-particle distributions in $\pi^{\pm}p$, $K^{\pm}p$ and $p^{\pm}p$

reactions at large x values are consistent to ∿10% accuracy with factorization of the Pomeranchukon contribution to the cross section. The triple Regge formalism appears to yield a surprisingly good description of the data [28].

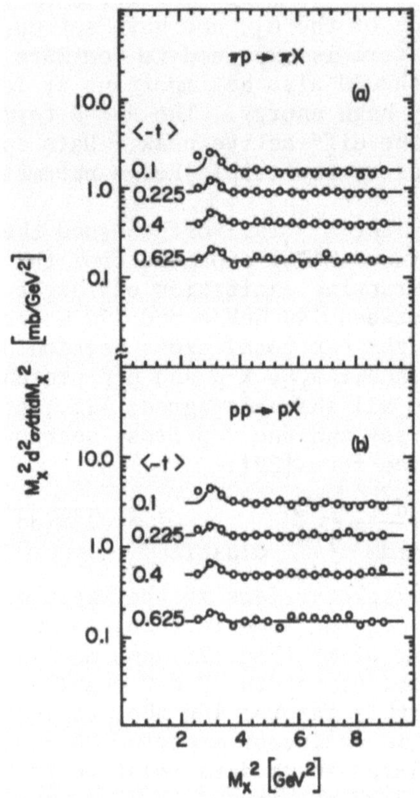

Figure 21: Test for the presence of the triple-Pomeranchukon contribution to leading-particle reactions.

III. INCLUSIVE PARTICLE PRODUCTION

Nuclear Targets

Multiparticle production off nuclear targets can, in principle, be used as a tool for examining the space-time development of hadronic processes. For example, if the asymptotic hadron final state in a hadron-nucleon collision evolves in a distance far shorter than the nuclear dimension, then in a hadron-nucleus collision we might expect that the final-state particles emitted in the initial interaction of the hadron with a particular nucleon will collide again with the downstream nucleons in that same nucleus, initiating a high-multiplicity cascade. On the other hand, if the initial collision involves the production of a resonant system which decays outside of the nucleus, there would then be very little cascading expected. Consequently, if the hadron-nucleon interaction time is short we expect, naively, that the charged-particle multiplicity in a hadron-nucleus collision will be a strong function of atomic number A and energy. On the contrary, the A dependence and s-dependence should be weak if the time scale for the development of the hadron final state is long.

Figure 22 presents the ratio (R_{NE}) of the inelastic charged-particle multiplicity in nuclear emulsion (average A~73) to that in pp collisions as a function of laboratory momentum [31]. After an initial increase of R_{NE} at energies below 100 GeV/c, the multiplicity in nuclear emulsion is observed to saturate at a value well below $R_{NE}=2$. This well-known result from cosmic ray studies, supplemented by new data from FNAL, speaks against a cascade model for production in nuclei and for a rather long time-constant for the development of the final state.

Busza et al [31] have performed a series of measurements of multiparticle production in $\pi^{\pm}$, $K^{\pm}$ and $p^{\pm}$ interactions with nuclear

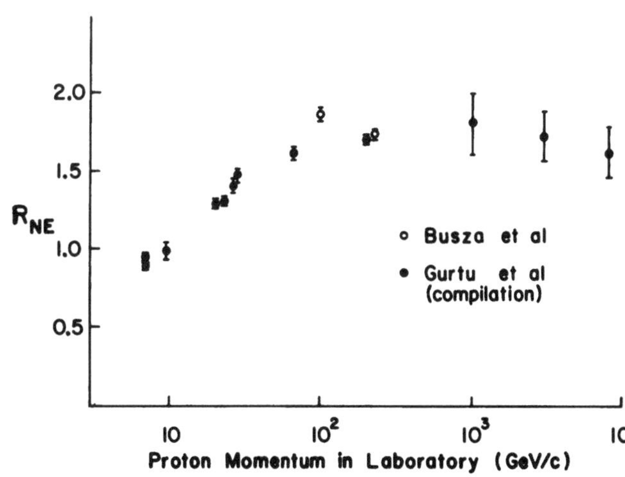

Figure 22: Ratio of inelastic multiplicities for particle production in nuclear emulsion relative to production in hydrogen.

targets ranging from hydrogen to uranium. The data consist of angular distributions of secondaries for 50 GeV/c, 100 GeV/c and 200 GeV/c beam particles. Figure 23 provides a global view of the overall richness of the data. The multiplicity in pA collisions is plotted for different intervals of the pseudo rapidity η, as a function of nuclear target material. ($\eta = \ln\cot(\theta/2)$, θ is the production angle in the laboratory. η is a good approximation to the laboratory rapidity variable $y = \frac{1}{2} \ln[(E+p_\ell)/(E-p_\ell)]$, where E and p_ℓ are, respectively, the energy and longitudinal momentum of any particle. η and y become indistinguishable when the transverse momentum is larger than the particle mass, $p_T^2 \gg m^2$.) Because of the difference in the inelastic πp, Kp and pp cross sections, and because even large nuclei are partly transparent, the average nuclear thickness of a target material depends on the type of incident particle. A parameter ($\bar{\nu}$) can be defined to take account of this difference in inelastic cross sections for different channels. The value of $\bar{\nu}$ for any nucleus A is given by $A\sigma_{hp}^{inelastic}/\sigma_{hA}^{inelastic}$, where h represents any incident hadron. Thus defined, the $\bar{\nu}$ parameter describes the average thickness of the nucleus in units of absorption mean free paths for hadron h. Also, $\bar{\nu}$ gives a measure of the average number of inelastic collisions h would make with the nucleons if subsequent to each collision it could interact anew.

The data presented in Fig. 23 indicate that the produced multiplicity is strongly dependent on η and on $\bar{\nu}$, except at large η (small angles, or forward production), where the multiplicity is independent of nuclear size. The authors indicate that although pion and proton-induced reactions exhibit different behavior for the same nucleus, the behavior is similar for target nuclei with identical values of $\bar{\nu}$. (The correction for $\bar{\nu}$ is important because, typically, pions have absorption lengths of $\sim$3 f in nuclear matter while protons have absorption length of only $\sim$2f.)

Figure 24 presents a plot of R_A (average multiplicity relative to hydrogen, integrated over η) as a function of $\bar{\nu}$ for $\pi^\pm$ and p data. There is no clear systematic variation of $R_A(\bar{\nu})$ with energy or with the type of incident beam particle. The dependence of R_A on $\bar{\nu}$ requires a quadratic term in $\bar{\nu}$ for an acceptable fit to the data. (For proton data R_A is approximately proportional to $A^{0.25}$.)

In Fig. 25 I show the η spectra for the proton data at 200 GeV/c. As $\bar{\nu}$ (or the nuclear-target size) increases, the multiplicity is observed to grow rapidly everywhere except in the very forward direction.

Although a variety of theoretical models have been suggested in order to explain all these remarkable features of the data [32], the Energy Flux Cascade model [33] appears to be in best overall

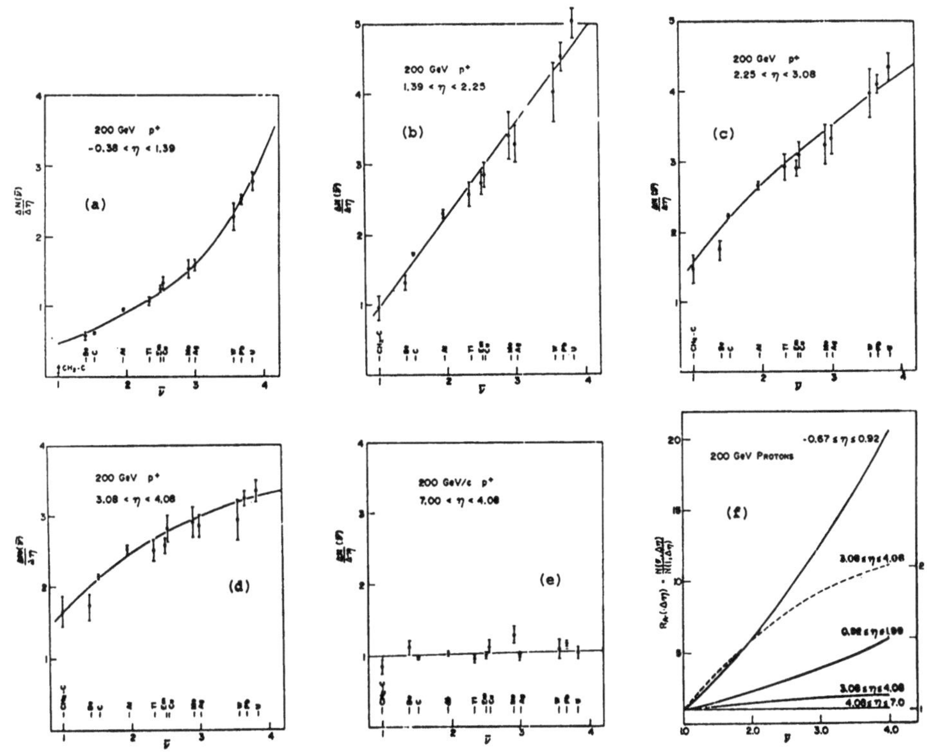

Figure 23: Inelastic multiplicities for particle production in proton-nuclear collisions at 200 GeV/c. The data are displayed for various angular regions (η) as a function of the parameter $\bar{\nu}$ (see text).

agreement with the measurements. In this model it is assumed that the initial collision of a hadron with one of the nucleons in a nucleus generates a distribution of energy which has the same rapidity content as that observed in hadron-nucleon collisions. This energy flux travels through the nucleus, expanding spatially with time, and behaves, as far as its interaction with nucleons in its path, as several systems of hadrons (one large-rapidity pack and perhaps several small-rapidity packs, any group being approximately 1 fermi wide, as measured in the rest frame of that pack). Although all the packs or slices of the original flux can collide with another nucleon, only the large rapidity component has enough energy to produce a new flux (of somewhat lower rapidity content). This process repeats $\bar{\nu}$ times yielding a weak muliplication of particles. Although the Energy Flux Model provides excellent qualitative agreement with the data, quantitative predictions, such as the asymptotic prediction $R_A \approx 1+ (\bar{\nu}-1)/3$, are

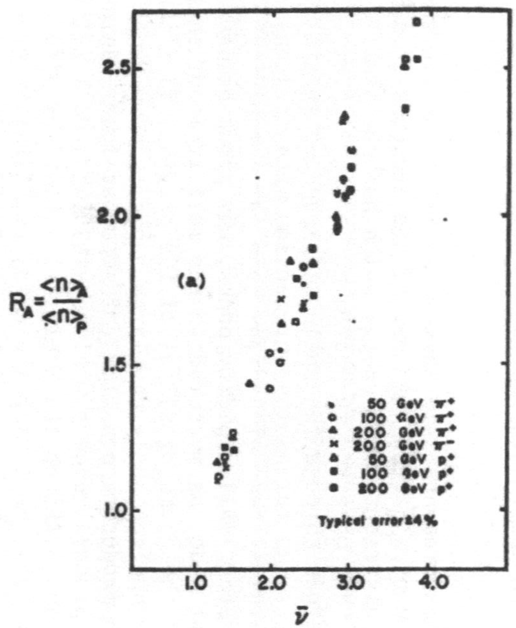

Figure 24: Ratio of inelastic multiplicities for particle production in proton-nuclear and pion-nuclear collisions relative to multiplicities on hydrogen for data between 50 GeV/c and 200 GeV/c, displayed as a function of $\bar{\nu}$.

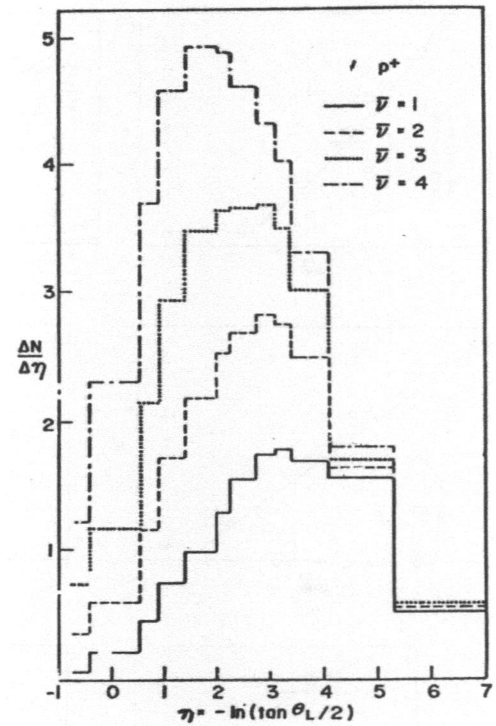

Figure 25: Differential multiplicity distributions as a function of the angular variable η for different target materials for 200 GeV/c incident-momentum protons.

clearly inconsistent with presently available measurements.

The Energy Flux idea certainly appears to come to grips with the problem of multiparticle production in nuclei, nevertheless, there is, perhaps, a more familiar way to understand the above physics. Figure 26 displays the $2\pi^-\pi^+$ and $2\pi^+3\pi^-$ mass distributions observed in inclusive pp collisions at 400 GeV/c. The data indicate that the mean value of the mass spectra are 2-3 Gev, depending somewhat on multiplicity. The widths of these distributions are 1.5-2.5 GeV, comparable to the mean values. Now, if as it is commonly believed, hadron clusters are produced in particle collisions, and the average number of particles in a hadron cluster is $\sim$3-4 (on the average we have about 4 clusters per pp collision at 400 GeV), we expect that the typical cluster of hadrons produced in high energy collisions will have a mass of $\sim$2.5 GeV and a width of $\sim$2 GeV. If the clusters are distributed uniformly in rapidity, then the forward-most cluster will have essentially all the beam momentum and the backward-most cluster will be essentially produced at rest in the laboratory. Because of the localization of the cluster mass, using an uncertainty-principle argument, the fast forward cluster would have an effective lifetime of $\gamma\hbar/\Gamma$ and a mean free path for decay of $\lambda \sim \gamma\hbar c/\Gamma \simeq (400/2)\ (2\times10^{-14})/2 \simeq 20$ fermi! Consequently, if the interaction occurs on a nucleon inside a nucleus, it is clear that the leading cluster will behave as a single massive object as it travels through nuclear matter. Thus time dilation will clearly diminish cascading effects within nuclear matter. If the leading cluster interacts again within the nucleus, the secondary interaction will produce new clusters distributed in rapidity in a manner similar to that in the first collision. Thus the cross section in nuclei at small y will

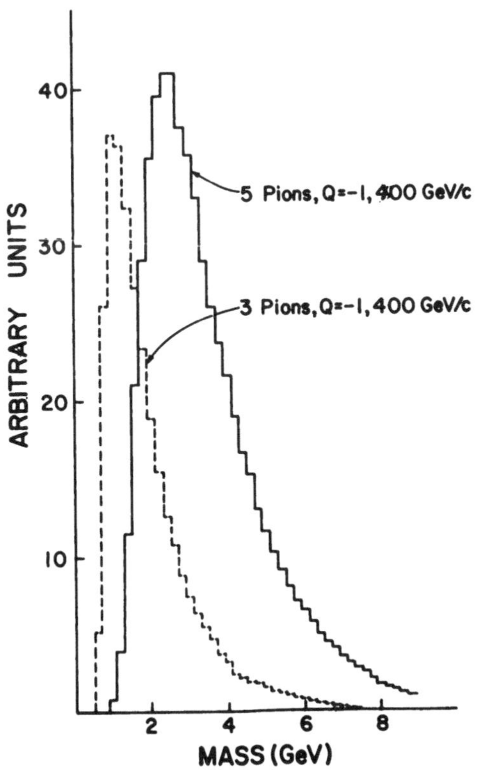

Figure 26: Mass distributions of $2\pi^+3\pi^-$ and $\pi^+2\pi^-$ systems produced in inclusive pp collisions at 400 GeV/c.

be increased as a result of the secondary collisions but the forward y component will not be grossly affected. (The slower secondary clusters will again tend not to have sufficient energy to multiply in the nucleus and therefore the cascading will not be severe.) Although the above remarks are rather qualitative, they may contain the essence of the physics of multiparticle production in nuclear matter [34].

Hydrogen Targets

Multiplicities. One of the general qualities to emerge from investigations of hadron collisions at Fermilab is the essential similarity of all hadronic matter. It appears that the total s-value of a system, and possibly the charge, determine all the major features of multiparticle production in hadronic collisions. Figure 27, taken from reference 5, displays the veracity of these remarks. The figure shows cross sections as a function of charged-particle multiplicities (n) observed in $p^{\pm}p$, $\pi^{\pm}p$ and $K^{+}p$ collisions at 100 GeV/c. Except for the absolute scales, the distributions are all quite similar. In detail, however, differences are apparent between the various channels. In particular, there is an unusually large excess of cross section in $\bar{p}p$ channels relative to pp reactions for $n \gtrsim 12$. The opposite may be true for n=2. It will be very interesting to compare $\bar{p}p$ and pp values of σ_n at $\gtrsim 300$ GeV/c, by which time the cross sections for the n=4 multiplicity may also reverse (i.e., $\sigma_{pp} > \sigma_{\bar{p}p}$). I expect this to happen because stronger final-state absorption in $\bar{p}p$ reactions should always tend to reduce $\bar{p}p$ cross sections relative to their analogous pp reactions (such as, for example, the low-multiplicity diffraction channels). In other words, as the $\bar{p}p$ and pp total cross sections become more equal, their small difference will appear, partially at the expense of low-multiplicity reactions, at ever increasing n-values. (I am assuming that the mean multiplicity in the annihilation channel, as in all processes, increases with s.)

Conclusions pertaining to the $\bar{p}p$ - pp difference similar to those reached above have been advanced previously by Eylon and Harari [6]. These authors suggest that on the basis of their model (mentioned earlier in the lecture) the quantity $R_n = [\sigma_n(\bar{p}p) - \sigma_n(pp)]/\sigma_n(pp)$, at fixed s, will approach the form $(3/2)^n$ for large n; also, for fixed n, $-R_n$ will approach the form $(1/2)^n$ at large s. Figure 28 displays a check of the dependence of R_n on n for several s-values [35]. The data have been fitted successfully to the form $R_n = \beta^n s^{-\alpha}$, where the s-dependent parameters β and α appear to approach "asymptotic" values of β=1.24 and α=0.73 at 100 GeV/c. In the specific model of Eylon and Harari β should approach 1.5 while α/2 should become equal to the difference between the intercepts of the exchanged meson and baryon trajectories.

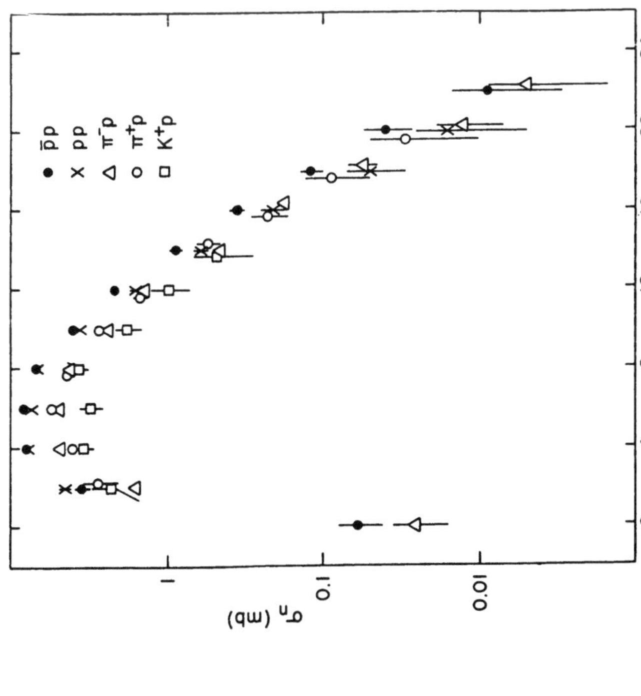

Figure 27: Cross section for producing n charged-particle inelastic final states in hadronic collisions at 100 GeV/c.

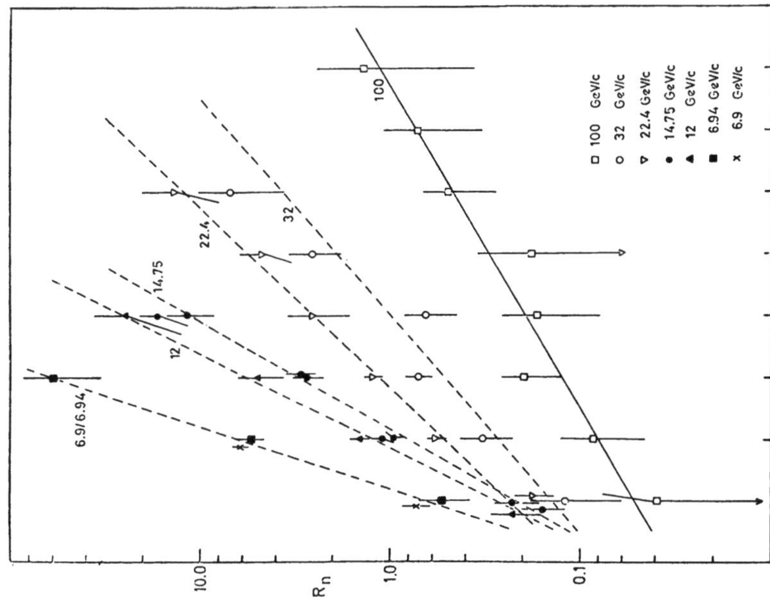

Figure 28: Linear fits to the ratios: $[\sigma_n(\bar{p}p)-\sigma_n(pp)]/[\sigma_n(pp)]$ at different momenta.

Both asymptotic predictions are in rough, albeit surprisingly good, agreement with the data. It is not possible as yet to test whether $-R_n$ approaches $(1/2)^n$. (This is related to the fact that $\Delta\sigma/\sigma(\bar{p}p\to\text{mesons})$ essentially saturates the Eylon-Harari bound of 1, which naively also implies that difference ratio R_n is due entirely to the annihilation contribution to the total cross section.) Nonetheless, as mentioned in the previous paragraph, higher energies may provide a confirmation of the essential prediction that, for fixed n and large s, $\sigma_n(pp) > \sigma_n(\bar{p}p)$.

The dependence of multiplicity distributions on target material is displayed in Figs. 29 and 30. Figure 29(a) compares $\bar{p}p$ with $\bar{p}n$ data at ∿15 GeV/c, and Fig. 29(b) compares pp with pn data at 300 GeV/c [36]. (The neutron data were obtained using deuterium targets. The distributions shown in Fig. 29 have been corrected for secondary rescattering and are somewhat model dependent.) Although in detail it appears that the inelastic mean charged-particle multiplicity $\langle n_c\rangle$ tends to be smaller in neutron-target data than in proton-target data, to an excellent approximation, the distributions in Fig. 29 simply interleave as a function of n_c:

$$\sigma_{n_c}(pp) = \tfrac{1}{2}\,[\sigma_{n_c-1}(np) + \sigma_{n_c+1}(np)]\ ,$$

again suggesting the presence of similar, target-independent, production processes.

Figure 30 displays multiplicity data for π^- on Ne at 200 GeV/c (solid circles), π^- Ne data at 10.5 GeV/c (open squares) and π^-p data at 205 GeV/c (solid triangles)[37]. The multiplicities are presented in terms of the Koba-Nielsen-Olesen (KNO) scaling parameters, [38] where n_π for the Ne data refers to pion shower tracks (i.e., nuclear break-up prongs have been excluded from consideration). Except for the fluctuations in the Ne data, due to enhanced coherent diffraction production at low multiplicities, the three distributions are again very similar, indicating the universality and target independence of multiparticle production processes.

It is well known that the dependence of $\langle n_c\rangle$ on s is not just simply logarithmic [5,26]. Several years ago I showed [39] that the dependence of $\langle n_c\rangle$ on s in pp collisions can be represented using the phenomenological expression $\langle n_c\rangle = A + B\,\ln s + C\,\ln s/s^{\frac{1}{4}}$. The fit was excellent for data from $\lesssim 10$ GeV/c up to ISR energies. This form is more appealing than an often used quadratic [5] dependence on $\ln^2 s$ because it provides a way to check whether, asymptotically, multiplicities grow as $\ln s$ and become independent of the incident channel. If all channels yield the same value for the parameter B, the result would suggest (although without a theoretical justification for the above form of $\langle n_c\rangle$) that the

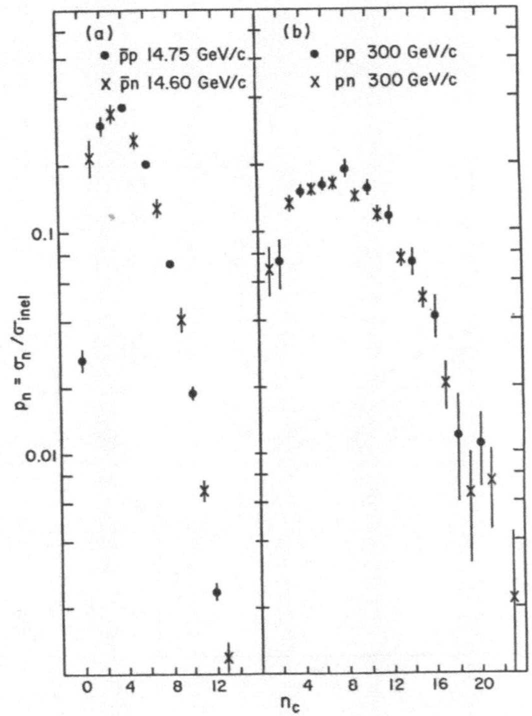

Figure 29: Comparison of $p^{\pm}$-hydrogen and $p^{\pm}$-neutron inelastic multiplicities.

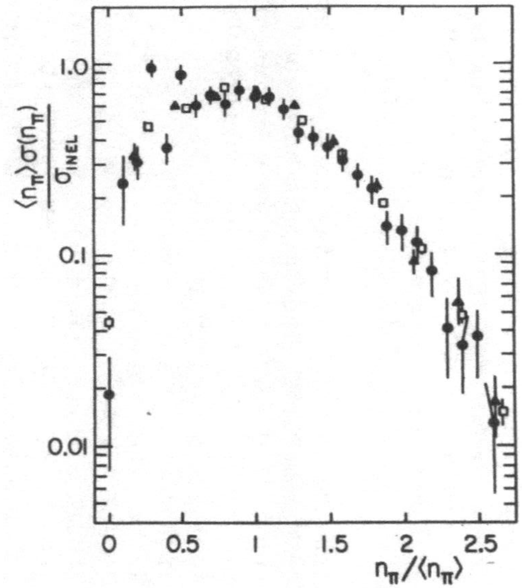

Figure 30: Comparison of π^--Ne data with π^-p multiplicities in terms of the KNO scaling parameters (see text).

small differences which exist in $\langle n_c \rangle$ are only low-energy effects and that at large-s multiplicities grow logarithmically and Pomeranchuk factorization holds. Figure 31 displays the highest energy $\langle n_c \rangle$ data available for γp, $p^{\pm}p$, $\pi^{\pm}p$ and $K^{\pm}p$ [40]. I have tried to restrict the fits to data for $\gtrsim$10 GeV/c incident momenta. Unfortunately, the only available γp measurements are from a low-energy SLAC experiment [41].

The $K^{\pm}p$ and $\pi^{+}p$ data each consist of only three points - just enough to determine the three parameters. The $\bar{p}p$ data have but four points and extend only up to 100 GeV/c. The best measurements are for pp and $\pi^{-}p$ collisions; these extend from ~20 GeV/c to ~400 GeV/c. The results of all the fits are given in Table I, and several of the curves are graphed in Fig. 31. Except for the relatively poor χ^2 for the fit to the sparse $\bar{p}p$ data, the fits and the agreement in the value of B for all the channels is quite impressive! We see that all data are consistent with Pomeranchuk factorization at large s-values to perhaps ~5% accuracy. It would be invaluable to have $K^{\pm}p$ and $\bar{p}p$ data points for momenta in excess of 200 GeV/c (and, of course, any high-energy γp data) to provide a better check of the form for $\langle n_c \rangle$ and of the universality of the B parameter.

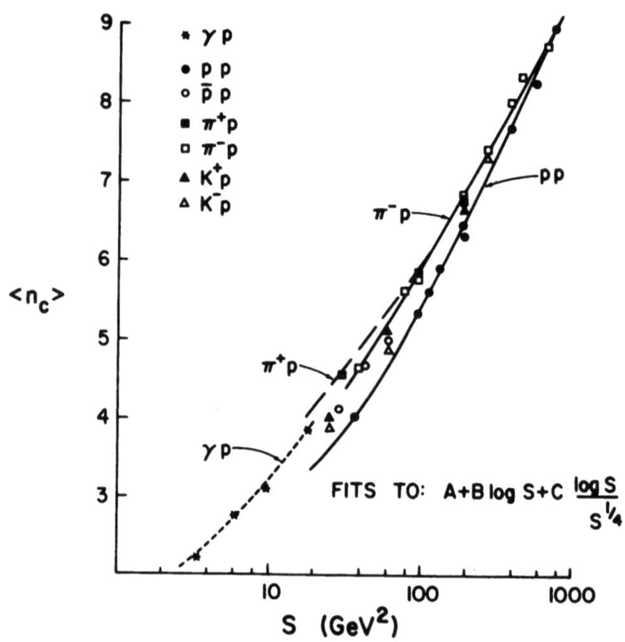

Figure 31: Inelastic average charged-particle multiplicities as a function of s for various hadronic channels. Curves represent fits to the data (see text and Table I).

TABLE I

Fits of Average Charged Particle Multiplicities to the Form:

$$A + B \ln s + \frac{C \ln s}{s^{\frac{1}{4}}}$$

CHANNEL	A	B	C	χ^2/d.f.
	PARAMETERS			
pp	5.35 ± 1.5	1.36 ± 0.05	-4.26 ± 0.92	1.07
$\bar{p}p$	5.44 ± 3.2	1.30 ± 0.07	-3.89 ± 2.0	4.18
K^+p	2.17 ± 5.0	1.29 ± 0.12	-1.57 ± 3.7	-
K^-p	6.38 ± 3.5	1.34 ± 0.10	-4.73 ± 2.3	-
π^+p	3.82 ± 6.3	1.19 ± 0.10	-2.26 ± 4.2	-
π^-p	0.67 ± 1.9	1.40 ± 0.10	-0.82 ± 1.0	1.15
γp	1.67 ± 0.9	1.28 ± 0.53	-1.09 ± 1.6	0.31

Inclusive Production Spectra and Limiting Behavior. Typical rapidity distributions for pions produced in high energy collisions are displayed in Fig. 32 for 102 GeV/c and 400 GeV/c pp data [42]. The data have been integrated over p_T, and no corrections have been applied for ∿10% $K^{\pm}$,∿1% $e^{\pm}$, and ∿2% $\bar{p}$ contamination. (The estimated corrections required for proton contamination of the π^+ spectra are indicated on the graph.) It is clear from these data that the cross section does not scale in the central region of pion production. In particular, for $y_{CM}=0$ ($y_{LAB}=2.69$ and 3.37 at 102 GeV/c and 400 GeV/c, respectively), the inclusive π^- cross section rises by 45±10%, while the π^+ cross section rises by 30±10% between the two energies. At 400 GeV/c the π^+ cross section is ∿15% greater than the π^- cross section near $y_{CM}=0$, indicating that asymptopia is still far away. For small values of y_{LAB}, where the π^+ yield is about three times the π^- yield, there is some indication that the cross section falls with energy [43]. The π^- data at 102 GeV/c and 400 GeV/c are compared in more detail in Fig. 33. The rise in the invariant cross section with increasing s is clearest at small x. Near $|x|$∿0.1 the cross section appears to scale; while at large $|x|$, and especially at small p_T, there is a small violation of the Hypothesis of Limiting Fragmentation [43].

Figure 34 displays invariant cross sections for pion spectra observed in $\pi^{\pm}p$ collisions [44]. The characteristics of these non-leading π-data are similar to those just discussed for pp collisions. Namely, near x=0 the cross section grows with energy,

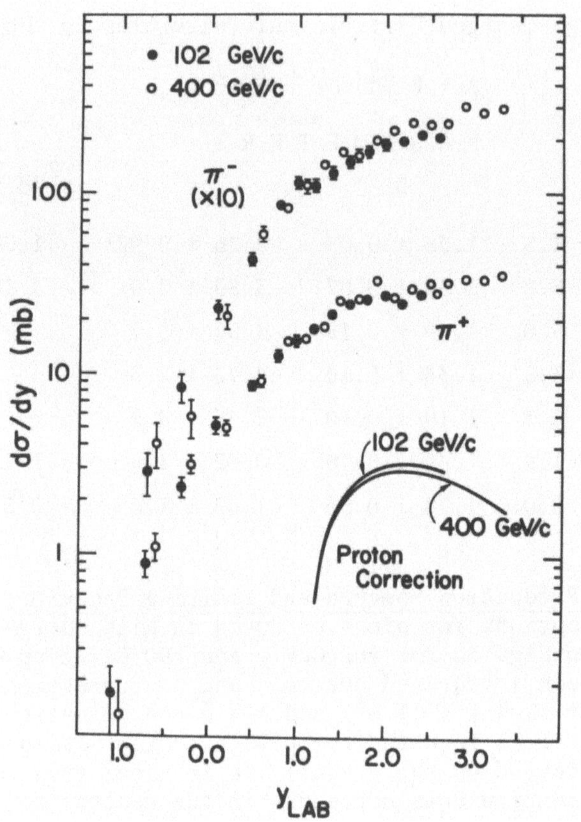

Figure 32: Rapidity distributions (integrated over p_T) of pions produced in pp collisions at 102 GeV/c and 400 GeV/c. Data have not been corrected for $K^{\pm}$, $e^{\pm}$ or $p^{\pm}$ contamination (see text).

while at large $|x|$-values cross sections tend to fall with increasing s.

Whitmore [5] has recently updated and refined an earlier study [45] of the energy dependence of inclusive particle-production cross sections at $y_{CM}=0$. Figure 35, taken from reference [5], summarizes the situation concerning pion production for various reactions. The data are presented as a function of $s^{-\frac{1}{4}}$ in terms of the invariant cross section, integrated over p_T, and normalized by the contributions of the Pomeranchukon to total cross sections for each of the incident channels [46]. (The solid line shown on the graph for π^- production represents recent ISR measurements.)

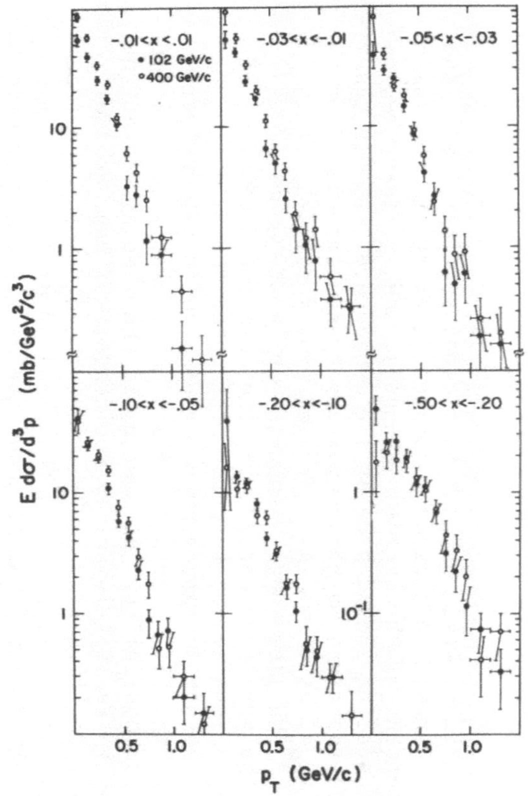

Figure 33: Invariant cross section for negative pions produced in pp collisions at 102 GeV/c and 400 GeV/c, as a function of p_T, for different regions of x. Data have not been corrected for small contamination from e^-, K^- and $\bar{p}$.

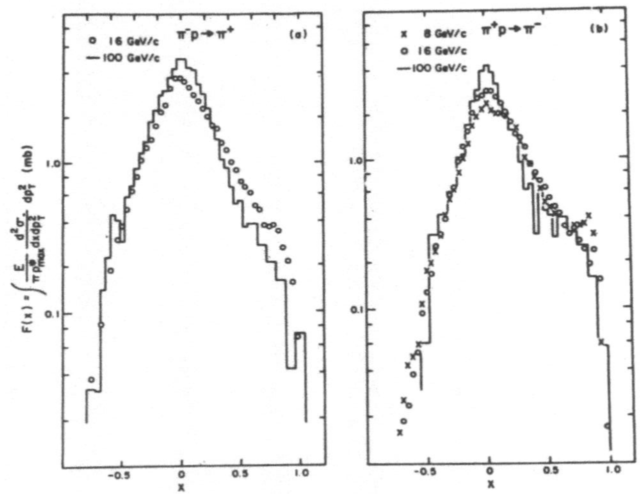

Figure 34: Invariant cross sections, integrated over p_T, for the inclusive production of π^+ and π^- in π^-p and π^+p reactions in figures (a) and (b), respectively.

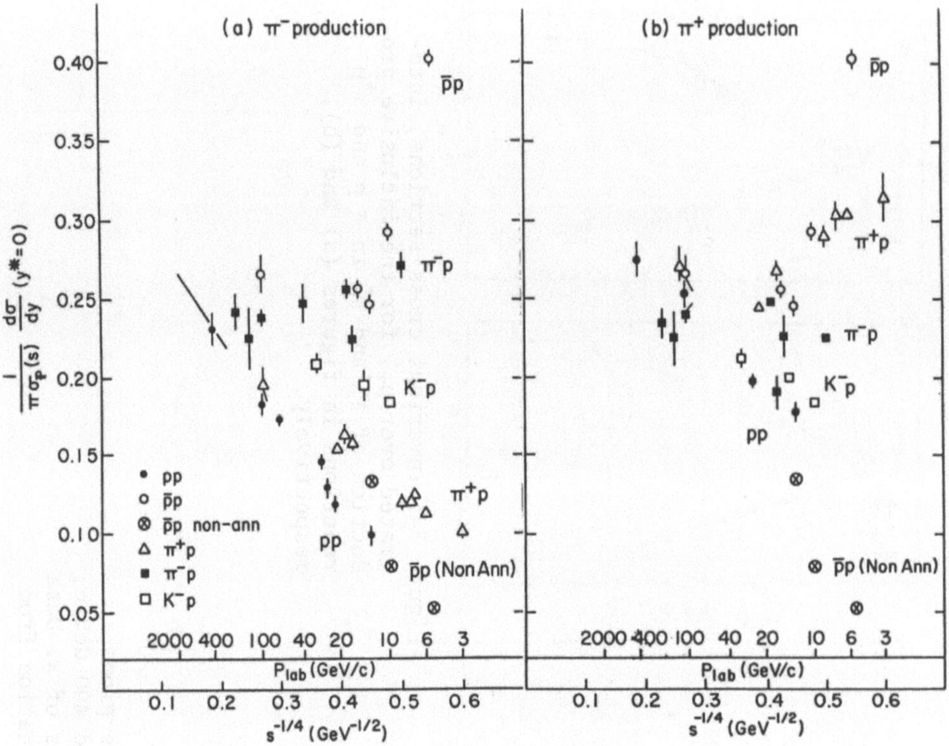

Figure 35: Density function for the production of π^- and π^+ mesons at $y_{CM}=0$, for various incident channels, as a function of $s^{-\frac{1}{4}}$. See text for explanation of the normalization.

Mueller-Regge ideas would suggest that cross sections for π^+ and π^- production should factorize at large s (converge to the same value independent of the incident channel), and, furthermore, the approach to scaling should have an $s^{-\frac{1}{4}}$ dependence. We see that data for all reactions may be converging to the same value of $d\sigma/dy_{CM}$ at $y_{CM}=0$ as $s\to\infty$. As indicated previously [39], this value in Mueller-Regge phenomenology should equal $B/2\pi$, where B is the parameter determining the asymptotic behavior of $\langle n_c\rangle$. From Table I we see that the best value for $B/2\pi$ is 0.21±0.01, a result in rather poor agreement with extrapolations suggested in Fig. 35. I am not certain how to interpret this discrepancy. One possibility, of course, is that the agreement for all B values in Table I is purely accidental. I can, however, take a rather positive view and say that the discrepancy between the extrapolation in Fig. 35 and the value of $B/2\pi$ is only off by $\sim$30% of the extrapolated value! Considering all the theoretical uncertainties regarding asymptopia, this is, in a sense, still a remarkable result [47].

Independent of the above discrepancy, the data displayed in Fig. 35, and the K_S^0 production data displayed in Fig. 36, suggest that, after leading-particle effects subside, the inclusive cross sections at x=0 increase asymptotically with increasing s, grossly consistent with expectations from Mueller-Regge ideas.

A comprehensive investigation of the approach to limiting behavior of pion production in the target-fragmentation regime has recently been reported in the literature [48]. Figure 37 displays the results of that compilation. The data are presented in terms of an integral of the inclusive cross section over p_T, and over a fixed interval of longitudinal momenta in the laboratory frame (p_L). Specifically, the function $G(p_L)$, which is defined as

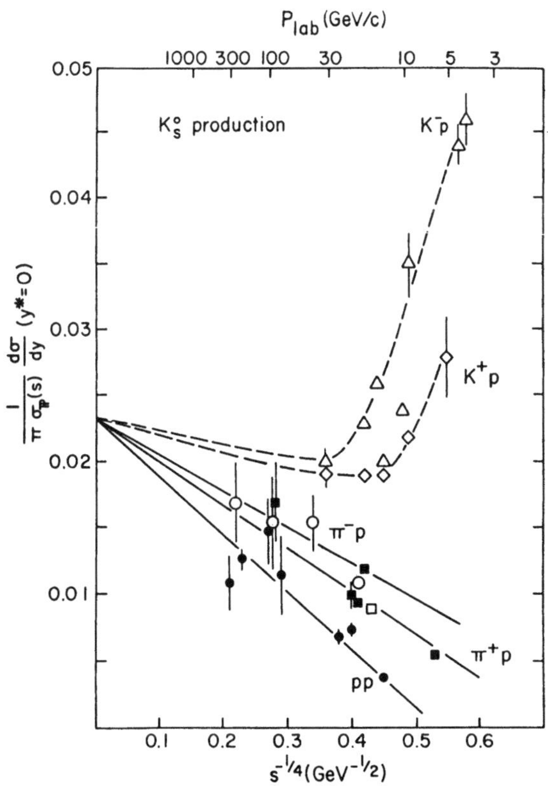

Figure 36: Density function for the production of K_S^0 mesons at $y_{CM}=0$, for various incident channels, as a function of $s^{-1/4}$.

$[\pi/\sigma_p(s)]\int E(d\sigma/d^3p)dp_T^2$, and is normalized as in Fig. 35 by the Pomeranchukon contribution to the total cross section [46], is examined as a function of $s^{-\frac{1}{2}}$. This sort of s-dependence is expected on the basis of application of Mueller-Regge ideas in the target-fragmentation regime for reactions studied in Fig. 37 [49]. Asymptotically all π^- production channels should factorize, and separately, all π^+ production channels should factorize. This appears to be happening at increasing s. At small s-values different channels can have contributions (in addition to the s-independent Pomeranchuk term) from different exchanges; furthermore, if Regge trajectories are not degenerate, and if leading trajectories do not have intercepts of $\alpha(0)=\frac{1}{2}$, then the approach to scaling can be fairly complicated and not just of the $s^{-\frac{1}{2}}$ form. Figure 37 suggests that, in fact, the picture is quite complicated. Nevertheless, the simple exoticity pattern of Chan et al [49] seems to be in qualitative agreement with the data. That is, reactions A+B→C+Anything, where the quantum numbers of $(AB\bar{C})$ are exotic, display least energy dependence in their inclusive cross sections.

Results similar to those compiled in Fig. 37 have recently been obtained in a counter experiment at BNL and Fermilab [50]. The data are for $K^{\pm}$, $p^{\pm}$ and $\pi^{\pm}$ projectiles incident on protons, in the momentum range between 4 GeV/c and 250 GeV/c. The variation with energy of $\pi^{\pm}$, $K^{\pm}$ and $p^{\pm}$ particle production was examined at a fixed $p_T \simeq 0.3$ GeV/c, and values of $y_{LAB}=0.6$, 0.4 and 0.2. These global studies indicate that the approach to asymptopia is, again, not just of the $s^{-\frac{1}{2}}$ form. I reproduce one of their graphs in Fig. 38, which summarizes their pion-production results. To achieve factorization of particle-proton and antiparticle-proton inclusive cross sections as $s\to\infty$, the authors require in addition to an energy dependent $s^{-\frac{1}{2}}$ term an s^{-1} term (from lower-lying trajectories) at low energies. Cross sections for meson-proton and baryon-proton channels here again appear to be in agreement with factorization at large s (when using normalizations akin to those suggested in reference [46]).

The SAS Group has also examined inclusive channels involving the exchange of quantum numbers [51]. The available data, which are for the region of projectile fragmentation, have been used to extract the effective Regge trajectories for production at large x values. The specific reactions studied are given in Fig. 39. The authors analyze their data using a semi-empirical one Regge-pole exchange formula, similar to our triple-Regge expression, keeping explicitly the total cross section for the scattering of the exchanged object R(t) on target B (see Fig. 20). The specific form used is:

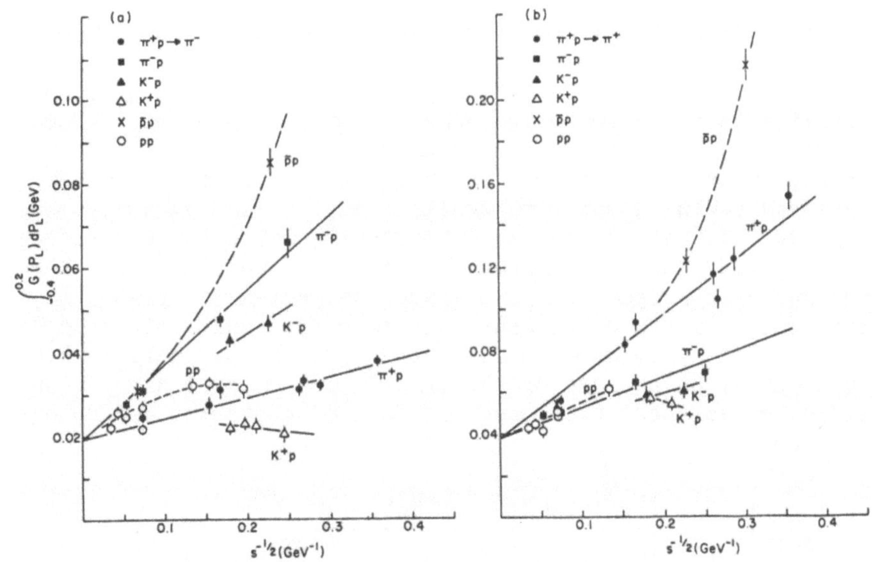

Figure 37: Density function for the production of π^+ and π^- mesons in the region of proton fragmentation in hadron-proton collisions. Data are displayed as a function of $s^{-1/2}$.

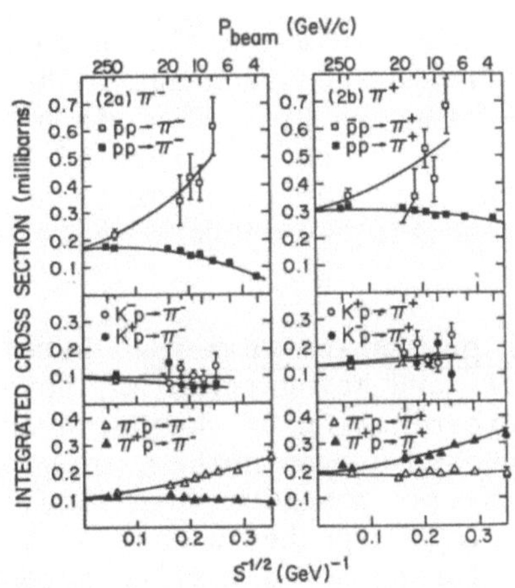

Figure 38: Cross sections for π^+ and π^- production in the region of proton fragmentation for hadron-proton collisions. The integration is over the following final-state momenta and approximate angles in the laboratory frame: $0.3\ GeV/c \leqslant |\vec{p}| \leqslant 0.6\ GeV/c$ and $50° \lesssim \theta \lesssim 60°$ (for details see ref. [50]).

$$\frac{d^2\sigma}{dtdx} \approx f(t)(1-x)^{1-2\alpha_{eff}} \sigma_{Rp}(M^2)$$

where f(t) contains all the t-dependence, α_{eff} are the effective values of the exchanged trajectories at the t-values specified in Fig. 39. The energy (i.e.,M^2) dependent cross section for Rp scattering is taken to be the same as for πp, Kp or pp scattering, depending on the nature of the exchanged object (for strange-meson exchange Kp is used, etc.). The data in all but the $\pi^+p \to p$+Anything reaction are consistent with scaling in s to ∿15% accuracy. Consequently, all data in the 50 GeV/c to 175 GeV/c range were combined to study the x-dependence of the cross section. For the $\pi^+p \to p$+Anything process the cross section decreases by ∿30% in the energy range of the experiment, and therefore only 140 GeV/c and 175 GeV/c data were used to extract α_{eff}. The ordinate in Fig. 39 is defined as an average of the cross section over s and t, normalized to the total cross section for Rp scattering at 200 GeV, as follows:

$$\left(\frac{d^2\sigma}{dtdx}\right)_{eff} = \left(\frac{\sigma_{Rp}(p_{in} = 200\ \mathrm{GeV/c})}{\sigma_{Rp}(M^2)} \frac{d^2\sigma}{dtdx}\right)_{average}$$

The values of α_{eff} for exchanges involving K* quantum numbers are $\alpha_{eff} \simeq 0.22$ at $\langle t \rangle \cong -0.3$ GeV^2, amazingly consistent with the K* trajectory. Similarly, the values of α_{eff} for baryon exchange processes typically equal $\alpha_{eff} \approx -0.55$, again in qualitative agreement with expectations from baryon exchange. The difference observed for the cross sections in Fig. 39(a) is presumably due to u-channel nucleon exchanges, which are expected to fall steeply with increasing s (as they, in fact, appear to be doing). The agreement observed between the C-conjugate reactions in Fig. 39(b) is impressive indeed.

Local Compensation of Quantum Numbers. Recent theoretical investigations by Krzywicki and Weingarten [52] have established a new industry for the determination of whether quantum numbers, such as charge, strangeness, baryon number, or even kinematic quantities such as transverse momentum, are locally compensated in rapidity space. The LCQN hypothesis asserts that any produced particle carrying a quantum number q must be accompanied nearby in rapidity space by a small group of particles carrying a total value of the quantum number -q. Local compensation of electric charge had previously been shown to hold [53], and now there is also evidence for the local compensation of transverse momentum in high energy collisions [54].

The best way to illustrate the essence of the LCQN hypothesis is through the idea of a zone graph. Let us consider for simplicity a ten-particle final state produced in a neutron-neutron collision. Two somewhat different possibilities for the distribution of the ten charges in rapidity space are shown in the sketch on the following page. A generalized charge-transfer variable Z(y) can be defined such that for a positive charge located at some particular y, Z(y) increases by one unit at that y value, and for a negative charge, Z(y) decreases by one unit at the position of that charge. The step function Z(y) for a complete event is known as a zone graph. Z=0 separates different zones. In general terms, the LCQM hypothesis requires that for large s-values the internal structure of zones (e.g., mean zone length, number of charges per zone) become s-independent. Furthermore, the properties of zones must be such that it is improbable to have a large number of particles per zone or a zone which is long in rapidity space. Finally, correlations between pairs of zones must decrease quickly with increasing separation in y. Thus stated, LCQM implies that the mean number of zones will grow with energy as ℓn s. These conditions are fulfilled in a qualitative way for data in the Fermilab energy regime. Figure 40 (from Bromberg et al, ref.[53]), displays the correlation function $D(y_1,y_2)=\langle Z(y_1)Z(y_2)\rangle - \langle Z(y_1)\rangle\langle Z(y_2)\rangle$ for 102 GeV/c and 400 GeV/c pp data as a function of $\Delta y=y_2-y_1$ for $y_1=0$ and $y_1=-1.2$. The curves were obtained by recalculating $D(y_1,y_2)$ after randomly reassigning the charges to

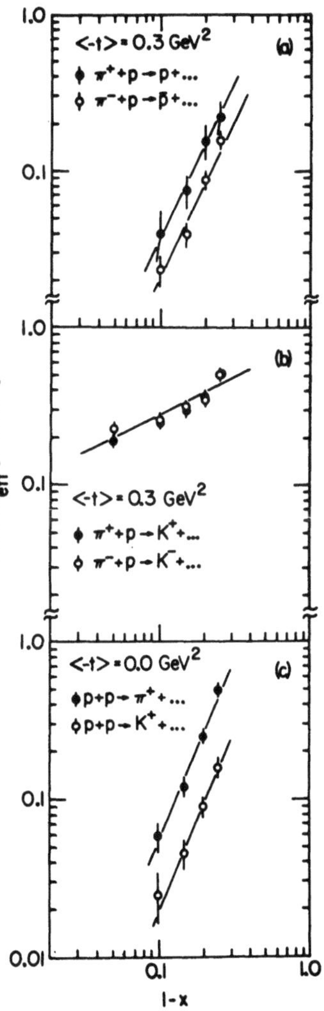

Figure 39: Cross sections for inclusive reactions involving the exchange of quantum numbers.

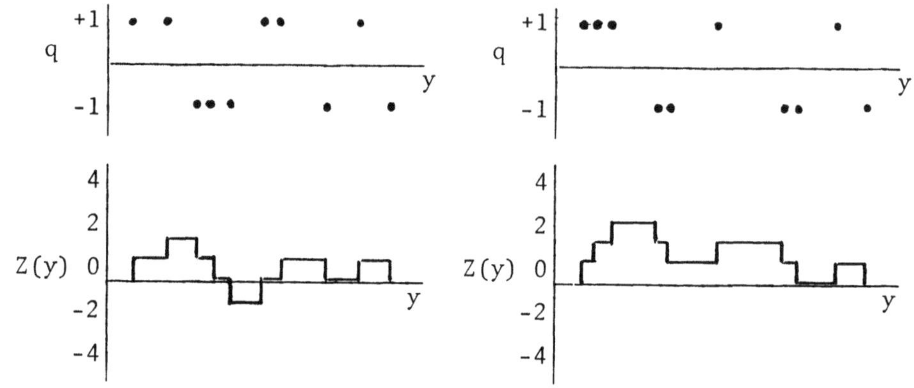

Two Possible Zone Graphs

the observed particle tracks. Hence the difference between the real and the randomized-charges data can be attributed to a mechanism which prevents large fluctuation in charge transfer. In particular, the small value of $D(y_1,y_2)$ reflects a tendency for zone multiplicities to be smaller than for the case of a random distribution of charge, and the rapid fall of $D(y_1, y_2)$ for large Δy, characterized by a correlation length of ~ 1.2 units in rapidity, shows a tendency for real zones to be typically shorter than in the randomized data, all consistent with expectations from the LCQN hypothesis.

The preliminary results on local compensation of transverse momentum [54] have been used to extract a lower bound on the value of the slope of the Pomeranchukon trajectory. The result is somewhat model dependent in its treatment of unobserved neutrals in the final state, but provides a stringent limit of $\alpha'_P(o) \gtrsim 0.2$ GeV^{-2}, which may be compared with phenomenological values of $\alpha'_P(o) \simeq 0.25$ GeV^{-2}. The fascinating implication of this result is that the dynamical mechanism responsible for the shrinkage of the elastic diffraction peak is the local compensation of transverse momentum [54].

Off-Shell Inclusive Scattering. We are all familiar with particle-exchange processes and with inclusive-production reactions. In the past few years interesting data have been presented which indicate that inclusive reactions initiated with off-shell particles bear a great similarity to real inclusive processes.

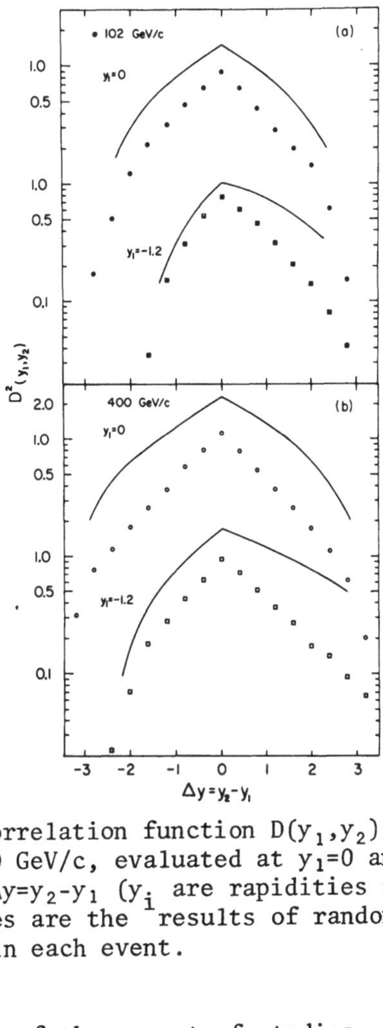

Figure 40: Zone correlation function $D(y_1,y_2)$ for pp data at 102 GeV/c and 400 GeV/c, evaluated at y_1=0 and y_1=-1.2, as a function of $\Delta y=y_2-y_1$ (y_i are rapidities in the center of mass). Curves are the results of randomizing the charge assignments in each event.

As an example of these sort of studies, I will discuss production of $\Delta^{++}(1236)$ in pp reactions. This process might be expected to proceed through pion exchange, however, this does not appear to be the case. In Fig. 41 I display the π^+p mass distribution for the reaction $pp\to\pi^+p$+Anything at 102 GeV/c and at 400 GeV/c [55]. A clear, energy independent peak at the $\Delta^{++}(1236)$ resonance is apparent in the data. The t-spectrum of the events in the Δ region is shown in Fig. 41(b). In Fig. 41(c) I show the x distribution of the π^+p events in the Δ mass region. The cross section is observed to scale with s. The curve on the data is from a calculation by R. Field. To fit the x spectrum, Field had to introduce a large off-shell damping term (form factor in t) for the pion-exchange contribution to the Δ^{++} process. This form factor was not required for the pion-exchange term in the

$pp \to p$+Anything reaction. In addition to pion exchange, a comparable contribution from ρ and A_2 exchange was required to fit the x distribution. Although it does not appear that pion exchange dominates the production, I will nevertheless compare the characteristics of the X^o system accompanying the production of Δ^{++} to $\pi^- p$ scattering at a center of mass energy $\sqrt{s} = M_{X^o}$.

Figure 42 demonstrates that the total charged-particle multiplicity (including "elastic" two-body $\pi^- p$ final states) of the X^o system has the same dependence on M^2 as real $\pi^- p$ data have on s. The f_2 moment $(\langle n^2\rangle - \langle n\rangle^2 - \langle n\rangle)$ for the X^o system produced in association with the Δ^{++} also has the same dependence on M^2 as real $\pi^- p$ data have on s. Taking this comparison one step further, in Fig. 42(c) I display the x distribution for the inclusive reaction $\pi^- p \to \pi^-$+Anything (smooth curve) and the analogous off-shell $R^- p \to \pi^-$ +Anything data for several regions of $M^2_{X^o}$. The distributions in Fig. 42(c) have been normalized in such a way that the integral of the data over the invariant phase space yields the values of

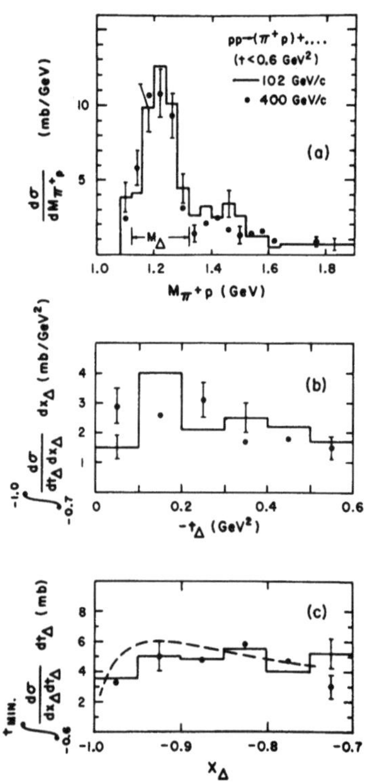

Figure 41: Characteristics of Δ^{++}(1236) production in pp collisions at 102 GeV/c and 400 GeV/c.

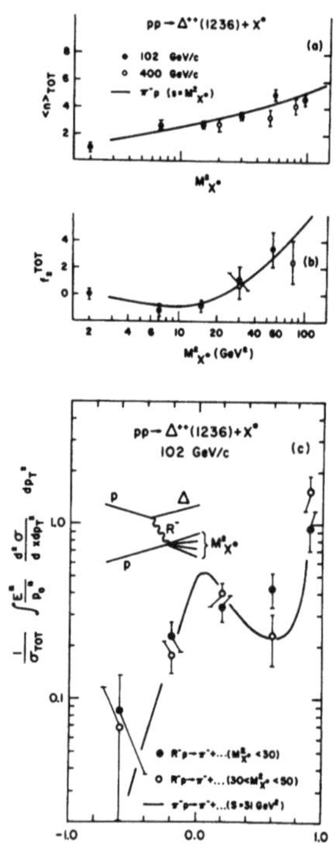

Figure 42: Comparison of off-shell Reggeon-proton (R^--p) scattering with real π^-p production data.

$\langle n \rangle_{TOT}$ given in Fig. 42(a). (The definition of x_{π^-} for R^-p scattering in such that positive x corresponds to π^- emission along the R^- "direction", where the usual Gottfried-Jackson t-channel direction is chosen as the R^-p collision axis.) The similarity between the reaction $\pi^-p \to \pi^-$+Anything and the off-shell process $R^-p \to \pi^-$+Anything at comparable s values is remarkable. This agreement is particularly puzzling considering the fits that Field made to the Δ production reaction. It appears therefore that the mass and charge of a hadronic system have the greatest bearing on the dynamics. The fact that R^- has admixtures of objects having different spin (but same isotopic spin) does not grossly affect the inclusive spectra.

A similar study to that discussed in Figs. 41 and 42 was also performed for the reaction pp→p+Anything [56]. Here off-shell $R^0p \to \pi^-$+Anything was compared to $\gamma p \to \pi^-$+Anything at values of $s_{\gamma p} \simeq M^2$. The inclusive spectra are shown in Fig. 43. The agreement is, again, impressive but quite surprising in view of the fact that for low values of M^2 the R object should mainly have properties characterizing the Pomeranchuk trajectory and certainly not a photon. Consequently, here again we see that gross features of inclusive production appear to be independent of all dynamic qualities, except for the charge and effective mass of the interacting system.

"If there is a question of importance, look to the masses" (Mao Tse Tung). This last part of my lecture can be summarized by the above quotation from Mao's Little Red Book, as paraphrased to me by Ed Berger. We have made a study of the p_T-dependence and the x-dependence of multiparticle systems produced in pp collisions [57]. Specifically, we have examined how the mean value, $\langle p_T \rangle$, and the full width at half maximum of the x distribution (Γ_x) vary with the mass, charge, and particle multiplicity of a produced system. The results of this investigation are summarized in Figs. 44 through 47. There is essentially no dependence on the charge of the system being examined (particularly for $\langle p_T \rangle$). At fixed mass, a weak variation with multiplicity can be discerned in the $\langle p_T \rangle$ data. A somewhat stronger variation is observed for Γ_x. The $\langle p_T \rangle$ dependence is essentially s-independent (between 102 GeV/c and 400 GeV/c) while the Γ_x values become smaller with increasing s (presumably due to approximate Feynman scaling). The variation of Γ_x and $\langle p_T \rangle$ with mass is very similar to that observed for stable hadrons (e.g., π,K,Λ etc.). In detail, the 3π systems appear to coincide in character with the properties of the long-lived objects [58]. The latter result may be due to the fact that clusters, which have typical multiplicities of $\sim$3, have production properties akin to the more stable particles. (This is not an entirely consistent picture because of the charge-independence of

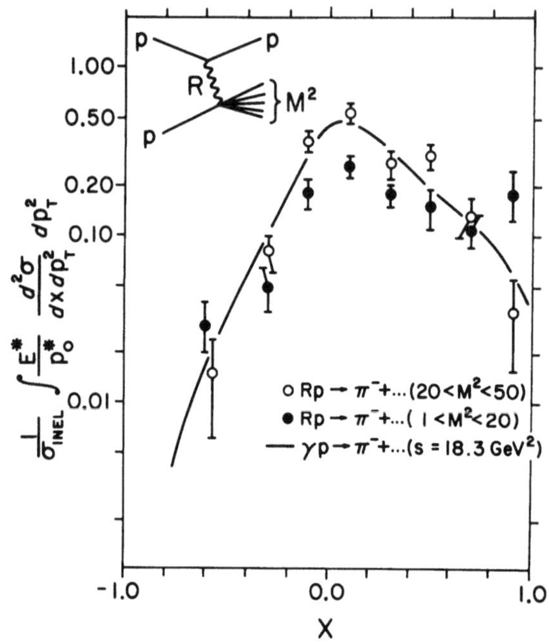

Figure 43: Comparison of the off-shell reaction $R^0+p\to\pi^-$+Anything with the inclusive data for $\gamma+p\to\pi^-$+Anything at $s\approx M^2$.

the result.) The rise in $\langle p_T\rangle$ with multiplicity at large masses can be understood, qualitatively, as due to a damped random-walk process. The unusual reversal at small masses may be due to a Bose effect (but here again there is inconsistency because of the lack of dependence of $\langle p_T\rangle$ on charge). The arrows labeled M.C. are the expected values of $\langle p_T\rangle$ (essentially independent of M) which our Monte Carlo model provides for the 400 GeV/c data. (This model contains proper single-particle distributions and multiplicities observed at 400 GeV/c but no explicit correlations among particles.) It is interesting that although the Monte Carlo predictions for $\langle p_T\rangle$ are completely incorrect, the predictions for Γ_x (not shown) appear to be in far better accord with the data. It is not clear to me whether this new result is important or whether it is just another way to study particle correlations. In any event, the mass of the produced system clearly has the largest effect on the dynamic properties of the hadron bundle, hence one should "look to the masses."

I thank J. Whitmore for providing me with an early version of his update of "Multiparticle Studies Utilizing Fermilab Bubble Chambers". In addition, I wish to thank E. L. Berger, D. Duke, P. Slattery, and C. Quigg for helpful discussions. Several of the topics in these lectures could not have been prepared without the excellent assistance I have received from our students J. Biel and

P. Stix. Finally, I wish to express my gratitude to D. Weingarten for a critical reading of the manuscript and for providing extensive suggestions as to content.

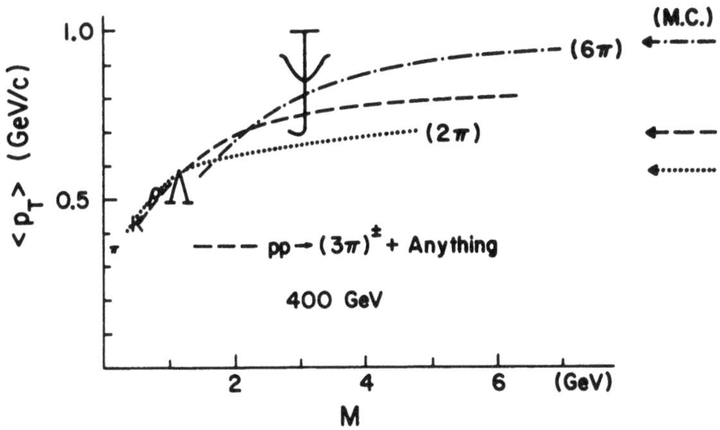

Figure 44: Comparison of $<p_T>$ for the production of long-lived particles in pp collisions with 400 GeV/c data (smoothed curves) for the reactions $pp \rightarrow (2\pi)^0$+Anything, $pp \rightarrow (3\pi)^{\pm}$ + Anything, and $pp \rightarrow (6\pi)^0$+Anything. The values of $<p_T>$ for the produced multipion systems are displayed as a function of the masses of the systems. (Only charged particles were used for the calculations.)

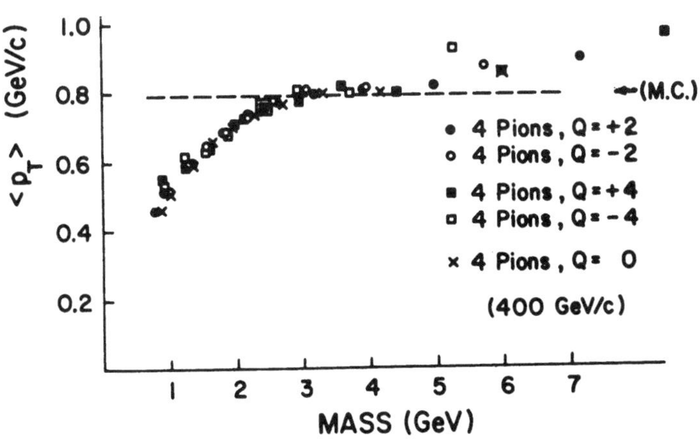

Figure 45: The value of $<p_T>$ for a (4π) system produced in the reaction $pp \rightarrow (4\pi)$+Anything at 400 GeV/c. This figure displays the variation of $<p_T>$ with the mass and the charge (Q) of the (4π) system. The dashed line represents the variation of $<p_T>$ with mass, expected on the basis of a Monte-Carlo calculation. (There is only a very weak dependence on charge in the Monte Carlo.)

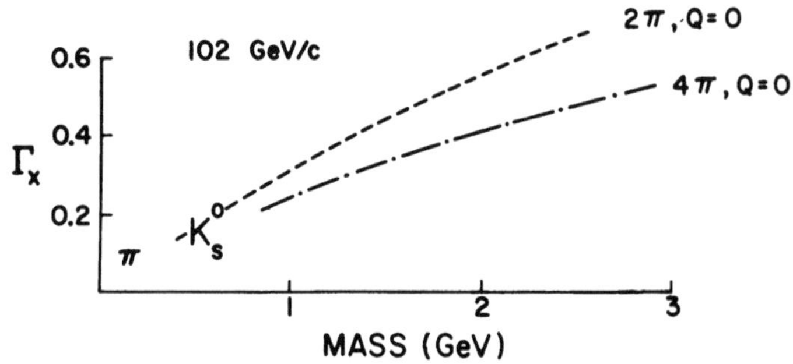

Figure 46: Comparison of the full width at half maximum (Γ_x) of the x-distribution ($d\sigma/dx$, integrated over p_T) for pions and K_s^o produced in pp collisions at 102 GeV/c with data (smoothed) for the reactions $pp\rightarrow(2\pi)^o$+Anything and $pp\rightarrow(4\pi)^o$+Anything, also at 102 GeV/c. The widths of the x-spectra are plotted as a function of the masses of the produced multipion systems.

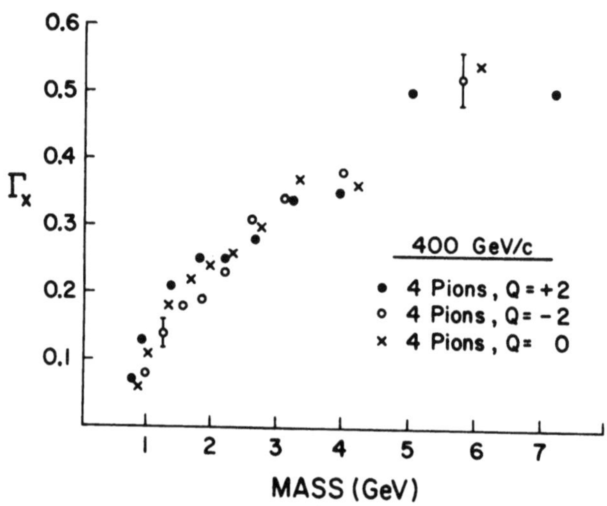

Figure 47: The dependence of Γ_x on the charge of a produced (4π) system in pp collisions at 400 GeV/c.

REFERENCES

1. J.J. Aubert et al, Phys. Rev. Lett. 33, 1404 (1974); J.E. Augustin et al, Phys. Rev. Lett. 33, 1406 (1974).

2. A. Carrol et al, Phys. Lett. 61B, 303 (1976). Also, private communication from W. Baker of the BNL-FNAL-Rockefeller Collaboration.

3. Private communication from H. Kobrak of the Chicago-San Diego-Wisconsin Collaboration.

4. A.V. Barnes et al, Phys. Rev. Letts. 37, 76 (1976). See also O.I. Dahl et al, Phys. Rev. Letts. 37, 80 (1976) for a study of $\pi^-p \to \eta n$.

5. J. Whitmore in Particles and Fields APS/DPF Meeting in Seattle (1975). H. Lubatti and P. Mockett eds. See also the comprehensive review of bubble chamber data by J. Whitmore, MSU Report (Feb. 1976) to appear in Physics Reports.

6. Y. Eylon and H. Harari, Nuclear Phys. B80, 349 (1974).

7. D. Ayres et al (FSAS Group), Phys. Rev. Lett. 35, 1195 (1975); see also the report of D. Cutts which appears in these proceedings. I have included the results of the Michigan-ANL-FNAL-Indiana Group in Fig. 5; see C. Akerlof et al, Phys. Rev. Lett. 35, 1406 (1975).

8. I. Ambats et al, Phys. Rev. D9, 1179 (1974); G. Brandenburg et al, Phys. Lett. 58B, 367 (1975).

9. R.L. Anderson et al (FSAS Group) Fermilab Report Pub-76/47-EXP/ANL-HEP-PR-76-28.

10. M. Davies and H. Harari, Phys. Lett. 35B, 239 (1971). See also R. Carnegie, Phys. Lett. 58B, 371 (1975).

11. Private communication from J. Butler (FSAS Group). See also results of Akerlof et al given in ref. (7). The latter measurements indicate that the ratios $\sigma_{elastic}$ to σ_{total} are 0.15/0.14/0.20 for $\pi^{\pm}p/K^{\pm}p/p^{\pm}p$ channels respectively. The small systematic discrepancies between the two sets of data have been commented on previously by J. Lach, Fermilab Report Conf -76/15-EXP (1976).

12. V. Barger and R. Phillips, Phys. Lett. 60B, 358 (1976).

13. A. Buras and J. Dias de Deus, Nucl. Phys. B71, 481 (1974).

14. See reference 13 for a comparison of pp data, and J. Butler, Ph.D. Thesis, MIT (1975) for a comparison of K^+p data.

15. C. Akerlof et al, Phys. Lett. 59B, 197 (1975). For the initial prediction of Chou and Yang see Phys. Rev. Letts. 20, 1213 (1968).

16. C. Ankenbrandt et al, FNAL-Yale Group, Fermilab-Conf-75/61-EXP (1975). See also J. Lach in reference (11).

17. R. Hendrick and B. Lautrup, Phys. Rev. D11, 529 (1975).

18. V. Bartenev et al, Phys. Rev. Lett. 31, 1367 (1973).

19. V.N. Bolotov et al, Nucl. Phys. B73, 365 (1974). See the lectures of E. Leader for a comprehensive discussion of charge-exchange data and their phenomenological implications.

20. H.R. Barton et al, OSU-MSU-Carleton-Alberta Collaboration, Preprint COO-1545-196. I thank N.W. Reay for a discussion of the data.

21. J. Biel, Ph.D. Thesis, University of Rochester (1976).

22. J. Biel, et al, Phys. Rev. Letters 36, 507 (1976); T. Ferbel, Proc. of Int'l School of Subnuclear Physics at Erice (1975), A. Zichichi ed.

23. This has been emphasized by V. Chaloupka in a preprint discussing similar data from SLAC.

24. A similar effect has been reported by a group at the ISR measuring the reaction $pp \rightarrow n\pi^+p$. See H. de Kerret et al CERN preprint (1976).

25. A. Minaka et al, Preprint KYUSHU-76-HE-6 (1976). Also, private communication from E. Berger.

26. For a review of the phenomenology and earlier data see H. Bøggild and T. Ferbel, Ann. Rev. Nuc. Sci 24, 451 (1974).

27. F. Winkelmann, Phys. Lett. 48B, 273 (1974).

28. R. Field and G. Fox, Nucl. Phys. B80, 367 (1974).

29. R.L. Anderson et al, FSAS Preprint (1976). I thank D. Ritson for providing me with this paper prior to general circulation.

30. M. Sogard, private communication from the FSAS Group. See also the paper of D. Cutts in these proceedings.

31. The data are from A. Gurtu et al, Proc. of Vth Int'l. Symp. on Many-Particle Hadrodynamics, Eisenach-Leipzig (1974), G. Ranft and J. Ranft, eds. and W. Busza et al, MIT/Carleton/ FNAL preprint (1976). See also W. Busza, VIth Int'l Conf. on Nuclear and Particle Physics, Los Alamos (1975). I wish to thank J. Elias for a helpful discussion of the Fermilab data.

32. See the review of L. Bertocchi, VIth Int'l Conf. on Nuclear and Particle Physics, Los Alamos (1975).

33. K. Gottfried, Phys. Rev. Letters 32, 957 (1974).

34. For an alternate description of multiparticle production on nuclear targets see the report of G. Berlad in these proceedings.

35. R.E. Ansorge et al, Phys. Letts. 59B, 299 (1975); J.G. Rushbrooke et al, Phys. Letts. 59B, 303 (1975).

36. The figure is from reference (5). The pn data are from A. Sheng et al, Phys. Rev. D12, 1219 (1975), and the $\bar{p}n$ data are from A. Fridman et al, Phys. Rev. D12, 3414 (1975).

37. J. Elliott et al, Phys. Rev. Letters 34, 607 (1975).

38. Z. Koba, H. Nielsen and P. Olesen, Nucl. Phys. B40. 317 (1972).

39. T. Ferbel, Phys. Rev. D7, 925 (1973).

40. I wish to thank J. Whitmore for providing me with as yet unpublished results from the latest exposures at the 30-inch bubble chamber.

41. K. Moffeit et al, Phys. Rev. D5, 1603 (1972).

42. C. Bromberg et al, Nucl. Phys. B107, 82 (1976).

43. This effect has been discussed previously in R. Schindler et al, Phys. Rev. Letters 33, 862 (1974).

44. Reference (5) and P. Bosetti et al, Nucl. Phys. B54, 141 (1973).

45. T. Ferbel, Phys. Rev. D8, 2321 (1973).

46. C. Quigg and E. Rabinovici, Phys. Rev. D13, 2525 (1976). The parameterization used by Whitmore for the cross sections are:

$\sigma(\pi N) = (15.18\ \mathrm{mb})p_{LAB}^{0.0755}$, $\sigma(KN) = (13.09\ \mathrm{mb})p_{LAB}^{0.0755}$,
$\sigma(NN) = (25.33\ \mathrm{mb})p_{LAB}^{0.0755}$.

47. It is uncertain, for example, how the latest parameterization of the Pomeranchukon contribution to total cross sections by Quigg and Rabinovici affects the asymptotic form of the theoretical expression for $\langle n_c \rangle$. Preliminary indications are that $\langle n_c \rangle$ will rise more rapidly than that indicated by the B $\ln s$ rise in Table I.

48. J. Whitmore et al, Phys. Lett. 60B, 211 (1976).

49. Chan, H-M., Hsue, C.S., Quigg, C. and Wang J-M, Phys. Rev. Letts. 26, 672 (1971).

50. E.W. Beier et al, University of Pennsylvania preprint, submitted to the Tbilisi Meeting.

51. R.L. Anderson et al, Inclusive Hadron Exchange Scattering at 50-175 GeV. I thank D. Ritson for providing me an early version of the preprint.

52. A. Krzywicki and D. Weingarten, Phys. Lett. 50B, 265 (1974).

53. J. Derré et al, French-Soviet Collaboration, preprint M-12 (1974); C. Bromberg et al, Phys. Rev. D12, 1224 (1975); D. Fong et al, Phys. Lett. 61B, 99 (1976); J. Lamsa et al, Phys. Rev. Lett. 37, 73 (1976).

54. D. Weingarten et al, Rochester-ISU-MSU-ANL-FNAL-Maryland Preprint (1976).

55. J.P. DeBrion et al, Phys. Rev. Letts. 34, 910 (1975).

56. J.P. DeBrion et al, Phys. Letters. 52B, 477 (1974).

57. P. Stix et al, Rochester Report (in preparation).

58. See the lecture of A.J.S. Smith in this volume for an interesting comparison of the latest results for $\langle p_t \rangle$ of $\mu^+\mu^-$ pairs produced in hadronic collisions as a function of the mass of the $\mu^+\mu^-$ system. It appears that $\langle p_t \rangle$ for $\mu^+\mu^-$ pairs is larger than for produced hadrons of equivalent invariant mass.

DISCUSSION

CHAIRMAN: Prof. T. Ferbel

Scientific Secretaries: A. Degré and K.L. Wernhard

DISCUSSION

- *LEADER:*

How can the parameter C in your formula for the average multiplicity be negative?

- *FERBEL:*

C has to be negative. To show this, we can rewrite the formula in the following way:

$$\langle n \rangle = A + B \ln s \left(1 + \frac{C/B}{s^{1/4}} \right) .$$

We assume that $\langle n \rangle$ is proportional to the plateau in the $d\sigma/dy$ distribution. The width of the plateau rises proportional to ln s. For $s \to \infty$ the height of the plateau is constant. Therefore

$$\langle n \rangle = B \ln s, \qquad \text{for } s \to \infty ,$$

but for lower energies, we have to correct for the increase in height with energy. This means that $\langle n \rangle$ reaches A + B ln s from the bottom; therefore, $C < 0$.

- *WIGNER:*

Does the A_1 show up in a variety of reactions or only in πp collisions?

- *FERBEL:*

To summarize the situation about the A_1, one can say the following: Initially, the A_1 was found in a πp experiment as a peak around 1.15 GeV in the 3π effective mass distribution. Secondly, it is also produced coherently on nuclei. Thirdly, the A_1 shows a ($\pi\rho$) substructure. Fourthly, a spin-parity analysis taking into account the production and the decay of the 3π-system -- Ascoli-analysis -- shows that 0^-, 1^+ and 2^- waves are present, 1^+ dominant, but the phase-shift does not go through 90° at the mass of the A_1.

- *RANFT:*

A double Regge model describes the mass distribution in the 3π (ρ,π) system in the A_1 range. It gives an s-wave but without the circle in the Argand diagram, thus the A_1 should not -- at least in this global manner, with respect to mass distributions -- be called a resonance.

- *BUCCELLA:*

To find the A_1 in $\rho\pi$ phase shift analysis is more difficult than to find the A_2. In fact, while J^P conservation allows only D-wave for the A_2, s-wave is also possible for the A_1. So, one has a two-channel problem, i.e., to diagonalize a 2×2 matrix.

- *MARCIANO:*

Caldi and Pagels have shown that SU(3) × SU(3) chiral symmetry does not require the existence of the A_1 meson. In their scheme, the B meson, a well-established resonance, is the chiral partner of the ρ.

- *ETIM:*

What happens to Weinberg's sum rules if you throw away the A_1 meson?

- *LIPKIN:*

SU(3) does not need the A_1 or the B, but if the Q states are there, SU(3) requires the rest of the octet to be there. Chiral symmetry requires an axial vector meson, but perhaps this could be the B as Marciano suggested. The quark model needs both the A_1 and B which are the $^3P_1$ and $^1P_1$ states. However, if the A_1 is not there, the quarkists can probably find a spin-orbit interaction to get rid of it.

A REVIEW OF THE ISR RESULTS

Giovanni Valenti

Istituto Nazionale di Fisica Nucleare

v. Irnerio 46, Bologna, Italy

I. INTRODUCTION

Operation of the CERN Intersecting Storage Rings (ISR) has provided a large variety of new results and insight in the study of high energy proton-proton interaction. This lecture will consist of a review of recent results on total cross-section, elastic scattering, large transverse momentum phenomena and direct lepton production.

II. TOTAL CROSS-SECTION

New measurements of the proton-proton total cross-section have been performed[1] with a precision of $\pm$ 0.6% over the energy range $23.5 < \sqrt{s} < 62.7$ GeV. Two different techniques were implemented by the two groups, CERN-ROME (CR), PISA-STONY BROOK (PBS): the PBS group derived the cross-section from

$$\sigma_{TOT} = \frac{R_{TOT}}{L}$$

where L is the Luminosity measured using the Van der Meer method[2], and R_{TOT} is the interaction rate recorded on a nearly 4π scintillation detector (Fig. 1).

To determine R_{TOT} from $R_{observed}$, the following corrections have to be considered:

i) loss of acceptance due to dead regions

ii) elastic events at very small angles escaping into the beam pipe.

This can be expressed by:

$$\sigma_{TOT} = \frac{R_{observed}\,(1+\varepsilon)}{L} + \Delta\,\sigma_{el}$$

where ε accounts for trigger losses and $\Delta\sigma_{el}$ corrects for losses of elastic events.

The CR group used the differential elastic cross-section at zero momentum transfer (t=0) to compute the total cross-section via the optical theorem

$$\sigma_{TOT} = \sqrt{\frac{16\pi}{1+\rho^2}\left(\frac{d\sigma}{dt}\right)_{t=0}}$$

ρ is the ratio of the real to the imaginary part of the forward elastic scattering amplitude, which at ISR energies is small and positive (Fig. 2)[3].

The forward elastic differential cross-section is given by:

$$\left(\frac{d\sigma}{d\omega}\right)_{\theta=0} = \left(\frac{d\sigma}{d\omega}\right)_{\theta} (1 - \varepsilon_{coul})\; e^{b|t|}$$

where $\left.\frac{d\sigma}{d\omega}\right._{\theta} = \frac{R_{el}}{\Delta\omega\cdot L}$, R_{el} = rate of elastic scattering,

$\Delta\omega$ = solid angle of the apparatus, ε_{coul} take into account coulomb scattering effects important at small t. From this one obtains:

$$\sigma_{TOT} = \sqrt{\frac{16\pi}{1+\rho^2}\,\frac{\pi}{\rho^2}\,(1 - \varepsilon_{coul})\; e^{b|t|}\,\frac{1}{\Delta\omega}\,\frac{R_{el}}{L}} \quad (*)$$

(*) In determining σ_{TOT}, the necessary values of ρ and b were input data. The CR group has also performed a coulomb nuclear interference measurement which should allow a simultaneous determination of σ_{TOT}, b, ρ.

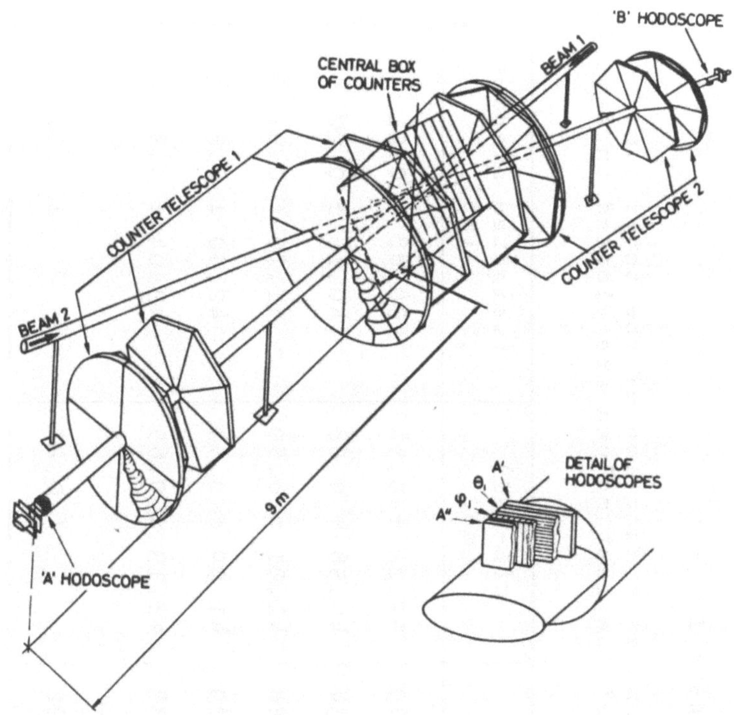

Fig. 1 ISR intersection I-8. General layout of the Pisa-Stony Brook large solid-angle secondary-particle detector, and the CERN-Rome hodoscope measuring elastic scattering in the vertical plane.

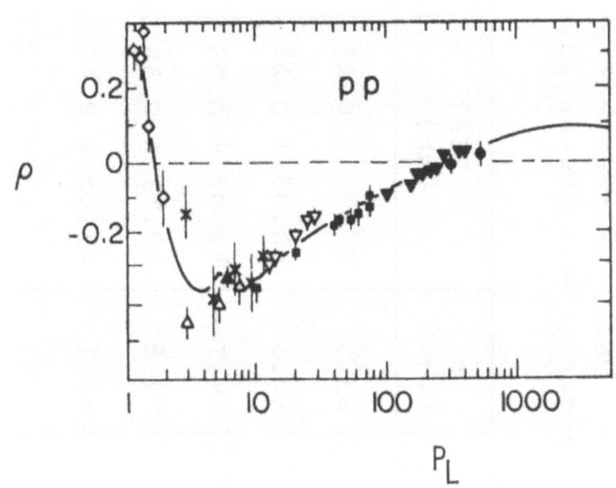

Fig. 2 Ratio of the real to the imaginary part of the forward amplitude plotted versus the laboratory momentum.

TABLE I

Total proton-proton cross-sections as a function of the total c.m. energy, $\sqrt{s}$. The weighted averages of the last column are plotted in fig. 2.

$\sqrt{s}$ (GeV)	σ(PSB) (mb)	σ(CR) (mb)	σ(L-ind) (mb)	σ(weighted average) (mb)
23.5	38.80 ± 0.25	39.01 ± 0.29	39.22 ± 0.54	38.88 ± 0.21
30.6	40.08 ± 0.24	40.35 ± 0.34	40.58 ± 0.68	40.16 ± 0.22
44.7	41.92 ± 0.25	41.45 ± 0.26	41.01 ± 0.50	41.70 ± 0.21
52.8	42.73 ± 0.34	42.38 ± 0.29	41.99 ± 0.54	42.50 ± 0.27
62.7	43.02 ± 0.40	43.05 ± 0.33	43.13 ± 0.65	43.04 ± 0.31
Scale error	± 0.36	± 0.22	± 0.25	± 0.28

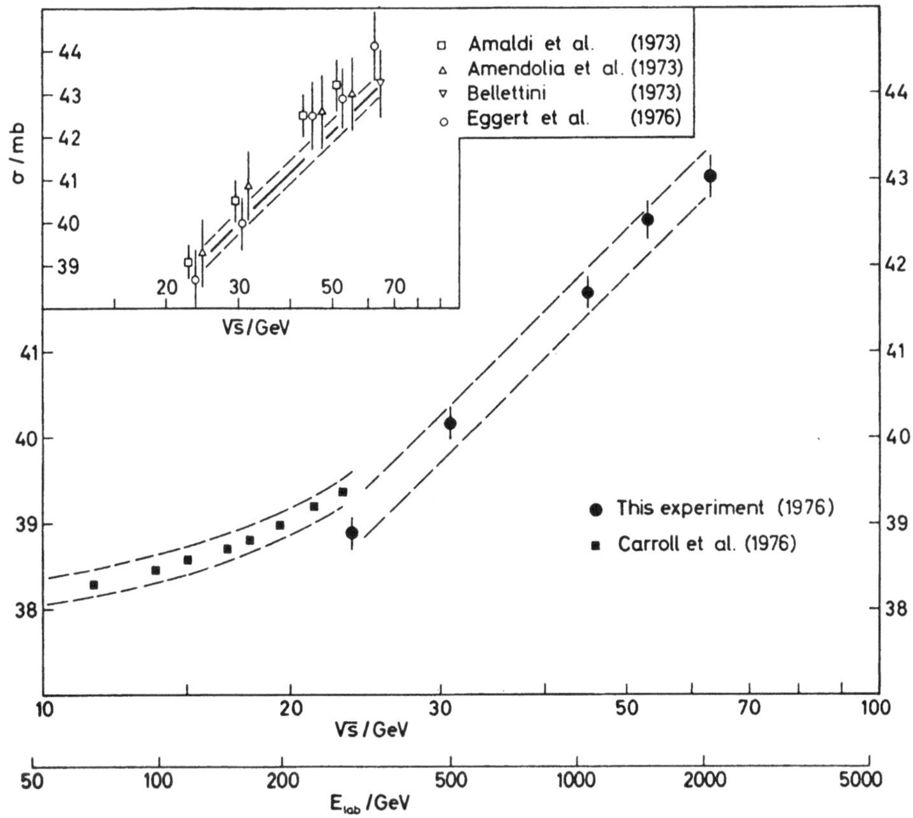

Fig. 3 The proton-proton total cross-section measurements obtained by the PBS-CR groups are plotted together with those from FNAL (Ref. 4). The broken lines represent in both cases the extended scale error. In the upper left corner, on a reduced scale, the results of all previous measurements at the ISR are shown (Ref. 5).

The two measurements were performed at the same time in the same interaction region of the ISR; this ensures that the luminosity is the same in both experiments so that an absolute (luminosity independent) measurement of the total cross-section could be obtained. Table I gives a comparison of the measurements obtained as well as weighted averages. The total cross-section compilation shown in Fig. 3 is in the range of energies above NAL up to 62.7 GeV in the c.m.s. and represent our best knowledge so far.

III. LARGE ANGLE ELASTIC SCATTERING

One of the key features of proton-proton elastic scattering is the dip in the differential cross-section $d\sigma/dt$ occurring at a $|t|$ value of 1.4 $(GeV/c)^2$. The existence and the characteristics of such a dip are important in the attempt to understand the shape of the proton. A detailed study of the dip position as a function of the total center-of-mass energy, $\sqrt{s}$, has been done by the CHOV collaboration (CERN-HAMBURG-ORSAY-VIENNA) at the Split Field Magnet[6]. The following features have emerged:

i) the t_{MIN} decreases with increasing s,

t_{MIN} $(\sqrt{s} = 23\ GeV) = -(1.44 \pm 0.02)\ (GeV/c)^2$

t_{MIN} $(\sqrt{s} = 62\ GeV) = -(1.26 \pm 0.03)\ (GeV/c)^2$

The slope parameter b seems to behave (as a function of $\sqrt{s}$) differently in different momentum transfer regions, that is:

ii) between $t = -0.25\ (GeV/c)^2$ to $t = -0.6\ (GeV/c)^2$

$b = (10.3 \pm 0.2)\ (GeV/c)^{2}$

between $t = -0.6\ (GeV/c)^2$ to $t = -1.1\ (GeV/c)^2$

$b\ (\sqrt{s} = 23\ GeV) = (9.1 \pm 0.1)\ (GeV/c)^{2}$

$b\ (\sqrt{s} = 62\ GeV) = (11.8 \pm 0.3)\ (GeV/c)^{2}$

iii) the peak value observed after the dip increases with s

$\sqrt{s} = 23$ GeV 45 ± 5 nbarns

$\sqrt{s} = 62$ GeV 73 ± 10 nbarns

iv) for $t < -2.0\ (GeV/c)^2$ $\frac{d\sigma}{dt}$ does not depend on $\sqrt{s}$.

These features are shown in Fig. 4. Fig. 5 gives a compilation for various values of $\sqrt{s}$. Notice that the ratio

$$\frac{t_{dip}\ (\sqrt{s} = 23\ GeV)}{t_{dip}\ (\sqrt{s} = 62\ GeV)} = 1.14$$

is in agreement with geometrical scaling; it should in fact be proportional, as σ_{TOT} and σ_{inel} to the square of the radius of the proton.

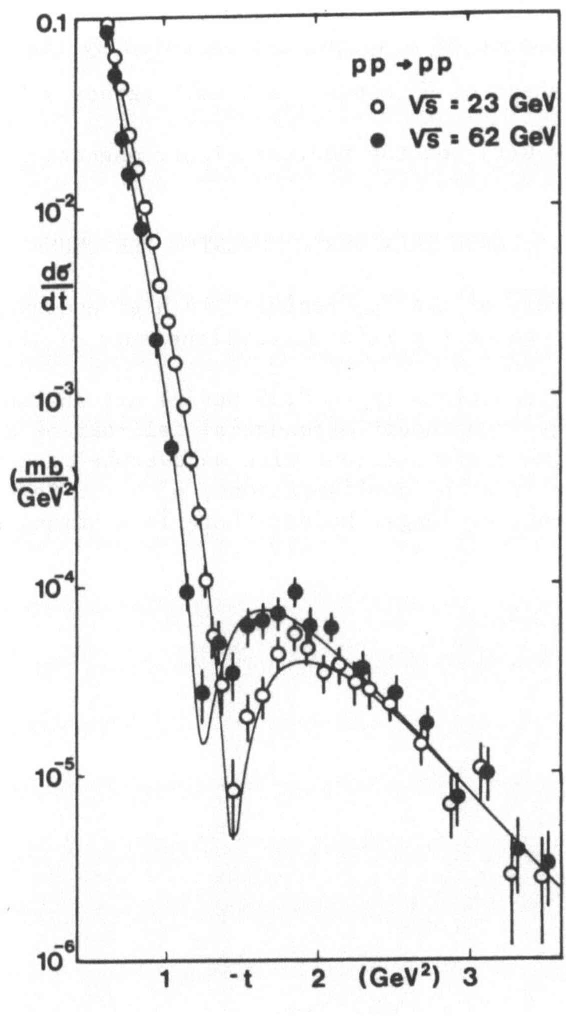

Fig. 4 Differential cross-section for momentum transfer $-t > 0.6$ GeV^2. The t-resolution in the vicinity of the minimum is ±0.013 GeV at 23 GeV and ±0.028 GeV^2 at 63 GeV. The solid lines represent the best fit result of a two-amplitude model (Ref. 8).

The same group has extended $\frac{d\sigma}{dt}$ measurements up to $-t = 9\ GeV^2$ at a center-of-mass energy $\sqrt{s} = 53$ GeV (Fig. 6)[7]. Two features emerge:

i) the absence of a second dip expected by the optical model

ii) the shape of $\frac{d\sigma}{dt}$ above $\sim 2.5\ GeV^2$ agrees with the fourth power $G(t)^4$ of the nuclear electromagnetic form factor.

IV. LARGE TRANSVERSE MOMENTUM PHENOMEMA

The discovery of two "p_T regimes" in high energy proton-proton interactions is one of the main accomplishements of the CERN ISR.

The low p_T reactions ($p_T < 0.35$ GeV/c) are characterized by a striking energy independent exponential fall-off of all single particle inclusive cross-sections with an overwhelming pion production[9]. In large p_T configurations, $p_T > 2.0$ GeV/c, the exponential cut-off no longer holds; there is a strong energy

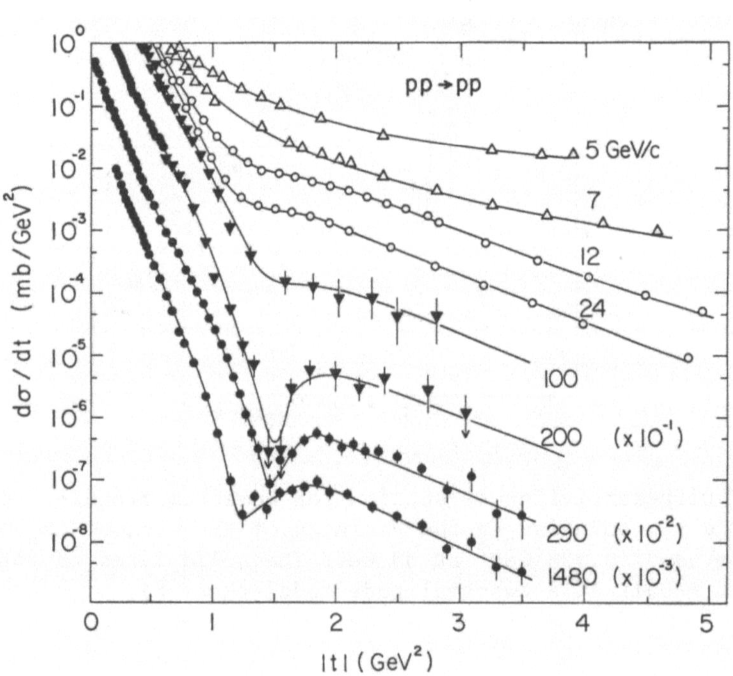

Fig. 5 Compilation of proton-proton differential cross-sections.

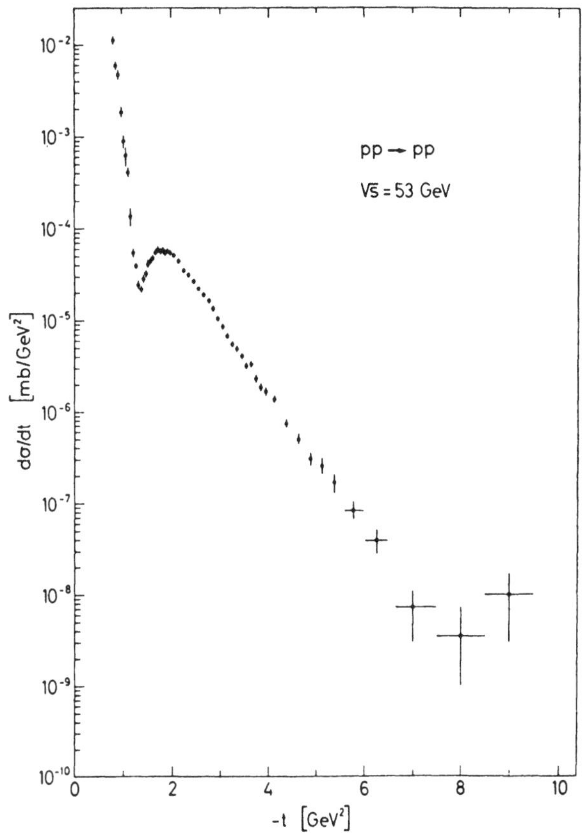

Fig. 6 Differential cross-section as a function of t. The error bars represent statistical errors and bin-to-bin systematical errors. An overall scale uncertainty of ±5% has to be added.

dependence in the production cross-sections, and heavier particles are more and more abundantly produced.

The growing interest in high p_T physics relies on the possibility of studying the small transverse distance structure of the proton. Here I will discuss recent ISR results only; more complete reviews of earlier experiments can be found in the literature[10].

Single particle inclusive distributions

Invariant cross-sections are observed to vary strongly with transverse momentum; data are usually parametrized according to

$$E \frac{d\sigma}{dt} = p_T^{-n} f(x_T,\theta)$$

where $x_T = 2p_T/\sqrt{s}$, $\sqrt{s}$ and θ are the total energy and production angle in the center-of-mass system (c.m.s.). Typical sets of data for π°, $\pi^\pm$ inclusive single particle distributions obtained by the CERN-Columbia-Rockefeller-Saclay (CCRS) Collaboration[11] are shown in Figs. 7, 8, 9. In Fig. 10 π° inclusive spectra at various center-of-mass energies are compared. Finally, Fig. 11 shows data obtained by the Aachen-CERN-Heidelberg-Munich (ACHM) Collaboration[12] and by the British-Scandinavian (BS) Collaboration[13] on π° and $\pi^\pm$ production respectively. Table II summarizes the values of the exponent n obtained by fits to these data.

The behaviour of $f(x_T,\theta)$ is known at the ISR up to ~ 0.35 and $\theta = 90°$ only; data are consistent with

$$f(x_T,\theta) = A\, e^{-Bx_T}$$

TABLE II

Single pion inclusive data

Data	Range of p_T	Range of $\sqrt{s}$	n
CCRS π°	> 2.5 GeV/c^2	23.5 to 62.4 GeV	7.2 ± 0.1
CCRS $\pi^\pm$	> 3.3 GeV/c^2	44.8 to 62.4 GeV	7.8 ± 0.2
ACHM π°	> 2.0 GeV/c^2	23.6 to 62.9 GeV	7.2 ± 0.2

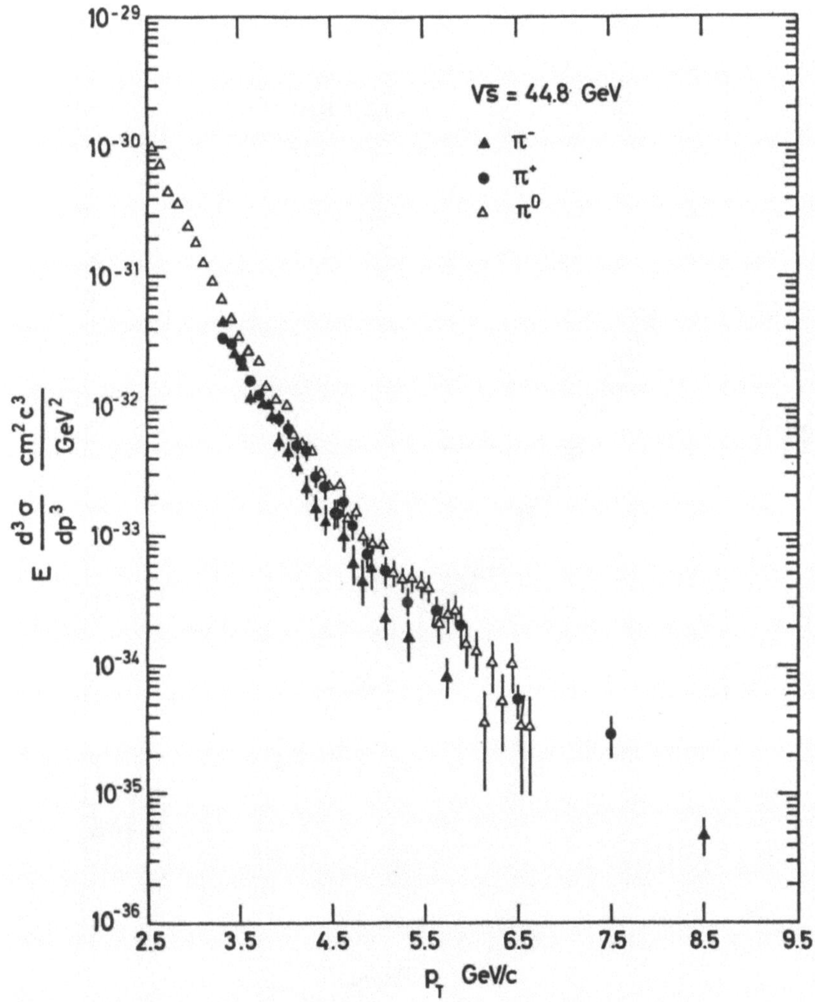

Fig. 7 Invariant cross-section for π^+, π^-, and π° production at $\theta = 90^\circ$ and $\sqrt{s} = 44.8$ GeV.

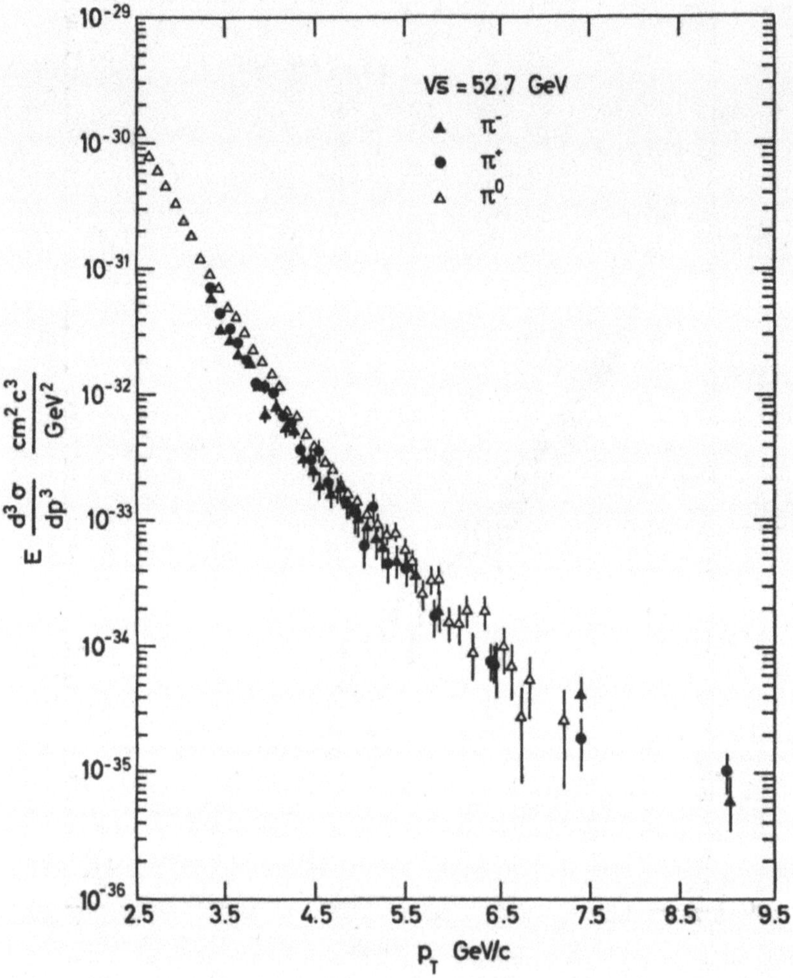

Fig. 8 Invariant cross-section for π^+, π^-, and π° production at $\theta = 90^\circ$ and $\sqrt{s}$ = 52.7 GeV.

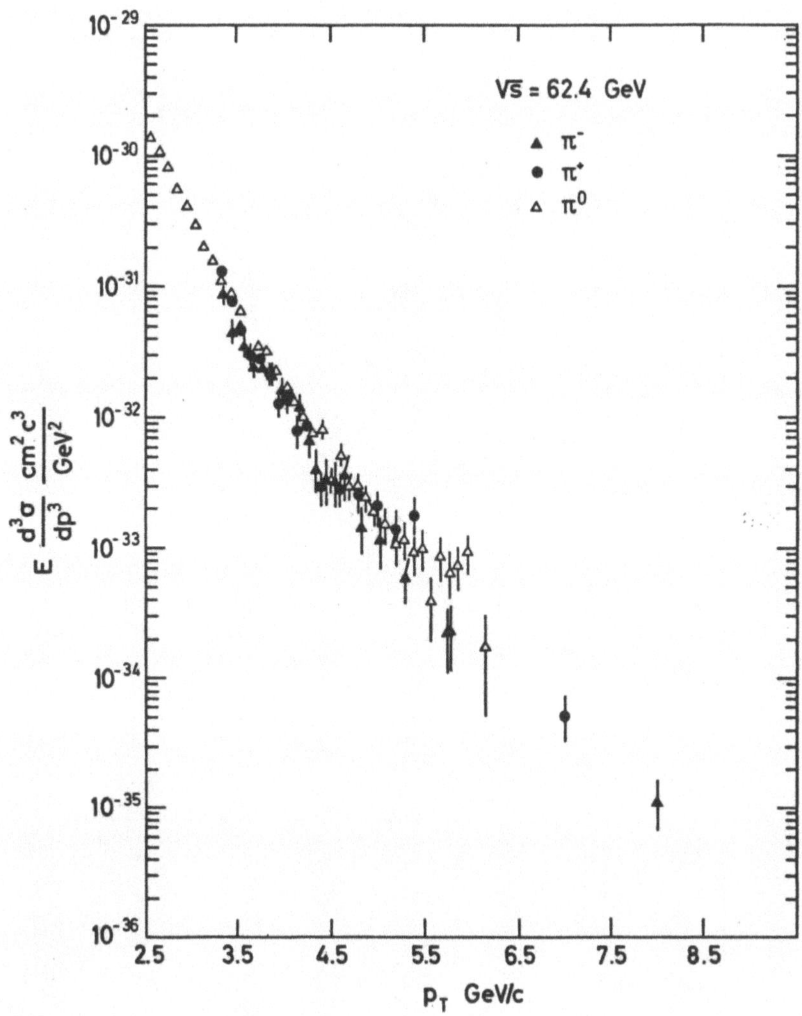

Fig. 9 Invariant cross-section for π^+, π^-, and π° production at $\theta = 90^\circ$ and $\sqrt{s} = 62.4$ GeV.

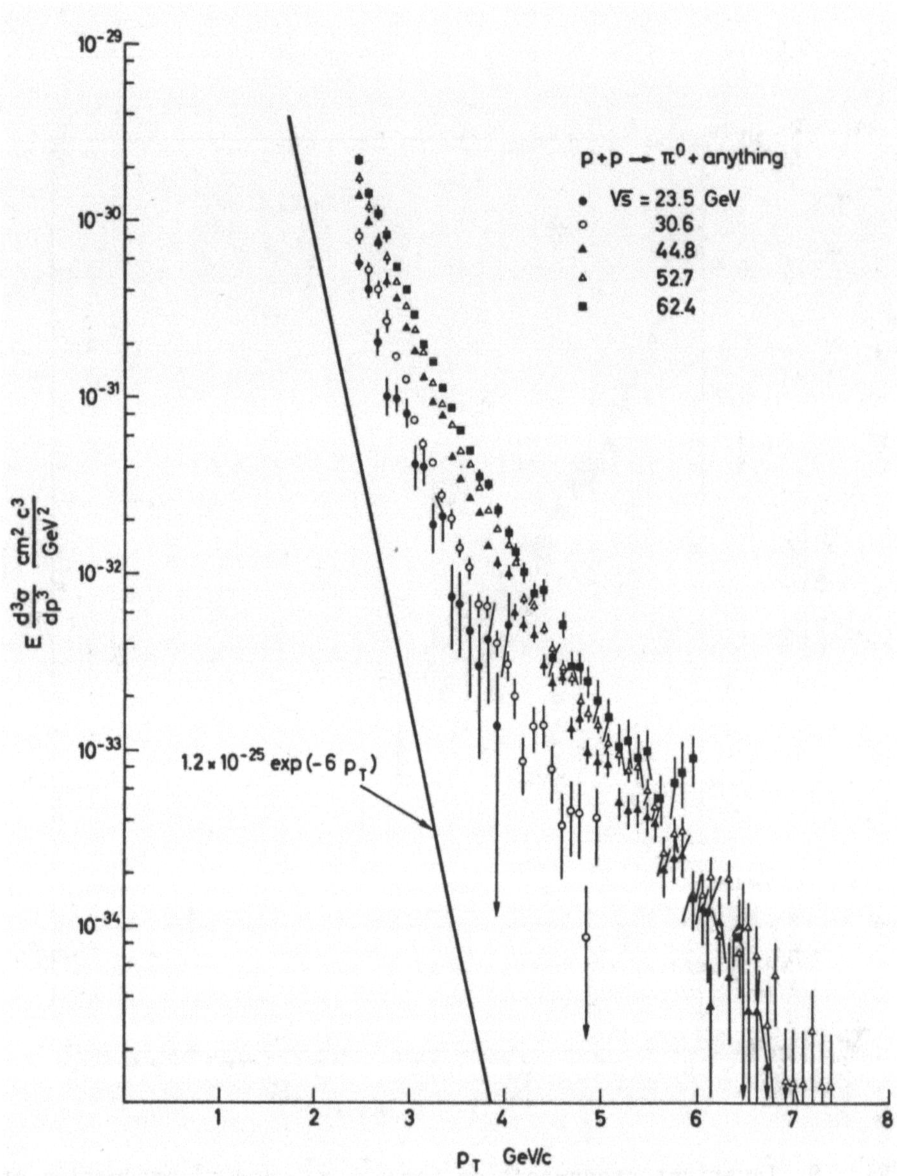

Fig. 10 Invariant cross-section for π° production at 90°, for five center-of-mass energies. The extrapolation from data with $p_T < 1$ GeV/c is shown for comparison.

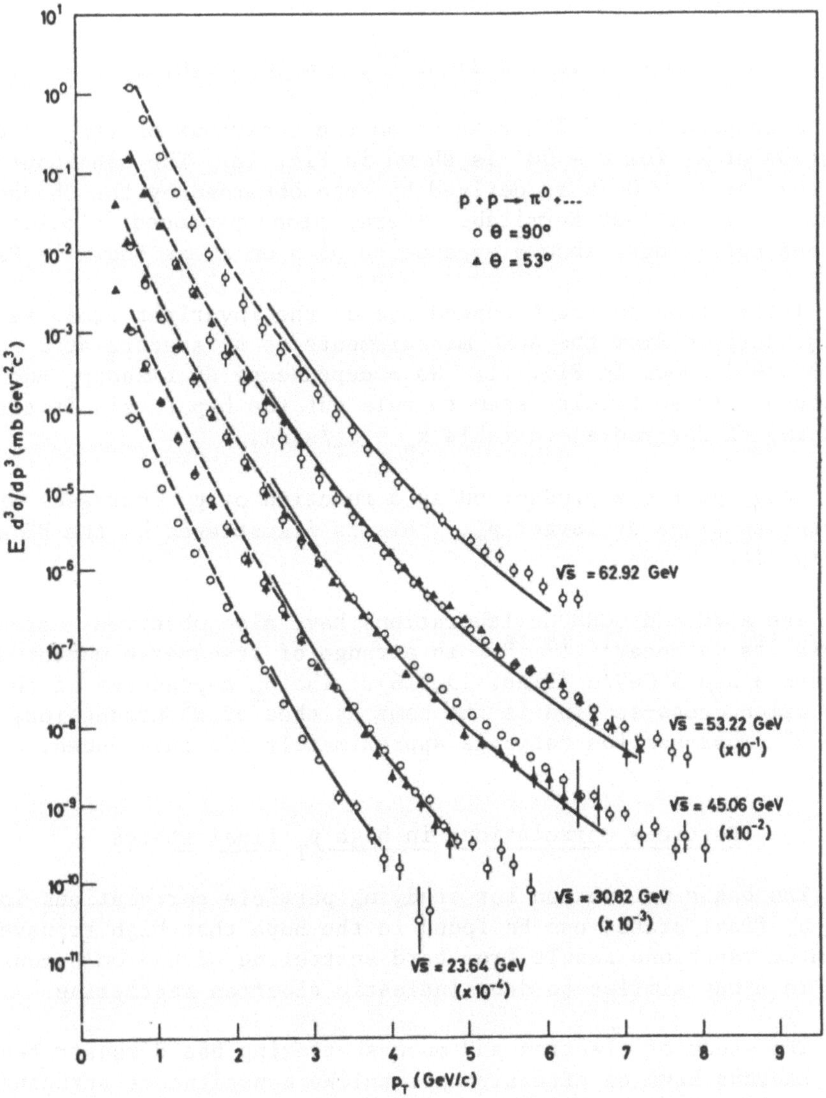

Fig. 11 Inclusive invariant cross-section for π° production for various values of θ and $\sqrt{s}$. The solid lines indicate the data of the CERN-Columbia-Rockefeller-Saclay Collaboration (Ref. 11) and the dashed lines those of the British-Scandinavian Collaboration (Ref. 13).

with $A = (14.8 \pm 0.6)\ 10^{-27}$ and $B = 12.6 \pm 0.2$.

A fit of the type $f(x_T,\theta) = C\ (1 - x_T)^m$ can also represent the data; in this case the best fit values are:

$$C = (12.9 \pm 0.2)\ 10^{-27},\ m = 10.5 \pm 0.2.$$

A compilation of ISR results on the behaviour of $f(x_T,\theta)$ as a function of x_T for $\theta = 90°$ is shown in Fig. 12. The behaviour of $f(x_T,\theta)$ for $x_T > 0.35$ is derived by data obtained by the Chicago-Princeton group[14] at Fermilab. There, pions produced in proton nucleus collisions, show a dependence of n on x_T as shown in Fig. 13.

Information on the θ dependence of the invariant cross-section can be derived from the ACHM measurements of π° spectra at $\theta = 53°$ and $\theta = 90°$ shown in Fig. 11. No θ dependence at fixed p_T and $\sqrt{s}$ is found. These results seem to rule out the hypothesis of scaling in terms of the radial variable $x_R = x_T/\sin\theta$.

Heavy particle production as a function of p_T increases with respect to pions at larger p_T. This is illustrated by the BS data in Fig. 14.

The ACHM and CCRS Collaborations have also observed η production via its $\gamma\gamma$ decay[11 15 16] in a range of transverse momentum between 3 and 5 GeV/c (Figs. 15, 16). The p_T dependence of the production cross-section is the same as that of π° production, the η to π° cross-section ratio is approximately 1/2 independent of $\sqrt{s}$.

Particle correlations in high p_T final states

The basic motivation for studying particle correlations in high p_T final states can be found in the hope that high transverse momentum reactions result from hard scattering of hadron's constituents in a way similar to deep inelastic electron scattering.

The study of electron and muon scattering has demonstrated that hadrons have an effective point-like constituents structure. The quark parton model interpretation leads one to expect that hadrons can scatter to large transverse momentum via hard, large angle scattering processes involving their constituents. The final state structure that should emerge from such a picture is characterized by the so called jets. One is brought to expect, in high energy proton-proton interactions, two coplanar jets, collinearity being spoiled by the internal mition of the hadron constituents. Each hadron of the jet, moreover, should have a limited momentum

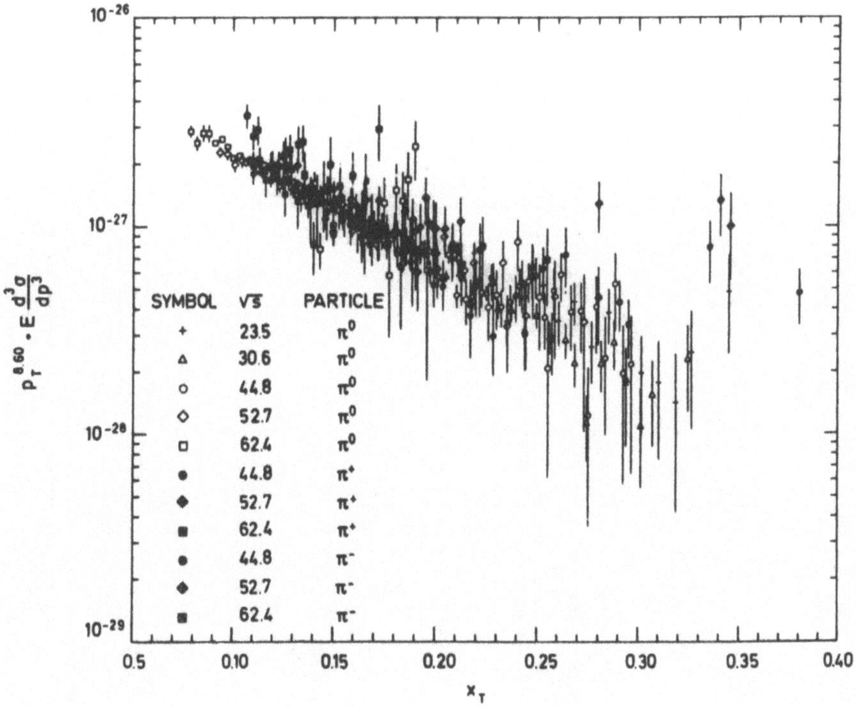

Fig. 12 Plot of the quantity $p_T^{8.6}$ $E(d^3\sigma/d^3p)$ for π^+, π^-, and π° production as a function of $x_T \equiv 2p_T/\sqrt{s}$.

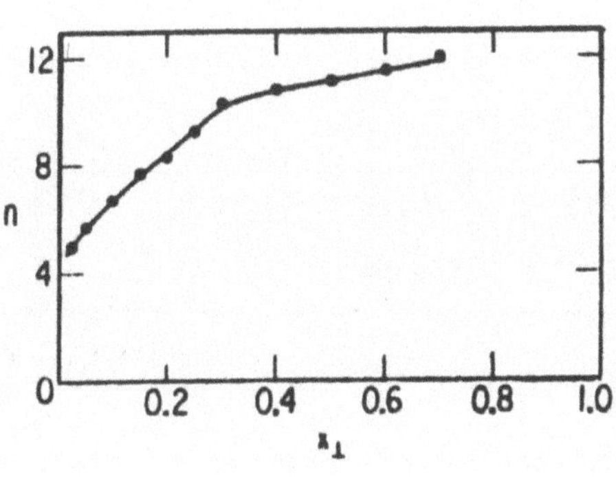

Fig. 13 Plot of the exponent "n" (see text) versus x_T for inclusive pion production at large $\sqrt{s}$.

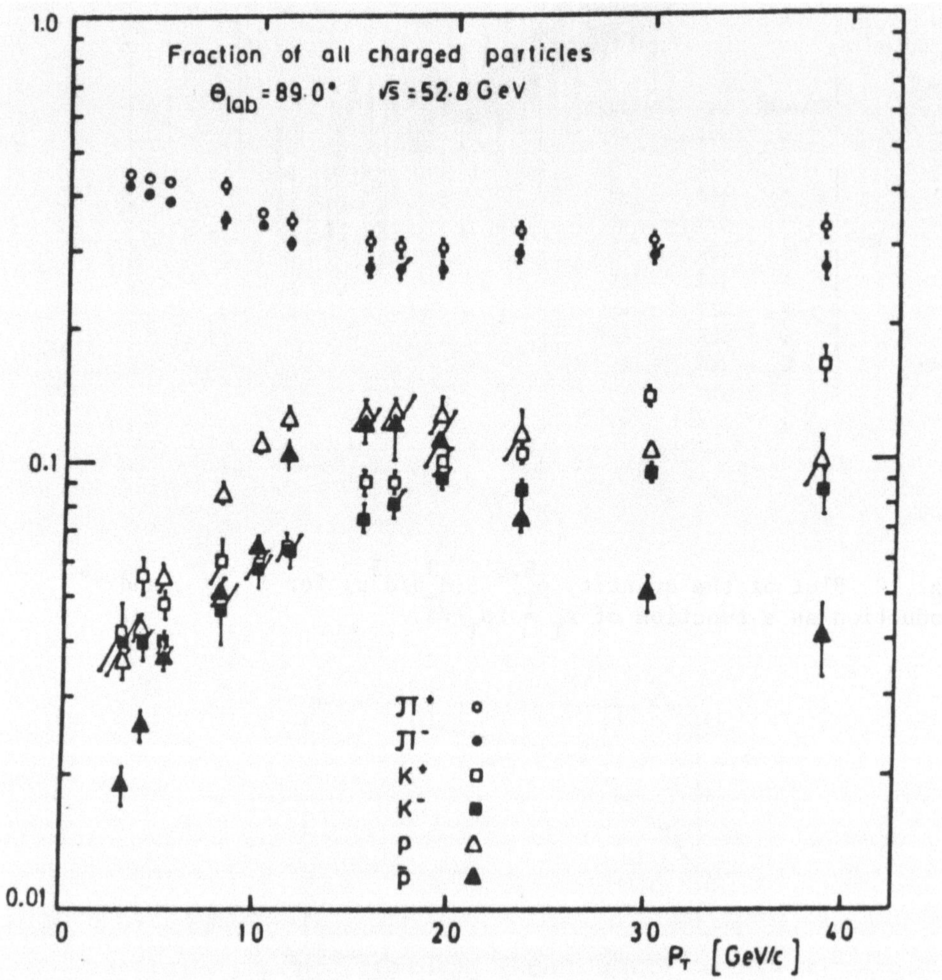

Fig. 14 the p_T dependence of the charged particle composition at $\sqrt{s}$ = 52.8 GeV, θ = 89°.

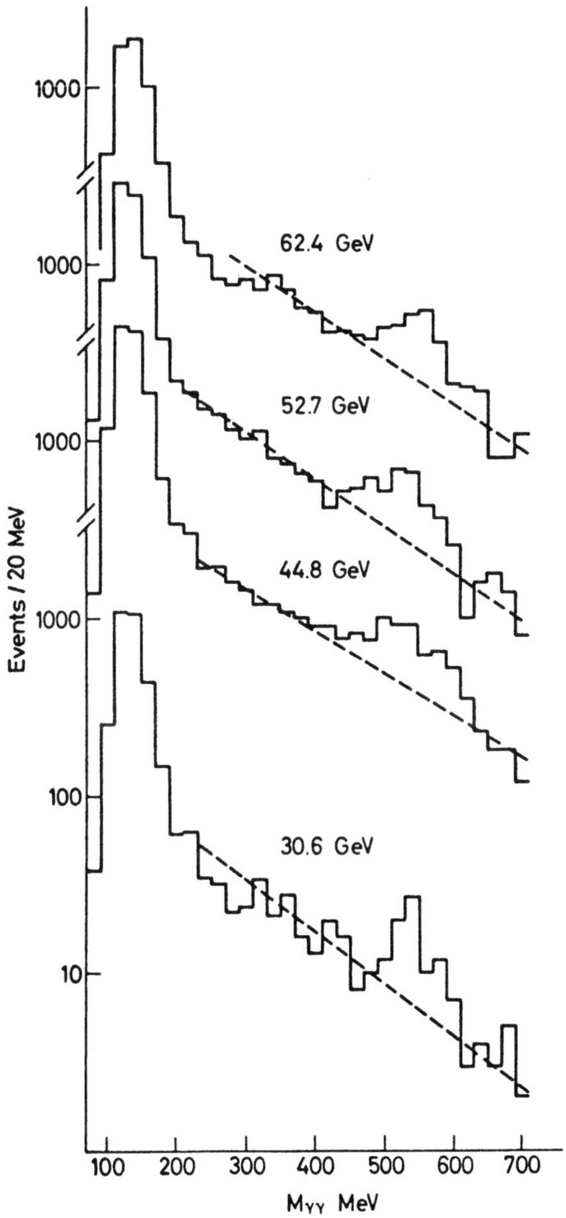

Fig. 15 Invariant mass distribution of two-photon events at four center-of-mass energies (CCRS).

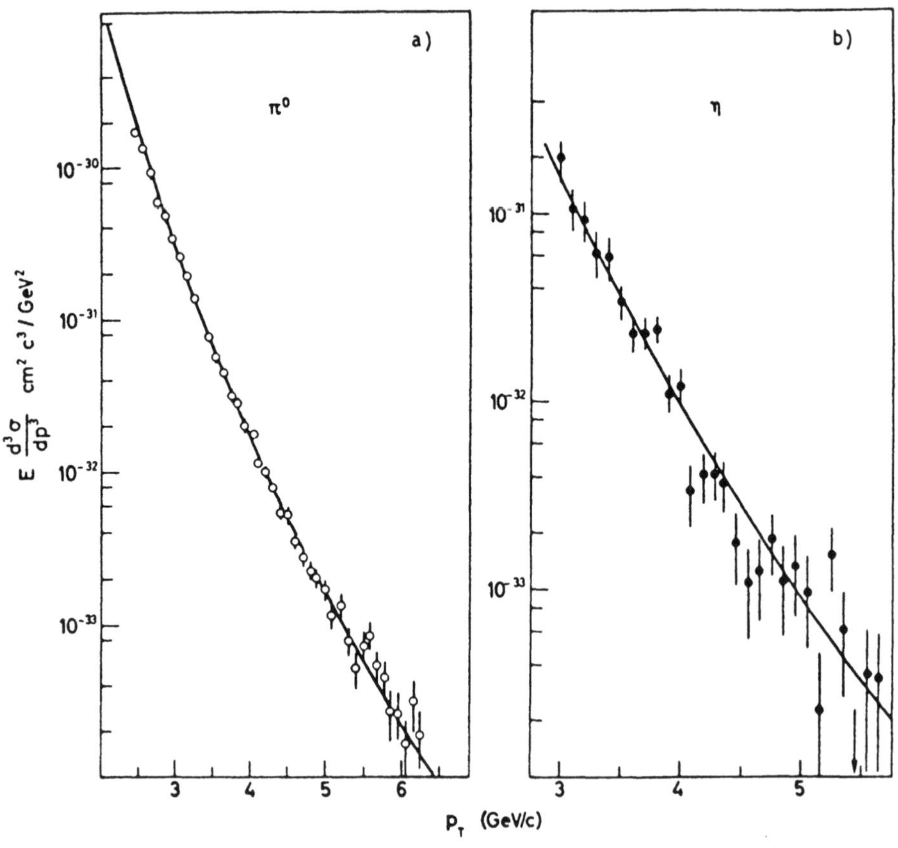

Fig. 16 Invariant cross-section for inclusive production of π°'s (Fig. 16a) and η's (Fig. 16b) at $\theta = 90^\circ$ and $\sqrt{s} = 52.7$ GeV. The curve shown in Fig. 16a is the best fit to the π° data; the curve of Fig. 16b is the best fit to the π° data, multiplied by 0.58.

component perpendicular to the jet axis.

Following P. Darriulat[17] it is possible to separate, in high p_T reactions, three components:

i) a component characteristic of a typical collision with small transverse momentum having

$$\sqrt{s}_{eff} < \sqrt{s} - 2\ p_T$$

ii) a component including the large transverse momentum particle itself and few others close in angle to it

iii) a component produced over a wide angular cone opposite to the high p_T particle.

The data of Fig. 17 from ACHM[12] show the features listed. In particular one can observe:

a) an enhancement in the same azimuthal region as the π° ($|\phi| < 30^\circ$) centered at the same pseudorapidity y as the π° ($\pm$ 0.7 units).

b) in a $\Delta\phi$ sector $\pm$ 30° around $\phi = 180^\circ$ the particle density peaks at $y = -y\ (\pi^\circ)$.

Two particle correlations are usually given in terms of the ratio of the two-particle invariant cross-section integrated over the π° transverse momentum, to the π° invariant cross-section integrated over the same momentum range, plotted versus the particle transverse momentum; that is

$$F = \frac{\int_{\Delta y_1} dy_1 \int_{p_{T1}}^{\infty} dp'_{T1} \frac{d^4\sigma}{dy_1 dp'_{T1} dy_2 dp_{T2}}}{\int_{\Delta y_1} dy_1 \int_{p_{T1}}^{\infty} dp'_{T1} \frac{d^2\sigma}{dy_1 dp'_{T1}}}$$

CCRS data on two particle correlations of Figs. 18, 19, 20 show a character analogous to the three-component picture previously described.

New data obtained by a CERN group[15] at the Split Field Magnet (SFM) are now available. Their study tries to disentangle the

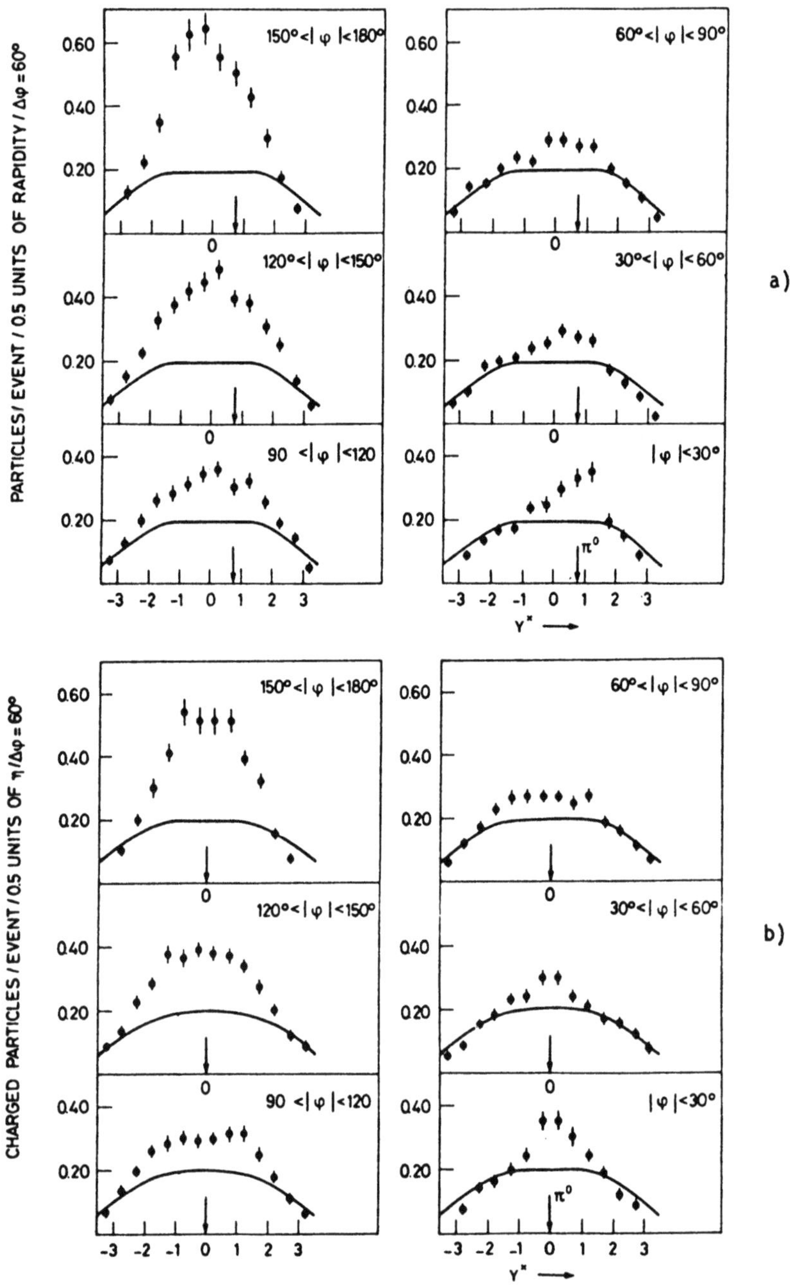

Fig. 17 Charged particle densities from ACHM (ref. 12) with $p_T(\pi^\circ) > 2$ GeV/c; $y^* = -\ln \mathrm{tg}(\theta/2)$ is the pseudorapidity. The solid lines show charged particle densities in minimum bias ($\sim$ inclusive) triggers: a) $\theta_{\pi^\circ} = 53^\circ$; b) $\theta_{\pi^\circ} = 90^\circ$.

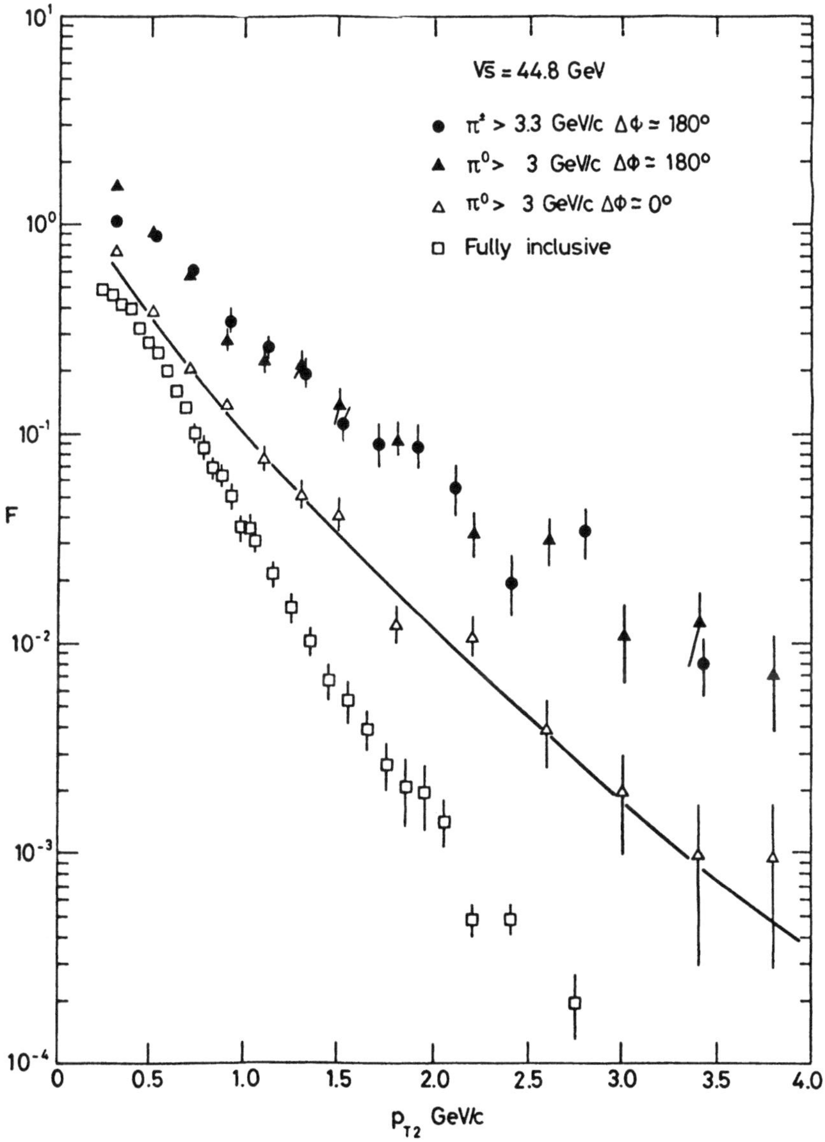

Fig. 18 The function F (see text) plotted versus the associated charged particle transverse momentum p_T, at $\sqrt{s}$ = 44.8 GeV. The probability per inelastic interaction is also shown.

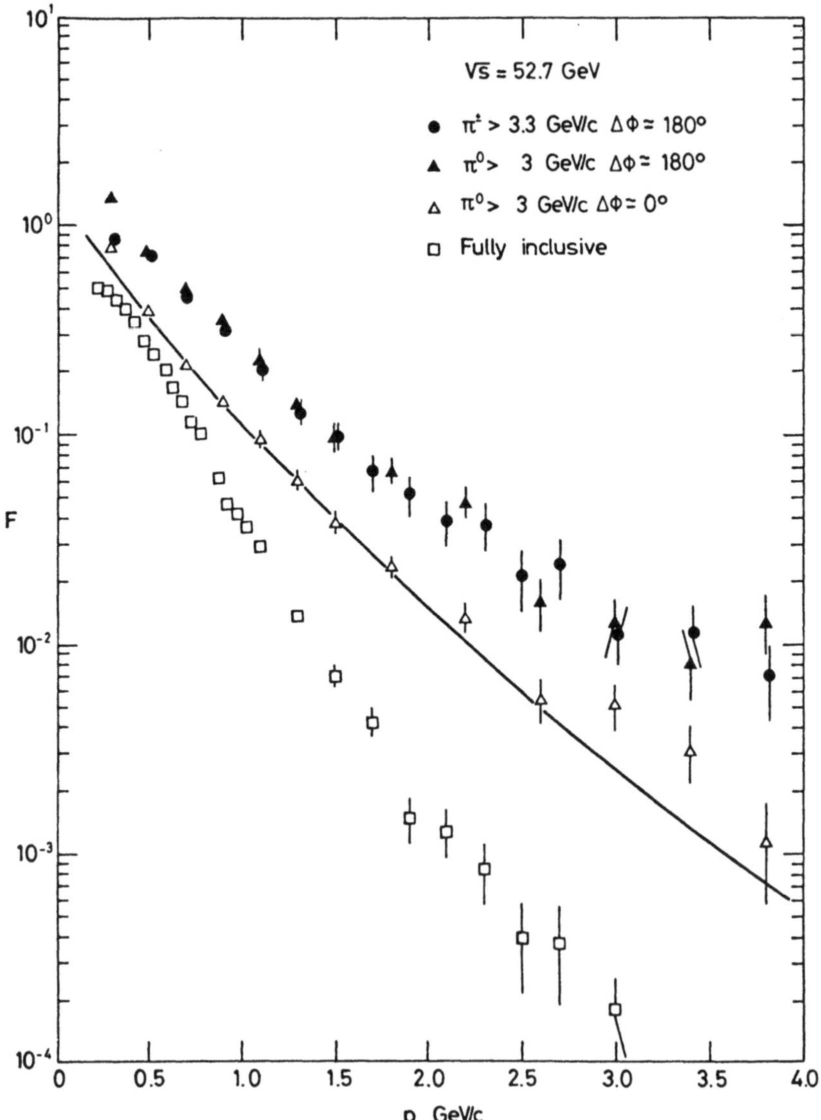

Fig. 19 The function F (see text) plotted versus the associated charged particle transverse momentum p_T, at $\sqrt{s}$ = 52.7 GeV. The probability per inelastic interaction is also shown.

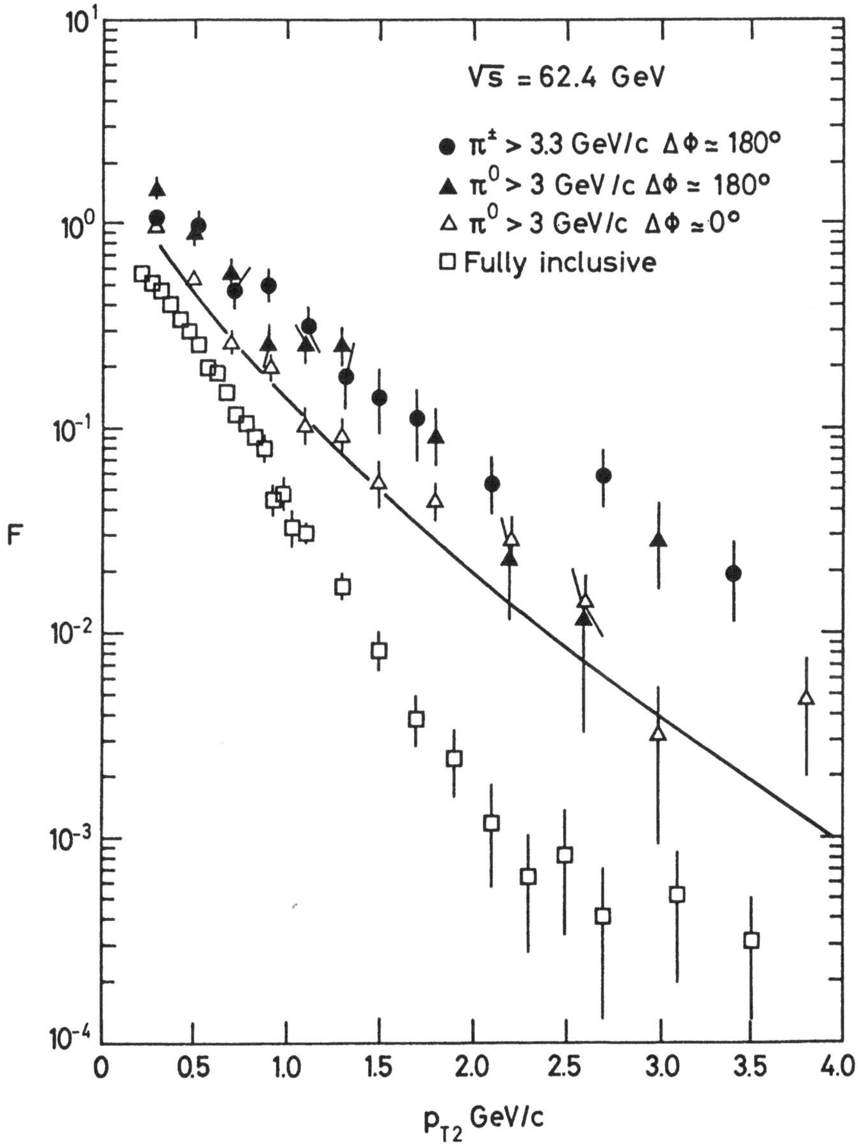

Fig. 20 The function F (see text) plotted versus the associated charged particle transverse momentum p_T, at $\sqrt{s}$ = 62.4 GeV. The probability per inelastic interaction is also shown.

"jets" pattern by studying the secondaries produced in high p_T triggered π° events. The kinematical definition used by the authors in describing the data is shown in Fig. 21.

Charged particles produced in the π° hemisphere

The average single particle distribution function integrated over ϕ in the region $|\phi| < 27°$ and over p_x for four bins of p_x between 0.4 and 1.7 GeV/c for 90° data is shown in Fig. 22. The relative minimum bias curve (dashed line) has errors of the order of 10 ÷ 20 %. The excess grows and shrinks in rapidity with increasing p_x; for $|y| < 2$, $|\phi| < 27°$, $(0.4 < p_x < 1.7)$ GeV/c the integral of the excess is:

$$0.13 \pm 0.02 \text{ charged particles/event.}$$

The associated invariant mass distribution for the $(\pi^\circ,\pi^\pm)$ system, in Fig. 22, shows a low mass excess over the background for uncorrelated minimum bias particle distribution. These features are consistent with correlated particles distributed symmetrically around the triggered π° with an average transverse momentum of 300 ± 100 MeV/c for all p_x.

The question whether this low mass excess could be explained by resonance production is answered by Fig. 23. The ρ peak is clearly visible but does not exhaust all the excess. Assuming that ρ and π° have the same p_T dependence $(2.7 < p_T < 5$ GeV/c) so that the ρ production cross-section can be expressed as a multiple of the π°'s, and using the following parametrization which has been found to describe correctly ρ production

$$\frac{d\sigma}{dp_T^2} = \frac{1}{(1 + p_T^2)^n} e^{-0.5\, p_T}$$

with $n = 4.0 \pm 0.6$ (in agreement with π production spectra), one obtains:

$$\frac{1}{2} (\rho^+ + \rho^-)/\pi^\circ = 0.9 \pm 0.2 \,^{(*)}.$$

(*) ρ decays have been substracted from π° cross-section.

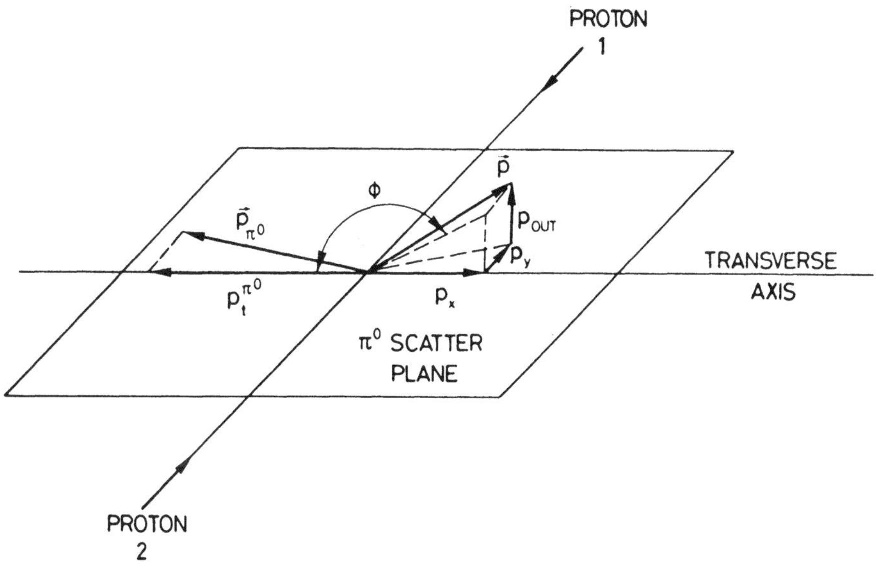

Fig. 21 Diagram showing the kinematical quantities used to describe the triggering π° and the secondary hadrons in the c. m. system. The hadron momentum, p, is broken into three components. The absolute value of the component along the π° transverse momentum is called p_x, the one out of the π° scatter plane is called p_{out}. The azimuthal angle of the hadron, ϕ, is referred to the π° scatter plane which is the plane containing the incoming protons and the π°.

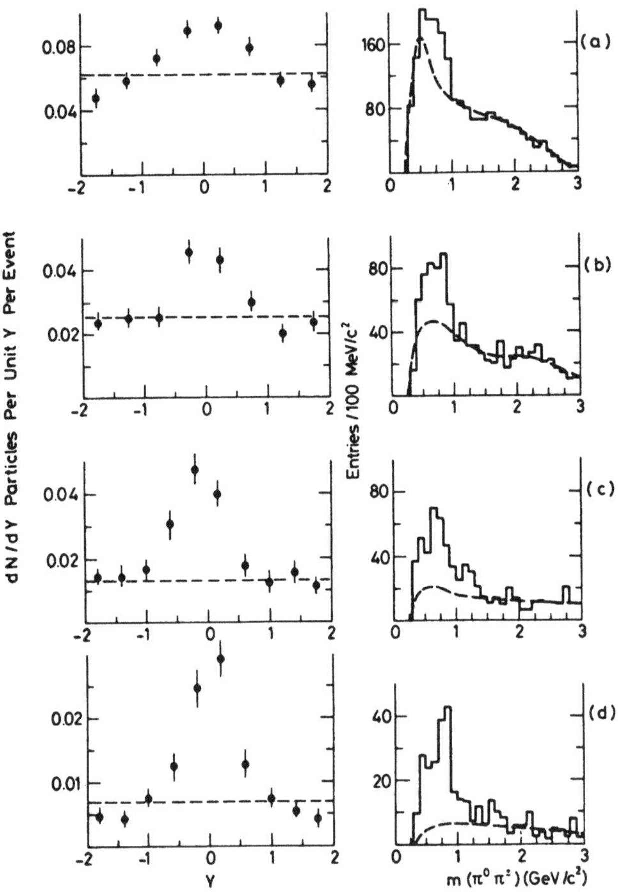

Fig. 22 Rapidity distributions and $\pi^{\circ} - \pi^{\pm}$ mass distributions for particles in the π° hemisphere with $|\phi| < 27°$. The data are sliced into different p_x intervals between 0.4 and 1.7 GeV/c: (a) $0.4 < p_x < 0.6$, (b) $0.6 < p_x < 0.8$ GeV/c, (c) $0.8 < p_x < 1.1$ GeV/c, (d) $1.1 < p_x < 1.7$ GeV/c. The mass distributions are plotted for $|y| < 2$. The dashed curves show the minimum bias rapidity distributions expected for uncorrelated particles.

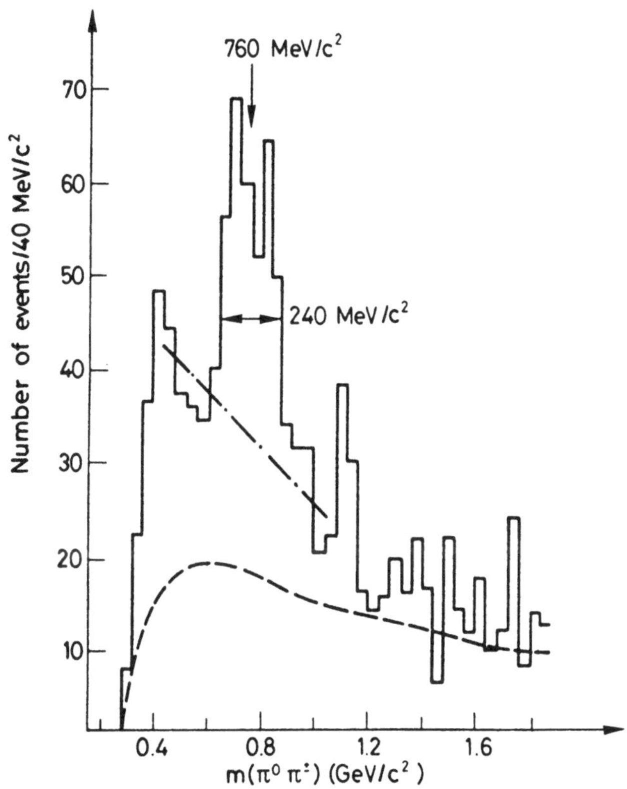

Fig. 23 The $\pi^\circ - \pi^\pm$ mass distribution for particles in the π° hemisphere, within the limits $|\phi| < 27°$, $|y| < 2$ and $0.7 < p_x < 1.7$ GeV/c. The dashed line is the mass distribution expected for uncorrelated particles and the dashed-dotted line is the background assumed to estimate the ρ signal.

In other words at $p_T = 2$ GeV/c only 16% of the π°'s come from decays.

The absence of $\omega \rightarrow \pi^+ \pi^- \pi^\circ$ and the presence of a strong three-particle correlation (at least as strong as the two-particle one), see Fig. 24, excludes that a common parent, if it exists, decays always into two particles.

Charged particles in the hemisphere opposite to the π°

A distribution of $|y|$ for charged particles in the hemisphere opposite to the π° with $|180^\circ - \phi| < 35^\circ$ is shown in Fig. 25. One observes a large positive/negative ratio at large p_x and large y, fairly independent of $p_T(\pi^\circ)$. Moreover, plot c) indicates an increase of particle density for high p_x restricted to $|y| < 1.5$. This increase is better evidentiated in Fig. 26 where charged multiplicity data having $|y| < 1$, $|180^\circ - \phi| < 30^\circ$ are plotted against $p_T(\pi^\circ)$; at $p_T(\pi^\circ) = 2.5$ GeV/c the charged multiplicity is twice the minimum bias one.

P_{out} distribution

The important point to emphasize, in order to test the jet structure in single particle distributions, is the coplanarity with the scatter plane defined by the trigger particle and the incoming protons. Fig. 27 shows $dN/d|p_{out}|$ for various intervals of p_x. A fit of this quantity in the form

$$\frac{dN}{d|p_{out}|} = A\, e^{-B\,|p_{out}|}$$

gives:

$$B = 2.1 \pm 0.3$$

fairly independent of p_x. Minimum bias data, on the contrary, show a $B = 1/\langle p_{out}\rangle$ with $\langle p_{out}\rangle \simeq 0.6\sqrt{p_x}$. To better understand the way in which transverse momentum is balanced, one can consider the single particle distribution in the variable $x_E = |p_x/p_T(\pi^\circ)|$. This is shown in Fig. 28 for $p_x > 1.2$ GeV/c; for comparison are also shown the distributions for x_F (Feynmann scaling) and z analogously defined for electron-proton and $e^+ - e^-$ scattering processes.

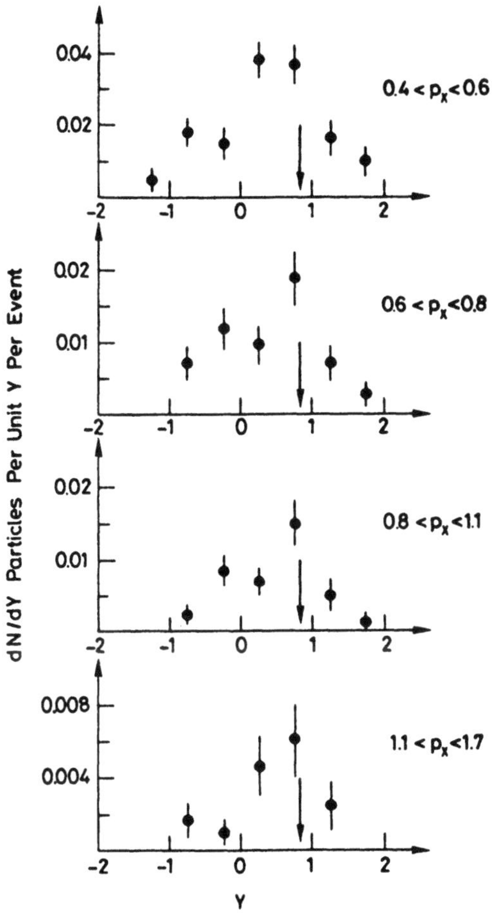

Fig. 24 Rapidity distribution of charged particles in the π° hemisphere from the 45° data. Secondaries from events in which a fast particle (p_x > 1 GeV/c, y > 0) occurs have been plotted; the fast particle is excluded from the plot.

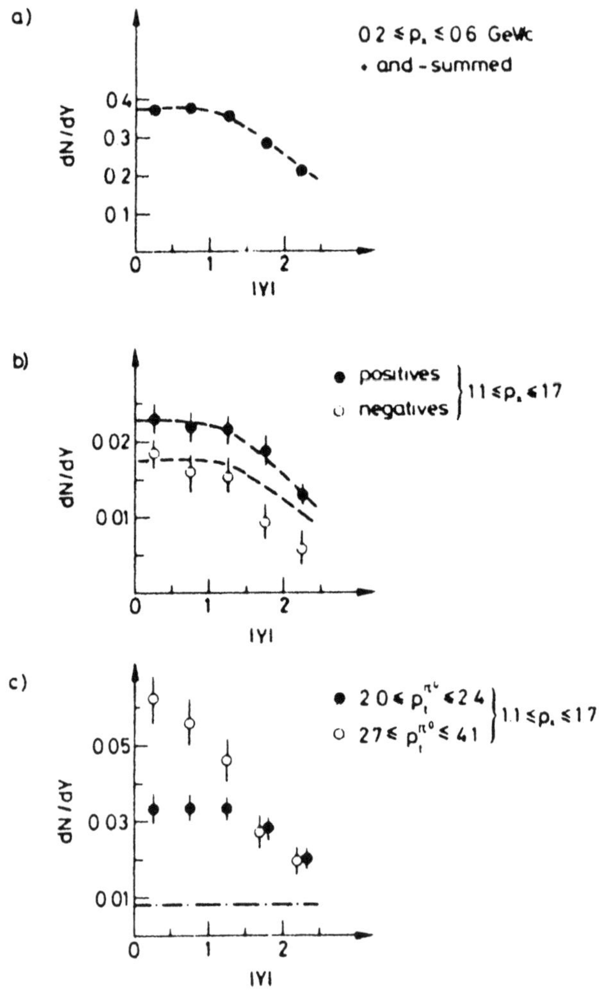

Fig. 25 Distribution of the absolute value of the rapidity for charged particles in the hemisphere opposite the π°, with $|180^\circ - \phi| < 35^\circ$. (a) Summed over +ve and −ve charges for $0.2 < p_x < 0.6$ GeV/c. The dashed line is drawn to guide the eye. (b) Broken into +ve and −ve charges for $1.1 < p_x < 1.7$ GeV/c. The dashed lines are the same curve as in (a) but scaled down to agree with the data in the $|y| < 1$ region. (c) Broken into $p_T(\pi^\circ)$ intervals, $2 < p_T(\pi^\circ) < 2.4$ GeV/c and $2.7 < p_T(\pi^\circ) < 4.1$ GeV/c, and summed over charges for $1.1 < p_x < 1.7$ GeV/c. The dashed-dotted line indicates the mean density from minimum bias events.

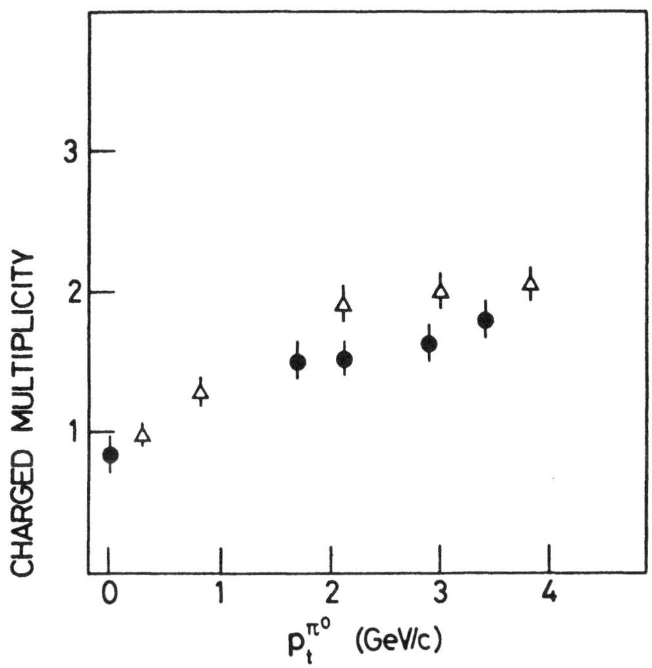

Fig. 26 Number of charged particles per event as a function of $p_T(\pi°)$ including the minimum bias value (shown at $p_T(\pi°) = 0$) for $|180° - \phi| < 30°$ and $|y| < 1$ (full dots). Also shown (open triangles) are data from ref. 12.

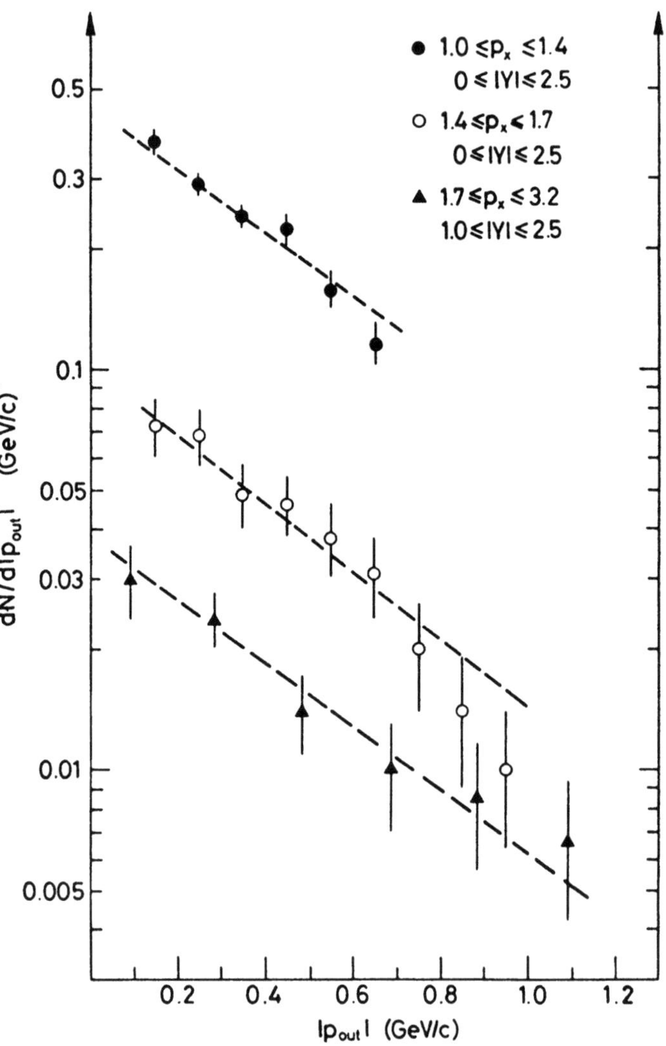

Fig. 27 Distribution of $|p_{out}|$ for different p_x intervals of the charged particles. The distributions are integrated over $|y| < 2.5$; for the highest p_x interval the y range is restricted to $1 < |y| < 2.5$. The dashed lines correspond to a function

$$dN/d|p_{out}| \propto \exp(-2|p_{out}|).$$

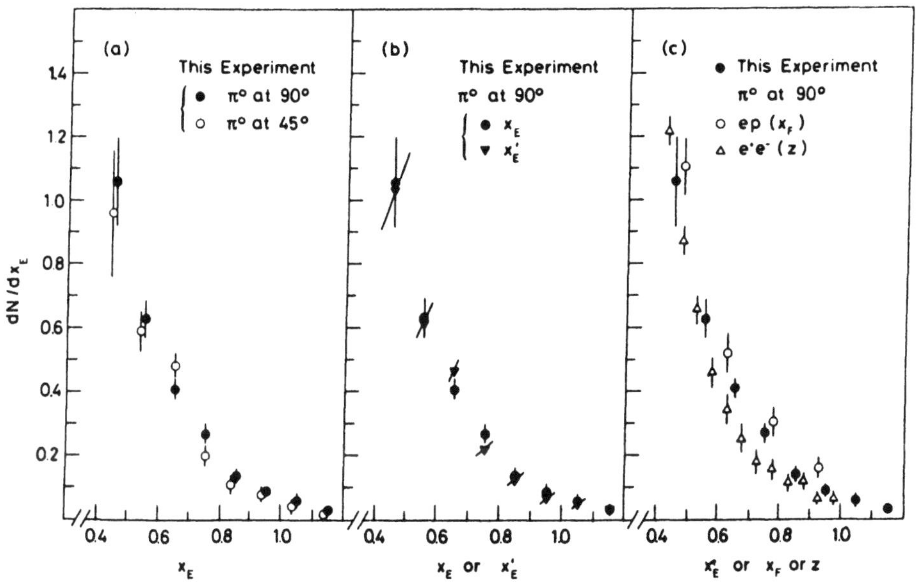

Fig. 28 a) The x_E distribution obtained from the 90° and from the 45° data. b) Comparison between the x_E and $x_{E'}$ distributions (90° data). c) The x_E distribution (90° data) is compared to the Feynmann x_F distribution observed in deep-inelastic $e^{\mp}$-p scattering (ref. 18) and with $\frac{1}{2}$dN/dz ($z = p/p_{max}$) observed in $e^+ - e^-$ annihilation at E_{cm} = 4.8 GeV/c (ref. 19).

V. DIRECT LEPTON PRODUCTION

The motivation for investigating direct lepton production in proton-proton interaction lies in the possibility of relating these leptons to the leptonic decays of new particles. Several measurements have been performed at the ISR in the last few years, contributing to the now well established feature of almost constant (e/π) ratio, for $\sqrt{s}$ from a few GeV up to the highest ISR center-of-mass energy $\sqrt{s} = 62.4$.

On the electron side two groups have published data on single electron production. The CCRS[20] Collaboration has measured the invariant production cross-section for single electrons with transverse momenta from 0.6 to 4.7 GeV/c and for five center-of-mass energies $\sqrt{s} = 23.5$, 30.6, 44.8, 52.7, 62.4 GeV. Fig. 29 gives, for fixed values of $\sqrt{s}$, the p_T (center-of-mass momentum) dependence of the charge averaged invariant differential cross-section $E\, d^3\sigma/d^3p$. Fit to the charge averaged pion data from the BS Collaboration[13] are superimposed to the data points. The agreement between pion and electron production seems to be good at all energies considered. A compilation of the e/π ratio for $\pi^{\pm}$ and π° (see Sec. III) data is given in Fig. 30. Finally the $\sqrt{s}$ dependence of the e/π ratio for the CCRS data is shown in Fig. 31, together with two possible fits.

The ACHMNR[21] Collaboration has extended the electron production data for $\sqrt{s} = 52.7$ to small p_T. While the direct electron production for $p_T > 1.5$ GeV/c seems to behave (except for the 10^{-4} factor) as the pion's one, below $p_T = 1.5$ GeV/c, it increases when p_T decreases something like $1/p_T$. Fig. 32 shows the e/π ratio in excess after all known background sources present in the experiment are taken into account. A compilation of e/π ratio versus p_T for $\sqrt{s} = 52.7$ is shown in Fig. 33.

In conclusion the e/π ratio emerging from these data appears to remain constant (or slowly varying) with s at a level of 10^{-4} from a few GeV up to 62 GeV (c.m.s). The new measurements reported at the time of the Palermo conference, furthermore, wash out a possible threshold effect around 25 GeV.

The recent discovery of the new family of particles (J ψ' ψ'')[23] decaying into lepton pairs (e^+e^-, $\mu^+\mu^-$) has suggested the possibility that the direct single electron yield could be explained as a decay product. Indeed such decays represent a sizeble contribution to the single electrons observed, but do not explain completely the e/π ratio data.

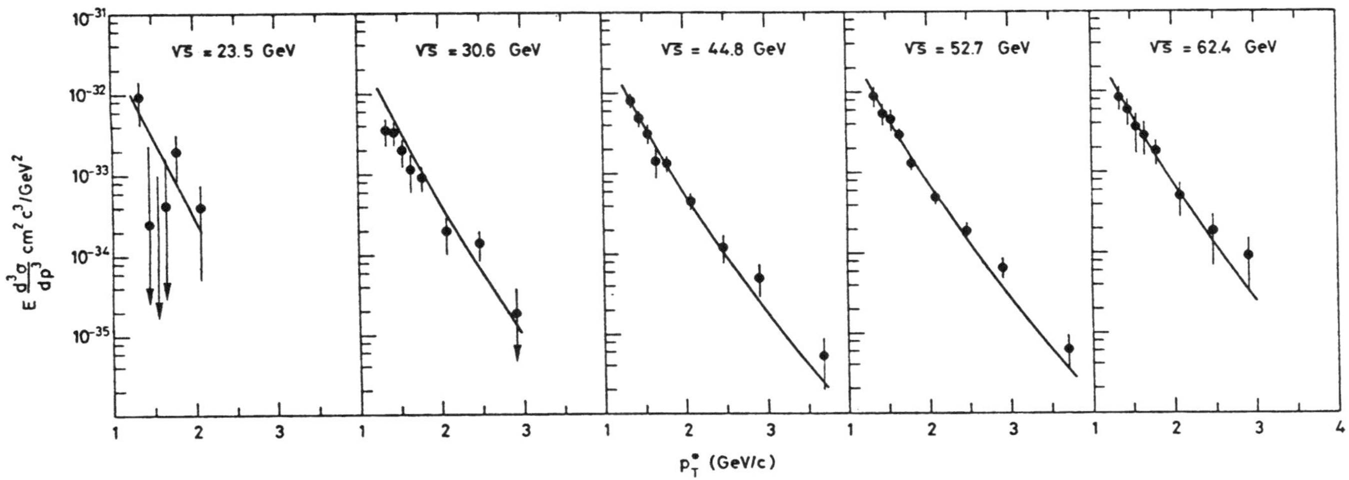

Fig. 29 The charge-averaged invariant cross-section for electron production plotted as a function of the center-of-mass momentum p_T for five values of $\sqrt{s}$. The curves represent a fit of the charge-averaged pion data of the BS Collaboration (ref. 13), and has been multiplied by 10^{-4}.

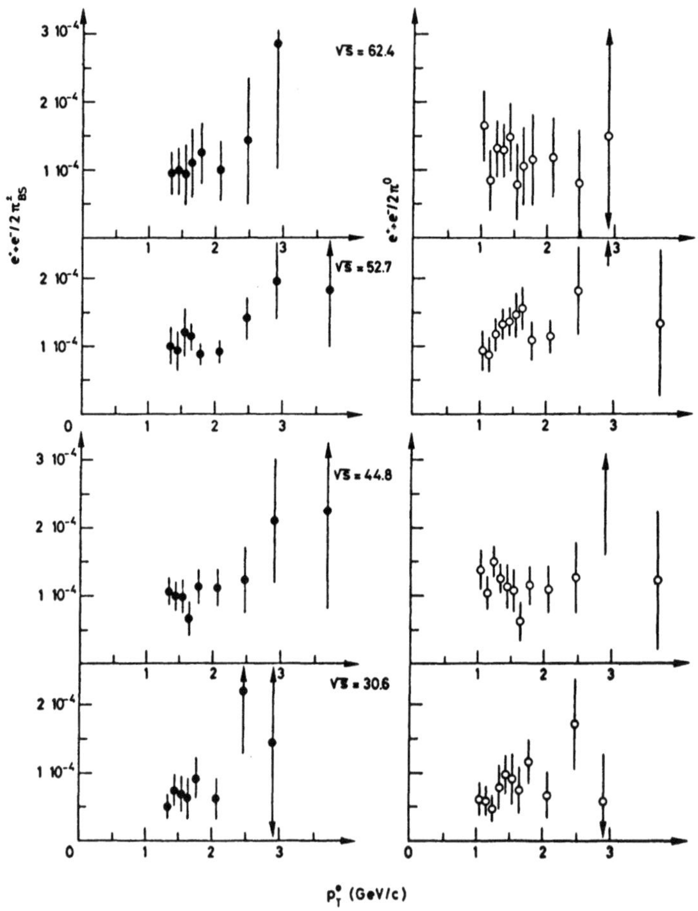

Fig. 30 a) The ratio of the charge-averaged invariant electron cross-section to the BS fit (ref. 13). b) The ratio of electrons to neutral pions plotted as a function of p_T for four values of $\sqrt{s}$.

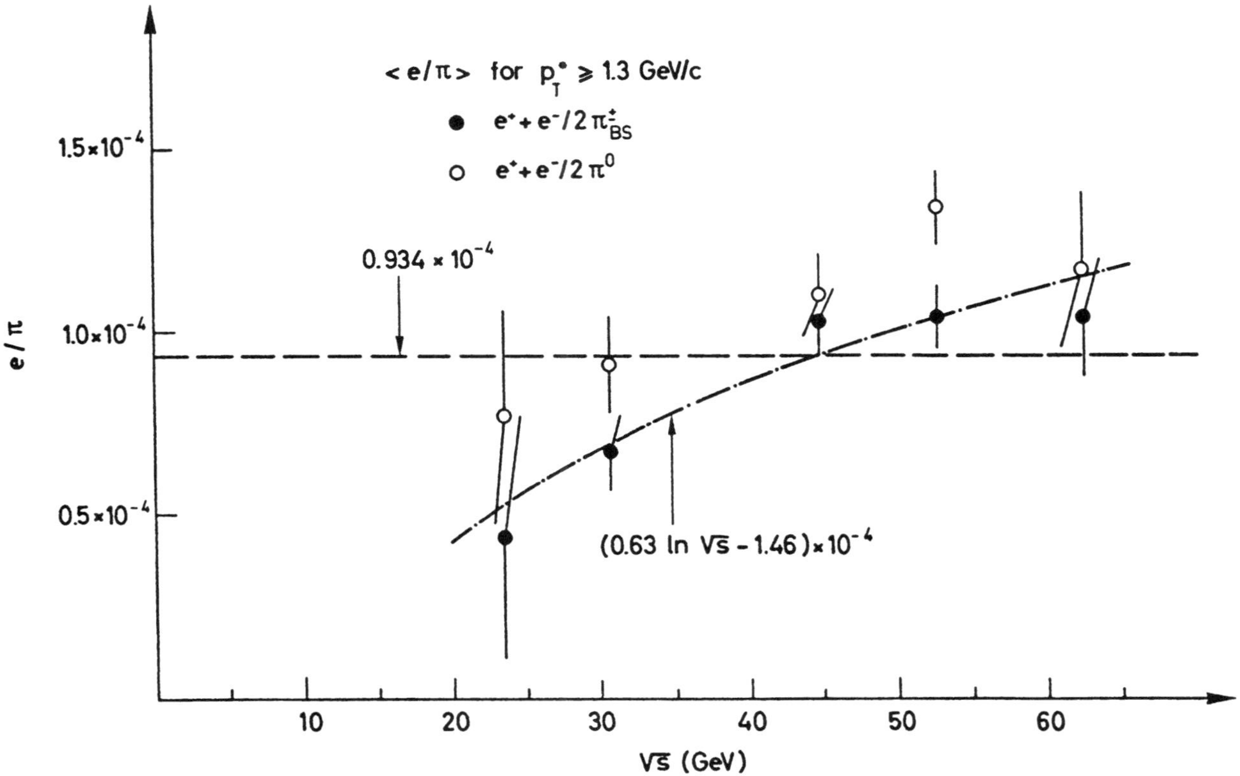

Fig. 31 The ratio of invariant cross-sections of electrons to pions of $p_T > 1.3$ GeV/c plotted as a function of c. m. s. energy $\sqrt{s}$. The two curves shown are fits to the full points.

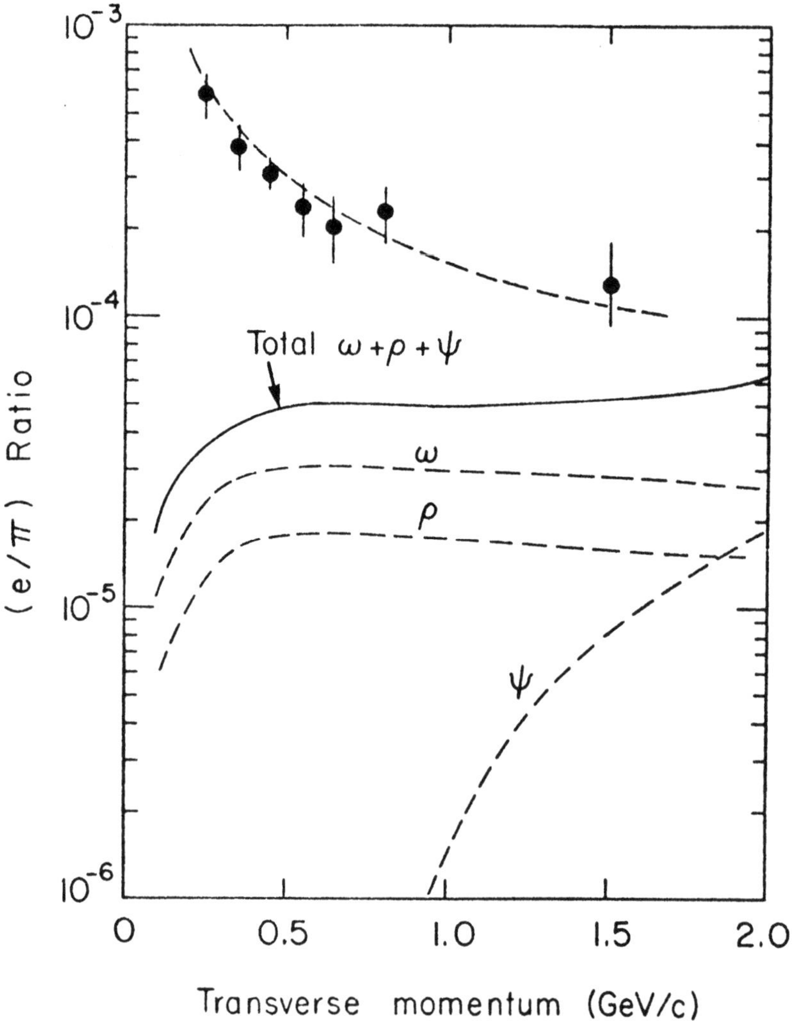

Fig. 32 Observed electron excess compared with the prediction of vector meson decays according to Bourquin and Gaillard (ref. 22).

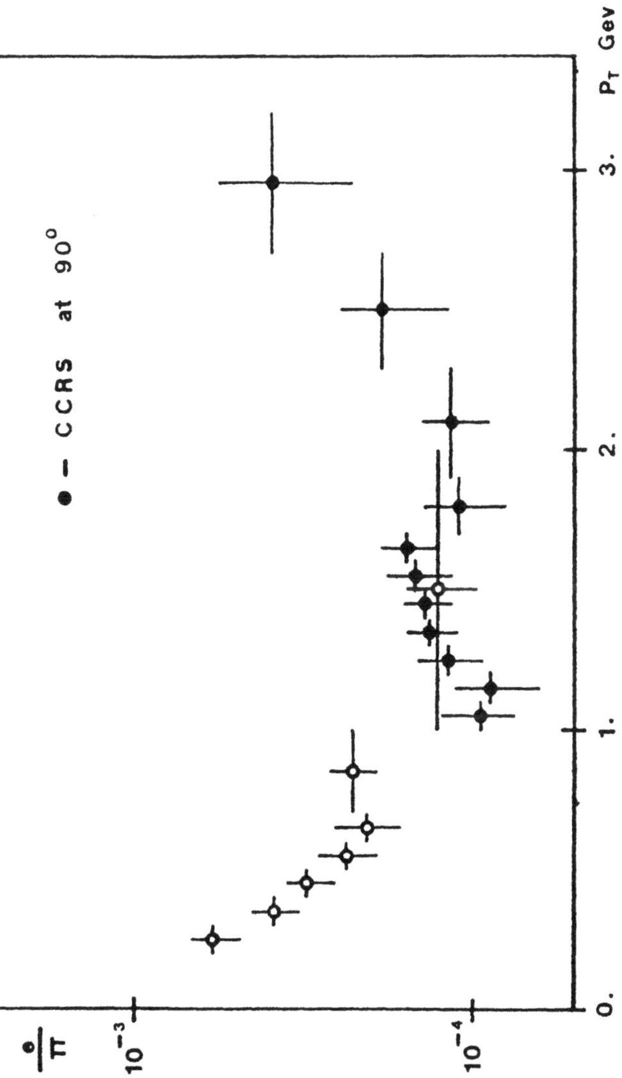

Fig. 33 P_T dependence of the e/π rate at ISR energies.

The J production in proton-proton collision at the ISR has been observed in both e^+e^- [24] and $\mu^+\mu^-$ [25]. The results obtained are:

$$B_{ee} \left. \frac{d\sigma_J}{dy} \right|_{y=0} = (7.5 \pm 2.5)\ 10^{-33}\ cm^2$$

for $\langle p_T \rangle = 0.67$ GeV/c, and

$$B_{\mu\mu} \left. \frac{d\sigma_J}{dy} \right|_{y=1.6} = (7.2 \pm 2.4)\ 10^{-33}\ cm^2$$

for $\langle p_T \rangle = 1.0 \pm 0.2$ GeV/c.

Electron-muon coincidences

Some very recent and preliminary results[26] seem to hint to the presence of electron-muon production in proton-proton collision. The existence of e-μ pairs is very relevant in the process of understanding the source of prompt leptons in hadron interactions.

The method implemented by the ACHMNR Collaboration is to compare muon yield in coincidence with identified electrons and hadrons. Electrons are identified using threshold gas Cherenkov counters, lead glass calorimeters and magnetic analysis; background due to electron pairs is detected via dE/dx measurements. Muons are identified by absence of visible interaction in a calorimeter and magnetic analysis.

Table III presents measured yield defined as:

$$\text{Yield} = \frac{\text{Arm1} \cdot \text{Arm2}}{\text{Arm1}}$$

where Arm1 and Arm2 are the electron and muon arm of the spectrometer respectively.

The excess of electron-muon with respect to hadron-muon coincidences is measured to be

$$(1.61 \pm 0.50)\ 10^{-4}.$$

A charge analysis of these e-μ events gives for the opposite charge excess yield

TABLE III

Fraction of events in Arm 1 (electron arm) with a valid count in Arm 2 (muon arm)

Selection in Arm 2 (muon arm)	Event selection in Arm 1 (electron arm)		
	Electrons	Hadrons	Electrons-Hadrons
Hadron absorber <u>in</u> <u>no</u> visible interaction in calorim. correct t.o.f.	$(3.72 \pm 0.47) \times 10^{-4}$ (63 events)	$(2.11 \pm 0.18) \times 10^{-4}$ (139 events)	$(1.61 \pm 0.50) \times 10^{-4}$ 3.2 s.d. from 0
Hadron absorber <u>in</u> visible interaction in calorim. or wrong t.o.f.	$(2.00 \pm 0.34) \times 10^{-4}$ (34 events)	$(1.87 \pm 0.17) \times 10^{-4}$ (123 events)	$(0.13 \pm 0.38) \times 10^{-4}$
Hadron absorber <u>out</u>	$(1.22 \pm 0.17) \times 10^{-2}$ (148 events)	$(1.02 \pm 0.04) \times 10^{-2}$ (596 events)	$(0.20 \pm 0.11) \times 10^{-2}$

$$\frac{e^{\pm}\mu^{\mp}}{e} = (1.07 \pm 0.43)\ 10^{-4}\ (*)$$

and for the equal charge configuration

$$\frac{e^{\pm}\mu^{\pm}}{e} = (0.23 \pm 0.44)\ 10^{-4}.$$

These results hint to an opposite charge e-μ pair production.

VI. DIRECTLY PRODUCED PHOTONS

A CERN group[27] has recently reported on a simultaneous measurement of large transverse momentum (1.6 to 3.8 GeV/c) yield of single photons and photon pairs at 90° production angle performed at the SFM.

The method implemented to extract the single photon production cross-section is the following:

i) measurement of the π° yield (π° is identified from the $\pi^\circ \to \gamma\gamma$ decay)

ii) calculation of the single photon yield from π° decay

iii) single photon yield measurement.

Several sources of systematic bias had to be considered in such a measurement; the most important are:

a) single photons do not have the photon pair (π°) signature, hence higher background

b) loss of photons by the apparatus affects single and pair rate differently

c) accurate knowledge of the energy response is necessary as one wants to compare rates at different detection energies (p_T fall-off).

The π° data, relative to $1.6 < p_T < 3.8$ GeV/c for $\sqrt{s} = 45$ and 53 GeV are compared to $\pi^{\pm}$ data from reference 28 in Fig. 34. Table IV gives the detailed single photon observed yield as well as the

(*) The data are not corrected for acceptances.

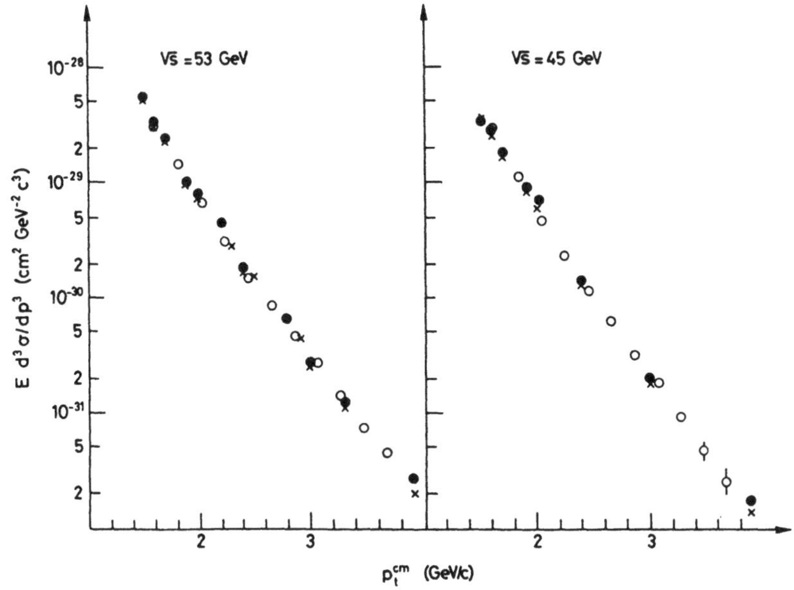

Fig. 34 Inclusive π° production cross-section at 90° for $\sqrt{s}$ = 45 and 53 GeV, as measured from resolved photon pairs (open circles). Also shown are the data of ref. 28 for π^{+} (full circles) and π^{-} (crosses).

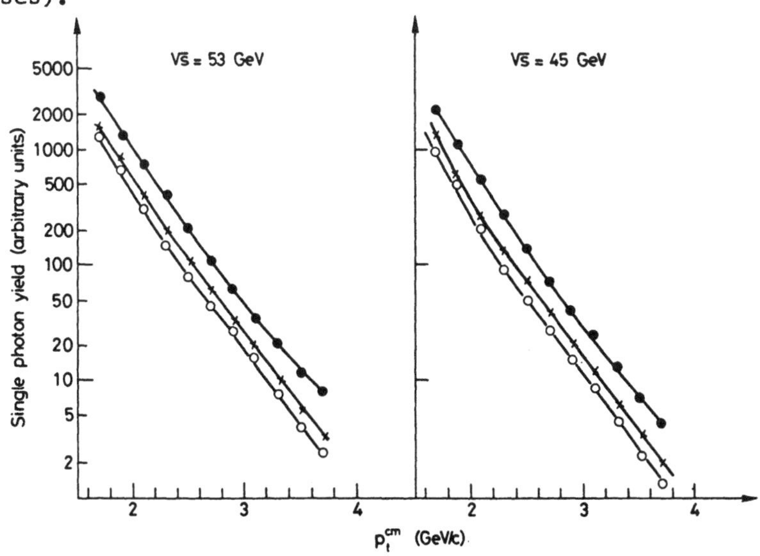

Fig. 35 Single photon yields uncorrected for antineutrons contamination at 90° for $\sqrt{s}$ = 45 and 53 GeV. The observed distributions (full dots) are compared to those inferred from the resolved photon pair data (open circles). The latter are also shown after addition of the contribution from $\eta \to \gamma\gamma$ decays (crosses).

TABLE IV

Single photon data

Data are integrated over both values of $\sqrt{s}$ (45 and 53 GeV) and over p_t^{cm} between 2.8 and 3.8 GeV/c. All numbers are given as fractions of the π^0 production cross-section in the same range of p_t^{cm}.

I - Observed single photon yield	
Single photon events	0.462 ± 0.008
Conversion losses	+ 0.032 ± 0.009
Events with an independent second photon	+ 0.037 ± 0.009
Antinucleon contamination	- 0.077 ± 0.046
	0.455 ± 0.053
II - Single photon yield from π^0 and η decays	
$\pi^0 \to \gamma\gamma$	0.192 ± 0.006
$\eta \to \gamma\gamma$	0.062 ± 0.011
Uncertainty in acceptance calculation	± 0.013
	0.255 ± 0.018
III - Excess of single photons	
γ/π ratio	0.20 ± 0.06
Uncertainty on energy response	± 0.07

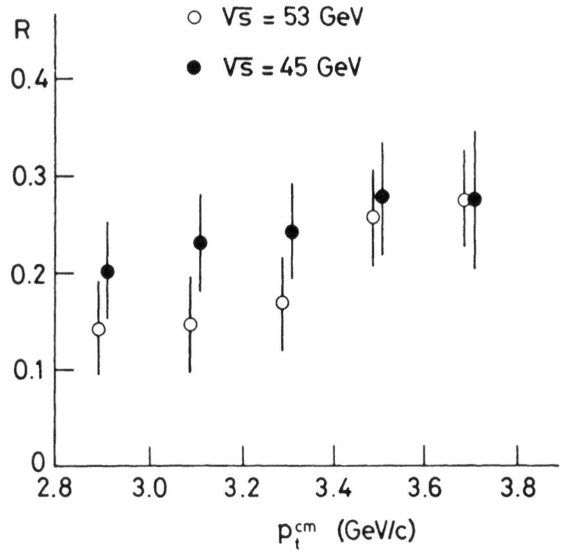

Fig. 36 The ratio R between the excess of single photons (corrected for antineutron contamination) not accounted for by π° and η decays and the π° yield measured.

contamination estimates. The data are integrated over both values of $\sqrt{s}$ and over $2.8 < p_T < 3.8$ GeV/c. Fig. 35 shows the p_T distribution of the single photon yield.

The ratio R between the excess of single photons not accounted for by π° and η decay and the π° yield as a function of the transverse momentum is shown in Fig. 36.

REFERENCES

1) - CERN-PISA-STONY BROOK Collaboration, Physics Letters 62B, 460 (1976).

2) - S. Van der Meer, CERN Internal Report ISR-PO/68-31 (1968).

3) - See for instance U. Amaldi et al., Annual Review of Nuclear Science vol. 26.

4) - A. S. Carrol et al., Physics Letters 61B, 303 (1976).

5) - U. Amaldi et al., Physics Letters 44B, 112 (1973),
S. R. Amendolia et al., Physics Letters 44B, 119 (1973),
S. R. Amendolia et al., Nuovo Cimento 17A, 735 (1973),
G. Bellettini, 5th International Conference on High Energy Collision, Stony Brook, 1973 (ed. C. Quigg) AIP Conf. Proc. No. 15, K. Eggert et al., Nuclear Physics B98, 93 (1975).

6) - CERN-HAMBURG-ORSAY-AUSTRIA Collaboration, Physics Letters 58B, 233 (1975).

7) - CERN-HAMBURG-ORSAY-AUSTRIA Collaboration, Physics Letters 62B, 363 (1976).

8) - R. J. N. Phyllips and V. Berger, Physics Letters 46B, 412 (1972).

9) - B. Alper et al., Physics Letters 44B, 521 (1973), M. Banner et al., Physics Letters 44B, 537 (1973), F. W. Büsser et al., Physics Letters 46B, 471 (1973).

10) - P. V. Landshoff, Large transverse momentum reactions, Plenary report at the 17th International Conference on High Energy Physics, London 1974,
D. Sivers, S. J. Brodsky and R. Blakenbecler, Large transverse momentum processes, Physics Reports 23, (1976).

11) - F. W. Büsser et al., High transverse momentum phenomena involving π and η mesons at the CERN ISR, paper presented at the Int. Symp. on lepton and photons, Stanford University, 21 - 27 August 1975,
F. W. Büsser et al., Physics Letters 51B, 306 (1974).

12) - K. Eggert et al., Nuclear Physics B98, 49 (1975).

13) - B. Alper et al., Nuclear Physics B87, 19 (1975).

14) - J. W. Cronin et al., Physical Review D11, 3105 (1975).

15) - P. Darriulat et al., Nuclear Physics B107, 429 (1976).

16) - F. W. Büsser et al., Nuclear Physics B106, 1 (1976).

17) - P. Darriulat, Hadronic collisions with large transverse momentum products, Rapporteur talk at the International Conference on High Energy Physics, Palermo 1975.

18) - P. Capiluppi et al., Nuclear Physics B79, 189 (1974).

19) - J. T. Dakin et al., Physical Review D10, 1401 (1974).

20) - B. G. Pope, Rapporteur talk at the International Conference on High Energy Physics, Palermo 1975.

21) - L. Baum et al., Physics Letters 60B, 5 (1976).

22) - M. Bourquin and J. M. Gaillard, Physics Letters 59B, (1975).

23) - See for example S. Ting, Rapporteur talk at the International Conference on High Energy Physics, Palermo 1975.

24) - F. Büsser et al., Physics Letters 56B, 1044 (1975).

25) - E. Nagy et al., Physics Letters 60B, 96 (1975).

26) - Aachen-CERN-Heidelberg-Munich-Northwestern-Riverside (ACHMNR) Collaboration private communication.

27) - P. Darriulat et al., Nuclear Physics B110, 365 (1976).

28) - B. Alper et al., Nuclear Physics B100, 237 (1975).

D I S C U S S I O N

CHAIRMAN: Prof. G. Valenti

Scientific Secretaries: M.A. Ichola and T. Wilkie

DISCUSSION

- *CRONIN:*

Where does the background estimate come from in the eμ charge correlation events? From hadronic measurements?

- *VALENTI:*

The estimate for the various eμ charge states is derived from the comparison of the rates of eμ events identified as background to the rates of good eμ candidates.

- *FERBEL:*

Can the eμ events be interpreted as charm?

- *ZICHICHI:*

There was a seminar at CERN about this. Gaillard has shown that these events are in the wrong phase-space region for charm decays since they are in the very forward direction.

What are you conclusions about Feynman scaling?

- *VALENTI:*

The conclusion is that Feynman scaling is broken at $x \sim 0$. Furthermore, the deviation seems to be stronger at small p_T.

- *WIGNER:*

I do not understand the significance of your single photon data. Could you elaborate please?

- *VALENTI:*

Experimentally, the comparison of the measured single photon yield to that expected from π^0 decays is done by first measuring the yield of π^0's -- identified as photon pairs with correct invariant mass -- then calculating the single photon yield expected from π^0 decays seen by the detectors, and finally by measuring the single photon yield itself. The result is that there is an excess of single photons not accounted for by π^0 decays, suggesting that high transverse momentum photons are directly produced in proton-proton collisions.

- *ZICHICHI:*

A bit of history: in the past, people estimated the number of π^0's produced in pp collisions simply by counting the number of photons observed. It was assumed that all these photons were produced by π^0 decay. One always found that the rate of neutral pions was greater than that of $\pi^{\pm}$. This is the first time the π^0 has been reconstructed from both photons, and this is the first time that the π^0 rate coincides with the $\pi^{\pm}$ rate.

HIGHLIGHTS OF THE TBILISI CONFERENCE

C.W. Fabjan

European Organization for Nuclear Research (CERN)

Geneva, Switzerland

The 18th International Conference on High-Energy Physics in the series of the so-called 'Rochester' Conferences was held in Tbilisi, the capital of Georgia, 15-21 July, 1976.

Georgia, which is one of the 15 republics of the Soviet Union, has produced men as unlike in character as C. Roustaveli, the great 12th century poet with his world-famous work "The Prince in the Lion's Skin" and Joseph Stalin. The chain of the Caucasus mountains, which surround and traverse this country, contrasts dramatically with the fertile plains, which ensured a privileged life to a succession of invaders over the last three millennia. Christianity found an early stronghold in this country, to which a number of monuments testify, illustrating eloquently the varied and highly advanced expressions of the architectural and decorative arts.

In selecting the "highlights" of the Conference I preferred not to present a catalogue of the many, very interesting results reported at Tbilisi, but rather to discuss only those issues which, during the last year, have most strongly affected our fundamental concepts. Some of the issues, according to common consensus, appeared to be settled, while others were still very much debated and promise exciting results in the near future.

LEPTON PRODUCTION IN HADRON COLLISIONS

In 1974 at the London Conference the first measurements of inclusive lepton production at FNAL and ISR energies were reported. The subsequent two years were marked by intense experimental activity, and work submitted to this conference permits us to draw preliminary conclusions.

Observation of a Dilepton Continuum with High Invariant Mass

Measurements from FNAL are summarized in Fig. 1, including both dimuon[1,2] and dielectron[1] data. The data are compared with one of many models proposed, namely the mechanism of parton-antiparton annihilation, as advocated by Drell and Yan[3]. The solid line in Fig. 1 shows the result of such a calculation, assuming a $(1-x)^{7/2}$ distribution for the coloured partons, where x is the ratio of the parton to nucleon momentum.

For some time it has been noticed that the p_T distribution of the dilepton pairs flattens with increasing mass of the pairs. This trend is clearly exhibited by the data[4], as shown in Fig. 2. Data for both proton-induced and pion-induced μ pairs are shown, permitting a comparison of cross-section and slope. This p_T behaviour, which differs considerably from the p_T distribution of normal hadronic events, was not well reproduced by Drell-Yan-type calculations;

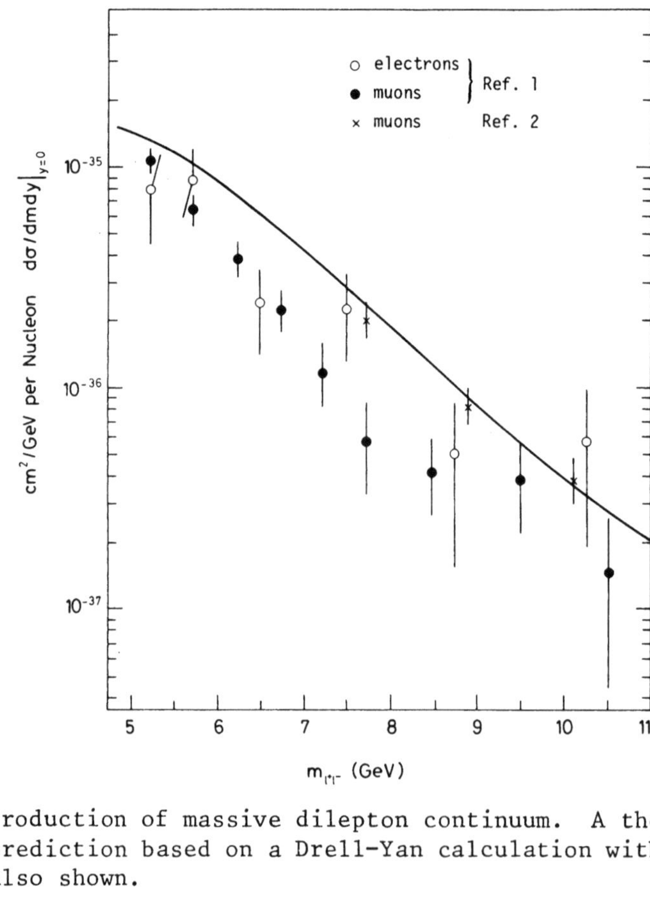

Fig. 1 Production of massive dilepton continuum. A theoretical prediction based on a Drell-Yan calculation with colour is also shown.

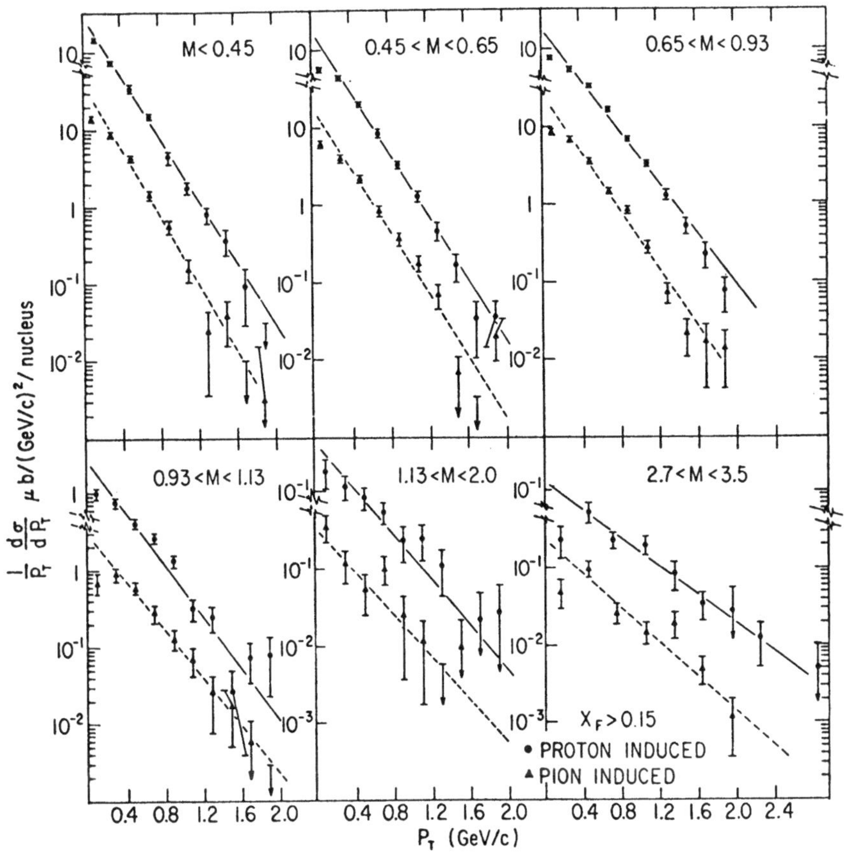

Fig. 2 The p_T dependence of pion- and proton-induced μ pairs as a function on the invariant mass range.

more recent recalculations of this effect, as discussed at the Conference[5], appear to have considerably better agreement with the p_T data.

An important result[3] of the Drell-Yan model is its prediction that for $s \to \infty$, $M \to \infty$, but s/M^2 finite, the cross-section for pair production should exhibit a scaling behaviour of the form

$$M^3\frac{d\sigma}{dM} = f\left(\frac{s}{M^2}\right) .$$

The observation of such a scaling behaviour at present machine energies would be of extreme importance, as, based on CVC arguments[6], connection can be made between the production cross-section at fixed s of a lepton pair of given invariant mass and the charged and neutral intermediate vector mesons of the same mass. The present status of

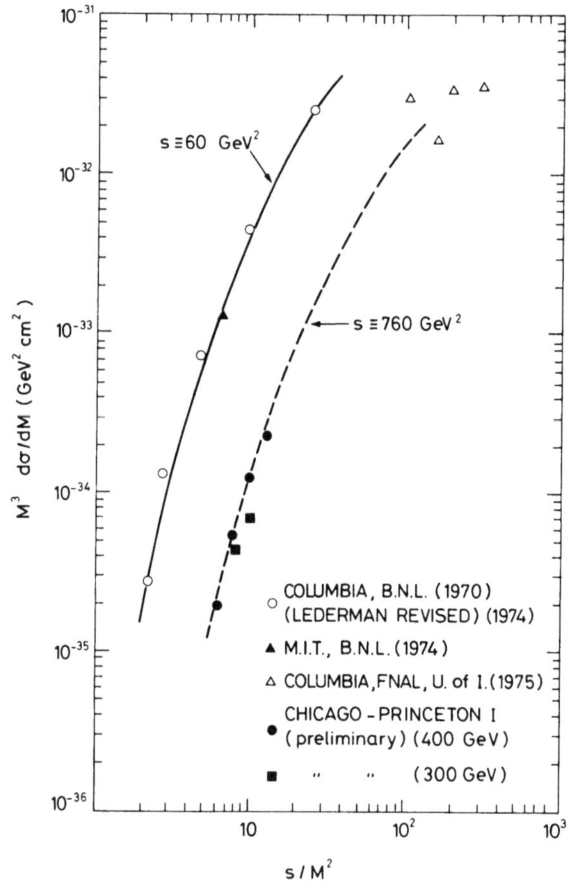

Fig. 3 The pair production cross-section $M^3(d\sigma/d\Gamma)$ as a function of scaling variable s/M^2.

the scaling behaviour of the data is summarized in Fig. 3. Whereas a large scaling violation between the BNL data and FNAL data is observed, there may be a hint in the data (compare the 300 GeV/c and 400 GeV/c points) that at FNAL energies incipient scaling may be reached. In the near future, new data from the CERN ISR will clarify this point.

Whereas the existence of a mass-continuum of dileptons has been clearly established experimentally, its theoretical interpretation requires further work. The experimental study of this continuum over a large range of invariant masses, and in particular at very low masses, and its dependence on $\sqrt{s}$, x_F, and p_T should allow a better distinction to be made between competing models of production[7].

Dileptons from Vector Meson Decay

New data on vector meson production were reported at the Conference. Precise knowledge of these cross-sections is of course necessary to evaluate the contribution of vector meson decays to the inclusive single lepton spectrum (see next subsection). Table 1 gives cross-sections for inclusive production, and Fig. 4 shows some of the relevant data, again for proton- and pion-induced production[4]. The dilepton continuum is also shown in Fig. 4, as the data have been corrected for background sources.

Table 1

Inclusive vector meson production by 150 GeV/c protons

	σ/nucleon
η	< 11 mb
$\rho = \omega$ (assumed)	8.6 ± 2.5 mb
ϕ	660 ± 220 μb
J/ψ	94 ± 31 nb

In Fig. 5 the measured dielectron spectrum is shown as reported by the Columbia-FNAL Group[1]. Refined analysis of the data resulted in improved energy resolution of the apparatus: the previously reported structure at 6 GeV [8] persists, containing 14 events. This structure is, however, not seen in a recently obtained data sample on $\mu^+\mu^-$ pairs by the same group; but, as the energy resolution for the μ pairs was considerably worse, the two experiments are not in

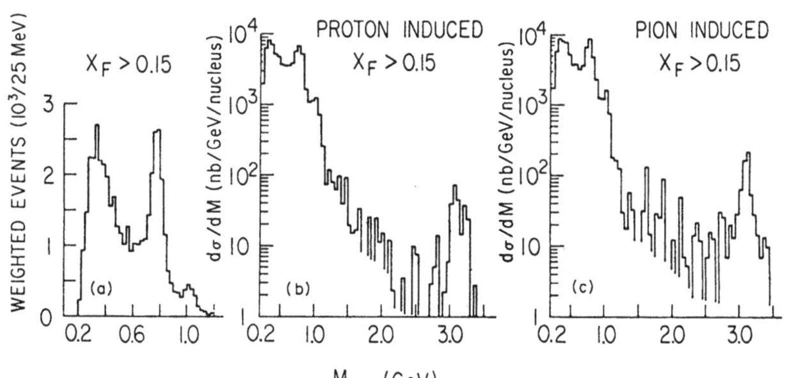

Fig. 4 The production of vector mesons and the low-mass μ-pair continuum.

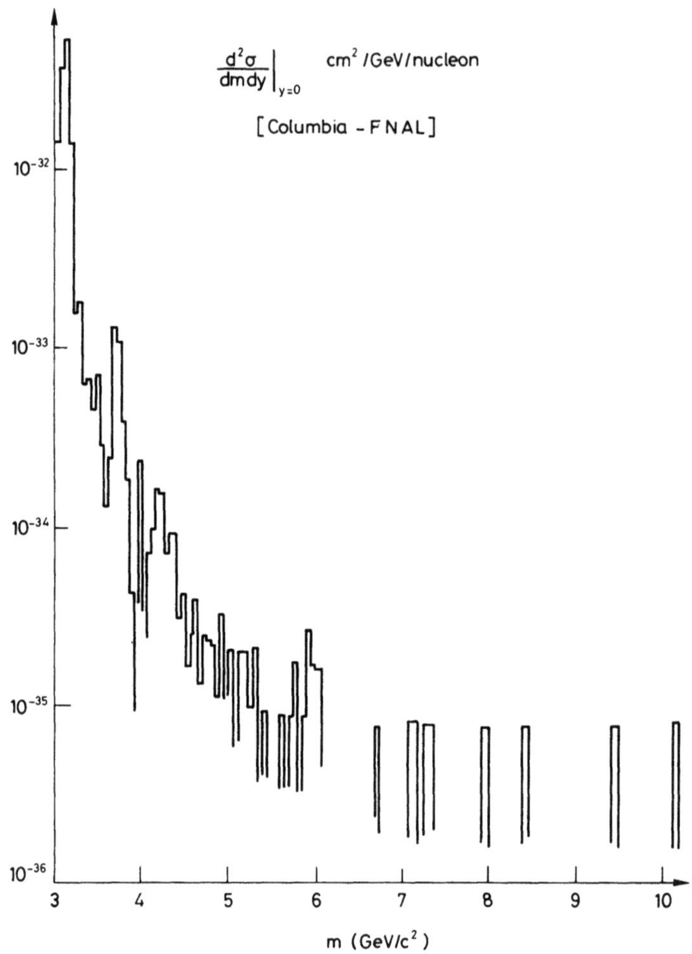

Fig. 5 The dielectron spectrum observed by the Columbia-FNAL Collaboration. With the clear evidence for the J/ψ and the ψ', structure is also observed at 4.3 GeV/c^2 and 6 GeV/c^2.

direct contradiction. The reanalysis of the electron data gave a somewhat sharper structure at 4.3 GeV and it is tempting to identify this region with the broad resonance structure observed at SLAC in σ_{tot} of e^+e^- annihilation. Further evidence for a resonance structure was reported by another group[9)], which observed through photoproduction not only the J/ψ and ψ', but also a clustering of eight events at $M \simeq 4.5$ GeV/c^2 above an estimated Bethe-Heitler background of 0.24 events.

Inclusive Single Lepton Production

The data obtained during the last two years up to the time of the Conference show the following global features[10]:

a) For $p_T > 1$ GeV/c and large production angle θ_{cm}, data have been obtained by the CERN-Columbia-Rockefeller-Saclay (CCRS), Chicago-Princeton II (CPII), and Columbia-FNAL Collaborations. The ratio R = lepton/pion is found to be $R = 10^{-4}$ with an indication that R may be rising with increasing values of $\sqrt{s}$ and p_T. Measurements at Serpukhov indicate a threshold for production of μ's around $\sqrt{s} = 8$ GeV reaching the plateau value of R^{μ} = muon/pion = 10^{-4} at $\sqrt{s} = 12$ GeV.

b) For $p_T < 1$ GeV/c and large production angle θ_{cm}, data have been reported by the Aachen-CERN-Harvard-Munich-Northwestern-Riverside Collaboration. The ratio R^e = electron/pion rises below $p_T = 1$ GeV with decreasing p_T up to values of $R^e = 3 \times 10^{-4}$ at $p_T \sim 250$ MeV/c ($\sqrt{s} = 52$ GeV). A similar behaviour is reported by the Pennsylvania-Stony Brook Collaboration at $\sqrt{s}$ values between 4.5 and 7 GeV, at variance with an old experiment, re-analysed by Winter, which resulted in $R^e = -(0.2 \pm 0.29) \times 10^{-4}$.

Three contributions dealt with the very important question of possible sources of this prompt lepton production. Based on the measured dilepton yield (continuum and vector mesons), the Chicago-Princeton II Group[4] computed the invariant cross-section for single inclusive lepton production at $\sqrt{s} = 17$ GeV (Fig. 6a). A parametrization of the data is necessary to extrapolate the measurements outside

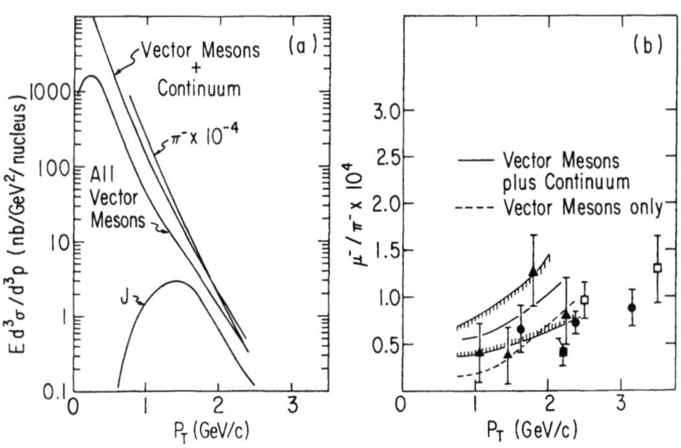

Fig. 6 a) Monte Carlo estimate of the single inclusive lepton yield, based on the measured dilepton spectrum.
b) Comparison of the Monte Carlo estimate of the μ^-/π^- ratio with various experimental results.

the limited kinematical region, which was covered in the measurements. In particular, the result for large p_T is a sensitive function of the p_T distribution of the high-mass pairs (J/ψ-particle, continuum). For comparison, the measured inclusive π-spectrum, downscaled by $\times 10^{-4}$, is also shown. The ratio $R = \mu/\pi$, computed in the above way, is compared with the data in Fig. 6b. Within the rather large errors, reflecting the uncertainties of the kinematical extensions and renormalization errors, the measured dilepton production accounts well for the observed ratio R. Consequently, electromagnetic processes are identified as a major, if not the only source of the observed single lepton spectrum.

The same conclusion is reached by the BNL-Yale-FNAL Group[11], which measured μ and μ-pair production at $p_T < 2$ GeV/c in the forward direction. Again, within errors, reflecting the assumptions necessary for kinematical extension of the data, the measured dimuon yield accounts for the observed single muon production. Furthermore, the group measured the polarization of the prompt μ's and found for the polarization parameter

$$P = -0.15 \pm 0.2 ,$$

consistent with zero. This measurement provides additional confirmation of the electromagnetic origin of the prompt leptons.

Perhaps contradictory evidence has been reported from a team working at Serpukhov[12]. Inclusive μ production was measured in the transverse momentum range $1.9 \leq p_T \leq 3.0$ GeV/c and at production angles $\theta_{cm} \simeq 90°$. A charge asymmetry was observed of

$$\frac{\mu^+}{\mu^-} = 1.2 \pm 0.1 ;$$

a measurement of the polarization yielded

$$P(\mu^+) = -(0.85 \pm 0.36) ,$$

whereas the μ^-'s were found to be unpolarized. As different kinematical regions were investigated in the Serpukhov and in the BNL-Yale-FNAL experiments, the two results are not necessarily contradictory; a planned remeasurement by the Serpukhov group should clarify this very important point.

THE J/ψ FAMILY AND ITS INTERPRETATION AS AN "-ONIUM" SYSTEM

Immediately after the discovery of the J/ψ and ψ′, an interpretation was suggested which identified these particles as bound states of a heavy spin ½ particle bound to its antiparticle. The level scheme[13] for such a bound state, closely resembling that of positronium, is shown in Fig. 7. At the conference, improved data on the intermediate states and their tentative quantum number assignment were presented. They provide crucial support for the bound-state interpretation; an impressive list of data can now be accounted for, if charmed quarks with $Q = \frac{2}{3}e$ and $M \sim 1.6$ GeV/c^2 are assumed as the constituents of the "-onium" system ("charmonium").

Further Evidence on Intermediate States

New data on radiative transitions of the form $\psi' \to \gamma + \chi$; $\chi \to \gamma + J/\psi$ were presented by two groups working at SLAC. The SLAC-LBL Group[14] reported evidence for states at 3.45, 3.50, and 3.55 GeV.

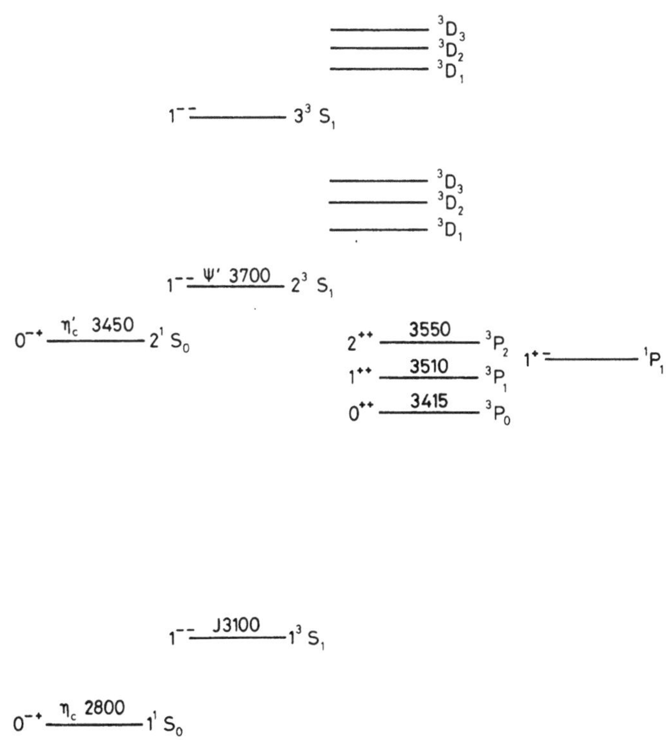

Fig. 7 The level scheme of "charmonium" with a tentative identification of the recently seen intermediate states.

The Pavia-Princeton-Maryland-UC San Diego-Stanford Group[15]) measured with considerable statistics the γ-lines at 3.415, 3.512, and 3.551 GeV. The information on these intermediate states is supplemented through the observation of the hadronic decay modes of χ states:

$$\psi' \to \gamma + \chi, \quad \chi \to \text{hadrons } (2\pi,\ 2K2\pi,\ 3\pi\pi,\ \ldots) .$$

Such data were reported by the SLAC-LBL Group[14]) and give further information on the 3.41, 3.50, and 3.55 states (Fig. 8). The angular distribution of the emitted photon with respect to the beam was also measured. For the $\chi(3415)$ it is consistent with $1 + \alpha \cos^2 \theta$, $\alpha = 1$, identifying this state as an S = 0 level. For the other states, the angular distribution was consistent with $\alpha < 1$.

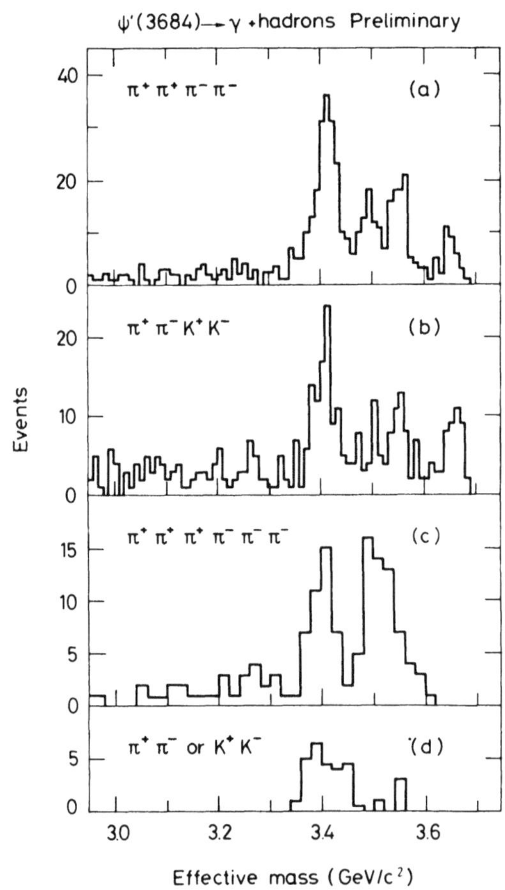

Fig. 8 Preliminary data on the hadronic decay modes of the intermediate states of the J/ψ family.

The Double-Arm Spectrometer (DASP) Collaboration[16] at DESY reported data on the decay channels

$$J/\psi \rightarrow m + \gamma, \quad m \rightarrow \gamma + \gamma$$

$$\psi' \rightarrow m + \gamma, \quad m \rightarrow \gamma + \gamma,$$

where $m = \pi^0, \eta, \eta', X$. The group reported 110 events of $J/\psi \rightarrow 3\gamma$; this sample contained decays to η, a hint of η' at $\sim$ 900 MeV, and 29 events clustered at 2.8 GeV above the background, estimated to be 14 events. Further evidence on this $\chi(2800)$ state is seen in decays of $\psi' \rightarrow \gamma + (2\gamma)$, where again five events above background are seen clustering at 2.8 GeV. The branching ratios as observed by the DASP Collaboration are given in Table 2.

Table 2

Radiative decays of J/ψ, ψ'

Decay mode	J/ψ	ψ'
$\eta\gamma$	$1.0 \pm 0.2 \times 10^{-3}$	$< 1.3 \times 10^{-3}$
$\eta'\gamma$	$< 4.4 \times 10^{-2}$	$< 1.4 \times 10^{-3}$
$\chi(2800)\gamma$	1.6×10^{-4}	$< 3.7 \times 10^{-4}$
$\pi^0\gamma$	$< 1.6 \times 10^{-4}$	

Based on this newly reported experimental evidence, the most plausible identification[17] with the states predicted by the charmonium model is given in Fig. 7.

J/ψ Family Interpreted by the Charmonium Model

In these calculations[13,17] the bound-state fermion is assumed to be a charmed quark of mass $M_C \sim 1.6$ GeV/c^2, of charge $Q = {}^2\!/_3 e$ and magnetic moment $\mu_C = e\hbar/2M_C c$. A phenomenological, strongly confining potential of the form

$$V = \alpha r + \beta/r + \gamma$$

is assumed; the problem is treated non-relativistically as typical momentum transfers in the system are of the order of hundreds of MeV.

The existence of the intermediate states provides one of the strongest suggestions for the validity of the charmonium picture. However, further high statistics work on these states and the experimental verification of those at present only tentatively assigned quantum numbers will be crucial. Besides a rather impressive list of correct predictions of the charmonium model, there are some open problems requiring further investigation[13]. A sample is given in Table 3.

Table 3

Some open problems with the charmonium model

Quantity	Experiment	Theory	Comments
$\Gamma(J/\psi \to \eta_c + \gamma)$	$\leq$ 7 keV	20-30 keV	
$\frac{M(^3P_2) - M(^3P_1)}{M(^3P_1) - M(^3P_0)}$	0.4-0.5	1.0-1.4	Relativistic effects may be important; a recent relativistic treatment gave 0.3-0.5
$(^3S-^1S)$ splitting	for n = 1: ∿ 300 MeV n = 2: ∿ 230 MeV	∿ 100 MeV	Only the ratio of (n=1) to (n=2) splitting is consistent with theory

EVIDENCE FOR CHARM

The mounting evidence for the "Hidden Charm" hypothesis as a possible explanation for the J/ψ family, and the observation of final states in deep inelastic ν scattering consistent with the production of hadrons carrying a new quantum number, found its culminating corroboration in the observation of hadrons with properties precisely predicted by the charm scheme.

Observation of Charmed Mesons by the SLAC-LBL Group

In a series of two different measurements[14] the group observed the decay of neutral and charged charmed mesons. The neutral state was observed through the decay modes

$$M_C \to K^{\pm} + \pi^{\mp} ,$$
$$\to K^{\pm} + 3\pi .$$

The data presented at the conference for these two decay modes are shown in Fig. 9. Particle identification was based on time-of-flight information.

From the data the mass of M_C was determined to be m = 1865 ± ± 15 MeV/c^2 with a width $\Gamma < 40$ MeV/c^2, compatible with the experimental resolution of the detector. Table 4 summarizes the observed production cross-sections for various intervals of W.

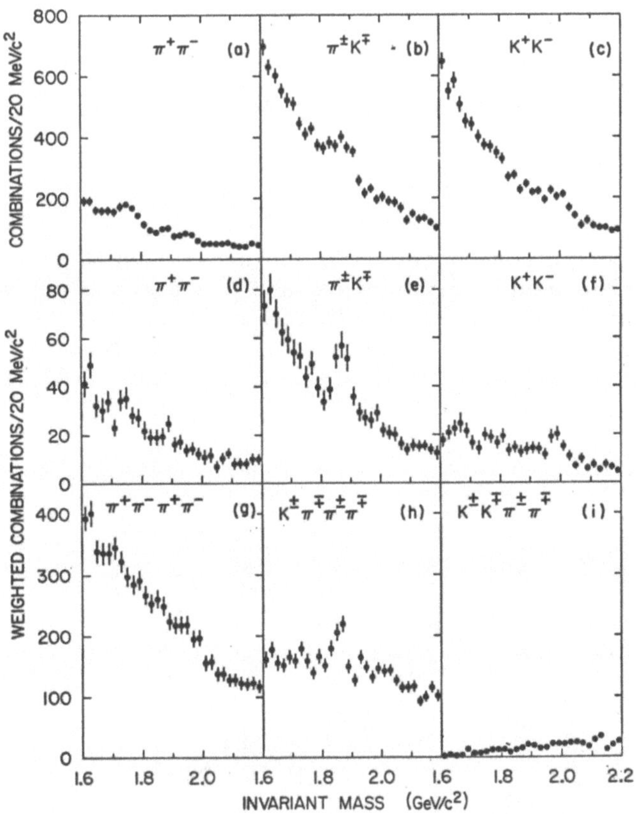

Fig. 9 Experimental evidence for decay of a neutral charmed meson. Graphs (a) to (c) show the invariant mass spectrum for an arbitrary assignment of particle identification. In graphs (d) to (f) the measured TOF was used to statistically improve the particle identification. A clear peak in the πK channel is observed. Graphs (g) to (i) show the resonance decaying into K3π.

Table 4

Cross-section for $M_C(1865)$ production

W	$\sigma_{hadronic}$ (nb)	BR•σ [nb]	
		Kπ	K3π
3.9-4.6	27 ± 3	0.20 ± 0.05	0.69 ± 0.15
4.8	18 ± 2	0.10 ± 0.07	
J/ψ,ψ′		< 0.02	< 0.04

In a subsequent running period, data were taken at W = 4.03 GeV, which corresponds to a local maximum in the multiply-peaked total cross-section around 4.0-4.3 GeV. These data (Fig. 10) provide clear evidence for a charged meson decay:

$$M_C' \to K^{\mp} + \pi^{\pm} + \pi^{\pm} .$$

Its mass was determined as m = 1875 ± 15 MeV/c^2 with a width of M ≤ 40 MeV/c^2, again consistent with the experimental resolution. The decay into the exotic final state and the absence of decays into non-exotic channels are expected for the weak decay of charmed mesons and constitute the strongest evidence for the existence of such states.

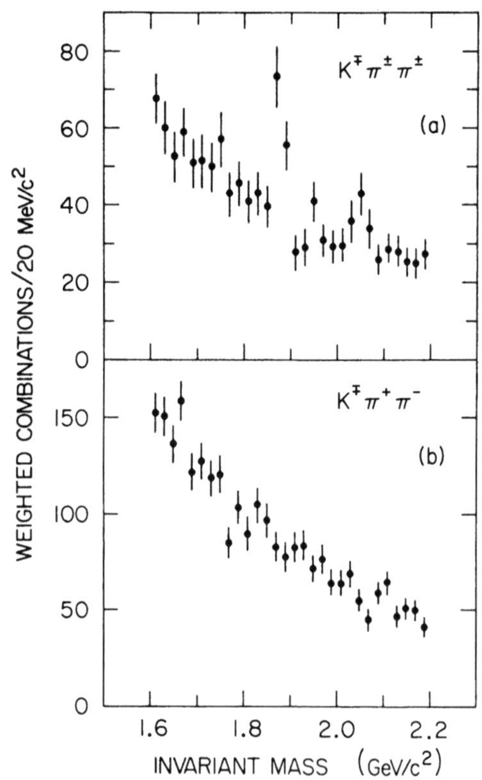

Fig. 10 Experimental evidence for a charged, charmed meson. The decay into the exotic channel $K^{\mp} + \pi^{\pm} + \pi^{\pm}$ is clearly seen, precisely as predicted by the charm scheme; it is absent in the non-exotic channel.

Evidence from Other Laboratories

Two groups working at DESY reported on the observation of anomalous lepton production[16]. One group, using the supraconducting solenoid "Pluto" for the momentum analysis of charged particles, studied the reaction

$$e^+ + e^- \rightarrow e + K_S^0 + x$$

at W = 4.1 GeV. They observed 27 such events above an estimated background of nine events, suggestive of a semileptonic decay of a charmed meson. They quote $BR \cdot \sigma \sim 1$ nb for this class of events. Similar evidence is reported by the DASP group, which used two magnetic spectrometers to study

$$e^+ + e^- \rightarrow e + x$$

at various values of W. A fairly strong signal was found at $W \simeq 4.4$ GeV. The signal shows a threshold effect, as no events were observed at W = 3.68 GeV. After careful analysis and subtraction of background sources, a total of 22 events was seen, of which nine are accompanied by K_S^0's. These events are characterized by a fairly high hadron multiplicity ($\langle n \rangle \simeq 5$), a relatively soft momentum spectrum of the leptons, and $BR \cdot \sigma \simeq 1$ nb. Again, this observation is consistent with the production and subsequent decay of a hadron pair carrying a new quantum number, conserved in strong and electromagnetic interactions.

The Dubna-Berlin-Budapest-Prague-Sofia Collaboration[18] reported very preliminary data on hadron production by neutrons of average momentum of 50 GeV/c on a carbon target. They observe 14 events in the invariant mass spectrum of $(\Lambda\pi^+\pi^-)$ clustering at 2090 MeV/c^2. The measured width of this cluster is Γ = 23 MeV/c^2, consistent with the experimental resolution and to be compared with the Λ(2100) having a width of Γ = 150-200 MeV/c^2. We should know shortly whether this effect represents the first evidence for a charmed baryon or whether it is merely a statistical fluctuation in the data.

The compelling evidence for charmed hadrons found at SPEAR was truly the highlight of this Conference; the hints of charm seen at other laboratories promise feverish activity and a rich harvest. Iliopoulos in his rapporteur's talk at the 1974 London Conference showed confidence that this 1976 Conference would be dominated by the discovery of charm. In a similar talk at Tbilisi, de Rujula proclaimed that charm will still occupy the centre of the stage at Tokyo in 1978.

HAVE WE SEEN A NEW HEAVY LEPTON?

The SLAC-LBL Group was not just content to announce the discovery of charmed hadrons. For a year this group has been observing anomalous lepton events, and more data permitting a more refined and suggestive analysis was presented at Tbilisi[14]. This group has now recorded 105 events of the form

$$e^+ + e^- \rightarrow e^\pm + \mu^\mp + \text{(missing energy)} ,$$

where no charged particle or photon was observed in the final state.

Whereas the experimental evidence for this class of events has been growing over the last year and is now well established, the interpretation of this effect is still being debated. The general features were presented, which characterize this phenomenon, and arguments were given, which, in principle at least, permit a discrimination between the production of a hadron pair carrying a new quantum number and the production of a pair of "sequential heavy leptons", given by the reaction

$$e^+ + e^- \rightarrow U^+ + U^- \quad \text{and}$$

the subsequent decay

$$U^- \rightarrow \nu_U + e^- + \bar{\nu}_e , \qquad U^- \rightarrow \nu_U + \mu^- + \bar{\nu}_\mu .$$

The cross-section for the observation of those events as a function of the centre-of-mass energy is shown in Fig. 11. The production shows a threshold around 3.5 GeV, rises to a maximum at 4.5 GeV, and decreases slowly at higher W. Indication of the nature of the decay can be obtained from the momentum distribution of the detected leptons, as shown in Fig. 12 as a function of the normalized momentum

$$\rho = (p - 0.65)/(p_{max} - 0.65)$$

and three ranges of W. The choice of the parameter ρ allows the comparison of the lepton spectra above the cut-off of 0.65 GeV/c taken at different values of W. In two of the three W-ranges, a three-body decay of a massive lepton is clearly favoured over a two-body decay of a boson.

The key to the discrimination between a new hadron and a heavy lepton is provided by the "missing energy" in the observed reaction, which for lepton decay would be carried by neutrinos. From the ρ-distribution of Fig. 12 it can be deduced with 95% confidence that the mass m of the unobserved particles must be

$$m < 0.7 \text{ GeV}/c^2 .$$

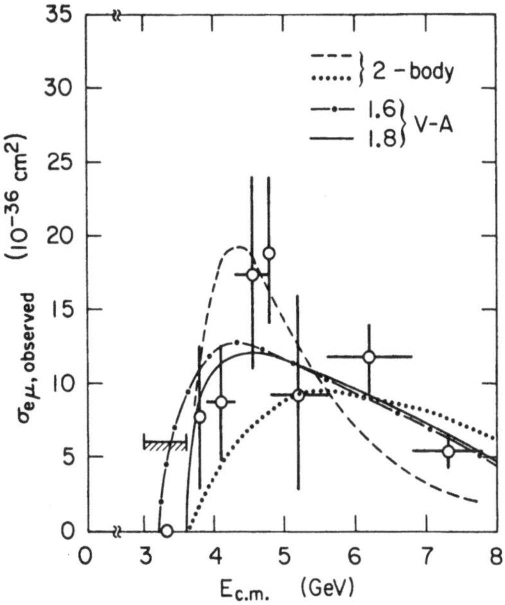

Fig. 11
The production cross-section of the "anomalous" lepton events as a function of E_{cm}. The solid and dash-dot curves are predicted for heavy lepton production, the dotted and dashed curves for a 1.8 GeV/c^2 boson and subsequent two-body decay.

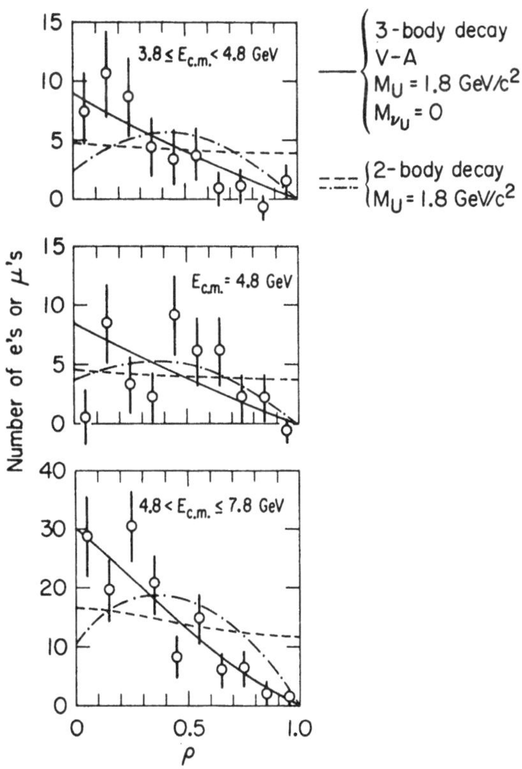

Fig. 12
The distribution in normalized momentum for the detected lepton. The solid curve corresponds to a heavy lepton, M_U = 1.8 GeV/c^2, $M_{\nu U}$ = = 0.0, and V-A coupling. The dashed and dashed-dotted curves are predictions for two-body decays of a 1.8 GeV/c^2 boson.

This excludes the possibility of neutrons, and hence, for example, the decay of a charmed baryon into $e^- + \bar{\nu}_e + n$. The possibility of unobserved K_L^0's can also be evaluated by studying the class of events

$$e^+ + e^- \rightarrow e + \mu + K_S^0 + x .$$

Similarly, the probability for missing π^0's can be evaluated by studying events with one or more visible photons. Such a careful analysis, evaluating also the effect of non-instrumented regions of the detector, indicated that at most 39% of the observed anomalous decays could contain undetected photons or charged particles.

The simplest hypothesis compatible with all the data is the production of a pair of heavy leptons, with a mass in the range 1.6 to 2.0 GeV/c^2.

CONCLUSION

This conference will be remembered by the announcement of the discovery of charmed hadrons. The evidence for anomalous lepton production in e^+e^- collisions and its interpretation as heavy lepton production will initiate a varied experimental activity and we may hope for more definitive conclusions soon. Deep inelastic lepton-hadron scattering continues to be a rich source for new information. The weak neutral currents have been studied by several groups with increased attention to detail. Several reports were presented at the conference, which indicate that parity is violated in these interactions in a way which is consistent with the Weinberg-Salam theory. With the increasing consensus that the electromagnetic and weak interactions are unified, the notion that gauge field theories ("quantum chromodynamics") are also underlying strong interactions seemed to be rapidly becoming accepted. One such consequence -- systematic scaling violation in deep inelastic lepton-hadron scattering -- was observed in μ scattering at FNAL and reported at the conference[19].

The range of activities in high-energy physics is perhaps best illustrated by the announcement[20] of the observation of a πμ bound state ("pionium"), produced in K_L^0 decays with a branching ratio of $\sim 10^{-7}$. Study of such atoms will not only permit a study of πμ interactions, but may also shed light on possible e-μ differences. On the more speculative side, an old idea of Pontecorvo[21] was discussed with new interest: although direct measurements of the rest mass of neutrinos slowly yield improved upper limits, a much more sensitive way to look for differences in the ν_μ and ν_e rest masses may be provided by the possible observation of "neutrino" oscillations. In such a case -- the well-known oscillations in a K^0-$\bar{K}^0$ beam may serve as a conceptual guide -- the intensity of, for example, an initially pure ν_e beam is predicted to vary in space with a characteristic length L, which depends on the mass difference and momentum of the two neutrinos. Observation of low-energy solar neutrinos may be the ultimate test, but meanwhile experiments at the highest energy accelerators are actively contemplated.

May this short 'tour d'horizon' have transmitted some of the excitement and importance of the new discoveries, given an indication of the experimental program with which some of the pivotal questions are being addressed and given a flavour of the breadth of the field. We are justified in our expectation to learn of similarly important discoveries at the next conference at Tokyo in 1978.

REFERENCES

Unless otherwise stated, all references are contributions to this Conference.

1) Columbia-FNAL Collaboration; D.C. Hom et al.

2) Chicago-Princeton I Collaboration; L. Klugberg et al.

3) S.D. Drell and T.M. Yan, Phys. Rev. Letters 25, 316 (1970).

4) Chicago-Princeton II Collaboration; K.J. Anderson et al.

5) M. Duong-van, private communication and SLAC-PUB 1726 (1976).

6) L.M. Lederman and B.G. Pope, Phys. Rev. Letters 27, 765 (1971).

7) M.B. Green et al., Nuovo Cimento 29A, 123 (1975).

8) D.C. Hom et al., Phys. Rev. Letters 36, 1236 (1976).

9) Columbia-FNAL-Univ. Illinois Collaboration; B. Knapp et al.

10) A. Staude, invited contribution.

11) BNL-Yale-FNAL Collaboration; R.K. Adair et al.

12) MPEI Moscow-IHEP Serpukhov Collaboration; N.Z. Anisimova et al.

13) T. Appelquist, invited contribution.

14) SLAC-LBL Collaboration; R. Schwitters, invited contribution.

15) Pavia-Princeton-Maryland-UC San Diego-Stanford Collaboration.

16) B. Wiik, invited contribution.

17) M. Chanowitz, invited contribution.

18) Dubna-Berlin-Budapest-Prague-Sofia Collaboration; I.A. Savin, invited contribution.

19) V.I. Zakharov, invited contribution.

20) M. Schwartz et al., BNL preprint (1976).

21) B. Pontecorvo, invited contribution.

HADRON NUCLEUS COLLISIONS IN THE COLLECTIVE* TUBE MODEL†

G. Berlad

Department of Physics

Technion, Haifa, Israel

Introduction

High energy particle production data measured with nuclear targets exhibit several outstanding features. These features appear to contradict the naive picture in which the projectile is assumed to propagate "on mass shell" through the nuclear medium by a succession of collisions with single nucleons in the target. We can distinguish two classes of such phenomena:

(1) A class of features that look like the corresponding particle nucleon scattering data. Among these we find the KNO ([1]) scaling function ([2]) and the normalized rapidity distribution

$$\frac{dN}{dy} \equiv \frac{1}{\sigma}\frac{d\sigma}{dy}$$

for large rapidity values ([3]). This would seem to indicate that particle nucleus scattering looks more like a single step collision of the projectile with chunks of nuclear matter, rather than a chain of successive collisions with individual nucleons inside the target.

(2) A second class of features exhibit production of particles with energies much higher than the kinematical limit set by a process where the projectile collides only with a single nucleon at one time. This is a big effect and cannot be explained by the Fermi motion of the nucleons inside the target nucleus. Such an energy excess has been observed by

A.M. Baldin ([4]) in an experiment at Dubna. Using various nuclear targets a large number of fast pions were detected in the backward direction. Their kinetic energies exceeded by factors of two and three the kinematical limit set by multistep collisions with single nucleons inside the target. Baldin called it "the cumulative effect".

On these observations we have based an intuitive picture of particle nucleus collisions. This plausible picture claims no deep understanding of the underlying detailed dynamics; however, when it is applied to the various phases of hadron nucleus interactions, all the main features of the experimental data are correctly reproduced. Some of its predictions will be discussed in detail hereafter.

The Collective Tube Model (CTM) ([2])

Let us consider a high energy projectile incident on a nuclear target of atomic number A . The CTM then assumes that:

(a) The projectile collides simultaneously with a tube of all nucleons situated along its path through the nucleus. For our present purposes it is sufficient to assume that there are, on the average, about $A^{1/3}$ nucleons in such a tube ([5]). The energy available for particle production in the projectile-tube collision, $E_{av}(A)$, is given by

$$E_{av}(A) = \sqrt{S} - m - mA^{1/3} \simeq A^{1/3} E_{lab} \qquad (1)$$

Here S is the C.M. energy squared, E_{lab} is the projectile lab. energy, and m is the mass of a nucleon. We shall refer to eq. (1) as the energy rescaling effect.

(b) In its respective CM-frame the projectile tube collision resembles a projectile-nucleon collision at the same available energy.

Assumptions (a) and (b) lead to the following results. Let us denote by σ^{i}_{hn} the cross section for a given final state "i" in hadron-nucleon scattering, and by σ^{i}_{hA} the respective cross section in hadron-nucleus scattering. Then,

$$\frac{\sigma^{i}_{hA}(E_{lab})}{\sigma^{inel}_{hA}} \simeq \frac{\sigma^{i}_{hn}(A^{1/3}E_{lab})}{\sigma^{inel}_{hn}} \qquad (2)$$

where σ^{inel} is the total inelastic cross section.

Using the experimental observation that the effective area of the nuclear target can be parametrized as

$$\frac{\sigma_{hA}^{inel}}{\sigma_{hn}^{inel}} \simeq K_h A^{\alpha_h} \qquad (3)$$

where K_h and α_h are projectile dependent constants [6], we derive two types of sum-rules [7,8]:

a relation connecting "hn" and "hA" processes,

$$\sigma_{hA}^{i}(E_{lab}) \simeq K_h A^{\alpha_h} \sigma_{hn}^{i}(A^{1/3} E_{lab}) \qquad (4)$$

and a relation connecting "hA_1" and "hA_2" processes,

$$A_1^{-\alpha_h} \sigma_{hA_1}^{i}(A_1^{-1/3} E_{lab}) \simeq A_2^{-\alpha_h} \sigma_{hA_2}^{i}(A_2^{-1/3} E_{lab}) \qquad (5)$$

These sum-rules allow us to compare experimental results from different nuclear targets at different energies.

Discussion of the High Transverse Momentum Data [8]

J.W. Cronin and his collaborators have observed that for large transverse momenta the A dependence of cross sections for inclusive production of hadrons from nuclear targets can be represented in the form [9]

$$\frac{d\sigma_{pA}}{dp_T} \propto A^{\alpha(p_T)}$$

where $\alpha(p_T) > 1$. Such a strong A-dependence is hard to understand unless collective effects are present. We claim that such a behaviour is natural in the CTM.

It has been observed both at the ISR [10] and FNAL [11] that at high energy and fixed but large transverse momenta the inclusive proton proton cross section for particle production shows a remarkable increase with energy. This increase is reflected onto hadron collisions by the energy rescaling effect, and may be

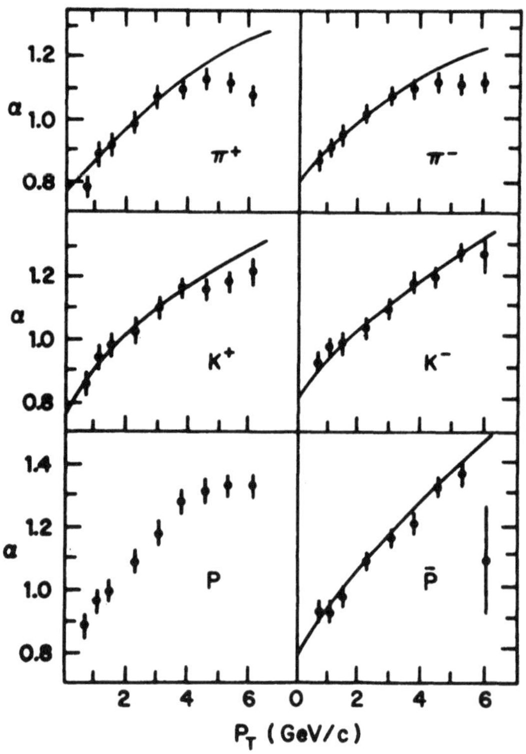

Fig. 1. - The effective A-exponent, $\alpha(p_T)$. Experimental points are from reference ([9]). The continuous lines are the CTM results. Predictions for proton production are not presented because part of the detected protons in the final state does originate from the fragmentation of the recoiling tube for which the CTM does not apply.

the cause of the steepened A-dependence observed in these processes. Let us see how this happens:

We first parametrize the pp data as follows:

$$\sigma^{i}_{pp}(E_{lab},p_T) \equiv \frac{d\sigma_{pp}}{dp_T}(E_{lab},p_T) \propto E_{lab}^{\beta(p_T)} \tag{7}$$

where $\beta(p_T)$ is implicitly energy dependent.

Then we insert eq. (7) into eq. (4) and obtain:

$$\sigma^{i}_{pA}(E_{lab},p_T) \propto A^{\alpha(p_T)} \quad ; \; \alpha(p_T) = \alpha_p + \beta(p_T)/3 \tag{8}$$

The effective A-dependence, $\alpha(p_T)$, is therefore determined by two contributions. One is the nuclear target area responsible for the α_p exponent, and the other, the $\beta(p_T)/3$ exponent, reflects, via the energy rescaling effect, the growth of the corresponding "pp" cross section with energy. The overall A-dependence is compared with experimental data of reference ([9]) in figure (1).

A most sensitive test for the necessity of energy-rescaling is provided by the high-p_T particle ratios ([12]). The trivial target geometry dependence (be it $A^{2/3}$ or A^1) is cancelled in these ratios, and, in the absence of energy rescaling, particle ratios from all nuclear targets at a given lab. energy and p_T should coincide. If, on the other hand, energy-rescaling is present, then, as particle-ratios from different nuclear targets at fixed p_T are plotted as a function of the available energy

$$E_{av} \simeq A^{1/3}E_{lab}$$

all points should follow a single universal line.

The failure of theories with no energy rescaling to explain the $\bar{p}/\pi^-$ ratio is clearly evident in figure (2). (Other particle ratios are not shown because, within experimental errors, they seem to be energy independent, thus they cannot distinguish energy rescaling from non-energy rescaling models.)

We now plot the same data as a function of the available energy (eq. (1)). This is shown in figure (3). In this representation, within experimental uncertainties, the data points do indeed show a tendency to align along a universal line.

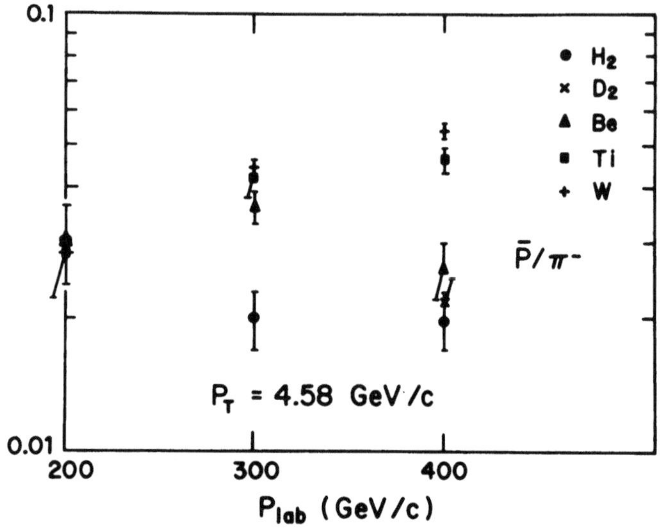

Fig. 2. - The $\bar{p}/\pi^-$ ratio ([12]). In the absence of energy rescaling all points of equal p_{lab}-values should reduce to a single point, regardless of atomic number. This is clearly incorrect.

The CTM and New Particle Production ([13])

The production of very massive particles can occur only above thresholds of extreme energy. However, if Hydrogen targets are replaced by nuclear targets and energy rescaling does take place, these thresholds are significantly lowered. To be explicit, if a certain threshold is at $E_{lab} = E_o$ in a hadron proton collision, the energy rescaling tells us that it should be lowered to $E_{lab} = A^{-1/3}E_o$ when a nuclear target of atomic number A is used. A reduction of almost an order of magnitude in the required energy can thus be gained by replacing a Hydrogen target by a Uranium target. This gain, superimposed on the enhancement due to

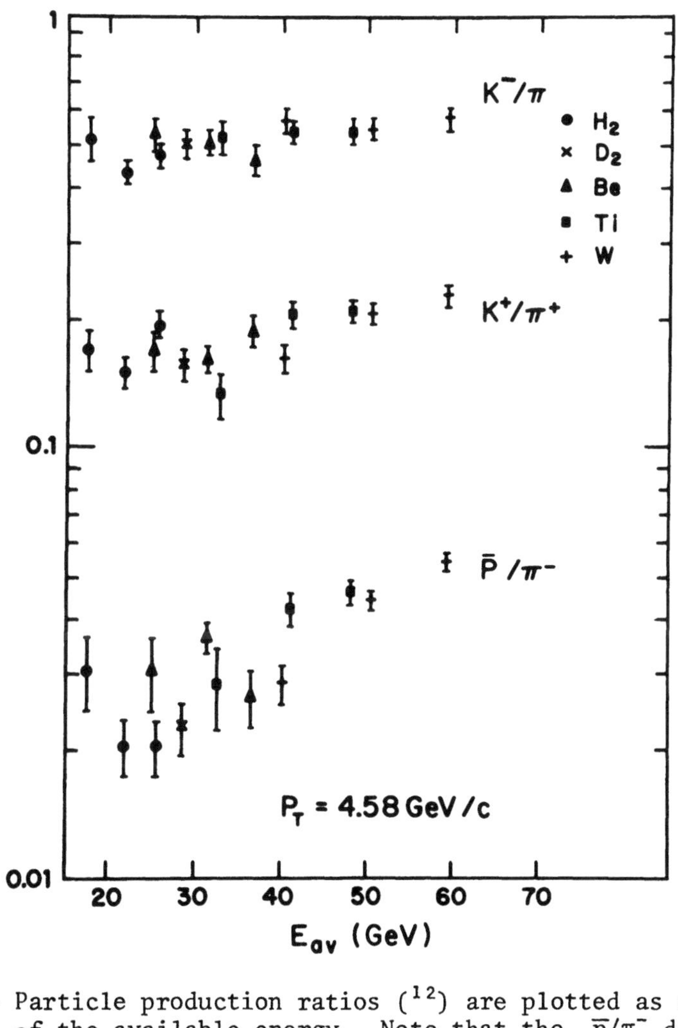

Fig. 3. - Particle production ratios ([12]) are plotted as a function of the available energy. Note that the $\bar{p}/\pi^-$ data (lowest data points) tend to align along a single line, in accord with the CTM prediction. All other particle ratios are energy independent, and thus insensitive to energy rescaling.

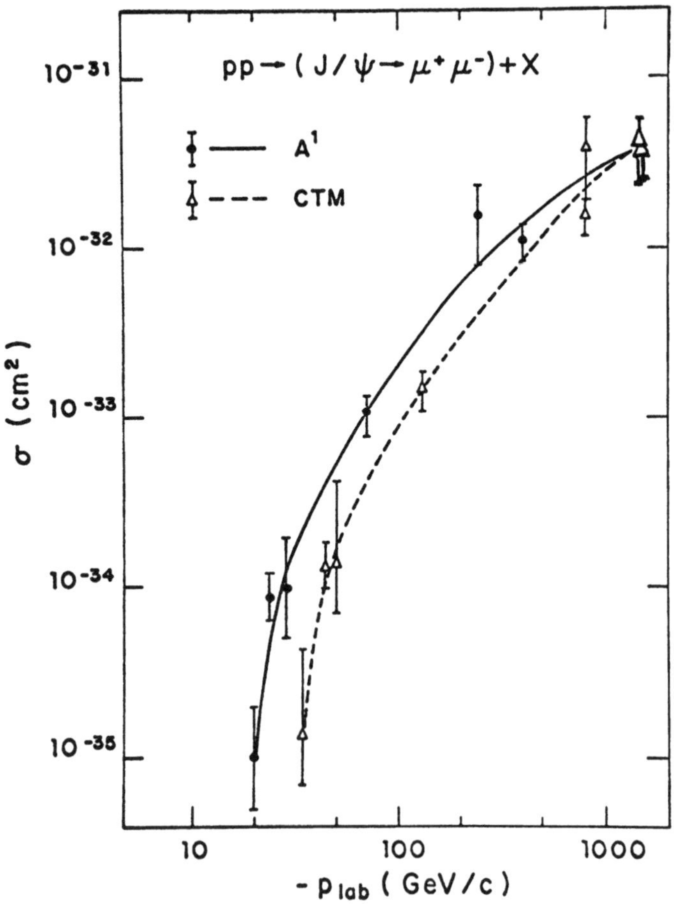

Fig. 4. - $\sigma(pp \rightarrow (J/\psi \rightarrow \mu^+\mu^-) + X)$ as a function of p_{lab} extracted from various nuclear targets ([14]). The circles correspond to an A^1-dependence with no energy rescaling. The triangles present the CTM predictions. The lines were drawn to guide the eye. Note the shift of the threshold in the CTM to higher p_{lab} - values.

the larger nuclear target area, makes it possible to measure processes with high thresholds and small cross section.

The effect of threshold shift for inclusive J/ψ production in the CTM is shown in figures (4) and (5). The experimental data of ref. ([14]) has been used as an input to equation (4) and (5) to predict, respectively, inclusive J/ψ production off Hydrogen and off Uranium. At large energies, far above the threshold, the gain is only marginal. Hence, at $E_{lab} = 20$ GeV/c, energy rescaling turns the "no go" process of J/ψ production on Hydrogen to a "go" process on Uranium, whereas above $E_{lab} = 100$ GeV/c it only enhances the effective A dependence from $A^{2/3}$ to A^1. This enhancement was confirmed by recent FNAL data ([15]), as demonstrated in Fig. (6).

Finally, what does the CTM tell us about the exotic W-boson production process? If we use as input a theoretical calculation of W-boson production in proton-proton collisions by E.A. Paschos and Ling-Lie Wang ([16]), then for a Uranium target we obtain from the CTM the cross section that is shown in figure (7). The threshold shift is enormous, and on the level of $\sigma_{pU \to W} \sim 10^{-33} - 10^{-34}$ cm^2 this cross section should be measurable with a 10^3 GeV/c proton beam (provided $m_W \simeq 70$ GeV).

What About Nucleus Nucleus Collisions?

There has been recently a continued improvement in the heavy atoms ionization techniques. Therefore, it is of interest to find out what could be gained by scattering high-energy, highly ionized nuclei on each other.

The answer is rather surprising. If in particle-nucleus collisions we gain a factor of $A^{1/3}$ in available energy, in a fully ionized nucleus (A_1) - nucleus (A_2) collision, the gain is approximately $(A_1 \cdot A_2)^{1/3}$. Moreover, whereas in particle-nucleus scattering only one tube is active in any given event while the rest of the nucleus acts as a spectator in nucleus-nucleus scattering, very many tube-tube collisions occur simultaneously, and particle production rates are strongly increased.

The results of a preliminary study of the nucleus-nucleus scattering problem, and the enormous effect of many simultaneous tube-tube collisions are presented in figures (8) and (9). Figure (8) shows the charged multiplicity ratio

$$R_{A_1 = 70, A_2} \equiv \langle n_c \rangle_{A_1 = 70, A_2} / \langle n_c \rangle_{pp}$$

For comparison one should recall that $R_{p,U(A = 239)} \simeq 2$ ([6]).

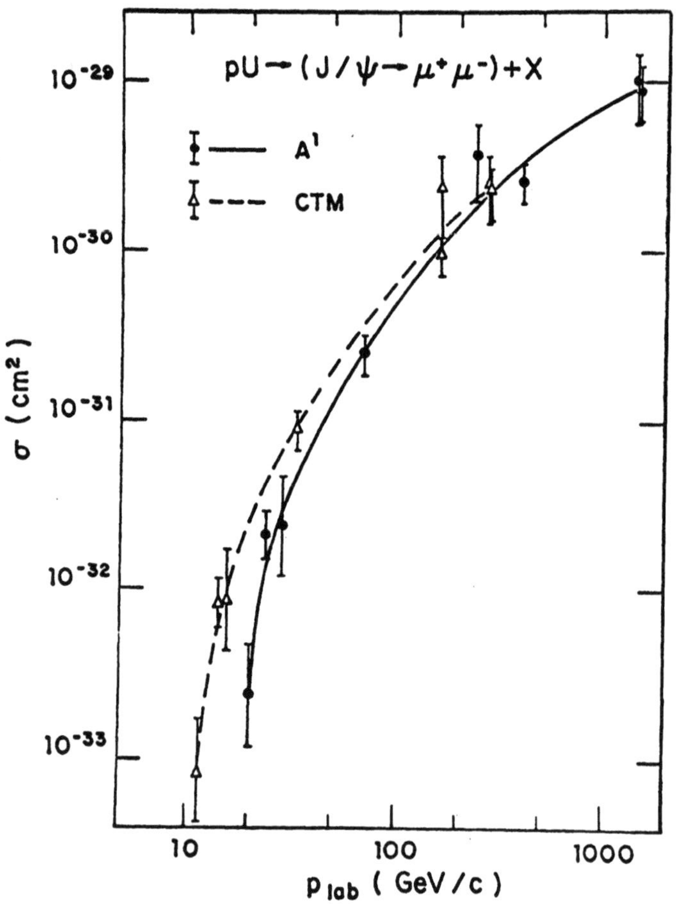

Fig. 5. - $\sigma(pU \to (J/\psi \to \mu^+\mu^-) + X)$ as a function of p_{lab}. The notation and input data are that of figure (4). Note the shift of the threshold in the CTM to lower p_{lab} - values.

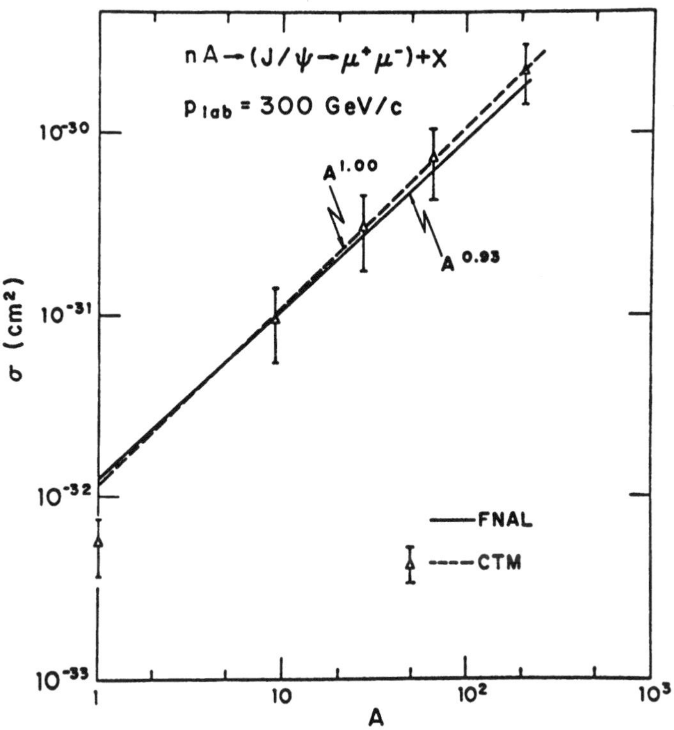

Fig. 6. - $\sigma(pA \to (J/\psi \to \mu^+\mu^-) + X)$ as a function of atomic number for $p_{lab} = 300$ GeV/c . The triangles represent the predictions of the CTM. Error bars reflect the experimental error of the input data taken from reference ([14]). The dashed line is the best fit to A^α (A > 9) of the CTM results. The solid line is the best fit to A^α of the experimental data ([15]). A closer value to the CTM result has been reported by Professor A.J.S. Smith at this school.

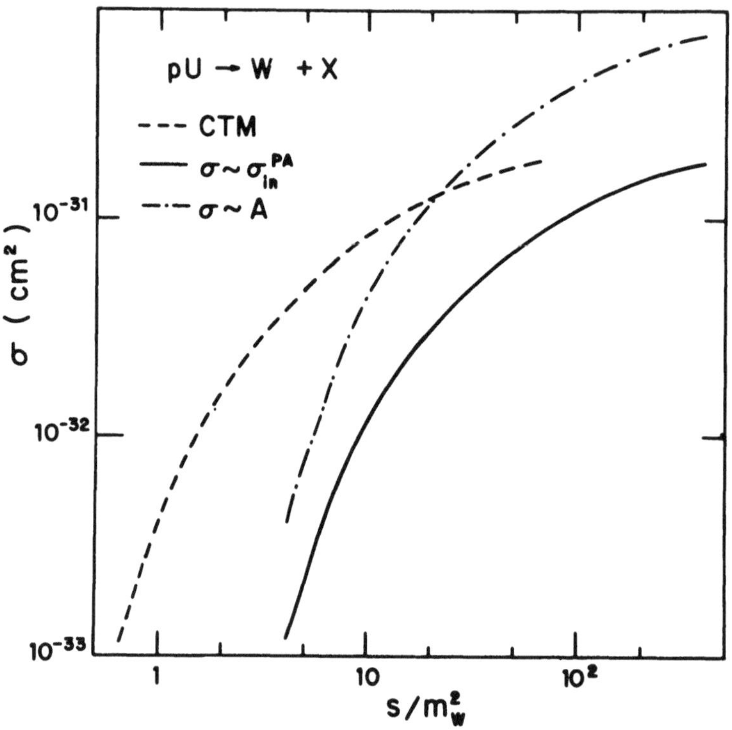

Fig. 7. - Prediction of the CTM for W-meson production. The input is a $pp \to W + X$ calculation of reference [16]. The L.H.S. curve represents the CTM result, the central curve the results of the assumption

$$\sigma^W_{pA} \sim A \cdot \sigma^W_{pp}$$

and the R.H.S. curve the results of

$$\sigma^W_{pA} \sim (\sigma^{inel}_{pA}/\sigma^{inel}_{pp})\sigma^W_{pp} \, .$$

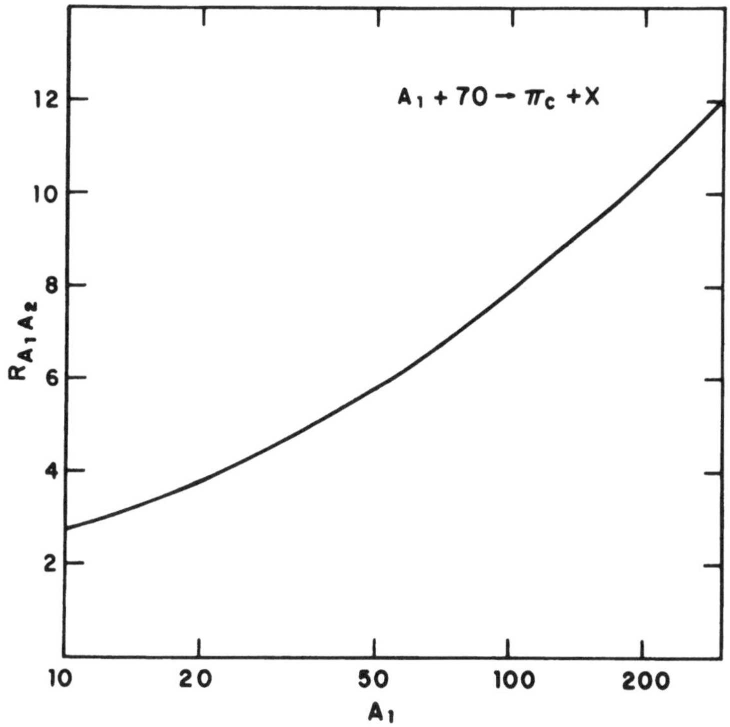

Fig. 8. - The normalized charged multiplicity for nucleus(A_1 = 70) - nucleus(10 < A_2 < 240) collisions as predicted by the CTM.

This tremendous growth in multiplicity that originates from all over the rapididty range is shown in figure (9) where the normalized rapidity distribution

$$\frac{dN}{dy} = \frac{1}{\sigma}\frac{d\sigma}{dy}$$

is plotted for "pp(1,1)", "pA(1,70)" and "AA(70,70)" collisions. Evidently, these results are stimulating enough to justify the channeling of at least a small fraction of the efforts aimed at constructing more powerful accelerators into advancing the techniques of heavy ion acceleration with existing machines.

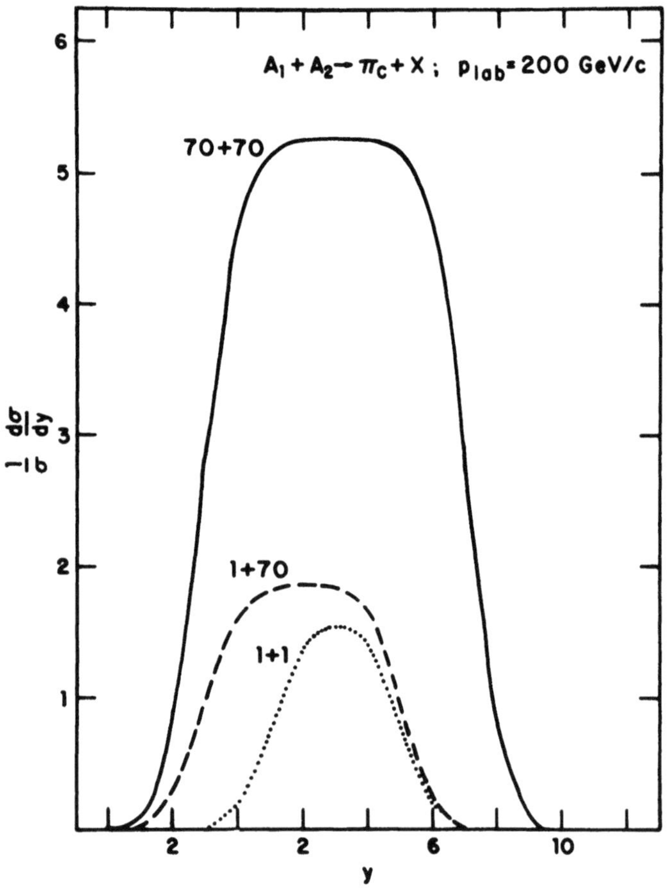

Fig. 9. - The normalized rapidity distribution as predicted by the CTM:
- lower curve, $A_1 = 1$ and $A_2 = 1$ (pp) is used as input;
- middle curve, $A_1 = 1$ and $A_2 = 70$ (p-Emulsion) ;
- upper curve, $A_1 = 70$ and $A_2 = 70$ (Emulsion-Emulsion).

A Comment on Inelastic Processes with Deuterium Targets ([17])

Deuterium targets have been used at high energies mainly for two purposes:

(i) to study neutron cross sections
(ii) to test theories of particle-nucleus collisions, where the target nucleus is relatively simple and well understood.

The Glauber Theory provides us with a good description of elastic scattering and diffraction dissociation of the Deuterium target. These processes, originating mainly from grazing collision of the projectile and the Deuterium, probe only the fringe of the target. In this peripheral region the tubes are "very short", and the probability of hitting both proton and neutron collectively is extremely small. CTM effects are, hence, negligible.

Inelastic collisions, on the other hand, do probe the central nuclear-matter distribution in the Deuteron. It can be shown ([17]) that for such processes there is a probability of about 20% to hit a tube containing both neutron and proton. This non-negligible collective effect has to be properly handled. There are experimental indications, e.g., that for some final states the data shows a positive shadowing. Such a behavior is unexplainable in a naive Glauber model where the projectile propagates "on shell" and the shadowing is always negative. In the CTM, however, it is a unique consequence of the model assumptions (see ref. ([17]) and fig. (1) therein). More complicated Glauber type models can be formulated by including inelastic intermediate states, but these are by no means unique. Different input assumptions will yield different predictions for the type and size of the shadowing effect. It has thus been proposed ([17]) that the CTM should be used to extract neutron cross sections from high energy inelastic processes with Deuteron targets.

Summary and Conclusions

There are many other effects which are successfully explained by the CTM. These, however, require a probabilistic treatment of the nucleon content of the struck tube ([5]).

In order of increasing subtlety they are:

(1) The dependence of the scaled charged multiplicity, R_A , on the target atomic number ([2]) .

(2) The dependence of the KNO scaling function on the target atomic number ([2]).

(3) The dependence of the scaled charged multiplicity and its dispersion on the number of fast proton (presumed to be fragments of the recoiling tube) ([2]).

(4) Shapes of rapidity and pseudo-rapidity distributions as functions of energy, atomic number and fast protons ([18]).

(5) Photon-Deuteron interactions ([17]).

It is rather amazing that such a naive picture of particle-nucleus scattering does explain so wide a field of high energy phenomena of extreme diversity. It is, therefore, our conviction that whatever the ultimate theory of high energy particle nucleus scattering will, eventually, turn out to be, collective interactions will certainly be there.

Acknowledgement

We would like to thank Professor A. Dar and Mr. Y. Afek for their valuable help in preparing this summary.

References

* We thank Professors O. Kofoed-Hansen and H. Lipkin for suggesting the use of the word "Collective" rather than "Coherent" in the present context.

† This research was performed in collaboration with Prof. A. Dar, Dr. G. Eilam and Mr. Y. Afek.

1. Z. Koba, H.B. Nielsen and P. Olesen, Nucl. Phys. B40, 317 (1972).

2. G. Berlad, A. Dar and G. Eilam, Phys. Rev. D13, 161 (1976).

3. W. Busza et al., FNAL preprint, June 1976.

4. A.M. Baldin et al., Yad. Fiz. 20, 1201 (1974) (Engl. Transl. 20, 629 (1975)).

5. A more sophisticated approach is possible, whereby a probability is calculated to hit a tube of exactly j nucleons. Physical quantities are then properly weighted with these probabilities and summed over the j's from 1 to A. See, e.g., reference ([2]).

6. W. Busza, Proceedings of the VI International Conference on High Energy Physics and Nuclear Structure, Santa Fe, June 1975, and references therein.

7. Due to the different CM systems of different projectile tube collisions, variables which are not invariant under boosts in the beam direction, such as e.g. the rapidity, must be treated with special care. This treatment is not displayed in eqs. (4) and (5). For details see reference ([8]).

8. Y. Afek, G. Berlad, G. Eilam and A. Dar, "Scaling Laws for Inclusive Production of Hadrons in High Energy Particle Nucleus Collisions", Technion Preprint PH-76-12.

9. J.W. Cronin et al., Phys. Rev. D11, 3105 (1975).

10. See, e.g., Professor Valenti's report in these proceedings.

11. See, e.g., Professor Cronin's report in these proceedings.

12. We would like to thank Professor Cronin for suggesting this test, and for providing us with his new data, which has been presented at this school.

13. Y. Afek, G. Berlad, G. Eilam and A. Dar, "Cumulative Enhancement of J/ψ Production in Hadron Nucleus Collisions", Technion Preprint PH-76-24.

14. S.C.C. Ting, Proceedings of the International Conference on High Energy Physics, Palermo, Italy (1975).

J.J. Aubert, Proceedings of the XIth Rencontre de Moriond, France (1976).

15. M. Binkley et al., "Dimuon Production on Nuclear Targets", FNAL Preprint, June 1976.

16. Emmanuel A. Paschos and Ling-Lie Wang, "The Quest for W's", BNL Preprint, BNL-21048.

17. A. Dar and Tran Thanh Van, "Inelastic Interactions with Deuterons at Very High Energies", Technion Preprint PH-76-61.

18. Y. Afek, G. Berlad, G. Eilam and A. Dar, "Rapidity Distributions in Hadron-Nucleus Collisions", Technion Preprint PH-76-48.

PRODUCTION OF DIMUONS BY PIONS AND PROTONS AT FERMILAB*

A. J. S. Smith

Department of Physics

Princeton University, Princeton, NJ 08540 U.S.A.

INTRODUCTION

In this lecture I shall describe an experiment in progress[1] at the Fermi National Accelerator Laboratory (FNAL) to study the production of $\mu^+\mu^-$ pairs in hadron-hadron collisions. So far, we have used beams of 225 - GeV π^+, π^- and protons incident on carbon and tin targets for most of the data. In an earlier run, a 150 GeV positive beam was used with a beryllium target. The results I report of the 150 GeV run are final; those at 225 GeV are preliminary.

The goal of this experiment is to carry out, as completely as possible, a measurement of μ-pair production over a wide range of kinematic variables and particle types, using a spectrometer with good resolution and very large acceptance. Only such a comprehensive experiment can measure differential cross sections without resorting to model-dependent assumptions. There are many interesting aspects of these measurements: Among them are the study of the production mechanism of the new particles[2] J/ψ and ψ', as well as a search for other states decaying to μ-pairs. I shall discuss in detail the features of J/ψ production by π^+, π^-, and protons, and compare them with the production of the more common vector mesons ρ^0, ω, ϕ, and with non-resonant μ-pairs in neighboring mass regions. The non-resonant pairs are also of great interest; for example, in 1970 Drell and Yan[3] proposed parton-antiparton ($q\bar{q}$) annihilations as their source, and since then many predictions of yields have been made by numerous authors. Virtual bremsstrahlung has been put forth as another possible source of non-resonant pairs.[4]

To produce μ-pairs with proton beams by quark-antiquark annihilation all antiquarks must come from the sea. On the other hand, pion beams contain valence antiquarks and hence present a more straightforward probe with which to test these models.[1] Also, because the valence antiquarks of π^+ and π^- have charge $\frac{1}{3}$ and $-\frac{2}{3}$ respectively, a cross-section ratio of $\frac{1}{4}$ is predicted for μ-pairs produced by π^+ mesons versus π^- on an isospin symmetric target. Thus by using π^+ and π^- beams of the same momentum incident on a carbon target, our experiment can test the Drell-Yan mechanism in a way not too dependent upon details of a particular model.

The experiment is also sensitive to processes resulting in "multimuon" final states (states having more than two muons). Such events could originate in associated production of J/ψ pairs, or of a J/ψ along with a charmed particle decaying promptly into a muonic final state. We have searched for multimuon events, finding none above background, as explained in what follows. This negative result puts quite stringent upper limits on associated production processes.

Finally, one of the most compelling motivations for this experiment was to ascertain the source of the single prompt leptons, first observed in high-energy p-nucleus collisions at Serpukhov[5] and FNAL.[6] (Prompt leptons are those not attributable to π or K decays.) These measurements were all made at high transverse momenta and near 90° in the center-of-mass system. Since then other experiments[7] have detected prompt leptons over a wide range of kinematic variables, and have found the ratio of the yields to be approximately $\mu/\pi \approx 10^{-4}$, and quite insensitive to the kinematic region of the experiment.

Our goal is to see if these single leptons can be accounted for by lepton pairs. (We actually study only muons -- there could be different sources of electrons.) We obtain the pair contribution to the inclusive single-muon yield by measuring the differential cross-sections for all μ-pairs, and then integrating over one muon in the pair. The conclusion, as described below, is that our measured μ-pairs at 150 GeV are sufficient to explain the prompt single muon measurements in the kinematic range of our experiment. When our 225 GeV data have been completely analyzed, they should give an even more comprehensive measurement of the pair contribution to the single-μ yield, because of greater statistical power and because of the wider range of acceptance for the higher momentum beam.

Before discussing the experiment in detail, it seems appropriate to define the variables of μ-pair production. Table I shows the variables we use, as well as the range of each accepted by the spectrometer. It is seen that we have virtually complete coverage for all μ-pairs produced in the forward hemisphere in the center of mass system.

Table I

Variables to Describe the Production of μ-Pairs

$M^2_{\mu\mu} \equiv (p_+ + p_-)^2$	$.211\ \text{GeV} < M_{\mu\mu} \lesssim 12\ \text{GeV}$
$x_F \equiv p^*_{/\!/}/p^*_{/\!/\,MAX} \simeq 2p^*_{/\!/}/\sqrt{s}$	$0 < x_F < 1$
$p_\perp \equiv$ Transverse momentum of pair	$0 < p_\perp \lesssim 4\ \text{GeV/c}$
$\Theta^* \equiv$ Helicity angle of μ's from decaying state	$0 < \lvert\cos\Theta^*\rvert < 0.8$

EXPERIMENTAL CONSIDERATIONS

To obtain a large acceptance detector for our experiment, we have modified the Chicago-Cyclotron spectrometer, constructed for μ-p scattering experiments by a Chicago-Harvard-Oxford group, whom we specially acknowledge. This detector, located in the muon laboratory at Fermi National Accelerator Laboratory, as shown in Fig. 1, is fed by a secondary beam from the "neutrino" target. The beam is very long (∿ 1 KM), a good feature for μ-p scattering, but less than optimal for hadron experiments, in that even at 225 GeV/c about 43% of the kaon flux is lost by decays.

For this experiment the highest possible beam momentum is desirable; not only is there more energy available to produce new objects, but the folding forward in the laboratory of the reaction products gives one a larger detection efficiency for a given magnet aperture. Unfortunately, the pion flux decreases drastically once the beam momentum exceeds roughly half the momentum of the primary protons. This fact determines the maximum useful beam momentum. For our first run, the accelerator operated at 300 GeV, so we chose p_{beam} = 150 GeV, obtaining a π^+/p ratio of $\sim \frac{1}{3}$. Last winter in our second run, when the primary energy was 400 GeV, we put p_{beam} = 225, at which $\pi^+/p \sim \frac{1}{5}$. So far we have used beryllium, carbon, and tin targets, with beams of π^+, π^-, and protons, the intensities being $\lesssim 2 \times 10^6$/burst for protons, $\lesssim 5 \times 10^5$ burst for π^+ and π^-.

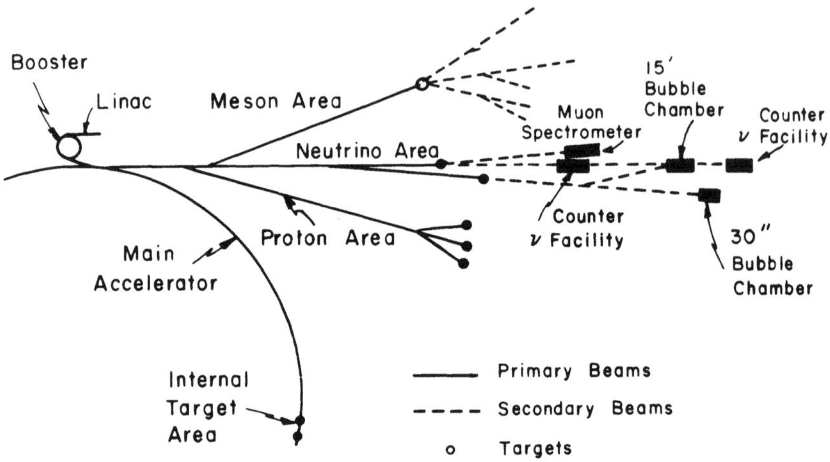

Fig. 1. Plan of the Fermi National Accelerator Laboratory. Our detector is shown as the muon spectrometer.

The beam was defined by three scintillation counters upstream of the detector, two of which were separated by a dipole magnet to sweep out debris from interactions along the beam. To identify pions and protons individually, three threshold gas Cerenkov counters averaging about 70 meters in length were placed in the beam. The construction was very simple: the existing beam pipes between transport elements were filled with helium gas to a pressure just below the threshold for Cerenkov radiation from protons. (The operating pressure was about .2 atmospheres absolute at 225 GeV/c.) In each counter, light from pions was focussed by a spherical mirror through a quartz window onto an RCA 31000-M phototube, which has excellent ultra-violet sensitivity. Information from the Cerenkov counters was recorded for each event, allowing us to accumulate π^+ and proton data simultaneously. This feature is very desirable in that many systematic effects, such as changes in magnet currents, beam intensity, detector efficiencies, to name a few, cancel out in a comparison of proton-induced and π^+-induced events.

The spectrometer is shown in Fig. 2. Interactions took place when the beam, typically 3cm x 3cm wide, struck the target, usually $\sim$ .25 interaction lengths thick. (The last beam-defining counter is shown as T_3 in the figure.) To detect the rare μ-pair events in

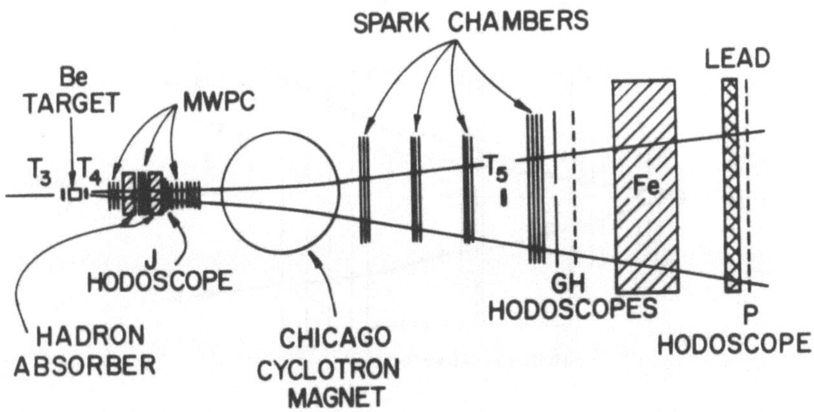

Fig. 2. The Spectrometer.

the presence of overwhelming hadronic processes, we absorbed the hadrons before a substantial fraction could decay by placing a 2.2 meter thick steel hadron absorber 1.3 meters downstream from the target. Thus only muons emerged into a large cylindrical magnet (the former Chicago-Cyclotron), 2.2 meters in radius and 1.27 meters in gap height. Downstream from the magnet, the muons had to penetrate an additional 2.5 meters of iron, after which they were detected by a large scintillation-counter hodoscope (shown as P in Fig. 2). The apparatus is charge symmetric - hence π^- - induced pairs were measured simply by reversing the polarity of all beam magnets.

The trigger requirements were as follows: 1) One and only one beam particle and no "halo" particles within a time interval of $\simeq$ 100 nsec. 2) $\geq$ 2 particles leaving the target, as determined by pulse-height in counter T_4 situated just downstream of the target. 3) $\geq$ 1 particle in the G hodoscope. 4) $\geq$ 2 hits in the P hodoscope. (To reject single μ's accompanied by low-energy electromagnetic showers, the P hodoscope was mounted flush against the downstream side of a 20cm-thick Pb wall. Additional rejection was accomplished by requiring the muon pulses to be in non-adjacent counters.) 5) No count in T_5, a 7.5cm-square counter downstream of the magnet to veto beam muons. 6) For most of the 225-GeV/c running, we also required two non-adjacent hits, either vertically or horizontally, in a hodoscope (shown as J in Fig. 2) placed at the downstream edge of the hadron absorber. This requirement, while reducing the trigger rate from background and low-mass pairs, had a completely negligible effect on the detection efficiency of pairs with masses > 1 GeV.

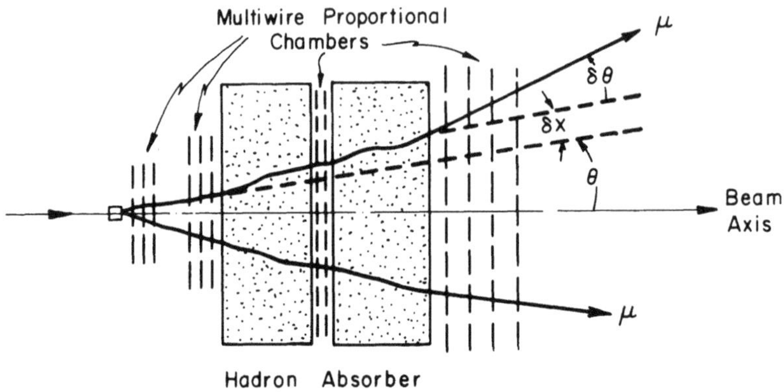

Fig. 3. Details of the upstream configuration of the detector. Variables are defined in the text.

Approximately 70% of all triggers were clean μ-pair events. The rest, mainly single muons showering into the P hodoscope, were easily eliminated by the analysis. The trigger rate was about 1 in 10^5 beam particles.

The particle trajectories were measured by multi-wire proportional chambers (MWPC) upstream of the magnet, and by wire spark chambers downstream, where instantaneous rates were always lower than 10^5/sec. The MWPC configuration is shown in Fig. 3. The unique feature of our experiment is the chambers in front of the absorber, which were used to measure the directions of the μ's before any significant multiple scattering occurred. These chambers enabled us to obtain good mass resolution, reconstruction efficiency, and background rejection even at the lowest μ-pair masses. For the data reported here, a 4-plane module 55cm x 55cm in area was used, consisting of an X- and a Y-measuring plane, and also a U and a V plane at 45° to X, Y. For data we are still analysing, an additional module with 3X and 3Y planes was used to determine more precisely the vertex in the target. To follow the μ's as they scattered through the absorber, another MWPC module (XYUV) was placed half way through the steel, where the hadron showers had subsided to an extent that the average hit multiplicity was about 4.

To eliminate the effect of the multiple scattering in the absorber from the momentum determination, we remeasured the muon trajectories before they entered the cyclotron magnet. Eight MWPC planes (4X, 4Y), each 1m by 1m in area, were used. To complete the momentum analysis, we measured the trajectories, after they had been deflected by the magnet, with 12 planes of capacitive read-out wire spark chambers 2m by 4m in area, and 8 even larger magnetostrictive planes 2m by 6m. The planes alternated U X X V, with U and V tilted ± 15° with respect to X, to provide narrow-angle stero "views".

Event reconstruction downstream of the absorber was straightforward, since all the chambers had in general better than 90% efficiency, and only about 10 percent of the events had an extra track, (usually an accidental) through the spectrometer. A typical mass resolution $\frac{\Delta M}{M}$ of $\approx$ 1.6% (standard deviation) was obtained. Accidental tracks, as well as any hadrons penetrating the absorber, were eliminated by requiring all μ tracks to point within resolution to struck counters in the P hodoscope. A small contamination from beam muons was eliminated, with no loss of good events, by rejecting any event in which the laboratory momentum of either muon exceeded 88% of the beam momentum.

The most difficult part of the analysis was reconstructing the events upstream of the hadron absorber. The procedure by which the tracks emanating from the target were matched to those downstream of the absorber was tested by a Monte Carlo simulation. Clean μ-pairs were generated and transported through the spectrometer. Hadron tracks and extra hits from real data were superimposed, and the reconstruction program then was asked to find the μ-tracks. For masses above 500 MeV a large and uniform efficiency was obtained for a typical x_F of 0.3. For a mass of 325 GeV the efficiency had only dropped by a factor of 2. The rejection of background from muons produced in the absorber was measured in a similar manner, this time the μ-pairs being generated within the first interaction length of the absorber. We found the rejection power to be greater than 15 at a mass of 300 MeV and to rise sharply with increasing pair mass to at least 100 by 1 GeV. (This factor is indpendent of the factor of 50 or more against primary interactions in the absorber obtained by demanding at least two particles in the counter T_4 in order to trigger.)

Dead-time in the readout electronics of the MWPC's upstream of the absorber limited their use to maximum incident beam fluxes of $\sim 10^6$/pulse. To study μ-pairs with masses $\gtrsim$ 1 GeV, we were able to turn these chambers off and run with fluxes as high as 3×10^6/ burst. This was possible because the MWPC's downstream of the absorber provided good enough direction measurements to separate background μ-pairs produced in the absorber from pairs of 1 GeV mass or more produced in the target. As shown in Fig. 3, because of multiple scattering the trajectory measured by these MWPC's alone differs from the true trajectory by an angle $\delta\Theta$ and a displacement δx. (A similar situation pertains to the y-coordinate.) Muon pairs produced in the target were separated from those produced in the steel absorber by a χ^2 procedure. Specifically, $\delta\Theta_x$, δx, $\delta\Theta_y$ and δy were varied to minimize the χ^2 of each of the two muon tracks for two assumptions: First, that the μ came from the center of the target; second, that it came from a point 1 interaction length into the absorber. Excellent separation was obtained even for masses below 1 GeV. In the expression for χ^2, the uncertainties in and correlations between $\delta\Theta_x$ and δx, $\delta\Theta_y$ and δy were calculated for the multiple scattering of the absorber, from well-known formulae in Rossi's book.[8] The mass resolution obtained by this method was typically $\frac{\Delta M}{M} \approx .04$, mainly due to the uncertainty in opening angle from multiple scattering. This is perfectly adequate to study the interesting features of both continuum and resonant μ-pairs.

FINAL RESULTS OF μ-PAIR PRODUCTION BY 150-GeV/c π^+ AND PROTONS

In this run, a beryllium target was used with positive beam to study the reactions $\binom{\pi^+}{p} + \text{Be} \to \mu^+\mu^- + X$. Figure 4 shows the mass spectra of all μ-pairs with $x_F \geq 0.15$, after correction has been made for the efficiency of the spectrometer. The acceptance calculation assumed a uniform angular distribution in $\cos\Theta^*$ for muons from the decaying pair. For the ρ-ω mass region, our analysis gives a distribution $dN/d(\cos\Theta^*) = 1 - (0.07 \pm 0.24) \cos^2\Theta^*$, consistent with the assumption used in the acceptance. In any case, the only effect of a 1 - standard deviation decay asymmetry would be to produce a uniforma 5% change in the over-all acceptance. The measured width of the ρ-ω peak is 100 MeV, narrow enough to indicate a substantial ω contribution.[9]

A striking feature of both pion-induced and proton-induced spectra is the prominent continuum present at very low masses, falling steeply by about 500 MeV. The observation of this continuum has important implications for the single direct lepton question, so it is important to be sure that is not caused by background of one form or another. The most obvious source of background, the decays of uncorrelated π's or K's, is eliminated because 99% of the observed pairs have muons of opposite charge ($\mu^+\mu^-$). Secondary processes in the shield are negligible because of trigger and reconstruction rejections. Other sources which have been estimated quantitatively include photoproduction of μ-pairs in target and absorber, secondary ρ-ω production and decay in the absorber, and even the associated production and subsequent decay of $K^+ K^-$ pairs. All background sources combined produce less than 10% of the observed low-mass pairs, so we conclude the signal is real.

There are some very plausible sources of this prompt low-mass continuum. The two most obvious are the Dalitz decays of η and ω, which, although they have not been observed experimentally, have reliably calculated branching ratios:[10]

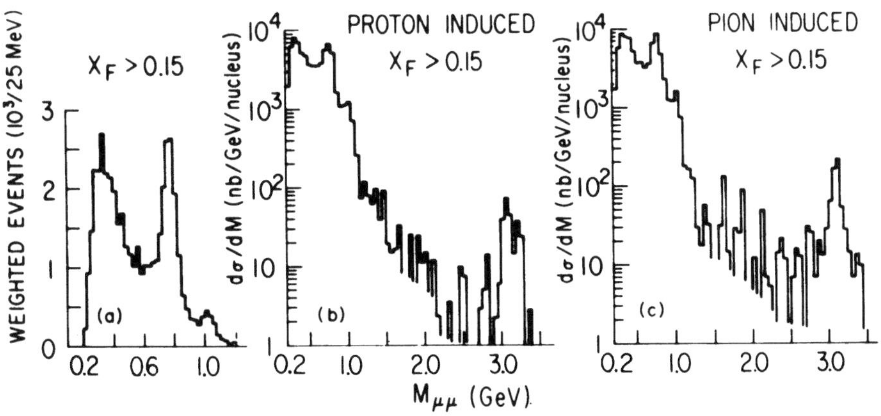

Fig. 4. Effective Mass Distributions, 150 GeV beam, corrected for detector efficiency, for $x_F > 0.15$. a) Combined π^+ and p induced events with $M_{\mu\mu} < 1.2$ GeV/c. b) Proton induced events, logarithmic scale. c) pion-induced events, logarithmic scale.

$\Gamma(\eta \to \mu^+\mu^- \gamma)/\Gamma(\eta \to \gamma\gamma) = 7.5 \times 10^{-4}$; $\Gamma(\omega \to \pi^0 \mu^+\mu^-)/\Gamma(\omega \to \pi^0 \gamma) = 5.5 \times 10^{-4}$. If we assumed all the low-mass continuum were due to η decay, the inclusive η production cross-section required would be 11 mb/nucleon, or about 10% of the π inclusive cross-section. This is entirely reasonable - although no measurements have been made at low $p_\perp$, at high $p_\perp$ the ratio of inclusive cross-sections[11] of η and π^0 is $\frac{\eta}{\pi^0} \sim 0.5$. If we assumed that ρ_0 and ω production are equal ($\sigma_{\rho_0} = \sigma_\omega$), then the process $\omega \to \mu^+\mu^-\gamma$ would explain 13% of the continuum.

The dependence of μ-pair production upon the kinematic variables of Table I was studied by first dividing the data into 6 mass intervals. (The dependence upon particle type, π^+ or p, is studied as well.) In Fig. 5 is shown the dependence upon longitudinal momentum, where $Ed\sigma/dx_F$ is plotted vs. x_F. In each mass interval π^+ mesons produce more μ-pairs at large x_F. It appears that the differential cross sections for π^+ and p approach equality at $x_F = 0$. Also, in Fig. 6 is shown the dependence of the differential cross sections upon $p_\perp$, in the form $p_\perp^{-1}\, d\sigma/dp_\perp$. Here no significant differences appear between π^+- and p-induced spectra. The data of Fig. 5 have been fitted to the forms $Ed\sigma/dx_F \propto (1-x)^c$ and $p_\perp^{-1}\, d\sigma/dp_\perp \propto e^{-bp_\perp}$ respectively. The lowest $p_\perp$ bin in Fig. 6 has not been included in the fit. The results of the fits are given in Table II, along with normalization constant obtained from the parametrization $E\, d^3\sigma/d^3p = A(1 - x_F)e^{-bp_\perp}$. As described in the pre-

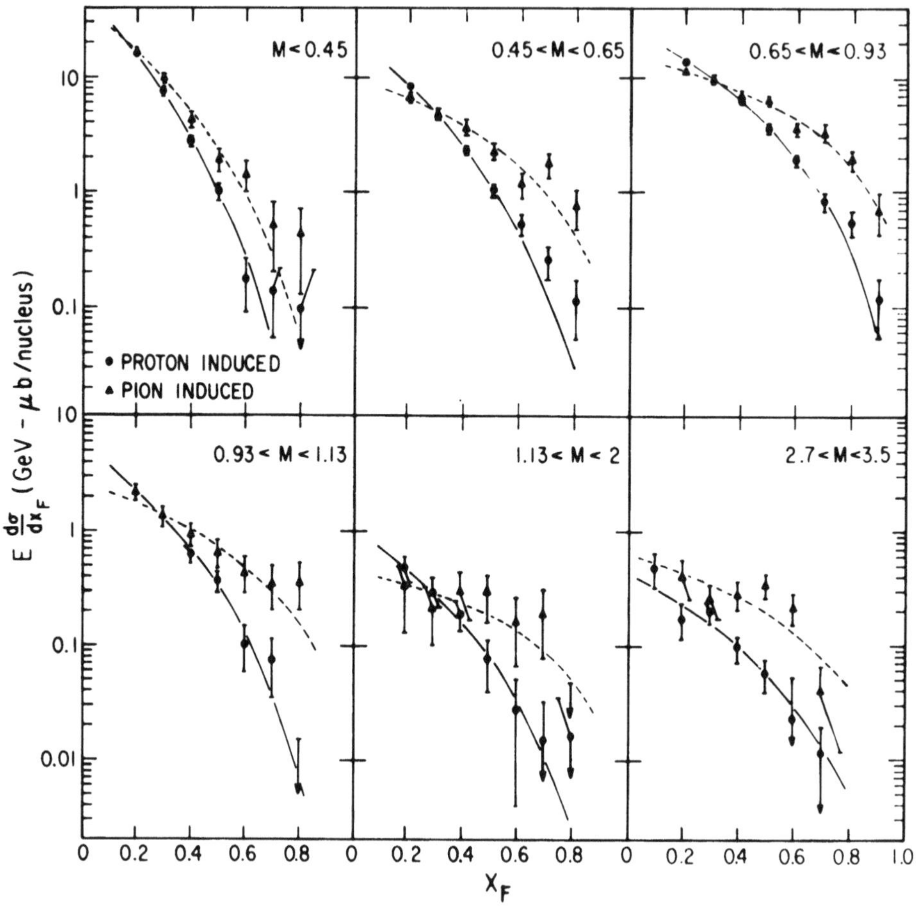

Fig. 5. Invariant cross-sections vs. Feynman X. The curves are the fits of Table II.

vious section, the high-mass data, mainly J/ψ production, were obtained without the front MWPC's. We verified that the normalization of this data and the rest, where the MWPC's were used, agreed to within 10%. Other results listed in the table include the cross-sections per nucleon obtained by using the dependence upon the atomic mass number A reported by Binkley et al.:[12] $\sigma(A) \propto A^{0.67}$ for $M_{\mu\mu} < 1.13$ GeV, $\propto A^{0.85}$ for $1.13 < M_{\mu\mu} < 2$, and $\propto A^{1}$ for $M_{\mu\mu} > 2$ GeV. As will be discussed below, we have confirmed

TABLE II - Results of fitting the Lorentz-invariant cross-section per Be nucleus to the form $Ed^3\sigma/d^3p = A(1 - x_F)^c \exp(-bp_\perp)$ for different mass regions. The x_F and $p_\perp$ projections of the data were fit independently. Calculation of the integrated cross-sections is described in the text.

Region	Mass GeV	Source	A nb/GeV^2/c^3	b (GeV/c)$^{-1}$	c	Cross Section /Nucleus $x_F > 0.15$ nb	Cross Section/Nucleon $x_F > 0.15$ nb	$x_F > 0$ nb
				Proton Production				
I	0.21 - 0.45	Continuum	2.67±.53 x 10^4	4.63±.15	6.03±.22	1470±300	340±70	1550±620
II	0.45 - 0.65	Continuum	9.03±1.8 x 10^3	4.58±.14	4.34±.21	800±160	185±37	620±220
III	0.65 - 0.93	ρ-ω	4.69±.95 x 10^3	3.79±.09	2.79±.12	960±190	220±44	510±150
		Continuum	1.83±.40 x 10^3			372±75	86±17	200±60
IV	0.93 - 1.13	φ	1.06±.21 x 10^3	3.93±.28	4.06±.40	127±38	29±9	83±25
		Continuum	.55±.11 x 10^3			66±20	15±5	43±13
V	1.13 - 2.0	Continuum	250±100	3.41±.85	3.78±.80	39±16	6.0±2.5	16±7
VI	2.7 - 3.5	J	36±12	2.08±.26	2.94±.32	14±5	1.6±.6	3.3±1.1
				Pion Production				
I	0.21 - 0.45	Continuum	2.20±.44 x 10^4	5.07±.25	4.30±.33	1660±330	383±76	1250±500
II	0.45 - 0.65	Continuum	4.02±.80 x 10^3	4.64±.24	1.92±.25	780±160	180±37	370±130
III	0.65 - 0.93	ρ-ω	3.41±.70 x 10^3	4.31±.16	1.34±.14	960±190	220±44	370±110
		Continuum	1.46±.30 x 10^3			410±82	94±20	160±50
IV	0.93 - 1.13	φ	4.90±1.5 x 10^2	3.61±.40	1.73±.44	160±48	37±11	70±21
		Continuum	1.47±.44 x 10^2			48±14	11±3	21±6.4
V	1.13 - 2.0	Continuum	89±53	3.2±2.4	1.33±1.0	40±24	6.2±3.7	10±6
VI	2.7 - 3.5	J	81±27	2.57±.36	1.72±.38	35±12	3.9±1.3	6.5±2.2

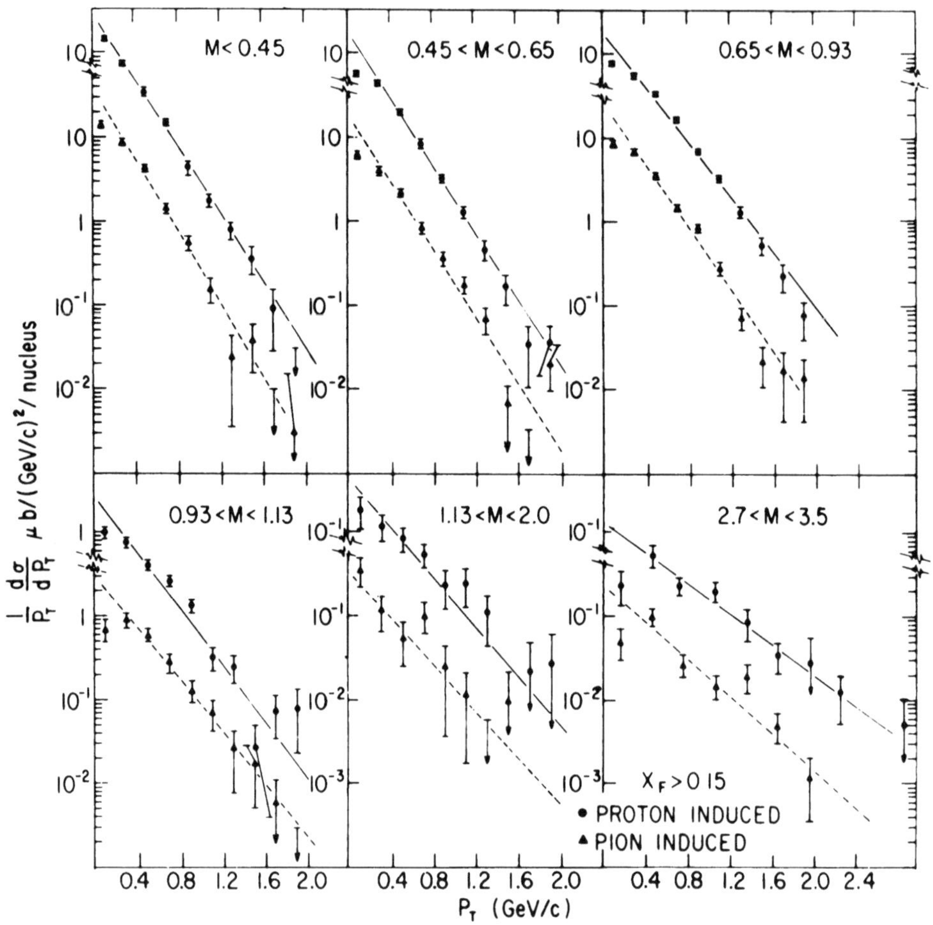

Fig. 6. Differential cross-section $d\sigma/dp_\perp/p_\perp$ vs. $p_\perp$. The curves are the fits of Table II.

in our 225 GeV experiment the linear A-dependence for high-mass μ-pairs. Finally we give total cross-sections for $x_F > 0$ for the various mass regions. Note that we cannot resolve ρ^0 and ω, so have assumed their production to be equal, as suggested by the width of the peak.

AN APPARENT SOLUTION OF THE SINGLE LEPTON PUZZLE

As mentioned in the introduction to this lecture, an important goal of our experimental program was to find out whether or not μ-pair production can account for the anomalous direct muons observed by numerous authors in p-nucleus collisions. In none of these previous experiments, however, was there a significant acceptance for detecting a second muon, should it have been there. By gathering up all the pairs, then, we can simply see how much they contribute to the yield of single μ's. It has already been estimated by several authors[13] that μ-pairs from the known resonances ρ^0, ω, φ, and J/ψ were insufficient, explaining at most about 30% of the single μ's. These estimates were made difficult by the fact that the data was obtained by various experiments differing vastly in almost every respect. By contrast, in our experiment we calculate the single μ yield from all pairs over our large kinematic region, and compare with single-μ measurements from the same target, beryllium. In this way many systematic uncertainties are made negligible.[14]

To include all significant μ-pairs in our calculation, we had to extrapolate our cross-sections down to $x_F = 0$ using a parametrization of the measured data for $x_F \geq 0.15$. First, we assumed the cross-section $E d\sigma/dx_F \propto d\sigma/dy$ to be flat in the region $0 < x_F < 0.15$. Second, we parametrized the data to the form $d\sigma/dx_F = A(1 - x_F)^c$ and then extrapolated to $x_F = 0$. These two methods agree to within 10%. If one used the (probably wrong) extrapolation $E d\sigma/dx_F = A(1 - x_F)^c$ down to $x_F = 0$ one obtains larger cross-sections because the rapidity plateau has been ignored. The parameters used in these fits are given in a recent publication,[15] and are omitted here. The contribution for each of the sources in Table II was calculated separately. The mass dependence of the continuum was parametrized as $\exp(-5M_{\mu\mu})$ within each mass interval. Two independent calculations, one by Monte Carlo techniques, the other by a numerical integration, agreed perfectly.

The sensitivity of our calculation to the $p_\perp$-dependence of the pair spectra was studied also. Because our pair measurements extended out past $p_\perp$ of 2 GeV/c, and virtually all single μ's of $p_\perp \leq 2$ GeV were found to come from pairs withint this $p_\perp$ region, we therefore do not depend on extrapolations.

Figure 7a shows our calculated single-μ yield at $x_F = 0$ as a function of $p_\perp$, indicating separately the contributions from vector mesons and the continuum. The π^- cross-section on beryllium,[16] multiplied by 10^{-4}, is shown for reference. It is seen that vector mesons cannot account for the observed single μ yield (which follows $\pi^- \times 10^{-4}$), especially as $p_\perp$ goes below 2 GeV/c. However, with the addition of the continuum, quite good agreement

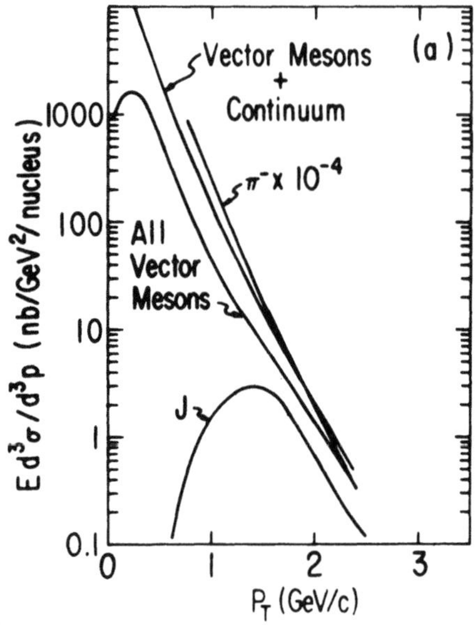

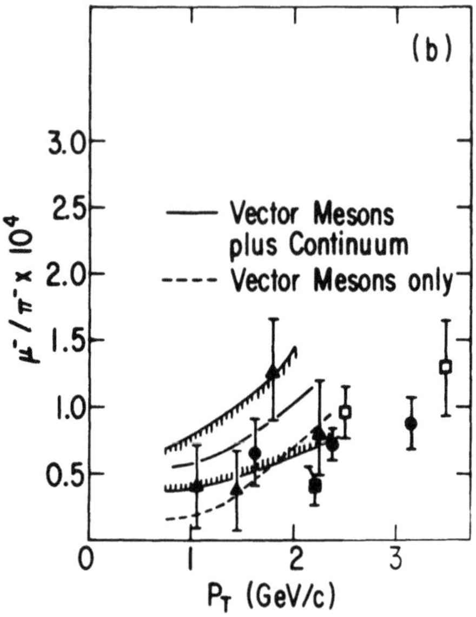

Fig. 7. a) Invariant cross-sections vs. $p_\perp$ for single μ's produced at $x_F = 0$ by the μ-pairs measured in 150 GeV p-Be collisions. The π^- cross-section is also shown, scaled by 10^{-4}. b) Comparison of existing single μ data with the μ^-/π^- ratio calculated from our pair measurements. (■ Denotes Ref. 5; ● and □, Ref. 6; ▲, Bintinger et al., Ref. 7).

is seen. Figure 7b shows a comparison of our calculated single μ's from pairs with other experimental measurements of μ^+/π^+ and μ^-/π^-. The agreement is good, indicating that pairs can account for the observed single μ yields. The hatch marks show the uncertainties in the calculation of μ^-/π^-, which are mainly due to the extrapolation of our pair cross-sections to $x_F = 0$. (In our 225 GeV data, the extrapolation, and hence the systematic uncertainty, will be significantly smaller.) The dashed curve in Fig. 7b shows the single-μ contribution from vector mesons alone. Single-μ yields for other values of x_F are shown in Fig. 8 as a function of $p_\perp$.

There have also been several measurements of single μ's in the forward direction.[17,18] In these experiments, all μ's are accepted which have $p_\perp \lesssim 600$ MeV/c because of multiple scattering. To compare with them, we have summed, at a given value of x_F, all single μ's with $p_\perp < 600$ MeV/c coming from pairs. The results are shown in Fig. 9. In particular, Fig. 9(b) shows once again that pairs can account for most of the single μ's and that for small values of x_F of the single μ, the continuum becomes the dominant source. An experiment by Kasha et al.[17] has also found that single

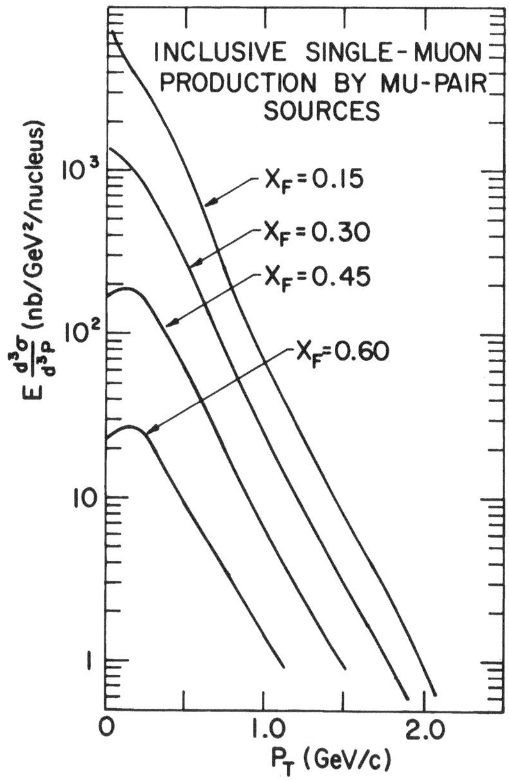

Fig. 8. Cross-section for inclusive single-μ production from all sources, as a function of $p_\perp$ for fixed x_F.

μ's in the forward direction arise mainly from pairs, but proposes a strong source of μ pairs (in addition to ρ^0, ω decays) with a mass of about 900 MeV. We definitely do not see such a source.

The basic conclusion from our measurements is, then, that μ-pairs can account for the yield of single μ's for $p_\perp \lesssim 2$ GeV at $x_F = 0$, and also in the forward direction. Particularly in the forward direction, the continuum pairs are a significant source of single μ's. It should be mentioned that there are no single-μ measurements, because of experimental difficulties, in the kinematic region ($x \simeq 0$, $p_\perp \lesssim 1$ GeV/c) where the largest contribution from decays of charmed particles is expected. Therefore it is still possible that a substantial contribution from the decays $D^0 \to K\mu\nu$ exists.[19]

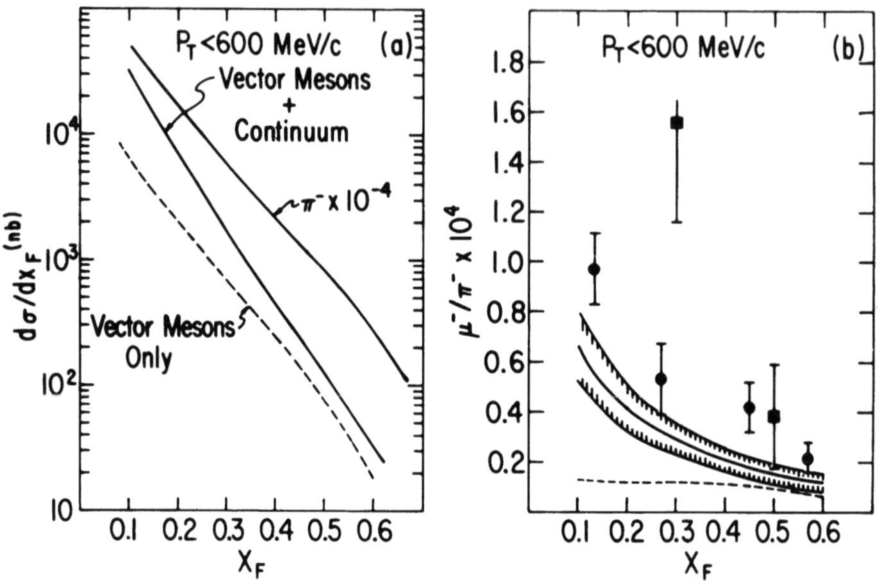

Fig. 9. a) Single-μ differential cross-section vs x_F in the forward direction (integrated over all $p_\perp$ < 600 MeV/c). b) Comparison of the μ/π ratio measured in Ref. 17 (●) and Ref. 18 (■) with the yield calculated from μ pairs. The solid line and hatch marks show the yield from μ-pairs of all sources. The dashed line is the yield from vector mesons only.

PRELIMINARY RESULTS: μ-PAIR PRODUCTION AT 225 GeV/c

At the higher beam momentum of 225 GeV we have a usefully large acceptance for all pairs produced in the forward hemisphere, and in our second run, which ended this April, we have collected about 20 times as much data as in the 150 GeV sample just described. In addition, we were able to use both positive and negative beams, which allows us to compare π^+ and π^- produced pairs. Except for a one-week run with a tin target to measure the A-dependence, we used exclusively a carbon target to exploit its isoscalar nature in interpreting the comparison of pairs produced by π^+ and π^- mesons. In all the results I shall now present, only the MWPC's downstream of the hadron absorber have been used in the analysis. Therefore I shall discuss only those pairs with masses > 1.5 GeV, where as explained above the upstream MWPC's are not needed. The complete analysis is in progress and will be reported within the next few months.

The general features of the high-mass μ-pair spectra are

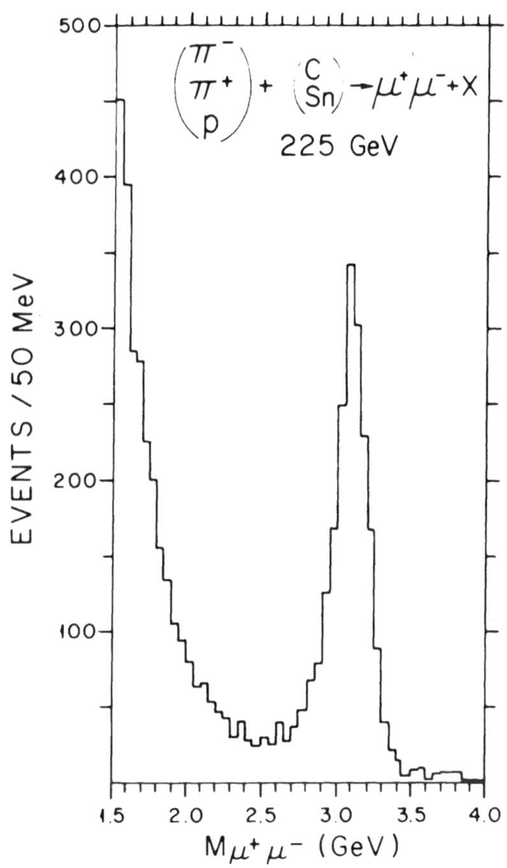

Fig. 10. Mass Spectrum of all $\pi^{\pm}$ and p induced events at 225 GeV, uncorrected for acceptance.

apparent from Fig. 10, in which all the data analyzed so far, about 85% of our exposure, are shown. We observe a sharply falling continuum out to about 2.7 GeV, where the large peak from the J/ψ sets in. The width of this peak shows the mass resolution of the spectrometer. A shoulder is seen around 3.7 GeV, the mass of the ψ'. In what follows, we shall discuss the properties of these events in detail, comparing and contrasting the resonant and non-resonant μ-pairs. Finally, the results of a search for events having 3 or more μ's will be given.

PRODUCTION OF THE J/ψ AND ψ'

The sample of J/ψ events obtained in our 225 - GeV run is the most comprehensive of any experiment to date. The more than 2000 events give adequate statistical power to determine precisely the

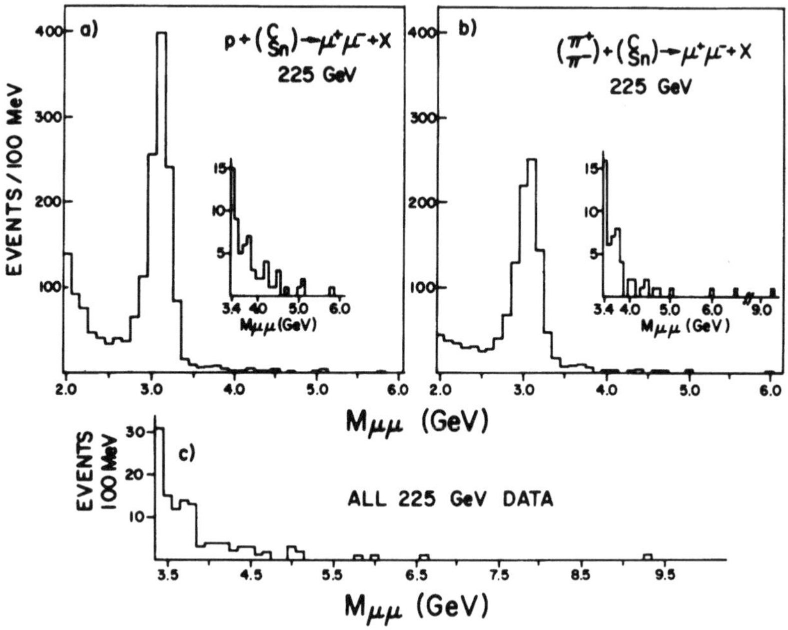

Fig. 11. Mass distributions uncorrected for efficiency. a) p-induced events, b) π^+ and π^- events combined, c) All events above 3.35 GeV. The inserts in a) and b) magnify the ψ' region.

dependence upon kinematical variables of the differential cross section for J/ψ production, over the wide range of spectrometer acceptance. The use of π^+, π^-, and proton beams gives additional insight into the production dynamics. The raw data for $M_{\mu\mu} > 2$ GeV are shown in Fig. 11. Part a) shows all p-induced events, part b) all $\pi^{\pm}$-induced events. In the inserts the scale is expanded for $M_{\mu\mu} \geq 3.35$ GeV, and in part c) of the figure all events from $\pi^{\pm}$ and p with masses $M_{\mu\mu} > 3.35$ have been combined. An apparent enhancement appears in all cases at M = 3.7 GeV, which we interpret as ψ' production. Although there are only a small number of ψ' candidates, they can be used effectively to get the approximate level of ψ' production relative to that of the J/ψ. To get as large a sample as possible we have combined carbon and tin data, assuming that J/ψ and ψ' production have the same dependence on A. We have performed a very simple background subtraction assuming a straight-line continuum, obtaining 6 π-induced events, and 4 p-induced events. Using the same x_F- and $p_\perp$-dependence for ψ' as for J/ψ we calculate the average detection efficiency and estimate roughly the ratio of ψ' to J/ψ production, finding:

$$\text{Proton-Induced:} \quad \frac{B\sigma(\psi')}{B\sigma(J/\psi)} \approx 3 \times 10^{-3}$$

$$\text{Pion-Induced:} \quad \frac{B\sigma(\psi')}{B\sigma(J/\psi)} \approx 7 \times 10^{-3}.$$

These values are lower than those reported by Snyder et al.,[20] who quoted, at $y = 0$ (or $x_F = 0$), for protons on beryllium, at 400 GeV, $[B\cdot d\sigma(\psi')/dy]/[B\cdot d\sigma(J/\psi)/dy] \approx 0.02$. This suggests that the ψ' cross section is increasing faster with energy than that of J/ψ in the range $20 < \sqrt{s} < 27$ GeV, although more work is needed to settle this issue.

Figure 11(c) shows that only a few events were observed in the mass range from 4 to 10 GeV, where the average detection efficiency was $\gtrsim 12\%$. There is thus no evidence for resonances to the level of this exposure, where each event corresponds to $\sim 10^{-35}$ cm^2/nucleon.

To obtain differential cross sections for J/ψ production, the events in the mass range $2.7 < M_{\mu\mu} < 3.5$ GeV were binned in x_F, $p_\perp$, and $\cos\Theta^*$, and then corrected for the spectrometer acceptance. This efficiency was calculated by a Monte Carlo simulation, which included effects of energy loss, multiple scattering and bremsstrahlung, as well as the transport of the particles through the spectrometer to the final hodoscope. To obtain absolute cross-sections we divided by the effective luminosity, corrected for absorption of the beam in the somewhat thick targets. We collected approximately half of the positive-beam carbon data with a 7.5 cm thick target, the other half with a 12.5 cm target. No evidence, either in absolute rate or in energy distribution of the events, was seen of secondary production, so we have combined the data from the two targets in what follows.

The differential cross-sections are shown in Fig. 12 as a function of x_F, in Fig. 13 of $p_\perp^2$, and in Fig. 14, of $\cos\Theta^*$. The cross-sections are expressed as nanobarns/nucleon, using a linear dependence upon the atomic mass number A, as observed by Binkley et al.[12] in neutron-nucleus interactions. (The result of our own measurement of the A-dependence of total cross-sections is given below.)

The x_F-dependence of the cross-section is, qualitatively, independent of target type, and much broader for π-induced events than for those induced by protons. The data have been fitted to the form $\frac{d\sigma}{dx_F} = A(1-x)^b$, as suggested by parton models; the results of

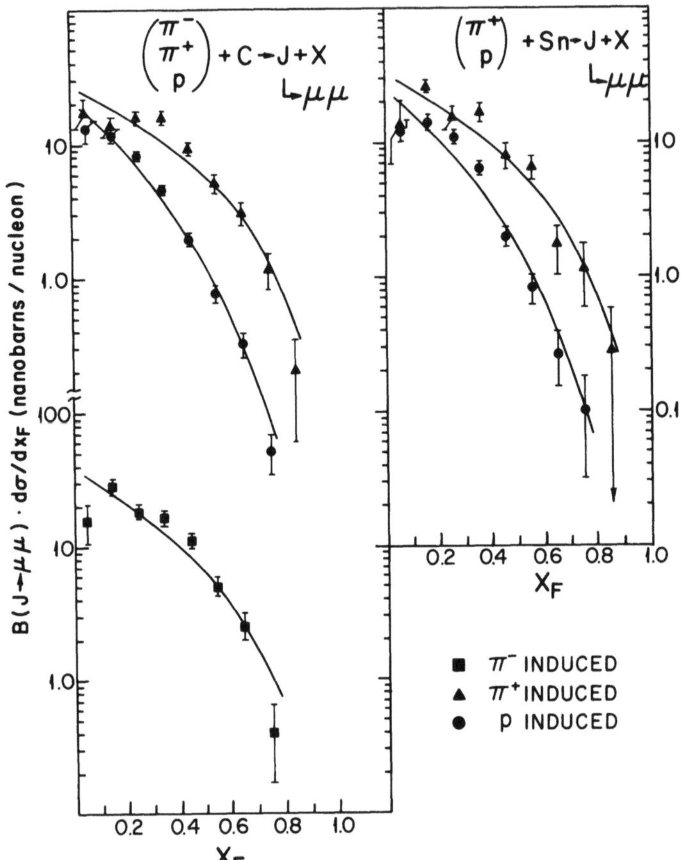

Fig. 12. Differential cross-sections $B(J\to\mu\mu)\cdot d\sigma/dx_F$ for J production. Fits have been made to the form $(1-x)^b$.

the fitting procedure are given in Table IIIa. The fit to the π^--C data is not good, being dominated by one high bin, but aside from that, the fits give a reasonable description. The values for π^+ and p induced events are in good agreement with our earlier results at 150 GeV/c.

Next, the distributions in $p_\perp^2$ were fitted to linear and quadratic forms, respectively:

$$B(J/\psi\to\mu\mu)\cdot\frac{d\sigma}{dp_\perp^2} = Ce^{-dp_\perp} \text{ and } B(J/\psi\to\mu\mu)\cdot\frac{d\sigma}{dp_\perp^2} = Fe^{-gp_\perp^2} .$$

As seen from the table, either parametrization is adequate, and there are no striking differences among the various reactions.

In Table IIIc are given the results of fitting the relative distributions in $\cos\Theta^*$ of Fig. 14 to the form

Table III - Properties of the Differential Cross-section for J/ψ Production at 225 GeV. A linear dependence on A has been used to obtain cross-sections per Nucleon.

a) Dependence on x_F: $B \cdot \frac{d\sigma}{dx_F} = A(1-x)^b$ nanobarns/nucleon.

Beam	Target	A	b	χ^2/dof
p	C	22.3±1.8	3.98±0.20	3.6
π^+	C	27.7±3.3	2.62±0.22	3.5
π^-	C	39.1±5.0	2.15±0.16	5.2
p	Sn	25.8±3.4	3.94±0.30	4.1
π^+	Sn	32.2±6.7	2.33±0.32	1.7

b) Dependence on $p_\perp$: (units of $\frac{d\sigma}{dp_\perp^2}$ are nanobarns/nucleon/(GeV/c)^2)

		LINEAR $B \cdot d\sigma/dp_\perp^2 = Ce^{-dp_\perp}$			QUADRATIC $B \cdot d\sigma/dp_\perp^2 = Fe^{-gp_\perp^2}$		
Beam	Target	C	d	χ^2/dof	F	g	χ^2/dof
p	C	9.1±0.9	1.97±.08	1.8	3.5±0.3	0.80±0.06	1.2
π^+	C	20.9±3.3	2.03±.15	.5	7.9±1.2	0.88±0.12	1.6
π^-	C	13.9±2.2	1.58±.13	1.5	6.2±0.8	0.59±0.06	1.0
p	Sn	9.8±1.5	1.86±.14	1.3	4.1±0.6	0.78±0.09	0.95
π^+	Sn	13.2±4.5	1.56±.25	0.95	6.2±1.4	0.64±0.14	1.0

c) Dependence on $\cos\theta^*_{\mu\mu}$: we have fitted the form $d\sigma/d(\cos\theta^*) \propto (1+\alpha\cos^2\theta^*)$

Beam	Target	α	χ^2/dof
p	C	-0.068±.24	3.14
π^+	C	0.181±.37	0.64
π^-	C	0.36 ±.44	1.3
p	Sn	-0.947±.31	2.0
π^+	Sn	-0.35 ±.45	1.45
p	C+Sn	-0.267±0.19	2.2
π^+,π^-	C+Sn	0.26 ±.25	1.34

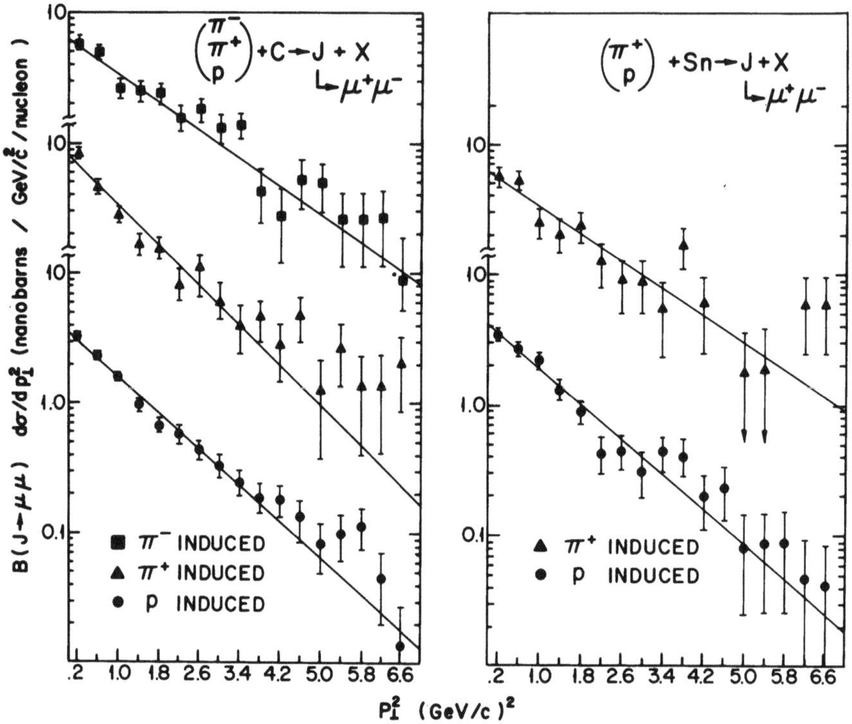

Fig. 13. Differential cross-sections $B(J\to\mu\mu)\cdot d\sigma/dp_\perp^2$ for J production as a function of $p_\perp^2$. The lines shown are fits to the form $Fe^{-gp_\perp^2}$.

$$\frac{d\sigma}{d(\cos\Theta^*)} = A(1 + \alpha\cos^2\Theta^*)$$

In no case is there compelling evidence of any significant polarization of the J/ψ.

To find more precise values of α, we have therefore combined all proton data (Fig. 14f) and all $\pi^\pm$ data (Fig. 14g), obtaining $\alpha_p = -0.26 \pm 0.19$ and $\alpha_\pi = 0.26 \pm .25$, respectively. Further study is required to see if the small polarization in the proton events is significant.

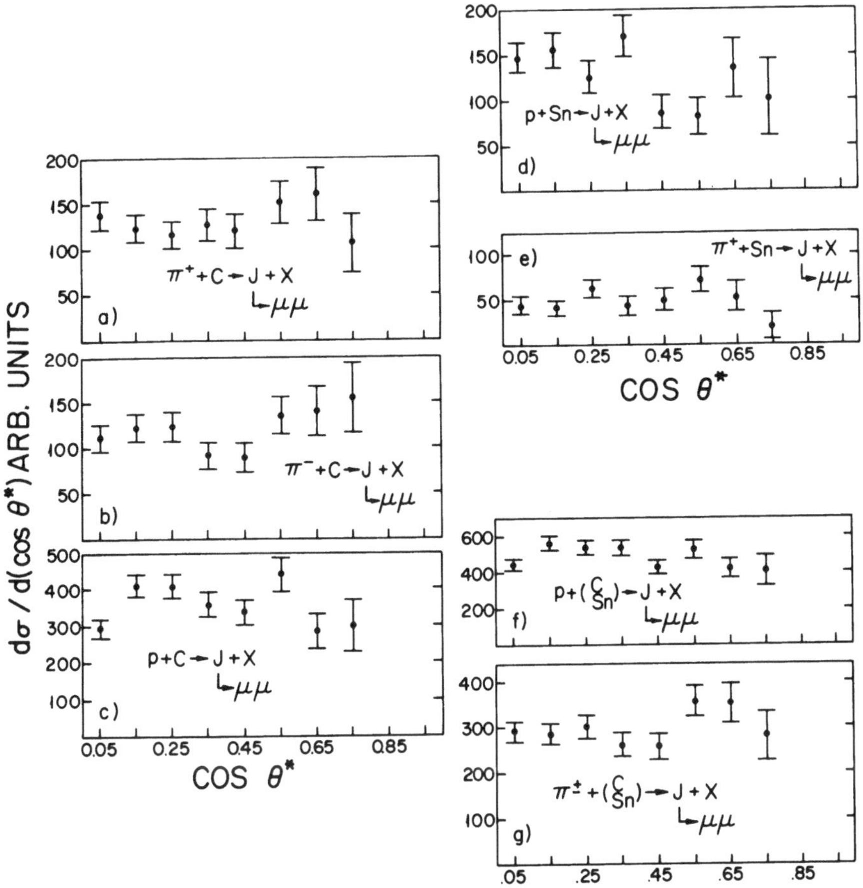

Fig. 14. Relative differential cross-sections $d\sigma/(\cos\Theta^*)$ for J/ψ production.

Values of the total cross-sections (times branching ratio) for J/ψ production are given in Table IV for the range $0 \leq x_F \leq 1$. The errors have been inflated somewhat arbitrarily because of the preliminary nature of the data, and should be about twice as small in the final analysis. To obtain the cross-section per nucleon, one must know the dependence on atomic mass number A, which is found by comparing the actually measured cross-sections per nucleus for carbon and tin. Assuming the form $\sigma(A) \propto A^{\alpha}$ we find $\sigma_{\pi} \propto A^{1.08 \pm 0.10}$ $\sigma_{p} \propto A^{1.03 \pm 0.15}$. This is consistent with a linear dependence, which we have therefore used in obtaining Table IV.

Table IV

Total Cross-Sections for J/ψ Production, 225 GeV Beam

Beam	Target	No. of Events	$\sigma(x_F>0)$ nb/nucleon
p	C	868	4.13±0.37
π^+	C	390	8.1 ±1.2
π^-	C	350	9.8 ±1.4
p	Sn	331	4.85±0.7
π^+	Sn	150	9.50±1.4

Note: A linear dependence on atomic mass number has been used.

The dependence of the cross section for J/ψ production on beam energy is quite interesting, as seen in Fig. 15, where $\left.\frac{d\sigma}{dx_F}\right|_{x_F=0}$ and $\left.\frac{d\sigma}{dy}\right|_{y=0}$ for proton-induced events are plotted versus $\sqrt{s} \simeq \sqrt{2M_p E_{Lab}}$. It would be better to plot the total cross-section σ vs $\sqrt{s}$, but most experiments[21] so far have only been sensitive near $x_F = 0$, where there is the simple relationship between x_F and the center-of-mass rapidity y: $d\sigma/dy\Big|_0 \simeq (2M_J/\sqrt{s})(d\sigma/dx_F)\Big|_0$. One sees a large increase from threshold up to $\sqrt{s} = 16.8$ GeV ($E_{Lab} = 150$ GeV) and then a virtual constancy of $\frac{d\sigma}{dy}$ for $\sqrt{s} > 20$ GeV. One infers from this a similar behaviour of the total cross section.

The results of Table IV provide a test of charge symmetry in inclusive J/ψ production by π^+ and π^-, and also a measure of the relative effectiveness of π and p in producing J/ψ's. First, we find that the ratio of the π^+/π^- cross-sections is consistent with unity:

$$\frac{\sigma^+(x>0)}{\sigma^-(x>0)} = 0.83 \pm .16.$$

In our final analysis, the error should be about half that given here, because in principle there are only very small systematic errors in this comparison. Also, as will be discussed below, the

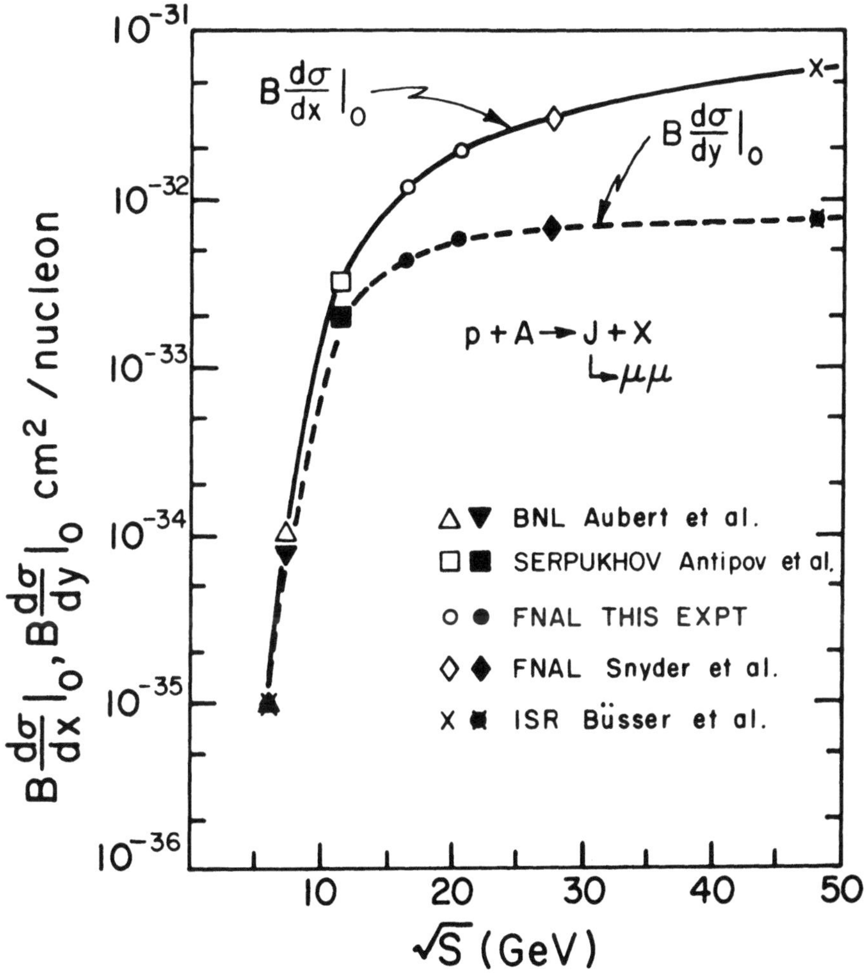

Fig. 15. Energy dependence of J production cross-sections at $x_F = 0$. The open points show $(d\sigma/dx_F)_0$, the solid points show $(d\sigma/dy)_0$, where y is the center-of-mass rapidity. For experiments using nuclear targets, a linear A dependence has been used to obtain cross-sections/nucleon.

continuum is smaller for π^+ events than for π^-, an effect which will push the above ratio closer to one. The analysis of this correction is not as yet complete. Secondly, the proton-induced events: averaging π^+ and π^- results, we have, for carbon,

$$\frac{\sigma_p(x>0)}{\sigma_\pi(x>0)} = \frac{4.13 \pm .37}{8.97 \pm .92} = 0.46 \pm .063.$$

Again, this error will be reduced somewhat in the final analysis.

To conclude the discussion of J/ψ production by hadrons, I summarize the most important qualitative features which have emerged:

A dependence	-	linear for both π^+ and p.
x_F dependence	-	independent of target material, much broader for π beams than proton beam.
$p_\perp$ dependence	-	independent of both beam and target material. Much broader than that of lower mass vector mesons.
Beam energy dependence	-	$d\sigma/dy\big|_{y=0}$ increases sharply from threshold, becoming virtually constant for $\sqrt{s} \gtrsim 20$ GeV. Dependence of total cross section has not been measured, except at Fermilab energy, and must be inferred.
Cross-section ratios	-	For x>0, π^+/π^- consistent with unity, π's give twice as large a cross section as protons.

STUDY OF NON-RESONANT μ-PAIRS

It is very important to learn about the process by which non-resonant, or continuum μ-pairs are produced. For example, the parton models make quantitative predictions for the cross sections, but say little about the transverse momenta expected for the pairs. So far, all previous experiments have used neutron or proton beams, so all the antiquarks are provided by the sea, leading to rather model-dependent predictions for the features of the resulting spectrum. This is the first experiment to use π^+ and π^- beams on an isoscalar target, a particularly attractive experimental situation, in that valence antiquarks are present in the beam. Annihilations of fractionally charged valence quarks in the target with the antiquark in the projectile should then show a charge assymetry

$$\sigma(\pi^+C)/\sigma(\pi^-C) = 1/4,$$

independent of details of the model. Unfortunately this large asymmetry is diluted by the presence of the isospin-symmetric quark-pairs in the sea. Using recent measurements of structure functions, Altarelli, Brandt, and Cabibbo,[22] and Farrar[23] have estimated the effect of the sea, finding the charge ratio to be close to 1 at low pair masses, and then to decrease slowly to the value 1/4 as M^2/s increases. For example, at $M_{\mu\mu} \simeq 2.5$ GeV the ratio is predicted to be ~ 0.6.

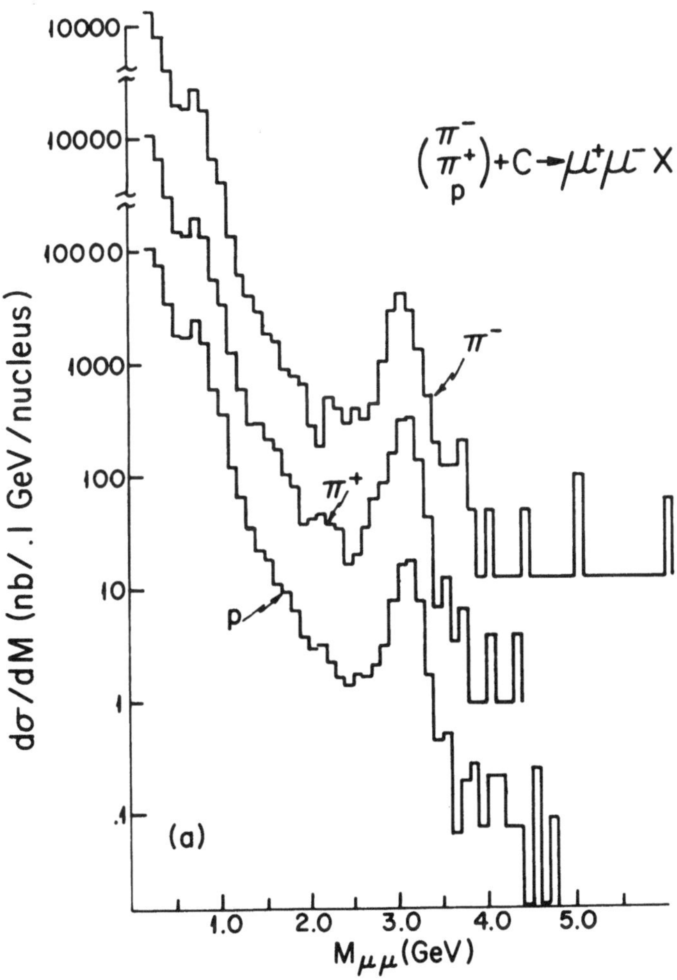

Fig. 16. Differential cross-sections $d\sigma/dM_{\mu\mu}$ versus $M_{\mu\mu}$. The decay background has been subtracted.

The mass spectra are shown on a logarithmic scale in Fig. 16. Decay background, as estimated from like-sign ($\mu^+\mu^+$ or $\mu^-\mu^-$) pairs has been subtracted. In four mass intervals below the J/ψ and two at higher masses, we now compare the production of dimuons by π^+ and p and by π^+ and π^- on carbon. Figures 17a and 17b show these cross-section ratios and equivalent ratios for the J/ψ mass region. Of particular interest is the π^+/π^- ratio, which appears to change from approximately unity at low mass to about .25 at the highest masses, except for the J/ψ mass region, where it returns, within errors, to unity. Within the limited statistics, this behaviour is in agreement with the prediction of Ref. 22.

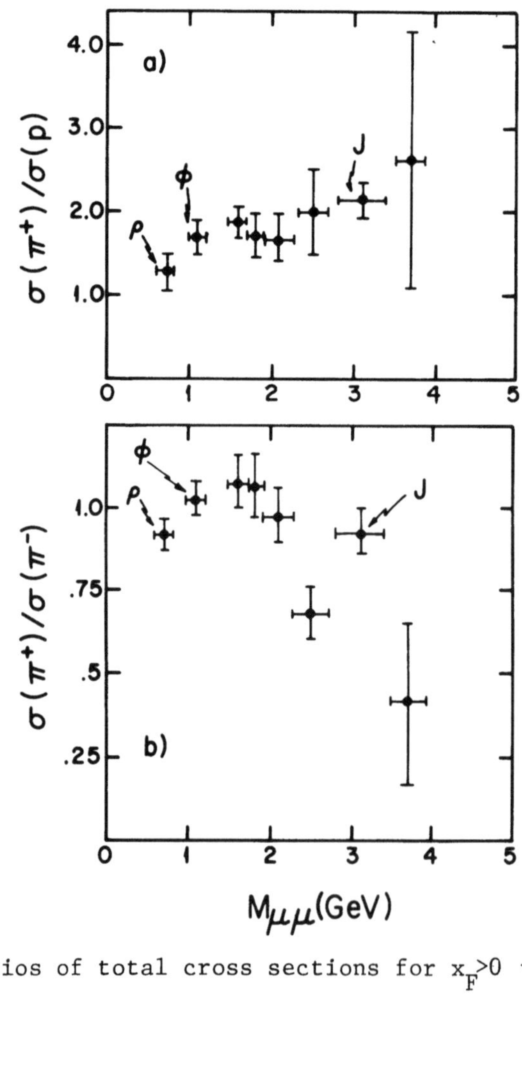

Fig. 17. Ratios of total cross sections for $x_F>0$ vs. pair mass.

When π^+ and proton induced pairs are compared, on the other hand (Fig. 17a), the J/ψ and neighboring continuum regions do not show any significant difference. In both Figs. 17a and 17b the ρ_o-ω and ϕ mass regions are also shown, where a charge ratio of 1 is expected from isospin conservation.

Other features of the data have been studied in a preliminary manner by selecting three mass regions having significant numbers of events, in a region relatively free of resonance signal. The regions are: 1.5-1.9 GeV/c^2, 1.9-2.3 GeV/c^2, and 2.3-2.7 GeV/c^2.

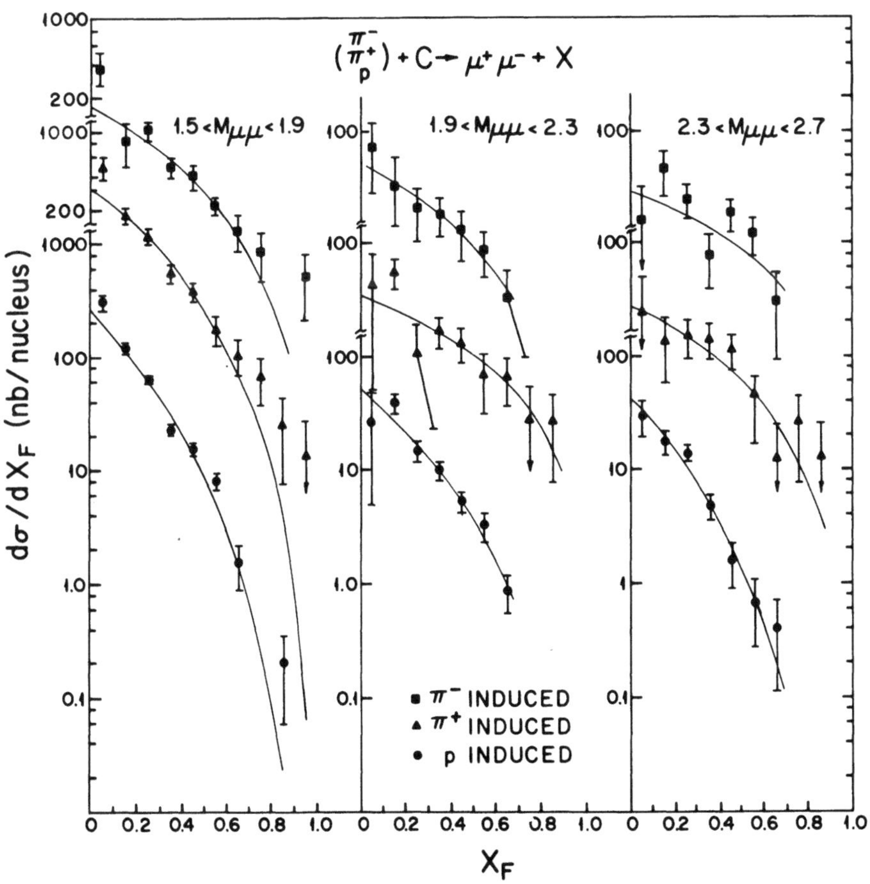

Fig. 18. Dependence of pair cross-sections on x_F for various intervals of pair mass.

Figure 18 shows the dependence of the cross-section in these bins on the variable $x_F \simeq \frac{2p_{//}^*}{\sqrt{s}}$, for p, π^+, and π^- incident on carbon. The basic feature of all these distributions, that the π-induced events have a slower fall-off with increasing x_F than those of p-induced events, is very similar to what we have observed in production of the resonances ρ, ω, ϕ, and J/ψ. The data have been fitted to the form $B \cdot d\sigma/dx_F = A(1 - x_F)^b$; the best fits, of somewhat limited statistical significance, are summarized in Table Va.

Figure 19 illustrates the dependence of the differential cross-sections $d\sigma/dp_\perp^2$ upon the transverse momentum $p_\perp$ of the dimuon. Exponentials both linear and quadratic in $p_\perp^2$ have been fit to these spectra, the results appearing in Table Vb. The quadratic fits are shown in the figure, although either hypothesis gives an adequate fit.

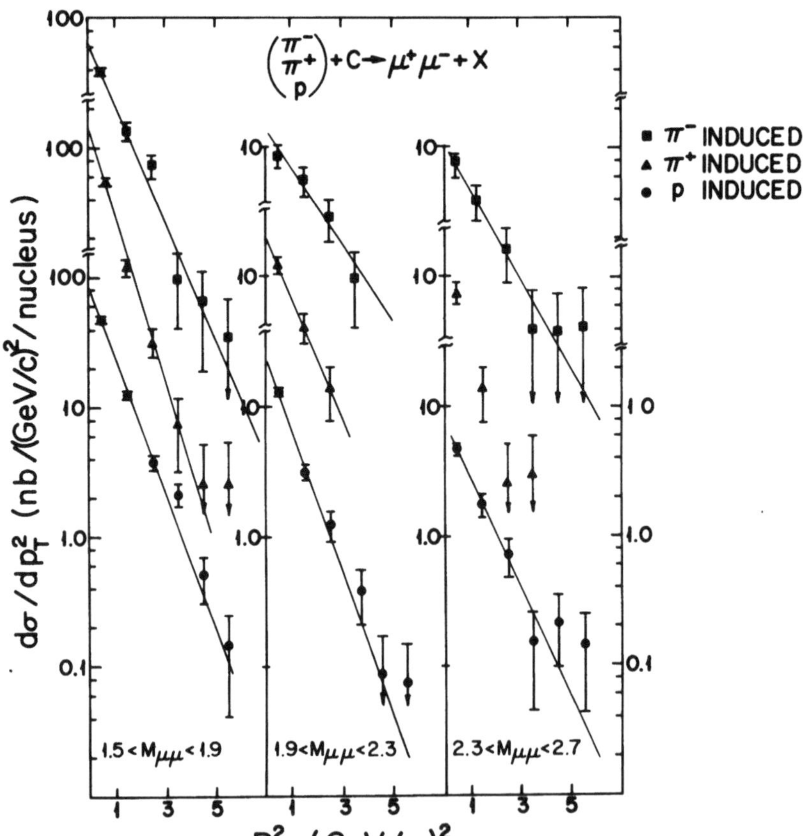

Fig. 19. Differential cross-section vs. $p_\perp^2$ for various intervals of pair mass.

Table V - Fits of Differential Cross-sections of Continuum Dimuons.

$$\begin{pmatrix} p \\ \pi+ \\ \pi- \end{pmatrix} + C \rightarrow \mu^+\mu^- + X$$

Note: All cross-sections are per Carbon Nucleus.

a) Dependence on x_F: $B \cdot \frac{d\sigma}{dx_F} = A(1-x)^b$ nanobarns/nucleon

Beam	Mass Region	A	b	χ^2/dof
p	1.5-1.9	260±32	4.9±.3	3.4
	1.9-2.3	51±21	3.8±.7	.82
	2.3-2.7	45±11	5.2±.6	.86
π^+	1.5-1.9	321±92	3.5±.6	1.4
	1.9 2.3	34±18	1.7±.8	.88
	2.3-2.7	28±29	2.3±1.6	.63
π^-	1.5-1.9	170±78	2.5±.7	1.5
	1.9-2.3	52±23	2.4±.8	.16
	2.3 2.7	29±11	1.7±.7	1.7

b) Dependence on p_T: (units of $\frac{Bd\sigma}{dp_T{}^2}$ are nanobarns/(GeV/c^2)/nucleus)

		Linear $B\cdot d\sigma/dp_T{}^2 = Ce^{-dp_T}$			Quadratic $B\cdot d\sigma/dp_T^2 = Fe^{-gp_T^2}$		
Beam	Mass Region	C	d	χ^2/dof	F	g	χ^2/dof
p	1.5-1.9	381±30	2.9±.1	3.9	84±4	1.2±.05	1.9
	1.9-2.3	110±33	3±.3	.75	24±3	1.3±.1	.55
	2.3-2.7	25±14	2.4±.5	.85	7.5±1.2	.98±.12	.75
π^+	1.5-1.9	586±106	3.3±.2	.88	110±11	1.4±.1	.3
	1.9-2.3	66±30	2.4±.5	.3	21±6	1.1±.25	.001
	2.3-2.7	89±60	3.5±.8	.44	17±6	1.7±.4	.5
π^-	1.5-1.9	269±60	2.7±.2	2.7	67±7	1.1±.1	1.1
	1.9-2.3	28±12	1.6±.4	1.2	13±3	.65±.15	.5
	2.3-2.7	32±18	2.0±.5	.5	11±3	.81±.18	.3

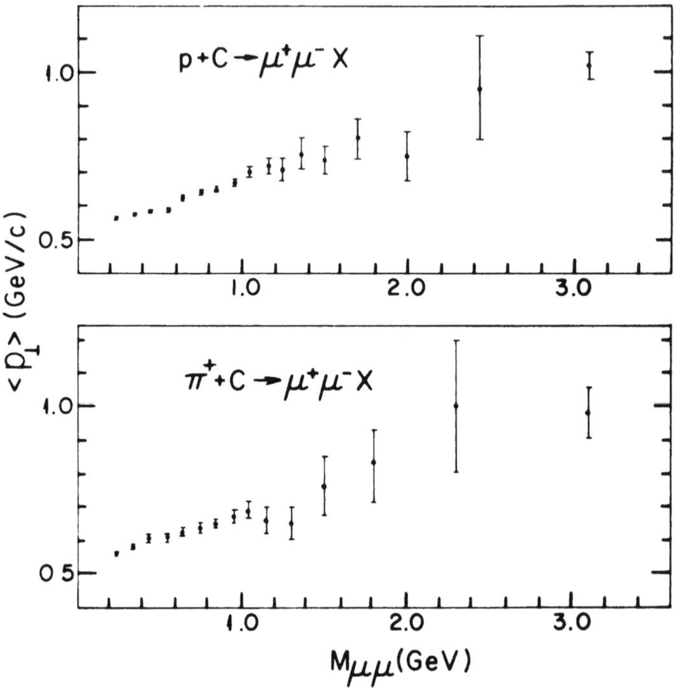

Fig. 20. Average $p_{\perp}$ of muon pairs vs. pair mass, uncorrected for spectrometer efficiency.

Figure 20 shows an interesting feature of μ-pair production. We have plotted the average transverse momentum $\langle p_{\perp} \rangle$ as a function of the dimuon mass, for π^{+} and p-induced events. (The analysis of the π^{-} sample is not yet completed.) No discontinuity at all is apparent between $\langle p_{\perp} \rangle$ of resonant and non-resonant dimuons. Instead, one sees a slow, smooth increase of $\langle p_{\perp} \rangle$ with increasing dimuon mass. As described in Professor Cronin's lecture, within rather large uncertainties this smooth increase continues out to μ-pair masses of 10 GeV.

We have also studied the differential cross-section as a function of $\cos\Theta^{*}$; a hint of polarization of the form $1 + \cos^{2}\Theta^{*}$ is suggested by the data, (see Fig. 21) but the statistics are insufficient for a firm conclusion to be reached. Finally, in Fig. 22 we show the dependence on $M_{\mu\mu}$ on a log log scale. Good agreement with a power law is seen.

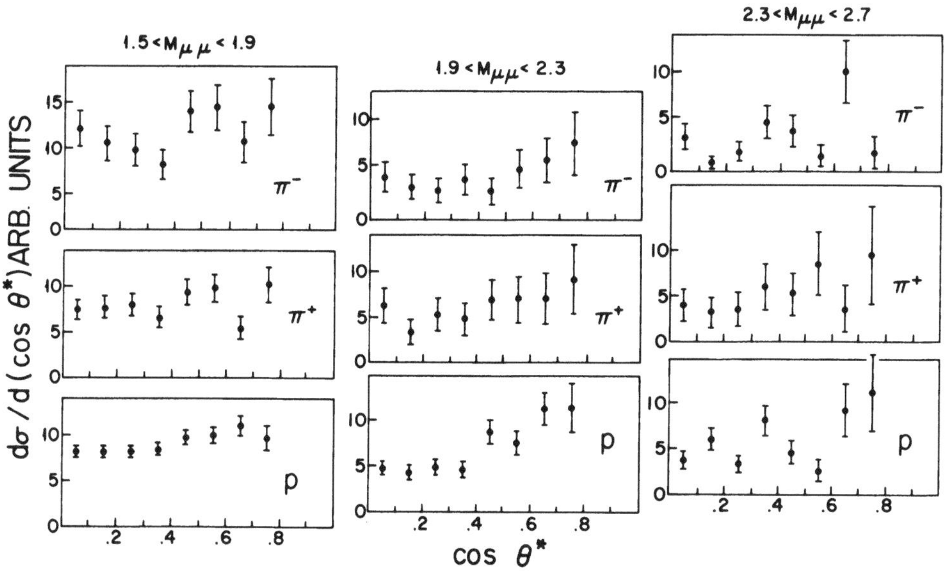

Fig. 21. Decay angular distribution for continuum dimuons. The angle θ^* is the helicity angle of the decay muons in the CM frame.

In an upcoming experiment to run at Fermilab in Spring '77, we shall measure these cross-sections to a level of 10^{-37} cm^2/event, approximately 100 times more sensitively than the present exposure. Because the results of our experiment and others are so far very suggestive of quark-anti-quark annihilation, one should perhaps take seriously another prediction, that the π^-/p cross-section ratio for $\mu\mu$ production rises sharply with increasing M^2/s, reaching a value > 100 for $M_{\mu\mu}$ = 10 GeV. Our new run will test this prediction conclusively.

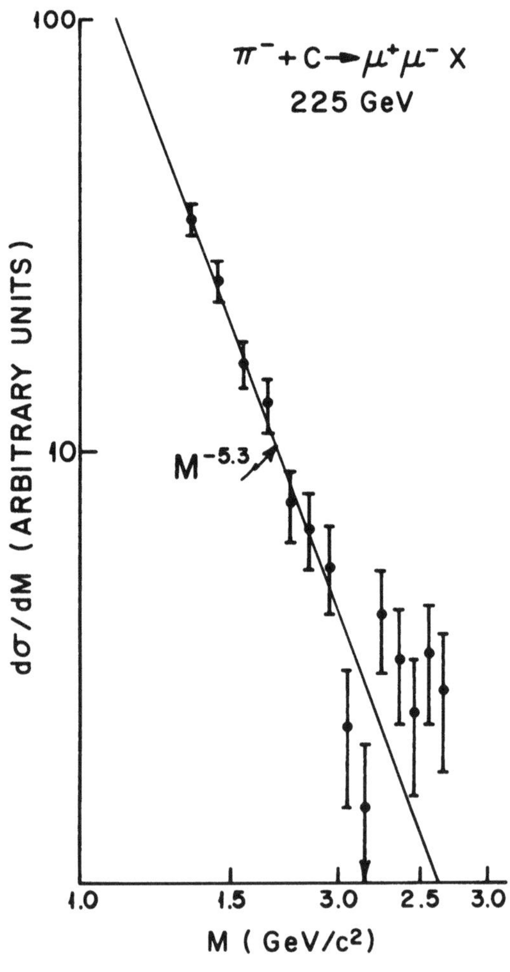

Fig. 22. Differential cross-section $d\sigma/dM_{\mu\mu}$ vs. $M_{\mu\mu}$, plotted on a log-log scale. An approximate agreement with a power law is suggested. All of the region $x_F>0$ is included in the cross-section.

SEARCH FOR MULTIMUON EVENTS

The efficiency of the spectrometer for the detection of an extra muon accompanying a μ-pair is very large. Such extra muons are expected if, for example, the J/ψ is produced in pairs, or in association with charmed mesons, as is required by the Okubo-Zweig-Iizuka rule.[24] We have searched all our data - about 850,000 μ-pair events in all - and 2100 J/ψ events for any having extra muons. After performing cuts on the data to reduce to a negligible level the backgrounds from secondary interactions and muons in the halo around the beam, we were left with 1229 three-muon events, and one event with 4 muons, the latter being a double $\rho \rightarrow \mu\mu$ event. Only 2 of the 2100 J/ψ events had an extra μ. This level of extra muons is completely consistent with the contamination expected from pi decay, namely $\approx 2 \times 10^{-3}$. We thus conclude there is no evidence at this level for extra muons accompanying either high or low mass pairs.

The computation of cross section limits is made simple by the fact that the major losses of single muons in the forward hemisphere are due to the 8 GeV range requirement imposed by the iron muon identifiers, to the geometrical aperture of the proportional chambers, and to a halo-suppressing requirement in the analysis that all muon laboratory momenta must be less than 100 GeV/c. The first eliminates an elliptical segment in the $x_F - p_\perp$ plane with intercepts at Feynman $x_F = .035$ and $p_\perp = 740$ MeV/c. The last eliminates muons with $x_F > .43$. Within these boundaries our acceptance is more than 50%. The two accompanied J/ψ events are shown by ■'s in Fig. 23, which shows that the extra μ's are low $p_\perp$, low x_F, as would be expected from decay muons. The cross section limit for J/ψ events accompanied by an extra muon is $\frac{\sigma(J + \mu)}{\sigma(J)} < 2.5 \times 10^{-3}$ with 90% confidence. This result rules out, for example, pair production of J/ψ's to a limit of 3.5%, within the virtually complete acceptance for detecting at least one of the 2 muons from the second J/ψ. The limits on associated production of charmed particles are of course dependent on the branching ratio assumed for their muonic decays, as well as the model of production. Eventually, we shall calculate limits for the popular models, but for now I simply state that the limit for a charmed particle accompanying a J/ψ is $\lesssim \frac{5 \times 10^{-3}}{BR(\text{meson} \rightarrow \mu)}$, which for the usual assumption of about 0.1 for the branching ratio, says that $\lesssim 5$ % of J/ψ's are produced in association with charm. Similar limits over a somewhat more restricted acceptance have been reported for neutron-nucleus collisions at Fermilab.[25]

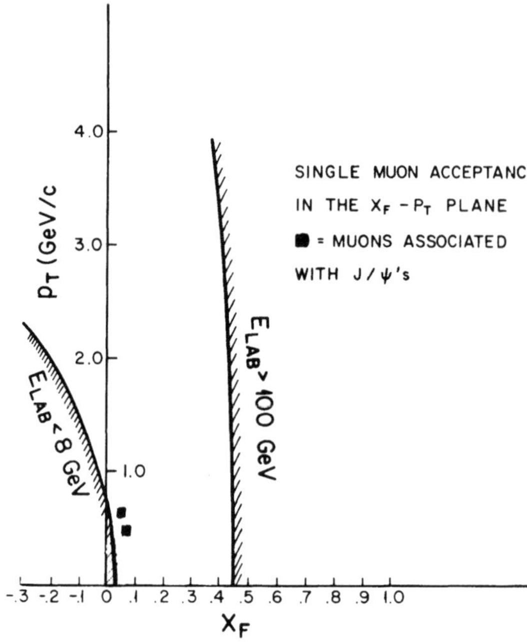

Fig. 23. Sensitive kinematic region for the search for multimuons.

CONCLUSION AND ACKNOWLEDGEMENT

The experimental results on dimuon physics are becoming quite precise. They are generally in agreement with the parton model, but there is a disturbing feature in that the high average transverse momenta are not easily explained. It is also puzzling that the differential cross sections of resonances and continuum are so similar in their dependence on the kinematic variables. Finally, we all wish to know why there is no evidence of muonic decays of charmed particles, either as evidenced in copious yields of single μ's or in extra muons accompanying J/ψ events.

I wish to thank Professor Zichichi for his kind invitation and superb hospitality. It is also a pleasure to acknowledge the staff of Fermilab and particularly the members of the Chicago-Harvard-Oxford Illinois μ-scattering group who have provided the spectrometer for our work.

REFERENCES

* This research was performed at Fermilab and supported in part by the NSF and ERDA.

1 The authors of this work are members of the Chicago-Princeton (II) collaboration: K.J. Anderson, G.G. Henry, J.E. Pilcher, E.I. Rosenberg, J.G. Branson, K.T. McDonald, G.H. Sanders, A.J.S. Smith, and J.J. Thaler. In addition G.E. Hogan and C. Newman participated in the analysis of the 225 GeV data. Preliminary results were reported in K.J. Anderson et al., Phys. Rev. Lett. 36, 237 (1976).

2 J.J. Aubert et al., Phys. Rev. Lett. 33, 1404 (1974); J.F. Augustin et al., op cit, p. 1406.

3 S.D. Drell and T.M. Yan, Phys. Rev. Lett. 25, 316 (1970).

4 G.R. Farrar and S.C. Frautschi, Phys. Rev. Lett. 36, 1017 (1976).

5 V.V. Abramov et al., reported in the Proceedings of the 17th Int. Conf. on High Energy Physics, London, 1974, p. V-53.

6 J.P. Boymond et al., Phys. Rev. Lett. 33, 112 (1974); J.A. Appel et al., Phys. Rev. Lett. 33, 722 (1974).

7 F.W. Büsser et al., Phys. Letters 53B, 212 (1974), D. Bintinger et al., Phys. Rev. Lett. 35, 72 (1975).

8 B. Rossi, High Energy Particles, Prentice Hall, 1956.

9 The ρ_o-width observed in $\rho\to\mu\mu$ decay is narrower than in $\rho\to\pi\pi$ decay because of the photon propagator in the former decay.

10 C. Jarlskog and J. Pilkuhn, Nucl. Phys. B1, 264 (1967); C. Quigg and J.D. Jackson, LBL Report UCRL-18487; C.H. Lai and C. Quigg, FNAL Preprint FN-296 (1976).

11 F.W. Büsser et al., Phys. Letters 55B, 232 (1975).

12 M. Binkley et al., Phys. Rev. Lett. 37, 574 (1976).

13 M. Bourquin and J.M. Gaillard, Phys. Letters, 59B, 191 (1975); J.W. Cronin, in Proceedings of the International School of Subnuclear Physics, Erice, 1975.

14 For example, the effects of the peculiar A-dependence of high $p_\perp$ processes reported by Cronin et al., Phys. Rev. D11, 3105 (1975) are avoided.

15 K.J. Anderson et al., Phys. Rev. Lett. 37, 803 (1976); K.J. Anderson et al., Phys. Rev. Lett. 37, 799 (1976).

16 T. Eichten et al., Nucl. Phys. B44, 333 (1972); W.F. Baker et al., Phys. Letters 51B, 303 (1974).

17 L.B. Leipuner et al., Phys. Rev. Lett. 35, 1613 (1975); H. Kasha et al., Phys. Rev. Lett. 36, 1007 (1976).

18 D. Buchholz et al., Phys. Rev. Lett. 36, 932 (1976).

19 This was emphasized by P.A. Piroué, Division of Particles and Fields Meeting, Upton, L.I., October 1976.

20 H.D. Snyder et al., Phys. Rev. Lett. 36, 1415 (1976).

21 F.W. Büsser et al., Phys. Letters 56B, 482 (1975); Y.M. Antipov et al., Phys. Letters 60B, 309 (1976); also References 2 (Aubert et al.) and 20.

22 G. Altarelli et al., Nucl. Phys. B92, 413 (1975).

23 G.R. Farrar, Nucl. Phys. B77, 426 (1974).

24 A. Okubo, Phys. Letters, 5, 165 (1963); G. Zweig, CERN Report TH-412 (1964); I. Iizuka, Prog. Theor. Phys., Suppl. No. 37-38, 21 (1966); D. Sivers, Phys. Rev. D11, 3253 (1975); R.M. Barnett and D. Silverman, Phys. Rev. D12, 2037 (1975).

25 M. Binkley et al., Phys. Rev. Lett. 37, 578 (1976).

DISCUSSION

CHAIRMAN: Prof. A.J.S. Smith

Scientific Secretaries: Y. Afek and G.J. Tarnopolsky

DISCUSSION

- *ETIM:*

What is the parametrization of the continuum which you added to the vector meson induced μ-pair production to finally compute the inclusive single-μ spectrum?

- *SMITH:*

We divide the continuum into mass regions, each parametrized with similar functional dependence. The results of the fit appear in the table shown.

- *ETIM:*

How sensitive to this particular parametrization is the computed inclusive muon yield?

- *SMITH:*

The main uncertainty comes from the extrapolation we made from the lower measured values at $x \triangleq 0.15$ down to $x \triangleq 0$. Fluctuations arising from the fit parameters are smaller than the ones introduced by the extrapolation.

- *PAULI:*

From these experiments, can you make any statements about heavy lepton production? Also, can any of the dimuon experiments be altered to get more definite statements on the properties or non-existence of heavy leptons in the 1-10 GeV range?

- *SMITH:*

Most dimuon experiments work because all particles but muons are absorbed. If the produced heavy leptons both decay into muons, their contribution would probably be buried under the dimuon yield from other sources. One should really try the $e^{\pm}\mu$ channel.

- *ZICHICHI:*

Heavy lepton production is also suppressed in hadronic reactions since it is not easy to produce a time-like photon of at least 4 GeV.

- *YOON:*

Can you explain why the ψ' bump in the mass spectrum you measured is so small?

- *SMITH:*

There is a ψ' signal, smaller than that observed by Lederman et al. Our mass resolution is just sufficient to see the bump. We then subtract a non-resonant background; the subtraction could be off by a factor of two, but the signal is small. One could ascribe this to the lower centre-of-mass energy of our experiment. Also, Lederman's sample is restricted to $x \triangleq 0$, while we get a wide range of x values. The signal might be diluted there. Comparing the J and ψ' yields, one should note that the branching ratio $B(\psi' \to \mu\mu)$ is only $(0.9 \pm 0.2)\%$.

- *LITTENBERG:*

Is your result inconsistent with the observation of a substantial signal of direct photons at the Intersecting Storage Ring (ISR)? I believe Farrar and collaborators have claimed they could explain all the direct muons on the basis of the amount of direct photons actually observed. Could you comment on this?

- *SMITH:*

I have not studied the photon experiment carefully enough to assess its implication. However, the ISR measurements are at a much higher energy than our experiment. Also, we do see a substantial continuum contribution throughout the resonance region. Perhaps this corresponds to the Farrar contribution. Clearly we shall study this. Anyway, I am sure our result is correct.

PHYSICS WITH THE SINGLE ARM SPECTROMETER AT FERMILAB

D. Cutts

Brown University, Department of Physics

Providence, Rhode Island 02912 USA

INTRODUCTION

I will describe some of the uses of a new spectrometer facility now operational at the Fermi National Accelerator Laboratory. Located in beam M6E of the Meson Lab, the Single Arm Spectrometer (SAS) is fully instrumented for high energy counter/wire chamber experiments. I will talk briefly about the spectrometer to give you a feel for the physics it makes accessible, but mainly I will concentrate on experiments we've done, are doing, or hope to do.

To build the Single Arm Spectrometer Facility, the large group of people indicated below contributed their time, some apparatus, or both. This group formed the collaboration for the initial experiment, a study of elastic scattering completed in

---

Fermilab Single Arm Spectrometer Group: D.S. Ayres, R. Diebold, and G.J. Maclay, Argonne National Laboratory; D. Cutts, R.E. Lanou, L.J. Levinson, and J.T. Massimo, Brown University; J. Litt, CERN and Daresbury Laboratory; R. Meunier, CERN; B. Gittelman, E. Loh, and M. Sogard, Cornell University; A.E. Brenner, J.E. Elias, and G. Mikenberg, Fermi National Accelerator Laboratory; L. Guerriero, P. Lavopa, G. Maggi, C. DeMarzo, F. Posa, G. Selvaggi, P. Spinelli, F. Walner, E.N. Anelli, Istituto di Fisica, INFN, Bari, Italy; D.S. Barton, J. Butler, J. Fines, J.I. Friedman, H.W. Kendall, B. Nelson, L. Rosenson, and R. Verdier, Massachusetts Institute of Technology; B. Gottschalk, Northeastern University; R.L. Anderson, D. Gustavson, K. Rich, D.R. Ritson, and G.A. Weitsch, Stanford Linear Accelerator Center.

1975. Much of these results are already published [1], and I will discuss them only briefly, dwelling on one particular application of the data: an impact parameter analysis of elastic scattering. In this same first experiment we obtained data on inelastic scattering in the diffractive and triple-Regge regions, and I will show some results for the inclusive reactions:

$$a + p \rightarrow a + X$$

and

$$a + p \rightarrow b + X$$

where $a = \pi^{\pm}, K^{\pm}, p^{\pm}$ and $b \neq a$. This inclusive study is being extended by the experiment now running in the Single Arm Spectrometer. A subgroup of the original collaborators, comprised of people from Bari, CERN, FNAL, MIT, and those of us at Brown, have added a detector around the target to measure multiplicity and pseudorapidity distributions in association with a forward particle detected by the spectrometer. I will discuss our plans and progress. Future experiments with SAS include a further inclusive study, by those of us now running, which will use nuclear targets, and an experiment to study the non-diffractive reaction $\pi^+p \rightarrow K^+\Sigma^+$, in the forward direction. (The collaboration for this latter experiment will include people from the original group not involved in the current run.) Finally, I will speak about a proposal which seems uniquely suited to some aspects of SAS, a search for heavy long-lived particles.

THE SINGLE ARM SPECTROMETER FACILITY

The SAS Facility is based on a forward arm spectrometer located at the end of the high momentum (to 200 GeV/c), high resolution secondary beam M6E at FNAL's Meson Lab. The spectrometer sits at 0° to the beam and functions essentially as a fourth stage to M6E, having focussing as well as bend magnets. A total of 8 Cerenkov counters are used in M6E and SAS for complete tagging of incident and scattered particle identifications (π, K, or p). Figure 1a shows the third stage of the beam line, with detectors for measuring the momentum, direction, and type of particles incident on the H_2 target. To choose the central scattering angle we pitch the beam in the vertical plane, as indicated in Figure 1b; the maximum pitch angle at present is $\sim$24 mrad, and about the pitch angle SAS has an angular acceptance $\simeq \pm 2$ mrad (vertically). The ability to run the beam at 0° into SAS proved to be very useful experimentally. Figure 1c illustrates the detectors in the spectrometer. Performance figures for SAS include the following:

---

[1] Fermilab Single Arm Spectrometer Group, Phys. Rev. Lett. 35, 1195 (1975); D.S. Ayres et al., to be published; R.L. Anderson et al., to be published.

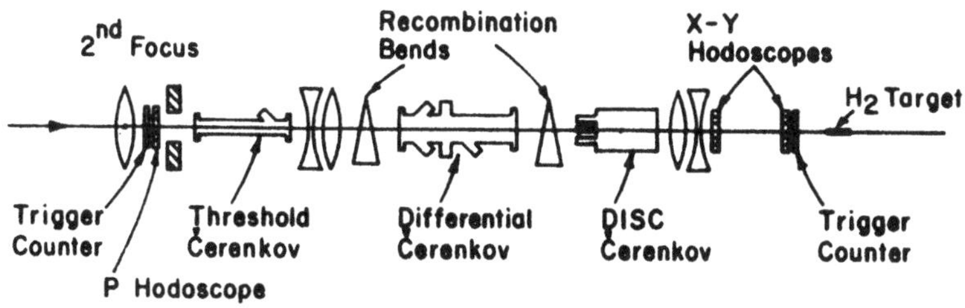

Figure 1a: Diagram of the third stage of beam line M6E.

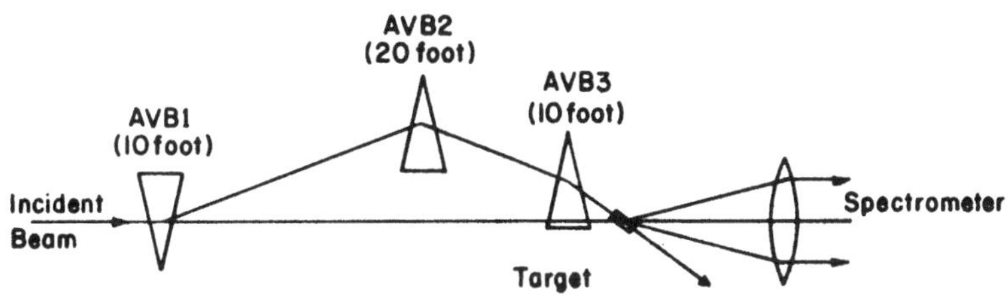

Figure 1b: Diagram of AVB system, determining scattering angle setting.

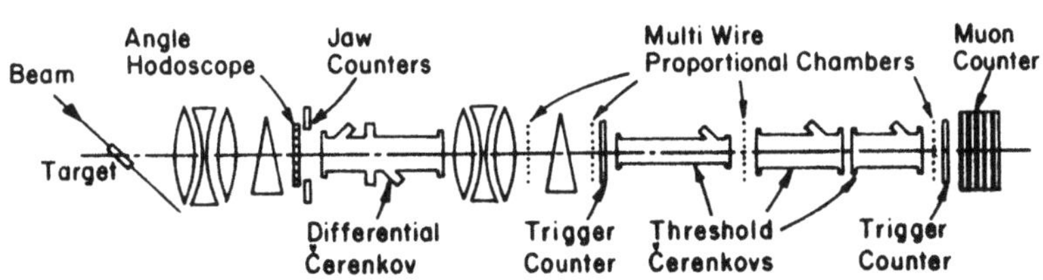

Figure 1c: Diagram of the Single Arm Spectrometer, showing detectors for particle identification, momentum, and angle determination.

full acceptance $\simeq$ 8 μstr., 4.5% $\Delta p/p$

r.m.s. resolution $\simeq$ 0.07% $\Delta p/p$, ~0.1 mrad.

With SAS one measures the momentum, scattering angle, and particle type of a single particle going forward in the lab. Its primary physics application then is to study high energy scattering reactions, involving a forward secondary. The obvious first experiment was elastic scattering: the good momentum resolution allowed a clean separation of elastic peaks--but more practically, the presence of easily observable peaks allowed the Facility to be built and tuned!

Sample results of the elastic scattering experiment are shown in the following pages. Our basic measurement was that of the differential cross section $d\sigma/dt$ for the six elastic reactions ($\pi^{\pm}p$, $K^{\pm}p$, $p^{\pm}p$) in the kinematic region:

incident momenta: 50, 70, 100, 140, 175 GeV/c

(s from 94 to 328 GeV^2)

momentum transfer: $.02 \lesssim |t| \lesssim 0.8\ (GeV/c)^2$

These data are plotted in Figures 2a and 2b. The following figure (3) gives our results for the slopes [2] of the elastic differential cross sections, as compared with measurements at lower energies, and with pp results from the ISR. Note that over an energy range the K^+p and pp cross sections exhibit shrinkage (b increases), $\pi^{\pm}p$ and K^-p are essentially flat, and $\bar{p}p$ antishrinks. We have integrated the differential cross sections to obtain the total elastic cross sections, plotted in Figure 4 together with the total cross sections from Carroll at al. [3], and the inelastic differences. The $\pi^{\pm}p$ elastic cross sections are essentially flat, K^+p rises slightly, and pp, $\bar{p}p$ elastic cross sections are falling. Both the K^+p and pp inelastic cross sections are clearly rising, as are their total cross sections in this energy range.

Elastic scattering is a fundamental hadron-hadron interaction, and the data are important to a number of models and methods of interpreting high energy strong interactions. In this talk I shall describe just one such application, the derivation of a picture of the scattering process in impact parameter space.

---

[2] The cross sections are well fit by the form e^{bt+ct^2}; we plot the effective slope b^{eff} at $t = -0.2\ (GeV/c)^2$.

[3] A.S. Carroll et al., Phys. Rev. Lett. 33, 928 and 932 (1974).

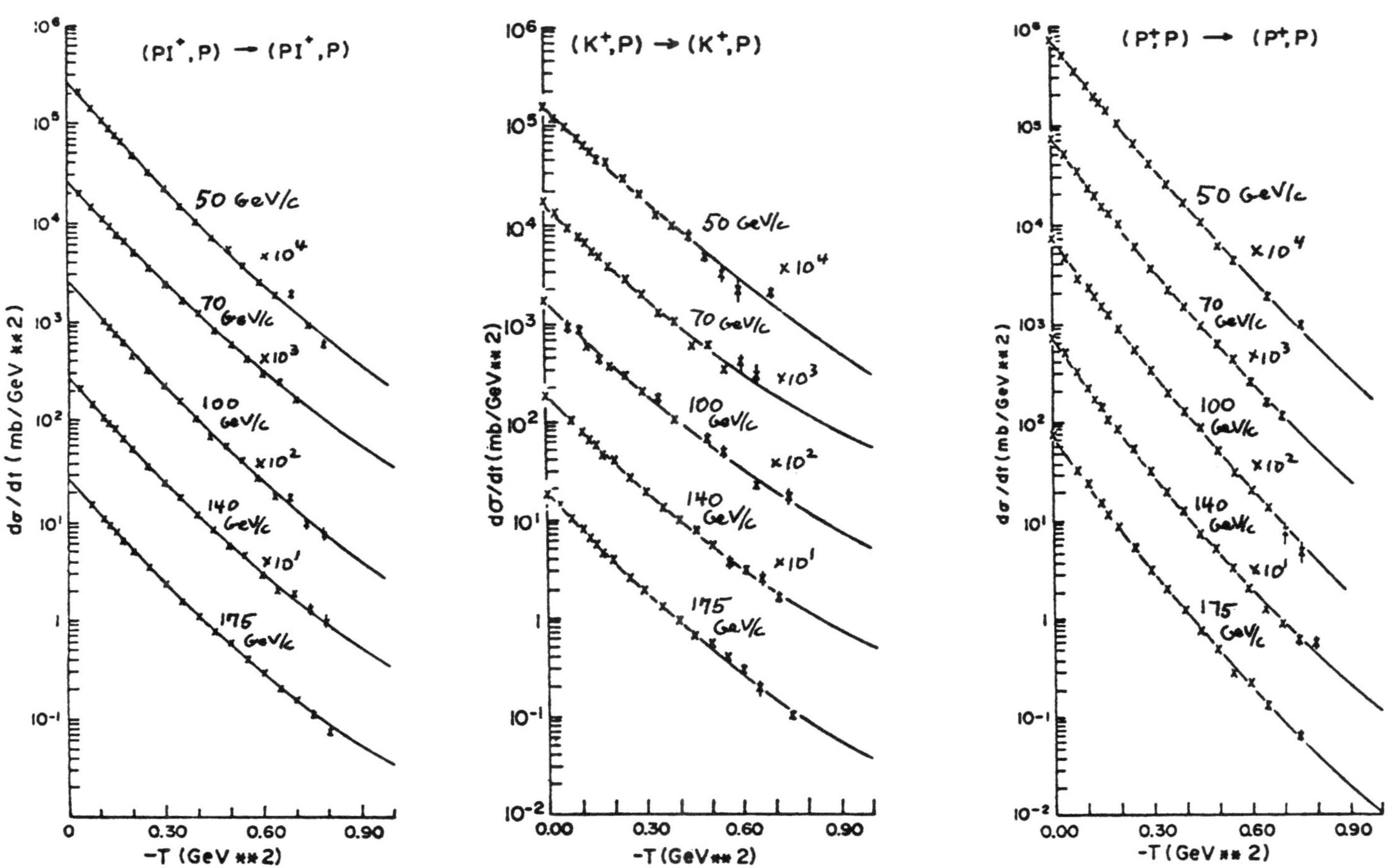

Figure 2a. Elastic differential cross section data for π^+p, K^+p, and pp.

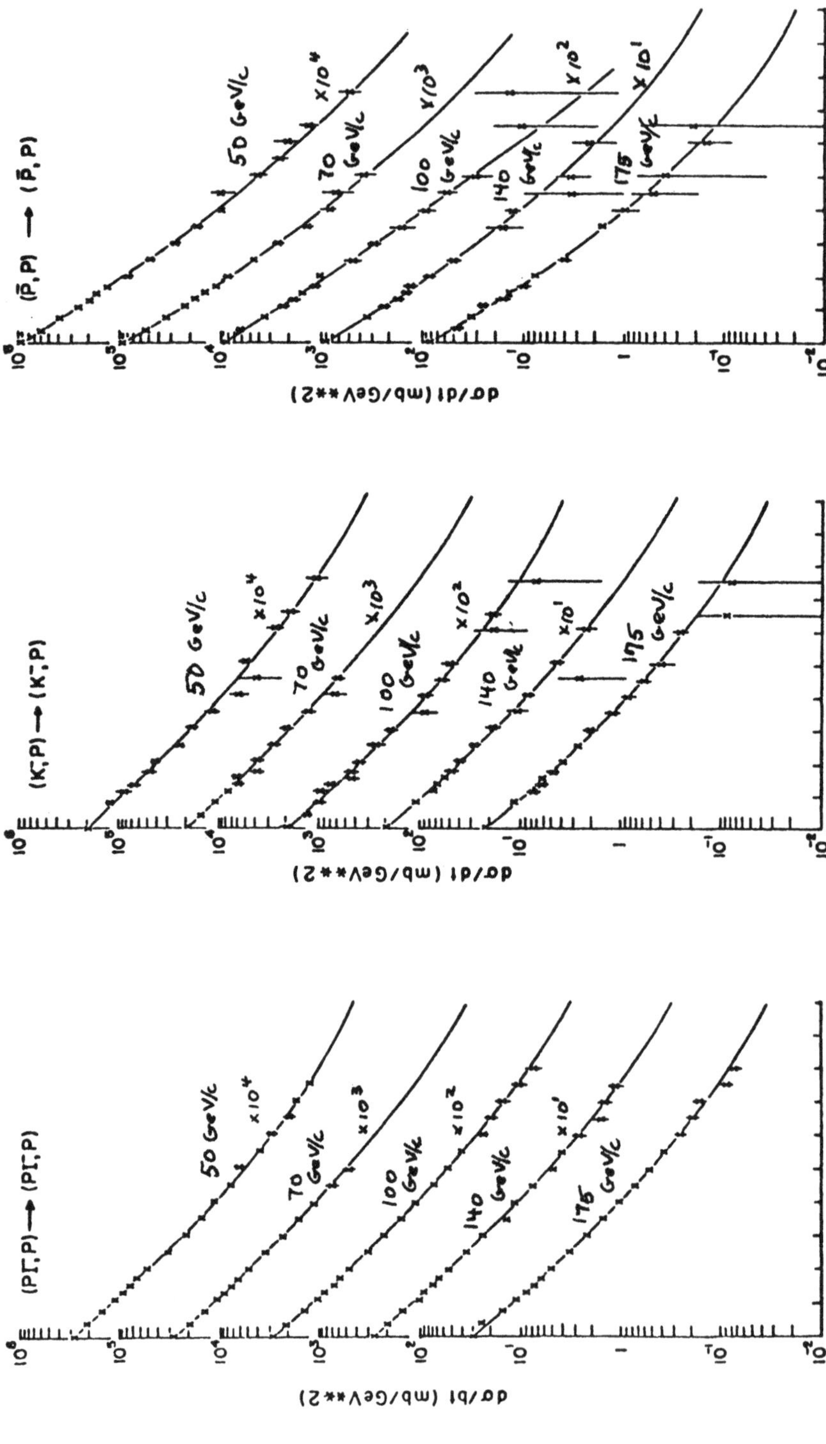

Figure 2b. Elastic differential cross section data for $\pi^- p$, $K^+ p$ and $\bar{p}p$.

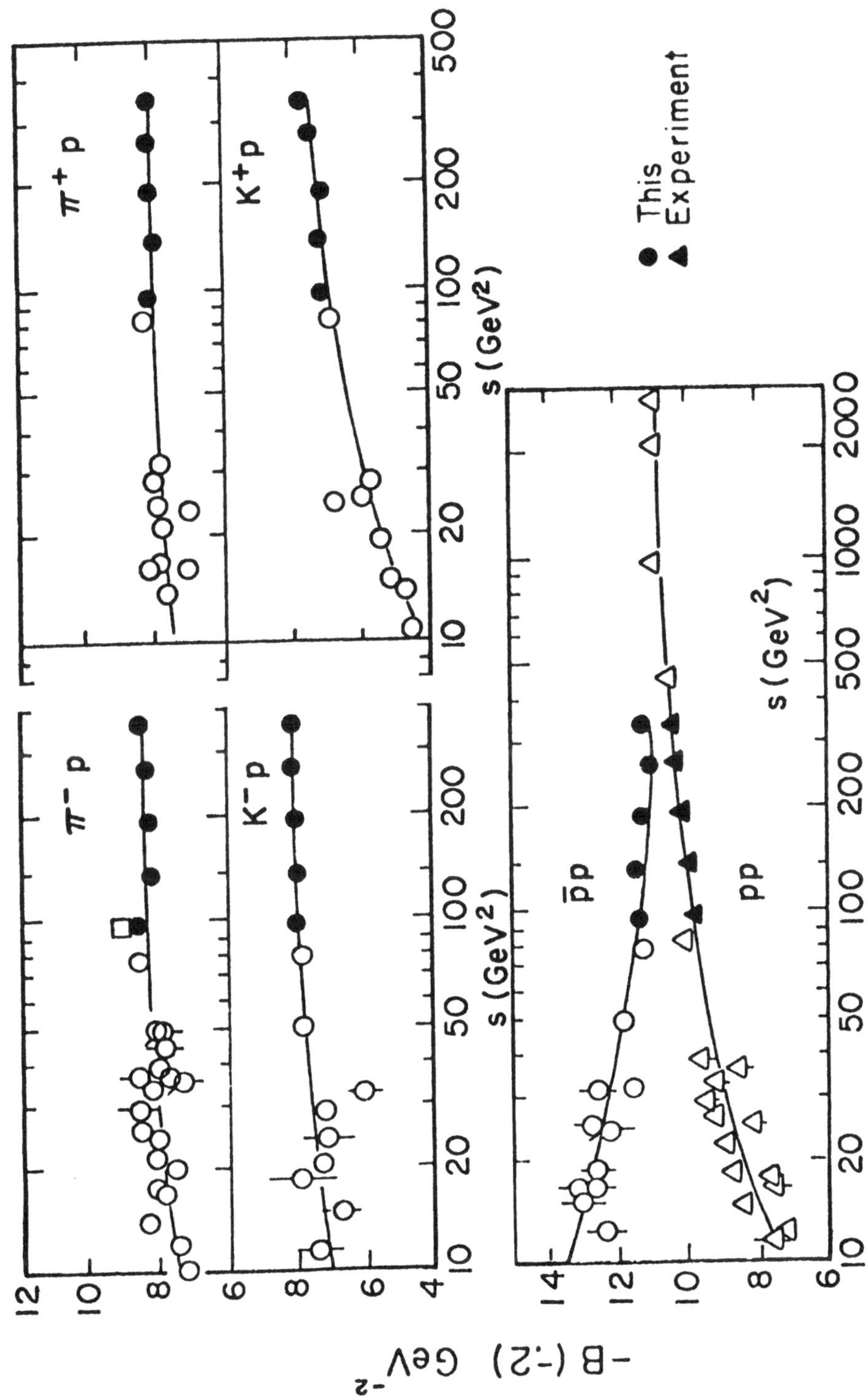

Figure 3. Slopes of the elastic differential cross section.

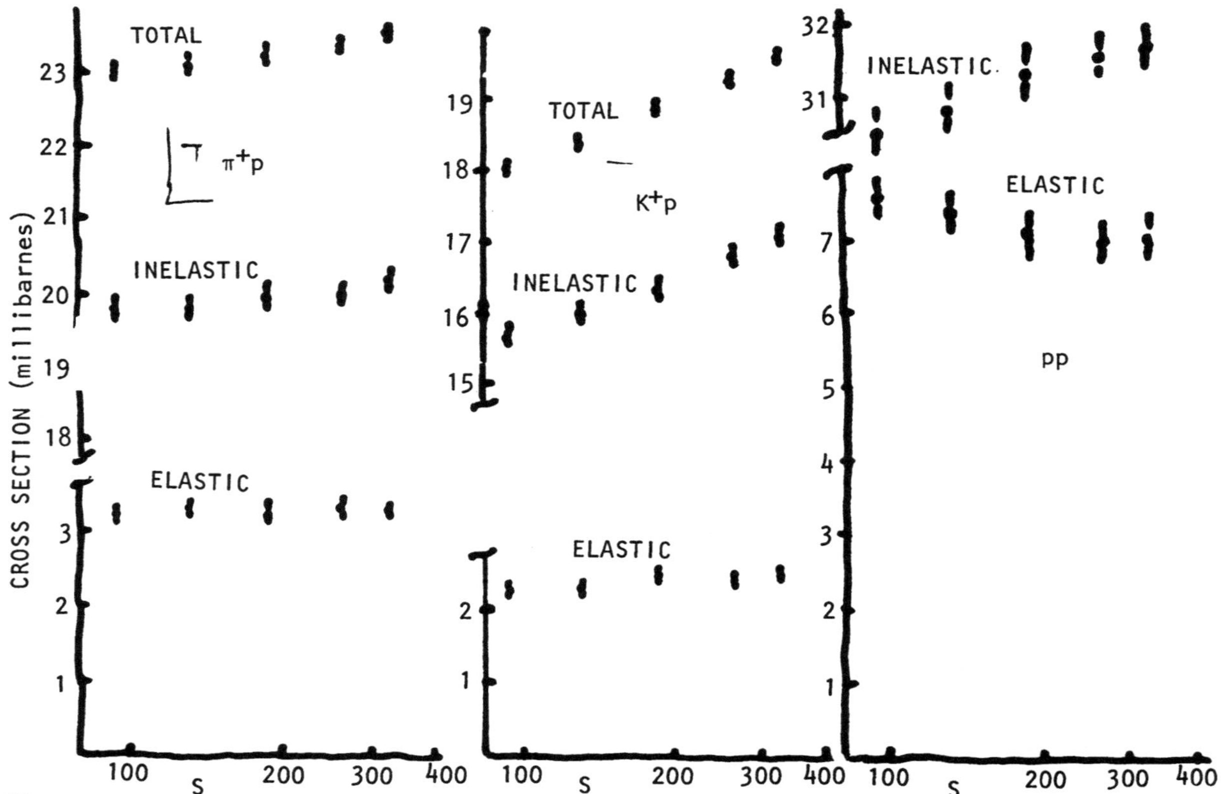

Figure

Figure 4a: Integrated cross sections in millibarnes for π^+p, K^+p, and pp.

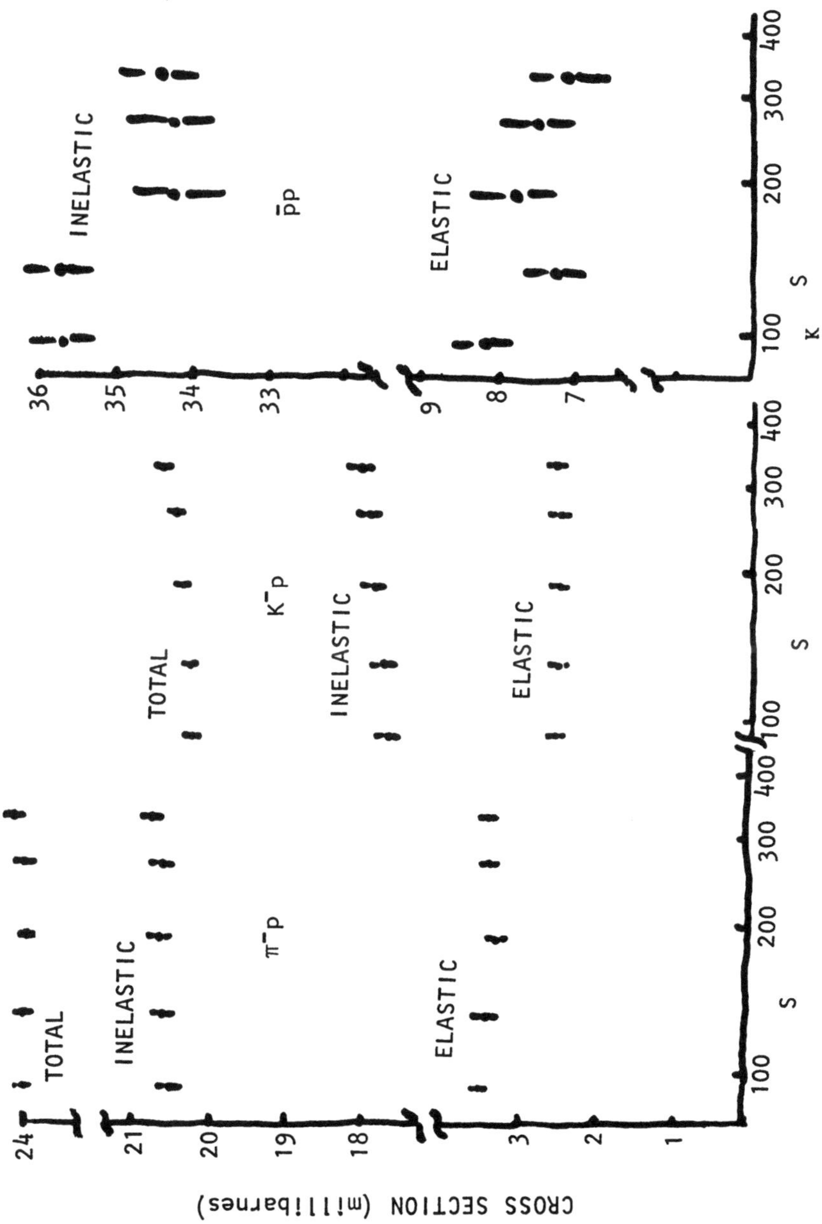

Figure 4b: Integrated cross sections in millibarnes for $\pi^- p$, $K^- p$, and $\bar{p}p$.

IMPACT ANALYSIS

From our elastic differential cross section data we make the numerical transform [4]

$$\mathrm{Im}\ h_{el}(s,b) = \frac{1}{\sqrt{4\pi}} \int_0^{\sqrt{-t_1}} \sqrt{\frac{d\sigma}{dt}}\, J_0(b\sqrt{-t})\ d(\sqrt{-t})$$

to find the elastic scattering amplitude h_{el} as a function of impact parameter b. The procedure has two minor problems: 1) the data is in the form $d\sigma/dt$ while what we really want is Im A(t) and 2) we have no data for $|t| > t_1$, where t_1 is generally 1.2 $(GeV/c)^2$. We assume Im A = A, and we apply a correction for the lack of high $|t|$ data by studying one energy where our data extends out to 1.4 $(GeV/c)^2$. Estimates of the ratio Re A(t)/Im A(t) indicate that our assumption of a purely imaginary amplitude introduces an error less than 0.5% at all b. Separate analyses using models of the high t behavior give the same results as those I will present, except for a small scale shift of the order of 3% at b = 0.

Having obtained $h_{el}(s,b)$ [$\equiv$ Im $h_{el}(s,b)$] as above we use the unitarity equation

$$\mathrm{Im}\ h_{el}(s,b) = \frac{1}{4}|h_{el}(s,b)|^2 + G_{inel}(s,b)$$

to find the differential cross sections in impact parameter space:

$$\frac{d\sigma_{tot}}{\pi db^2} = \mathrm{Im}\ h_{el}(s,b)$$

$$\frac{d\sigma_{el}}{\pi db^2} = \frac{1}{4}|h_{el}(s,b)|^2$$

$$\frac{d\sigma_{inel}}{\pi db^2} = G_{inel}(s,b)$$

The inelastic "overlap function" G_{inel} we find simply by subtracting the elastic from the total cross sections, both of which come directly from the transform of our data. Thus from our measured elastic $d\sigma/dt$ we have a full description of the b-space scattering.

[4] H.I. Miettinen, CERN preprint TH.1864 (1974).

The first results one obtains in this analysis are the amplitudes Im $h_{el}(s,b)$ for the six elastic reactions. All are nearly gaussian in b, as expected for transforms of nearly exponential t-space data. Over our energy range (s from 94 to 328 GeV^2) we find Im h_{el} ∿ independent of energy for all b, for the reactions $\pi^{\pm}p$ and K^-p. For K^+p we observe a slight increase with energy, at all b; and for $\bar{p}p$ we see a slight decrease with energy, at all b. Finally, we find the amplitudes Im h_{el} for pp decreasing at small b and increasing at larger b, as s increases. These results are expected, as the amplitude integrated over b gives the total cross section:

$$\sigma_{tot} = \int \text{Im}\, h_{el}(s,b)\ \pi db^2$$

The absorption probability for a head-on collision ("opacity") is given by the function $G_{inel}(s,b = 0)$; our results are shown in Figure 5. The striking feature is that the mesons are more transparent than the protons, with $K^{\pm}$ slightly less opaque than $\pi^{\pm}$; $\pi^{\pm}p$, K^-p opacities are approximately constant while the K^+p opacity is increasing by ∿2% over our energy range. Note the $\bar{p}p$ and pp opacities are decreasing to the apparently constant values of 0.94 found for pp at the ISR. As mentioned before, the absence of large t data in our transform could cause a systematic shift in these results, such that the opacities shown are $\lesssim$ 2.5% low; this effect, however, wouldn't change the general comparison between reactions. The following Figure 6 shows for the six reactions the calculated r.m.s. interaction distances defined by

$$\sqrt{R^2} = \left[\frac{\int b^2 \frac{d\sigma}{db^2}\, bdb}{\int \frac{d\sigma}{db^2}\, bdb}\right]^{1/2}$$

the size of the K^+p and pp interaction is increasing with energy decreasing in $\bar{p}p$, and essentially constant for $\pi^{\pm}p$ and K^-p. Thus the change with energy in the total cross section is reflected directly in the growth in the interaction size rather than, for pp, in an increase in central opacity.

To examine more closely the energy dependence of the K^+p and pp cross sections, we have plotted the differential total cross section differences, defined by

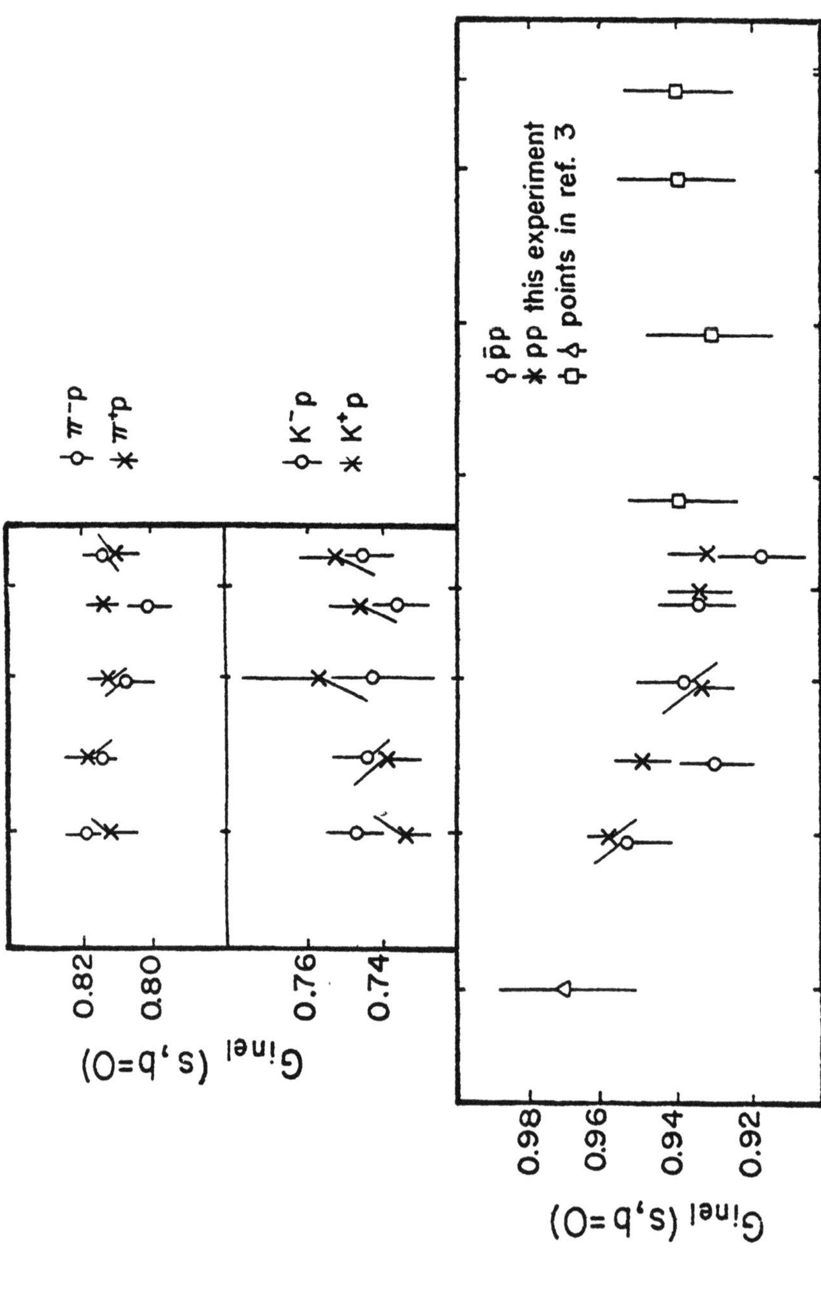

Figure 5. Inelastic overlap function $G_{inel}(s,b = 0)$.

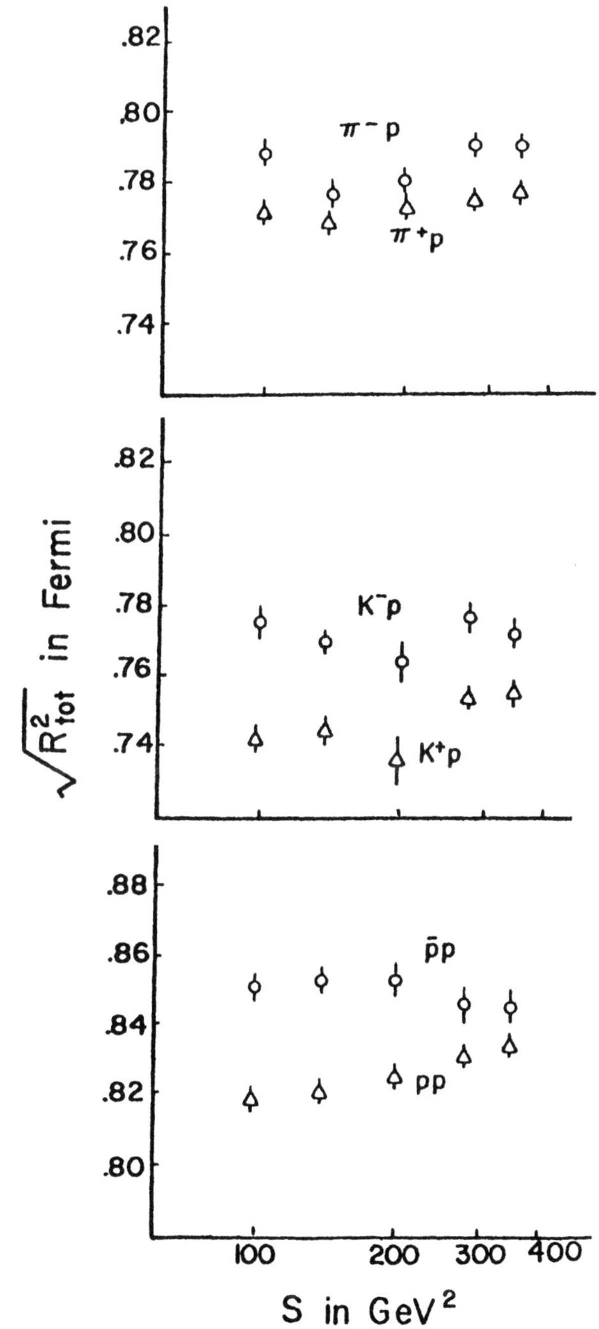

Figure 6. Calculated r.m.s. interaction distances, in Fermi, as a function of energy.

$$\Delta_{tot}(b) = \left(\frac{d\sigma_{tot}}{\pi db^2}\right)_{175} - \left(\frac{d\sigma_{tot}}{\pi db^2}\right)_{70}$$

A function of b, this quantity gives the difference between the differential cross sections measured at 175 GeV/c and at 70 GeV/c. We have also calculated the same expression subtracting 50 GeV/c from 140 GeV/c data. Since the range in energy covered with each pair of data (175-70 and 140-50) was similar, a comparison of the two results is sensitive to the presence of systematic errors. The values for $\Delta_{tot}(b)$ we obtained are shown in Figure 7, for pp and K^+p. Errors plotted are only statistical, but there seems to be good agreement between the two independent pairings of data. We show also the differential cross section differences calculated for the "ϕ"p reaction, where to isolate Pomeron exchange we define the amplitude [5]

$$h(\phi p) \equiv h(K^+p) + h(K^-p) - \frac{1}{2}\,h(\pi^+p) - \frac{1}{2}\,h(\pi^-p) \quad .$$

Note that Δ_{tot} is negative at small b, for the pp reaction; in other words, the pp differential cross section decreases at small b and increases at larger b, while both K^+p and "ϕ"p increase at small b. [In looking at these graphs remember the total cross section is gotten by integrating with πdb^2; the pp total cross section rise with energy is in full agreement with such an integration of the data shown.] The following page (Figure 8) shows the same relations for the inelastic cross section differences rather than the total. The K^+p and "ϕ"p are more or less the same, but there is a dramatic difference in the small b results for pp. Thus the decrease in the total cross section in pp at small b is related to the decrease in the elastic cross section, while the rise at large b (and overall rise) comes from the inelastic cross section increase.

We have studied the cross-over effects by examining $d\sigma/dt$ elastic scattering data for $K^{\pm}p$ and $p^{\pm}p$ [1]. The idea is that differences in the particle-antiparticle cross sections isolate the imaginary non-flip ω-exchange contribution. We have examined the quantity

$$\Delta(xp) \equiv \frac{d\sigma_{tot}(K^-p)}{\pi db^2} - \frac{d\sigma_{tot}(K^+p)}{\pi db^2}$$

[5] C. Quigg and E. Rabinovici, Phys. Rev. 13, 2525 (1976).

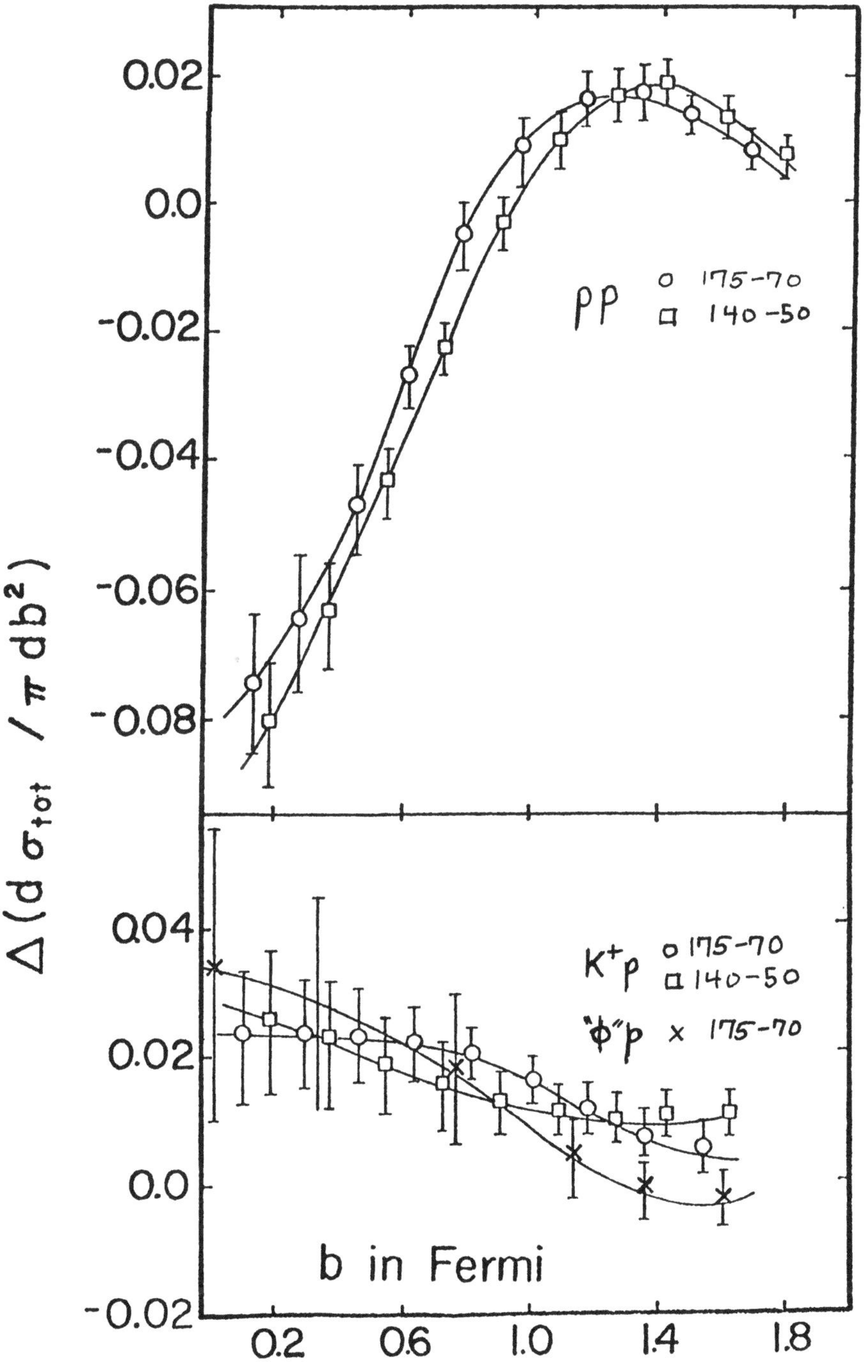

Figure 7. Differential total cross section differences, as a function of impact parameter.

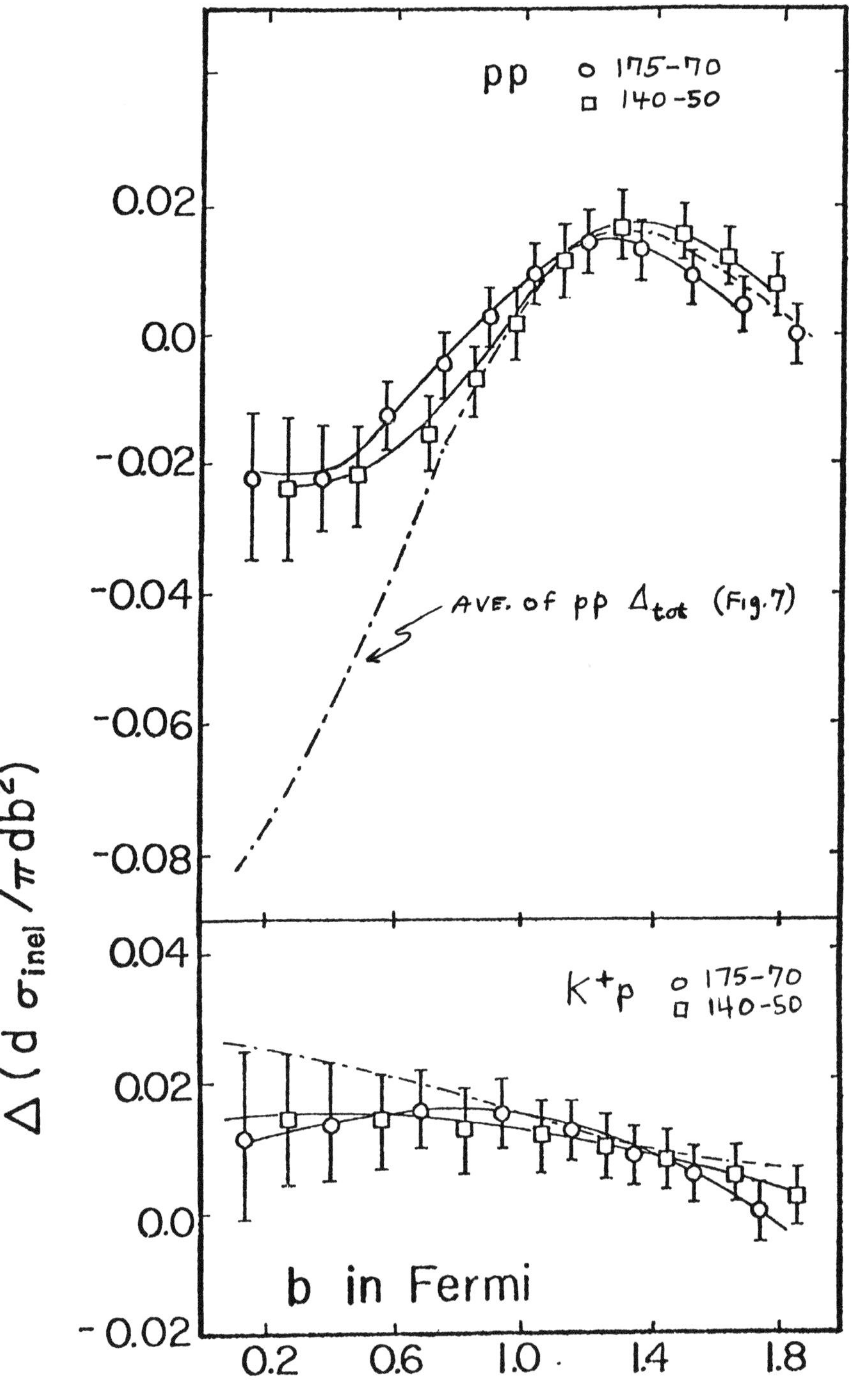

Figure 8. Differential inelastic cross section differences, as a function of b.

as a function of impact parameter b. The results are shown in Figure 9: data on the particle-antiparticle differences are plotted for two energies, as well as the corresponding expressions for pp. For both reactions we find the difference to be peripheral centered roughly around 1 Fermi, and decreasing rapidly with energy fully consistent with a simple Regge-pole interpretation.

The impact analysis gives us a measure of the overlap of two hadronic density distributions; from the pp data we can extract the proton distribution and then apply the result to πp and Kp to find the hadronic density distributions for π and K. Figure 10 shows the comparison of our results for the Fourier transform of the hadronic densities with the Fourier transform of the hadronic charge distribution, as suggested by Chao and Yang [6]. Our results are the lines, and the data points are measurements of the form factors F_1 and G_E, for pp [7], and the pion form factor [8], as a function of t. Our agreement with the model is good, for pp and πp, and we make a prediction for the Kp form factor.

We have tested in impact parameter space quark model predictions for the amplitudes as a function of b. The comparisons are generally ≈20% off, as is the amplitude relation proposed by Lipkin [9]:

$$A(\pi p) = \frac{1}{2} A(Kp) + \frac{1}{3} A(pp) \quad .$$

To test the relation, one picks particular combinations of the charge states to satisfy Regge exchanges. Two such combinations give the following relations:

$$\text{Rule 1:} \qquad \frac{\frac{2}{3} a(\pi^- p) + \frac{1}{3} a(\pi^+ p)}{\frac{1}{2} a(K^+ p) + \frac{1}{3} a(\bar{p}p)} - 1 = 0$$

$$\text{Rule 2:} \qquad \frac{a(K^- p) - \frac{1}{2} a(K^+ P) + \frac{1}{3} a(pp)}{\frac{1}{2} a(\pi^- p) + \frac{1}{2} a(\pi^+ p)} - 1 = 0$$

Our results are shown in Figure 11. The relations are good in the region of 1 Fermi and are worst at large b.

[6] T.T. Chao and C.N. Yang, Phys. Rev. 170, 1591 (1968).
[7] L.E. Price et al., Phys. Rev. D8, 2063 (1973).
[8] C.S. Bebek et al., Phys. Rev. D13, 25 (1976), and G.T. Adylov et al., Phys. Letters 51B, 402 (1974).
[9] H.J. Lipkin, Phys. Letters B56, 76 (1975).

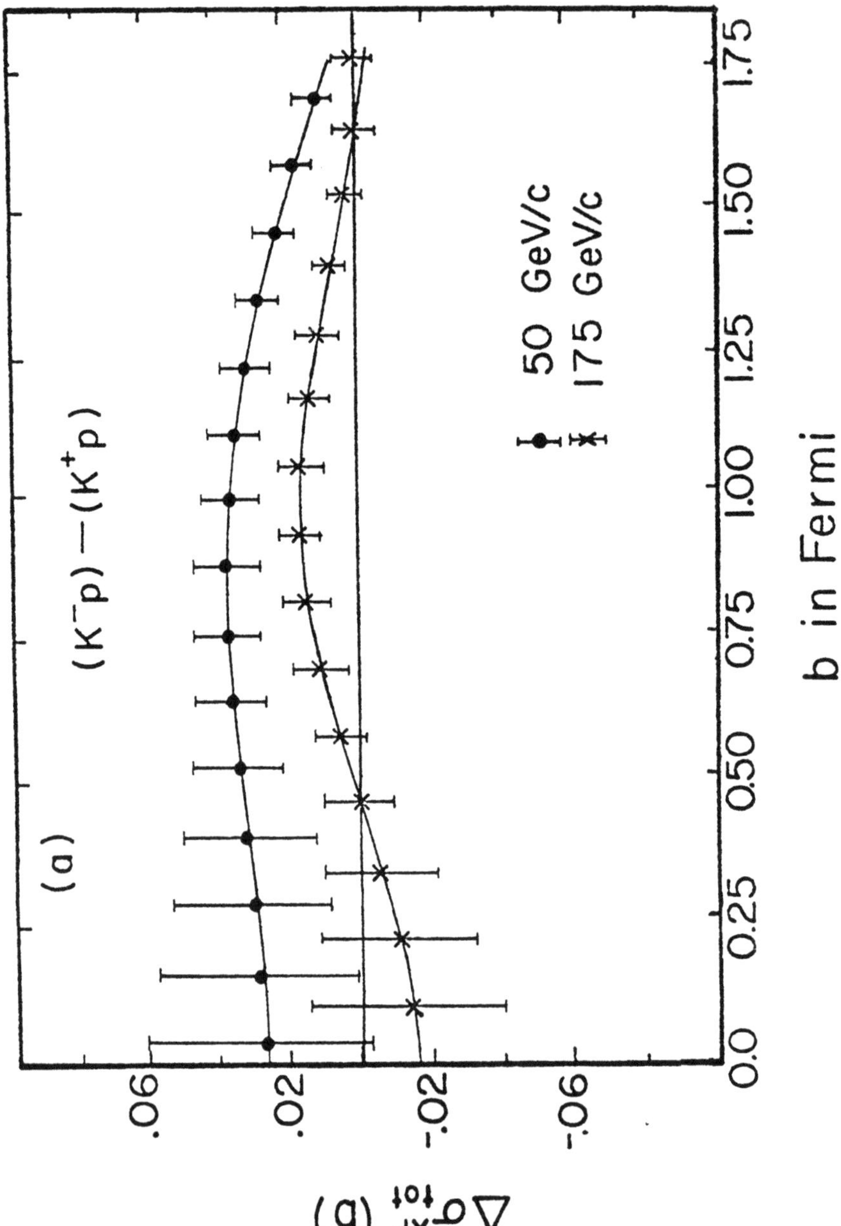

Figure 9. Particle-antiparticle differences as a function of impact parameter.

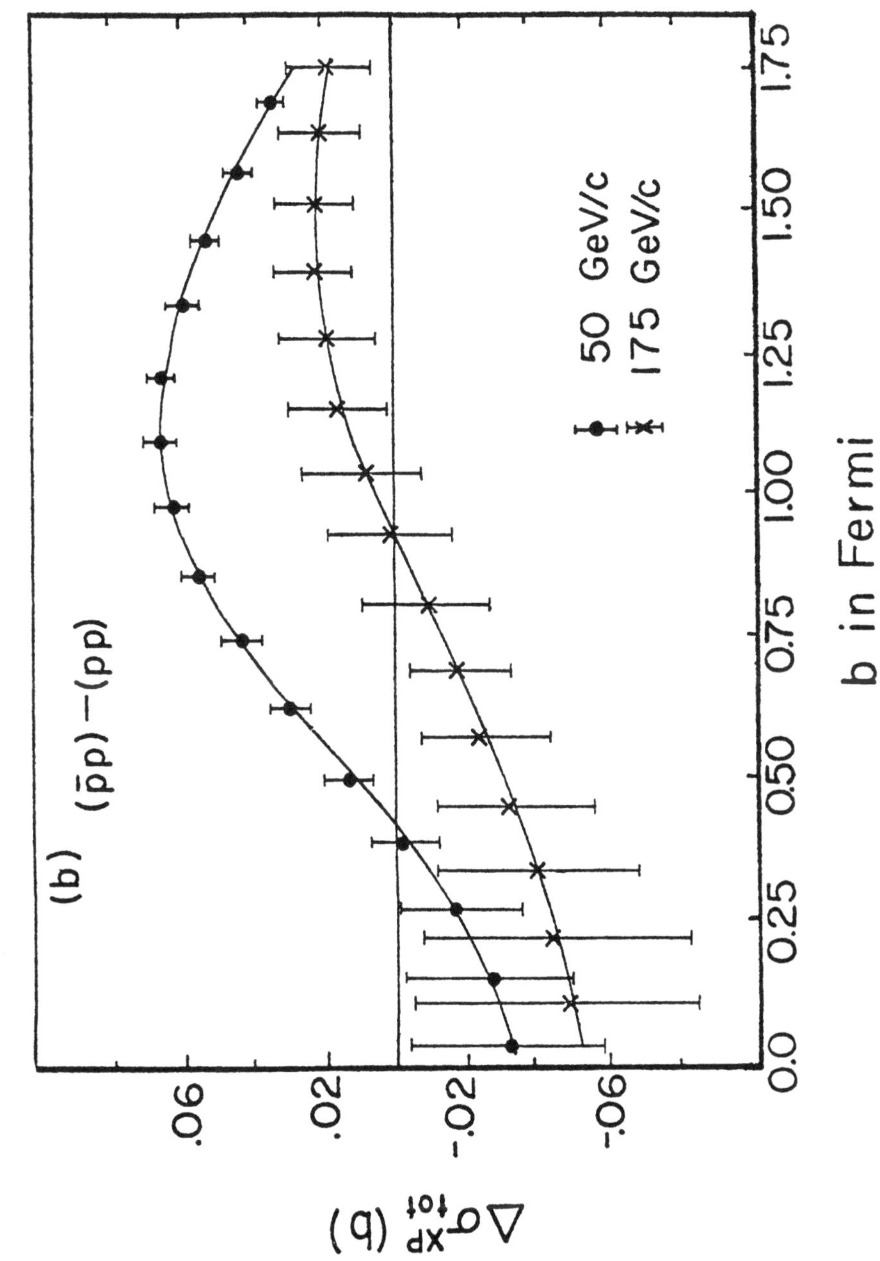

Figure 9. Particle-antiparticle differences as a function of impact parameter.

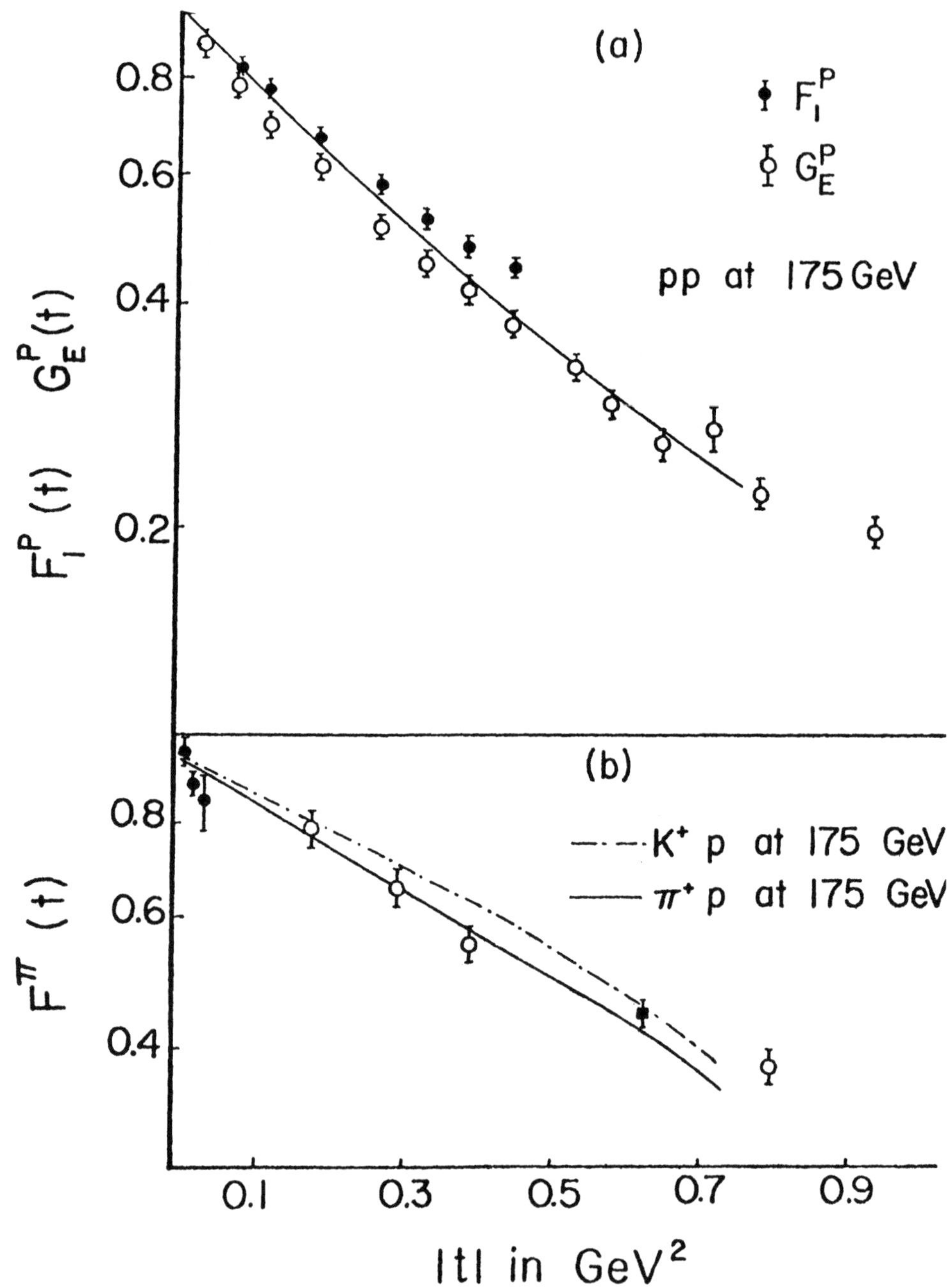

Figure 10. Comparison of proton form factor (a) and pion form factor (b) measurements (data points) with calculated transforms of proton, pion, and kaon hadronic density distributions (lines).

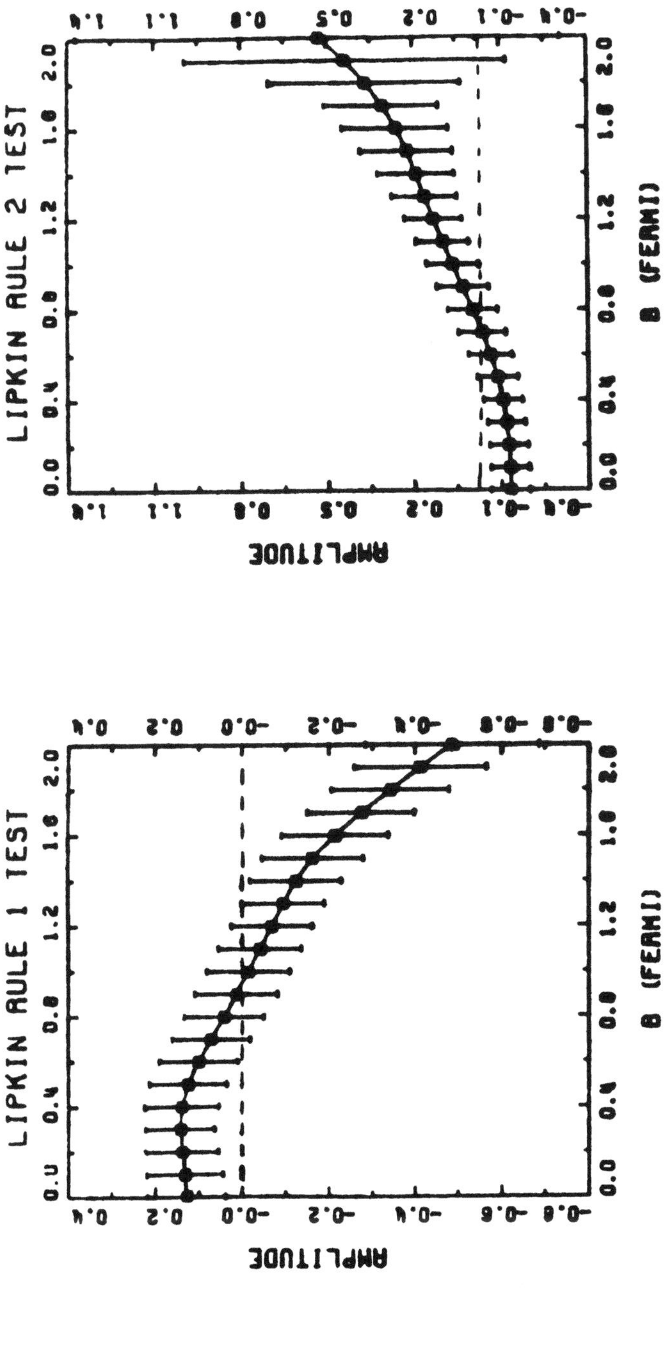

Figure 11. Quark model amplitude relations in impact parameter space.

INELASTIC RESULTS

As part of the initial data taking with the Single Arm Spectrometer, we recorded events for which the momentum of the particle detected in SAS was below the elastic peak. Some of these inclusive events came "for free" in the elastic data taking; and at most of the elastic settings (a given scattering angle, determining t) we also took additional inelastic data by lowering the momentum of the spectrometer in steps up to 25% below the beam momentum. This data, then, covered the kinematic range

$$1.0 > x > 0.75$$

$$.03 < |t| < 0.7\ (\mathrm{GeV/c})^2 \quad \text{(or up to 1.2 in some cases)}$$

for the five incident momenta

50, 70, 100, 140, and 175 GeV/c.

Note the Feynman variable x is simply determined (to an excellent approximation, at our energies) by the momenta measured in the lab:

$$x \simeq 1 - \frac{Mx^2}{s} \simeq \frac{p_{\text{spectrometer}}}{p_{\text{beam}}}$$

Since we tagged for each event the incident particle type (π, K, or p) and the particle type in the spectrometer, we took data simultaneously on 9 reactions:

$$a + p \rightarrow b + X \quad ,$$

where a,b = π, K, or p. Together with two beam polarities (in this first experiment, sign of b = sign of a) we have data for 18 reactions at 5 energies and a number of kinematic points in the ranges indicated above -- generally referred to as the "projectile fragmentation" region.

How do we study this data? We have used the "triple Regge" picture (see reviews by Boggild and Ferbel [10] and Fox [11]) as a guide. In this model one describes the single particle inclusive cross section by a sum of terms involving different combinations of the possible exchanges i, j, k, each with residue

[10] H. Boggild and T. Ferbel, Ann. Rev. Nuc. Sci., 24, 451 (1974).
[11] G.C. Fox, p. 180, in High Energy Collisions, AIP Conference Proceedings #15 (1973).

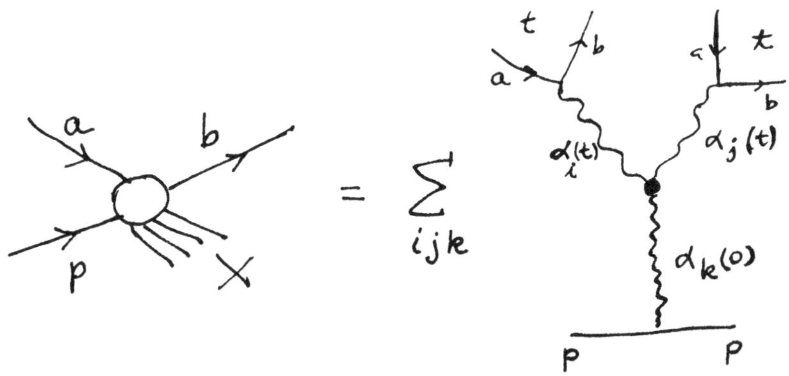

Triple Regge picture.

functions and having t, M^2, s dependences given by the appropriate $\alpha(t)$'s. With R = Reggeon and P = Pomeron, one writes the terms expected to be dominant as: PPP, PPR, RRP; their contributions to the cross section are:

$$\frac{d\sigma}{dtdM^2}: \quad \sim \frac{1}{M^2} \quad \text{(PPP)}, \qquad \sim \frac{1}{M^3} \quad \text{(PPR)}, \qquad \sim \frac{1}{s} \quad \text{(RRP)}$$

or, using the relation $M^2 \simeq (1 - x)s$:

$$s\frac{d\sigma}{dtdM^2} \quad \sim (1 - x)^{-1} \quad \text{(PPP term)}, \qquad \sim (1 - x)^{-3/2}\, s^{-1/2} \quad \text{(PPR term)}, \qquad \sim \text{const.} \quad \text{(RRP term)}$$

We have not as yet made a proper Triple Regge Analysis, using all terms and the full t, M^2, s dependences; what we have done so far is to use the above picture to help parametrize the data.

We have divided our inelastic events into three categories: diffractive, inclusive (without quantum number exchange), and exchange (of strangeness or baryon number). The data grouped as "diffractive" are of the type:

$$a + p \rightarrow a + X \qquad (a = \pi^{\pm}, K^{\pm}, p^{\pm})$$

with kinematics close to the elastic, specifically:

$$0.975 < x < 0.995$$

$$2.4 \text{ GeV}^2 < M^2 < 9 \text{ GeV}^2$$

I will say only that this data has the characteristics that:

$$M^2 \frac{d\sigma}{dtdM^2} \text{ is independent of } M^2\text{, for fixed } t$$

in all six channels; and, further that for all channels the data "factorizes" -- namely, the expression above is the same for all reactions, when normalized by the appropriate elastic cross section. These results are just those expected if the processes are dominated by a single term, PPP, common to all channels.

Data of a similar type (particle a → particle a) but with a larger missing mass ($M^2 > 4$ GeV^2) we studied separately, including events over the full range of x ($x > 0.75$) and $.03 < |t| < 0.7$ (GeV/c)^2. There is an enormous amount of data (over 500 graphs) of the type shown in Figure 12, and at present our analysis has only fit events to a simple relation suggested by the triple Regge picture and partly by the data itself. We suppose the invariant cross section depends only on two terms, one that varies as 1/1-x (like PPP) and one that varies as (1-x). The latter term may be some combination of PPR or RRP terms, perhaps, but describes the data well. For each term we allow s and t dependence, writing

$$s \frac{d^2\sigma}{dtdM^2} = A_1 e^{B_1 t + C_1 t^2} (1 + D_1(E^{-1/2} - 0.1))(1 - x)$$

$$+ A_2 e^{B_2 t + C_2 t^2} (1 + D_2(E^{-1/2} - 0.1)(\frac{1}{1 - x}) \quad ,$$

where E is the incident laboratory beam energy. We have separated parameters depending on s and t, writing a form suggested by our elastic data ($e^{Bt + Ct^2}$) and typical energy dependence expected in the Regge picture.

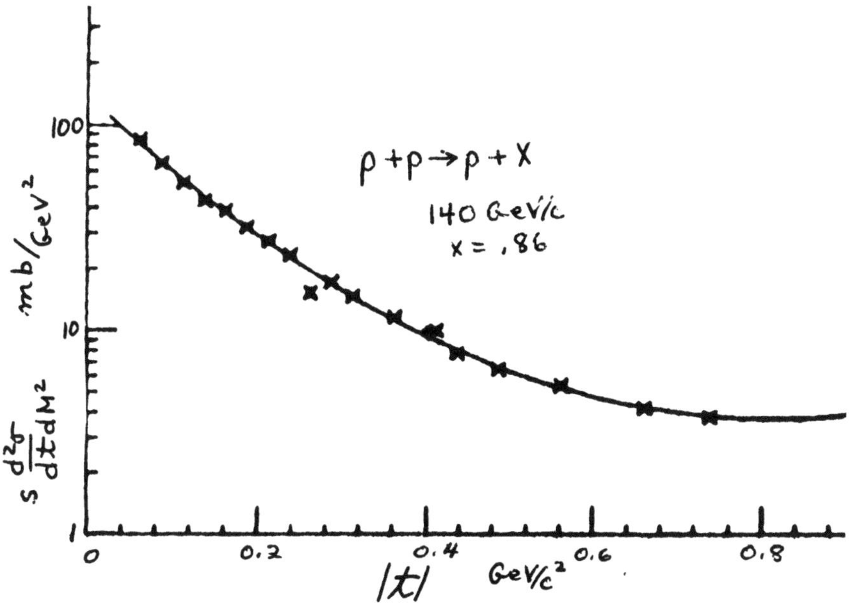

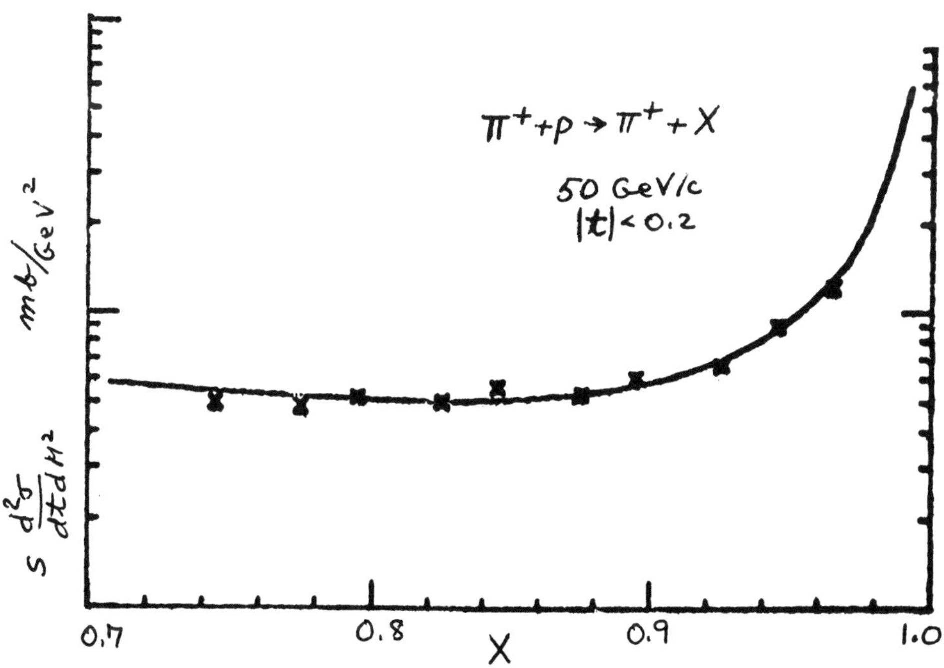

Figure 12. Sample inelastic data.

TABLE I

Channel	A_1 (mbs/GeV^2)	B_1 GeV^{-2}	C_1 GeV^{-4}	D_1 GeV$^{1/2}$
π^+	113 ± 7	-8.00 ± 0.30	6.21 ± 0.50	-0.68 ± 0.60
π^-	120 ± 7	-7.76 ± 0.32	5.33 ± 0.50	1.15 ± 1.00
p^+	612 ± 30	-8.32 ± 0.16	5.24 ± 0.24	-0.28 ± 0.40
p^-	609 ± 50	-8.42 ± 0.50	6.25 ± 0.80	-2.15 ± 1.50
K^+	69 ± 8	-7.66 ± 0.80	6.85 ± 1.00	-1.29 ± 1.73
K^-	65 ± 14	-7.75 ± 1.30	5.90 ± 1.50	-3.81 ± 4.00

Channel	A_2 (mbs/GeV^2)	B_2 GeV^{-2}	C_2 GeV^{-4}	D_2 GeV$^{1/2}$
π^+	2.30 ± 0.12	-5.21 ± 0.15	1.72 ± 0.20	3.64 ± 0.40
π^-	2.50 ± 0.14	-5.04 ± 0.16	1.20 ± 0.22	3.39 ± 0.46
p^+	6.08 ± 0.36	-7.24 ± 0.14	2.58 ± 0.22	2.69 ± 0.28
p^-	6.09 ± 0.70	-6.53 ± 0.60	-0.66 ± 1.50	8.16 ± 2.00
K^+	1.30 ± 0.10	-3.64 ± 0.32	0.83 ± 0.44	5.51 ± 2.00
K^-	1.56 ± 0.14	-3.85 ± 0.40	0.64 ± 0.60	7.31 ± 2.50

Our results for this fit are shown in the table above, from which we can make the following observations:

1. The parameter $A_2 \neq 0$; all channels exhibit a diffractive peak at $x \sim 1$.

2. We find $C_2 \sim 0$ or small: the $(1 - x)^{-1}$ contribution has a simple exponential t dependence.

3. In contrast, C_1 is large: the $(1 - x)$ term has a strong change in slope with t.

4. At high x, there is large energy dependence (the parameter D_2 is large), but

5. we see very little energy dependence away from $x = 1$ ($D_1 \sim 0$).

From the energy dependence in the $(1 - x)^{-1}$ term, we conclude the contribution is not completely from PPP but has some other part, possibly PPR (which goes as $s^{-1/2}$). These terms must have roughly the same relative size for the different reactions to explain the observed factorization at high x, as shown in Figure 13. Here we plot the ratios of the invariant cross sections as normalized by the corresponding elastic cross sections, as:

$$R_\pi/R_p \equiv \frac{\dfrac{d\sigma_\pi}{dtdM^2}\Big/\dfrac{d\sigma_\pi^{el}}{dt}}{\dfrac{d\sigma_p}{dtdM^2}\Big/\dfrac{d\sigma_p^{el}}{dt}}$$

Note in the region near $x = 1$ all reactions factorize, while down to $x = .75$ the π and K ratios remain together but increasingly fall below proton ratio. The implication in a simple Regge picture is that the same exchange dominates both π and K, but that an additional contribution to $p + p \rightarrow p + X$ enters away from the diffractive limit.

A third grouping of our inelastic data was made by selecting events for which the particle identified in the spectrometer indicated a change of strangeness or baryon number relative to the incident beam particle. Because of the low statistics on these inelastic "exchange" reactions, we concentrated our analysis on the six channels with high incident beam flux:

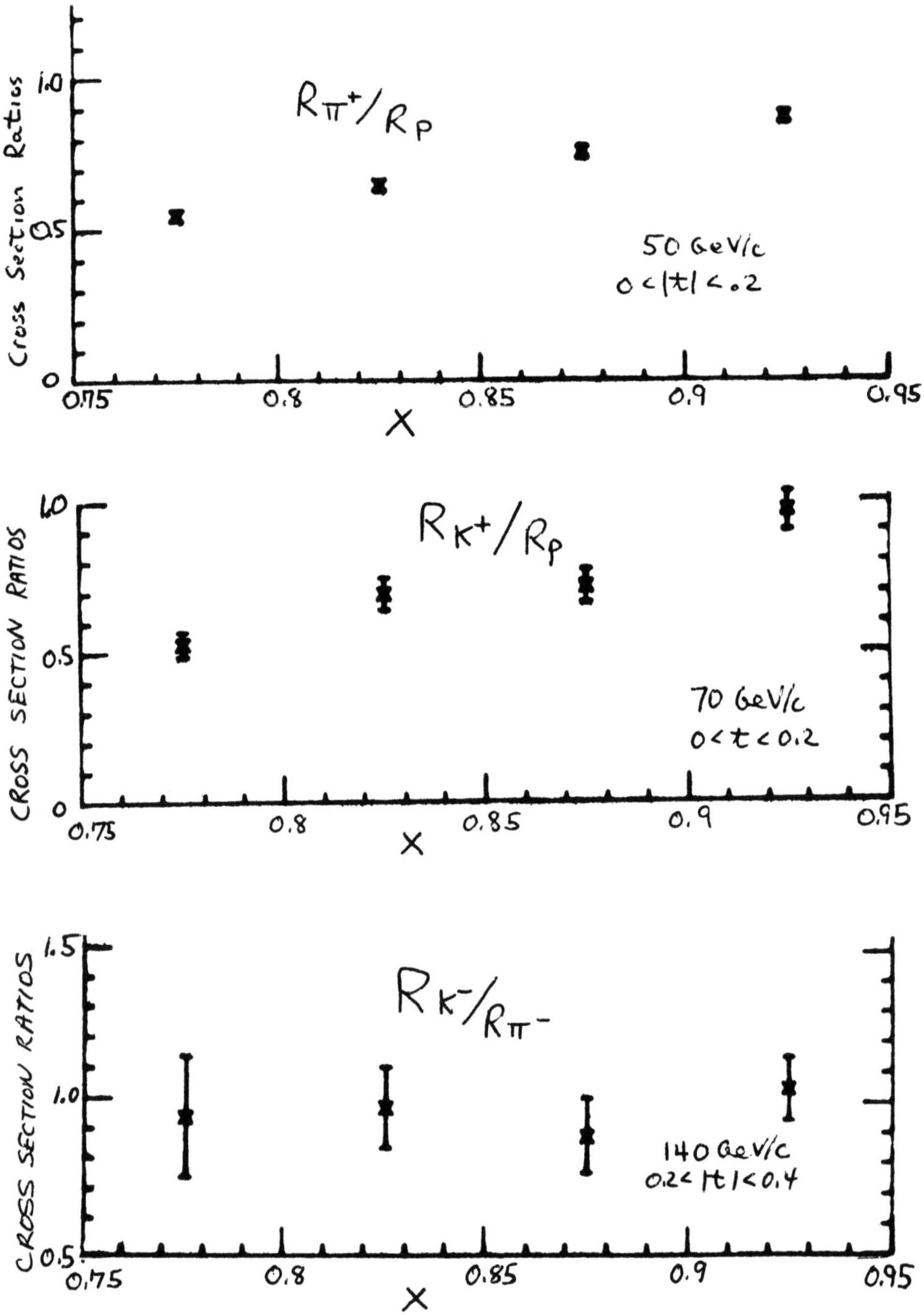

Figure 13. Ratios of invariant cross sections normalized by elastics, as a function of x.

reaction	expected exchange
$\pi^+ + p \to K^+ + X$ $\pi^- + p \to K^- + X$	$\alpha_{K*} \sim 0.3 + t$
$\pi^- + p \to \bar{p} + X$ $\pi^+ + p \to p + X$ $p + p \to \pi^+ + X$	$\alpha_N \sim -0.5 + t$
$p + p \to K^+ + X$	$\alpha_\Lambda \sim -0.7 + t$

We parametrized the cross section explicitly according to the Triple Regge Model outlined above; writing the invariant cross sections for these reactions ($a + p \to b + X$) as:

$$\frac{d^2\sigma}{dtdx} = \frac{|\beta|^2}{16\pi S_0^2}(1 - x)^{1-2\alpha_{eff}}\,\sigma_{tot}^{bp}(M^2)\,e^{b_{eff}t}$$

Our results are given in the table below. We find the effective trajectory $\alpha_{eff} \sim -.6$ for reactions probably dominated by nucleon exchange, α_{eff} small and positive for those involving K exchange, and α_{eff} large and negative for Λ exchange reactions. These values are in good agreement with those expected by the exchange picture. To show the data and this $(1 - x)^{1-2\alpha_{eff}}$ dependence explicitly, we plot in Figure 14 the logarithm of a normalized cross section

$$\ln\left[\frac{\sigma_t^{bp}(200)}{\sigma_t^{bp}(M^2)}\frac{d^2\sigma}{dtdx}\right] \equiv \ln\left[\left(\frac{d^2\sigma}{dtdx}\right)_{eff}\right]$$

vs. $\ln(1 - x)$. The slopes of channels expected to have similar exchanges are clearly in good agreement. The relative normalization (the residue function β) agrees less well, except for the $\pi \to K$ reactions.

TABLE II

Channel	Effective -t Range in GeV^2	Laboratory Energy Range in GeV	$\|\beta\|^2/16\pi s_0^2$ (GeV^{-2})	α_{eff}	b_{eff} (GeV^{-2})
$\pi^+ + p \to K^+ + \ldots$	0.1 - 0.65	50 - 140	0.155 ± 0.015	0.20 ± 0.06	3.7 ± 0.3
$\pi^- + p \to K^- + \ldots$	0.1 - 0.65	50 - 140	0.135 ± 0.014	0.25 ± 0.06	3.7 ± 0.3
$\pi^- + p \to \bar{p} + \ldots$	0.2 - 0.65	50 - 175	0.128 ± 0.012	-0.55 ± 0.15	2.0 ± 0.3
$\pi^+ + p \to p + \ldots$	0.2 - 0.65	140 - 175	0.240 ± 0.022	-0.50 ± 0.10	3.1 ± 0.5
$p + p \to \pi^+ + \ldots$	0.2 - 0.20	50 - 175	0.270 ± 0.025	-0.65 ± 0.15	4.4 ± 0.5
$p + p \to K^+ + \ldots$	0.2 - 0.20	50 - 175	0.108 ± 0.015	-0.65 ± 0.20	3.0 ± 1.0

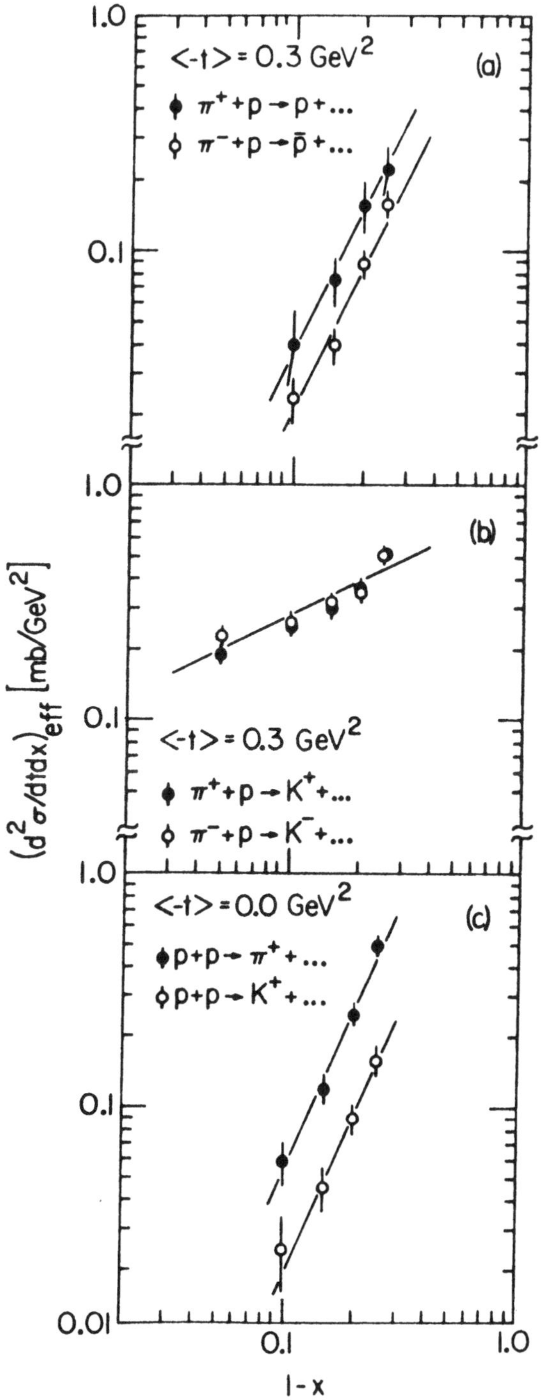

Figure 14. Comparison of inelastic exchange reactions.

PHYSICS IN PROGRESS

The Single Arm Spectrometer as a detector provides excellent particle identification with good momentum and angle resolution, for forward scattered particles. Although its aperture is relatively small it can be easily set at a particular value of (x, p_T) by selecting the spectrometer momentum and angle; with successive runs at different settings it can span a very wide region in the forward single particle inclusive kinematics, as illustrated below.

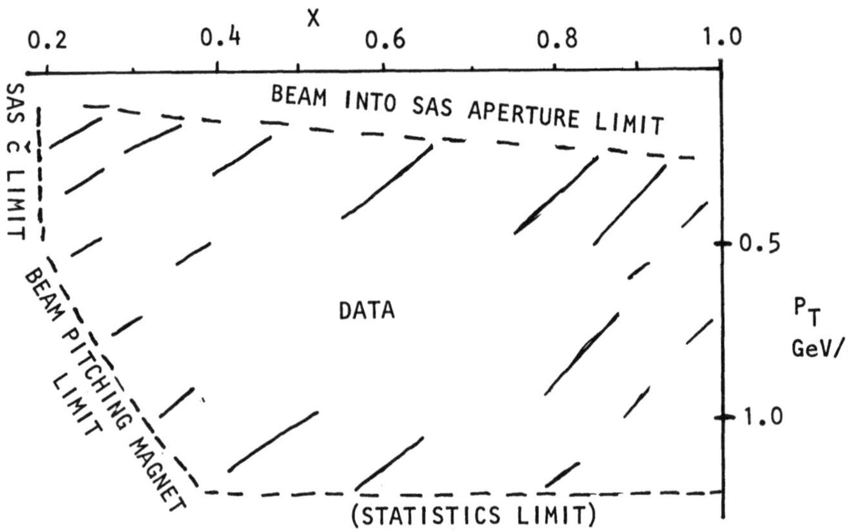

A subset (Bari, Brown, CERN, Fermilab, MIT) of the original SAS collaboration is taking advantage of this feature. We are engaged in a broad study of inclusive scattering, extending the single particle coverage begun in the earlier experiment to as wide a kinematic region as possible and including data with opposite as well as the same sign particles detected in the beam and spectrometer. Additionally, we have added detectors around the target (see Figure 16) capable of measuring the charged multiplicity and pseudorapidity distributions on an event-by-event basis, in association (or not) with a forward particle detected by the spectrometer.

This inclusive study is currently in progress. We have data already at 100 GeV/c and 175 GeV/c with positive beam (π^+, K^+, p data taken simultaneously) filling most of the kinematic area

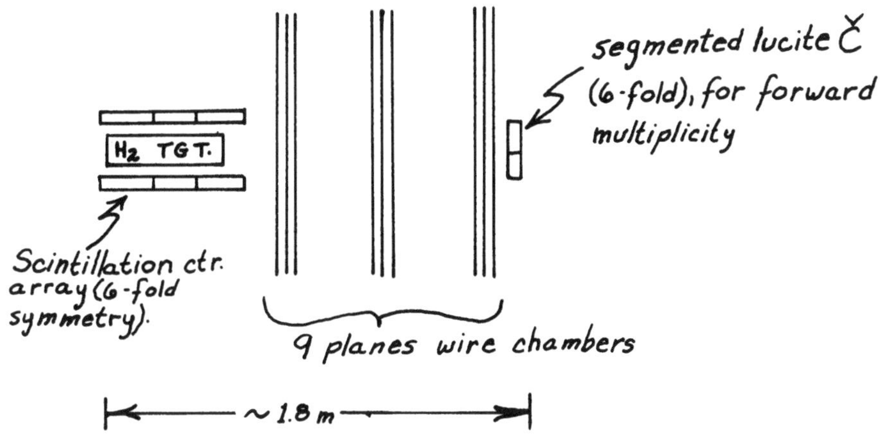

Figure 16. Multiplicity detector.

shown, for SAS set on positives, and some data with the spectrometer set on negative particles. The run is resuming shortly; we will finish data taking at 175 GeV/c positive beam, take data with negative beam at 175 GeV/c and 100 GeV/c, and possibly explore 50 GeV/c for the maximum extent in s. The analysis is progressing well. The software to compute single particle cross sections from SAS detector measurements was developed for the earlier experiment, so the main problem has been to handle information from the multiplicity detector. Our analysis is now able to unravel the wire chamber hits and Cerenkov pulse heights into reasonable multiplicity distributions, and we are proceeding to examine the data already on tape.

An interesting feature of the apparatus is that, by setting the spectrometer at 0° (pitching magnets off), we can use SAS in veto to make a total cross section trigger. Thus SAS can be used to measure the integrated cross section as well as map out the differential cross sections, and -- with the detector around the target -- measure multiplicity distributions associated with either. "SAS-veto" data has already proved useful in allowing us to compare our results with well known pp or πp charged multiplicity and prong distributions. Another check on our data has come from on-line "scaler" cross sections for the single particle inclusive reactions; where there is overlap there seems to be good agreement with earlier measurements.

CONCLUSION: FUTURE EXPERIMENTS WITH SAS

I will finish my discussion of physics with the Single Arm Spectrometer by mentioning several experiments due to be run in the next year. The subgroup now taking data has been approved to study "A-dependence of Inclusive Scattering and Associated Multiplicity." We will explore similar kinematic regions as indicated in Figure 15 using Be, Al, Cu, Sn, Pb targets, recording as before the multiplicity and pseudorapidity distributions for each event, and with added information from a new detector for π°'s at large p_T. A separate subgroup (ANL, Cornell, Fermilab, SLAC) of the original collaboration will make a "Study of $\pi^+p \rightarrow K^+\Sigma^+$ and $\pi^+p \rightarrow K^+Y^{*+}$." This experiment will study these non diffractive reactions at $|t| < 0.6$.

As a final topic I will mention a proposed "Search for Heavy Long-Lived Particles." Those of us currently doing the inclusive studies have noted the unique suitability of the M6-SAS (beam, spectrometer) combination for a mass search. Set at 0° the spectrometer is essentially a 4th stage of the beam; together with M6 there are 8 Cerenkov counters (3 differential) available, as well as long flight paths. We have proposed to use combined Cerenkov and time-of-flight techniques to search for heavy stable ($\tau \gtrsim 10^{-8}$ sec.) particles. Three $\check{C}$ counters can be used as a light particle veto, for $\sim 10^{-10}$ rejection; and others can be spread throughout the mass interval up to 5 GeV (mass), above which time-of-flight measurements become accurate. With this scheme the entire mass region can be sampled simultaneously with one beam setting. We will measure pulse heights in the Cerenkov counters, for better mass determination, and an existing muon detector (at the end of SAS) would tag muon-like events. The experiment as a whole would improve sensitivity over previous searches by a factor of $\sim$200, and could set an upper limit for heavy particle production of $\sim 4 \times 10^{-5}$ μb/GeV/c str., per nucleus. This corresponds to sampling $\sim 10^{10}$ incident beam particles, of both signs, and would take in all about 2 weeks. This mass search seems an exciting possibility to us and a good example of the varied and interesting physics accessible with the Single Arm Spectrometer.

DISCUSSION

CHAIRMAN: Prof. D. Cutts

Scientific Secretary:

DISCUSSION

- *PAULI:*

Is there some special reason why the predictions of the simple quark model and the Lipkin quark model should diverge at large b, impact parameter?

- *CUTTS:*

Actually, I think that we would expect the data to blow up compared to quark model predictions at large t - small b.

- *LIPKIN:*

It is rather surprising that the quark model gives you results at all for this. We have ignored the fact that the pion radius is different from the proton radius. We should put in form factors here but if we do, we wind up with a theory with more parameters than data.

AZIMUTHAL CORRELATIONS IN PARTICLE PRODUCTION AT LOW $p_\perp$

Gisela Ranft

Sektion Physik, Karl-Marx-Universität

DDR - 701 Leipzig

Summary

The inclusion of Bose-Einstein statistics in an independent cluster model explains the experimentally observed differences in the production of like or unlike mesons at small separation in rapidity and small azimuthal angle. This effect accounts for an important part of the observed short range correlation. It also shows up in the azimuthal asymmetry parameter B as a function of the mean transverse momentum of the two produced particles under study. This azimuthal asymmetry parameter as function of rapidity B (Δy) basically allows to discriminate between local or non-local conservation of transverse momenta. Together with single particle spectra and rapidity correlations the azimuthal dependence explains the observed band structure in two-particle inclusive production in a conventional way.

This lecture is based on several papers done together with J. Ranft.

1. Introduction

When the study of inclusive reactions became fashionable it was soon realized that a wide variety of models is able to explain single particle spectra $d^3N/d^3p = E\, d^3\sigma/d^3p$ at least in their general trend, but that correlation data - in particular in differential form - allow to discriminate between various proposed models for the production mechanism. The investigation of two-particle correlations in rapidities $d^2N/dy_1\,dy_2$ revealed a short range nature in the non-diffractive component which could be explained by intermediate cluster production /1/. From the y_1, y_2 correlation data if followed that in the non-diffractive component of particle production the number of clusters increases with energy, $\langle m\rangle \sim \ln s$, the average cluster mass is independent of energy, around $\langle M\rangle \simeq 1.5$ to 2 GeV, and the average decay multiplicity of a cluster into charged particles is $\langle \nu_{ch}\rangle \simeq 3$ to 5. These numbers rely strongly on the value of the correlation function R

$$R = \frac{\dfrac{d^6N}{d^3p_1\,d^3p_2}}{\dfrac{d^3N}{d^3p_1}\cdot\dfrac{d^3N}{d^3p_2}} - 1 \qquad (1)$$

in the centre of the rapidity plot; i.e. at $y_1 \simeq 0$, $y_2 \simeq 0$. There also the diffractive component contributes. Moreover, at small rapidity difference $\Delta y = y_1 - y_2$ of the two observed particles the Bose effect of identical particles simulates a short range effect. Studied also as function of the transverse momenta $p_{\perp 1}$ and $p_{\perp 2}$ and their relative angle ϕ, this effect enhances the two-particle correlation at $\phi \simeq 0°$ and $\Delta p_\perp = |p_{\perp 1} - p_{\perp 2}| \simeq 0$. This effect reduces the amount of short range attraction which has to be explained by clusters; it makes cluster masses and decay multiplicities smaller such that only

$$\langle M\rangle \sim 1.2 \text{ to } 1.5 \text{ GeV}$$

and

$$\langle \nu_{ch}\rangle \sim 2 \text{ to } 2.3 \text{ per cluster.}$$

This suggests even more that clusters are a superposition of actually identified resonances.

In Section 2 we shortly describe how the Bose effect is incorperated in the cluster model in the volume element of creation of the considered two like particles. In Section 3 we stress that the Bose effect acting at the birth of the two like particles and the second order interference effect (à la Hanbury Brown-Twiss) are not identical but rather act in the same direction. Examples are quoted how to discriminate between these two effects. Sections 4 and 5 deal with applications of azimuthal correlations, such as the use of the azimuthal asymmetry parameter B (Δy) to distinguish between local or non-local conservation of transverse momenta and the explanation of the observed band structure in invariant two-particle masses as a particular form of two-particle correlations in y, $p_\perp$ and azimuthal angle ϕ.

2. The Bose effect for like particles and the cluster model

2.1. The data

The difference between like and unlike two-particle correlations - attributed to the Bose effect - is seen when both particles 1 and 2 are in the same momentum state, $\vec{p}_1 = \vec{p}_2$. It is however so strong as to show up also in partially integrated distributions such as in

- $dN/d\phi$ for small Δy (integrated over $p_{\perp i}$; see Fig. 2.1). It is $\phi = \cos^{-1} \frac{\vec{p}_{\perp 1} \cdot \vec{p}_{\perp 2}}{|\vec{p}_{\perp 1}| |\vec{p}_{\perp 2}|}$.
- the azimuthal asymmetry as function of Δy

$$B(\Delta y) = \frac{dN/d\phi\,(\phi > \frac{\pi}{2}) - dN/d\phi\,(\phi < \frac{\pi}{2})}{dN/d\phi\,(\phi > \frac{\pi}{2}) + dN/d\phi\,(\phi < \frac{\pi}{2})} \tag{2.1}$$

  for small Δy (integrated over $p_{\perp i}$; see Fig. 2.2)
- the correlation function $C(\Delta y, \phi)$ for small Δy and $\phi \simeq 0$ (integrated over $p_{\perp i}$; see Fig. 2.3)
- the azimuthal asymmetry B ($\Delta p_\perp$) for small Δy and small $\Delta p_\perp = |\vec{p}_{\perp 1}| - |\vec{p}_{\perp 2}|$ (see Fig. 2.4)
- the azimuthal asymmetry B ($q_\perp$) for small Δy and $\Delta p_\perp$ with $q_\perp = (|\vec{p}_{\perp 1}| + |\vec{p}_{\perp 2}|)/2$ (see Fig. 2.5).

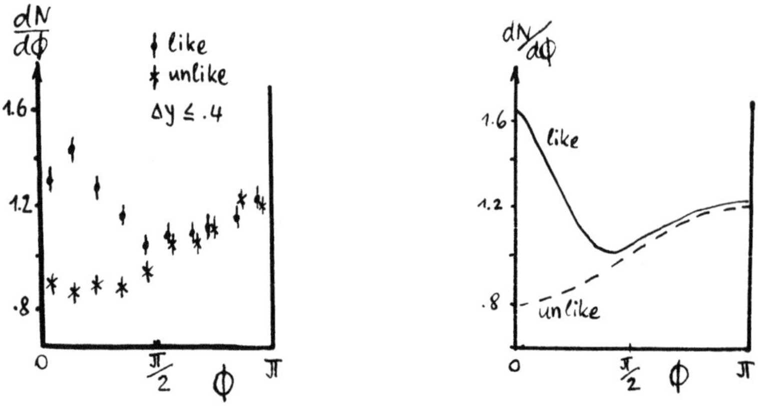

Fig. 2.1 a) b)
The experimental distribution $dN/d\Phi$ for like and unlike pion pairs at $\Delta y < 0.4$ in the reaction $\pi^- p \to \pi\pi X$ at 40 GeV/c. It is $2 \leq n_{ch} \leq 20$. a) Experiment. b) Theory. From ref. /2/.

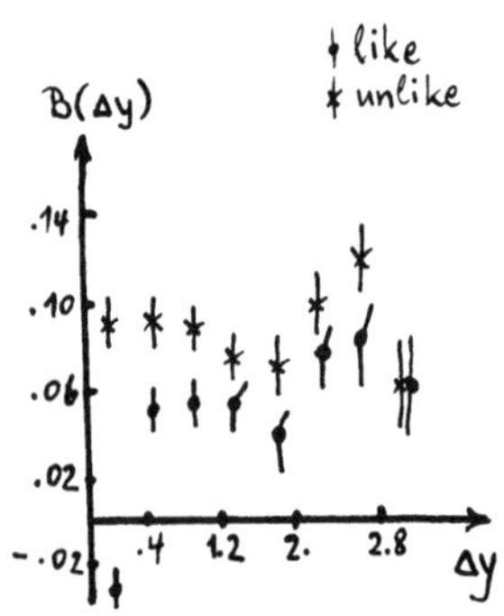

Fig. 2.2 a)

Fig. 2.2
The asymmetry parameter B as function of Δy for like and unlike pion pairs without restriction in y_1 and y_2
a) in the reaction $\pi^- p \to \pi\pi X$ at 40 GeV/c for $6 \leq n_{ch} \leq 20$. From ref./2,
b) in the reaction $pp \to \pi\pi X$ at 205 and 102 GeV/c from the Rochester - Michigan - ANL - FNAL - SUNY colla- boration as given by T. Ferbel, Proc. V-th Internat. Symposium on Many Par- ticle Production, Leipzig 1974, p.38!

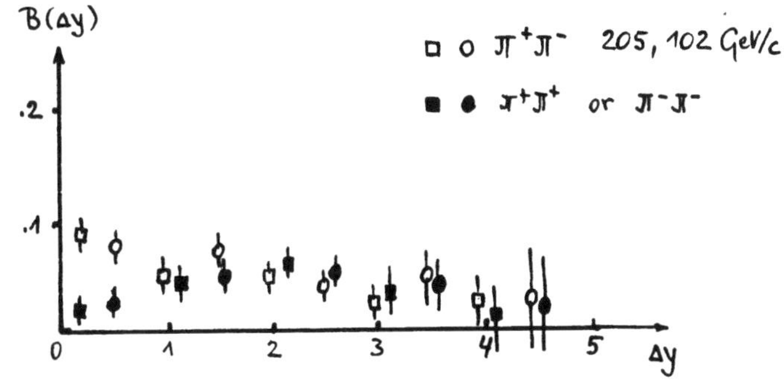

Fig. 2.2 b)

2.2. The cluster model and Bose effect

We describe the observed short range rapidity dependence within a model with neutral clusters; in this model the two observed particles can originate from one cluster (one-cluster term, denoted by 1F) or from two clusters (two-cluster term, denoted by 2F).
The short range rapidity correlation of two particles of the one-cluster term is contained in the function

$$Q_2(y_1, y_2, \sqrt{s}) \simeq \exp\left(-\frac{(y_1-y_2)^2}{2L^2}\right) \tag{2.2}$$

with the correlation length L. The long range rapidity correlation of two particles is due to both particles emerging from different clusters, thus

$$\simeq Q_1(y_1, \sqrt{s})\, Q_1(y_2, \sqrt{s}) \tag{2.3}$$

where $Q_1(y, \sqrt{s})$ describes the rapidity dependence in the decay of one particle. Both functions Q_1 and Q_2 are calculated from the mass and the rapidity distribution of the clusters as well as their decay distribution in rapidity and multiplicity /6/. They are normalized to 1. For small y, $Q_1(y, \sqrt{s})$ is constant representing the rapidity plateau.
Further rapidity dependence is contained in the Bose effect, see eq. (2.7) below.
The clusters have the average charged decay multiplicity $\langle v \rangle$, its second moment is $\langle v(v-1) \rangle$. The charged multiplicity in the collision is n, + or - denote the observed kind of secondaries in the cluster decay (v_+, v_-) or the collision (n_+, n_-). We assume

that their average values $\langle \nu \rangle_n$, $\langle \nu(\nu-1) \rangle_n$ are independent of n; the average values $\langle n \rangle$ and $\langle n(n-1) \rangle$ are functions of s.
The inclusive rapidity distribution is /2/ for like particles

$$\mathcal{N}_2^{like}(y_1, y_2, s) = \frac{1}{\sigma_{in}} \frac{d\sigma_{in}^{like}}{dy_1 dy_2}$$

$$= Q_2(y_1, y_2, s) \langle n \rangle \frac{\langle \nu(\nu-1) \rangle}{\langle \nu \rangle}$$

$$+ Q_1(y_1, s)\, Q_1(y_2, s) \left[\langle n(n-1) \rangle - \langle n \rangle \frac{\langle \nu(\nu-1) \rangle}{\langle \nu \rangle} \right] \qquad (2.4)$$

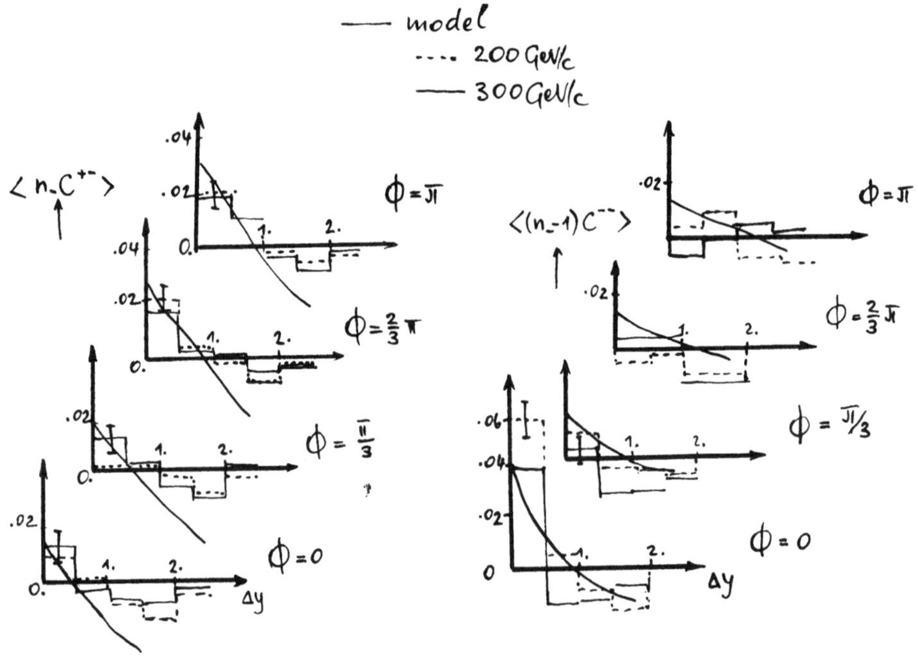

Fig. 2.3
Comparison of ($\Delta y, \phi$) correlations in the cluster model with data on pp $\rightarrow \pi^+\pi^- X$ and pp $\rightarrow \pi^-\pi^- X$ at 200 and 300 GeV/c by Oh et al., ref. /4/. In the model the peak in the like pion combination is due to Bose statistics. From ref. /5/.

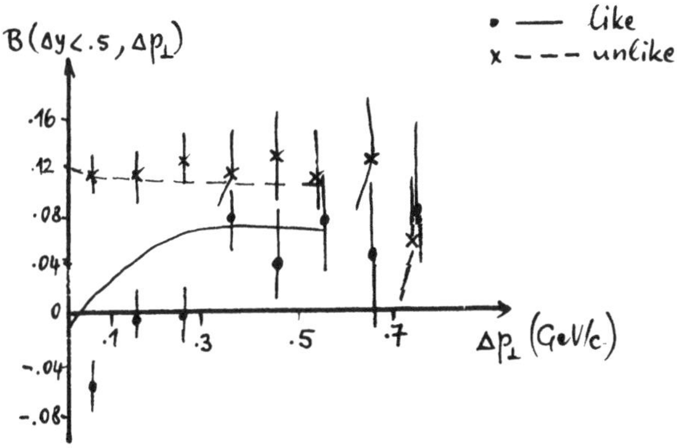

Fig. 2.4
The azimuthal asymmetry parameter B($\Delta y < .5$, $\Delta p_\perp$) as function of $\Delta p_\perp = |\vec{p}_{\perp 1}| - |\vec{p}_{\perp 2}|$ for $\pi^- p \rightarrow \pi^\pm \pi^\pm X$ and $\pi^- p \rightarrow \pi^+ \pi^- X$ at 40 GeV/c. The theoretical curves are from the cluster model described in the text. From ref. /6/.

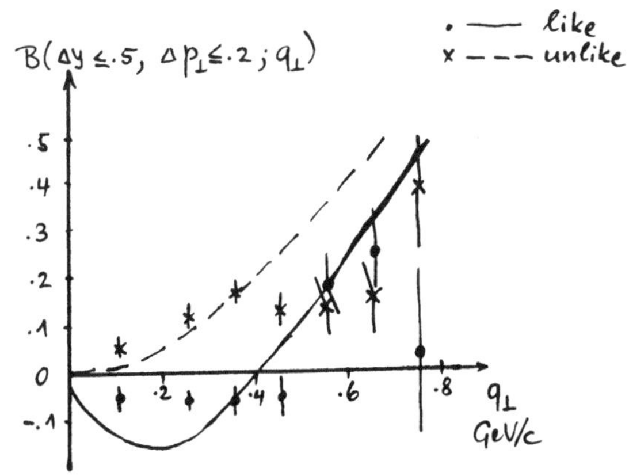

Fig. 2.5
The azimuthal asymmetry parameter $B(\Delta y < .5, \Delta p_\perp < .2, q_\perp)$ as function of $q_\perp = (|\vec{p}_{\perp 1}| + |\vec{p}_{\perp 2}|)/2$ for $\pi^- p \rightarrow \pi^\pm \pi^\pm X$ and $\pi^- p \rightarrow \pi^+ \pi^- X$ at 40 GeV/c. The theoretical curves are from the cluster model described in the text. From ref. /6/.

for unlike particles

$$\mathcal{N}_2^{unlike}(y_+, y_-, s) = \frac{1}{\sigma_{in}} \frac{d\sigma_{in}^{unlike}}{dy_+ \, dy_-} = Q_2(y_+, y_-, s) \frac{\langle \nu_+ \nu_- \rangle}{\langle \nu_+ \rangle} \langle n_+ \rangle$$

$$+ Q_1(y_+, s)\, Q_1(y_-, s) \left[\langle n_+ n_- \rangle - \langle n_+ \rangle \frac{\langle \nu_+ \nu_- \rangle}{\langle \nu_+ \rangle} \right] \qquad (2.5)$$

Here the first term is the one-cluster term, the second term is the two-cluster term.
There are three sources of transverse momentum and azimuthal angle dependence

i) phase space for not too large transverse momenta $p_{\perp i}$ behaves as /8/

$$g(\phi, p_{\perp 1}, p_{\perp 2}) \simeq 1 - \frac{\pi}{2} B \frac{p_{\perp 1} p_{\perp 2}}{\langle p_\perp \rangle^2} \cos\phi \qquad (2.6)$$

with the azimuthal asymmetry parameter $B \simeq \frac{1}{n-1}$ depending on the number of produced particles. This n dependence suggests the use of a semi-inclusive description. In general B is also a function of the cluster mass M and the cluster transverse momenta $\langle k_\perp \rangle$ /3/.

ii) the $p_\perp$ and ϕ dependence of the Bose effect for like particles, for which the parametrisation /2,3/

$$f(\vec{p}_1, \vec{p}_2, \overline{\Delta p_\parallel}, \overline{\Delta p_\perp}) = 1 + \exp\left[- \frac{(p_{\parallel 1} - p_{\parallel 2})^2}{2(\overline{\Delta p_\parallel})^2} - \frac{(\vec{p}_{\perp 1} - \vec{p}_{\perp 2})^2}{2(\overline{\Delta p_\perp})^2} \right] \qquad (2.7)$$

can be adopted. For two particles having equal momenta $\vec{p}_1 \equiv \vec{p}_2$, f reproduces the factor 2 demanded by statistics /2/; it gives a Gaussian-like smearing around $\vec{p}_1 = \vec{p}_2$ and for very different momenta $\vec{p}_1 \neq \vec{p}_2$ gives the factor 1. Via the uncertainty principle, the parameters $\overline{\Delta p_\parallel}, \overline{\Delta p_\perp}$ are related to the transverse $\Delta x_\perp \simeq 1/\Delta p_\perp$ and longitudinal extension $\Delta x_\parallel \simeq 1/\Delta p_\parallel$ of the interaction region.

iii) the transverse momentum cut-off. Adopting a thermodynamic law for the cluster decay this is contained in the term

$$\sim \exp(-E/T) = \exp(-m_\perp \cosh y / T) \qquad (2.8)$$

$m_\perp = (p_\perp^2 + m^2)^{1/2}$ is the particle's transverse mass.

The Δy, $p_{\perp i}$ and ϕ dependence due to the short range term and the Bose effect for the one-cluster contribution to the two-particle distribution is given by

$$Q_2(\Delta y, s)\, I_{1F}(\Delta y, \phi) = N_1 \int dy_1 dy_2 dp_{\perp 1} dp_{\perp 2}\, \delta(\Delta y - (y_1 - y_2))\, d\eta$$

$$p_{\perp 1} p_{\perp 2} \exp\left(-\frac{m_{\perp 1}\cosh(y_1-\eta)}{T}\right) \exp\left(-\frac{m_{\perp 2}\cosh(y_2-\eta)}{T}\right)$$

$$\left\{1 + \exp\left[-\frac{1}{2(\overline{\Delta p})^2}\left(m_{\perp 1}\sinh y_1 - m_{\perp 2}\sinh y_2\right)^2 \right.\right.$$

$$\left.\left. + p_{\perp 1}^2 + p_{\perp 2}^2 - 2 p_{\perp 1} p_{\perp 2} \cos\phi\right]\right\} \qquad (2.9)$$

and for the two-cluster contribution to the two-particle distribution is given by

$$I_{2F}(\Delta y, \phi) = N_2 \int dy_1 dy_2 dp_{\perp 1} dp_{\perp 2}\, p_{\perp 1} p_{\perp 2}\, \delta(\Delta y - (y_1 - y_2))$$

$$\exp\left(-\frac{m_{\perp 1} + m_{\perp 2}}{T}\right)\left[1 + \exp\left(-\frac{1}{2(\overline{\Delta p_\parallel})^2}\left\{m_{\perp 1}\sinh y_1 - m_{\perp 2}\sinh y_2\right\}^2\right.\right.$$

$$\left.\left. - \frac{1}{2(\overline{\Delta p_\perp})^2}\left\{p_{\perp 1}^2 + p_{\perp 2}^2 - 2 p_{\perp 1} p_{\perp 2}\cos\phi\right\}\right)\right] \qquad (2.10)$$

We obtain the inclusive two-particle distribution for unlike particles (+,-)

$$\mathcal{N}_2^{unlike}(y_+, y_-, \phi, p_{\perp +}, p_{\perp -}) = \frac{1}{\sigma_{in}} \frac{d\sigma_{in}}{dy_+ dy_- d\phi\, dp_{\perp +} dp_{\perp -}} =$$

$$N_1\, p_{\perp +} p_{\perp -} \exp\left[-\frac{m_{\perp +}\cosh y_+}{T}\right] \exp\left[-\frac{m_{\perp -}\cosh y_-}{T}\right]$$

$$\frac{\langle \nu_+ \nu_- \rangle}{\langle \nu_+ \rangle} \langle n_+ \rangle \left(1 - \frac{\pi}{2} B_{1F}^{+-} \frac{p_{\perp +} p_{\perp -}}{\langle p_\perp \rangle^2}\cos\phi\right) +$$

$$+ N_2\, p_{\perp +} p_{\perp -} \exp\left(-\frac{m_{\perp +} + m_{\perp -}}{T}\right) Q_1(y_+, s)\, Q_1(y_-, s)\,.$$

$$\cdot\left[\langle n_+n_-\rangle - \frac{\langle \nu_+\nu_-\rangle}{\langle \nu_+\rangle}\langle n_+\rangle\right]\left(1 - \frac{\pi}{2}B_{2F}^{+-}(s)\,\frac{p_{\perp +}p_{\perp -}}{\langle p_\perp\rangle^2}\cos\phi\right) \tag{2.11}$$

for like particles (1,2)

$$N_2^{like}(y_1, y_2, \phi, p_{\perp 1}, p_{\perp 2}) = \frac{1}{\sigma_{in}}\,\frac{d\sigma_{in}}{dy_1\,dy_2\,d\phi\,dp_{\perp 1}\,dp_{\perp 2}} =$$

$$= \int d\eta\; N_1\, p_{\perp 1} p_{\perp 2} \exp\left[-m_{\perp 1}\frac{\cosh(y_1-\eta)}{T}\right]\exp\left[-m_{\perp 2}\frac{\cosh(y_2-\eta)}{T}\right]$$

$$\left[1 + \exp\left\{-\frac{1}{2(\overline{\Delta p})^2}\left(m_{\perp 1}\sinh y_1 - m_{\perp 2}\sinh y_2\right)^2 + p_{\perp 1}^2 + p_{\perp 2}^2\right.\right.$$

$$\left.\left. - 2p_{\perp 1}p_{\perp 2}\cos\phi\right\}\right]\langle n\rangle\frac{\langle \nu(\nu-1)\rangle}{\langle \nu\rangle}\left(1 - \frac{\pi}{2}B_{1F}^{\pm\pm}\frac{p_{\perp 1}p_{\perp 2}}{\langle p_\perp\rangle^2}\cos\phi\right)$$

$$+ N_2\, p_{\perp 1}p_{\perp 2}\exp\left(-\frac{m_{\perp 1}+m_{\perp 2}}{T}\right)\left[1 + \exp\left(-\frac{1}{2(\overline{\Delta p_\parallel})^2}\left\{m_{\perp 1}\sinh y_1 -\right.\right.\right.$$

$$\left.\left.\left. - m_{\perp 2}\sinh y_2\right\}^2 - \frac{1}{2(\overline{\Delta p_\perp})^2}\left\{p_{\perp 1}^2 + p_{\perp 2}^2 - 2p_{\perp 1}p_{\perp 2}\cos\phi\right\}\right)\right]$$

$$\cdot Q_1(y_1,s)\,Q_1(y_2,s)\left[\langle n(n-1)\rangle - \langle n\rangle\frac{\langle \nu(\nu-1)\rangle}{\langle \nu\rangle}\right]\cdot$$

$$\cdot\left(1 - \frac{\pi}{2}B_{2F}^{\pm\pm}(s)\,\frac{p_{\perp 1}p_{\perp 2}}{\langle p_\perp\rangle^2}\cos\phi\right) \tag{2.12}$$

Here the transverse momentum dependence is made explicit. η is the rapidity of the cluster. The normalisation constants N_1 and N_2 are given by the normalisations of $Q_1(y,s)$ and $Q_2(y_1,y_2,s)$ to 1.
The data can be described with the parameters of the cluster decay

$$\frac{\langle \nu(\nu-1)\rangle}{\langle \nu\rangle} = 0.4 \quad \text{and} \quad \frac{\langle \nu_+\nu_-\rangle}{\langle \nu_+\rangle} = 1.3\ ;$$

whose numerical values are suggested by data on $\bar{p}p$ annihilation and the statistical bootstrap model, the long range asymmetry parameter

$$B_{2F}\ (\Delta y > 2) = \begin{cases} 0.08 & \text{at } 40\ \text{GeV/c} \\ 0.06 & \text{at } 100\ \text{GeV/c} \\ 0.04 & \text{at } 200\ \text{GeV/c} \end{cases}$$

and the asymmetry parameter B_{1F} due to one-cluster decay, which is related to $\langle v \rangle$, the cluster mass M and the cluster transverse momentum $\langle k_\perp \rangle$

$$B_{1F} = \frac{1}{\langle v-1 \rangle} + \left\langle \frac{k_\perp^2}{M^2 + k_\perp^2} \right\rangle \sqrt{1 + \frac{m^2}{\langle p_\perp^2 \rangle}} \; . \qquad (2.13)$$

Good numerical values are

$$B_{1F}^{+-} = 0.12 \quad \text{and} \quad B_{1F}^{\pm\pm} = 0.08 \; .$$

The comparison of the theory with the data is shown in Figs. 1.1 to 1.5. Here we have defined in Fig. 1.4

$$B(\Delta y, q_\perp, \Delta p_\perp < \delta) = \frac{A(\Delta y, p_{\perp 1}, \delta) - C(\Delta y, p_{\perp 1}, \delta)}{A(\Delta y, p_{\perp 1}, \delta) + C(\Delta y, p_{\perp 1}, \delta)}$$

and in Fig. 1.5 (2.14)

$$B(\Delta y, \Delta p_\perp > \delta) = \frac{D(\Delta y) - E(\Delta y)}{D(\Delta y) + E(\Delta y)} \qquad (2.15)$$

with

$$A(\Delta y, p_{\perp 1}, \delta) = \int_{\pi/2}^{\pi} d\phi \int_{p_{\perp 1} - \delta/2}^{p_{\perp 1} + \delta/2} dp_{\perp 2} \, \mathcal{N}_2(\phi, \Delta y, p_{\perp 1}, p_{\perp 2}) \qquad (2.16)$$

and

$$D(\Delta y) = \int_{\pi/2}^{\pi} d\phi \int_0^\infty dp_{\perp 1} \left[\int_0^{p_{\perp 1} - \delta/2} dp_{\perp 2} \mathcal{N}_2(\ldots) + \int_{p_{\perp 1} + \delta/2}^{\infty} dp_{\perp 2} \mathcal{N}_2(\ldots) \right] \qquad (2.17)$$

and $C(\Delta y, p_{\perp 1}, \delta)$ $(E(\Delta y))$ being defined similarly to $A(\Delta y, p_{\perp 1}, \delta)$ $(D(\Delta y))$ for ϕ in the range 0 to $\pi/2$. In eq. (2.14) (corresponding to Fig. 1.4) we integrate over events inside a ring of width δ around the collision axis for fixed $|\vec{p}_{\perp 1}| = p_{\perp 1}$. In eq. (2.15) (corresponding to Fig. 1.5) we integrate over all events besides of those inside the ring of width δ. This ring shifts with $p_{\perp 1} = |\vec{p}_{\perp 1}|$. The parameters of cluster extension in eq. (2.10) have the values

$$\overline{\Delta p_\perp} = 0.3 \pm 0.05 \text{ GeV/c}$$

$$\overline{\Delta p_\parallel} = 0.15 \pm 0.05 \text{ GeV/c}$$

suggested by the data.

3. Bose effect and second order interference

The difference between like and unlike particle correlation also has been explained by the analogy to the Hanbury-Brown-Twiss effect /8/. An analysis was proposed in terms of the so called $\Delta\Delta$ plot which studies the two-particle distribution over the $(q_0, \Delta q_\perp^2)$ plane with $q_0 = |E_1 - E_2|$ ($\vec{p}_1$, E_1 and $\vec{p}_2$, E_2 are the momenta and energies of the two particles) and $\Delta q_\perp$ is given in the following way:
If

$$\vec{n} = \frac{(\vec{p}_1 + \vec{p}_2)}{|\vec{p}_1 + \vec{p}_2|} \quad \text{and} \quad \vec{w} = \vec{p}_1 - \vec{p}_2 ,$$

than

$$w_{\parallel} = \vec{w} \cdot \vec{n}$$

and

$$\Delta\vec{q}_\perp = \vec{w} - w_{\parallel} \cdot \vec{n} \tag{3.1}$$

This analysis has been carried through in a number of experiments; it leads to a peak for small q_0 and small $\Delta q_\perp^2$, from which according to the probability to observe like meson pairs

$$P(q_0, q_\perp) \sim 1 + \frac{I(q_\perp R)}{1 + q_0^2 \tau^2} \tag{3.2}$$

(I is the Bessel function) the interaction radius R and the interaction time τ has been determined by fit /9/. This effect is due to the path ambiguity of two identical bosons being emitted at two different space time points, which demands a symmetrisation in the co-ordinates of the two particles. This effect amounts to intensity correlations which are constructive for bosons; it will be present even with randomly emitted particles (similarly to the astronomical case, where this second order interference serves to determine the radius of stars, uses randomly emitted photons). The effect described in Section 2 however is related to a dynamical effect (i.e. a correlated emission) which can be understood via statistics.
A clear discussion of the distinction of these two effects was given by Kripfganz /10/ who also proposes examples in which these effects can be studied separately. In particular a quantum statistical treatment including full isospin conservation predicts correlations of unlike bosons, such as $\pi^+\pi^-$. These correlations are not present if only charge conservation is

taken into account. Here no second order interference can build up. Although these correlations are in the order of a few per cent only in most cases (and therefore undetectable), they are big and negative in the decay

$$\Psi_{3.1}\ (I=0,\ G=-1) \longrightarrow \pi^+\pi^- + (N\pi^0)\ .$$

It should be possible to search for this effect. Another example is the Bose peak in $\pi^0\pi^0$; in the decay $\pi^0 \rightarrow 2\gamma$ second order interference is superimposed to the effects present in the $\pi^0\pi^0$ distribution, it makes the peak in $\gamma\gamma$ correlations more narrow.

4. Azimuthal correlations and local conservation of transverse momentum

The hypothesis of local conservation of transverse momentum in non-diffractive multi-particle production processes was proposed by Krzywicki and Weingarten /11/. It was used to derive a bound on the Pomeron slope. Due to difficulties in detecting all neutral particles a direct experimental test of this hypothesis is rather difficult.
However, it seems possible to test the concept of local $p_\perp$ conservation in experiments detecting only charged particles if the azimuthal asymmetry parameter (2.1) is used. In order to exclude diffractive processes, only large multiplicity events should be considered.
In ref. /12/ performed two Monte Carlo calculations producing particles in an UJM like fashion; intermediate cluster production is not important for the argument. The two calculations are

i) transverse momentum of each particle is compensated by all other particles regardless of their distance in rapidity, as in the usual UJM.
ii) transverse momentum of each particle is compensated locally in rapidity space. In events with n particles with rapidities y_i (i = 1, ..., n) we define weights

$$w_{ij} \sim \exp\left(-\frac{y_i - y_j}{\lambda}\right)\ ,\quad \lambda \simeq 1 \qquad (4.1)$$

and balance the transverse momentum of particle j by all other particles i = j according to the weights w_{ij} .

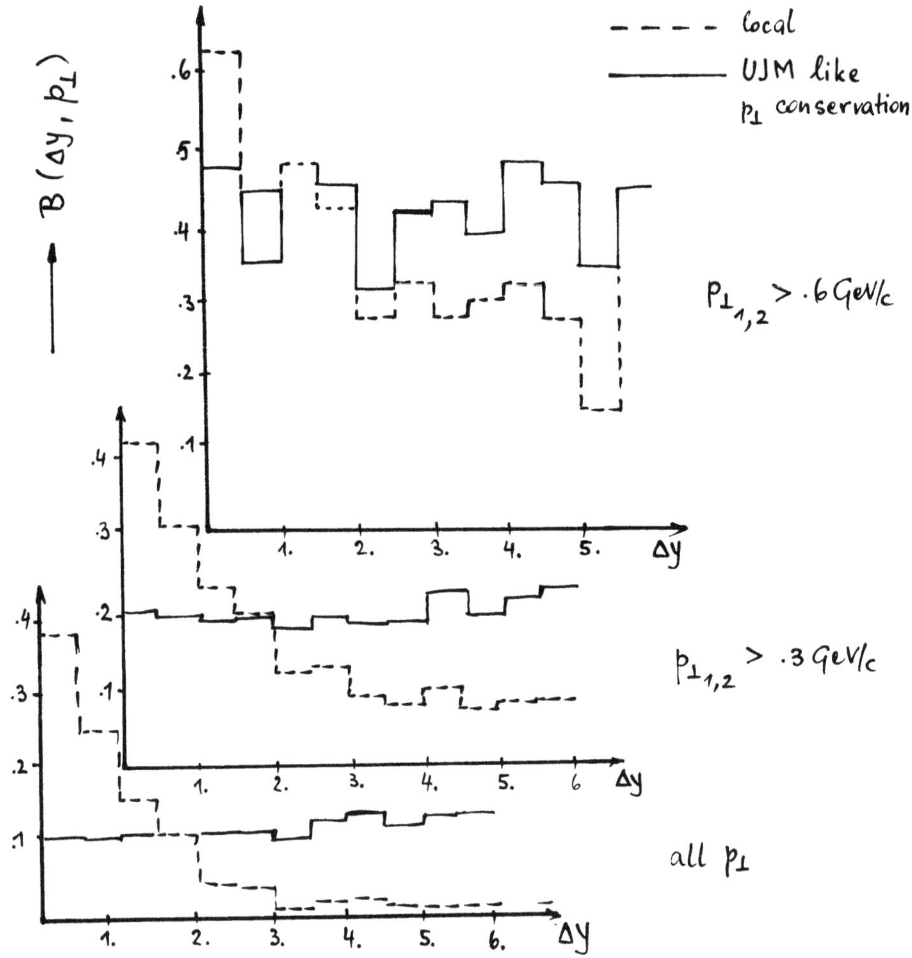

Fig. 4.1
The azimuthal asymmetry parameter $B(\Delta y, p_\perp)$ for charged particles plotted as function of Δy for three different $p_\perp$ cut-offs in the two models
(i) with UJM like $p_\perp$ conservation, ———
(ii) with local $p_\perp$ conservation, — — —

In both calculations the transverse momentum compensation is done for all particles produced in the event, but we calculate the asymmetry parameter defined in eq. (2.1) only by considering the charged particles.
In Figure 4.1 we plot $B(\Delta y)$ as function of Δy

a) without $p_\perp$ cut-off
b) only for particle pairs with $p_{\perp i}$ and $p_{\perp j} > 0.3$ GeV/c
c) as case b) but with $p_{\perp i}$ and $p_{\perp j} > 0.6$ GeV/c.

In all three cases we find:

i) in the uncorrelated jet-like model, $B(\Delta y)$ is independent of Δy and
ii) in the model with local p conservation strong short range behaviour is present.

No data is available on $B(\Delta y)$ of charged particles for large multiplicities only, excluding diffractive events savely. In the data on $B(\Delta y)$ shown in Figure 1.2 b) leading particles have been subtracted and in this way also some diffractive events have been eliminated. These data show very strong long range behaviour rather similar to our calculation i) (i.e. non-local $p_\perp$ conservation). It is difficult to believe that all of this is due to the diffractive component, however, no firm conclusion can be drawn. Weingarten /12/ also has presented analytical arguments that it should be possible to test local compensation of transverse momentum in experiments detecting only charged particles. His conclusions are similar to ours but we here are more specific in proposing to use the quantity $B(\Delta y)$. These conclusions are also reached by Graßberger et al. /14/.
It also was studied in the Monte Carlo calculation /15/ whether by rotation the secondaries by an angle Θ (around an axis perpendicular to the collision plane) the observes long range correlations in $B(\Delta y)$ (see Figure 1.2 b)) could be manufactured from Monte Carlo events produced with local $p_\perp$ conservation. This would have demonstrated that in particle production even at low $p_\perp$ there is an axis of preference (different from the collision axis), which would minimize the correlation length. The non-local $p_\perp$ conservation - suggested by the data - then would be a result of the rotation of the particle jet relative to the collision axis. This way the peak of $B(\Delta y)$ at small Δy in the case of local $p_\perp$ conservation is reduced, but even rotations by rather large angles do not produce the long range effect present in the data (e.g. for $\sqrt{s} = 56$ GeV, the peak is reduced from $B(\Delta y \simeq 0) \simeq 0.53$ to $B(\Delta y \simeq 0) \simeq 0.38$ for $\Theta \simeq 1^\circ$ without a significant change for $\Delta y \gtrsim 2$.

5. Two-particle correlations and band structure

The band structure observed in the invariant mass spectrum of two hadrons i and j (i and j being p, $\bar{p}$, π^+, π^- or K^-) in the reaction $p + Be \rightarrow h_i + h_j + X$ /16/ can be considered as a special kind of a two-particle correlation, since the invariant mass is $M_{ij} = (p_i + p_j)^2$. The measurements are done at $x_i = x_j = 0$. It is /17/

$$\left.\frac{dN}{dM_{ij}}\right|_{x_i = x_j = 0} = \left(1 + R^{ij}(0,0)\right) f_1^i(0,p_{\perp i}^2) f_1^j(0,p_{\perp j}^2) \cdot \frac{2\mu_i \mu_j}{M_{ij}} \left(1 - \frac{\pi}{2} B \frac{p_{\perp i} p_{\perp j}}{\langle p_\perp \rangle^2} \cos\phi\right) \tag{5.1}$$

where $f_1^i(x,p_\perp^2)$ and $R^{ij}(y_i=0,y_j=0)$, $\mu_{\perp i}$, B and ϕ are the single particle spectrum of particle i and the rapidity correlation, the transverse mass, the azimuthal asymmetry and the azimuthal angle of the two particles i and j. In the experiment the main contribution is due to $\cos\phi = -1$. With thermodynamic single particle spectra $f_1(x,p_\perp)$ for $p_\perp < 1.5$ GeV/c and a fit of the constituent interchange model to spectra at higher $p_\perp$, as well as with $R^{ij}(0,0) \simeq R^{\pi_{ch}\pi_{ch}}(0,0) \simeq 0.6$ used for all combinations of secondaries, the observed band structure at 30 GeV/c can be understood (Figure 5.1). However, due to the rise with s of the single particle spectra at x = 0, which is different for the various pairs of secondaries, the band structure is predicted to change drastically with energy. As an example, Figure 5.2 shows the change of the band structure at M_{ij} = 2 GeV from 30 GeV/c to 300 GeV/c.

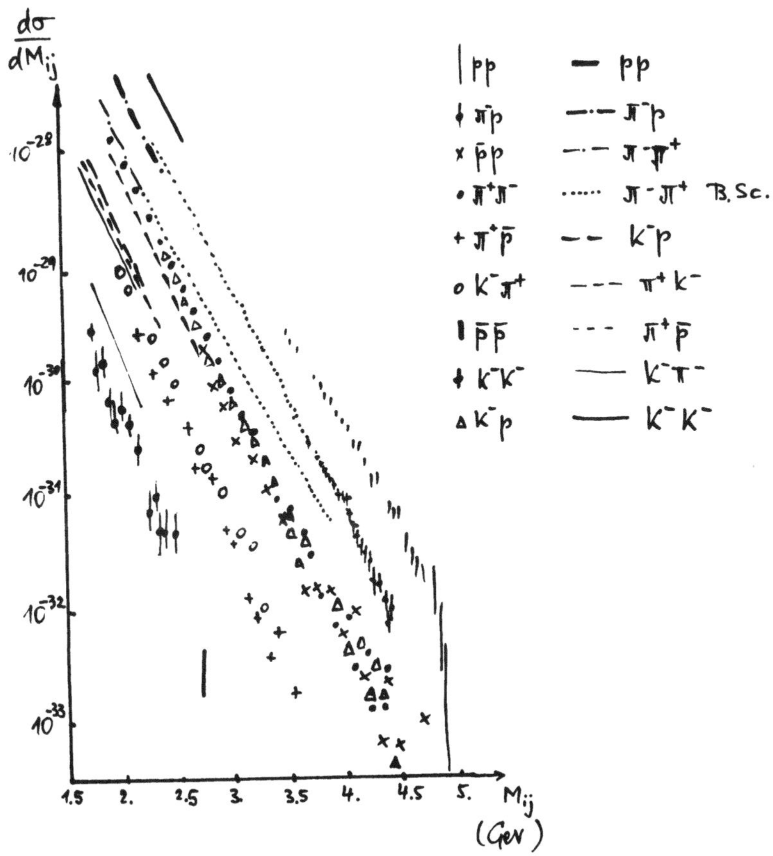

Fig. 5.1
Comparison of the observed (ref./16/) two-particle mass distributions at 30 GeV/c with eq. (5.1), see text. From ref. /17/.

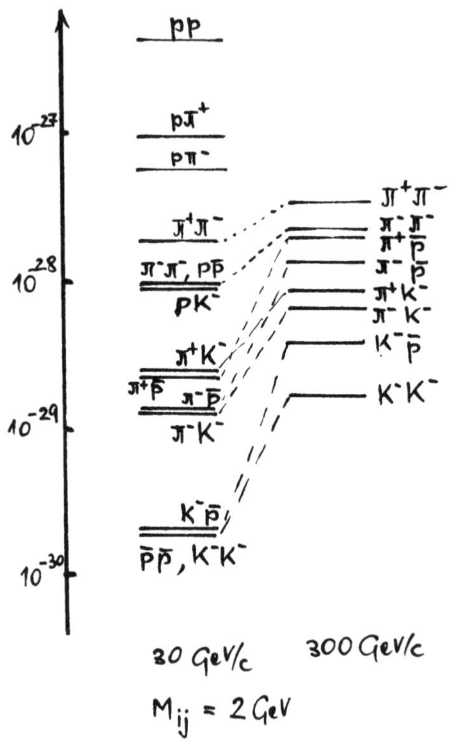

Fig. 5.2
Change of the band structure of the two-particle invariant mass at M_{ij} = 2GeV from 30 to 300 GeV/c as predicted by eq. (5.1) with thermodynamic single particle spectra. From ref. /17/.

References

/ 1/ G. Ranft and J. Ranft, Nucl. Phys. B 53 (1973) 217.
/ 2/ G. Ranft, J. Ranft, M. Sabau, K.G. Fadeev, and A.N. Solomin, Nucl. Phys. B 86 (1975) 63.
/ 3/ G. Ranft and J. Ranft, Nucl. Phys. B 92 (1975) 207.
/ 4/ B.Y. Oh et al., Phys. Letters (1975)
/ 5/ G. Ranft and J. Ranft, Phys. Letters 57B (1975) 373.
/ 6/ G. Ranft, J. Ranft, R. Kirschner, N.G. Fadeev, S. Lutov, and R.M. Nazargulov, Nucl. Phys. B, to appear.
/ 7/ G. Ranft and J. Ranft, Nucl. Phys. B 83 (1974) 285.
/ 8/ G.I. Kopylov and M.I. Podgoretzki, Jad. Fiz. 18 (1973) 656.
E.V. Shurjak, Phys. Letters 44B (1973) 387.
G. Cocconi, Phys. Letters 49B (1974) 459.
/ 9/ G.I. Kopylov, " Δ Δ plot", Dubna preprint E2-8549 (1975).
A. Para, Talk at the IV Internat. Seminar on High Energy Physics Problems, Dubna, June 1975.
/10/ J. Kripfganz, "Quantum statistical correlations in the frame work of an uncorrelated jet model", Karl-Marx-University preprint KMU-HEP 7607 (1976).
/11/ A. Krzywicki, Nucl. Phys. B 86 (1975) 296.
D. Weingarten, Phys. Rev D 11 (1975) 1924.
/12/ J. Ranft and G. Ranft, "Azimuthal asymmetry and local p conservation", Karl-Marx-University preprint KMU'HEP 7514 (1975).
/13/ D. Weingarten, Rochester preprint 1975.
/14/ P. Grassberger and J.L. Meunier, private communication (1976).
/15/ J. Ranft, unpublished.
/16/ J.J. Aubert et al., Phys. Rev. Lett. 35 (1975) 639.
/17/ J. Ranft and G. Ranft, Phys. Rev. Lett. 36 (1976) 988.

DISCUSSION

CHAIRMAN: Prof. G. Ranft

Scientific Secretary: F. Bopp

DISCUSSION

- *FERBEL:*

I just wish to stress that the latest measurements on zone-zone correlations from the 30 inch bubble chamber for 200 GeV/c and 300 GeV/c pp collisions indicate that data obey the sort of features expected from local compensation of transverse momentum -- as pointed out by Weingarten et al. Using these results, Weingarten has extracted a lower limit of 0.2 GeV^{-2} for the slope of the Pomeranchuk trajectory. The value of α'(Pomeranchuk) is $\sim$ 0.25 GeV^{-2}, which implies that local compensation of transverse momentum in multiparticle production is responsible for a large fraction of the shrinkage of the elastic scattering distribution with increasing energy.

- *RANFT:*

So far, most analysis leads to different conclusions. Pure local transverse momentum conservation could not be established.

- *WEILL:*

What is the connection between the correlation described here and the "Goldhaber" effect studies in $\bar{p}p$ pionic annihilation? In this one, if one restricts Δp -- the difference of the momenta of the pions -- one observes no effect for like-pion pairs; but one observes an increase in the negative correlations for unlike pions.

- *RANFT:*

The considered effect should be of the same nature. The reported observations could arise from the use of wrong variables or insufficient narrow intervals.

- *BASILE:*

To what extent do diffractive processes contribute to the $\pi\pi$ correlations?

- *RANFT:*

Diffractive processes are negligible, as events with high multiplicity ($n_c > 6$) have been selected.

- *BASILE:*

If particles were not identified, how can you obtain the Bose effect?

- *RANFT:*

The particles are predominantly pions. The influence of other particles is, in this respect, negligible.

- *BERLAD:*

Is one to understand the rise of low mass $\bar{p}p$ pairs from 30 GeV to 300 GeV as growth of exotic cluster production with energy?

- *RANFT:*

No, they are completely explained in terms of single particle distributions, which rise at x = 0, and two-particle rapidity correlations.

- *BERLAD:*

Do the correlations at small Δy and $\Delta\phi$ appear in the same form in low and high multiplicity inclusive data?

- *RANFT:*

Yes, the effect is consistent at different multiplicities.

MONOPOLES

Patrizio Vinciarelli*

CERN

Geneva, Switzerland

1. INTRODUCTION

In these lectures I would like to discuss some recent work done in collaboration with Walter Troost at CERN.[1)] Under very general conditions, we are led to predict the existence of magnetic monopoles. To put our work in perspective, I will review briefly what I regard as key developments in the history of monopoles.

The symmetry of the free Maxwell equations under transformations of electric into magnetic fields led Dirac to suggest their existence.[2)] In his approach, monopoles are introduced in the theory in analogy with electrons, via independent field degrees of freedom. Though motivated by symmetry requirements, Dirac's extension of electrodynamics was a matter of choice, as for the introduction of a muon field. The inclusion of monopoles had however immediate implications of considerable theoretical interest, since it poses very stringent constraints for the consistency of the theory. I have in mind:

a) Dirac's famous "quantization condition," which could account for the universality of the electric charge;

b) gauge invariance of the second kind, which could account for the masslessness of the photon;

c) "duality" of the electromagnetic field, which could account for our living in 4-dimensional space-time.

On the other hand, Dirac's extension was faced by what then appeared to be a technical nuisance: the description of the electromagnetic field in the presence of a magnetic monopole in terms of the e.m. potential. Dirac adopted a potential with a singularity line, the (in-)famous "Dirac string", and modified the connection of the e.m. field tensor to the potential:

$$F_{\mu\nu} = (\partial \times A)_{\mu\nu} \quad - \quad \text{fictitious string field.} \qquad (1)$$

To many people this "trick" is not esthetically pleasing. Objectively, it led to the introduction of dynamical coordinates and momenta to describe the motion of the string, a fictitious notion. It took a real "tour de force" of Dirac to derive the theory from an action principle.[3] Also, it has been a source of ambiguities, as witnessed by controversies about the precise character of the "quantization condition."[4]

In spite of these and other difficulties, in light of the simplicity of the idea and, undoubtedly, the reputation of his author, much effort has been put over the years in (unsuccessful) experimental searches, yielding ever increasing lower bounds on the monopole mass and decreasing upper bounds on its production cross section. (Un-)fortunately the mass of the monopole is a free parameter in Dirac's theory, which can be adjusted to accomodate any lower bound.

The disillusionment caused by continuing experimental failures was enhanced in recent years by theoretical developments. The advent of unified gauge theories of weak and electromagnetic interactions modified our visions of symmetries, while prodiving an alternative explanation for the universality of the electric charge. It is amusing to note that these developments weakened the case for Dirac's theory, but eventually suggested a conceptually deeper and more beautiful realization of Dirac's original suggestion.

In unified gauge theories, the non-abelian gauge group is spontaneously broken and electromagnetic gauge invariance remains as a left-over unbroken U(1) subgroup. Then, as demonstrated by Polyakov and 't Hooft,[5)] particles carrying a magnetic charge with respect to this subgroup appear as soliton solutions to the classical field equations. These solutions do not require the introduction of external sources. They correspond to bona fide bound states of the theory with a non-vacuum quantum number, a magnetic charge, which may be identified with a topological charge.[6)] The energy of the solutions is finite. The mass of the bound states, which for small values of the gauge coupling constant is given by the energy of the classical solutions, is therefore calculable and was estimated to be $E \simeq M_W/\alpha \simeq 10^4$ GeV. This large value would account for the failure of past searches in accelerator experiments. It would also predict the failure of similar searches in the foreseeable future.

The existence of magnetic monopoles is therefore compelling within the theoretical framework of unified gauge theories. However the framework itself is special, e.g. for its reliance of the Higgs-Kibble mechanism for spontaneous symmetry breakdown. It is then of interest to abstract from previously known examples and inquire about the most general conditions leading to the existence of magnetic monopoles as solitons.

2. - THE MODEL OF 'T HOOFT AND POLYAKOV

The model of 't Hooft and Polyakov[5] is based on the SU(2) gauge Lagrangian:*)

$$\mathcal{L} = -\tfrac{1}{4}F^{a\mu\nu}F^a_{\mu\nu} + \tfrac{1}{2}(D^\mu\phi^a)(D_\mu\phi^a) - V(\phi) \ , \qquad (2)$$

$$F^a_{\mu\nu} = \partial_\mu A^a_{\ \nu} - \partial_\nu A^a_{\ \mu} + e\varepsilon^{abc}A^b_{\ \mu}A^c_{\ \nu} \ , \qquad (3)$$

$$D_\mu\phi^a = \partial_\mu\phi^a + e\varepsilon^{abc}A^b_{\ \mu}\phi^c \ . \qquad (4)$$

Here ϕ^a and $V(\phi)$ denote a triplet of scalar (Higgs) fields and

*) Our conventions are $g^{\mu\nu}$ = diag. (1,-1,-1,-1); ε^{abc} is completely antisymmetric with $\varepsilon^{123} = 1$.

its self-interacting potential leading to a non-vanishing (vacuum expectation) value of ϕ^a in the ground state solution:

$$\phi^a = \delta^{3a}F, \qquad A^a_\mu = 0 \quad , \tag{5}$$

with the usual choice of gauge. The 1,2 components of the vector field will then acquire a mass (eF) whereas the third component will describe the surviving Abelian electromagnetic interaction.

The field equations admit also a soliton solution of the form[5)]

$$\phi^a = \frac{r^a}{r}\,\phi(r), \quad A^a_i = \varepsilon^{aij}\,\frac{r^j}{r}\,A(r), \quad A^a_o = 0 \quad , \tag{6}$$

with

$$\phi(r) \xrightarrow[r\to\infty]{} F, \qquad A(r) \xrightarrow[r\to\infty]{} \frac{1}{er} \quad . \tag{7}$$

To interpret it, 't Hooft proposed to use the following gauge invariant definition of the e.m. field:

$$F_{\mu\nu} = \frac{1}{|\phi|}\,\phi_a F^a_{\mu\nu} - \frac{1}{e|\phi|^3}\,\varepsilon_{abc}\phi^a(D_\mu\phi^b)D_\nu\phi^c \tag{8}$$

which reduced to the familiar one in the usual (vacuum) gauge:

$$F_{\mu\nu} = \partial_\mu A^3_\nu - \partial_\nu A^3_\mu, \quad \phi^a = \delta^{a3}F \quad . \tag{9}$$

Expression (8), evaluated for the solution (6), yields

$$F_{ij} = -\frac{1}{er^2}\,\varepsilon_{ijk}\,\frac{r^k}{r} \quad , \tag{10}$$

representing the radial magnetic field generated by a magnetic pole of strenght 1/e located at the origin:

$$\frac{1}{4\pi}\ \frac{1}{2}\,\partial_i\ \varepsilon^{ijk}\ F_{jk} = \frac{1}{e}\,\delta(\underline{x}) \tag{11}$$

Hence, the interpretation of the solution (6) as describing a magnetic monopole carrying two Dirac units of magnetic charge g, corresponding to n=2 in the quantization condition:

$$eg = \frac{1}{2}\,n \quad . \tag{12}$$

It is tempting to wonder about the "origin" of this charge. Arafune, Freund and Goebel[6)] noted that $F_{\mu\nu}$, as defined in Eq. (8), may be rewritten in the form

$$F_{\mu\nu} = M_{\mu\nu} + H_{\mu\nu} \quad , \tag{13}$$

where

$$M_{\mu\nu} = \partial_\mu B_\nu - \partial_\nu B_\mu \ , \qquad B_\mu = \frac{\phi^a}{|\phi|}\ A^a_\mu \quad , \tag{14}$$

$$H_{\mu\nu} = \frac{1}{e}\ \varepsilon_{abc}\hat{\phi}^a\ \partial_\mu\hat{\phi}^b\ \partial_\nu\hat{\phi}^c \ , \qquad \hat{\phi}^a = \frac{\phi^a}{|\phi|} \quad . \tag{15}$$

They observe that the magnetic current

$$K_\mu = \frac{1}{2}\ \varepsilon_{\mu\nu\rho\sigma}\ \partial^\nu\ F^{\rho\sigma} \tag{16}$$

for the solution (6) reduces to

$$K_\mu = \frac{1}{2}\,\varepsilon_{\mu\nu\rho\sigma}\partial^\nu H^{\rho\sigma} = \frac{1}{2e}\ \varepsilon^{\mu\nu\rho\sigma}\varepsilon_{abc}\partial^\nu\hat{\phi}^a\partial^\rho\hat{\phi}^b\partial^\sigma\hat{\phi}^c \quad , \tag{17}$$

leading to an expression for the magnetic charge $\tilde{Q}$ which involves only the asymptotic behavior as $R \to \infty$ of the triplet of Higgs fields:

$$\tilde{Q} = \frac{1}{4\pi} \int d^3x\, k_o = \lim_{R\to\infty} \frac{1}{8\pi e} \int_{S^2_R} (d^2\sigma)_i\, i\varepsilon^{ijk} \varepsilon_{abc} \hat{\phi}^a \partial_j \hat{\phi}^b \partial_k \hat{\phi}^c \ . \qquad (18)$$

They then proceed to identify the magnetic charge of the solution (6) with (equivalent) topological characters of the Higgs fields configurations (Kronecker index, Brouwer degree, homotopy class, Poincaré-Hopf index).

One is thus led to believe that the following conditions are necessary for the existence of monopoles as solitons:

a) the imbedding of U(1) in larger gauge groups;

b) the spontaneous breakdown of such groups to U(1) via the Higgs mechanism, in the presence of multiplets of spinless fields.

Is this really necessary?

3. - MODEL NO. 1 IN THE RADIAL GAUGE

Let us consider the SU(2) gauge Lagrangian

$$\mathcal{L} = -\frac{1}{4} F^{a\mu\nu} F^a_{\mu\nu} + \frac{1}{2} (D^\mu \eta^a) D_\mu \eta^a, \ \vec{\eta}^{\,2} = \frac{M^2}{e^2} \qquad (19)$$

where $\vec{n}$ denotes a real spinless isovector field subject to the constraint $\vec{n}^2 = M^2$ [7]. This constraint may be implemented with the addition of a Lagrange multiplier term

$$\mathcal{L}' = \frac{1}{2} \lambda(x) \left(\vec{\eta}^{\,2} - \frac{M^2}{e^2} \right) \qquad (20)$$

to the Lagrangian (19). The constraint eliminates one of the three field degrees of freedom of $\vec{n}$. The model which is obtained may be understood as a suitable strong coupling limit of models defined by the Lagrangian (2).

The equations of motion are

$$D_\nu F^{a\mu\nu} = e\varepsilon^{abc} n^b D^\mu n^c , \tag{21}$$

$$D_\nu D^\nu n^a = \lambda(x)\, n^a . \tag{22}$$

In analogy with (6), we look for static soliton solutions of the form

$$n^a = \frac{M}{e}\frac{r^a}{r} , \quad A_i^a = \varepsilon^{aij}\frac{r^j}{r}A(r) , \quad A_0^a = 0 \tag{23}$$

In contrast to (6), this ansatz involves only one reduced function A(r), which controls the amplitude of the gauge fields. The amplitude for the isovector field is constrained to M. The orientation of $\vec{n}$ in isospin space is not defined at the origin, which is therefore an exceptional point. To avoit ambiguities we shall carefully proceed to: a) solve the Lagrange equations with the ansatz (23) in any domain which excludes the origin; b) extend the solution to the whole of space in the sense of distributions, checking for the absence of external sources.

Substitutions of the ansatz (23) in Eqs. (21-22) for any domain excluding the origin leads to the reduced field equations:

$$K''(r) - \frac{K(K^2-1)}{r^2} = M^2 K(r), \qquad r>0, \tag{24}$$

$$-2\frac{K^2(r)}{r^2} = \lambda(r), \qquad r>0, \tag{25}$$

where

$$K(r) = 1 - erA(r). \tag{26}$$

Equation (25) determines the Lagrange multiplier function λ(r). The dynamical equation (24) corresponds to the reduced Hamiltonian:

$$\mathcal{H} = \frac{4\pi}{e^2}\int_0^\infty dr\left[K'^2 + \frac{(K^2-1)^2}{2r^2} + M^2K^2\right]. \tag{27}$$

It is immediately seen that a solution of Eq. (24) will have a finite energy only if the following boundary conditions are satisfied:

$$K(r) \underset{r\to\infty}{\sim} o(r^{-\frac{1}{2}}), \quad K(r) \underset{r\to 0}{\sim} 1 + o(r^{\frac{1}{2}}). \tag{28}$$

Equation (24) actually implies:

$$K(r) \underset{r\to\infty}{\sim} \not{c}_2 e^{-Mr}, \quad K(r) \underset{r\to 0}{\sim} 1 - \frac{1}{3}M^2r^2 \ln(\not{c}_1 r). \tag{29}$$

With the help of standard mechanical analog techniques, one may convince oneself that indeed there exists a solution of Eq. (24) in the domain $r > 0$ with these asymptotic behaviors. In Fig. 1 the solution, constructed numerically, is displayed (for later reference, this is denoted as the $\mathcal{N} = 1$ curve).

We must now extend this solution to the whole of space in the sense of distributions. We shall proceed carefully since the singularity in the n-field at the origin, though very mild, might induce some people to suspect the presence of an external source in the n-field equation. The problem is well posed if the variations $\frac{\delta \mathcal{L}}{\delta n^a}$ and $\frac{\delta \mathcal{L}}{\delta \partial_i n^a}$ [where $\mathcal{L}$ is the Lagrangian including the multiplier term (20)] are uniquely defined as distributions, which is the case if they are locally summable. As implied by Eq. (28), corresponding to the presumed solution,

$$\frac{\delta \mathcal{L}}{\delta n^a} \underset{r\to 0}{\sim} -2\frac{M}{e}\frac{r^a}{r^3}, \quad \frac{\delta \mathcal{L}}{\delta \partial_i n^a} \underset{r\to 0}{\sim} \frac{m}{er}\left(\delta_{ai} - \frac{r^a r^i}{r^2}\right) \tag{30}$$

and they are indeed locally summable. We must then check that, for any test function T(x), the functional

$$J(T) = \int d^3x\, T(x)\left[\frac{\delta \mathcal{L}}{\delta n^a} - \partial^\mu \frac{\delta \mathcal{L}}{\delta \partial_\mu n^a}\right] \tag{31}$$

vanishes. To this end, we break up the volume integral into two parts extending to the interior and exterior of a sphere of radius R, J_R and $J_{\bar{R}}$ respectively. Clearly $J_{\bar{R}} = 0$. By partial integration,

$$J_R = \int_{V_R} d^3x \left[T(x) \frac{\delta \mathcal{L}}{\delta n^a} + (\partial^\mu T) \frac{\delta \mathcal{L}}{\delta \partial_\mu n^a} \right] + - \int_{S_R^2} (d^2\sigma)^i \, T(x) \frac{\delta \mathcal{L}}{\delta \partial_i n^a} \,, \tag{32}$$

in obvious notations. Since T and $\partial_i T$ are test functions and therefore bounded, Eqs. (30) imply that the volume integrals can be made as small as desired by a proper choice of R. The real <u>test</u> is with the surface integral. Now since,

$$\left| \frac{m}{e r} \left(\delta_{ia} - \frac{r_i r_a}{r^2} \right) \right| \leq \frac{2m}{er} \,, \tag{33}$$

this integral is bounded by

$$\frac{8\pi m R}{e} \left[\max_{|x| \leq R} T(x) \right] , \tag{34}$$

and can therefore be made infinitesimally small as well. Thus $J_R = 0$ and $J = 0$, proving the absence of external sources.

The e.m. field tensor for the present model may be defined, in analogy with Eq. (8), to be:

$$\mathcal{F}_{\mu\nu} = \frac{e}{m} n_a F^a_{\mu\nu} - \frac{e^2}{m^3} \varepsilon^{abc} n_a (D_\mu n_b) D_\nu n_c \,. \tag{35}$$

For our solution it takes the same value as in Eq. (10) and implies a magnetic charge:

$$\tilde{Q} = \frac{1}{8\pi} \int d^3x\, \varepsilon^{ijk}\, \partial_i \mathcal{F}_{jk} = \frac{1}{e}\,. \tag{36}$$

Thus we claim to have found bona fide monopole-solitons in theories defined by the Lagrangian (19).

This result suggests that the answer to the questions raised at the end of the previous section is: no. In fact, making a conventional choice of gauge in the Lagrangian (19), so that

$$\eta^a(x) = \frac{m}{e}\,\delta^{a3} \tag{37}$$

we obtain

$$\mathcal{L} = \frac{1}{4} F^{a\mu\nu} F^a_{\mu\nu} + m^2 W^+_\mu W^{-\mu}\,, \tag{38}$$

where

$$W^\pm_\mu = \frac{1}{\sqrt{2}} (A^1_\mu \pm A^2_\mu). \tag{39}$$

We recognize this as a theory of two massive charged vector mesons interacting with a U(1) (massless) gauge field. Spinless excitations (Higgsons) are not present. Since the Lagrangian (38) is expected to be equivalent to the Lagrangian (19), this theory should exhibit the same monopole-solitons. We will show that this is indeed the case and proceed from there to demonstrate the existence of monopole-solitons in much wider a class of theories.

4. - THE DESCRIPTION OF MONOPOLES IN ABELIAN GAUGES

The transformation connecting "abelian gauges", as defined by Eq. (37), to "radial gauges", as defined by Eq. (23), is singular. This singularity, of topological origin, will reflect itself in drastic differences of language in the description of the same physical systems, monopoles in particular, in the two classes of gauges. We prefer the description in terms of abelian gauges because, as we shall see, it will enable us to present a simple physical interpretation of the binding forces which are responsible for the existence of monopole-solitons solutions. Thus we will understand *why* such solutions exist, carry a magnetic charge, have a finite energy, etc. ...
From another point of view, the description in terms of abelian gauges is also more fundamental, since it is in this class of gauges that the vacuum state and perturbation theory are formulated. Finally, only the description in terms of abelian gauges will be available to us in extending our analysis to theories where U(1) is not imbedded in larger gauge groups. Such imbeddings are in fact *inessential* for the generation of monopole-solitons. They play a cosmetical role, via the use of radial gauges, in allowing a singularity-free description of

monopole solutions in terms of ordinary functions with a domain which extends to the whole of space. This is not possible in abelian gauges, where the description will involve the use of "sections" instead of functions, a notion borrowed from the mathematics of fiber bundles and deeply connected to the gauge principle, which was recently suggested by Wu and Yang.[8)] This notion is not yet familiar among physicists and some were tempted to conclude that "sectional monopoles" are not the same as "radial monopoles". In fact there is no intrinsic difference between them.

The apparent difference is similar to that experienced in the description of the surface of a sphere in terms of spherical (θ,ϕ) and cartesian (x,y,z) coordinates. The surface of the sphere (the analogue of the monopole) does not change with the coordinate system, but in the first case its description involves points of singularity (the analogue of the Dirac string) or a split into "sections",[*]) whereas in the second coordinate systems, which used more variables and is redundant, the description is automatically singularity-free. The imbedding of U(1) into larger gauge groups, such as SU(2), SU(3), etc. ... serves a similar purpose, that of providing a special singularity-free "coordinate system" for the monopole, at the

*Overlapping regions with associated singularity-free coordinate systems and coordinate transformations between different systems in their overlap.

expense of some redundancy. The difference is only one of language.

We will adopt the language of sections. While referring the reader to the original articles of Wu and Yang[8] for a comprehensive discussion, we recall here the basic ideas which are necessary to follow our subsequent developments. Space in the presence of a monopole is divided into two regions, R_a and R_b. With the monopole located at the origin and in terms of spherical coordinates, these may be chosen to be:

$$R_a : 0 \leq \theta < \frac{\pi}{2} + \delta\ , \quad 0 < r\ , \quad 0 \leq \varphi < 2\pi\ , \tag{40}$$

$$R_b : \frac{\pi}{2} - \delta < \theta \leq \pi\ , \quad 0 < r\ , \quad 0 \leq \varphi < 2\pi\ , \tag{41}$$

where $0 < \delta \leq \frac{\pi}{2}$. The e.m. field is described by a potential in each region. For example:

$$(A_0)_a = (A_r)_a = (A_\theta)_a = 0\ , \ (A_\varphi)_a = \frac{g}{r} \tan\frac{\theta}{2}\ , \tag{42}$$

$$(A_0)_b = (A_r)_b = (A_\theta)_b = 0\ , \ (A_\varphi)_b = -\frac{g}{r} \operatorname{ctan}\frac{\theta}{2}\ , \tag{43}$$

in obvious notations. Both Eqs. (42) and (43) define Dirac potentials with a string singularity which is, however, outside their respective domains. In the overlap region

$$R_{ab} : \frac{\pi}{2} - \delta < \theta < \frac{\pi}{2} + \delta\ , \quad 0 < r\ , \quad 0 \leq \varphi < 2\pi \tag{44}$$

the two potentials are related by a gauge transformation

$$(A_\mu)_a = (A_\mu)_b + \frac{i}{e} S_{ab} \partial_\mu S_{ab}^{-1} , \tag{45}$$

where

$$S_{ab} = \exp(i 2 e g \varphi) . \tag{46}$$

The requirement that the gauge transformation be univalued yields Dirac's quantization condition for the monopole charge, Eq. (12). The pair of potentials then defines a "section" which is analytic everywhere, and describes the true field of a monopole without the need to subtract off any fictitious term:

$$\mathcal{F}_{\mu\nu}(x) = \partial_\mu (A_\nu)_a - \partial_\nu (A_\mu)_a , \quad x \in R_a , \tag{47}$$

$$\mathcal{F}_{\mu\nu}(x) = \partial_\mu (A_\nu)_b - \partial_\nu (A_\mu)_b , \quad x \in R_b , \tag{48}$$

$$\frac{1}{8\pi} \partial_i \epsilon^{ijk} \mathcal{F}_{jk} = g \delta(\underline{x}) , \tag{49}$$

via Gauss' theorem. Charged (gauge dependent) matter fields are also defined as sections:

$$\Psi = (\Psi_a, \Psi_b), \quad \Psi_a(x) = S_{ab} \Psi_b(x), \quad x \in R_{ab}. \tag{50}$$

Equations of motion for e.m. potentials and matter fields are written down for each element of a section in the corresponding domain. They are compatible in virtue of Eqs. (45) and (50). When not explicitly carried out, this splitting of equations is to be understood in the remainder of this paper.

5. - A PHYSICAL PICTURE OF MONOPOLE-SOLITONS

So far little has been said in the way of understanding why monopole-solitons exist. Arguments based on topological considerations in the radial gauge can hardly be regarded as a satisfactory explanation. Furthermore, such arguments necessitate the imbedding of U(1) in larger gauge groups, which we would like to repudiate as inessential. In this section we hope to fill the gap by advancing a simple physical interpretation of the origin of monopole-solitons. In the next section this simple physical picture will serve as a guide in constructing the ansatz which will exhibit monopole-solitons in the model defined by the Lagrangian (38). Then our intuition will be confirmed.

Let us perform the following conceptual experiment. Given a point monopole charge, represented for the moment by an external source, we place it in a medium described by a matter field carrying a spin and a magnetic moment, which we may think of as an infinite set of magnetic dipoles. Via the interaction with the e.m. field, the medium will polarize:

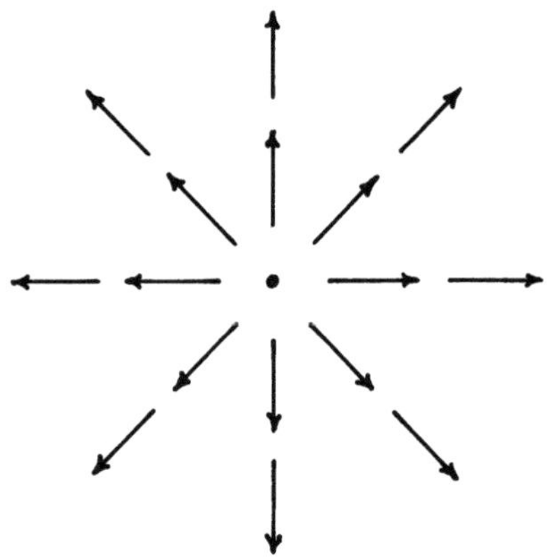

Because of the singularity in the e.m. field at the location of the external point source, we expect the matter field to become singular at the same point. If this singularity has the right strength, then:

a) the infinite accumulation of oriented magnetic dipoles at one point will build up the monopole charge self-consistently, in absence of external sources, via a phenomenon which is converse to charge screening;*)

b) the dipoles will reinteract with their coherently generated field and be attracted to the center of the monopole; the ensuing negative interaction energy will compensate for the infinite part of the energy required to create the e.m. field of the monopole.

Thus particles carrying a magnetic charge could bootstrap themselves into the theory in the form of solitons.

We shall see that this indeed occurs in a wide class of models characterized by the following general conditions:

1) U(1) gauge invariance;

2) the presence of charged matter fields with an intrinsic spin and a positive anomalous magnetic moment.

The second requirement may be understood by exploring in more detail proposition b) within the framework on non-linear field theories. In fact the interaction with the normal component of the magnetic moment arises in such theories as part of the gauge invariant kinetic term of the matter field. This term contributes

---

*) The fact that a (monopole) charge can be constructed by superimposing dipoles can be understood in ordinary linear electromagnetic theory as follows. The equation for the e.m. potential is $\nabla^2 A(\vec{r}) = \rho(\vec{r})$, where $\rho(\vec{r})$ is the source term. For a point charge located at $\vec{a}$ we have $\rho(\vec{r}) = \delta(\vec{r}-\vec{a})$ and, for a dipole moment $\vec{d}$, $\rho(r) = \vec{d}\cdot\vec{\nabla}\delta(\vec{r}-\vec{a})$. Consider now a distribution of dipoles, centered around the origin and radially oriented with density $1/a^2$. The source term then becomes $\rho(\vec{r}) = \int d^3a(\vec{a}/a^3)\cdot\vec{\nabla}\delta(\vec{r}-\vec{a}) = \int d^3a(\vec{\nabla}\cdot a/a^3)\delta(\vec{r}-\vec{a})$, by partial integration. Now $\vec{\nabla}\cdot\vec{a}/a^3 = \delta(\vec{a})$, so that $\rho(\vec{r}) = \delta(\vec{r})$, i.e., the source corresponding to a point charge at the origin. Note the essential role played by the $1/r^2$ singularity in the dipole distribution at the origin. Thus, by accumulating an infinite number of dipoles, a charge may be obtained.

a positive definite quantity to the Hamiltonian and cannot therefore compensate for infinities in other terms, such as the self-energy of the e.m. field. This cancelling role can be taken up by the anomalous component of the magnetic moment.

6. - MODEL NO. 1 IN AN ABELIAN GAUGE

Armed with the intuition just developed and the section formalism, we can now go back to the model Lagrangian (38) and construct an ansatz for monopole-soliton solutions. Our physical picture suggests that the polarization vectors of the charged vector mesons ($\vec{W}^{\pm}$) be aligned along the radial direction with respect to the location of the point monopole charge. This requirement essentially determines the structure of the presumed solution (g = 1/e):

$$x \in R_a: \begin{cases} \vec{A}^3(x) = \frac{1}{er} \tan\frac{\theta}{2} \vec{1}_\varphi , & (51) \\ \vec{W}^+(x) = \frac{K(r)}{er} e^{-i\varphi} \frac{1}{\sqrt{2}} (\vec{1}_\varphi + i \vec{1}_\theta) , & (52) \end{cases}$$

$$x \in R_b: \begin{cases} \vec{A}^3(x) = \frac{-1}{er} \operatorname{ctan}\frac{\theta}{2} \vec{1}_\varphi , & (53) \\ \vec{W}^+(x) = \frac{K(r)}{er} e^{i\varphi} \frac{1}{\sqrt{2}} (\vec{1}_\varphi + i \vec{1}_\theta) , & (54) \end{cases}$$

where the regions R_a and R_b are as specified by Eqs. (40-41). Notice that the consistency conditions (45), (46) and (50) are satisfied. The sections $\vec{A}^3$, $\vec{W}^+$, and $\vec{W}^-$ are everywhere analytic. The fields, in vector notations, are given by

$$\vec{F} = \vec{\partial} \times \vec{A}^3 = \frac{1}{er^2} \vec{1}_r , \tag{55}$$

$$\vec{F}^3 = \vec{\partial} \times \vec{A}^3 + ie \vec{W}^+ \times \vec{W}^- = \frac{1}{er^2} (1-K^2) \vec{1}_r , \tag{56}$$

$$\vec{F}^+(x) = (\vec{\partial} + ie\vec{A}^3) \times \vec{W}^+ \tag{57}$$

$$x \in R_a : = \frac{iK'}{er} e^{-i\varphi} \frac{1}{\sqrt{2}} (\vec{1}_\varphi + i \vec{1}_\theta) , \tag{58}$$

$$x \in R_b : = \frac{iK'}{er} e^{i\varphi} \frac{1}{\sqrt{2}} (\vec{1}_\varphi + i \vec{1}_\theta) . \tag{59}$$

The equations of motion which derive from the Lagrangian (38) are:

$$\partial F^3_{\mu\nu} = ie (W^{-\nu} F^+_{\mu\nu} - W^{+\nu} F^-_{\mu\nu}) , \tag{60}$$

$$(\partial^\nu + ie A^{3\nu}) F^+_{\mu\nu} = m^2 W^+_\mu + ieW^{+\nu} F^3_{\mu\nu} . \tag{61}$$

Our ansatz satisfied Eqs. (60) trivially, while reducing Eqs. (61) to the single non-linear equation:

$$K''(r) = M^2 K(r) + \frac{K(K^2-1)}{r^2} \quad . \tag{62}$$

Not surprisingly, this is the same as the corresponding dynamical equation obtained from our analysis in the radial gauge, Eq. (24). Needless to say, the effective Hamiltonian

$$\mathcal{H} = \int d^3x \left[\vec{F}^+\cdot\vec{F}^- + \frac{1}{2}\vec{F}^3\cdot\vec{F}^3 + M^2\vec{W}^+\cdot\vec{W}^-\right], \tag{63}$$

$$= \frac{4\pi}{e^2}\int_0^\infty dr\left[K'^2 + \frac{(K^2-1)^2}{2r^2} + M^2K^2\right], \tag{64}$$

also coincides with the corresponding expression obtained in the radial gauge, Eq. (27). Thus, following the steps of section 3), the g = 1/e monopole-soliton implied by the Lagrangian (38) may be easily constructed: it is physically undistinghishable from its radial gauge ancestor.

At this point, before proceeding to obtain monopole-solitons in wider and more interesting classes of models, we should demonstrate explicitly the claims made on the basis of our intuition in the previous section, in particular concerning the role of the anomalous magnetic moment of the W and the origin of the magnetic charge.

The Lagrangian (38) describes charged massive vector mesons carrying an anomalous magnetic moment:

$$\vec{\mu}_a = ie\vec{W}^+ \times \vec{W}^- \tag{65}$$

in addition to the normal component implied by minimal coupling of the e.m. field to the vector mesons. As implied by Eqs. (55-56) and (29), it is the anomalous component (65) which reduces in $\vec{F}^3$, Eq. (56), the Coulomb singularity of the magnetic field $\vec{\mathfrak{F}}$, Eq. (55), at the origin to:

$$\vec{F}^3 \underset{r\to 0}{\sim} \frac{2M^2}{3e} \ln(\not{c} r)\, \vec{1}_r . \tag{66}$$

This logarithmic singularity is sufficiently mild and does not cause any problem. In particular the energy density in Eq. (64) in integrable: as we predicted in the previous section, this may be interpreted by saying that the attraction exerted by the magnetic charge on the anomalous magnetic moments of the W's offsets the infinity in the Coulomb self-energy, making it possible for a monopole to exist as a finite energy soliton solution.

It is with the anomalous moment (65) that we must find also the mechanism for the origin of the magnetic charge carried by the soliton. For the purpose of this identification, let us consider the Bianchi identities:

$$\varepsilon^{ijk}\left[\partial_i F^a_{jk} + e\epsilon^{abc} A^b_i F^c_{jk}\right] = 0 . \qquad (67)$$

In particular for a = 3 we have, in vector notations:

$$\vec{\partial}\cdot\vec{\mathcal{F}} = -ie\,\vec{\partial}\cdot(\vec{\omega}^+ \times \vec{\omega}^-) - ie(\vec{\omega}^+\cdot\vec{F}^- - \vec{\omega}^-\cdot\vec{F}^+) \qquad (68)$$

This equation enables us to express the magnetic charge in terms of vector meson fields:

$$\tilde{Q} = \frac{1}{4\pi}\int d^3x\, \vec{\partial}\cdot\vec{\mathcal{F}} \qquad (69)$$

$$= \frac{-ie}{4\pi}\int d^3x \left[\vec{\partial}\cdot(\vec{\omega}^+ \times \vec{\omega}^-) + \vec{\omega}^+\cdot\vec{F}^- - \vec{\omega}^-\cdot\vec{F}^+\right] \qquad (70)$$

$$= \lim_{R\to 0}\left\{-\frac{ie}{4\pi}\int_{S_R} d\vec{\sigma}\cdot(\vec{\omega}^+ \times \vec{\omega}^-)\right\}$$
$$+ \lim_{R\to 0}\left\{\frac{-ie}{4\pi}\int_{V_{\bar{R}}} d^3x \left[\vec{\partial}\cdot(\vec{\omega}^+ \times \vec{\omega}^-) + \vec{\omega}^+\cdot\vec{F}^- - \vec{\omega}^-\cdot\vec{F}^+\right]\right\}, \qquad (71)$$

where S_R denotes the surface of the sphere of radius R and $V_{\bar{R}}$ the exterior of this surface. Since the second term vanishes identically, we obtain:

$$\tilde{Q} = \lim_{R\to 0}\left\{\frac{-ie}{4\pi}\int_{S_R} d\vec{\sigma}\cdot(\vec{\omega}^+ \times \vec{\omega}^-)\right\} . \qquad (72)$$

Now, for our soliton solution, we find

$$\vec{\mu}_a = ie\,(\vec{W}^+ \times \vec{W}^-) \underset{r\to 0}{\sim} \frac{-1}{er^2}\,\hat{r}_r \tag{73}$$

leading to

$$\tilde{Q} = 1/e\;. \tag{74}$$

Thus we may concisely say that "monopole-solitons are made from dipoles" in the sense that the latter, in addition to playing a crucial role in the binding of the soliton, are responsible for the generation of its magnetic charge.

7. - MODEL NO. 2: ELECTRODYNAMICS OF VECTOR MESONS

In the previous section we demonstrated that it is possible to dispense of Higgs fields and still retain monopole-soliton solutions. This was done in the context of a model with an underlying SU(2) gauge invariance of the Lagrangian, explicitly broken down to an exact U(1) gauge invariance by mass terms. The latter invariance is obviously indispensable for the existence of monopoles, but there is no obvious need for the imbedding of U(1) in larger gauge groups or the invariance of the interaction Lagrangian with respect to any such group. The physical picture presented in section 5 suggests that such imbeddings are inessential to the formation of the soliton and of its magnetic charge. Now we will demonstrate that this is indeed the case, a conclusion which will appear as a byproduct of the extension of our analysis to a wider class of models.

Let us consider the Lagrangian

$$\mathcal{L} = -\frac{1}{4}\left[\mathcal{F}_{\mu\nu} + i\varkappa e\left(W_\mu^+ W_\nu^- - W_\nu^+ W_\mu^-\right)\right]^2 + -\frac{1}{2} F_{\mu\nu}^+ F^{-\mu\nu} + M^2 W_\mu^+ W^{-\mu} , \tag{75}$$

where

$$\mathcal{F}_{\mu\nu} = \partial_\mu A_\nu - \partial_\nu A_\mu , \tag{76}$$

$$F_{\mu\nu}^+ = (\partial_\mu + ieA_\mu) W_\nu^+ - (\partial_\nu + ieA_\nu) W_\mu^+ . \tag{77}$$

This Lagrangian describes two massive charged vectors mesons, with an unspecified anomalous magnetic moment proportional to $\varkappa$, in interaction with the photon field. The quartic self-interaction of the charged mesons incorporated into (75) is dictated by very general physical requirements on the classical theory, such as causality of propagation. From another view-point the presence of this term ensures the positivity of the Hamiltonian. In summary, the Lagrangian (75) is the most general one describing three vector mesons with an exact U(1) gauge invariance, parity conserving interactions and involving only dimensionless coupling constants.[9)]

For $\varkappa = 1$ the Lagrangian (75) reduces to the Lagrangian (38) with the vertices of an SU(2) Yang-Mills gauge model and for which we already exhibited monopole-soliton solutions. However, in the light of our physical picture, there appears to be nothing that requires the anomalous magnetic moment of the W's to take on its gauge theory value $\varkappa = 1$: we expect monopole-soliton solutions to persist, though with different characters, for all values of $\varkappa > 0$. A generalization of the ansatz which we have applied successfully to the model with $\varkappa = 1$ immediately suggests itself:

$$x \in R_a \begin{cases} \vec{A}(x) = \frac{1}{er} \tan\frac{\theta}{2}\, \vec{1}_\varphi\,, & (78) \\[2ex] \vec{W}^+(x) = \frac{1}{\sqrt{\varkappa}} \frac{K(r)}{er} e^{-i\varphi} \frac{1}{\sqrt{2}} (\vec{1}_\varphi + i\vec{1}_\theta)\,, & (79) \end{cases}$$

$$x \in R_b \begin{cases} \vec{A}(x) = \frac{-1}{er} \operatorname{ctan}\frac{\theta}{2}\, \vec{1}_\varphi\,, & (80) \\[2ex] \vec{W}^+(x) = \frac{1}{\sqrt{\varkappa}} \frac{K(r)}{er} e^{i\varphi} \frac{1}{\sqrt{2}} (\vec{1}_\varphi + i\vec{1}_\theta)\,, & (81) \end{cases}$$

where the regions R_a and R_b are again as specified by Eqs. (40-41). The fields, in vector notations, are given by:

$$\vec{\mathcal{F}} = \vec{\partial} \times \vec{A} = \frac{1}{er^2} \vec{1}_r \,, \tag{82}$$

$$\vec{F}^0 = \vec{\partial} \times \vec{A} + i\kappa e \vec{W}^+ \times \vec{W}^- = \frac{1}{er^2}(1-K^2)\vec{1}_r \,, \tag{83}$$

$$\vec{F}^+(x) = (\vec{\partial} + ie\vec{A}) \times \vec{W}^+ \,, \tag{84}$$

$$x \in R_a : = \frac{1}{\sqrt{\kappa}} \frac{iK'}{er} e^{-i\varphi} \frac{1}{\sqrt{2}} (\vec{1}_\varphi + i\vec{1}_\theta) \,, \tag{85}$$

$$x \in R_b : = \frac{1}{\sqrt{\kappa}} \frac{iK'}{er} e^{i\varphi} \frac{1}{\sqrt{2}} (\vec{1}_\varphi + i\vec{1}_\theta) . \tag{86}$$

The equations of motion which derive from the Lagrangian (75) are:

$$\partial^\nu F^0_{\mu\nu} = ie(W^{-\nu} F^+_{\mu\nu} - W^{+\nu} F^-_{\mu\nu}) \,, \tag{87}$$

$$(\partial^\nu + ieA^\nu) F^+_{\mu\nu} = M^2 W^+_\mu + i\kappa e W^{+\nu} F^0_{\mu\nu} . \tag{88}$$

Our ansatz satisfies Eqs. (87) trivially and reduces Eqs. (88) to the single non-linear equation:

$$K''(r) = MK(r) + \kappa \frac{K(K^2-1)}{r^2} . \tag{89}$$

This is similar, though not equivalent, to Eq. (62). The effective Hamiltonian is:

$$\mathcal{H} = \int d^3x \left[\vec{F}^+ \cdot \vec{F}^- + \frac{1}{2} \vec{F}^0 \cdot \vec{F}^0 + M^2 \vec{W}^+ \cdot \vec{W}^- \right], \tag{90}$$

$$= \frac{4\pi}{\varkappa e^2} \int_0^\infty dr \left[K'^2 + \varkappa \frac{(K^2-1)}{2r^2} + M^2 K^2 \right]. \tag{91}$$

The boundary conditions which must be met by a solution of Eq. (89) for finiteness of the energy are the same as in Eq. (28). However, Eq. (89) now leads to asymptotic behaviors near the origin which differ for $\varkappa \lessgtr 1$:

$$K(r) \underset{r\to 0}{\sim} 1 - |\alpha|\, r^\alpha + \cdots, \quad \alpha = \frac{1+\sqrt{1+8\varkappa}}{2}, \quad \varkappa < 1, \tag{92}$$

$$K(r) \underset{r\to 0}{\sim} 1 - \frac{M^2}{2(\varkappa-1)} r^2 + \cdots, \qquad \varkappa > 1. \tag{93}$$

The asymptotic behavior at infinity remains as in Eq. (29). Mechanical analogue techniques indicate that a solution of Eq. (89) with these asymptotic behaviors indeed exists for any $\varkappa > 0$. Numerical solutions corresponding to $\varkappa = 0.5$ and $\varkappa = 2$ are displayed in Fig. 1. The e.m. field for our solutions, given by Eq. (82), is of course independent of $\varkappa$ and corresponds to a point magnetic charge $g = 1/e$.

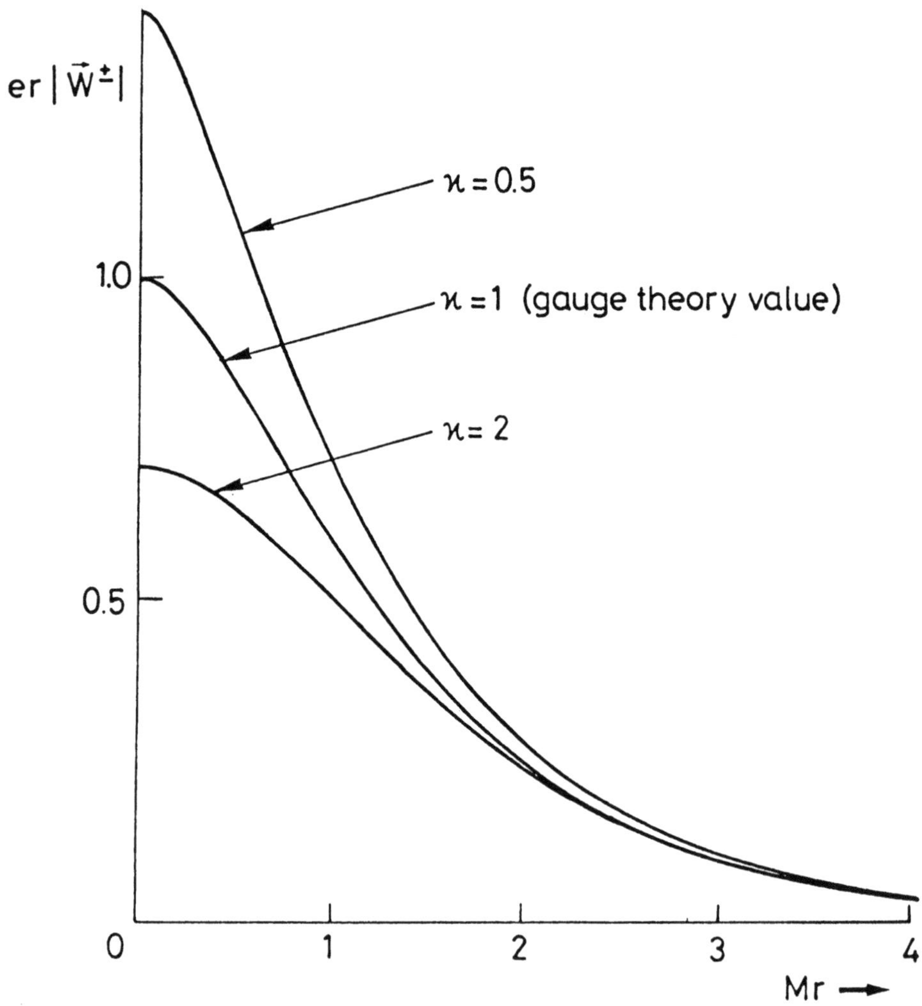

Figure 1

It is interesting to compare the monopoles arising in models with different values of $\varkappa$. Figure 2 shows their masses E (classical energies) and radii R:

$$E = \frac{4\pi}{\varkappa e^2} \int_0^\infty dr \left[K'^2 + x \frac{(K^2-1)}{2r^2} + m^2 K^2 \right], \tag{94}$$

$$= 2 \frac{4\pi}{\varkappa e^2} \int_0^\infty dr\, K^2(r) \quad , \tag{95}$$

$$R^2 = \frac{1}{E} \frac{8\pi}{\varkappa e^2} \int_0^\infty dr\, r^2 K^2(r) , \tag{96}$$

where we made use of an equipartition theorem implied by a trivial condition of monimum of the functional (91) under the scaling operation $K(r)\to K(\lambda r)$. Notice that, although for $\varkappa = 1$, the gauge theory value, monopoles are relatively heavy, as $\varkappa \to \infty$ their mass becomes arbitrarily small. On the other hand as $\varkappa \to 0$ the same diverges: in fact we would then be approaching electrodynamics of minimally coupled vector mesons in the presence of a Dirac monopole*).

*) Notice however that as the limit $\varkappa \to 0$ is taken our solution, while acquiring an infinite energy, remains sourceless, which is not true for a Dirac monopole solution.

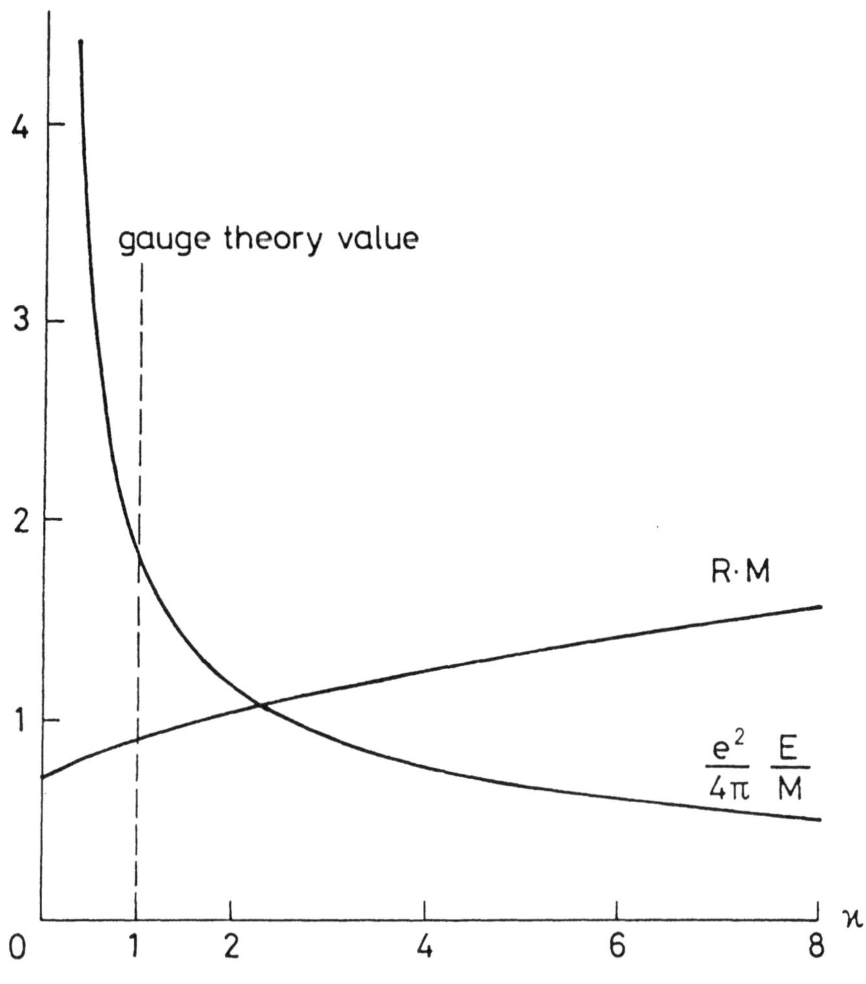
4
3
2
1
0
gauge theory value
R·M
$\frac{e^2}{4\pi}\frac{E}{M}$
1
2
4
6
8
$\varkappa$

Figure 2

8. - MODEL NO. 3: ELECTRODYNAMICS OF NON RELATIVISTIC SPIN 1/2 PARTICLES

So far we have discussed models where the "constituents" of monopole solitons are spin 1 particles. However our physical picture of "monopoles made from dipoles" does not require that the latter be carried by spin 1/2 fields. Unfortunately at the classical level relativistic c-number spinor field theories are ill-defined (e.g., the Hamiltonian is not bounded below). To circumvent these difficulties and enable ourselves to expose in a simple and unambiguous way interesting aspects of the problem, we will adopt for the moment Galileian c-number spin 1/2 fields. Of course, in general this will amount to more than a non-relativistic (low velocity) approximation, as the Pauli principle will also be ignored.

With this reservation, we introduce the following Hamiltonian density:

$$\mathcal{H} = \frac{1}{2m}\left\{\phi^+[\vec{\sigma}\cdot(-i\vec{\partial}-e\vec{A})]^2\phi + \chi^+[\vec{\sigma}\cdot(-i\vec{\partial}+e\vec{A})]^2\chi\right\}$$
$$+ m(\phi^+\phi+\chi^+\chi) + \frac{1}{2}\left[\vec{\mathcal{F}} - \frac{\varkappa e}{2m}(\phi^+\vec{\sigma}\phi - \chi^+\vec{\sigma}\chi)\right]^2, \tag{97}$$

involving two oppositely charged spin 1/2 (two-component) fields with the same mass, ϕ and χ, in interaction with the magnetic field $\vec{\mathcal{F}}$. Notice that in addition to the "normal" magnetic moment coupling, as generated by a non-relativistic expansion of the Dirac equation, an "anomalous" magnetic moment interaction is also included by a term of the form (magnetic field - spin density)^2, implying also a spin-spin self-interaction. This is done in analogy with our models 1 and 2 for vector fields. In the vector case such a form is needed to meet general physical requirements[9] on the classical theory, whereas in the present case we do not know of similar restrictions. Note however that the suppression of the spin-spin term in Eq. (97) would destroy positivity.

Motivated by our physical picture of monopole-solitons which suggests

$$\vec{r}\cdot\vec{\sigma}\,\phi = \pm r\phi, \quad \vec{r}\cdot\vec{\sigma}\,\chi = \mp r\chi, \tag{98}$$

we make the following ansatz:

$$x \in R_a: \begin{cases} \vec{A}(x) = \dfrac{1}{2er}\tan\dfrac{\theta}{2}\,\vec{1}_\varphi, & (99) \\[2ex] \phi(x) = \begin{pmatrix} \cos\theta/2 \\ \sin\theta/2\, e^{i\varphi} \end{pmatrix} \dfrac{K(r)}{er}\sqrt{\dfrac{m}{2\varkappa}}, & (100) \\[2ex] \chi(x) = \begin{pmatrix} \sin\theta/2\, e^{-i\varphi} \\ -\cos\theta/2 \end{pmatrix} \dfrac{K(r)}{er}\sqrt{\dfrac{m}{2\varkappa}}, & (101) \end{cases}$$

$$x \in R_b: \begin{cases} \vec{A}(x) = -\frac{1}{2er} \operatorname{ctan} \theta/2 \, \vec{1}_\varphi \,, & (102) \\ \phi(x) = \begin{pmatrix} \cos \theta/2 \, e^{-i\varphi} \\ \sin \theta/2 \end{pmatrix} \frac{K(r)}{er} \sqrt{\frac{m}{2\kappa}} \,, & (103) \\ \chi(x) = \begin{pmatrix} \sin \theta/2 \\ -\cos \theta/2 \, e^{i\varphi} \end{pmatrix} \frac{K(r)}{er} \sqrt{\frac{m}{2\kappa}} \,. & (104) \end{cases}$$

Notice that

$$\vec{\mathcal{F}} = \vec{\partial} \times \vec{A} = \frac{1}{2er^2} \vec{1}_r \tag{105}$$

corresponding to the Dirac unit of magnetic charge, g = 1/2e.*)
The Euler equations implied by the Hamiltonian (97) are reduced by this ansatz to

$$K''(r) = 2m^2 K(r) + \frac{\kappa}{2} \frac{K(K^2-1)}{r^2} \,, \tag{106}$$

*) This is to be contrasted with the value g = 1/e for which we exhibited solutions in our previous models with vector matter fields. The justification for this difference is the existence of a connection between the possible values of the magnetic charge for monopole-solitons and the spins of their constituent matter fields: g e = s.

which, apart from a redefinition of parameters, is the same as Eq. (89). At this point monopole-soliton solutions may be constructed as in our model No. 2.

9. - MISCELLANEOUS CONSIDERATIONS

Thus monopole-solitons exist in theories which contain:

a) the (massless) photon field;

b) charged matter fields carrying a positive anomalous magnetic moment.

Note that:

1) The constituent matter fields may carry spin $s = 1/2, 1, \ldots$ and correspondingly there will exist monopole-solitons with magnetic charges $g = \pm 1/2e, \pm 1/e, \ldots$;

2) As it is evident in the section formalism, Eq. (46), each value of g defines a homotopy class of U(1);

3) The imbedding of U(1) in larger groups is inessential;

4) The mechanism (Higgs or otherwise) by which such larger gauge groups are broken down to U(1) is a fortiori also inessential;

5) The monopole-solitons that we have constructed have zero electric charge. Solutions exist for arbitrary values of the electric charge and may be constructed by a simple extension of our ansatz;

6) The monopole-solitons which we have constructed have zero angular momentum.

Let us now come to the mass of our monopoles, which for small values of the coupling constant e may be calculated in good approximation ignoring quantum effects. If we identify the W fields of our model No. 1 or 2 with the intermediate vector boson of the weak interactions having $\varkappa = 1$, the gauge theory value, the monopole mass would be predicted to be very large ($E \simeq 10^4$ GeV), as in the case of 't Hooft and Polyakov. A more easily accessible value ($E \simeq 10^2$ GeV) would be obtained if the same models could be applied to describe the interactions of particles such as the $\rho^{\pm}$(750 MeV) mesons. It is unlikely that lower mass monopoles with spin 1 matter fields, as constituents, exist because of the apparent lack of light constituents of this type.

We know however of light spin 1/2 charged particles: the electron and the positron. They do not carry an anomalous magnetic moment at the Lagrangian level. However, they acquire one by 2nd order radiative corrections and it is positive: $\varkappa = \alpha/2\pi$. Let us then assume that we can apply the Hamiltonian (97) to provide an effective description of the electron-positron-photon system. Since $\varkappa$ is very small, a solution of Eq. (106) is approximately given by

$$K(r) = \exp(-\sqrt{2}\, m r) \tag{107}$$

where m is the electron mass. Our speculative application of model No. 3 then leads to a monopole of mass:

$$E = \sqrt{2}\,\frac{\pi}{\alpha^2}\,m_e \simeq 4\,3\ \mathrm{GeV}. \tag{108}$$

For technical reasons we have not yet assessed the accuracy of this estimate.

REFERENCES

1) W. Troost and P. Vinciarelli, "Monopoles from dipoles", CERN preprint TH.2195 (1976); and other unpublished work.

2) P.A.M. Dirac, Proc. Roy. Soc. London A133, 60 (1931).

3) P.A.M. Dirac, Phys. Rev. 74, 817 (1948).
See also: J. Schwinger, Particles, Sources and Fields, Vol. 1 and 2 (Addison-Wesley, 1970 and 1973).

4) B. Zumino, Strong and Weak Interactions, 1966 Int. School of Physics, Erice, edited by A. Zichichi (Acad. Press, New York and London), p. 711.

5) A. M. Polyakov, JEPT Letters 20, 194 (1974);
G. 't Hooft, Nuclear Phys. B79, 276 (1974).
See also: T. T. Wu and C. N. Yang, in Properties of Matter under Unusual Conditions, edited by H. Mark and S. Fernback (Interscience, New York, 1969), pp. 349-354.

6) See, e.g., J. Arafune, P. G. O. Freund and C. J. Goebel, J. Math. Phys. 16, 433 (1975).

7) The Lagrangian (19) has been proposed as the basis for a model of weak and e.m. interactions by L. D. Faddeev, Sov. Phys. Dokl. 18, 382 (1973).

8) T. T. Wu and C. N. Yang, Nuclear Phys. B107, 365 (1976); Stony Brook preprint ITP SB 76-11 (1976).

9) G. Velo, Nuclear Phys. B65, 427 (1973).

D I S C U S S I O N

CHAIRMAN: Prof. P. Vinciarelli

Scientifiec Secretaries: A. Audrito and A. de la Torre

DISCUSSION

- *KLEINERT:*

In what sense is your monopole charge built from magnetic dipole moments?

- *VINCIARELLI:*

Not in the sense that, starting from dipoles, you or I can make a monopole. We cannot make a monopole because there is an infinite potential barrier associated with the generation of the charge. The sense of "monopoles from dipoles" was made precise by the Bianchi identity, which I have discussed and for which I have given a physical interpretation: the monopole charge is built from an infinite accumulation of magnetic dipoles at one point.

- *KLEINERT:*

In your third model, you need an anomalous magnetic moment κ in order to have a monopole. Thus, within a fundamental theory, you will have to invoke radiative corrections to create $\kappa = 0$. Are you not afraid the same radiative corrections will make your monopole "puff up" into many mesons?

- *VINCIARELLI:*

The danger is not the possibility of decay into ordinary particles, which is excluded because of the charge of the monopole, but rather that the finiteness of the energy in the one monopole sector, which is implied by our effective Lagrangian, will not be retained by the full-blown quantum field theory.

- *KLEINERT:*

Let me suggest using the technique that is standard in superconductivity: first, make a meson Lagrangian out of your fermion one, analogous to the Ginzburg-Landau equation, as I have explained in my lecture -- then you have certain radiative corrections included. After this, solve for a monopole in the meson theory.

- *VINCIARELLI:*

We are presently investigating a similar approach.

- *ROSSI:*

Do you think you could obtain a derivation of your effective Lagrangian by summing up the quantum contributions to the effective action, at least to first order in $\hbar$ -- loop expansion?

- *VINCIARELLI:*

We are now making an attempt along similar lines.

- *YANG:*

In your second example, the fact that you have a finite total energy is very interesting; but I do not believe you can claim that, for your solution, $\nabla \cdot H = 0$ is valid everywhere. If you accept my statement, then your variational equations are, for your solutions, not valid at the origin. The solution is still interesting because it has finite energy. It does not, however, satisfy the field equation everywhere. I do not believe you can cure this problem with the idea of sections.

- *VINCIARELLI:*

I never said that the magnetic field of our solutions is divergenceless everywhere in space. In fact

$$\nabla \cdot H = \frac{1}{2} \varepsilon^{ijk} \partial_i F_{jk} = \delta(\underline{x})$$

if the point-like magnetic charge is located at the origin.

Our equations are valid and satisfied everywhere by our solutions if these are formulated in terms of sections. We need sections, or alternative formalisms which also exploit U(1) gauge invariance, only for the purpose of describing unambiguously the electromagnetic field in the presence of a monopole charge. This charge is, in our case, self-consistently generated from within the -- conventional -- field theories we study, without the need to introduce external sources or new degrees of freedom carrying magnetic charge.

QUARKS, COLOR AND OCTONIONS

F.BUCCELLA

Istituto di Fisica "G.Marconi" - ROMA

The enigmatic properties of quarks have been described by introducing for them a new SU(3) degree of freedom, which is an exact symmetry+, with the additional constraint that only states scalar under this new group, named $SU_c(3)$ to distinguish it from the SU(3) of isospin and strangeness, can be observed[2]. This assumption implies that quarks, which transform under $SU_c(3)$ as the fundamental representation, cannot be observed alone but only in pairs $q\bar{q}$ (mesons) or in triplets (baryons). This new degree of freedom accounts for the symmetry in the others quantum numbers of the baryon wave function and succesfully explains ($\pi^0 \to 2\gamma$) or reduces (R value) previous discrepancies. The main purpose of this talk is to show that the octonion algebra supplies a natural framework both for the SU(3) character of the new degree of freedom and for the non observability of non singlet states[3].

There have been already attempts to connect isospin to the quaternion algebra[4]; in fact its three immaginary units e_i have the product rule:

$$e_i e_j = -\delta_{ij} + \varepsilon_{ijk}\, e_k$$

which is evidently invariant under rotations in the three-dimensional space of the e_i's : one has then an SU(2). However one would like to identify one of the e_i's , for instance e_3 with the ordinary unit of quantum mechanics $i = \frac{1}{\hbar}[q,p]$

Therefore one should consider only the transformations which leave e_3 invariant, namely the rotations around the third axis: this

+ This is equivalent to assume that quarks are parafermions of order three[1].

reduces SU(2) to U(1). In conclusion the quaternion algebra is connected to SU(2), but if it is considered an extension of the complex numbers of quantum mechanics one is left with U(1).

Let us consider now octonions with their seven imaginary units multiplying as [5]

$$e_i e_j = -\delta_{ij} + h_{ijk}\, e_k \qquad 1)$$

where h is an antisymmetric tensor equal to one for (123) and the others six triplets obtained from it with a cyclic permutation on the cycle (1243657) (for the others triplets $h_{ijk}=0$)
It is useful to introduce the so called split octonions [5]:

$$u_1 = \frac{1}{2}(e_1 + ie_4) \qquad u_1^* = \frac{1}{2}(e_1 - ie_4)$$

$$u_2 = \frac{1}{2}(e_2 + ie_5) \qquad u_2^* = \frac{1}{2}(e_2 - ie_5)$$

$$u_3 = \frac{1}{2}(e_3 + ie_6) \qquad u_3^* = \frac{1}{2}(e_3 - ie_6)$$

$$u_\circ = \frac{1}{2}(1 + ie_7) \qquad u_\circ^* = \frac{1}{2}(1 - ie_7)$$

In the basis of the u's eq.1) reads:

$$\begin{aligned} &u_i u_j = \varepsilon_{ijk} u_k^* \\ &u_i u_j^* = -\delta_{ij} u_\circ \\ &u_\circ u_i = u_i \\ &u_\circ^* u_i = 0 \\ &u_i u_\circ = 0 \\ &u_i u_\circ^* = u_i \\ &u_\circ u_\circ = u_\circ \\ &u_\circ u_\circ^* = 0 \end{aligned} \qquad 2)$$

Eqs.2) show the intimate connection between octonions and SU(3); indeed if the u_i's and the u_i^*'s are classified respectively in a 3 and 3^* representations of SU(3) and $u_\circ$ and $u_\circ^*$ in a 1, the products just written show invariance under SU(3) (in fact two 3's combine antisymmetrically to give a 3^*, the product of a 3 times a 3^* gives a singlet and the multiplication by $u_\circ$ and $u_\circ^*$ does not change the transformation properties under SU(3)). All together the seven immaginary units of the octonions build up a 7 representation of G_2, which is the symmetry algebra of the octonions,[5] while the SU(3) considered is the subalgebra of G_2 which leaves e_7 invariant. Beyond being SU(3) invariant eqs. 2) enjoy the property that no 6 is present in the product of two 3's and no octet in the product of a 3 times a 3^*: therefore three 3's combine to give a singlet as well as a pair 3 and 3^*; this is just what is expected for the $SU_c(3)$ degree of freedom. In fact the split octonions algebraic structure arises if one defines in a six-dimensional vector space, which is a $3 \oplus 3^*$ representation of SU(3), an SU(3) invariant product which closes in a space which contains only 1, 3 and 3^* representations.[6]

The property, which makes octonions a particularly appealing framework for $SU_c(3)$, is the fact that they supply a rather natural explanation for the non-observability of non-singlet states. In fact the quarks, which are assumed to transform as a 3 of $SU_c(3)$, are described by the u_i's; but these numbers do not associate ($(u_1u_2)u_3 \neq u_1(u_2u_3)$), which is an undesired property for an observed quantity; however the subspace of the $SU_c(3)$ singlets, spanned by $u_\circ$ and $u_\circ^*$, is just the ordinary complex algebra.[3]

The identification of $SU_c(3)$ as the subalgebra of G_2, which does not act on one (e_7) of the immaginary octonion units brings to a non trivial connection between the color and flavor groups. In fact, given the flavor group F , the color-flavor group $SU_c(3) \otimes F$, has to be included in a larger one, which contains G_2, where F does not commute with the six generators which complete $SU_c(3)$ to G_2. To this extent one employs e_7, which is unaffected by $SU_c(3)$, in representing F; the action of the additional generators of G_2 enlarges e_7 to $O_\circ$ (the space of the purely immaginary octonions). Let us consider the simple case where F is the SU(3) of isospin and strangeness. The SU(3) algebra may be represented by the Gell-Mann matrices λ_a times the immaginary unit i (the commutator of two $i\lambda_a$ is a combination of the others). The antihermitian traceless 3x3 complex matrices may be decomposed into the real antisymmetric (called A) and the purely immaginary symmetric ones (called $iS_\circ$). In order to establish a connection with $SU_c(3)$ one identifies the immaginary unit which multiplies $S_\circ$ with e_7; the SU(3) of flavor is spanned by $e_7 S_\circ \oplus A$.

The six generators needed to complete $SU_c(3)$ to G_2 rotate e_7 into the others six immaginary units; so one has $G_2 \oplus O_\circ \otimes S_\circ \oplus A$ which shows symmetry between the seven octonion immaginary units.

The products $O_o \otimes S_o$ transform as a $7 \otimes 5$ under $G_2 \otimes SO(3)$ (SO(3) is the Lie algebra generated by A) and, with an appropriate definition of their Lie product, build up, together with G_2 and A, the exceptional algebra F_4(7); F_4 in fact can be obtained from SU(3) by generalizing the complex numbers to octonions in the 3x3 traceless antihermitian matrices and by adding the symmetry group of octonions G_2(8) (the complex numbers have no continuous symmetry).

The generators needed to complete $SU_c(3) \otimes SU(3)$ to F_4 transform as a $(3,6^*)\oplus(3^*,6)$ under the two SU(3)'s (the $SU_c(3)$ content is easily understood since both the transverse octonions $e_{i\neq 7}$ and the six generators which act on e_7 transform as a $3\oplus 3^*$ under $SU_c(3)$). The special role plaid by the left-handed currents in weak interactions makes rather appealing to consider, in the framework of unified theories(9), the chiral groups. The main point is to represent the operator which interchanges the corresponding vector and axial vector charges in terms of e_7; since this operator has square +1, while $e_7^2 = -1$, the simplest thing to do is to write it as ie_7 where i is an immaginary unit; so $SU_L(3)\otimes SU_R(3)$ is spanned by:

$$e_7 S_o \oplus A \oplus i S_o \oplus i e_7 A$$

The extension of $SU_c(3)$ to G_2 brings to:

$$G_2 \oplus O_o \otimes (S_o + iA) \oplus (A + iS_o).$$

The generators just written build up, again with an appropriate definition of their Lie product, the exceptional algebra E_6,(7) which consists of the generators of the three SU(3)'s and of a $(3,3^*,3^*)\oplus(3^*,3,3)$ representation of $SU_c(3) \otimes SU_L(3) \otimes SU_R(3)$. In connection with the conjectured existence of two others quarks of charge -1/3(10), SU(6) may be proposed as a possible flavor group. An extension of $SU_c(3)\otimes SU(6)$ can be obtained by identifying the generators, previously employed to build chiral $SU(3) \otimes SU(3)$, as a basis of the collinear subalgebra A $[\lambda_a(1\pm\sigma_x)/4]$; if one adds to i, corresponding to the x direction, the immaginary units j and k, corresponding to the y and z directions, one gets the complete set Q_o of the immaginary quaternionic units; $A(\vec{\sigma}/2)$ may be identified with the symmetry of quaternions $SU_Q(2)$. In conclusion a basis for the SU(6) algebra is given by: $e_7 S_o \oplus A \oplus Q_o \otimes S_o \oplus Q_o \otimes e_7 A \oplus SU_Q(2)$. With the previous extension method one is lead to:

$$G_2 \oplus O_o \otimes (S_o + Q_o \otimes A) \oplus (A + Q_o \otimes S_o) \oplus SU_Q(2)$$

which give rise to E_7(7), consisting of the generators of $SU_c(3)\otimes SU(6)$ and of a $(3,15^*)\oplus(3^*,15)$ representation of this group.

The partitions performed for the generators of the exceptional algebras considered are rather similar; in fact the extension of the real 3x3 matrices to the complex and to the quaternionic ones brings from F_4 to E_6 and E_7 respectively: in E_7 there is also the symmetry of quaternions $SU_Q(2)$; the extension of $Q_\circ$ to $O_\circ$ and of $SU_Q(2)$ to the symmetry of octonions G_2 brings from E_7 to E_8 (7). The groups E_6 (11) and E_7 (12) have been avocated in connection with the attempts of building up a unified theory of strong and weak and electromagnetic interactions.

In this framework the fundamental fermions, quarks and leptons, are classified in the fundamental representation of these groups. Since the exceptional groups may be characterized as the symmetries of the Jordan algebras on the real (F_4), complex (E_6), quaternionic (E_7) 3x3 hermitian octonionic matrices (7), it is not surprising that these mathematical objects play a role in describing the fundamental representations of these groups.

In particular the smallest non-trivial representation of F_4 is described by a traceless hermitian octonionic matrix; the basis of this representation can be written as: $S_\circ \oplus O_\circ \otimes A$.

Under $SU_c(3)\otimes SU(3)$ $S_\circ$ and e_7A transform as the (1,8) while $e_{i\neq 7}A$ as a $(3,3)\oplus(3^*,3^*)$ representation. The reducible representation $27\oplus 27^*$ of E_6 is described by a complex 3x3 octonionic hermitian matrix: under F_4 this object transforms as a $26\oplus 1\oplus 26\oplus 1$, which makes easy to understand that the $SU_c(3)\otimes SU_L(3)\otimes SU_R(3)$ content of the 27 is $(3,3,1)\oplus(3^*,1,3)\oplus(1,3^*,3)$. Finally the 56 representation of E_7 consists of a complex 3x3 octonionic matrix and of a complex number: from its $SU_c(3)\otimes SU(3)$ content one can extrapolate, grouping together the SU(3) multiplets with the same behaviour under $SU_c(3)$, the $SU_c(3)\otimes SU(6)$ content of the 56 to be $(1,20)\oplus(3,6)\oplus\oplus(3^*,6^*)$ (12).

Quarks (antiquarks) are identified with the triplets (antitriplets) of color and leptons with the singlets.

A remarkable property of this approach is the proper electric charge universality between quarks and leptons: this is easily seen at the F_4 level, where they belong respectively to the 3 and 8 representations of SU(3), and is a consequence of the fact that the transverse (associated to quarks) and longitudinal (associated to leptons) octonions anticommute or commute respectively with the imaginary unit e_7 involved in the definition of the generators of SU(3) (8).

REFERENCES

1) O.W.Greenberg; Phys.Rev.Lett. 13,598 (1964).

2)-H.Fritzsch and M.Gell-Mann; Proc. XVIth International Conference on High Energy Physics, NAL, vol.2, p.135.
-M.Gell-Mann; Acta Phys.Austriaca, Suppl.IX, 733 (1972).

3)-M.Günaydin and F.Gürsey; Phys.Rev. D9, 3387 (1974).
-F.Gürsey; The John Hopkins Univ., Workshop on Current Problems in High Energy Particle Theory, p.15 (1974).

4)-D.Finkelstein, J.M.Jauch, S.Schiminovitch and D.Speiser; Journ. Math.Phys. 3,207 (1962).
-D.Finkelstein, J.M.Jauch and D.Speiser; Journ.Math.Phys. 4,136 (1963).
-G.G.Emch; Helv.Phys.Acta 36,739,770 (1963).

5) M.Günaydin and F.Gürsey; Lettere al Nuovo Cimento 6,401 (1973) and Journ.Math.Phys. 14,1651 (1973).

6) R.Casalbuoni, G.Domokos and S.Kövesi-Domokos; Nuovo Cimento 31A, 423 (1976).

7)-See for instance R.D.Shafer; Introduction to Non Associative Algebras (Acad.Press,1966).
-N.Jacobson, Exceptional Lie Algebras (M.Dekker 1971).

8) F.Gürsey; Kyoto International Symposium on Mathematical Problems in Theoretical Physics, ed H.Araki, p.189 (Springer 1975).

9)-J.Pati and A.Salam; Phys.Rev. D8,1240 (1973).
-H.Georgi and S.L.Glashow; Phys.Rev.Letters 32,433 (1974).
-H.Fritzsch and P.Minkowski; Ann. of Phys. 93, 193 (1975).
-H.Fritzsch, M.Gell-Mann and P.Minkowski; Phys.Lett. 59B,256 (1975).

10) R.M.Barnett; Phys.Rev.Lett. 34,41 (1975).

11) F.Gursey, P.Ramond and P.Sikivie; Phys.Lett. 60B,177 (1976).

12)-F.Gürsey, P.Ramond and P.Sikivie; Phys. Rev. D12,2166 (1975).
-F.Gürsey and P.Sikivie; Phys.Rev.Lett. 36,775 (1976).
-P.Ramond; Nucl.Phys. B110,214 (1976).

FIELD THEORY APPROACH TO THE STATISTICAL BOOTSTRAP

Etim-Etim

CERN, Geneva[+]

1. INTRODUCTION

There is currently a new trend in particle physics as is evidenced by the rapidly growing interest in the dynamics of extended structures and the mass spectrum they sustain.

It is assumed that these structures can be described by quantum field theory in which an infinite summation over some suitable collective modes takes into account the essential non-perturbative features of the many-body problem. The new approach therefore goes beyond canonical perturbation theory.

The Statistical Bootstrap Model was devised to describe precisely the kind of extended objects (fireballs = hadrons) for which the above field theory description is assumed to apply. From this point of view it is natural to enquire if a consistent field theory

[+] On leave of absence from Laboratori Nazionali Frascati, Italy; Present Address: Z.i.F., University of Bielefeld, Germany

underlies the bootstrap idea. This is indeed the case. The bootstrap constraint can infact be respected by a non--local field theory. We therefore propose to generalize the statistical bootstrap model to a bootstrap field theory of which we shall construct here the simplest non--trivial element - the two-point function.

2. HAGEDORN-FRAUTSCHI AND YELLIN REPRESENTATIONS

It was first hinted by Yellin that there is a formal analogy between the statistical bootstrap model and a non-polynomial Lagrangian field theory. Unfortunately for the simplest and most studied non-polynomial Lagrangian

$$\mathcal{L}_{int}(x) = \Lambda e^{i\lambda\varphi_0(x)} \tag{1}$$

where Λ is the major and λ the minor coupling constant and $\varphi_0(x)$ a free scalar field, the spectral function of the "superpropagator"

$$\Delta(p^2) = e^{\lambda^2\Delta_F(p^2)} \tag{2}$$

grows less than exponentially for large $m = \sqrt{p^2}$

$$\mathrm{Im}\left[\Delta(p^2)\right]_{REG} \xrightarrow[p^2\to\infty]{} \exp(a\,m^{2/3}) \tag{3}$$

There is no doubt that one can find a non-polynomial Lagrangian with a spectral function which grows exponentially as required or at worst one that over-shoots it. However quite apart from the inherent complexity of working with non-polynomial Lagrangians, the main difficulty is in the approach itself - the specification of an interaction Lagrangian and consequently an equation of motion. In the bootstrap approach, the bootstrap constraint is the "equation of motion".

Its most familiar formulation is in the feed-back

A HADRON

→ is a composite of an undetermined number of all kinds of HADRONS, each of which in turn ──┘

from which it is clear that our business is with the way hadrons are assumed to be constituted. An infinite hierarchy belies this structure. In order to appreciate this point of view it is instructive to examine briefly how one would construct a hadron otherwise.

It is enough for our purposes to compare the phase space density of the composite system with the density of mass states of the constituents.

The density of states of N independent particles in a box of volume V_0 + and total energy E is given by

$$\sigma_N(E, V_0) = \left[\frac{V_0}{(2\pi)^3}\right]^N \int \delta\left(E - \sum_{i=1}^{N} E_i\right) \delta^{(3)}\left(\sum_{i=1}^{N} \vec{p}_i\right) \prod_{i=1}^{N} d^3\vec{p}_i \quad (4)$$

If the N particles are identical a factor of 1/N! is also required in the RHS.

For simplicity we ignore quantum numbers, charge, spin, isospin, etc.

+ is necessary for "confinement"

(a) For a meson considered as a quark-antiquark pair, $q\bar{q}$, N = 2 and

$$\sigma_2(E, V_0) \xrightarrow[E\to\infty]{} E^2 \tag{5a}$$

while the mass spectrum is just a delta function.

$$\rho(m) = \delta(m - m_q) \tag{5b}$$

(b) For a baryon as a quark triplet N = 3 and

$$\sigma_3(E, V_0) \xrightarrow[E]{} E^5 \tag{6a}$$

The mass spectrum on the other hand is a sum of delta functions

$$\rho(m) = \sum_{i=1}^{3} \delta(m - m_i) \tag{6b}$$

(c) For the slightly more interesting case of a meson formed of a gas of an elementary input boson b, $N = \infty$, ie meson $\equiv$ bb, bbb, bbbb,.. Nb, ($N \to \infty$) one finds

$$\sigma_\infty(E, V_0) = \sum_{N=2}^{\infty} \frac{1}{N!} \sigma_N(E, V_0)$$

$$\xrightarrow[E\to\infty]{} \exp(aE^{3/4}) \tag{7a}$$

$$\rho(m) = \delta(m - m_b) \tag{7b}$$

From these simple examples it is evident that infinite compositeness is necessary if the density of all states of motion $\sigma(E, V_0)$ must grow fast enough to approach an exponential. This condition is however far from sufficient to ensure the bootstrap equality

$$\sigma(E, V_o) = \rho(E=m) \tag{8}$$

as the last example clearly shows. The special strong interaction dynamics responsible for (8) does not therefore rely only on the wealth of elementary constituents available. Secondly eq. (8) cannot be valid for arbitrary volume V_o . In fact if the volume V_o is very small the contribution of the N-particle cluster to the LHS of (8) is suppressed by the factor V_o^N as against the linear V_o dependence of the states contributing to $\rho(m)$ (cf. eq.(4)). If on the other hand V_o is very large the N-particle clusters dominate overwhelmingly in $\sigma(E, V_o)$ and the equality in (8) becomes impossible. There is thus a critical volume V_o at which the density of all states of motion of all possible non-interacting N-clusters $\sigma(E, V_o)$ becomes equal to the dynamical density of mass states $\rho(m)$. V_o is a crucial parameter which sets the mass scale. At such a volume eq. (8) is an extremely revealing and interesting equality between the many intrinsic degrees of freedom of a massive resonance as counted by the mass spectrum and the equally large number of dynamical states of motion which can be generated just by clustering.

Such equality presupposes that no hadron is elementary. Consequently if the input boson in example (c) above is composite one can write symolically

$$b \equiv bb, \quad bbb, \quad bbbb \; .... \; Nb, ... \; (N \to \infty)$$

Hence if for fixed N (N = 2, 3, 4 ....) one iterates eq. (7a) for each factor b the density of states of motion $\sigma(E, V_o)$ becomes

$$\sigma(E,V_o) = \sum_{N=2}^{\infty} \frac{1}{N!}\left[\frac{V_o}{(2\pi)^3}\right]^N \int \delta(E-\sum_{i=1}^{N} E_i)\,\delta^{(3)}(\sum_{i=1}^{N} \vec{p}_i) \times \prod_{i=1}^{N} \sigma(m_i,V_o)\,dm_i\, d^3\vec{p}_i \tag{9}$$

which on making use of eq.(8) can also be rewritten as

$$\rho(m) = \sum_{N=2}^{\infty} \frac{1}{N!}\left[\frac{V_o}{(2\pi)^3}\right]^N \int \delta(m-\sum_{i=1}^{N} E_i)\,\delta^{(3)}(\sum_{i=1}^{N} \vec{p}_i) \times \prod_{i=1}^{N} \rho(m_i)\,dm_i\, d^3\vec{p}_i \tag{1o}$$

Eq.(1o) is the mathematical formulation of the constitutional assumption of the feed-back. It was first translated into this form by Frautschi. It has been observed that eq.(1o) is a reformulation of Hagedorn's hadronic thermodynamics in the language of the microcanical ensemble. This is essentially correct if one is satisfied with taking Laplace transforms. But from the point of view of Hagedorn's original arguments for the equilibrium of hadronic matter it goes very much beyond thermodynamics. According to Hagedorn equilibrium of hadronic matter, taking place in the short time of about $1o^{-23}$ sec., is not brought about by a large number of collisions between particles, and therefore requires no relaxation time. It is an equilibrium between the enormous number of competing decay channels of an excited hadron. The probability weights of these decay channels are given directly by the S-matrix. There is thus no implication for the application of classical statistical thermodynamics nor the assumption that S-matrix elements should have no symmetry properties and no momentum dependence. Consequently in the approximation in which it is valid eq.(1o) represents not so much the Laplace transform of a partition function but an important information on the constitution of a hadron which the S-matrix must contain. We shall call it the cluster or Hagedorn-Frautschi representation.

Eq.(1o) can be solved exactly for $\rho(m)$. It was first done by Yellin. To exhibit the solution explicitly in covariant form let us make use of the fact that the critical volume V_0 in eqs. (8) and (1o) sets a mass scale in the theory to introduce a lowest mass discrete state in the mass spectrum at $m = m_0$ where for later convenience we write

$$V_0 = \frac{4\pi}{3}(4\pi B)^{3/2}$$
$$B = 1/4\pi m_0^2 \qquad (11)$$
$$\delta_0(p^2 - m_0^2) = \theta(p_0)\,\delta(p^2 - m_0^2)$$

Eq. (1o) can now be rewritten in explicitly covariant form as

$$B\rho(p^2) = B\delta_0(p^2 - m_0^2) + \sum_{N=2}^{\infty} \frac{1}{N!} \int \delta^{(4)}\left(p - \sum_{i=1}^{N} q_i\right) \times \qquad (1o)$$
$$\times \prod_{i=1}^{N} B\rho(q_i^2)\, d^4 q_i$$

The presence of the term $\delta_0(p^2 - m_0^2)$ in the RHS of eq.(1o) is not strictly faithful to the idea that no hadron is elementary. However from the way it was introduced the "elementarity" of the particle with mass m_0 reflects the length scale to which one probes. It is not overtly inconsistent with the bootstrap philosophy, for what is "elementary" at one length scale may be "composite" when one probes to smaller distances.

Yellin's solution of eq. (1o') is

$$B\rho(p^2) = \sum_{N=1}^{\infty} g_N \, \Omega_N(p^2) \; ; \; g_1 = 1 \tag{11}$$

where

$$\Omega_N(p^2) = \int \delta^{(4)}\left(p - \sum_{i=1}^{N} p_i\right) \prod_{i=1}^{N} B\delta_o(p_i^2 - m_o^2)\, d^4 p_i \tag{12}$$

is the N-body phase space and the g_N are numerical coefficients which can be computed iteratively by substituting (11) into (1o') or more straightforwardly by using their Laplace transforms. We shall refer to eq. (11) as the granular or Yellin representation.

3. BOOTSTRAP FIELD THEORY

In the statistical bootstrap model (SBM) eq.(11) is nothing more than an interesting analytic solution of eq. (1o'). The point of view of bootstrap field theory (BFT) is radically different. According to it eqs.(1o') and (11) provide two different but physically equivalent ways of interpreting the complex structure of a hadron. For this reason it is argued that the two representations should be unitarily equivalent. In other words, if in the granular picture a hadron is described by the field $\varphi(x)$ and in the cluster picture by $\Phi(x)$ then there should exist a unitary transformation $\mathcal{U}$ such that

$$\Phi(x) = \mathcal{U}\, \varphi(x)\, \mathcal{U}^{\dagger} \tag{13}$$

Actually much less will be required; as U will be implemented, the fact that it is an isometry ($U^+U = 1$, $(UU^+)^2 = UU^+$) will suffice. Armed with U the SBM follows as a very special version of bootstrap field theory, obtained by making use of an interesting decomposition of phase space (totally ordered manifold)

$$\Omega_N(p^2) = \delta_{1N} B \delta_0(p^2 - m_0^2) + \frac{1}{d_N - 1} \sum_{\ell=2}^{N} \sum_{n_1+n_2+\cdots+n_\ell=N} \times$$
$$\times \int \delta^{(4)}\Big(p - \sum_{i=1}^{\ell} q_i\Big) \prod_{i=1}^{\ell} \Omega_{n_i}(q_i^2)\, d^4 q_i \qquad (14)$$

where d_N is the number of partitions of N.

Eq.(14) is a bootstrap equation, indeed the bootstrap equation. The Frautschi bootstrap expresses the form invariance of (14) under the (order-preserving) defornations

$$\begin{aligned} \Omega_N(p^2) &\longrightarrow \hat{\Omega}_N(p^2) \\ \Omega_{n_i}(q_i^2) &\longrightarrow \hat{\Omega}_{n_i}(q_i^2) \end{aligned} \qquad (15)$$

i.e.

$$\hat{\Omega}_N(p^2) = \delta_{1N} B \delta_0(p^2 - m_0^2) + \sum_{\ell=2}^{N} \sum_{n_1+n_2+\cdots+n_\ell=N} \gamma_\ell(n_1 n_2 \cdots n_\ell) \times$$
$$\times \int \delta^{(4)}\Big(p - \sum_{i=1}^{\ell} q_i\Big) \prod_{i=1}^{\ell} \hat{\Omega}_{n_i}(q_i^2)\, d^4 q_i \qquad (16)$$

in the special case of dilatations

$$\hat{\Omega}_N(p^2) = g_N \Omega_N(p^2)$$
$$\hat{\Omega}_{n_i}(q_i^2) = g_{n_i} \Omega_{n_i}(q_i^2) \tag{15'}$$

The problem represented by eqs.(14)-(16) belongs to Algebraic Topology. The dynamics in BFT is not in eqs.(1o') and (16). These equations are nothing more than an interesting way of expressing a particular normalization condition (quantization). The dynamics of the mass spectrum is completely specified by an abstract operation, called the cluster product, which is defined so as to simulate strong interactions.

3.1 CLUSTER PRODUCT

The three main features of strong interactions we wish to simulate are: -

(1) the state of two free pions is different from that of a ρ -meson and that of three free pions from that of ω. Symbolically

$$\begin{aligned} &\pi\pi \neq \rho \\ &\pi\pi\pi \neq \omega \neq \rho\pi \neq A_1 \end{aligned} \tag{17a}$$

etc.

Introducing formally a bracket symbol e.g.

$$\begin{aligned} \rho &= (\pi\pi) \\ \omega &= (\pi\pi\pi) \\ A_1 &= (\rho\pi) \end{aligned} \tag{17b}$$

etc.

to indicate that ρ is formed from two pions, ω from three, A_1 from ρ,π and so on, we can rewrite eq.(17a) as

$$\pi\pi \mp (\pi\pi)$$

$$\pi\pi\pi \mp (\pi\pi\pi) \mp (\pi\pi)\pi \mp ((\pi\pi)\pi)$$

etc. (17c)

The bracket operation is thus non-associative. Mathematically bracketing is a combinatorial problem, and as such has been known for a long time. Given a sample of size N e.g. the product

$$M_N = \pi_1 \pi_2 \cdots\cdots \pi_N \tag{17d}$$

the number of different ways of putting brackets between various factors in the product is called the bracketing coefficient b_N. The difference between one bracketing problem and another arises from the type of addtional constraints imposed on the formation of brackets e.g. bracketing only two numbers at a time, and that difference is reflected in b_N.

Bracketing is our mechanism for resonance formation starting with a given input. It consists in associating with each monomial M_N a Hilbert space H_N of degenerate states of dimension b_N. Each of the N factors in (17d) carries a basic unit of an additive "quantum number".

(2) The square of hadron masses are quantized i.e.

$$m^2 = m^2(N) \tag{18}$$

(3) The mass spectrum of hadrons grows exponentially; in other words the degeneracy at the mass level $m^2=m^2(N)$ grows exponentially with $m(N)$. An important property of the bracketing coefficient is that it too grows exponentially with some algebraic function of N. Hence if the mass spectrum is arranged to be equal to b_N the function (18) is known, at least asymptotically.

Consider then a set[+)]

$$\pi^{\pm}(p_1)\,,\ \rho^{\pm}(p_2)\,,\ \omega^{\pm}(p_3)\,,\ \cdots\cdots \tag{19}$$

of creation (+) and annihilation (-) operators in momentum space. Their (generalized) Wick product is

$$W(\pi^{\pm},\rho^{\pm},\omega^{\pm},\ldots;p)=\int\delta^{(4)}(p-\sum_{i=1}p_i)\,\pi^{\pm}(p_1)\rho^{\pm}(p_2)\omega^{\pm}(p_3)\cdot \times\prod_{i=1}d^4p_i \tag{20}$$

Introduce an abstract operation $\mathbb{C}$, which transforms Wick products into the creation and annihilation operators of particles.

+) $\pi^{\pm}$, $\rho^{\pm}$ do not stand for the corresponding particles in the Rosenfeld tables.

$$\mathbb{C} : W(\pi^{\pm}, \rho^{\pm}, \omega^{\pm}, \ldots; p) \longrightarrow (\pi^{\pm}, \rho^{\pm}, \omega^{\pm}, \ldots; p)$$

$$(\pi^{+}, \rho^{+}, \omega^{+}, \ldots; p) = (\pi^{-}, \rho^{-}, \omega^{-}, \ldots; p)^{+} \qquad (21)$$

and satisfies the following conditions:-

$\mathbb{C}$ 1. For all single-narticle onerators e.g. $\pi^{\pm}(p)$, $\rho^{\pm}(p)$, $\omega^{\pm}(p)$etc.

$$(\pi^{\pm}; p) = \pi^{\pm}(p) \qquad (22)$$

$$(\rho^{\pm}; p) = \rho^{\pm}(p), \text{ etc}$$

Hence on the vacuum of the onerators in (19)

$$\mathbb{C}|0\rangle = |0\rangle \qquad (23)$$

Because ($\pi^{\pm}, \rho^{\pm}, \omega^{\pm}, \ldots; p$) is by definition a single--narticle operator

$$\mathbb{C}(\mathbb{C}(\pi^{\pm}, \rho^{\pm}, \omega^{\pm}, \ldots; p)) = (\pi^{\pm}, \rho^{\pm}, \omega^{\pm}, \ldots; p) \qquad (22')$$

imnlving

$$\mathbb{C}^{2} = \mathbb{C} \qquad (24)$$

$\mathbb{C}$ has therefore no inverse $\mathbb{C}^{-1}$. The decay of a resonance is therefore not the inverse of $\mathbb{C}$. It is much more com-nlicated and will not be considered here.

$\mathbb{C}$ 2. For all local operators, $\pi^{\pm}$, $\rho^{\pm}$, $\omega^{\pm}$ .... irrespective of their number, $(\pi^{\pm}, \rho^{\pm}, \omega^{\pm}, \ldots; p)$ is local and the commutator

$$\left[(\pi^{-}, \rho^{-}, \omega^{-}, \ldots; p), (\pi^{+}, \rho^{+}, \omega^{+}, \ldots; q)\right] = c\text{-number} \neq 0 \tag{25}$$

is finite for all p^2.

The combination $\mathbb{C}W = W^{*}$ will be called a bracket or cluster or Wick star product in the set (19). The particles of which $(\pi^{\pm}, \rho^{\pm}, \omega^{\pm}, \ldots; p)$ are the creation and annihilation operators will be called clusters or simply resonances.

A few properties of W^{*} are that

(i) it is, as expected, non-associative

$$(\pi(\rho\omega)) \neq ((\pi\rho)\omega) \neq ((\omega\pi)\rho) \neq (\pi\rho\omega)$$

(ii) it is commutative

$$(\pi\rho) = (\rho\pi)$$

(iii) it is distributive over addition

$$(\pi\,[\rho+\omega]) = (\pi\rho) + (\pi\omega)$$

3.2 SBM OR PHASE SPACE BOOTSTRAP AS SPECIAL CASE OF BFT

Let

$$\varphi_0(x) = \int d^4p\, \delta_0(p^2 - m_0^2)\left(\bar{a}_0(p)e^{ipx} + \bar{a}_0^{+}(p)e^{-ipx}\right) \tag{26}$$

with

$$\left[\bar{a}_0(p), \bar{a}_0^{+}(q)\right] = 2E\delta^{(3)}(\vec{p} - \vec{q})\,(2\pi)^{-3} \tag{27}$$

be a free scalar field of mass m_0. Define the new operators

$$a_0(p)=\frac{\sqrt{2\pi^2}}{m_0}\delta_0(p^2-m_0^2)\bar{a}_0(p)$$
$$a_0^+(p)=\frac{\sqrt{2\pi^2}}{m_0}\delta_0(p^2-m_0^2)\bar{a}_0^+(p) \tag{28}$$

with commutation relation

$$\left[a_0(p), a_0^+(q)\right]=B\delta_0(p^2-m_0^2)\,\delta^{(4)}(p-q) \tag{29}$$

For the W* product of the a_0's we use the more convenient notation

$$a(p,N)=\mathbb{C}\int\delta^{(4)}(p-\sum_{i=1}^{N}q_i)\prod_{i=1}^{N}a_0(q_i)d^4q_i$$
$$a(p;n_1,n_2,\cdots n_\ell)=\mathbb{C}\int\delta^{(4)}(p-\sum_{i=1}^{\ell}q_i)\prod_{i=1}^{\ell}a(q_i;n_i)d^4q_i \tag{3o}$$

According to conditions $\mathbb{C}$1 and $\mathbb{C}$2 the operators in (3o) are local and have commutators

$$\left[a(p,N)\,;\,a^+(q,N')\right]=\delta_{NN'}\,\delta^{(4)}(p-q)N!\,\Omega_N(p^2)$$

$$\left[a(p,n_1,n_2,\cdots n_\ell), a^+(q,n_1',n_2',\cdots n_{\ell'}')\right]= \tag{31}$$
$$=\delta_{\ell\ell'}\,\delta^{(4)}(p-q)\sum_{\mathcal{P}}\Big(\prod_{i=1}^{\ell}\delta_{n_i n'_{\mathcal{P}(i)}}n_i!\Big)\times$$
$$\times\,\Omega_{n_1+n_2+\cdots+n_\ell}(p^2)$$

where $\mathcal{P}\equiv$ permutation of (1,2,...l). Note that eq.(3o) is a rewriting of the physics contained in eq.(17b). Eq. (31) is then the most naive generalisation of eq.(29) and gives rise to the following mapping

$$H_N \equiv \left\{ a^+(p;n_1,n_2,\cdots n_\ell)\,;\; N=\sum_{i=1}^{\ell} n_i\,;\; \ell \geq 2 \right\} \longrightarrow \Omega_N(p^2) \qquad (32a)$$

$$\frac{1}{\sqrt{N!}} J a(p,N) J^+ = \delta_{1N}\, a_0(p) + \frac{1}{\sqrt{d_N - 1}}\, \mathbb{C} \sum_{\ell=2}^{N} \frac{1}{\sqrt{\ell!}} \sum_{N=n_1+n_2+\cdots+n_\ell} \times$$

$$\times \int \delta^4\Big(p-\sum_{i=1}^{\ell} q_i\Big) \prod_{i=1}^{\ell} \frac{1}{\sqrt{n_i!}}\, a(q_i,n_i)\, d^4 q_i \qquad (32b)$$

between the Hilbert space H_N of dimension b_N and phase space, $b_N = (d_N - 1)^+$ is the number of different bracketings of first (.......) and second ((.....) (....) (...)) order for a sample of size N and J is an isometry ($J^+J=1$, $(JJ^+)^2 = JJ^+$).

The local space-time fields corresponding to the operators in (3o) are defined by

$$\varphi_N(x) = \frac{1}{\sqrt{2\pi^2}} \int d^4p \sqrt{p^2}\left(a(p,N)\, e^{ipx} + a^+(p,N)\, e^{-ipx} \right)$$

$$\varphi_1(x) \equiv \varphi_0(x) \qquad (33)$$

with $a(p,1) = a_0(p)$, and similarly for $\varphi_{n_1 n_2 \cdots n_\ell}(x)$ in terms of $a(p,n,n_2\ldots n_1)$. Consider next the operators defined by

---

$^+$ It is $d_N - 1$ and not d_N because $a(p,N) = a(p,\underbrace{1,1\ldots 1}_{N})$ by definition.

$$A(p) = a_0(p) + \sum_{N=2}^{\infty} \frac{\alpha_N}{\sqrt{N!}} a(p,N) \tag{34}$$

$$A(p,\ell) = \mathbb{C} \int \delta^{(4)}(p - \sum_{i=1}^{\ell} q_i) \prod_{i=1}^{\ell} A(q_i)\, d^4 q_i$$

$$= \sum_{n_1, n_2 \dots n_\ell = 1}^{\infty} \frac{\alpha_{n_1} \alpha_{n_2} \dots \alpha_{n_\ell}}{\sqrt{n_1!\, n_2! \dots n_\ell!}} a(p, n_1 n_2 \dots n_\ell) \tag{35}$$

$$C(p) = a_0(p) + \sum_{\ell=2}^{\infty} \frac{\gamma_\ell}{\sqrt{\ell!}} A(p,\ell) \tag{36}$$

with space-time fields

$$\varphi(x) = \sum_{N=1}^{\infty} \frac{\alpha_N}{\sqrt{N!}} \varphi_N(x) \ ; \quad \alpha_1 = 1 \tag{34'}$$

$$\Phi_\ell(x) = \sum_{n_1, n_2 \dots n_\ell = 1}^{\infty} \frac{\alpha_{n_1} \alpha_{n_2} \dots \alpha_{n_\ell}}{\sqrt{n_1!\, n_2! \dots n_\ell!}} \varphi_{n_1 n_2 \dots n_\ell}(x) \tag{35'}$$

$$\Phi(x) = \varphi_0(x) + \sum_{\ell=2}^{\infty} \frac{\gamma_\ell}{\sqrt{\ell!}} \Phi_\ell(x) \tag{36'}$$

With no restrictions on the coefficients $\alpha \equiv \{\alpha_N\}$ and $\gamma \equiv \{\gamma_\ell\}$ these fields are not local. From eqs.(31), (34) - (36) the vacuum expectation value of their commutators are

$$\langle 0|[\varphi(x), \varphi(y)]|0\rangle = 4\pi i B \int ds\, \rho(s;\alpha)\Delta(x-y;s) \tag{37a}$$

$$\langle 0|[\Phi(x), \Phi(y)]|0\rangle = 4\pi i B \int ds\, \sigma(s;\alpha,\gamma)\Delta(x-y;s) \tag{37b}$$

where $\Delta(x-y,s)$ is the free field commutator for a particle of mass $\sqrt{s}$ and

$$B\rho(p^2;\alpha) = B\delta_c(p^2-m_0^2) + \sum_{N=2}^{\infty} |\alpha_N|^2 \Omega_N(p^2) \tag{38a}$$

$$B\sigma(p^2;\alpha,\gamma) = B\delta_c(p^2-m_0^2) + \sum_{\ell=2}^{\infty} |\gamma_\ell|^2 \int \delta^{(4)}(p-\sum_{i=1}^{\ell} q_i) \times \prod_{i=1}^{\ell} B\rho(q_i^2,\alpha)\, d^4q_i \tag{38b}$$

Eq.(13) is thus sufficient to obtain the bootstrap equation

$$B\rho(p^2;\alpha) = B\delta_c(p^2-m_0^2) + \sum_{\ell=2}^{\infty} |\gamma_\ell|^2 \int \delta^{(4)}(p-\sum_{i=1}^{\ell} q_i) \times \prod_{i=1}^{\ell} B\rho(q_i^2,\alpha)\, d^4q_i \tag{39}$$

and implies

$$\alpha = \alpha(\gamma) \tag{4o}$$

Eq.(39) determines (4o) only in modulus. In fact from

$$C(p) = \mathcal{U} A(p) \mathcal{U}^{+} \tag{41}$$

and

$$\langle 0|[C(p), C^{+}(q)]|0\rangle = \langle 0|[A(p), A^{+}(q)]|0\rangle$$

one gets

$$|\alpha_N|^2 = \delta_{1N} + \sum_{\ell=2}^{N} |\gamma_\ell|^2 \sum_{n_1+n_2+\cdots+n_\ell=N} |\alpha_{n_1}|^2 |\alpha_{n_2}|^2 \cdots |\alpha_{n_\ell}|^2 \tag{42}$$

which, with $g_N = |\alpha_N|^2$, $g_{n_i} = |\alpha_{n_i}|^2$,

$|\gamma_\ell|^2 = \gamma_\ell(n_1, n_2, \ldots n_\ell)$, is exactly what one would obtain by substituting (15') in (16). From eq.(41) and the commutation relations (31) it is easy to see that (42) follows from the operator equation

$$\frac{\alpha_N}{\sqrt{N!}} \mathcal{U} a(p,N) \mathcal{U}^{+} = \delta_{1N}\, a_0(p) + \mathbb{C} \sum_{\ell=2}^{N} \frac{\gamma_\ell}{\sqrt{\ell!}} \sum_{n_1+n_2+\cdots+n_\ell=N} \times$$
$$\times \int \delta^4\Big(p - \sum_{i=1}^{\ell} q_i\Big) \prod_{i=1}^{\ell} \frac{\alpha_{n_i}}{\sqrt{n_i!}}\, a(q_i, n_i)\, d^4 q_i$$

Comparing (32b) and (43) we see that if phase space bootstrap (the Frautschi equation) is all one needs U can very well be just an isometry. Eq.(43) is then consistent with (32b) if

$$\alpha_N = \delta_{1N} + \sum_{\ell=2}^{N} \gamma_\ell \sum_{n_1+n_2+\cdots+n_\ell=N} \alpha_{n_1}\alpha_{n_2}\cdots\alpha_{n_\ell} \qquad (44)$$

In turn (44) is consistent with (42) if

$$\mathrm{Re}(\alpha^*_{\mathbb{P}(N)}\alpha_{\mathbb{P}'(N)}) = 0; \; \mathbb{P}(N) \neq \mathbb{P}'(N) \qquad (45)$$

where $\mathbb{P}(N)$ is a partition of N

$$\mathbb{P}(N) : N = n_1 + n_2 + \cdots + n_\ell ; \; \ell \geq 2$$

and

$$\alpha_{\mathbb{P}_\ell(N)} = \alpha_{n_1}\alpha_{n_2}\cdots\alpha_{n_\ell}\,\delta(N - \sum_{i=1}^{\ell} n_i)$$

$$\alpha_N = \sum_{\ell=2}^{N} \gamma_\ell\, \alpha_{\mathbb{P}_\ell(N)} \qquad (46)$$

From eqs.(15), (16), (32b), (39), (42), (43) and (44) we see that many algebraic systems, e.g. functions, real and complex numbers, operators, the subspaces of a totally ordered topological space, obey bootstrap equations of the same general form. Thus one has only to verify the existence of a bootstrap relationship in a simple system (in our case the category of N-body (N=1,2...) phase space) and then reproduce it in other algebraic systems by suitable homomorphisms. Given this fact the Frautschi bootstrap equation is clearly nothing more than an interesting way of expressing a particular normalization condition (eq. (31)) for the states in H_N. Actually one need not go beyond eq.(34) to derive the Frautschi equation: for fixed N substitute $\hat{a}(q_i, n_i) = \alpha_{n_i}\, a(q_i, n_i)$ into eq.(32b) and change $1/(d_N-1)$ to γ_ℓ then

Phase space
→ consists of phase spaces
each of which in turn ⟲

yields the bootstrap equation upon taking commutators and summing over N.

3.3 CORRELATION BETWEEN MASS AND SIZE OF CLUSTER

By definition, the cluster operators $a(p, n, n_2 \ldots n_l)$ ($n_i, \ell \geq 1$) are local and have finite commutators for all p^2. This definition actually stipulates a definite correlation between the mass $m_N = \sqrt{p^2}$ and the size $N = n_1 + n_2 + \cdots + n_l$ of a cluster. Consider, for simplicity, the case l=1 and take the limit $N \to \infty$ in eq.(31). Condition $\mathbb{C}$ 2 then implies the inequality

$$0 < \lim_{N\to\infty} \left[N!\, \Omega_N(m_N^2) \right] < \infty \tag{47}$$

and hence the said correlation. If for small m_0 and $m_N \gg m_0$, $\Omega_N(m_N^2)$ is approximated by

$$\Omega_N(m_N^2) \simeq \frac{\pi B^2}{2(N-1)!\,(N-2)!} \left(\frac{\pi B}{2} m_N^2\right)^{N-2} \tag{48}$$

then (47) is satisfied only if

$$m_N^2 \xrightarrow[N\to\infty]{} \left(\frac{2}{\pi B e}\right) N = \left(\frac{8\,m_o^2}{e}\right) N \tag{49}$$

where $e \simeq$ 2.78.... If the mass-size correlation is the same for all clusters then the square of cluster masses are quantized

$$m_N^2 = m_o^2 + b(N-1)\,;\quad b = 8m_o^2/e \tag{5o}$$

4. BFT IN GENERAL

BFT is concerned not with the structure of phase space but how to formulate dynamics in the Hilbert space

$$H = \bigoplus_{N=1}^{\infty} H_N$$

$$H_N \equiv \left\{ a^+(p, n_1, n_2, \cdots n_\ell)|0\rangle,\ N = \sum_{i=1}^{\ell} n_i\,,\ \ell \geq 1 \right\} \tag{51}$$

The definition of the cluster products in eq.(3o) with only phase space correlation was done for the sake of simplicity and to get directly to the results of SBM. In general we have

$$a_{\lambda n_1 n_2 \cdots n_\ell}(p) = \frac{1}{\sqrt{n_1!\,n_2!\cdots n_\ell!}}\, \mathbb{C}\int \delta^{(4)}\Big(p - \sum_{i=1}^{\ell} q_i\Big)\, \psi_{\lambda n_1 n_2 \cdots n_\ell}(p, q_1 \cdots q_\ell) \prod_{i=1}^{\ell} a(q_i; n_i)\, d^4 q_i \tag{52}$$

in terms of wave functions $\psi_{\lambda n_1 n_2 \cdots n_\ell}(p, q_1 \cdots q_\ell)$ which may depend on additional quantum numbers represented collectively by λ. With (52) the normalization of the states in H_N is no longer given by (31). At this point we can incorporate the information that mass and size of clusters are correlated directly in the normalization of states and quantize as follows

$$\left[a_{\lambda n_1 n_2 \cdots n_\ell}(p), a^{+}_{\lambda' n'_1 n'_2 \cdots n'_{\ell'}}(q) \right] = \delta_{\lambda\lambda'} \delta_{\ell\ell'} \times \delta_o(p^2 - m^2(n_1 n_2 \cdots n_\ell)) \delta^4(p-q) \tag{53}$$

Eq.(53) is true for $l = n_1 = 1, n_i = o$ $i \geqslant 2$ (eq.(29)) and is satisfied by (52) if

$$\int \delta^{(4)}(p - \sum_{i=1}^{\ell} q_i) \psi^{*}_{\lambda' n_1 n_2 \cdots n_\ell}(p, q_1 q_2 \cdots q_\ell) \psi_{\lambda n_1 n_2 \cdots n_\ell}(p, q_1 q_2 \cdots q_\ell) \times \prod_{i=1}^{\ell} \Omega_{n_i}(q_i^2) d^4 q_i = \delta_{\lambda\lambda'} B \delta_o(p^2 - m^2(n_1 n_2 \cdots n_\ell)) \tag{54}$$

Note that if $\psi_{\lambda N} \neq \psi_{\lambda \underbrace{11 \cdots 1}_{N}}$ then $a_{\lambda N} \neq a_{\lambda \underbrace{11 \cdots 1}_{N}}$ and H_N in eq.(51) has dimension d_N

The operators

$$A_\lambda(p) = a_{\lambda o}(p) + \sum_{N=2}^{\infty} \sum_{\ell=1}^{N} \sum_{n_1 + n_2 + \cdots + n_\ell = N} a_{\lambda n_1 n_2 \cdots n_\ell}(p) \tag{55}$$

are not local. Their mass spectrum is given by the vacuum expectation value of the commutator

$$\langle 0|[A_\lambda(p), A^+_{\lambda'}(q)]|0\rangle = \delta_{\lambda\lambda'}\delta^4(p-q)\,B\rho(p^2) \tag{56}$$

with

$$\rho(p^2) = \delta_0(p^2-m_0^2) + \sum_{N=2}^{\infty} d_N\,\delta_0(p^2-m_N^2) \tag{57}$$

Making use of $m_N^2 = M_0^2 + b\,(N-1)$ and the recursion relation

$$d_N = \frac{1}{N}\sum_{\ell=0}^{N-1} \sigma(N-\ell)\, d_\ell \tag{58}$$

one gets

$$b\rho(p^2) = \sum_{N=0}^{\infty} d_N\,\omega_N(p^2) \tag{59}$$

where

$$\omega_N(p^2) = \sigma\left(\frac{p^2-m_0^2}{b}+1-N\right)\Big/\left(\frac{p^2-m_0^2}{b}+1\right) \tag{60}$$

and $\sigma(n)$ is the divisor function, that is the sum of all divisors of n. Eq.(59) is the "Yellin expansion" in this case and (58) the "bootstrap" equation analogous to

(42). Eq.(58) is a group theory (linear) decomposition with respect to the quantum number(N-1), that is the mass squared operator.

We are unfortunately not yet fully equipped to go beyond the mass spectrum and define the S-matrix. We still have to define an operation corresponding to decay and consider in some more detail the field theory properties of this non-local model. However what has been done is clear and can be concisely stated: Given the C^*-algebra generated by the creation and annihilation operators of a "particle" carrying a basic unit of an additive "quantum number" a multi-nary non-associative bracket product can be defined which maps the monomials

$$M_N = \underbrace{a_1 \, a_1 \, \cdots\cdots a_1}_{N\text{-times}} \tag{17d}$$

into N-degenerate vector spaces. The degeneracy of the subspaces H_N consisting of products with at most two overall bracketings is exponential in $\sqrt{N}$ as $N \to \infty$. The "quantum number" N can be anything provided it is additive. In the statistical bootstrap model it is the number of particles in phase space. In the dual resonance model it is spin.

As far as the mass spectrum is concerned the difference between these two models is in the different normalization of the states.

In fact eq.(46)

$$\alpha_{\mathbb{P}_\ell(N)} = \alpha_{n_1} \alpha_{n_2} \cdots \alpha_{n_\ell} \delta(N - \sum_{i=1}^{\ell} n_i); \; \ell \geq 1 \tag{46}$$

can also be used to generate H_N. In the dual model the α_{n_i} are commuting operators and the states $\alpha^+_{\mathbb{P}(N)}|0\rangle$, $\alpha^+_{\mathbb{P}'(N)}|0\rangle$ are orthogonal for $\mathbb{P}(N) \neq \mathbb{P}'(N)$. The need for bracketing does not arise because the states $\alpha^+_n|0\rangle$, $n \geq 1$ are assumed given from the beginning. If these states are normalized with respect to phase space and eq.(44) is imposed the statistical bootstrap model is obtained. The relationship between the two models ends here.

ACKNOWLEDGEMENTS

The work reported here was done in collaboration with Prof. Hagedorn. He is however not responsible for any inaccuracies in this report. I would like to thank G.C. Rossi for valuable discussions on combinatorics and the SBM, to F. Gliozzi for very helpful information on the point of view of the dual model and to W. Nahm for explaining his own point of view of SBM and the dual model.

REFERENCES

1. R. Hagedorn; Suppl. Nuovo Cimento 3 (1965) 147
CERN Lecture Notes, CERN 71-12 (1971)

2. R. Hagedorn and I. Montvay; Nucl. Phys. B59 (1973) 45

3. S. Frautschi; Phys. Rev. D3 (1971) 2821

4. S. Frautschi and C.J. Hamer; Phys.Rev. D4 (1971) 2125

5. J. Yellin; Nucl. Phys. B52 (1973) 583

6. W. Nahm; Nucl. Phys. B45(1972) 525

7. L. Comtet; Advanced combinatorics, D. Reidel Publishing Company, Dordrect (Holland, 1974)

8. R.D. Schafer; An Introduction to Non-Associative Algebras, Academic Press, New York (1966)

9. E.H. Spanier; Algebraic Topology, McGraw-Hill (1966) Chs. 1 and 7

1o. J. Riordan, An Introduction to Combinatorial Analysis, John Wiley, New York (1967)

D I S C U S S I O N

CHAIRMAN: Prof. E. Etim

Scientific Secretary: M.A. Ichola

DISCUSSION

- *LITTENBERG:*

How do you verify the form of $\rho(m)$ at high energies when you cannot identify individual resonances?

- *ETIM:*

It is not necessary to resolve individual resonances at very high energies in order to test the form of $\rho(m)$ there. In fact, the theoretical mass spectrum is a continuous function in the energy region where the resonances merge into a continuum.

- *WILKIE:*

The model is similar to the generalized Veneziano model and gives a spectrum of bosons of higher and higher spins. Is it possible to put in internal quantum numbers?

- *ETIM:*

It is not possible to give a proper treatment of angular momentum in this model, and this should be done before any attempt is made to include isospin.

- *FERBEL:*

I have often heard the statement you have just made -- namely that the mass spectrum grows exponentially. Could you explain what that means?

- *ETIM:*

That the spectrum grows exponentially! It means that if the mass is allowed to go to infinity the mass spectrum $\rho(m)$ behaves as an exponential exp (bm) where b is a constant.

- *FERBEL:*

I know that, but what does it mean and what does it imply?

- *ETIM:*

It is based on the bootstrap assumptions I drew attention to in the lecture. It implies that at very high energies, the number of open channels in a given hadronic reaction increases with energy and does so exponentially. Implications for cosmology have been discussed in various papers by Hagedorn, Frautschi, Huang, and Weinberg.

FIFTY YEARS OF SYMMETRY OPERATORS

E.P. Wigner

Princeton University

A BIT OF HISTORY

The title of my address, so kindly provided by the leadership of this conference, clearly indicates that I should be principally concerned with the role which the symmetry and invariance principles play in quantum mechanics, with their applications and effectiveness - it was precisely 50 years ago that these were recognized. Nevertheless, I like to say a few words about the role of symmetry principles in pre-quantum theory because the comparison of this role with the role played by the same principles at present seems to me very interesting.

Well, the first application of symmetry principles in physics is almost 150 years old. In 1830, J.F.C. Hessel determined 32 crystal classes[1)], those finite subgroups of the three-dimensional rotation-reflection group which have only elements of the order 1, 2, 3, 4, or 6. This was soon followed by the determination, simultaneously by Schönflies[2)] and by Fedorov[3)], of the 230 space groups, that is the 230 discrete subgroups of the euclidean group which contain three non-coplanar displacements. Let me admit at this point that it always fills me with admiration that neither Schönflies nor Fedorov have missed a single one of the 230 space groups - neither did Hessel miss any of the 32 crystal classes. They must have been very careful workers. Of course, their interest in the problems they solved was motivated by the early ideas on crystal structure, dating back many years to Steno[4)]. Indeed, according to classical mechanics, if the very few atoms the positions of which are not determined by crystal symmetry, occupy equilibrium positions, the same will be true also for those atoms - infinite in number in an infinite crystal - the positions of which *are* determined by the crystal symmetry, rotational or displacement. In this sense then, the concept of crystal symmetry is rigorous in classical physics, as is also the wealth of consequences for the properties of the crystal which follow from the crystal's symmetry. As will be mentioned later, none of this is true if classical mechanics is replaced by quantum mechanics - the classical crystal symmetry then becomes an approximate concept and can be best motivated by the Born-Oppenheimer approximation. It was rigorous, though, in the world picture prevalent at the time its founders established it - a fact worth remembering.

In classical physics, the theory of crystal symmetry was the most colourful consequence of the symmetry properties of space-time but it was not the only remarkable consequence. The other remarkable consequence was the validity of the various conservation laws for energy, linear and angular momentum, motion of the centre of mass. Actually, these were known before their connection with the invariance properties of space-time (Galilei invariance originally) was discovered by Hamel, Klein, and Nöther[5]. Again, these are rigorous consequences of the invariance properties of space-time, valid for isolated systems, i.e. closed physical systems, not subject to forces due to objects not included in the system. The derivation of the conservation laws from the invariance principles required great skill - its discoverers also deserve our admiration.

The last and most important pre-quantum mechanical application of symmetry principles is, as we all know, due to Einstein. He was also the one who first emphasized the fundamental importance of these principles - the special theory of relativity is a direct consequence of them. The general theory of relativity also seems to be founded on a symmetry principle but, as will be discussed further below and as was already recognized 60 years ago by Kretschman[6], the way the symmetry principle is applied in this connection is rather different from its application in the cases mentioned before. As I said, I will return to this point later.

Before sketching the role the symmetry principles played in the last 50 years, in particular in quantum mechanics, it may be good to insert a few remarks on the essence of these principles, on their role in the structure of physics. It will be my endeavour to be brief in this because I have already had the opportunity and pleasure to discuss this question at a meeting similar to the present one[7].

EVENTS, LAWS OF NATURE, INVARIANCE PRINCIPLES, KINEMATIC AND THEN DYNAMIC

The physicist endeavours to find the laws of nature, that is the correlations, between events. We know that if a heavy object is at the time t_0 at a distance z_0 from the ground and is at rest, it will be at time t at the distance $z = z_0 - g(t-t_0)^2/2$ from the ground. The two events between which this law established a correlation are the positions and velocities at the times t_0 and t. As we all know, the law just given can be greatly generalized - it is given in the simple form only as an illustration of what "law of nature" and "correlation between events" mean. We also know that the events between which physics establishes correlations has undergone drastic changes in the course of our discipline's history: it was, originally, the occupation of definite positions by the objects, it became the magnitude of various kinds of field strengths at all points of space later - in quantum mechanics it seems to be the outcome of an observation. Naturally, in order to make a prediction of a future event on the basis of the knowledge of past events, the system the events of which we are considering must be free of unknown outside influences, it must be "isolated" at least to such an extent that the unknown part of the outside influences is negligible. This can create serious problems, both practical and theoretical. However, if we disregard these problems, the laws of nature become verifiable and they describe truly amazing correlations between events, in amazingly simple and attractive mathematical language. It is well to emphasize, nevertheless, that the laws of nature only give correlations between events, they do not describe or predict all the events we experience - they would not even if they were perfect. This fact appears in the usual mathematical formulation of the laws of nature as the needed input of the initial conditions, the position and velocity of the falling object in the example mentioned.

The preceding characterization of the laws of nature, and of the role of initial conditions, may be short and incomplete but we are all cognizant of their essence. The same applies to the concept of the symmetry or invariance principles, at least the kinematic ones - in contrast to the laws of nature these describe correlations between the laws of nature, that is correlations between the correlations between events. They postulate the equivalence of several ways to describe the events - the correlations are postulated to be the same in all these descriptions. If we can translate from one description of the events to the other equivalent descriptions of these events, the postulate of the identity of the correlations provides a great deal of information about these correlations.

Let me illustrate this point on a very simple, in fact trivial, example. It is well known that Newton's first law as formulated already by Galileo "any velocity once imparted to a body will be rigidly maintained as long as there are no causes of acceleration or retardation" is invariant under Galilei transformations (and also under Lorentz transformations). It is not commonly observed that the converse is also true if it is also assumed that the motion is fully determined by the initial position and velocity, i.e. if the position at time t, to be denoted by $X(t;x,v)$ is a uniquely defined function of t and the initial position x and velocity v

$$X(0;x,v) = x \qquad X'(0;x,v) = v \tag{1}$$

X' denoting the derivative of X with respect to the first variable. This, of course, is not true for bodies of asymmetric nature - in fact for such bodies the definition of the "position" may be quite complex. In the following, however, the existence of a uniquely defined $X(t;x,v)$ will be assumed and proved, on the basis of Galilei or Poincaré invariance that

$$X(t;x,v) = x + vt \ . \tag{2}$$

In order to prove (2), we first consider $X(t;0,0)$, i.e. a body at rest and at the origin at $t = 0$. If this were different from 0, rotational invariance would be violated. Hence,

$$X(t;0,0) = 0 \ . \tag{3}$$

If we now describe this body's motion from the point of view of a coordinate system which is at the position $-x$ at time 0, and which has a uniform velocity $-v$, the body will have, in this coordinate system, at time t the position

$$X(t;x,v) = X(t;0,0) + x + vt \tag{3a}$$

which, together with (3), indeed gives (2) - if we assume that X is uniquely determined by its variables.

It may be worth observing that the preceding argument cannot be carried out if the space is one-dimensional. In this case, in fact,

$$X(t;x,v) = \frac{1}{2} gt^2 + x + vt \tag{4}$$

is consistent with Galilei or Poincaré invariance - though not if reflection invariance is also present - g being a number characteristic of the body, i.e. independent of x,v,t but possibly different for different bodies. A similar situation holds in quantum mechanics also: the derivation of the representations of the Poincaré group (to be considered later) yields less far-reaching results if there is only one space-like dimension. The limitation to symmetric bodies can, though, be eliminated.

What the preceding argument shows is that Newton's first law is not only compatible with Galilei invariance, it is a consequence thereof - or of Poincaré invariance.

This example was worked out in some detail because it renders it less surprising, at least in the case of elementary systems, that the equations of motion can be derived from invariance principles also in quantum mechanics. Altogether it is perhaps the most important function of invariance considerations to eliminate assumed laws of nature which are incompatible with them. Another important function, to be discussed later, is the derivation of some of the consequences of these laws - many more in quantum mechanics than in classical theory where this is restricted to the consequences of the conservation laws. Important as these are, they are quite restricted.

The preceding considerations refer to kinematic invariance principles and what I would like to call *active* ones. They postulate correlations between different sets of events, sets which can be transformed into each other by an invariance transformation. They actually increase our knowledge of correlations between events. If we find that hydrogen peroxide reacts with a permanganate and discolours it now and here in Erice, we can infer that this will be the case also at another time, at another place, and will be true also on a rapidly moving airplane. In contrast, there are invariance laws which postulate a variety of descriptions between the same events. The oldest example for this is the description of the same electromagnetic field by a variety of potentials, obtainable from each other by the addition of an arbitrary gradient. All these postulate the same correlations between "events" which are, in this case, the magnitudes and directions of the electric and magnetic fields associated with the points of space-time. Unimportant as this may appear to be, it does unquestionably have a certain mathematical attraction to have a variety of descriptions of the same events. The most important non-active invariance is the basis of the general theory of relativity - the same field, that is the same set of events, can be described by all the coordinate systems which can be transformed into each other. All the true consequences of the theory, all the correlations, can be obtained by means of coordinate systems chosen in particular ways, as characterized most concretely by Fock[8]. Surely, ordinary mechanics can also be formulated not only in terms of Cartesian coordinates but in terms of any type of coordinates - in fact Lagrange's equations do that. But the physical content of the equations is not increased by the more general formulation - only the mathematical attraction of the theory is. As was mentioned before, this was first recognized by Kretschman[6]. Nevertheless, the mathematical attraction of the more general formulation is so great that the condition of the invariance of the gravitational equations under arbitrary coordinate transformations has not been seriously questioned. Similar invariances under "gauge transformations" have been introduced also in modern quantum mechanics and they play an important role in particular in particle theory.

Let me now come to the last kind of invariances I want to mention - the dynamic invariances. These are also of increasing importance. They refer to particular types of interactions which show definite and often surprising regularities. The oldest such regularity is the very closely equal interaction between a pair of protons, a pair of neutrons, and between a proton and a neutron pair. This equality of the interactions is approximate - it has nevertheless many interesting consequences which will be a bit further discussed below. Even more interesting is the SU(3) symmetry between particles, the "eight-fold way", discovered by Gell-Mann and Ne'eman[9] though it is even less accurate. I am sure all of us here are familiar with it.

Let me herewith conclude the general discussion of the various types of symmetries and go over, first to a discussion of the fundamental differences in their application in classical and quantum physics, then to a more specific discussion of the conclusions they led to in the various areas of our discipline.

WHY DO THE SYMMETRY PRINCIPLES CONTRIBUTE MORE TO QUANTUM THAN TO CLASSICAL MECHANICS?

Two of the reasons for the increased importance of symmetry principles are rather obvious. The description of a state of the system is more involved mathematically in quantum than in classical theory. The latter describes the state of n particles by 6n numbers (positions and velocities), the first one by functions of 3n variables - a much more complex concept. If the symmetry principles make in both theories the same relative contribution to the solution of the equations, their actual contribution is much greater in quantum than in classical theory. The second obvious reason is that quantum mechanics has been applied in much more widely extended areas than were the domains of classical physics. In particular, the latter took the properties of its objects, size of the atoms, density of the substances, as given from the outside, whereas quantum theory tries to derive them from its basic principles.

There are, however, in addition to the foregoing, other reasons for the increased effectiveness of the symmetry postulates. The first and most general one of these is that the states of a quantum mechanical system form a linear manifold. In particular, the superposition of two or more states with the same definite energy*) (or the same momentum or the same value of any other quantity) is still a state with the same definite energy (or the same momentum, etc.). It further follows from the rotational invariance of the energy that if we subject a state with a definite energy to a rotation, it still will have the same definite energy. This is, of course, true also in classical theory. However, in quantum mechanics it is also possible to make a superposition of the initial and all kinds of rotated states and each such superposition will still have the same definite energy. If we superpose with equal weights states subjected to all possible rotations, we obtain either 0 - that is no state - or a spherically symmetric state. Hence, the existence of spherically symmetric states - except for single particles non-existent in classical theory - is, in

*) By this we mean that its state vector is a characteristic vector of the energy operator.

quantum mechanics, a consequence of the rotational invariance of the energy. Even if the procedure sketched above gives 0, i.e. no spherically symmetric state, it is possible to construct finite sets of states, each member of which, if subjected to a rotation, becomes a linear combination of the members of the finite set to which it belongs. This is a mathematical theorem, known by now for a long time. Hence, the linear character of the state vectors, the possibility to form superpositions of them, guarantees that the states with a definite energy have simple properties with respect to rotations. This plays an important role in the theory of simple systems and leads to the theory of angular momentum J. This has an important role in the theory of small systems - atoms and molecules - but is not very relevant in the macroscopic domain. After all, no one wants to construct a spherically symmetric state of a crystal - even if it does follow from the theory that such states exist.

The linear character of the states in quantum mechanics has other interesting consequences. If we subject the states not only to rotations but to all possible Poincaré transformations (Lorentz transformations and displacements) we can form sets of states - infinite sets though - with remarkable properties. Each set remains "closed" under any Poincaré transformation, i.e. if we subject any member of any set to such a transformation, it becomes a linear superposition of the original members of the same set - a rather simple set characterized by two numbers, the rest energy of the states of the set (which is the same for all members of the set) and the angular momentum of the states at rest (called the spin, also the same for all members of the set). Since the Poincaré transformations include time displacement, the time development of each member is also determined. What is sketched here is the theory of the irreducible representations of the Poincaré group[10]) - surely not a new subject. It follows from what I said, conversely, that every state of any system can be considered as a linear combination of states each of which belongs to an irreducible representation of the Poincaré group so that its time development is well known. This seems wonderful, but the trouble is that we do not know, in general, how to decompose the state vector of an arbitrary system into vectors which transform by such a representation-the state vectors are usually given in terms of other quantities. Only in the case of elementary particles, all the states of which transform to superpositions of each other by an irreducible representation of the Poincare group, does the existence of this symmetry give a complete description of the behaviour of all states. Even this is, however, remarkable - it also reminds us of the similar situation in classical mechanics where the admittedly much simpler character of the motion of the particle also follows from the Galilei or Lorentz invariance.

The preceding discussion emphasizes the importance of the linear character of the quantum mechanical states as facilitating the application of the invariance principles. In fact, this character has a decisive role in the application of the symmetry relations in quantum mechanics. Another circumstance which increases the role of the symmetry relations is the importance of the response of systems to small perturbations of the total Galilei or Poincaré symmetry. Such a perturbation is caused, for instance, by a weak homogeneous magnetic or electric field or some other outside interaction which destroys part of the Galilei or Poincaré symmetry, leaves some of it valid and has a simple behaviour even under those symmetry operations which it destroys. It is a consequence of the effect of a weak, homogeneous magnetic field that it splits an energy level with angular momentum J into 2J+1 equidistant levels, the splitting being proportional to the intensity of the magnetic field. There are many other kinds of perturbations about the consequences of which the symmetry principles furnish equally far-reaching information[11].

The last reason for the increased effectiveness of the symmetry principles in quantum mechanics, perhaps the most important one, is the existence of approximate symmetries in virtually every area of application of this theory. The oldest example herefore is the Russell-Saunders coupling in atomic spectra. It is based on the fact that the velocities of the electrons in the atoms, particularly of those on the outside, are small as compared with light velocity. As a result, the magnetic fields they create are small and have little effect on the energy levels, either by influencing the orbits, or by interacting with the electronic spins. The consequences of the assumption that the effect of these interactions is small can fill books, they include Landé's interval rule, his g formula, and many other conclusions. Naturally, all of them are approximate.

It may be good, in conclusion, to recall the case in which the symmetry relation is less effective in quantum than in classical mechanics. Crystal symmetry is, at least at low temperature, a rigorous concept in classical theory - in quantum mechanics it is an approximate one, just as are those mentioned in the last paragraph. It would be valid if the nuclear masses were infinitely larger than the electronic mass - since this is not truly the case, it is approximate. Born and Oppenheimer have developed an approximation scheme[12] which brings this out clearly, but the effect of the departure from the approximation has not been analysed for crystals, even though a number of cases are known where it breaks down badly. This presents a very interesting problem, but a fully logical solution, pointing to all kinds of deviations from the classical theory, does not seem to be easy.

This concludes what I wanted to say about the reasons of the greater effectiveness of the symmetry principles in quantum rather than in classical theory. A very short review of the application of these principles will follow a few critical remarks of a general nature on the invariance principles.

SOME ADVERSE REMARKS

It is natural to ask at this point: how sure are we that the "rigorous symmetries", those of the Poincaré group, are really valid? Do we have any reason to doubt them?

There is no question that they are valid to a very high degree of accuracy. They may not be absolutely accurate because every actual wave function extends to infinity and, of course, the true physical space is not flat, it has a gravitational curvature interfering with the Poincaré symmetry. If one makes an estimate of the inaccuracy of the symmetry on this basis, one arrives at a fantastically low number, perhaps unobservably low. But perhaps not quite. Anyway, there is at present no experimental reason to attribute weight to this observation.

It should be mentioned though, that what we now believe to be the true rigorous symmetry is smaller than what we thought of before 1956. Before that, it was believed that space reflection is also an accurate symmetry, i.e. that if a succession of events is possible, i.e. compatible with the laws of nature, the set of events which, if reflected in a mirror looks like these, is also compatible with the laws of nature. This symmetry was first suspected to be absent in the process of β-decay by Lee and Yang[13)] and their surmise was confirmed by Wu and collaborators[14)]. Thus space reflection, although a highly accurate symmetry in processes not affected by the interaction responsible for β-decay, is not a truly accurate symmetry. The same seems to be true of "time reflection" or reversal of the direction of motion.

Neither of these observations affect the remaining part of the Poincaré group, the so-called proper group. Also, for most phenomena, the effect of these "symmetry breaks" is negligibly small. Nevertheless, the fact that some symmetries in which we had full confidence are broken - violently under appropriate conditions - makes one less certain about the absolute validity of the remaining symmetries.

Are there any other reservations about the present concepts of symmetry and invariance? The connection between the invariance principles and the conservation laws was mentioned before. Can all conservation laws be derived from symmetry principles? No - the conservation laws for electric charge, for baryon and lepton numbers, do not follow from any accurate symmetry known at present. Are these going to undergo modifications? We do not know.

Let me now conclude by saying a few words on the application of the symmetry principles in the various principal areas of quantum mechanics - the role of the different symmetries which are recognized to be approximate providing the most interesting conclusions.

THE ROLE OF THE SYMMETRY PRINCIPLES IN SPECIAL AREAS OF PHYSICS

The symmetry principles' first application in quantum mechanics was to the theory of *atomic spectra*. Their effectiveness in this area was most eloquently stated by Max von Laue as quoted in the Preface of the book published on this subject[11)] by the present writer: "almost all rules of spectroscopy follow from the symmetry of the problem".

Some of these symmetries are essentially rigorous, such as the aforementioned equality of the energy of 2J+1 states, or the splitting of these energy levels into 2J+1 equidistant levels by a weak homogeneous magnetic field. The same applies to the formulae of Hönl and Kronig giving the relative strengths of the optical transitions originating from the members of such a split level. Some other rules, such as those of Landé also mentioned before, are approximate, the approximation being based on the assumption that the average velocities of the electrons are small as compared with light velocity. It is not possible to go into detail on these questions at this occasion - books can be, and have been, written on them.

The role of symmetry principles is less prominent in *molecular physics*. Surely, the rules given before which follow rigorously from Poincaré invariance are valid here also but they are less important because the density of energy levels is, as a rule, so great that one does not closely investigate individual energy levels. Of the approximate symmetries, the most important one is valid in the Born-Oppenheimer approximation[12)] - in this approximation the motion of the nuclei can be treated essentially classically, the potentials acting between them being given by the electronic energy levels which correspond to the specified positions of the nuclei. It follows that in this approximation the symmetry theory of classical mechanics is valid, and indeed, many molecules show a classical symmetry - the H atoms of CH_4 are practically always close to the corners of a regular tetrahedron. In classical theory, at very low temperature, they would always be exactly located at the corners of such a tetrahedron.

There is one exception to the very close accuracy of the classical theory: if a molecule contains several identical atoms, as does CH_4 or even H_2, the symmetry or antisymmetry of the wave function in the variables of these atoms influences the rotational structure drastically[15)] and renders also certain transformations very unlikely: we have orthohydrogen and parahydrogen and they transform into each other very slowly[16)].

Most of the remarks made above on the theory of molecular systems apply also in *solid state physics*. The energy levels of a solid are, in fact, practically infinitely dense on the energy scale. There is, however, an approximation which goes beyond that used for molecules.

Once the positions of the nuclei are fixed - and since the Born-Oppenheimer approximation is used and since this gives results close to those of classical physics, they do form a regular crystal lattice - the potential field acting on the electrons is a field with a definite crystal symmetry. The further approximation which is made is to attribute an individual wave function to each electron[17]. In atoms, this is called the Hartree-Fock approximation. The electrons then move in a field with the crystal symmetry and this has many interesting consequences[18]. In particular, the energy levels form zones, called Brillouin zones, and the electrons can jump from one level of such a zone to one close by. The electric conductivity and many other properties can be calculated from this picture and the agreement with the experimental findings is truly surprising. According to the picture used, if all Brillouin zones are either empty or fully occupied, an electric field of normal intensity cannot cause any electron to change its position in any of the zones and the electric field thus cannot induce a current. This is indeed confirmed by the fact that materials, such as diamond, in which, according to the theory, all Brillouin zones are filled or empty, are insulators - their electric conductivity is billions times smaller than that of metals. This agreement between the very approximate theory and experiment is truly surprising - no real explanation seems to be known therefor.

In *nuclear physics*[19], at least the basic rules of the coincidence of the energies of 2J+1 states, where $J\hbar$ is the angular momentum in the rest frame, is verified. The magnetic splitting of the lines has not been observed because its magnitude is inversely proportional to the mass of the constitutents, and hence about 2000 times smaller than in atomic spectra - unobservable for realizable magnetic fields. However, the other consequences of the Poincaré invariance appear to hold.

The very close equality of the proton-proton, neutron-proton, neutron-neutron interactions leads to a new approximate symmetry. If it were complete, isobars such as ^{7}Li and ^{7}Be would have identical energy spectra, and the two spectra are indeed very similar in this and in all similar cases. In the case of isobars such as ^{14}C, ^{14}N, ^{14}O, the spectra would be identical only for ^{14}C and ^{14}O - many of the levels of ^{14}N cannot be realized in ^{14}C and ^{14}O because of the exclusion principle. Indeed, the spectra of ^{14}C and ^{14}O are very similar and there is an energy level of ^{14}N which corresponds to each of the energy levels of ^{14}C and ^{14}O - the three levels form what is called an isospin triplet. A similar situation prevails throughout the whole nuclear table and several rules that can be derived for the properties of the members of isospin multiplets are well satisfied. Their energies differ, but the dependence of the difference is, as expected, to a very good approximation only a quadratic function of the neutron-proton difference. The β transitions between them satisfy the symmetry rule also very closely. The isospin concept turned out to be a very useful and rather accurate one.

If one further assumes that the nuclear forces are independent of the spin - this corresponds to the Russell-Saunders model for atoms discussed above - one obtains further regularities. These are a good deal less accurate than the isospin regularities because the nuclear forces are spin-dependent. In spite of this, as has been shown particularly by Radicati, the resulting "supermultiplet theory" has several interesting confirmations.

Let me now come to the last area I will mention, *particle physics*[20] Since our whole meeting was devoted to it, I will not discuss it in detail. The approximate symmetry, the SU(3) symmetry, leading to the "eightfold way" is not very closely realized. Nevertheless, it has led to *many* very interesting conclusions which were confirmed experimentally. In particular, it led to predicting the discovery of new particles to complete the multiplet postulated by the SU(3) symmetry. It also led to a rather simple expression, the Gell Mann-Okubo formula, for the mass differences between the members of a multiplet. The basis of the symmetry, on the other hand, is, in my opinion, not as clear as in the case of the other approximate symmetries Several attempts were made to derive them; very interesting ones, but not fully convincing.

Well, it was a pleasure to review the problems and accomplishments of the symmetry theories, even though the review had to be, naturally, very superficial. I often wonder how long our division of information into the three groups of initial conditions, laws of nature, and invariances will persist. At present it is interesting and useful to use these concepts.

REFERENCES

1) J.F.C. Hessel, Ostwald's Klassiker der exakten Naturwissenschaften, No. 89 (Leipzig, 1897), p. 91.

2) A. Schönflies, Kristallsysteme and Kristallstruktur (Leipzig, 1891).

3) E.S. Fedorov, Zap. Min. Obsk. 38, 1 (1891) (Trans.: Min. Soc.).

4) N. Steno, De solido intra solidem naturaliter contento dissertationis prodromus (Florence, 1669).

5) G. Hamel, Z. Math. Phys. 50, 1 (1904).
F. Engel, Nach. Ges. Wiss. Göttingen, 1918, p. 375; E. Nöther, *ibid.* p. 235.

6) A. Kretschman, Ann. der Physik 53, 575 (1917).

7) The rôle of invariance principles in natural philosophy, Rendiconti della Scuola Internazionale di Fisica Enrico Fermi, Corso XXIX (Academic Press, New York, 1964).

8) V. Fock, The theory of space, time, and gravitation (Pergamon Press, London, 1957).

9) M. Gell-Mann, Phys. Rev. 125, 1067 (1962).
Y. Ne'eman, Nuclear Phys. 26, 222 (1961).

10) E.P. Wigner, Ann. of Math. 40, 149 (1939).

11) E.P. Wigner, Gruppentheorie und ihre Anwendung auf die Quantenmechanik der Atomspectren (Vieweg, Braunschweig, 1931). Somewhat updated English translation: Academic Press, New York, 1959.
M. Hamermesh, Group theory and its application to physical problems (Addison Wesley Publ. Co., Reading, Mass., 1962).

12) M. Born and J.R. Oppenheimer, Ann. der Physik 84, 457 (1927).

13) T.D. Lee and C.N. Yang, Phys. Rev. 104, 254 (1956).

14) C.S. Wu, E. Ambler, R.W. Hayward, D.D. Hoppes and R.P. Hudson, Phys. Rev. 105 1413 (1957).

15) G. Herzberg's books, in particular his Spectra of Diatomic Molecules (Van Nostrand, New York, 1939 and 1950) describe very vividly both the applicability of the classical picture and the deviations from it in the case of identical atoms.
I. Kovacs' Rotational Structure in the Spectra of Diatomic Molecules (Hilger, London, 1969) gives even further details.

16) K.F. Bonhöffer and P. Harteck, Z. für Physikalische Chemie B4, 113 (1929).

17) J. Callaway, Electron band theory (Academic Press, New York, 1964).

18) M.J.O. Strutt, Ann. der Physik 85, 129 (1928).
F. Bloch, Z. für Physik 52, 555 (1928).
R. Peierls, Ann. der Physik 4, 121 (1930).
L.P. Bouckaert, R. Smoluchowski and E. Wigner, Phys. Rev. 50, 58 (1936).

19) A very interesting discussion on this subject was presented at the 14th Solvay Conference: Symmetry properties of nuclei (Gordon and Breach, 1974).

20) M. Gell-Man and Y. Ne'eman, The eightfold way (W.A. Benjamin, New York, 1964).
Highlights in particle physics (ed. A. Zichichi) (Editrice Compositori, Bologna, 1973).

CLOSING CEREMONY

The Closing Ceremony took place on Saturday, 7 August 1976. The Director of the School presented the prizes and scholarships as specified below.

PRIZES AND SCHOLARSHIPS

Prize for *Best Student* - awarded to William J. MARCIANO - Rockefeller University - New York, NY, USA.

Thirteen scholarships were open for competition among the participants. They were awarded as fellow:

Patrick M.S. Blackett Scholarship - awarded to William J. MARCIANO, Rockefeller University, New York, NY, USA.

Gunnar Källen Scholarship - awarded to Barbara YOON, MIT, Cambridge, MA, USA.

James Chadwick Scholarship - awarded to Nigel H. PARSONS, Glasgow University, Glasgow, UK.

Amos-de Shalit Scholarship - awarded to Gideon BERLAD, TECHNION, Haifa, Israel.

André Lagarrigue Scholarship - awarded to Myron R. PAULI, MIT, Cambridge, MA, USA.

Giulio Racah Scholarship - awarded to Giora J. TARNOPOLSKY, ETH, Zürich, Switzerland.

Giorgio Ghigo Scholarship - awarded to Arturo GARCIA, Oxford University, Oxford, UK.

Enrico Persico Scholarship - awarded to Fred POSNER, Harvard University, Cambridge, MA, USA.

Peter Preiswerk Scholarship - awarded to Thomas D.B. WILKIE, University of Durham, Durham, UK.

Gianni Quareni Scholarship - awarded to Junko SHIGEMITSU, Cornell University, Ithaca, USA.

Antonio Stanghellini Scholarship - awarded to Alberto C. DE LA TORRE, Gesamthochschule, Wuppertal, Germany.

Alberto Tomasini Scholarship - awarded to Hung PHAM QUANG, CERN, Geneva, Switzerland.

Ettore Majorana Scholarship - awarded to Paolo ROSSI, Scuola Normale Superiore, Pisa, Italy.

Prize for *Best Scientific Secretary* - awarded to Myron R. PAULI - MIT, Cambridge, MA, USA.

Prize for *Honorary Best Student* - awarded to Eugene P. WIGNER, Princeton University, Princeton, NJ, USA.

The following participants gave their collaboration in the scientific secretarial work:

Yachin AFEK
Orlando ALVAREZ
Andrea AUDRITO
Dario BISELLO
Fritz BOPP
Franco CERVELLI
André DEGRE'
Alberto C. DE LA TORRE
Massimo FALCIONI
Barry A. FREEDMAN
Arturo GARCIA
Paolo GIUSTI
M. Alimi ICHOLA
Bernard JANCEWICZ
David R.T. JONES
William J. MARCIANO
Bruno MATHIS
Nigel H. PARSONS
Myron R. PAULI
Hung PHAM QUANG
Fred POSNER
Gabriele PUGLIERIN
Paolo ROSSI/Pisa
Junko SHIGEMITSU
Pasquale SODANO
Giora J. TARNOPOLSKY
Karl-Ludvig WERNHARD
Thomas D.B. WILKIE
Barbara YOON

PARTICIPANTS

Yachin AFEK — TECHNION - Israel Institute of Technology
Department of physics
HAIFA 32000, Israel

Orlando ALVAREZ — Harvard University
Department of Physics
CAMBRIDGE, MA 02138, USA

Ivan ANDRIC — Universität Bielefeld
Abteilung Theoretische Physik
Universität Strasse
48 BIELEFELD, D

Andrea AUDRITO — Istituto di Fisica dell'Università
Corso Massimo d'Azeglio, 46
10125 TORINO, Italy

Milla BALDO-CEOLIN — Istituto di Fisica dell'Università
via Marzolo, 8
35100 PADOVA, Italy

Maurizio BASILE — CERN
EP Division
1211 GENEVA 23, Switzerland

René BERGER — Musée des Beaux Arts
Palais de Rumine
1005 LAUSANNE, Switzerland

Gideon BERLAD — TECHNION - Israel Institute of Technology
Department of Physics
HAIFA 32000, Israel

Dario BISELLO	Istituto di Fisica dell'Università via Marzolo, 8 35100 PADOVA, Italy
Volker BLOBEL	Universität Hamburg II. Institut für Experimentalphysik Notkestieg 1 HAMBURG 52, D
Fritz BOPP	Gesamthochschule Siegen Fachbereich Mathematik- Naturwissenschaften-Physik Hölderlinstrasse 3 59 SIEGEN 21, D
Franco BUCCELLA	Istituto di Fisica dell'Università piazzale delle Scienze, 5 00185 ROMA, Italy
Elio CALLIGARICH	INFN - Sezione di Pavia via A. Bassi, 6 27100 PAVIA, Italy
Giovanni CARAROMEO	CERN EP Division 1211 GENEVA 23, Switzerland
Roberto CASALI	INFN - Sezione di Pisa via Livornese 56010 SAN PIERO A GRADO (Pisa), Italy
Franco CERVELLI	INFN - Sezione di Frascati Casella Postale 70 00044 FRASCATI (Roma), Italy
Reinhold CHRISTIAN	Institut für Theoretische Physik der Universität Wien Boltzmanngasse 5 1090 WIEN, Austria
Luisa CIFARELLI	CERN EP Division 1211 GENEVA 23, Switzerland
Jean CLEYMANS	Universität Bielefeld Abteilung Theoretische Physik Universität Strasse 48 BIELEFELD, D

James W. CRONIN	The University of Chicago The Enrico Fermi Institute 5630 Ellis Avenue CHICAGO, IL 60637, USA
David CUTTS	Brown University Department of Physics PROVIDENCE, RI 02912, USA
Luigi DADDA	Politecnico di Milano piazza Leonardo da Vinci, 32 20133 MILANO, Italy
Richard H. DALITZ	The University of Oxford Department of Theoretical Physics 12 Parks Road OXFORD OX1 3PQ, UK
André DEGRE	Centre de Recherches Nucléaires Labo PNPP-HE B.P. 20/CRO 67037 STRASBOURG CEDEX, France
Alberto C. DE LA TORRE	Gesamthochschule Wuppertal Fachbereich 8 - Physik Hòfkamp 82-64 56 WUPPERTAL, D
Bernard D'ESPAGNAT	Université Paris XI LPTPE Bâtiment 210 91 ORSAY, France
John C. ECCLES	"Ca' a la Gra'" 6611 CONTRA (Locarno) TI Switzerland
Etim ETIM	CERN Th Division 1211 GENEVA 23, Switzerland
Christian W. FABJAN	CERN EP Division 1211 GENEVA 23, Switzerland
Massimo FALCIONI	Istituto di Fisica dell'Università piazzale delle Scienze, 5 00185 ROMA, Italy

Thomas FERBEL	The University of Rochester River Campus Station Department of Physics ROCHESTER, NY 14627, USA
Harm FESEFELDT	Max-Planck-Institut für Physik und Astrophysik Institut für Physik Föhringer Ring 6 8 MÜNCHEN 40, D
Barry A. FREEDMAN	Massachusetts Institute of Technology Room 6-416 Center for Theoretical Physics CAMBRIDGE, MA 02139, USA
William F. FRY	University of Wisconsin-Madison Department of Physics 1150 University Avenue MADISON, WI 53706, USA
Arturo GARCIA	The University of Oxford Department of Theoretical Physics 12 Parks Road OXFORD OX1 3PQ, UK
Paolo GIUSTI	CERN EP Division 1211 GENEVA 23, Switzerland
Michel GOURDIN	Université Paris VI LPTHE Tour 16 - 1er étage 4 place Jussieu 75230 PARIS CEDEX 05, France
M. Alimi ICHOLA	Collège de France Laboratoire de Physique Corpusculaire 11 place Marcelin-Berthelot 75231 PARIS CEDEX 05, France
Bernard JANCEWICZ	University of Wroclaw Institute for Theoretical Physics ul. Cybulskiego 36 50205 WROCLAW, Poland
Peter D. JARVIS	Imperial College of Science and Technology The Blackett Laboratory Prince Consort Road LONDON SW7 2BZ, UK

Kjell JOHNSEN
CERN
ISR Division
1211 GENEVA 23, Switzerland

David R.T. JONES
The University of Oxford
Department of Theoretical Physics
12 Parks Road
OXFORD OX1 3PQ, UK

Hagen KLEINERT
Freie Universität Berlin
FB 20, WE 4
Arnimallee 3
1000 BERLIN 33, D

Otto KOFOED-HANSEN
Danish Atomic Energy Commission
Research Establishment RISØ
4000 ROSKILDE, Denmark

Johann KÜHN
Max-Planck-Institut für Physik und Astrophysik
Institut für Physik
Föhringer Ring 5
8 MÜNCHEN 40, D

Elliot LEADER
University of London
Westfield College
Kidderpore Avenue
LONDON NW3 7ST, UK

Endre LILLETHUN
University of Bergen
Department of Physics
Allegt. 55
5014 BERGEN, Norway

Harry LIPKIN
The Weizmann Institute of Science
Department of Nuclear Physics
REHOVOT, Israel

Laurence S. LITTENBERG
Brookhaven National Laboratory
Physics Department
UPTON, NY 11973, USA

William J. MARCIANO
Rockefeller University
Physics Department
NEW YORK, NY 10021, USA

Bruno MATHIS
Freie Universität Berlin
Institut für Theoretische Physik
Arnimallee 3
1000 BERLIN 33, D

Andrew McPHERSON
Rutherford Laboratory
CHILTON, Didcot
Oxon., OX11 OXQ, UK

Ezio MENICHETTI
Istituto di Fisica dell'Università
Corso Massimo d'Azeglio, 46
10125 TORINO, Italy

Adriana MINGUZZI RANZI
Istituto di Fisica dell'Università
via Irnerio, 46
40126 BOLOGNA, Italy

Claudio ORZALESI
CERN
Th Division
1211 GENEVA 23, Switzerland

Oliver E. OVERSETH
The University of Michigan
The H.M. Randall Laboratory of Physics
ANN ARBOR, MI 48104, USA

Nigel H. PARSONS
Glasgow University
Department of Natural Philosophy
GLASGOW G12 8QQ, UK

Myron R. PAULI
Massachusetts Institute of Technology
Room 6-411A
Department of Physics
CAMBRIDGE, MA 02139, USA

Dubravko PEVEC
Ruder Boskovic Institute
P.O. Box 1016
41001 ZAGREB, Croatia, Yugoslavia

Hung PHAM QUANG
CERN
Th Division
1211 GENEVA 23, Switzerland

Fred POSNER
Harvard University
Department of Physics
CAMBRIDGE, MA 02138, USA

Giuliano PREPARATA
CERN
Th Division
1211 GENEVA 23, Switzerland

Gabriele PUGLIERIN
Istituto di Fisica dell'Università
via F. Marzolo, 8
35100 PADOVA, Italy

Gisela RANFT	Karl-Marx-Universität Sektion Physik Linnestrasse 5 701 LEIPZIG, DDR
Leonardo ROSSI	INFN - Sezione di Genova viale Benedetto XV, 5 16132 GENOVA, Italy
Paolo ROSSI	Istituto di Fisica dell'Università via F. Marzolo, 8 35100 PADOVA, Italy
Paolo ROSSI	Scuola Normale Superiore piazza dei Cavalieri, 7 56100 PISA, Italy
Jun John SAKURAI	CERN Th Division 1211 GENEVA 23, Switzerland
Dieter SCHLATTER	CERN EP Division 1211 GENEVA 23, Switzerland
Stuart J. SHARROCK	University College of London Department of Physics and Astronomy LONDON WC1, UK
Junko SHIGEMITSU	Cornell University Laboratory for Nuclear Studies ITHACA, NY 14853, USA
A.J. Stewart SMITH	Princeton University Department of Physics Joseph Henry Laboratories Jadwin Hall P.O. Box 708 PRINCETON, NJ 08540, USA
Pasquale SODANO	University of Alberta Department of Physics EDMONTON T6G 2J, Canada
Martin SOHNIUS	Max-Planck-Institut für Physik und Astrophysik Institut für Physik Föhringer Ring 6 8 MÜNCHEN 40, D

Giora J. TARNOPOLSKY
L.H.E.
Eidgenössische Technische Hochschule
ZÜRICH, Switzerland

Val L. TELEGDI
The University of Chicago
The Enrico Fermi Institute
5630 Ellis Avenue
CHICAGO, IL 60637, USA

Jean-Michel THENARD
CERN
EP Division
1211 GENEVA 23, Switzerland

Sam C.C. Ting
Massachusetts Institute of Technology
Department of Physics
CAMBRIDGE, MA 02139, USA

Giovanni VALENTI
CERN
EP Division
1211 GENEVA 23, Switzerland

Patrizio VINCIARELLI
CERN
Th Division
1211 GENEVA 23, Switzerland

Guy von DARDEL
University of Lund
Sölvegatan 14
223 62 LUND, Sweden

Gustav Adolf VOSS
DESY
Notkestieg 1
2000 HAMBURG 52, D

Raymond WEILL
Institut de Physique Nucléaire
B.S.P. - Dorigny
1015 LAUSANNE, Switzerland

Steven WEINBERG
Harvard University
Lyman Laboratory of Physics
CAMBRIDGE, MA 02138, USA

Karl-Ludvig WERNHARD
CERN
EP Division
1211 GENEVA 23, Switzerland

Gian Carlo WICK
Columbia University
Physics Department
P.O. Box 132
NEW YORK, 10027, USA

Eugene P. WIGNER
Princeton University
Department of Physics
Joseph Henry Laboratories
Jadwin Hall
P.O. Box 708
PRINCETON, NJ 08540, USA

Thomas D.B. WILKIE
University of Durham
Department of Physics
Science Laboratories
South Road
DURHAM DH1 3LE, UK

Catherine WILQUET
Inter-University for High Energies
ULB-VUB
avenue de la Plaine, 2
1050 BRUXELLES, Belgium

Chen Ning YANG
SUNY
Institute for Theoretical Physics
STONY BROOK, NY 11794, USA

Barbara YOON
Massachusetts Institute of Technology
Room 6-415
Department of Physics
CAMBRIDGE, MA 02139, USA

Thomas YPSILANTIS
CCSEM
91016 ERICE, Italy

INDEX

MIX
Papier aus verantwortungsvollen Quellen
Paper from responsible sources
FSC® C105338

If you have any concerns about our products,
you can contact us on
ProductSafety@springernature.com

In case Publisher is established outside the EU,
the EU authorized representative is:
Springer Nature Customer Service Center GmbH
Europaplatz 3, 69115 Heidelberg, Germany

Printed by Libri Plureos GmbH
in Hamburg, Germany